KT-546-000

HARRAP'S
ILLUSTRATED DICTIONARY OF
SCIENCE

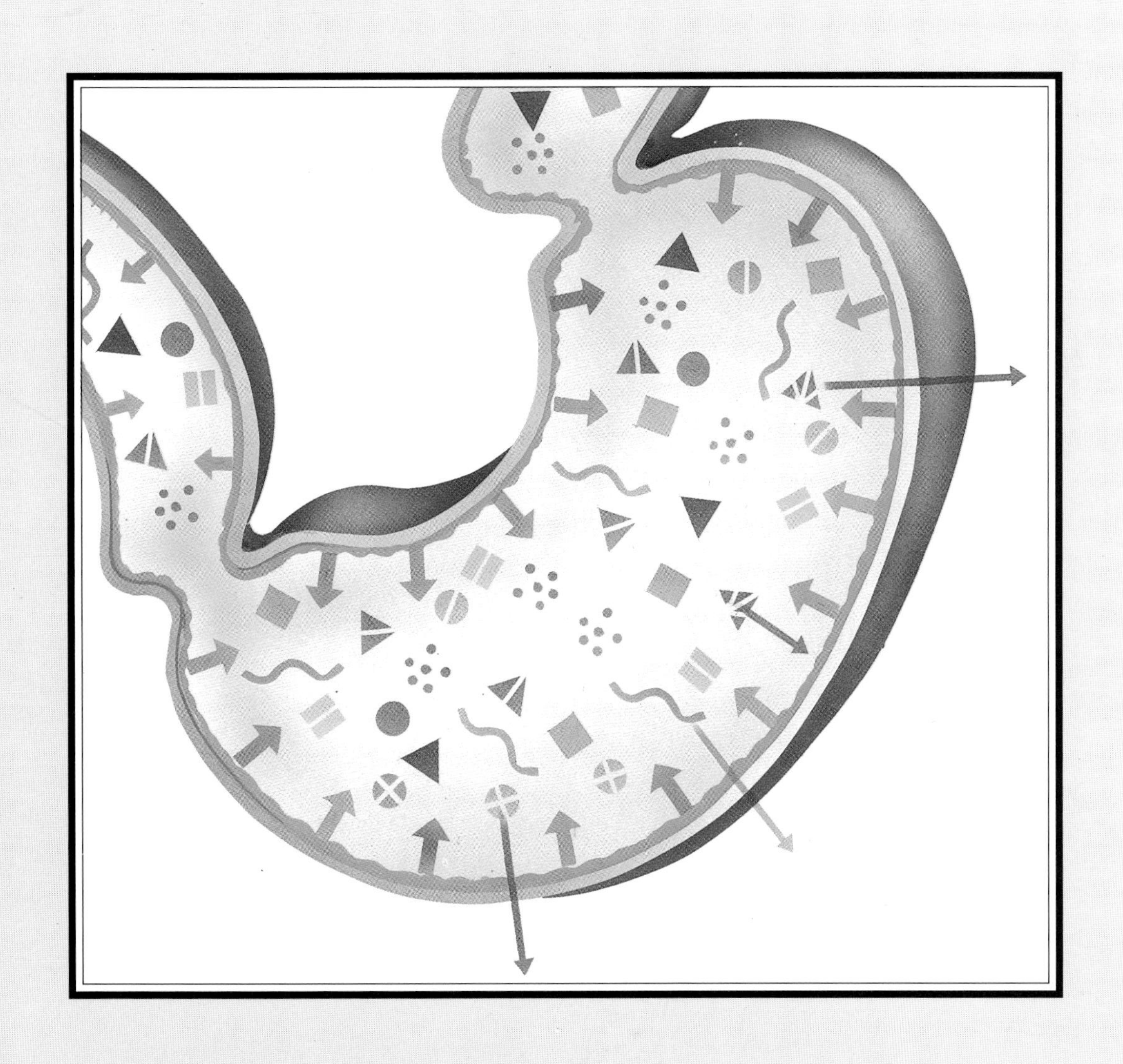

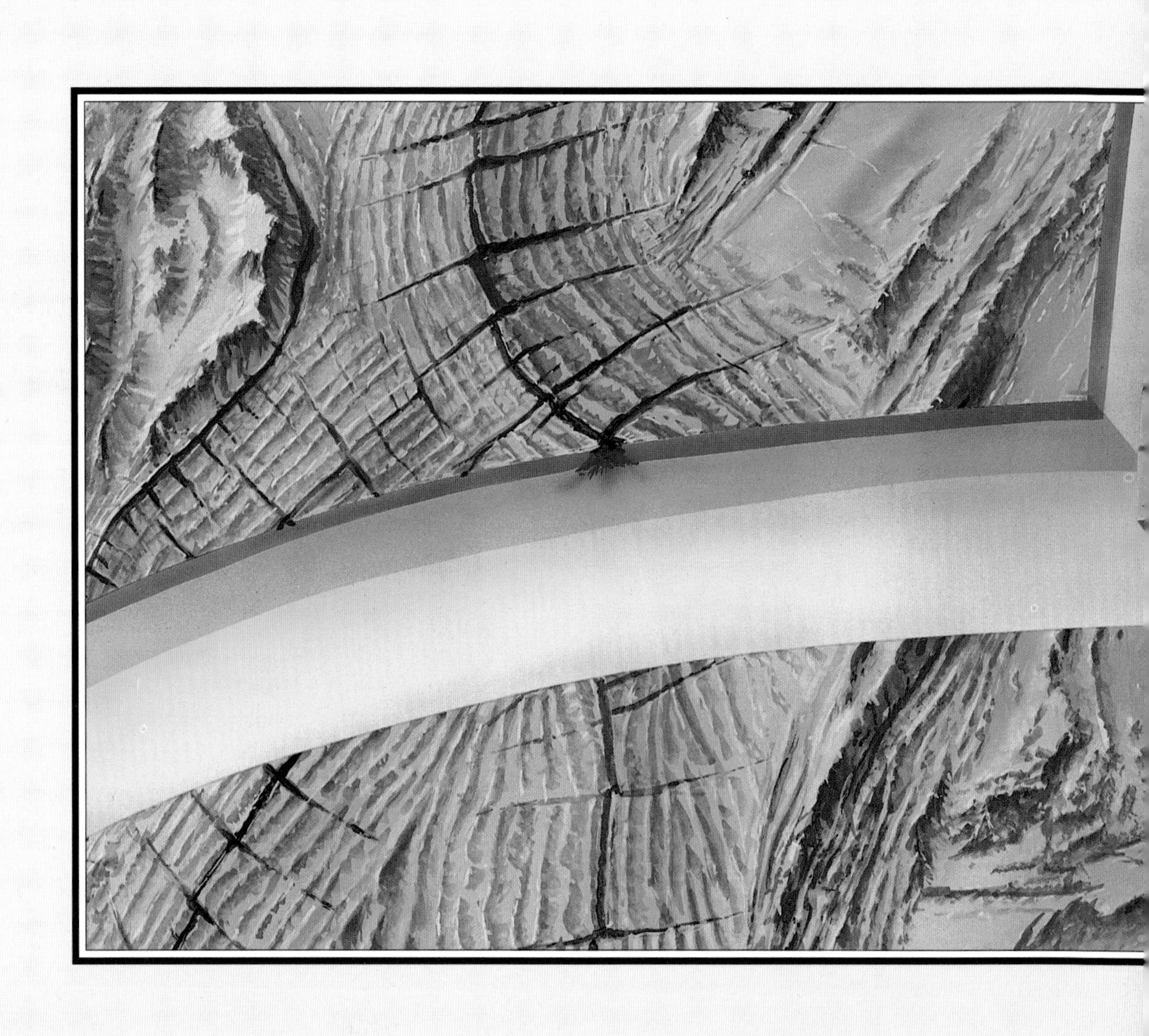

HARRAP'S
ILLUSTRATED DICTIONARY OF
SCIENCE

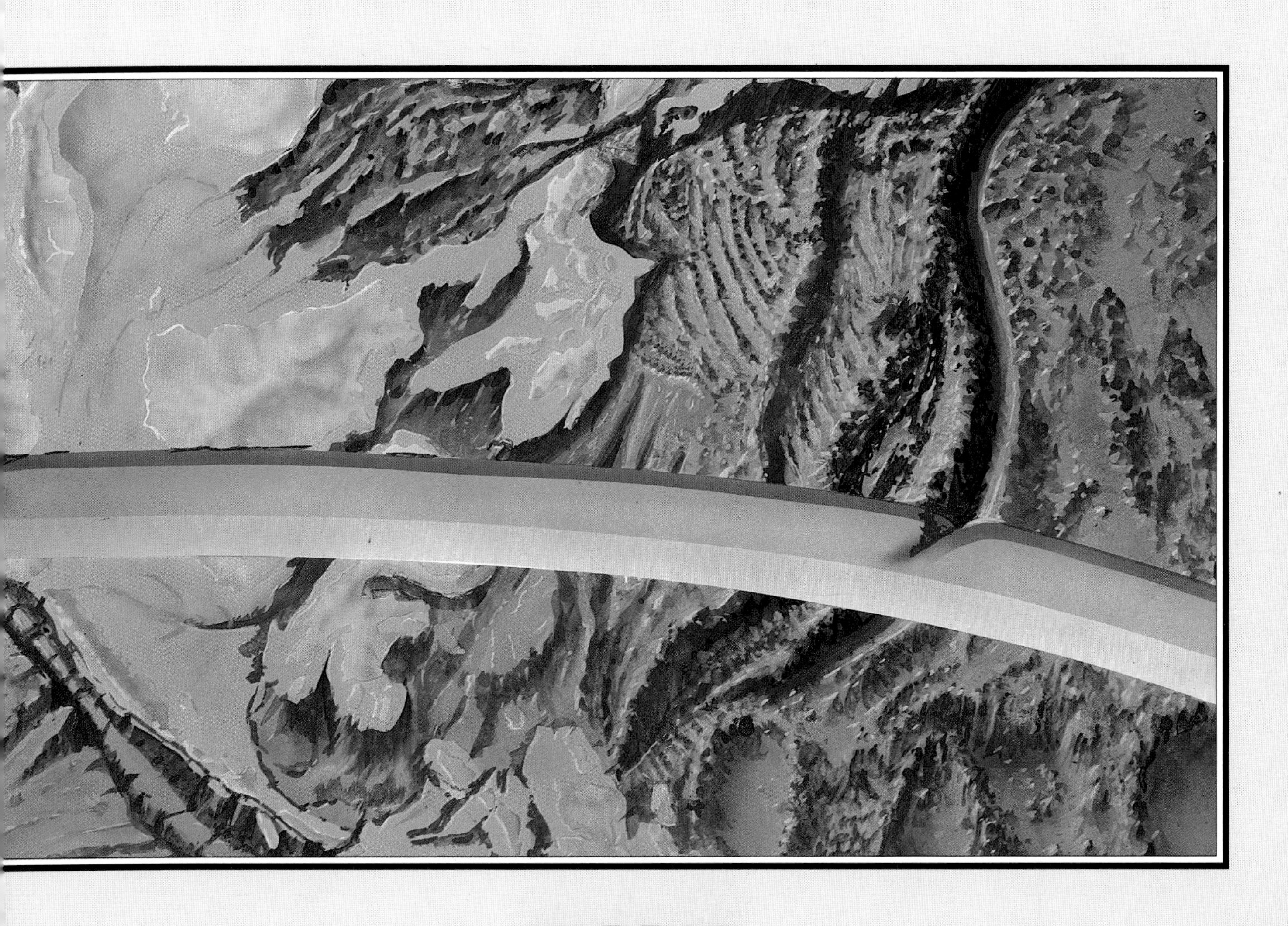

HARRAP

London

Editor Candida Hunt
Assistant Editor Monica Byles
Designer Nicholas Rous

Design Consultant John Ridgeway
Project Editor Lawrence Clarke

Contributing Consultants
Bernard Dixon
Dougal Dixon
Linda Gamlin
Iain Nicolson
Martin Sherwood
Christine Sutton
John-David Yule

Additional Contributors
P. le P. Barnett J.G. Bateman
M.G. Desebrock P.C. Gardner
B. Gibbs G.S. Harbinson
P.Hutchinson A.J. Pinching
R.P. Revel C. Richmond
P.R. Robinson W.O. Saxton
M. Scott Rohan

AN EQUINOX BOOK

Planned and produced by:
Equinox (Oxford) Ltd
Musterlin House, Jordan Hill, Oxford, England

Published in this edition 1988 by
Harrap Ltd,
19–23 Ludgate Hill,
London EC4M 7PD

ISBN 0 245 – 54780 – 0

Printed in Spain by Heraclio Fournier S.A., Vitoria
10 9 8 7 6 5 4 3 2 1

Introductory pictures
1 Detail of Digestion (page 41)
2-3 Detail of Plate Tectonics (page 142)
4-5 Detail of Stars (page 228)
7 Detail of Cell Division (page 32)

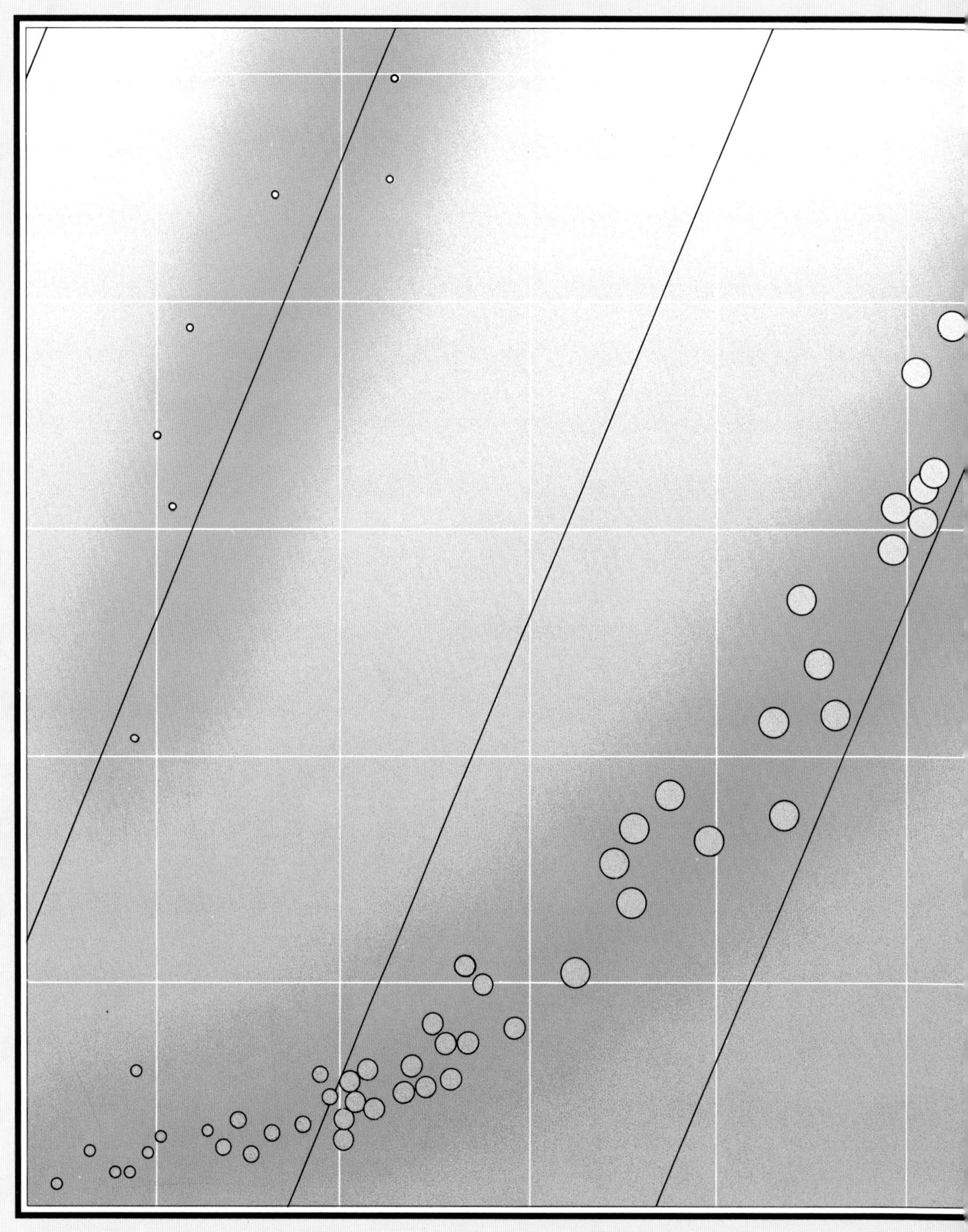

Illustrations

Units of Measurement

In general the International System of Units – SI units – is used throughout this book. This system is founded upon seven empirically defined *base units* (eg ampere), which can be combined, sometimes with the assistance of two geometric *supplementary units* (eg radian) to yield the *derived units* (eg cubic meter) which together with the base units constitute a coherent set of units capable of application to all measurable physical phenomena. Some of the derived units have special names (eg volt). For the sake of convenience smaller and larger units, the multiples and submultiples of the SI units, can be formed by adding certain prefixes (eg milli-) to the names of the SI units. In any instance only one prefix can be added to the name of a unit (thus nanometer, not millimicrometer). Of the other units in common scientific use, it is recognized that several (eg hour) will continue to be used alongside the SI units, although combinations of these units with SI units (as in kilowatt hour) are discouraged. However, other units (eg angstrom unit) are redundant if the International System is fully utilized, and it is intended that these should drop out of use.

Base units

Quantity	Unit	Symbol
length	meter	m
mass	kilogram	kg
time	second	s
electric current	ampere	A
temperature	kelvin	K
luminous intensity	candela	cd
amount of substance	mole	mol

Derived units

Quantity	Name	Symbol
frequency	hertz	Hz
force	newton	N
work, energy	joule	J
power	watt	W
pressure	pascal	Pa
quantity of electricity	coulomb	C
potential difference	volt	V
electric resistance	ohm	Ω
capacitance	farad	F
conductance	siemens	S
magnetic flux	weber	Wb
flux density	tesla	T
inductance	henry	H
luminous flux	lumen	lm
illuminance	lux	lx

Supplementary units

Quantity	Name	Symbol
plane angle	radian	rad
solid angle	steradian	sr

Non-SI units in common use

Quantity	Name	Symbol
plane angle	degree	°
plane angle	minute	′
plane angle	second	″
time	minute	min
time	hour	h
time	day	d
volume	liter	l
mass	tonne	t
energy	electronvolt	eV
mass	atomic mass unit	u
length	astronomical unit	AU
length	parsec	pc

Index notation

Very large and very small numbers are often written using powers of ten. The American system in the table below names large numbers according to the number of groups of three zeros which follow 1,000 when they are expressed in numerals – eg 1 billion (bi- meaning two) is 1,000 followed by two groups of three zeros.

Name	Numeral	Value in powers of ten	SI prefix	SI symbol
one	1	10^0	—	—
ten	10	10^1	deka	da
hundred	100	10^2	hecto	h
thousand	1,000	10^3	kilo	k
million	1,000,000	10^6	mega	M
billion	1,000,000,000	10^9	giga	G
trillion	1,000,000,000,000	10^{12}	tera	T
tenth	0.1	10^{-1}	deci	d
hundredth	0.01	10^{-2}	centi	c
thousandth	0.001	10^{-3}	milli	m
millionth	0.000001	10^{-6}	micro	μ
billionth	0.000000001	10^{-9}	nano	n
trillionth	0.000000000001	10^{-12}	pico	p

Abbreviations not already given above

A	mass number
Å	angstrom unit
AC	alternating current
AF	audio frequency
asb	apostilb
atm	atmosphere
AW	atomic weight
bhp	brake horse power
bp	boiling point
Btu	British thermal unit
C°	centigrade degree
°C	degree Celsius
Cal	see Kcal
cal	calorie
ccp	cubic close-packed
CGS	centimeter-gram-second (system)
Ci	curie
dB	decibel
DC	direct current
EHF	extremely high frequency
emf	electromotive force
emu	electromagnetic unit
esu	electrostatic unit
f	femto-
fcc	face-centered cubic
FM	frequency modulation
G	universal constant of gravitation
g	gram
g	acceleration due to gavity
Gb	gilbert
gr	grain
Gs	gauss
h	Planck constant
ha	hectare
hcp	hexagonal close-packed
HF	high frequency
hp	horse power
IF	intermediate frequency
ir	infrared
k	kilo-
k	Boltzmann constant
kcal	kilocalorie
kgf	kilogram force
L	lambert
LW	long wave
ly	light year
M	mega-
mbar	millibar
MF	medium frequency
MKSA	meter-kilogram-second- ampere (system)
mp	melting point
mya	million years ago
MW	medium wave
MW	molecular weight
Mx	maxwell
$\mathcal{N}$	Avogadro number, neutron number
nmr	nuclear magnetic resonance
Oe	oersted
O.N.	oxidation number
P	poise
pH	hydrogen ion concentration
ph	phot
ppm	parts per million
R	röntgen
R	universal gas constant
°R	degrees Rankine or (Reaumur)
rad	radian
rd	rad (dose)
rpm	revolutions per minute
sb	stilb
sg	specific gravity
SHF	superhigh frequency
sr	steradian
STP	standard temperature and pressure
subl	sublimation point
UHF	ultrahigh frequency
uv	ultraviolet
VHF	very high frequency
yr	year
$\mathcal{Z}$	atomic number

Introduction

We live in a world where scientific terms are part of everyday life. To make sense of our world we need to understand the vocabulary of science - from acid rain to the exploration of the Universe. The entries in this book encompass all the principal fields of science. It explains matter and energy (physics); the composition of substances and how they affect each other (chemistry); the science of living things (biology) and particularly of humans (medicine); our understanding of the Earth (geology and physical geography) and of the Universe (astronomy); and the practical apparatus of science (technology). Scientific understanding today is presented against its historical background, and biographical entries provide details about the key figures in the development of scientific knowledge.

The illustrations in the book stand on their own as a pictorial sequence of key scientific topics. Cross-references are provided (eg ➧ page 13) so that they can also be used with relevant text entries. There is also a cross-referencing system between entries (using SMALL CAPS), which helps to set a particular item within a wider context.

American spellings have been used; entries are listed with their English spelling and given the appropriate cross-reference when they might otherwise be difficult to find (eg **Haemoglobin** see HEMOGLOBIN).

Many scientific terms contain frequently occurring word stems. With a knowledge of these, technical terms can be more easily understood.

aero- relating to air (eg aerodynamics)
ante- before (eg antenatal)
anti- against (eg antibiotics)
arche- ancient (eg archeology)
arthr- relating to joints (eg arthritis)
auto- self (eg autolysis)
-blast bud (eg osteoblast)
brachi- arm (eg brachiopod)
bronch- windpipe (eg bronchitis)
bryo- moss (eg bryophytes)
-cardi- relating to the heart (eg pericardial cavity)
cephalo- relating to the head (eg cephalopod)
cerebro- of the brain (eg cerebrospinal)
chemo- drug (eg chemotherapy)
chrom- colored (eg chromatography)
cortico- outer part (eg corticoid)
cyto-, -cyte relating to cells (eg cytoplasm)
-derm skin or outer covering (eg dermatology)
dent-, -dont- relating to teeth (eg dentist)
ecto- outside (eg ectoderm)
-ectomy removal (eg hysterectomy)
endo- inside (eg endoderm)
-emia blood condition (eg anemia)
epi- upon or on the outside of (as in epidermis)
eu- true (eg eusocial animal)
-fer, -fera bearing (eg conifer)
gam- breeding, mating (eg gamete)
gastro- relating to the stomach (as in gastroenteritis)
-gen producing, growing (eg pathogen)
-germ relating to reproduction (as in germination)
geo- to do with the Earth (eg geography)
gyne- relating to women (eg gynecology)
hemo- to do with blood (eg hemoglobin)
herbi- relating to plants (eg herbivore)
hetero- others (eg heterotroph)
homeo- similar (eg homeostasis)
hydro- to do with water (eg hydrology)
hyper- excessive (eg hyperbaric chamber)
hypo- below, deficient (eg hypothalamus)
-itis inflammation (eg neuritis)
-karyo- cell (eg prokaryote)
laparo- abdomen (eg laparoscopy)
lact- to do with milk (eg lactation)
litho- stone (eg lithosphere)
meso- middle (eg mesoderm)
micro- small (eg microscope)
mono- one (eg monocotyledon)
morpho- shape (eg morphology)
myo- muscle (eg myocardium)
necro- death (eg necrosis)
neuro- nerve (eg neurotransmitter)
oculo- of the eye (eg binocular vision)
oligo- few (eg oligochaete)
oo- relating to egg or ovum (eg oomycete)
ophthalm- of the eye (eg opthalmology)
ornith- of birds (eg ornithology)
ortho- correct (eg orthodontics)
osteo- of bone (eg osteomalacia)
paleo- ancient (eg paleomagnetism)
path- relating to disease (eg pathology)
pedi- juvenile (eg pedomorphosis)
peri- around (eg perihelion)
phage- consumer or destroyer (eg bacteriophage)
photo- to do with light (eg photometry)
-plasm, -plasty growth, living matter (eg protoplasm)
phyto-, -phyte of plants (epiphyte)
poly- many (eg polyethylene)
pro- before (eg prokaryote)
proto- first (eg protozoa)
-rrhage, -rrhea flow (eg hemorrhage)
-sclero- hard (eg atherosclerosis)
-scope view (eg endoscope)
-stomy hole (eg colostomy)
sym-, syn- alike, together (eg symbiosis)
tachy- fast (eg tachycardia)
-tomy cut (eg tonsillectomy)
therm-, thermo- temperature, heat (eg thermodynamics)
tox- poison (eg toxemia)
-troph, -trophy of feeding, growing (eg autotroph)
tele- far off (eg telescope)
vaso- tube (eg vasodilation)
-vore eater (eg carnivore)
xero- dry (eg xerophyte)

NOBEL PRIZES IN SCIENCE 1901–1987

Year	Chemistry	Physics	Physiology or Medicine
1901	J. H. van't Hoff *Neth.*	W. C. Röntgen *Ger.*	E. A. von Behring *Ger.*
1902	Emil Fischer *Ger.*	H. A. Lorentz *Neth.* Pieter Zeeman *Neth.*	Ronald Ross *UK*
1903	S. A. Arrhenius *Swe.*	A. H. Becquerel *Fra.* Pierre Curie *Fra.* Marie S. Curie *Fra.*	N. R. Finsen *Den.*
1904	William Ramsay *UK*	J. W. S. Rayleigh *UK*	Ivan P. Pavlov *Russia*
1905	Adolf von Baeyer *Ger.*	Philipp Lenard *Ger.*	Robert Koch *Ger.*
1906	Henri Moissan *Fra.*	Joseph Thomson *UK*	Camillo Golgi *It.* S. Ramón y Cajal *Spain*
1907	Eduard Buchner *Ger.*	A. A. Michelson *USA*	C. L. A. Laveran *Fra.*
1908	Ernest Rutherford *UK*	Gabriel Lippman *Fra.*	Paul Ehrlich *Ger.* Élie Metchnikoff *Russia*
1909	Wilhelm Ostwald *Ger.*	Guglielmo Marconi *It.* C. F. Braun *Ger.*	Emil T. Kocher *Swi.*
1910	Otto Wallach *Ger.*	J. D. van der Waals *Neth.*	Albrecht Kossel *Ger.*
1911	Marie S. Curie *Fra.*	Wilhelm Wien *Ger.*	Allvar Gullstrand *Swe.*
1912	Victor Grignard *Fra.* Paul Sabatier *Fra.*	N. G. Dalén *Swe.*	Alexis Carrel *Fra.*
1913	Alfred Werner *Swi.*	Heike Kamerlingh Onnes *Neth.*	C. R. Richet *Fra.*
1914	T. W. Richards *USA*	Max von Laue *Ger.*	Robert Barany *Austria*
1915	Richard Willstätter *Ger.*	William H. Bragg *UK* William L. Bragg *UK*	———
1916	———	———	———
1917	———	C. G. Barkla *UK*	———
1918	Fritz Haber *Ger.*	Max Planck *Ger.*	———
1919	———	Johannes Stark *Ger.*	Jules Bordet *Bel.*
1920	Walther Nernst *Ger.*	C. E. Guillaume *Swi.*	S. A. S. Krogh *Den.*
1921	Frederick Soddy *UK*	Albert Einstein *Ger./Swi.*	———
1922	F. W. Aston *UK*	N. H. D. Bohr *Den.*	A. V. Hill *UK* Otto Meyerhof *Ger.*
1923	Fritz Pregl *Austria*	Robert A. Millikan *USA*	F. G. Banting *Can.* J. J. R. Macleod *Can.*
1924	———	K. M. G. Siegbahn *Swe.*	Willem Einthoven *Neth.*
1925	Richard Zsigmondy *Ger.*	James Franck *Ger.* Gustav Hertz *Ger.*	———
1926	T. Svedberg *Swe.*	J. B. Perrin *Fra.*	Johannes Fibiger *Den.*
1927	Heinrich Wieland *Ger.*	A. H. Compton *USA* C. T. R. Wilson *UK*	Julius Wagner-Jauregg *Austria*
1928	Adolf Windaus *Ger.*	O. W. Richardson *UK*	C. J. H. Nicolle *Fra.*
1929	Arthur Harden *UK* Hans von Euler-Chelpin *Swe.*	L. V. de Broglie *Fra.*	Christian Eijkman *Neth.* F. G. Hopkins *UK*
1930	Hans Fischer *Ger.*	Chandrasekhara V. Raman *India*	Karl Landsteiner *Austria*

Year	Chemistry	Physics	Physiology or Medicine
1931	Carl Bosch *Ger.* Friedrich Bergius *Ger.*	———	Otto H. Warburg *Ger.*
1932	Irving Langmuir *USA*	Werner Heisenberg *Ger.*	E. D. Adrian *UK* C. Sherrington *UK*
1933	———	P. A. M. Dirac *UK* E. Schrödinger *Austria*	Thomas H. Morgan *USA*
1934	Harold C. Urey *USA*	———	G. H. Whipple *USA* G. R. Minot *USA* W. P. Murphy *USA*
1935	Frédéric Joliot-Curie, Irène Joliot-Curie *Fra.*	James Chadwick *UK*	Hans Spemann *Ger.*
1936	P. J. W. Debye *Neth.*	C. D. Anderson *USA* V. F. Hess *Austria*	Henry H. Dale *UK* Otto Loewi *Austria*
1937	Walter N. Haworth *UK* Paul Karrer *Swi.*	C. J. Davisson *USA* George P. Thomson *UK*	Albert von Szent-Gyorgyi *Hung.*
1938	Richard Kuhn *Ger.*	Enrico Fermi *It.*	Corneille Heymans *Bel.*
1939	Adolf Butenandt *Ger.* Leopold Ružička *Swi.*	E. O. Lawrence *USA*	Gerhard Domagk *Ger.*
1940	———	———	———
1941	———	———	———
1942	———	———	———
1943	Georg von Hevesy *Hung.*	Otto Stern *USA*	E. A. Doisy *Den.* Henrik Dam *USA*
1944	Otto Hahn *Ger.*	I. I. Rabi *USA*	Joseph Erlanger *USA* H. S. Gasser *USA*
1945	A. I. Virtanen *Fin.*	Wolfgang Pauli *Austria*	Alexander Fleming *UK* E. B. Chain *UK* Howard W. Florey *UK*
1946	J. B. Sumner *USA* J. H. Northrop *USA* W. M. Stanley *USA*	P. W. Bridgman *USA*	H. J. Muller *USA*
1947	Robert Robinson *UK*	Edward V. Appleton *UK*	C. F. Cori *USA* Gerty T. Cori *USA* B. A. Houssay *Arg.*
1948	Arne Tiselius *Swe.*	P. M. S. Blackett *UK*	Paul H. Müller *Swi.*
1949	W. F. Giauque *USA*	Hideki Yukawa *Jap.*	W. R. Hess *Swi.* Egas Moniz *Port.*
1950	Otto Diels *FRG* Kurt Alder *FRG*	C. F. Powell *UK*	Philip S. Hench *USA* Edward C. Kendall *USA* Tadeus Reichstein *Swi.*
1951	E. M. McMillan *USA* G. T. Seaborg *USA*	John D. Cockcroft *UK* Ernest T. S. Walton *Ire.*	Max Theiler *S. Africa*
1952	A. J. P. Martin *UK* R. L. M. Synge *UK*	Felix Bloch *USA* E. M. Purcell *USA*	S. A. Waksman *USA*
1953	Hermann Staudinger *Ger.*	Frits Zernike *Neth.*	F. A. Lipmann *USA* Hans A. Krebs *UK*
1954	Linus C. Pauling *USA*	Max Born *UK* Walther Bothe *FRG*	J. F. Enders *USA* F. C. Robbins *USA* T. H. Weller *USA*
1955	Vincent du Vigneaud *USA*	Willis E. Lamb, Jr. *USA* Polykarp Kusch *USA*	A. H. T. Theorell *Swe.*

Chemistry	Physics	Physiology or Medicine
1956		
Cyril N. Hinshelwood *UK* Nikolai N. Semenov, *USSR*	W. B. Shockley *USA* W. H. Brattain *USA* John Bardeen *USA*	D. W. Richards, Jr. *USA* A. F. Cournand *USA* W. Forssmann *FRG*
1957		
Alexander R. Todd *UK*	Tsung-Dao Lee *China* Chen Ning Yang *China*	Daniel Bovet *It.*
1958		
Frederick Sanger *UK*	P. A. Cherenkov *USSR* Igor Y. Tamm *USSR* Ilya M. Frank *USSR*	Joshua Lederberg *USA* G. W. Beadle *USA* E. L. Tatum *USA*
1959		
Jaroslav Heyrovsky *Czech.*	Emilio Segrè *USA* Owen Chamberlain *USA*	Severo Ochoa *USA* Arthur Kornberg *USA*
1960		
W. F. Libby *USA*	D. A. Glaser *USA*	F. M. Burnet *Australia* P. B. Medawar *UK*
1961		
Melvin Calvin *USA*	Robert Hofstadter *USA* R. L. Mössbauer *FRG*	Georg von Bekesy *USA*
1962		
M. F. Perutz *UK* J. C. Kendrew *UK*	L. D. Landau *USSR*	J. D. Watson *USA* F. H. C. Crick *UK* M. H. F. Wilkins *UK*
1963		
Giulio Natta *It.* Karl Ziegler *FRG*	Eugene P. Wigner *USA* Maria G. Mayer *USA* J. Hans D. Jensen *FRG*	John Carew Eccles *Australia* Alan Lloyd Hodgkin *UK* Andrew F. Huxley *UK*
1964		
Dorothy Crowfoot Hodgkin *UK*	Charles Hard Townes *USA* Nikolai G. Basov *USSR* Alexander M. Prokhorov *USSR*	Konrad E. Bloch *USA* Feodor Lynen *FRG*
1965		
Robert Burns Woodward *USA*	Richard P. Feynman *USA* Shinichiro Tomonaga *Jap.* Julian S. Schwinger *USA*	François Jacob *Fra.* André Lwoff *Fra.* Jacques Monod *Fra.*
1966		
Robert S. Mulliken *USA*	Alfred Kastler *Fra.*	Francis P. Rous *USA* Charles B. Huggins *USA*
1967		
Manfred Eigen *FRG* Ronald G. W. Norrish *UK* George Porter *UK*	Hans Albrecht Bethe *USA*	Ragnar Granit *Swe.* Haldan Keffer Hartline *USA* George Wald *USA*
1968		
Lars Onsager *USA*	Luis W. Alvarez *USA*	Robert W. Holley *USA* H.G. Khorana *USA* Marshall W. Nirenberg *USA*
1969		
Derek H. R. Barton *UK* Odd Hassel *Nor.*	Murray Gell-Mann *USA*	Max Delbrück *USA* Alfred D. Hershey *USA* Salvador E. Luria *USA*
1970		
Luis Federico Leloir *Arg.*	Louis Eugène Néel *Fra.* Hans Olof Alfvén *Swe.*	Julius Axelrod *USA* Bernard Katz *USA* Ulf von Euler *Swe.*
1971		
Gerhard Herzberg *Can.*	Dennis Gabor *UK*	Earl W. Sutherland *USA*
1972		
Stanford Moore *USA* William H. Stein *USA* Christian B. Anfinsen *USA*	John Bardeen *USA* Leon N. Cooper *USA* John Robert Schrieffer *USA*	Gerald M. Edelman *USA* Rodney R. Porter *UK*
1973		
Ernst Otto Fischer *FRG* Geoffrey Wilkinson *UK*	Leo Esaki *Jap.* Ivar Giaever *USA* Brian D. Josephson *UK*	Konrad Lorenz *Austria* Nikolaas Tinbergen *UK* Karl von Frisch *FRG*

Chemistry	Physics	Physiology or Medicine
1974		
Paul J. Flory *USA*	Martin Ryle *UK* Antony Hewish *UK*	Albert Claude *Bel.* George E. Palade *USA* Christian de Duve *Bel.*
1975		
John W. Cornforth *Australia/UK* Vladimir Prelog *Swi.*	Aage Bohr *Den.* Ben Mottelson *Den.* James Rainwater *USA*	David Baltimore *USA* Renato Dulbecco *USA* Howard M. Temin *USA*
1976		
William N. Lipscomb *USA*	Burton Richter *USA* Samuel C. C. Ting *USA*	Baruch S. Blumberg *USA* D. Carleton Gajdusek *USA*
1977		
Ilya Prigogine *Bel.*	Philip W. Anderson *USA* Nevill F. Mott *UK* John H. van Vleck *USA*	Roger Guillemin *USA* Andrew V. Schally *USA* Rosalyn Yalow *USA*
1978		
Peter D. Mitchell *UK*	Peter L. Kapitsa *USSR* Arno A. Penzias *USA* Robert W. Wilson *USA*	Werner Arber *Swi.* Daniel Nathans *USA* Hamilton O. Smith *USA*
1979		
Herbert C. Brown *USA* Georg Wittig *FRG*	Sheldon L. Glashow *USA* Abdus Salam *Pak.* Steven Weinberg *USA*	Allan M. Cormack *USA* Godfrey N. Hounsfield *UK*
1980		
Paul Berg *USA* Walter Gilbert *USA* Frederick Sanger *UK*	James W. Cronin *USA* Val L. Fitch *USA*	Baruj Benacerraf *USA* Jean Dausset *Fra.* George D. Snell *USA*
1981		
Kenichi Fukui *Jap.* Roald Hoffmann *USA*	Nicolaas Bloembergen *USA* Arthur L. Schawlow *USA* Kai M. Siegbahn *Swe.*	Roger W. Sperry *USA* David H. Hubel *USA* Torsten N. Wiesel *Swe.*
1982		
Aaron Klug *UK*	Kenneth G. Wilson *USA*	Sune K. Bergström *Swe.* Bengt I. Samuelsson *Swe.* John R. Vane *UK*
1983		
Henry Taube *USA*	S. Chandrasekhar *USA* William A. Fowler *USA*	Barbara McClintock *USA*
1984		
Robert Bruce Merrifield *USA*	Carlo Rubbia *It.* Simon van der Meer *Neth.*	Niels K. Jerne *Den.* Georges J. F. Köhler *FRG* César Milstein *UK/Arg.*
1985		
Herbert A. Hauptman *USA* Jerome Karle *USA*	Klaus von Klitzing *FRG*	Michael S. Brown *USA* Joseph L. Goldstein *USA*
1986		
Dudley R. Herschbach *USA* Yuan T. Lee *USA* John C. Polanyi *Can.*	Ernst Ruska *FRG* Gerd Binnig *FRG* Heinrich Rohrer *Swi.*	Stanley Cohen *USA* Rita Levi-Montalcini *It./USA*
1987		
Charles Pedersen *USA* Donald Cram *USA* Jean-Marie Lehn *Fra.*	Georg Bednorz *Swi.* Alex Müller *FRG*	Susumu Tonegawa *Jap.*

Abbreviations

Arg. Argentina *Bel.* Belgium *Can.* Canada *Czech.* Czechoslovakia *Den.* Denmark *Fin.* Finland *Fra.* France *FRG* Federal Republic of Germany *Ger.* Germany *Hung.* Hungary *Ire.* Ireland *It.* Italy *Jap.* Japan *Neth.* Netherlands *Nor.* Norway *Pak.* Pakistan *Port.* Portugal *Swe.* Sweden *Swi.* Switzerland *UK* United Kingdom *USA* United States of America *USSR* Union of Soviet Socialist Republics

Aa
Block LAVA.
Abbe, Cleveland (1838-1916)
Nicknamed "Old Probabilities", US astronomer and meteorologist. In 1869, as director of the Cincinnati Observatory, he published the first daily weather forecasts in the US, becoming in 1871 the first chief meteorologist of the US Weather Service.
Abbe, Ernst (1840-1905)
German physicist, research director of and partner in the optical firm of CARL ZEISS and founder of the Carl Zeiss Foundation (1891). He invented an apochromatic condenser LENS for use in a MICROSCOPE.
Abbot, Charles Greeley (1872-1973)
US astrophysicist, noted for his studies of solar radiation, and director of the Smithsonian Astrophysical Observatory from 1907 to 1944.
Abdomen
In VERTEBRATES, the part of the body between the CHEST and the PELVIS. In humans, it contains most of the GASTROINTESTINAL TRACT (from the stomach to the colon), together with the LIVER, GALL BLADDER and SPLEEN in a cavity lined by PERITONEUM. The KIDNEYS, ADRENAL GLANDS and PANCREAS lie behind this cavity, with the abdominal AORTA and inferior VENA CAVA. It is separated from the chest by the DIAPHRAGM. In insects, crustaceans and arachnids the abdomen forms the rear division of the body, behind the THORAX.
Abel, John Jacob (1857-1938)
US pharmacologist who first isolated the hormone ADRENALINE (1897) and prepared INSULIN in crystalline form (1926).
Aberration of light
In astronomy, a displacement between a star's observed and true position caused by the Earth's motion about the Sun and the finite nature of the velocity of light. The effect is similar to that observed by a man walking in the rain: though the rain is in fact falling vertically, because of his motion it appears to be falling at an angle. The maximum aberrational displacement is 20.5″ of arc; stars on the ECLIPTIC appear to move to and fro along a line of 41″; stars 90° from the ecliptic appear to trace out a circle of radius 20.5″; and stars in intermediate positions ellipses of major axis 41″. (➧ page 115, 237)
Aberration, Optical
The failure of a lens to form a perfect image of an object. The commonest types are chromatic aberration, where DISPERSION causes colored fringes to appear around the image; and spherical aberration, where blurring occurs because light from the outer parts of the lens is brought to a focus at a shorter distance from the lens than that passing through the center. Chromatic aberration can be reduced by using an ACHROMATIC LENS, and spherical aberration by separating the elements of a compound lens.
Abiogenesis see SPONTANEOUS GENERATION.
Ablation
In glaciology, the loss of snow and ice from the surface of a GLACIER by melting EVAPORATION or SUBLIMATION; also the quantity so lost.
Abortion
Ending of PREGNANCY before the fetus is able to survive outside the womb. It can occur spontaneously (in which case it is often termed miscarriage) or it can be artificially induced. Spontaneous abortion may occur as a result of maternal or fetal disease or of faulty implantation in the WOMB. Induction may be mechanical, chemical or using HORMONES, the maternal risk varying with fetal age, the method used and the skill of the physician.
Abraham, Karl (1877-1925)
German psychoanalyst whose most important work concerned the development of the LIBIDO, particularly in infancy. He suggested that various PSYCHOSES should be interpreted in terms of the interruption of this development.
Abreaction see CATHARSIS.
Abscess
An accumulation of PUS, usually representing one response of the body to bacterial infection. Abscesses may occur in any tissue or organ of the body, producing pain, redness and swelling.
Abscisic acid
A plant HORMONE that inhibits plant growth, induces bud dormancy and promotes leaf ABSCISSION (shedding).
Abscission
In botany, the process whereby plants shed leaves, flowers and fruits. Controlled by plant HORMONES such as AUXINS and ABSCISIC ACID, leaf drop occurs in many plants through the formation of an abscission layer of corky cells in the leaf stem. This restricts sap flow to the leaf.
Absolute
An adjective frequently encountered in scientific terminology, used to imply a fundamental theoretic or physical significance, as opposed to one merely empirical or practical (as in absolute SPACE-TIME, TEMPERATURE scale, UNITS), to define an actual, as opposed to an apparent, relative or comparative, measurement (as in absolute HUMIDITY; MAGNITUDE), or to describe a limiting case (as in ABSOLUTE ZERO or, for an aircraft, absolute ceiling).
Absolute zero
The TEMPERATURE at which all substances have zero thermal ENERGY and thus, it is believed, the lowest possible temperature. Although many substances retain some nonthermal ZERO-POINT ENERGY at absolute zero, this cannot be eliminated and so the temperature cannot be reduced further. Originally conceived as the temperature at which an ideal GAS at constant pressure would contract to zero volume, absolute zero is of great significance in THERMODYNAMICS, and is used as the fixed point for ABSOLUTE temperature scales. In practice the absolute zero of temperature is unattainable, though temperatures within a few millionths of a KELVIN of it have been achieved in CRYOGENICS laboratories.

$$0K = -273.16°C = -459.69°F$$

Absorption
Any process by which a substance incorporates another substance into itself, or takes in radiant or sound ENERGY. In chemistry it is distinguished from ADSORPTION, in which the adsorbed substance merely adheres to the surface of the adsorbent. In AIR POLLUTION control and the chemical industries gas absorption is a key process, while the extent to which different surfaces absorb SOUND is an important factor in ACOUSTIC design. According to QUANTUM THEORY, an atom or MOLECULE can absorb a PHOTON of ELECTROMAGNETIC RADIATION only if the quantum of energy so received raises it exactly from its present energy level to a higher one, a fact of great significance for SPECTROSCOPY and for the theory of DYES and COLOR vision. (➧ page 21)
Abundance ratio see ISOTOPES.
Abyssal fauna
The animals inhabiting the oceans at depths greater than 1km. Here, temperatures range between 5°C and 1°C, the pressure approaches 600 atmospheres (at 6km depth) and daylight is absent. Abyssal animals are all highly specialized, some being filter-feeders, others scavengers or predators. Some of these animals are blind; some possess BIOLUMINESCENT lures and body panels for prey attraction, prey detection or species recognition. In general the abyssal animals are widely scattered, but some live clustered around geothermal (hydrothermal) vents, where superheated, mineral-laden water erupts from the seabed. The bacteria living around these vents provide the food supply for these communities (see CHEMOSYNTHESIS).
Abyssal hills
Small hills on the OCEAN floor, commonly flanking or projecting through the ABYSSAL PLAINS. They are thought to represent once volcanic seamounts now largely buried by sediment.
Abyssal plain
A large, flat area of the OCEAN floor, lying usually between 4km and 6km below the surface. These plains, which together cover some 40% of the Earth's surface, are floored by thick layers of mud and other sediments.
Acanthocephala (spiny-headed worms)
A phylum of worms parasitic on vertebrates such as mammals, fish and birds. Named for their proboscis, which bears tiny hooks that anchor them to the intestinal walls of their hosts, these worms are so degenerate that most have little more than a reproductive system and simple brain. They often cause fatal infections.
Acanthodii
A group of fish known only as fossils. They occur in rocks from the Silurian to the PERMIAN period, 438-248 million years ago. The Acanthodii are characterized by spines that supported the front edges of the fins. They were the earliest fish to have jaws.
Acceleration
The rate at which the VELOCITY of a moving body changes. Velocity, a VECTOR quantity, is speed in a given direction, a body can accelerate both by changing its speed and by changing its direction. The units of acceleration, itself a vector quantity, are those of velocity per unit time, eg meters per second, per second (m/s^2).

According to NEWTON's second law of MOTION, acceleration is always the result of a force acting on a body; the acceleration (**a**) produced in a body of mass (m) by a force (**F**) is given by $\mathbf{a}=\mathbf{F}/m$. The acceleration due to gravity (g) of a body falling freely near the Earth's surface is about $9.81 m/s^2$. In the aerospace industry the accelerations experienced by men and machines are often expressed as multiples of g. Headward (vertical) accelerations of as little as $3g$ can cause pilots to black out (see SPACE MEDICINE).
Accelerators, Particle
Atomic research tools used to accelerate SUBATOMIC PARTICLES to high velocities. Their power is rated according to the kinetic energy they impart, measured in ELECTRON VOLTS (eV). Linear accelerators use electrostatic and electromagnetic fields to accelerate particles in a straight line, but greater energies are obtained by leading the particles around a spiral or circular path (see CYCLOTRON; SYNCHROCYCLOTRON; SYNCHROTRON). (➧ page 49)

Accelerometer
A device used to measure ACCELERATION, usually consisting of a heavy body of known mass free to move in only one dimension, in which it is supported by springs. Any acceleration experienced along that line is computed from electrical measurements of the resulting distortion in the springs. Three such instruments set at mutual right angles are needed to measure accelerations in three dimensions.
Acclimatization
The process of adjustment that allows an individual organism to survive under changed conditions. In a hot, sunny climate humans acclimatize by eating less and drinking more, and by developing more pigment in the skin. At higher altitudes humans can adjust to the diminished oxygen by increased production of red blood corpuscles.
Accommodation
The capacity of the eye to move the position of its lens to focus on objects at a range of different distances. It takes place because the ciliary muscles contract to flatten the lens when the eye is focusing on near objects. The lens becomes stiffer with increasing age, so a person's near vision is lost.
Accumulator (storage battery) see BATTERY.
Acetabulum
Hemispherical socket at the side of the pelvis into which the head of the femur (thighbone) fits and rotates.
Acetaldehyde (ethanal, CH_3CHO)
A colorless, flammable liquid, an ALDEHYDE, made by catalytic oxidation of ETHENE. An important reagent, it is used in the manufacture of dyes, plastics and many other organic chemicals. In the presence of acids it forms the cyclic polymers paraldehyde, $(CH_3CHO)_3$, and metaldehyde $(CH_3CHO)_4$. The former is used as an anticonvulsant, and the latter as a solid fuel for portable stoves and as a poison for snails and slugs.
Acetates
Compounds in which the acid hydrogen atom in ACETIC ACID is replaced either by a metal (forming soluble acetate SALTS containing the ion CH_3COO^-) or by an organic radical (giving a covalent ESTER such as ethyl acetate, $CH_3COOC_2H_5$). Acetate esters are of great commercial importance, being used as solvents and for the manufacture of plastics and fibers. The various cellulose acetates are used in sheet form as a base for photographic films and also for fibers.
Acetic acid (ethanoic acid, CH_3COOH)
Most important of the CARBOXYLIC ACIDS, a pungent, colorless liquid used to make ACETATES. The ACETATE anion is important in BIOSYNTHESIS. Acetic acid is produced by bacterial action on alcohol in air (yielding vinegar), and industrially by the oxidation of HYDROCARBONS, ACETALDEHYDE or ETHANOL. Pure ("glacial") acetic acid solidifies to icelike crystals at 17°C; it is corrosive. MW 60.85, mp 16.7°C, bp 118°C.
Acetone (2-propanone, CH_3COCH_3)
The simplest KETONE, a fragrant, colorless liquid used in industry to make methyl methacrylate, solvent C and for organic syntheses. It is prepared industrially by dehydrogenation of 2-propanol (see ALCOHOLS) and as a co-product with phenol in the cumene-phenol process. Large amounts of acetone occur in diabetics (see DIABETES MELLITUS; ACIDOSIS). MW 58.08, mp −95°C, bp 56°C.
Acetylcholine (ACh)
An important NEUROTRANSMITTER.
Acetylene (ethyne, CHCH)
The simplest ALKYNE; a colorless, flammable gas prepared by reaction of water and calcium carbide (or acetylide; see CARBON); it is a very weak ACID. Acetylene may explode when under pressure, so is stored dissolved in acetone. It is used in the oxyacetylene torch for cutting and welding metals, in lamps, and also in the synthesis of ACRYLIC ACID and various fine chemicals. MW 26.04, subl −84°C.
Acetylsalicylic acid see ASPIRIN.
Achene
In biology, a dry indehiscent FRUIT with one seed, as in the buttercup.
Acheson, Edward Goodrich (1856-1931)
US inventor who discovered the powerful abrasive, CARBORUNDUM, and devised a method for producing high-quality GRAPHITE. Earlier he had assisted EDISON in developing the incandescent filament lamp.
Achromatic lens (achromat)
A compound LENS designed to minimize the effects of chromatic ABERRATION, which is due to the dispersion of light in REFRACTION. This is done by employing lens elements made of glasses of different dispersive powers, traditionally crown glass and flint glass. In practice a given lens can be correctly balanced only for a few wavelengths. (♦ page 237).
Acid
A substance capable of providing HYDROGEN ions (H^+) for chemical reaction. In an important class of chemical reactions (acid-base reactions) a hydrogen ION (identical to the physicist's PROTON) is transferred from an acid to a BASE, this being defined as any substance that can accept hydrogen ions. The strength of an acid is a function of the availability of its acid protons (see pH). Free hydrogen ions are available only in SOLUTION, where the minute proton is stabilized by association with a solvent molecule. In aqueous solution it exists as the hydronium ion (H_3O^+).

Chemists use several different definitions of acids and bases simultaneously. In the Lewis theory, an alternative to the Brönsted-Lowry theory outlined above, species that can accept ELECTRON pairs from bases are defined as acids.

Many chemical reactions are speeded up in acid solution, giving rise to important industrial applications (acid-base CATALYSIS). Mineral acids including SULFURIC ACID, NITRIC ACID and HYDROCHLORIC ACID find widespread use in industry. Organic acids, which occur widely in nature, tend to be weaker. CARBOXYLIC ACIDS (including ACETIC ACID and OXALIC ACID) contain the acidic group −COOH; aromatic systems with attached hydroxyl group (PHENOLS) are often also acidic. AMINO ACIDS, constituents of proteins, are essential components of all living systems. (♦ page 130)
Acid anhydrides
Class of organic compounds derived formally (but not usually in practice) from CARBOXYLIC ACID by elimination of water. They have the general formula RCOOCOR′, and are often prepared by reaction of an ACID CHLORIDE with a carboxylate. Chemically they resemble acid chlorides, but are less violently reactive. They are used in the FRIEDEL-CRAFTS and DIELS-ALDER reactions, and to make ESTERS.
Acid chlorides (acyl chlorides)
Class of organic compounds of general formula RCOCl; prepared by reacting CARBOXYLIC ACIDS with phosphorus (III or V) chloride or thionyl chloride ($SOCl_2$). They are volatile, fuming, pungent liquids, corrosive and very reactive. They react with ALCOHOLS to give ESTERS, with AMMONIA and AMINES to give AMIDES, and with water to give carboxylic acids. They are reduced to KETONES by GRIGNARD REAGENTS and in the FRIEDEL-CRAFTS REACTION. Other acid HALIDES are similar. (See also ALDEHYDES; ACID ANHYDRIDES.)
Acidosis
Medical condition in which the acid-base balance in the blood PLASMA is disturbed in the direction of excess acidity, the pH falling below 7.35. It may cause deep sighing breathing, drowsiness or coma.
Acid rain
Rainfall rendered acidic by SULFUR dioxide and NITROGEN oxides, which mix with atmospheric moisture and fall to Earth as ACIDS in rain, snow or fog. It is caused by nitrogen oxide from car exhausts and sulfur dioxide emitted by fossil-fueled power stations and large industry, and is partly due to tall chimney stacks, built to decrease local AIR POLLUTION. However, the longer sulfur dioxide is in the ATMOSPHERE, the more likely it is to be oxidized to sulfuric acid, to be deposited kilometers away, damaging buildings, acidifying lakes and destroying forests. Normal rain has a pH of 5.6; in parts of N America and Europe, the average pH is now 4-4.5.
Acne vulgaris
A common pustular SKIN disease of the face and upper trunk, most prominent in adolescence and occasionally in the menopause, as a result of hormonal changes. Blackheads become secondarily inflamed because of either local production of irritant FATTY ACIDS by BACTERIA or bacterial infection itself. In severe cases, with secondary infection and picking of spots, scarring may occur. Acne rosacea is an unsightly but harmless skin disease, giving a red-faced appearance.
Acorn worms
A group of marine invertebrates belonging to the phylum HEMICHORDATA.
Acoustics
The science of SOUND, dealing with its production, transmission and effects. Engineering acoustics deals with the design of sound-systems and their components, such as microphones, headphones and loudspeakers; musical acoustics is concerned with the construction of musical instruments, and ULTRASONICS studies sounds having frequencies too high for humans to hear them.

Architectural acoustics gives design principles of rooms and buildings having optimum acoustic properties. Anechoic chambers, used for testing acoustic equipment, are completely surfaced with diffusing and absorbing materials so that reverberation is eliminated. Noise insulation engineering is a further increasingly important branch of acoustics. (♦ page 225)
Acquired characteristics
Features that an organism acquires during its lifetime as a result of environmental influence, way of life, accidents, etc, rather than those determined by its genes. Acquired characteristics contribute to the organism's PHENOTYPE, but cannot be inherited (see CENTRAL DOGMA; LAMARCK).
Acquired Immune Deficiency Syndrome see AIDS.
Acromegaly (gigantism)
Abnormally large stature, starting in childhood. It may be caused by a constitutional trait or by HORMONE disorders during growth. The latter are usually excessive secretion of growth hormone or thyroid hormone before the EPIPHYSES have fused.
Acrophobia see PHOBIA.
Acrylic acid (propenoic acid, $CH_2{:}CHCO_2H$)
An unsaturated carboxylic acid used in the manufacture of acrylic resins. Its derivatives, acrylonitrile and methyl methacrylate, are also used widely to produce synthetic fibers and resins. MW 72.06, mp 13°C, bp 141.6°C.

ACTH (Adrenocorticotrophic hormone or corticotropin)
A HORMONE secreted by the PITUITARY GLAND that stimulates the secretion of various STEROID hormones from the cortex of the ADRENAL GLANDS.

Actinides
The 15 elements with atomic numbers (see ATOM) 89-103, beginning with ACTINIUM, analogous to the LANTHANUM SERIES, though rather more diverse in properties. The elements through uranium occur in nature; except actinium, they show higher valencies than +3. The TRANS-URANIUM ELEMENTS are primarily synthetic although traces of some of them may still occur naturally; the +3 valence state becomes progressively more stable, and higher valencies less stable. (See also PERIODIC TABLE.) (➧ page 130)

Actinium (Ac)
Radioactive TRANSITION ELEMENT in Group IIIB of the PERIODIC TABLE, resembling LANTHANUM; it occurs in URANIUM ores, and Ac^{227} (half-life 22yr) is formed by irradiation of RADIUM. It is the prototypical member of the ACTINIDES. AW 227, mp 1230°C, bp 3200°, sg 10. (➧ page 130)

Actinomycetes
A large group of filamentous bacteria, most of which are found in soil. They help to maintain soil fertility by their action in breaking down organic matter, and are valuable as a source of ANTIBIOTICS such as Streptomycin. A few of the actinomycetes cause disease.

Actinopterygii
Ray-finned fish, the major subclass of the OSTEICHTHYES, which have fins supported by bony rays; unlike the SARCOPTERYGII, they lack lobes in the bases of the paired fins, and most have a SWIMBLADDER instead of lungs. Primitive actinopterygians are included in the Infraclass Chondrosteri and have asymmetrical tails, heavy scales and an upper jawbone that is fused to the bones of the cheek. They were abundant during the Carboniferous period, 360-286 million years ago, but today are represented only by the sturgeons and paddlefish.

Holosteans (the Infraclass Holostei) are also known mainly from fossils, their only living representatives being the gars and the bowfin. The earliest holosteans replaced chondrosteans during the Triassic period, about 220 million years ago, and show characteristics intermediate between those of chondrosteans and teleosts (the Infraclass Teleostei). The latter group make up the vast majority of present-day fish. Teleosts have symmetrical tails, thin scales and an upper jawbone that is free from the cheek. They are adapted to a great variety of feeding habits and number over 20,000 species.

Activation energy
In a chemical reaction, the amount of additional energy required by the reactant molecules for them to reach the transition state through which products are formed. The transition state is always at a higher energy level than either reactants or products. CATALYSTS act by lowering the activation energy of a reaction.

Active galaxy see GALAXY.

Acupuncture
An ancient Chinese medical practice in which fine needles are inserted into the body at specified points, used for relieving pain and in treating a variety of conditions including RHEUMATISM. Although it is not yet understood how acupuncture works, it is still widely practiced in China and increasingly in the West, mainly as a form of ANESTHESIA, for which it is often accompanied by the passage of an electric current.

Acute
In diseases, intense but of short duration. Chronic diseases are usually less severe but often persist for the rest of the patient's life.

Adams, John Couch (1819-1892)
British astronomer who, independently of LEVERRIER, inferred the existence of the planet NEPTUNE from the PERTURBATIONS it induced in the orbit of URANUS.

Adaptation
The process by which a living organism becomes more suited to its environment. The word is most often used in an evolutionary sense to mean the very gradual adaptation of a species to its environment by means of random variation and NATURAL SELECTION. The products of this process – such as the thick fur of an Arctic mammal, or the tendrils of a climbing plant – are also commonly referred to as "adaptations".

"Adaptation" is also used to describe the process by which sensory organs adjust themselves to different light levels, sound intensities, etc, allowing them to operate with equal sensitivity over a range of different conditions.

Adaptive radiation
The evolution, from a single ancestor, of a number of descendants with a great variety of adaptations to different NICHES. The phenomenon is best seen on oceanic islands that very few species have been able to colonize. Those that do manage to reach the island encounter a great range of empty niches and tend to undergo rapid adaptive radiation until these are filled. Classic examples are the finches of the Galapagos Islands and the Hawaiian honeycreepers.

Addiction see DRUG ADDICTION.

Addison's disease
Failure of STEROID HORMONE production by the ADRENAL GLAND cortex, first described by English physician Thomas Addison (1793-1860). Its features include brownish skin pigmentation, loss of appetite, nausea and vomiting, weakness, and malaise.

Adenine see NUCLEIC ACIDS; NUCLEOTIDES.

Adenoids
Lymphoid tissue (see LYMPH) draining the nose, situated at the back of the throat. They are normally largest in the first five years and by adult life have undergone ATROPHY. Excessive size resulting from repeated nasal infection may lead to mouthbreathing, middle-ear diseases, sinusitis and chest infection. Surgical removal of the adenoids may be needed.

Adenosine triphosphate (ATP) see NUCLEOTIDES.

Adhesion
The force of attraction between contacting surfaces of unlike substances, such as glue and wood or water and glass. Adhesion is due to intermolecular forces of the same kind as those causing COHESION. Thus the force depends on the nature of the materials, temperature and the pressure between the surfaces. A liquid in contact with a solid surface will "wet" it if the adhesive force is greater than the cohesive force within the liquid. (See also ADHESIVES; SOLUTION.)

Adhesives
Substances that bond surfaces to each other by mechanical ADHESION (the adhesive filling the pores of the substrate) and in some cases by chemical reaction. Thermoplastic adhesives (including most animal and vegetable glues) set on cooling or evaporation of the solvent. Thermosetting adhesives (including the epoxy resins) set on heating or when mixed with a catalyst.

Adiabatic process
In THERMODYNAMICS, a change in a system without transfer of HEAT to or from the environment. An example of an adiabatic process is the vertical flow of air in the atmosphere; air expands and cools as it rises, and contracts and grows warmer as it descends. The generation of heat when a gas is rapidly compressed, as in a piston engine or SOUND waves, is approximately adiabatic.

Adipose tissue
Specialized fat-containing connective TISSUE, mainly lying under the skin and within the ABDOMEN, whose functions include insulation and the storage of fat, which is broken down to provide energy when required. In individuals its distribution varies with age, sex and OBESITY.

Adler, Alfred (1870-1937)
Austrian psychiatrist who broke away from SIGMUND FREUD to found his own psychoanalytic school, "individual psychology", which saw AGGRESSION as the basic drive. Adler emphasized the importance of feelings of inferiority in individual maladjustments to society.

Adolescence
In humans, the transitional period between childhood and adulthood. Physical changes such as the appearance of pubic hair, the onset of periods in girls and the breaking of a boy's voice, are caused mostly by the secretion of adult sex hormones. Adolescence is often also a time of emotional turmoil.

Adonis
Asteroid about one kilometer in diameter with a highly eccentric ORBIT. Its perihelion is within the orbit of Venus, its aphelion beyond that of Mars. It was discovered in 1936.

Adrenal glands (suprarenal glands)
Two ENDOCRINE GLANDS, one above each kidney. The inner portion (medulla) produces the hormones ADRENALINE and noradrenaline and is part of the autonomic NERVOUS SYSTEM. The outer portion (cortex) produces a number of STEROID HORMONES that control sexual development and function, glucose metabolism and electrolyte balance. Adrenal cortex damage causes ADDISON'S DISEASE. (➧ page 111)

Adrenaline (epinephrine)
A HORMONE secreted by the ADRENAL GLANDS, together with smaller quantities of noradrenaline. The nerve endings of the sympathetic NERVOUS SYSTEM also secrete both hormones, noradrenaline in greater quantities. These mediate the "fight or flight" response to stress situations: blood pressure is raised, smaller blood-vessels are constricted, heart rate is increased, METABOLISM is accelerated, and levels of blood glucose and FATTY ACIDS are raised. (➧ page 111)

Adrian, Edgar Douglas, 1st Baron Adrian of Cambridge (1889-1977)
English physiologist who shared the 1932 Nobel Prize for Physiology or Medicine with CHARLES SHERRINGTON for their work in elucidating the functioning of the neurons of the NERVOUS SYSTEM.

Adsorption
The ADHESION of molecules of a fluid (the adsorbate) to a solid surface (the adsorbent); the degree of adsorption depends on temperature, pressure and the surface area – porous solids such as CHARCOAL being especially suitable. The forces binding the adsorbate may be physical or chemical; chemical adsorption is specific, and is used to separate mixtures (see CHROMATOGRAPHY). Adsorption is used in gas masks and to purify and decolorize liquids. (See also ABSORPTION.)

Aerial photography see PHOTOGRAMMETRY.

Aerobe
An organism that needs oxygen for its survival. The term is usually applied to certain kinds of BACTERIA, as some bacteria are ANAEROBES. Among higher organisms almost all are aerobic apart from a few internal parasites. An obligate aerobe is one that cannot live without oxygen; a facultative aerobe is one that can also tolerate anaerobic conditions.

ANIMAL AND PLANT CELLS

▼ ***The cytoskeleton has two major components: microtubules and microfilaments. The latter produce cytoplasmic streaming – an important feature of many eukaryotes, especially amebae and fungi. These filaments are made of actin, one of the muscle proteins. Since myosin, the other muscle protein, is present in the cell, microfilaments may contract in a similar way.***

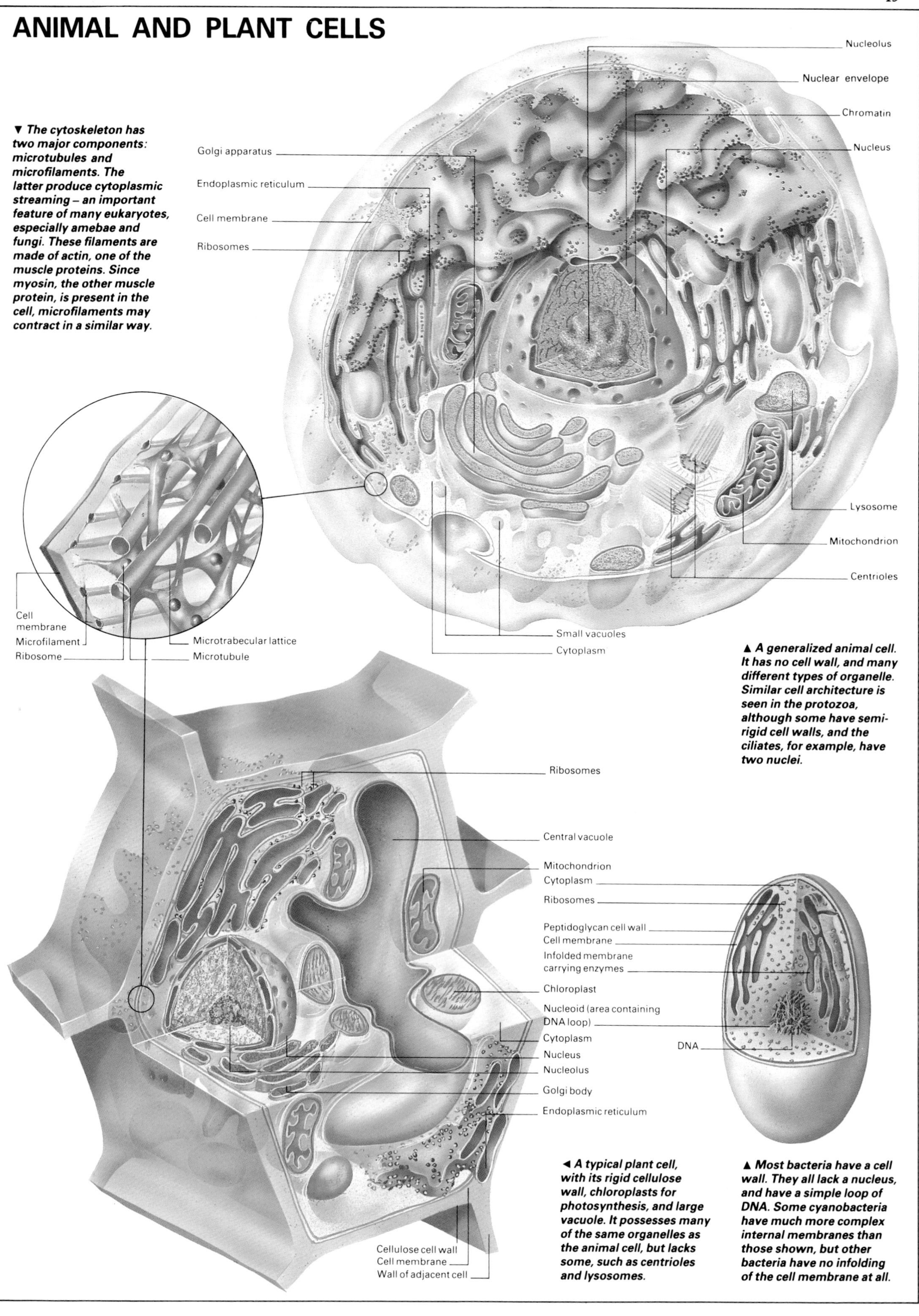

▲ ***A generalized animal cell. It has no cell wall, and many different types of organelle. Similar cell architecture is seen in the protozoa, although some have semi-rigid cell walls, and the ciliates, for example, have two nuclei.***

◀ ***A typical plant cell, with its rigid cellulose wall, chloroplasts for photosynthesis, and large vacuole. It possesses many of the same organelles as the animal cell, but lacks some, such as centrioles and lysosomes.***

▲ ***Most bacteria have a cell wall. They all lack a nucleus, and have a simple loop of DNA. Some cyanobacteria have much more complex internal membranes than those shown, but other bacteria have no infolding of the cell membrane at all.***

Aerodynamic efficiency
The EFFICIENCY with which an airfoil uses the AERODYNAMIC forces acting on it: in particular the ratio of lift to drag.
Aerodynamics
The branch of physics dealing with the flow of air or other gas around a body in motion relative to it. Aerodynamic forces depend on the body's size, shape and velocity; and on the density, compressibility, VISCOSITY, temperature and pressure of the gas. At low velocities, flow around the body is streamlined or laminar, and causes low drag; at higher velocities TURBULENCE occurs, with fluctuating eddies, and drag is much greater. "Streamlined" objects, such as airfoils, are designed to maintain laminar flow even at relatively high velocities. Pressure impulses radiate at the speed of SOUND ahead of a moving body; at SUPERSONIC velocities these impulses pile up, producing a shock wave – the "sonic boom" (see DOPPLER EFFECT). In airplane design all of these factors must be considered. (See also WIND TUNNEL; REYNOLDS NUMBER.)
Aeroembolism
Presence of air in the blood circulation. Direct entry of air into veins may occur through trauma, cannulation or surgery, and a large air embolus reaching the HEART may cause death. In acute decompression (as with sudden surfacing after deep diving) bubbles of air come out of solution. These may block small blood vessels causing severe muscle pains ("bends"), tingling and choking sensations and occasionally PARALYSIS or COMA. Recompression and slow decompression is the correct treatment.
Aerosol
A suspension of small liquid or solid particles (0.1-100 μm diameter) in a gas. Examples include smoke (solid particles in air), FOG and CLOUDS. Aerosol particles can remain in suspension for hours, or even indefinitely. Commercial aerosol sprays are widely used for insecticides, air fresheners, paints, cosmetics, etc. (See also COLLOID; ATMOSPHERE.)
Aestivation
In animals, a period of inactivity during the hottest or driest part of the year. It is similar to HIBERNATION, and generally involves a lowering of the metabolic rate to allow survival without food for a prolonged period. Lungfish, and some amphibians that inhabit semi-arid regions, show aestivation, entombing themselves in burrows lined with mucus to conserve moisture. The term aestivation is also used in botany, to refer to the way in which the parts of a flower are folded in the flower-bud; this is an important characteristic in plant classification.
Aetiology see ETIOLOGY.
Affective disorders
Psychiatric illnesses characterized by changes in "affect" or mood. The main conditions are DEPRESSION (both neurotic and psychotic) and MANIC-DEPRESSIVE PSYCHOSIS.
Afterbirth
The material, primarily the PLACENTA, expelled from the mother's body after child delivery.
Afterglow
The radiant coloration of the western sky following sunset. It arises from the SCATTERING of sunlight by dust particles (see COLLOID) in the upper atmosphere. The term is also used as an alternative name for phosphorescence (see LUMINESCENCE).
Afterimage
The illusory persistence of a visual image after the object that produced the image has gone. It may be positive (the same COLOR or shade as the original) or negative (the complementary color) depending on the background color.
Agar (agar-agar)
Gelatinous product prepared from the ALGAE *Gracilaria* and *Gelidium*. It dissolves in hot water, the solution gelling on cooling (see COLLOID). Its main use is as a thickener in nutrient media used for growing bacteria and fungi.
Agassiz, Jean Louis Rodolphe (1807-1873)
Swiss-American naturalist, geologist and educator, who first proposed (1840) that large areas of the northern continents had been covered by ice sheets (see ICE AGE) in the geologically recent past. He is also noted for his studies of fish. Becoming natural history professor at Harvard in 1848, he founded the Museum of Comparative Zoology there in 1859. On his death he was succeeded as its curator by his son, Alexander Agassiz (1835-1910).
Agassiz, Lake
A large prehistoric lake that covered parts of North Dakota, Minnesota, Manitoba, Ontario and Saskatchewan in the PLEISTOCENE epoch, named for LOUIS AGASSIZ. It was formed by the melting ice sheet as it retreated (see ICE AGE). When all the ice had melted, the lake drained northward, leaving fertile silt.
Agate
A gem stone, a variety of CHALCEDONY streaked with bands of color, formed by intermittent depositions of concentric SILICA on the walls of cavities in volcanic rock. It is often used in jewelry and ornamental work, usually after being dyed.
Agglomerate
Rock made up of angular fragments of lava in a matrix of smaller, often ashy, particles. It is a result of volcanic activity (see VOLCANO). (See also CONGLOMERATE.)
Agglutinins
ANTIBODIES found in BLOOD plasma that cause the agglutination (sticking together) of antigens such as foreign red blood cells and bacteria. Each agglutinin acts on a specific antigen, removing it from the blood. An agglutinin is produced in large quantities after immunization with its corresponding antigen. Agglutinins that agglutinate red blood cells are called isohemagglutinins, and the blood group of an individual is determined by which of these are present in his blood. Group O blood contains isohemagglutinins anti-A and anti-B; group A contains anti-B; group B contains anti-A, and group AB contains neither.
Aggregation
In physics, the clustering of particles into larger groups or aggregates such as those forming rigid bodies.
Aggression
Behavior adopted by animals, especially vertebrates, in the defense of their TERRITORY and in the establishment of social hierarchies. An animal's aggressive behavior is usually directed towards members of its own species, and is commonly ritualized, the combatants rarely inflicting serious wounds upon one another. Ritual fighting has become established by the evolution of a language of signs, such as the threat posture, by which animals make known their intentions. Equally important are submission or appeasement postures, which acknowledge defeat.
Aging
A process of deterioration that ends with the death of the organism. In biology, the process of aging is considered to begin at the very start of the organism's life – the moment when the egg is fertilized. The process may be caused by the gradual accumulation of mistakes in the genetic material (the DNA) at each cell division, or by other processes that interfere with the control of DNA expression.
Agnatha
A small group of vertebrates that are distinguished from the other vertebrates (the Gnathostomata) by their lack of jaws. They are also known as the jawless fish, and include two living groups, the lampreys and hagfish, as well as a number of fossil forms that were common during the Ordovician, Silurian and Devonian periods, 505-360 million years ago. The lampreys are eel-like, and feed on fish by attaching themselves to their prey and sucking their body fluids. Hagfish live by feeding on dead or dying fish and are, unlike the lampreys, exclusively marine. (➧ page 17)
Agonistic behavior
Behavior resulting from conflict between an animal's instinct to fight and its instinct to flee when it is confronted by a member of its own species at the border of its TERRITORY. Agonistic behavior often results in would-be combatants presenting their flanks to one another, thus avoiding both threat and submissive postures.
Agoraphobia see PHOBIA.
Agranulocytosis
A serious condition characterized by a marked decrease in BLOOD levels of granulocytes, part of the immune system (see IMMUNITY). It is a rare complication of treatment with certain drugs, or may follow a serious infection.
Agricola, Georgius (Georg Bauer, 1494-1555)
German physician and scholar, "the father of mineralogy". His pioneering studies in geology, metallurgy and mining feature in his *De natura fossilium* (1546) and *De re metallica* (1556).
Agronomy
The branch of agricultural science dealing with production of field crops and management of the SOIL. The agronomist studies crop diseases, selective BREEDING, crop rotation and climatic factors, tests the soil, investigates SOIL EROSION and designs LAND RECLAMATION and IRRIGATION schemes.
AIDS (Acquired Immune Deficiency Syndrome)
A condition caused by the HIV VIRUS, which kills T-cells, an essential component of the immune system (see IMMUNITY). This leads to immunosuppression and vulnerability to disease. Common manifestations of AIDS include infections such as *Pneumonocystis carnii* pneumonia and CRYPTOSPORIDOSIS and rare forms of cancer such as KAPOSI'S SARCOMA.
Air see ATMOSPHERE.
Airglow
A faint reddish or greenish light, similar to the AURORA, visible in night skies at low and middle latitudes. It is caused by the reforming of molecules split by the Sun's ultraviolet light.
Air mass see METEOROLOGY.
Air plant see EPIPHYTE.
Air pollution
The contamination of the atmosphere by harmful vapors, aerosols and dust particles, resulting principally from human activities but to a lesser extent from natural processes. Natural pollutants include pollen particles, salt-water spray, wind-blown dust and fine debris from volcanic eruptions. Most man-made pollution involves the products of COMBUSTION: smoke, which often includes dioxin and other carcinogens (from burning wood, coal and oil, and from refuse incineration); carbon monoxide, oxides of nitrogen, hydrocarbons and lead (from automobiles), and oxides of nitrogen and sulfur dioxide (mainly from burning coal). Other industrial processes, crop-spraying, the use of aerosols, industrial accidents and atmospheric nuclear explosions also contribute. Many pollutants undergo further chemical changes in the atmosphere; for example, oxides of nitrogen and hydrocarbons, when exposed to sunlight, produce ozone, which damages plants and precipitates asthma attacks in humans. Automobile emission control is a key area for

current research, exploring avenues such as lean-burn engines, the thorough OXIDATION of exhaust gases by catalytic converters, and the use of low-lead or lead-free GASOLINE. On the industrial front, flue-gas cleansing using catalytic conversion (see CATALYSIS) or centrifugal, water-spray or electrostatic precipitators is becoming increasingly widespread. The matching of smokestack design to local meteorological and topographic conditions is important for the efficient dispersal of remaining pollutants. Domestic pollution can be reduced by restricting the use of high-pollution fuels, as has been achieved in "smokeless zones". (See also ACID RAIN; POLLUTION).

Air pressure see ATMOSPHERE.

Air pump see PUMP.

Air sacs
Small respiratory cavities: in birds, leading off the lungs and entering the bones; in many insects, expansions in the TRACHEAE. (See also ALVEOLI.)

Airsickness see MOTION SICKNESS.

Air turbulence
Irregular eddying in the ATMOSPHERE, such as that encountered in gusts of wind. Turbulence disperses water vapor, dust, smoke and other pollutants through the atmosphere, and is important in transferring heat energy upward from the ground. There is little turbulence in the upper atmosphere except in developing thunderclouds.

Alabaster
Fine-grained, massive form of GYPSUM, usually translucent and white; used ornamentally for centuries, being easily carved.

Albedo
The ratio between the amount of light reflected from a surface and the amount of light incident upon it. The term is usually applied to celestial objects within the SOLAR SYSTEM: the Moon reflects about 7% of the sunlight falling upon it, and hence has an albedo of 0.07.

Albertus Magnus, St (1193-1280)
German scholastic philosopher and scientist; the teacher of St Thomas Aquinas. Albertus's main significance was in promoting the study of ARISTOTLE and in helping to establish Aristotelianism and the study of the natural sciences within Christian thought. In science he did important work in botany, and was possibly the first to isolate ARSENIC.

Albinism
Deficiency of pigment (usually MELANIN) in an organism. The skin and hair of albino animals (including humans) is white while the irises of their eyes appear pink. Albinism, which may be total or only partial, is generally inherited. Albino plants contain no CHLOROPHYLL and unless they are parasitic rapidly die, as they are unable to perform PHOTOSYNTHESIS.

Albite
Common mineral found in igneous rocks, consisting of sodium aluminum silicate ($NaAlSi_3O_8$); often forms vitreous crystals of various colors. It is one of the three end-members (pure compounds) of the FELDSPAR group.

Albumins
Group of PROTEINS soluble in water, present in both animals and plants. Ovalbumin is the chief protein in egg white. Serum albumin occurs in blood plasma; it controls osmotic pressure.

Alchemy
A blend of philosophy, mysticism and chemical technology, originating before the Christian era, seeking variously the conversion of base metals into gold, the prolongation of life and the secret of immortality. The late medieval period saw the discovery of NITRIC, SULFURIC and HYDROCHLORIC acids and ETHANOL (*aqua vitae*, the water of life) in the alchemists' pursuit of the "philosopher's stone" or elixir which would transmute base metals into gold. In the early 16th century PARACELSUS set alchemy on a new course, towards a chemical pharmacy (IATROCHEMISTRY). Having strong ties with ASTROLOGY, interest in alchemy, particularly in the Hermetic writings, has never quite died out.

Alcoholism
Compulsive drinking of alcohol in excess, one of the most serious problems in modern society. Many people drink for relaxation and can stop drinking without ill effects; alcoholics cannot give up drinking without great discomfort: they are dependent on alcohol, both physically and psychologically.

Alcohol is a DEPRESSANT that acts initially by reducing activity in the higher centers of the BRAIN, resulting in loss of inhibitions and a sense of freedom from responsibility and anxiety. This is the basis for initial psychological dependence. With further alcohol intake, thought and body control are impaired (see also INTOXICATION). Many of the diseases associated with alcoholism are in part due to MALNUTRITION and VITAMIN deficiency: CIRRHOSIS, NEURITIS and dementia. Drinking bouts alternating with abstinence are sometimes called DIPSOMANIA. Prolonged alcohol withdrawal leads to DELIRIUM TREMENS.

Alcohols
A class of organic compounds, of general formula ROH, containing a hydroxyl group bonded to a carbon atom. They are classified as monohydric, dihydric, etc., according to the number of hydroxyl groups; and as primary, secondary or tertiary according to the number of hydrogen atoms adjacent to the hydroxyl group. Alcohols occur widely in nature, and are used as solvents and antifreezes and in chemical manufacture. They are obtained by fermentation, oxidation or hydration of ALKENES from petroleum and natural gas, and by reduction of fats and oils. The alcohol of alcoholic drinks is ethyl alcohol or ETHANOL.

Aldebaran
Alpha Tauri, the 14th brightest star in the night sky and the brightest star in TAURUS. At a distance of 21pc, it has an absolute magnitude varying about −0.3 and an apparent magnitude varying about 0.85.

Aldehydes
Class of organic compounds of general formula RCHO, containing a carbonyl group (see also KETONES). They are highly reactive, and find many uses in industry in the preparation of solvents, dyes, resins and other compounds. Many aldehydes occur in nature and are often responsible for the flavor and scent of animals and plants. The simplest aldehydes are FORMALDEHYDE and ACETALDEHYDE. Aromatic aldehydes, such as BENZALDEHYDE and vanillin, are used in dyes and as perfumes and food flavorings. Aldehydes can be prepared by dehydrogenation or oxidation of primary ALCOHOLS, or by reduction of ACID CHLORIDES. Aldehydes may be reduced to primary alcohols, or oxidized to CARBOXYLIC ACIDS. They undergo addition reactions with BASES such as AMMONIA and CYANIDES; and condensation reactions with HYDRAZINE, alcohols, ACID ANHYDRIDES and other reactive compounds.

Alder, Kurt (1902-1958)
German organic chemist who shared the 1950 Nobel Prize for Chemistry with OTTO DIELS for demonstrating the usefulness of the diene synthesis (DIELS-ALDER REACTION) in forming ALICYCLIC COMPOUNDS.

Alembert, Jean Le Rond d' (1717-1783)
French philosopher, physicist and mathematician, a leading figure in the French Enlightenment and coeditor with DIDEROT of the renowned *Enyclopedia*. His early fame rested on his formulation of D'ALEMBERT'S PRINCIPLE in mechanics (1743). His other works treat calculus, music, philosophy and astronomy.

Alembic
An early type of still, popularly associated with the experiments of alchemists (see ALCHEMY); the term strictly refers only to a particular form of DISTILLATION head.

Alexandrite
Rare variety of the mineral CHRYSOBERYL, found in the Ural Mts. A valuable GEM, it has a brilliant luster, and appears dark green in daylight but red in artificial or transmitted light.

Alfvén, Hannes Olof Gösta (b. 1908)
Swedish physicist who shared the 1970 Nobel Prize for Physics with LOUIS NEEL for contributing to the development of PLASMA physics. Alfvén himself introduced the study of MAGNETOHYDRODYNAMICS.

Algae
A large and extremely diverse group of photosynthetic organisms. They are mostly aquatic, and range in size from microscopic single-celled organisms living on trees, in snow, ponds and the surface waters of oceans, to strands of seaweed several meters long. Some algae are free-floating, some grow attached to a substrate.

Algae are separated into several major divisions, primarily on the basis of pigmentation. Green algae (Division Chlorophyta) are found mostly in fresh water and may be single-celled (as in *Chlamydomonas*), form long filaments (like *Spirogyra*) or a flat, leaf-like layer of cells (like the sea lettuce, *Ulva kactyca*). Other green algae are more substantial, being built up of intermeshed filaments, or pseudoparenchyma. This is also found in the brown algae (Division Phaeophyta), which includes the familiar seaweeds, such as the wracks found on rocky shores. The largest, the kelps, can grow to enormous lengths. Red algae (Division Rhodophyta) are found mostly in warmer seas, and include forms with hard, chalky "skeletons". Various unicellular organisms are classed with the algae by some taxonomists, but not by others; they include the diatoms, desmids and dinoflagellates, important constituents of marine PLANKTON. *Euglena*, which is both photosynthetic and motile, was once included with the algae, but is now placed with the PROTOZOA. The group once known as blue-green algae are now classed as CYANOBACTERIA.

Algae in both marine and freshwater habitats are important as the basis of food chains (see ECOLOGY). Some of the larger algae are economically important; for example, the red algae *Porphyra* and *Chondrus crispus* are used as foodstuffs. *Gelidium*, another red alga, is a source of AGAR, and the kelps (such as the giant kelp *Macrocystis*) produce alginates, one use of which is in the manufacture of icecream. Other uses of algae are in medicine and as manure.

The algae are often included in the PLANT KINGDOM, but some taxonomists consider that their simple level of organization separates them from the plants, and they classify them in the Kingdom PROTOCTISTA. (➧ page 139)

Algol
Beta Persei, second brightest star in the constellation PERSEUS. It is a multiple star of at least three but probably four components, two of which form an eclipsing binary (see VARIABLE STAR) causing a 10h diminution of brightness every 69h. (➧ page 228)

Alicyclic compounds
Class of organic compounds in which carbon atoms are linked to form one or more rings. AROMATIC COMPOUNDS are excluded because of their special properties. In general, alicyclic compounds resemble analogous ALIPHATIC COM

ANIMAL EVOLUTION

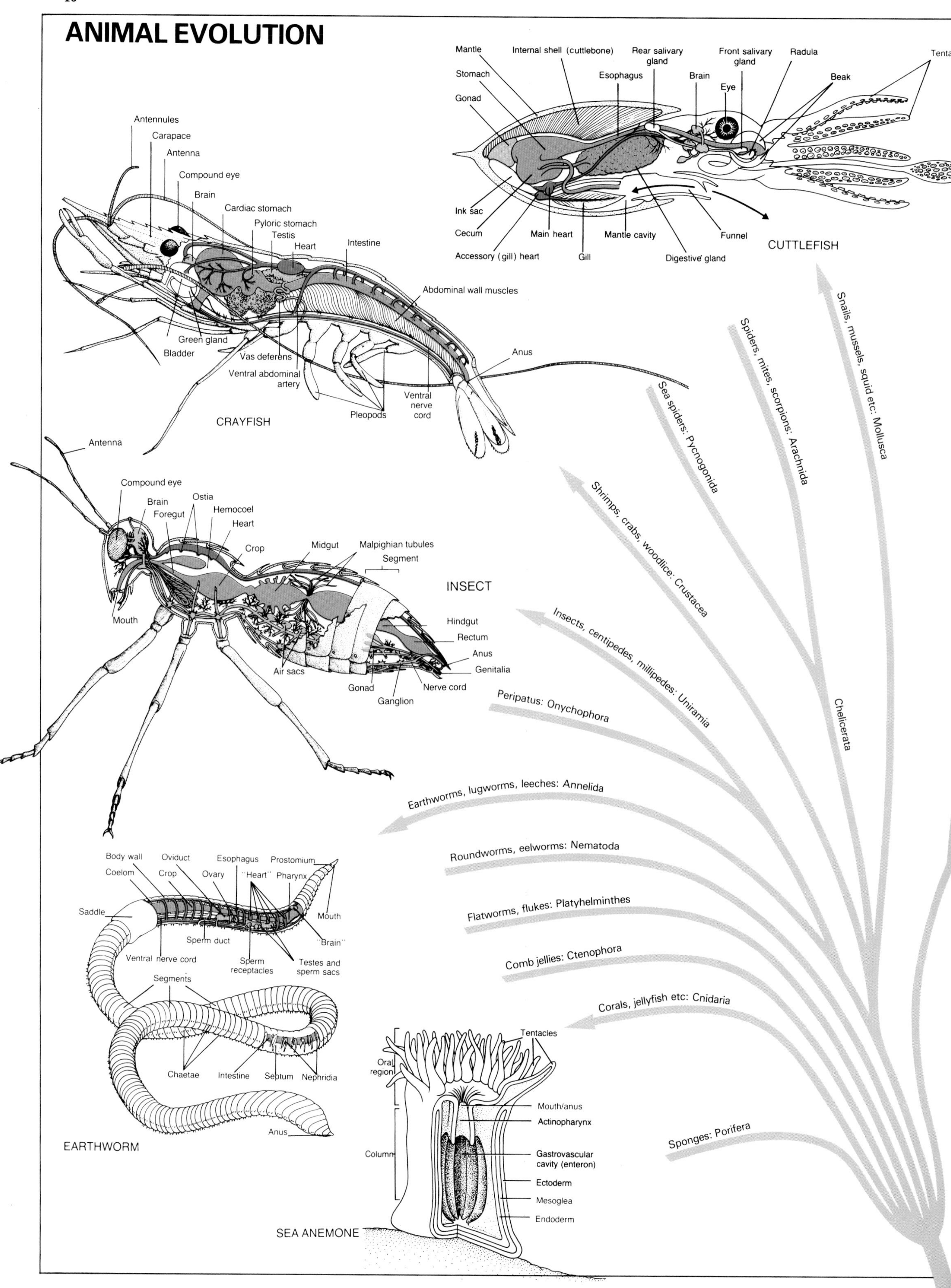

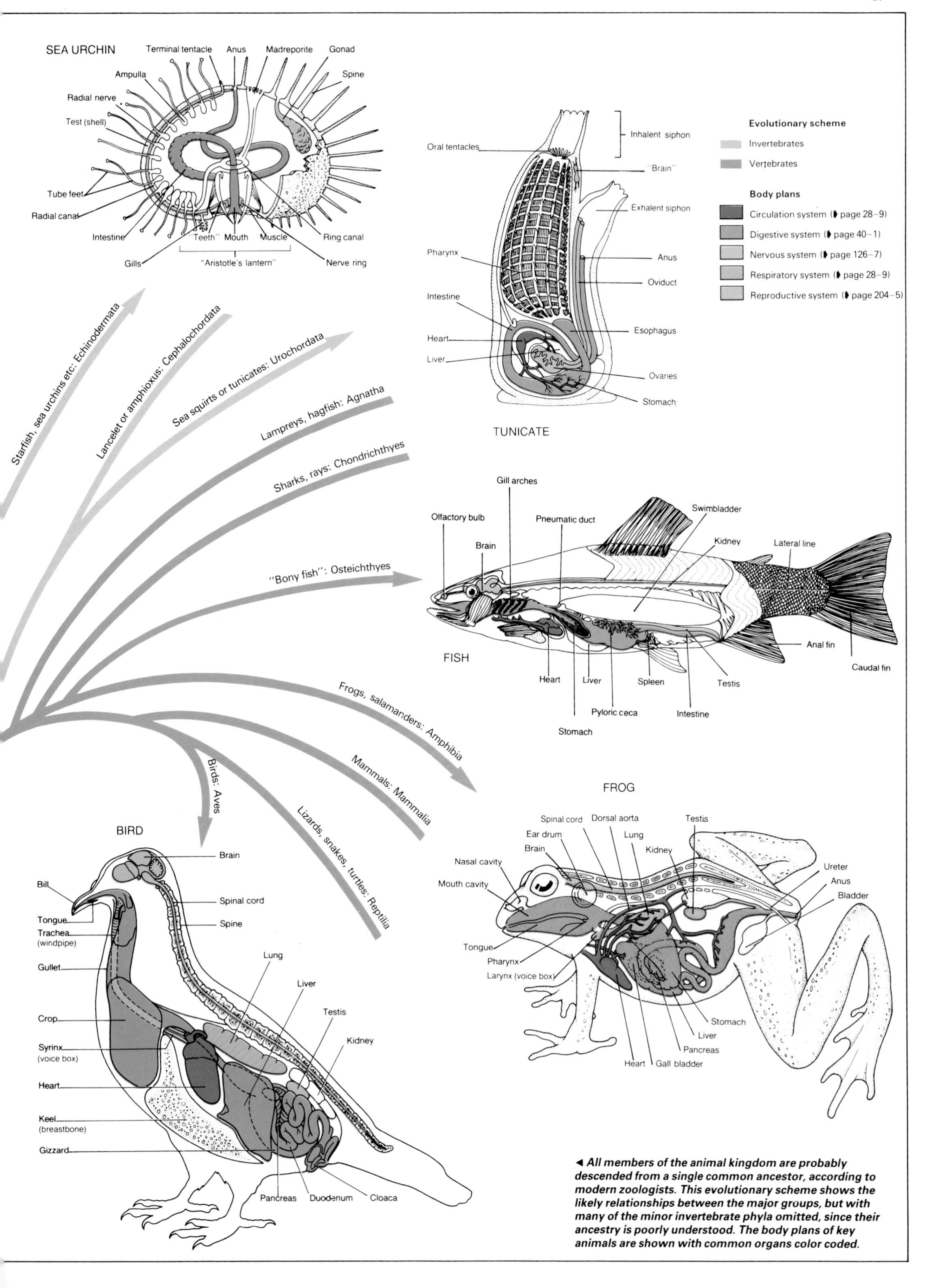

◀ *All members of the animal kingdom are probably descended from a single common ancestor, according to modern zoologists. This evolutionary scheme shows the likely relationships between the major groups, but with many of the minor invertebrate phyla omitted, since their ancestry is poorly understood. The body plans of key animals are shown with common organs color coded.*

POUNDS. However, strain occurs in small rings (with three, four or five members) because the angles between adjacent bonds are less than the preferred angle of 109° 28′, and these compounds are less stable and more reactive. Larger rings are nonplanar and unstrained. Many TERPENES, such as MENTHOL, are alicyclic. (See also HETEROCYCLIC COMPOUNDS.)

Alimentary canal see GASTROINTESTINAL TRACT.

Aliphatic compounds
Major class of organic compounds that includes all those with carbon atoms linked in straight or branched open chains. The other classes are ALICYCLIC, HETEROCYCLIC and AROMATIC compounds.

Alkali
A metal or ammonium salt which dissolves in water to give a solution with a pH>7 (see BASE). Alkalis neutralize acids to form salts and turn red litmus paper blue. Common alkalis are sodium hydroxide (NaOH), ammonium hydroxide (NH_4OH), sodium carbonate (Na_2CO_3) and potassium carbonate (K_2CO_3). They have important industrial applications in the manufacture of glass, soap, paper and textiles. Caustic alkalis are corrosive and can cause severe burns.

Alkali flats
Level, barren areas in dry regions covered with EVAPORITES, mainly salts of the ALKALI METALS and ALKALINE-EARTH METALS. Alkali flats are formed by the repeated periodic evaporation of shallow lakes lacking outlets.

Alkali metals
Highly reactive metals in Group IA of the PERIODIC TABLE, comprising LITHIUM, SODIUM, POTASSIUM, RUBIDIUM, CESIUM and FRANCIUM. They are soft and silvery white with low melting points. Alkali metals react with water to give off hydrogen, and so much heat is generated that spontaneous combustion may occur. Because of this extreme reactivity they never occur naturally as the metals, but are always found as monovalent ionic salts. (➧ page 130)

Alkaline-earth metals
Gray-white metals in Group IIA of the PERIODIC TABLE, comprising BERYLLIUM, MAGNESIUM, CALCIUM, STRONTIUM, BARIUM and RADIUM. They never occur naturally in an uncombined state, but are usually found as carbonates or sulfates. Except for beryllium, they are highly reactive and inflammable, readily dissolving in acids to form divalent ionic salts. The hydroxides of the four heaviest elements are alkalis. (➧ page 130)

Alkaloids
Narcotic poisons found in certain plants. They have complex molecular structures and are usually heterocyclic nitrogen-containing BASES. Many, such as coniine (from hemlock) or atropine (deadly nightshade), are extremely poisonous. Others, such as morphine, nicotine and cocaine, can be highly addictive, and some, such as mescaline, can cause hallucinations. But in small doses alkaloids are often powerful medicines, and are used as analgesics, tranquilizers, and cardiac and respiratory stimulants. Caffeine (found in coffee and tea) is a stimulant.

Alkalosis
Medical condition in which the blood PLASMA becomes excessively alkaline (the pH rises above 7.45) with resulting nausea, anorexia or TETANY.

Alkanes (paraffins)
The homologous series of saturated HYDROCARBONS of general formula C_nH_{2n+2}. The lowest members, which are gases, are METHANE, ETHANE, PROPANE and BUTANE; higher members are named for the number of carbon atoms in the molecule. From pentane (C_5H_{12}) to heptadecane ($C_{17}H_{36}$) they are liquids, and above that waxy solids. Alkanes with four or more carbon atoms have several ISOMERS, the straight-chain isomers being called normal alkanes (*n*-alkanes). Branched alkanes are named as derivatives of the longest straight chain in the molecule. Alkanes are obtained from petroleum and natural gas either directly or by catalytic reforming. They are soluble in most organic solvents, but not in water. Typical reactions include combustion in air, decomposition and rearrangement on heating, isomerization and condensation with alkenes (with acid catalyst), nitration, sulfonation, and halogenation by fluorine, chlorine and bromine (with heat or light). The monovalent radicals C_nH_{2n+1} derived from alkanes by loss of one hydrogen atom are called alkyl groups (methyl, ethyl, etc), usually denoted by the symbol R. (See also OCTANE.)

Alkenes (olefins)
The homologous series of unsaturated HYDROCARBONS having one or more double bonds (see BOND, CHEMICAL) between adjacent carbon atoms. The monoalkenes (one double bond) have general formula C_nH_{2n}; their systematic names are derived from those of the corresponding ALKANES by replacing the suffix -ane by -ene. They are prepared by thermal cracking of alkanes (from petroleum or natural gas), dehydration of ALCOHOLS, or base-catalyzed elimination of hydrogen halides from ALKYL HALIDES. Alkenes physically resemble the corresponding alkanes, but chemically their properties are due mainly to the double bond. Many reagents add across the double bond: hydrogen (with nickel or platinum catalyst), halogens, hydrogen halides and sulfuric acid. Alkenes are oxidized by permanganate or hypochlorite to GLYCOLS, and by OZONE to ozonides which readily decompose to ALDEHYDES and KETONES. Alkenes may readily be polymerized (by catalysts) to give plastics and resins. (See also BUTADIENE; ETHENE; ISOPRENE.)

Alkylation
The introduction of an alkyl group (see ALKANES) into a compound by substitution or addition, usually done by reacting the compound with an ALKENE or an ALKYL HALIDE. (See FRIEDEL-CRAFTS REACTION.) Specifically, in petroleum refining alkylation refers to the thermal or catalytic process in which branched alkanes are reacted with alkenes to yield highly branched products of high OCTANE rating.

Alkyl halides
Organic compounds consisting of an alkyl group (see ALKANES) bonded to a HALOGEN atom; polyhalogen derivatives of alkanes are similar. They are prepared by direct halogenation of alkanes (except for the iodides), addition of hydrogen halide to ALKENES, or halogenation of ALCOHOLS by hydrogen halides, phosphorus (III) halides, etc. Alkyl halides are used as solvents and as intermediates in chemical manufacture. The halogen atom is readily replaced by other NUCLEOPHILES such as hydroxide or cyanide. Elimination of hydrogen halides (base-catalyzed) yields alkenes.

Alkynes (acetylenes)
The homologous series of unsaturated HYDROCARBONS having one or more triple bonds (see BOND, CHEMICAL) between adjacent carbon atoms. The monoalkynes (one triple bond) have general formula C_nH_{2n-2}; their systematic names are derived from those of the corresponding ALKANES by replacing the suffix -ane by -yne. Except for ACETYLENE they are prepared by elimination of two hydrogen halide molecules from a dihaloalkane. Alkynes physically resemble the corresponding alkanes, but chemically their properties are due mainly to the triple bond, and are similar to those of the ALKENES. Addition reactions take place in two stages, forming first a substituted alkene and then a substituted alkane. The triple bond being nucleophilic (see NUCLEOPHILES), alkynes add to unsaturated compounds such as aldehydes or ketones. Alkynes readily polymerize to various products, including AROMATIC and ALICYCLIC compounds.

Allele
In genetics, an alternative form of a GENE found at a particular locus. For example, there may be a gene that determines eye color: one allele of that gene produces blue eyes, while another allele produces brown eyes. In diploid organisms, where there are two copies of every chromosome (other than the sex chromosomes), there will be two alleles of each gene in every organism. If these are identical the organism is said to be homozygous for that gene; if they are different the organism is said to be heterozygous. In a heterozygous organism, usually only one allele of the pair is expressed – this is described as the dominant allele (see DOMINANCE), while the other is the recessive. (➧ page 32)

Allergy
A hypersensitive state acquired through exposure to a particular material (allergen), reexposure bringing to light an altered capacity to react. Susceptibility is often inherited, but manifestations vary with age. Skin manifestations include ECZEMA and urticaria (see HIVES). In the nose and eyes HAY FEVER results, and in the GASTROINTESTINAL TRACT diarrhea may occur. In the LUNGS a specific effect leads to spasm of bronchi, which gives rise to ASTHMA. In most cases, the route of entry determines the site of the response; but skin rashes may occur regardless of route and asthma may follow eating allergenic material. If the allergen is injected, ANAPHYLAXIS may occur. Common allergens include drugs (penicillin, aspirin), foods (shellfish), plant pollens, animal furs or feathers, insect stings and the house dust mite.

Allopathy
A term sometimes used to distinguish orthodox medicine from homeopathy.

Allotropy
The occurrence of some elements in more than one form (known as allotropes) which differ in their crystalline or molecular structure. Allotropes may have strikingly different physical or chemical properties. Notable examples include DIAMOND and GRAPHITE, OXYGEN and OZONE, and SULFUR. (See also POLYMORPHISM.)

Alloy
A combination of metals with each other or with nonmetals such as carbon or phosphorus. They are useful because their properties can be adjusted as desired by varying the proportions of the constituents. Very few metals are used today in a pure state. Alloys are formed by mixing their molten components. The structures of alloys consisting mainly of one component may be substitutional or interstitial, depending on the relative sizes of the atoms. The study of alloy structures in general is complex. (See also PHASE EQUILIBRIA.)

The commonest alloys are the different forms of STEEL, which all contain a large proportion of iron and small amounts of carbon and other elements. BRASS and BRONZE, two well-known and ancient metals, are alloys of copper, while PEWTER is an alloy of tin and lead. The very light but strong alloys used in aircraft construction are frequently alloys of aluminum with magnesium, copper or silicon. Solders contain tin with lead and bismuth; type metal is an alloy of lead, tin and antimony. Among familiar alloys are those used in coins: modern "silver" coinage in most countries is an alloy of nickel and copper. Special alloys are used for such purposes as die-casting, dentistry, high-temperature use, and for making thermocouples, magnets and low-expansion materials. (See also AMALGAM; BABBITT METAL; GERMAN SILVER; INVAR; MONEL METAL.)

Allport, Gordon Willard (1897-1967)
US psychologist, important figure in the study of personality, who stressed the "functional autonomy of motives". Among his many works, *The Nature of Prejudice* (1954) has become a classic in its field.

Alluvium
Material such as GRAVEL, SILT and SAND deposited, mainly near their mouths, by streams and rivers. Alluvium makes rich agricultural soil, and the earliest civilizations originated as farming communities centered on alluvial flood plains.

Alopecia
Loss of hair, usually from the scalp, resulting from disease of hair follicles. Male pattern baldness is an inherited tendency, often starting in the twenties. Alopecia areata is a disease of unknown cause usually producing patchy baldness, though it may be total. It can be caused by certain drugs and poisons.

Alpha Centauri
Multiple star in the constellation CENTAURUS, comprising a DOUBLE STAR around which orbits at a distance of 10,000AU a red dwarf, Proxima Centauri, which is the nearest star to the Solar System, being 1.33 parsecs distant. (➧ page 228, 240)

Alpha particles
HELIUM nuclei ($_2He^4$) emitted at velocities of about 1.6Mm/s from radioactive materials undergoing alpha disintegration (see RADIOACTIVITY). Alpha particles, discovered by RUTHERFORD in 1899, carry a double positive charge and are strongly absorbed by air, thin paper and metal foils. (➧ page 147)

Altair
Brightest star in the constellation AQUILA and the twelfth brightest in the night sky (apparent magnitude +0.77). It has an extremely rapid rotation and is 5.1 parsecs from the Earth. (➧ page 228)

Alternating current (AC) see ELECTRICITY.

Alternation of generations
In plants, the alternation of two distinct types in the life-cycle. It is best seen in the mosses and liverworts (Bryophytes) and in the ferns, horsetails and clubmosses (Pteridophytes), but it also occurs in all higher plants, and in some of the seaweeds and freshwater algae. The basic cycle is the same in all plants. One stage, the sporophyte, has two sets of chromosomes (ie it is diploid) and produces spores. The spores germinate to give the second stage, known as the gametophyte. This has only one set of chromosomes (ie it is haploid), and produces eggs and sperm (gametes). Upon fertilization the egg gives rise to a new sporophyte plant. Meiosis, or reduction division, occurs during spore-formation, so the spores are haploid. Fertilization restores the diploid condition in the sporophyte.

In a few algae the sporophyte and gametophyte generations are identical in external appearance, but in other plants one stage is usually much larger and longer lived than the other. In algae this may be either the haploid or the diploid phase. In bryophytes it is the haploid stage that is dominant, this being the familiar moss or liverwort plant (thallus). The diploid sporophyte is represented by a stalked, spore-producing head growing from the thallus, and largely dependent on it for food. In pterodophytes the situation is reversed, the sporophyte being the familiar fern plant, while the gametophyte is a small, inconspicuous thallus that lives for less than a year. In higher plants – the conifers and their relatives (gymnosperms) and the flowering plants (angiosperms) – the gametophyte stage is reduced to just a few cells, and occurs within the tissues of the sporophyte – that is, within the cone or the flower. A parallel development has taken place within certain seaweeds, notably those of the genus *Fucus*, the kelps and wracks. Here the dominant phase is diploid, and the haploid gametophyte stage has been suppressed.

The term "alternation of generations" was at one time also applied to various animal life-cycles, such as those of the Cnidaria ("coelenterates"), which generally show two distinct phases.

Alternative medicine
Also known as complementary medicine; medical treatment not usually practiced by conventional doctors, eg acupuncture, homeopathy.

Altitude sickness
A condition caused by lack of OXYGEN in blood and tissues due to low atmospheric PRESSURE. Symptoms include breathlessness, headache, and faintness. At 5000m mental changes occur including indifference, euphoria and faulty judgment, but complete ACCLIMATIZATION is possible up to those heights. At very high altitude (6000m to 7000m), CYANOSIS, COMA and death rapidly supervene.

Altocumulus see CLOUDS.

Altostratus see CLOUDS.

Alum
A double salt comprising sulfates of a trivalent metal and a monovalent metal or ammonia combined with 12 molecules of water of crystallization:

$$M^IM^{III}(SO_4)_2.12H_2O.$$

The monovalent CATION is commonly potassium, sodium or ammonium; the trivalent cation may be aluminum, chromium or ferric iron. Alums are soluble in water and are usually acid. They are used as astringents (styptic pencils), as a mordant in dyes, and in the manufacture of baking powder, antiperspirants and fire extinguishers. Potash alum (potassium aluminum sulfate, $KAl(SO_4)_2.12H_2O$) is used in the sizing of paper and in water purification.

Alumina (aluminum oxide) see ALUMINUM.

Aluminum (Al)
Silvery white metal in Group IIIA of the PERIODIC TABLE, the most abundant metal, comprising 8% of the Earth's crust. It occurs naturally as BAUXITE, CRYOLITE, FELDSPAR, clay and many other minerals, and is smelted by the Hall-Héroult process, chiefly in the US, USSR and Canada. It is a reactive metal, but in air is covered with a protective layer of the oxide. Aluminum is light and strong when alloyed, so that aluminum ALLOYS are used very widely in the construction of machinery and domestic appliances. It is also a good conductor of electricity and is often used in overhead transmission cables where lightness is crucial. AW 27.0, mp 660.5°C, bp 2520°C, sg 2.6989 (20°C). (➧ page 130)

Aluminum compounds are trivalent and mainly cationic (see CATION), though with strong bases aluminates are formed. (See also ALUM.) Aluminum oxide (Al_2O_3), or alumina, is a colorless or white solid occurring in several crystalline forms, and is found naturally as CORUNDUM, EMERY and BAUXITE. Solubility in acid and alkali increases with hydration. mp 2045°C, bp 2980°C. Aluminum chloride ($AlCl_3$) is a colorless crystalline solid, used as a catalyst (see FRIEDEL-CRAFTS REACTION). The hexahydrate is used in deodorants and as an astringent.

Alvarez, Luis Walter (b. 1911)
US physicist awarded the 1968 Nobel Prize for Physics for work on SUBATOMIC PARTICLES, including the discovery of the transient resonance particles. He helped develop much of the hardware of NUCLEAR PHYSICS.

Alveoli
In air-breathing vertebrates, the numerous tiny globular sacs that make up the lung; they are the end-point of the many-branched tubes (the bronchi and bronchioles) through which air passes into the lung. Through the thin walls of the alveoli, oxygen passes into the blood from the air, and carbon dioxide passes into the lung from the blood. The term alveolus is also used more generally for any cavity in a living body, for example the socket in the jaw bone that holds the teeth. (➧ page 28)

Alzheimer's disease
A rapidly progressing form of DEMENTIA occurring in old age, first described by the German neurologist Alois Alzheimer (1864-1915).

AM (amplitude modulation) see RADIO.

Amalgam
An ALLOY of MERCURY with other metals. Most metals except iron will form amalgams; those with high mercury content are liquid, but most are solid. Amalgams of some NOBLE METALS occur naturally: SILVER and GOLD are extracted from their ores by forming amalgams. Dental amalgam, containing silver, copper, zinc and tin, is used to fill teeth. Various amalgams may be used as ELECTRODES. (See also MIRROR.)

Amber
Fossilized RESIN from prehistoric EVERGREENS. Brownish, it is highly valued and can be easily cut and polished for ornamental purposes. Its chief importance is that FOSSIL insects up to 20 million years old have been found embedded in it. The main source of amber is along the shores of the Baltic Sea.

Ambergris
Waxy solid formed in the intestines of sperm whales, perhaps to protect them from the bony parts of their squid diet. When obtained from dead whales it is soft, black and evil-smelling, but on weathering (as when found as flotsam) it becomes hard, gray and fragrant. Ambergris is used as a perfume fixative and, in the East, as a spice.

Ambidexterity see HANDEDNESS.

Amebae
A large order (Amoebida) of the Class Sarcodina (Rhizopoda) of PROTOZOA. They are unicellular with one or more nuclei. They move by extending pseudopodia, into which they flow; and feed by surrounding and absorbing organic particles (see AMEBOID MOVEMENT, PHAGOCYTOSIS). Reproduction is almost always asexual, generally by binary FISSION, though sometimes by multiple fission of the nucleus: a tough wall of cytoplasm forms about each of these small nuclei to create cysts, which can survive considerable rigors, returning to normal ameboid form when conditions improve. Certain amebae can reproduce sexually. Amebae are found wherever there is moisture, some parasitic forms living within other animals: *Entamoeba histolytica*, for example, causes amebic DYSENTERY in humans. The best-known of the group is the giant ameba, *Amoeba proteus*, which can achieve a diameter of 1mm. Many of the smaller amebae have shells, either secreted or built up from sand grains. (➧ page 13)

Amebiasis
A severe form of DYSENTERY caused by a PROTOZOAN, *Entamoeba histolyca*.

Ameboid movement
A form of movement seen in some single-celled organisms, and in some free-moving cells of multicellular organisms, such as the white blood cells of the human immune system. It is typical of the AMEBAE, a group of protozoans. They move by extending the cell in one direction, forming a

"false foot" or pseudopodium, then expanding into the pseudopodium while contracting the cell at the rear. This movement is achieved by the outer layer of the cell's protoplasm, the jelly-like ectoplasm, contracting or expanding as required. The inner part of the protoplasm, known as the endoplasm, is more liquid, and flows into the pseudopodium in response to the contractions of the ectoplasm. These contractions are brought about by the protein molecules actin and mysosin, which also produce contraction in muscles. The movement is sustained by the continuous conversion of endoplasm to ectoplasm at the forward-moving point of the cell, and the conversion of ectoplasm to endoplasm at the rear. Most cells that show ameboid movement also exhibit PHAGOCYTOSIS.

Amenhorrhea
Absence or abnormal stoppage of menstrual flow. Primary amenhorrhea is the failure of menstruation to begin at puberty. Secondary amenhorrhea is the cessation of established periods, and is usually the result of pregnancy or breastfeeding. It sometimes also results from frequent strenuous exercise, MALNUTRITION or ANOREXIA NERVOSA.

American Association for the Advancement of Science (AAS)
The largest US organization for the promotion of scientific understanding. Founded in Boston in 1848 but now centered in Washington, it has over 100,000 individual and 300 corporate members. Its publications include the weekly *Science* and it publishes a number of technical and popular periodicals.

American Philosophical Society
The oldest surviving US learned society, based in Philadelphia where it was founded by BENJAMIN FRANKLIN in 1743. The US counterpart of the ROYAL SOCIETY OF LONDON (1660), it currently has approaching 600 US and foreign members. It has an extensive library, much relating to early American science. Its own regular publications were initiated in 1769 with its *Transactions*.

Americium (Am)
Silvery white radioactive TRANSURANIUM ELEMENT, one of the ACTINIDES. It is prepared by NEUTRON irradiation of PLUTONIUM. Am^{241}, the most readily available ISOTOPE (half-life 458yr) emits GAMMA RAYS and is used in industrial density and thickness gauges. (➧ page 130)

Amethyst
Transparent violet or purple variety of QUARTZ, colored by iron or manganese impurities. The color changes to yellow on heating. Amethysts are semiprecious GEMS. The best come from Brazil, Uruguay, Arizona and the USSR.

Amia (bowfin)
A North American fish that is a survivor of the fossil group Holostei. (See also SWIMBLADDER.)

Amides
Class of ALIPHATIC COMPOUNDS, of general formula $RCONH_2$, derived from CARBOXYLIC ACIDS and AMMONIA by replacing the acid hydroxyl group by the amino group (NH_2). N-substituted amides are derived from primary or secondary AMINES instead of ammonia. Other amides are derived analogously from inorganic OXYACIDS or from SULFONIC ACIDS (see also SULFA DRUGS). Amides are prepared by reaction of ACID CHLORIDES, ACID ANHYDRIDES or ESTERS with AMMONIA or AMINES, or by partial HYDROLYSIS of NITRILES. Most simple amides are low-melting solids with strong HYDROGEN BONDING, soluble in water. Formamide and N-substituted amides are liquids widely used as solvents. Amides are both weak ACIDS and weak BASES. They may be hydrolyzed to form CARBOXYLIC ACIDS, and dehydrated to form NITRILES. Metallic HYDRIDES convert them to AMINES, and treatment with bromine and sodium hydroxide (the Hofmann degradation) yields amines with one fewer carbon atom. Polymeric amides (such as NYLON) are used as SYNTHETIC FIBERS, and similar amide linkages join AMINO ACIDS in PROTEINS and PEPTIDES. (See also IMIDES; UREA.)

Amines
Class of organic compounds derived from AMMONIA by replacing one or more hydrogen atoms by alkyl groups (see ALKANES) or aryl groups (see AROMATIC COMPOUNDS). Primary amines have general formula RNH_2; secondary R_2NH; and tertiary R_3N. HETEROCYCLIC nitrogen bases (including the ALKALOIDS and PYRIDINE) are tertiary amines. Amines may be formed by reduction of AMIDES, NITRILES or nitro compounds, or by reaction of ammonia with organic HALIDES, ALCOHOLS or SULFONIC ACIDS. Simple amines are pungent liquids that are strong BASES and LIGANDS; many occur naturally in decaying organic matter. They give AMIDES with acid derivatives. Amines have many uses, including the manufacture of dyes, drugs and synthetic fibers. (See also AMINO ACIDS; ANILINE.)

Amino acids
An important class of organic compounds containing a carboxyl group ($-COOH$), one or more amino ($-NH_2$) groups (see AMINES) and a side-chain (R). Twenty or so α-amino acids ($RCH[NH_2]COOH$) are the building blocks of the PROTEINS found in all living matter. When purified chemically, amino acids are white, crystalline solids, soluble in water; they can act as ACIDS or BASES depending on the chemical environment (see pH). In neutral solution they exist as ZWITTERIONS. An amino acid mixture may be analyzed by CHROMATOGRAPHY. All α-amino acids contain at least one asymmetric carbon atom, to which are attached the carboxyl group, the amino group, a hydrogen atom and the side-chain, which differs for each amino acid and determines its character. Because of this asymmetry, amino acids can exist in two mirror-image forms (see STEREOISOMERS). Generally only L-isomers occur in nature, but a few bacteria contain D-isomers. Humans synthesize most of the amino acids needed for nutrition, but depend on protein foods for eight "essential amino acids" that they cannot produce. As each amino acid contains both an acid and an amino group, they can form a long chain of amino acids bridged by PEPTIDE BONDS and called PEPTIDES or polypeptides. Peptide synthesis from constituent amino acids is a stage in PROTEIN SYNTHESIS. Proteins may be broken down again into their constituent amino acids, as in digestion. When amino acids are deaminated (the amino group removed), the nitrogen passes out as UREA, uric acid or ammonia gas. The remainder of the molecule enters the CITRIC ACID CYCLE, being broken down to provide energy. Scientists have produced amino acids and simple peptide chains by combining carbon dioxide, ammonia and water vapor under the sort of conditions (including electric discharges) thought to exist on Earth millions of years ago. This may provide a clue to the origin of LIFE. (➧ page 40)

Ammeter
An instrument used to measure electric currents greater than 1 μA. Most direct-current ammeters are similar in design to the moving-coil GALVANOMETERS used for smaller currents, though they differ in passing most of the test current through a low "shunt" RESISTANCE (thus bypassing the coil) and in using a pointer fixed to the coil assembly to indicate the reading on the linearly calibrated scale. For alternating currents either a rectifier can be used with a moving-coil instrument or the less sensitive hot-wire or moving-iron instruments can be used. (See also ELECTRICITY; VOLTMETER.)

Ammonia (NH_3)
Colorless acrid gas, made by the HABER PROCESS; a covalent HYDRIDE. The pyramidal molecule turns inside out very rapidly, which is the basis of the ammonia clock (see ATOMIC CLOCK). Ammonia's properties have typical anomalies due to HYDROGEN BONDING; liquid ammonia is a good solvent. Ammonia is a BASE; its aqueous solution contains ammonium hydroxide, and is used as a household cleaning fluid. It forms ammine (NH_3) LIGAND complexes with transition metal ions, and yields AMIDES and AMINES with many organic compounds. Ammonia is used as a fertilizer, a refrigerant, and to make ammonium salts, UREA, and many drugs, dyes and plastics. mp −78°C, bp −33°C.

On reaction with acids, ammonia gives ammonium salts, containing the NH_4^+ ion, which resemble ALKALI METAL salts. They are mainly used as fertilizers. The analogous quaternary ammonium salts, NR_4^+, are made by alkylation of tertiary AMINES and are used as ANTISEPTICS. Ammonium chloride (NH_4Cl), or sal ammoniac, a colorless crystalline solid used in dry cells and as a flux, is formed as a byproduct in the SOLVAY PROCESS. subl 340°C. Ammonium nitrate (NH_4NO_3), a colorless crystalline solid, is used as a fertilizer and in explosives. mp 170°C. (See also HYDRAZINE.) (➧ page 37, 233)

Ammonites
Extinct order of mollusks (Class: CEPHALOPODA), extant between 200 and 70 million years ago. Typically spiral-shelled, of diameter 0.01-2m, (though helical – see HELIX – shells have been found), they evolved rapidly and their FOSSILS are thus of use in dating geological strata.

Amnesia
The total loss of MEMORY for a period of time or for events. In cases of CONCUSSION, retrograde amnesia is the permanent loss of memory for events just preceding a head injury, while post-traumatic amnesia applies to a period after injury during which the patient may be conscious but incapable of recall, both at the time and later.

Amniocentesis
Diagnostic test in pregnancy in which cells from the AMNION are removed by BIOPSY and examined for CHROMOSOME abnormalitites.

Amnion
A tough membrane surrounding the embryo of reptiles, birds and mammals and containing the AMNIOTIC FLUID. All land-laid eggs contain amnions and an outer shell; those of fish and amphibians do not, and therefore must be laid either in moist surroundings or in water. (See also AMNIOTES.)

Amniotes
A subgroup of the VERTEBRATES, consisting of the mammals, reptiles and birds, characterized by the development of an AMNION to protect the embryo.

Amniotic fluid
The fluid contained within the AMNION, which provides a moist, aquatic environment for the embryo. In reptiles and birds it is within the egg. In mammals the amnion encloses the embryo within the uterus. The "breaking of the waters" shortly before giving birth is the release of the amniotic fluid

Amoebae see AMEBAE.

Ampere (A)
The SI base unit of electric current, named for A.M. AMPERE and defined as the constant current that, if maintained in two straight parallel conductors of infinite length, of negligible circular cross-section, and placed 1m apart in vacuum, would produce between these conductors a force equal to 2×10^{-7} newton per meter of length. (See also ELECTRICITY.)

ATOMIC SPECTRA

Photoelectric effect

► *In the photoelectric effect, light of certain wavelengths, notably in the ultraviolet, knocks electrons out of a material. The variation according to wavelength can be explained only if light consists of photons, localized wave "packets", as opposed to the continuous waves of classical electromagnetic theory. Changing the intensity of the light, however, changes the energy of the photons and therefore the energy of the electrons emitted.*

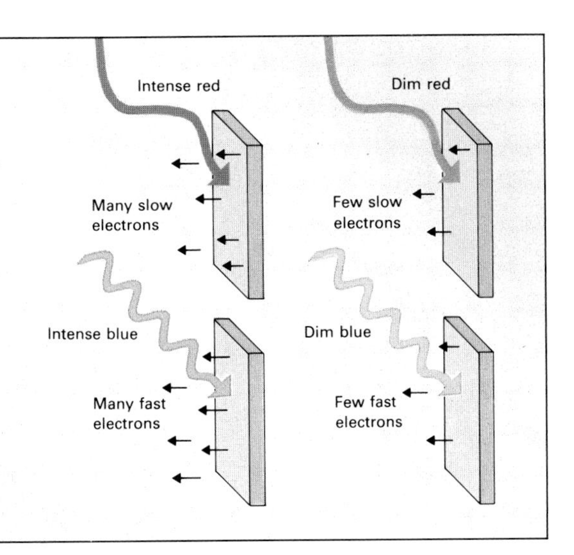

Absorption

Incoming photon
Electron knocked into higher orbit
Ground orbit
Outer orbit

◄ *Just as a material has its unique emission spectrum, so it also has a characteristic absorption spectrum, or specific wavelengths at which light is absorbed by its atoms. These wavelengths correspond to the amount of energy required to "knock" the electrons into a higher, or more energetic, orbit. Only a few specific orbits are possible: intermediate positions are never found.*

Energy levels of an atom

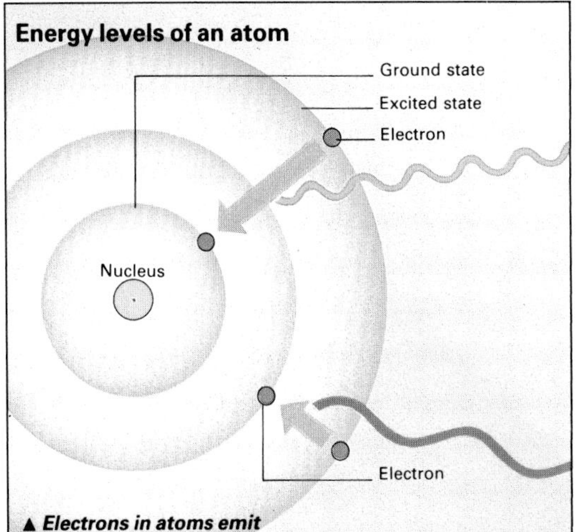

▲ *Electrons in atoms emit light when they change energy, often after being excited to higher energy levels than usual, perhaps by heating. With each jump from a higher energy level to a lower one, an electron emits a single photon. The energy (frequency) of the photon depends on the size of the jump: the bigger the jump, the greater the energy of the emitted photon, and the higher its frequency.*

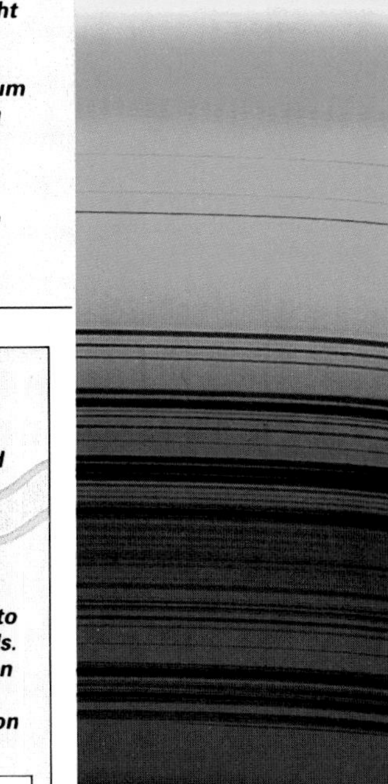

► *The frequencies of light emitted or absorbed by atoms provide their "fingerprints". A spectrum of the light from the Sun shows dark lines where photons of particular frequencies have been absorbed as electrons in atoms jump to higher energy levels.*

Lasers

► *The acronym laser stands for "light amplification by stimulated emission of radiation". One photon stimulates emission of another of identical wavelength, and in phase with the first. These photons are reflected back and forth, stimulating more photons until a powerful beam of "coherent" light has been formed.*

Stimulated emission

Incoming photon
Second photon emitted in phase with incoming photon
Electron knocked into lower orbit

◄ *When an atom absorbs light, electrons are excited by photons of the appropriate energy. They can be stimulated to fall back to the ground level orbit by a photon of frequency corresponding to the change in energy levels. This is stimulated emission – the emitted photon and the original incident photon move away in step.*

Gas laser

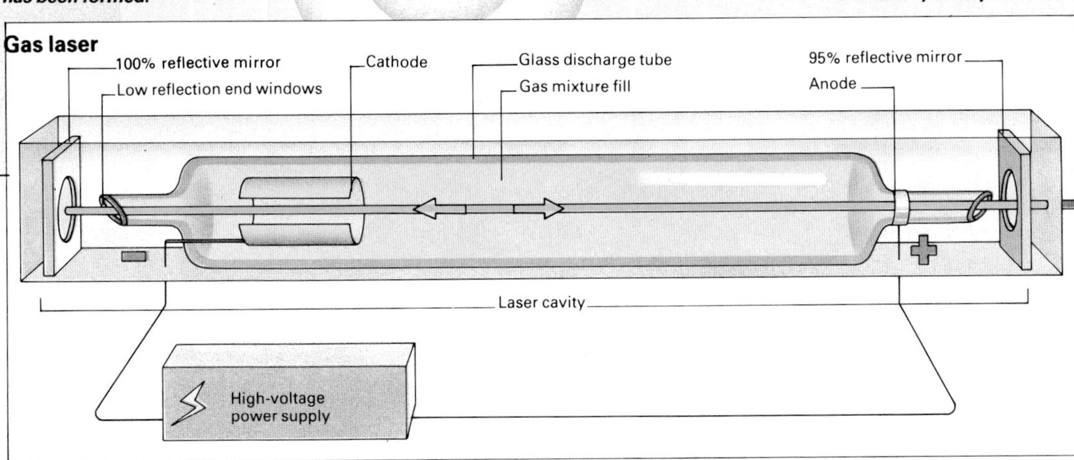

◄ *In a typical gas laser, a high voltage excites the atoms. On descending to its original condition a gas atom releases a photon which collides with nearby atoms, inducing them to emit also. Photons reflect back and forth, and laser light is emitted.*

Ampère, André Marie (1775-1836)
French mathematician, physicist and philosopher best remembered for many discoveries in electrodynamics and electromagnetism. In the early 1820s he developed OERSTED's experiments on the interaction between magnets and electric currents and investigated the forces set up between current-carrying conductors.

Amphetamines
A group of STIMULANT drugs, including benzedrine and methedrine, now in medical disfavor following widespread abuse and addiction. They counteract fatigue, suppress appetite, speed up performance (hence "Speed") and give confidence, but pronounced DEPRESSION often follows; thus psychological and then physical addiction are encouraged. A paranoid PSYCHOSIS (resembling SCHIZOPHRENIA) may result from prolonged use. While no longer acceptable in treatment of OBESITY, they are useful in narcolepsy, a rare condition of abnormal sleepiness.

Amphibians
A group of vertebrates that includes two fossil groups, the Lepospondyla and Labyrinthodontia and three living groups, the Anura (frogs and toads), Urodela (newts and salamanders) and the Apoda (caecilians). Amphibians typically spend part of their life in water and part on land. Many living members have soft, moist skin through which they breathe, but also have gills and/or lungs. The group is composed of about 3000 species and is widespread throughout the world. Amphibians of temperate regions commonly hibernate because they are POIKILOTHERMS and become sluggish at low temperatures. (♦ page 17)

Amphiboles
A class of SILICATE minerals found in IGNEOUS ROCKS and metamorphic SCHIST and GNEISS. They contain infinite double-chains of SiO_4 tetrahedra, and have a cleavage of about 56°. Amphiboles include HORNBLENDE, JADE and certain ASBESTOS minerals.

Amphioxus
A small, fish-like invertebrate, also known as a lancelet. (See CEPHALOCHORDATES.) (♦ page 17)

Amphipods
One of the largest orders of CRUSTACEANS, with over 3600 species, found in both fresh and salt water. Typical amphipods include freshwater shrimps and sandhoppers.

Amphisbaenids
A family of burrowing reptiles, also known as worm lizards. Most have no limbs, but some species have tiny vestigial forelimbs. All have hard, chisel-like snouts with which they burrow through the soil. They feed on insects, worms and small vertebrates. The amphisbaenids are not true lizards, though they are distantly related to the lizards and snakes.

Amplitude
In WAVE MOTIONS, the maximum displacement from its MEAN value of the oscillating property; thus for a PENDULUM half the extent of its swing. Amplitude modulation (AM) is a common method of encoding a carrier wave in RADIO. (♦ page 118, 225)

Amputation
The surgical or traumatic removal of a part or the whole of a limb or other structure. It is surgically necessary for severe limb damage, GANGRENE, loss of BLOOD supply and certain types of CANCER.

Amyl compounds
Organic compounds containing the amyl group C_5H_{11} (see ALKANES), which has eight ISOMERS. They include the amyl ALCOHOLS, synthesized from HYDROCARBONS or extracted from FUSEL OIL, and used as a solvent. Isoamyl nitrite ($C_5H_{11}ONO$) is a vasodilator used to give relief in ANGINA PECTORIS. Amyl acetate (which has a pleasant, fruity smell and is known as banana oil) is used as a solvent for NITROCELLULOSE and as a flavoring for candy.

Anabolism
Biochemical processes whereby larger molecules are synthesized from smaller ones, and the body's tissues built up. In CATABOLISM, by contrast, molecules are broken down to yield energy or to be excreted as waste. Together, anabolism and catabolism make up METABOLISM. Anabolic steroids are hormones that promote the buildup of body tissue. (See also ATP.)

Anaerobe
Any organism that can live without oxygen. Many BACTERIA and PARASITES are facultative anaerobes (that is, they can survive without oxygen for short or long periods), and a few, notably the ARCHEBACTERIA, are obligate anaerobes (unable to use oxygen in respiration). Most obligate anaerobes are killed by exposure to oxygen.

In breaking down food molecules to obtain energy, anaerobes rely only on GLYCOLYSIS, to produce ATP. Thus they do not break down sugars complexly to carbon dioxide and water as AEROBES do in aerobic RESPIRATION, which uses oxygen as an electron acceptor. Instead, anaerobes break sugars down to a 3-carbon molecule, pyruvate, which they then convert to lactic acid, ethanol or some other product (see FERMENTATION).

Anaesthesia see ANESTHESIA.

Analgesics
Drugs used to relieve pain. Aspirin and paracetamol are mild but effective. Piroxicam, indomethacin and ibuprofen are, like aspirin, useful in treating ARTHRITIS by reducing INFLAMMATION as well as relieving pain. NARCOTIC analgesics derived from OPIUM ALKALOIDS range from the milder codeine and dextropropoxyphene, suitable for general use, to the highly effective and addictive MORPHINE and HEROIN. These are reserved for severe acute pain and terminal disease, where addiction is either unlikely or unimportant.

Analog
In a plant or animal, an organ performing the same function as one in another species but which differs from it in origin and structure (ie is not a HOMOLOG). Thus a bird's wing and a bee's wing are analogous, whereas a bird's wing and a monkey's arm are homologous, both being derived from the forelimb of the ancestral reptiles.

Analysis see PSYCHOANALYSIS.

Analysis, Chemical
Determination of the compounds or elements comprising a chemical substance. Qualitative analysis deals with what a sample contains; quantitative analysis finds the amounts.

Anaphylaxis
A severe allergic reaction (see ALLERGY) caused by foreign material to which the person is sensitized, and mediated by HISTAMINE and KININS. In humans, sudden severe breathlessness – due to spasm in bronchi and larynx – and circulatory collapse (SHOCK) occur.

Anastomosis
Anatomically, a vessel that connects two arteries or veins, or an artery with a vein, without going through capillaries. Surgically, any passage that connects two tubes that are not normally connected, for example between two ends of gut where a section has been removed from the middle.

Anatomy
The structure and form of biological organisms. The subject has two main divisions: gross anatomy, dealing with components visible to the naked eye, and microscopic anatomy, dealing with microstructures seen only with the aid of a microscope. As structure is closely related to function, anatomy is related to PHYSIOLOGY.

The study of anatomy is as old as that of MEDICINE, though for many centuries physicians' knowledge of anatomy left much to be desired. ANAXAGORAS had studied the anatomy of animals, and anatomical observations can be found in the Hippocratic writings (see HIPPOCRATES), but it was ARISTOTLE who was the true father of comparative anatomy. Human dissection (the basis of all systematic human anatomy) was rarely practiced before the era of the Alexandrian School and the work of HEROPHILUS and ERASISTRATUS. The last great experimental anatomist of antiquity was GALEN. His theories, as transmitted through the writings of the Arab scholars Rhazes and AVICENNA, held sway throughout the medieval period. Further progress had to await the revival of the practice of human dissection by SERVETUS and VESALIUS in the 16th century. The latter founded the famous Paduan school of anatomy; its members also included FALLOPIUS and FABRICIUS, whose pupil WILLIAM HARVEY reunited the studies of anatomy and physiology in postulating the circulation of the BLOOD in *de Motu Cordis* (1628). This theory was confirmed some years later when MALPIGHI discovered the capillaries linking the arteries with the veins. Important developments in the late 18th century included the foundation of HISTOLOGY by BICHAT and of modern comparative anatomy by CUVIER.

The rise of microscopic anatomy has of course depended on the development of the microscope; one of its early successes was SCHWANN's cell theory in 1839.

Anaxagoras (*c*500-*c*428 BC)
Greek philosopher of the Ionian school, resident in Athens, who taught that the elements were infinite in number and that every thing contained a portion of every other thing. He also discovered the true cause of ECLIPSES, thought of the Sun as a blazing rock and demonstrated that air has substance.

Anaximander (610-*c*545 BC)
Greek philosopher of the Ionian school who taught that the cosmos was all derived from one primordial substance by a process of the separating out of opposites. He was probably the first Greek to attempt a map of the whole known world, and thought of the Earth as a stubby cylinder situated at the center of all things. Animal life, he thought, had begun in the sea.

Anaximenes of Miletus (*c*570-*c*500 BC)
Greek philosopher of the Ionian school who held that all things were derived from air; this becoming, for instance, fire on rarefaction, water and finally earth on condensation.

Anderson, Carl David (b. 1905)
US physicist who shared the 1936 Nobel Prize for Physics for his discovery of the positron (1932). Later he was codiscoverer of the first meson (see SUBATOMIC PARTICLES).

Anderson, Philip Warren (b. 1923)
US physicist who shared the 1977 Nobel Prize for Physics with N.F. MOTT and J.H. VAN VLECK, for their contributions to solid state physics.

Andrews, Roy Chapman (1884-1960)
US naturalist, explorer and author. From 1906 he worked for the American Museum of Natural History (later becoming its director, 1935-41) and made important expeditions to Alaska, the Far East and Central Asia.

Androecium see FLOWER.

Androgens
STEROID HORMONES that produce secondary male characteristics such as facial and body hair and a deep voice. They also develop the male reproductive organs. The main androgen is TESTOSTERONE, produced in the TESTES; Small amounts occur in women in addition to the ESTROGENS and may produce some male characteristics. (See also ENDOCRINE GLANDS; PUBERTY.) (♦ page 111)

Andromeda
Constellation in the N Hemisphere. The Great Andromeda Nebula (M31) is the most distant object visible to the naked eye in N skies. It is the nearest external GALAXY to our own, a larger spiral (49kpc across), and about 670kpc away.

Androsterone
An androgenic steroid (male sex hormone), derived from testosterone.

Anechoic chamber see ACOUSTICS.

Anemia
Condition in which the amount of HEMOGLOBIN in the BLOOD is abnormally low, thus reducing the blood's oxygen-carrying capacity. Anemic people may feel weak, tired, faint and breathless, have a rapid pulse and appear pale.

Of the many types of anemia, five groups can be described. In iron-deficiency anemia the red blood cells are smaller and paler than normal. Usual causes include inadequate diet (especially in PREGNANCY), failure of iron absorption, and chronic blood loss (as from heavy MENSTRUATION, HEMORRHAGE or disease of the GASTROINTESTINAL TRACT). In megaloblastic anemia the red cells are larger than normal. This may be due to nutritional lack of VITAMIN B_{12} or folate, but the most important cause is pernicious anemia, in which patients cannot absorb from the gut the B_{12} necessary for red-cell formation. In aplastic anemia inadequate numbers of red cells are produced by the bone MARROW. In hemolytic anemia the normal life-span of red cells is reduced, either because of ANTIBODY reactions or because they are abnormally fragile. A form of the latter is sickle-cell anemia, a hereditary disease in which abnormal hemoglobin is made. Finally, many chronic diseases, including RHEUMATOID ARTHRITIS, chronic infection and UREMIA, suppress red cell formation and thus cause anemia.

Anemometer
Any instrument for measuring wind speed. The rotation type, which estimates wind speed from the rotation of cups mounted on a vertical shaft, is the most common of mechanical anemometers. The pressure-tube instrument utilizes the PITOT-TUBE effect, while the sonic or acoustic anemometer depends on the velocity of sound in the wind. For laboratory work a hot-wire instrument is used: here air flow is estimated from the change in RESISTANCE it causes by cooling an electrically heated wire.

Anesthesia
Absence of sensation, which may be of three types: general, local or pathological. General anesthesia is a reversible state of drug-induced unconsciousness with muscle relaxation and suppression of REFLEXES; this facilitates many surgical procedures and avoids distress. While ETHANOL and NARCOTICS have been used for their anesthetic properties for centuries, modern anesthesia dates from the use of diethyl ETHER by WILLIAM MORTON in 1846 and of CHLOROFORM by SIR JAMES SIMPSON in 1847. Nowadays injections of short-acting barbiturates such as pentothal sodium are frequently used to induce anesthesia; rapidly inhaled agents, including halothane, ether, nitrous oxide, trichlorethylene and cyclopropane, are used for induction and maintenance. Local and regional anesthesia are the reversible blocking of pain impulses by chemical action of COCAINE derivatives (eg procaine, lignocaine). Nerve trunks are blocked for minor surgery and dentistry, and more widespread anesthesia may be achieved by blocking spinal nerve roots, useful in obstetrics and patients unfit for general anesthesia. Pathological anesthesia describes loss of sensation following trauma or disease, eg leprosy.

Aneurysm
A pathological ballooning of a blood vessel. This may occur in the heart after coronary thrombosis, or in the aorta and arteries due to ATHEROSCLEROSIS, high blood pressure, congenital defect, trauma or infection (specifically syphilis). They may rupture, causing hemorrhage, which in the heart or aorta is rapidly fatal. Again, their enlargement may cause pain, swelling or pressure on nearby organs; these complications are most serious in the arteries of the brain.

Anfinsen, Christian Boehmer (b. 1916)
US biochemist, corecipient of the 1972 Nobel Prize for Chemistry for research into the structure of the ENZYME ribonuclease (see also NUCLEIC ACIDS).

Angina pectoris
Severe, short-lasting chest pain caused by inadequate blood supply to the myocardium, often due to coronary artery disease. It is precipitated by exertion or other stresses that demand increased heart work. Pain may spread to nearby areas, often the arms, and sweating and breathlessness may occur.

Angiosperms (flowering plants)
Large and very important class of plants, characterized by having seeds that develop completely enclosed in the tissue of the parent plant, rather than unprotected as in the only other seedbearing group, the GYMNOSPERMS. The tissue surrounding the seed develops from the ovary wall or carpel, which, like other parts of the FLOWER (stamens, petals, sepals) is derived from a modified leaf. These complex flowers, many of which produce nectar to attract insect pollinators, are also a distinctive feature of the group, as are the FRUITS that develop from them. Containing about 250,000 species distributed throughout the world, and ranging in size from tiny herbs to huge trees, angiosperms are the dominant land flora of the present day. They have sophisticated mechanisms to ensure that pollination and fertilization take place and that the resulting seeds are readily dispersed and able to germinate. There are two subclasses: MONOCOTYLEDONS (with one seed-head) and DICOTYLEDONS (with two). (See also SPOROPHYTE.) (◆ page 53, 139)

Ångström, Anders Jonas (1814-1874)
Swedish physicist who was one of the founders of SPECTROSCOPY and was the first to identify hydrogen in the solar spectrum (1862). The ANGSTROM UNIT is named in his honor.

Ångström unit (Å)
The CGS UNIT used to express optical wavelengths (see LIGHT), and equal to 0.1nm.

Anhydrides
Compounds derived from others by reversible DEHYDRATION. Most inorganic anhydrides are soluble OXIDES that dissolve in water to give ALKALIS or OXYACIDS. (See also ACID ANHYDRIDES.)

Anhydrite
Mineral form of anhydrous calcium sulfate (see CALCIUM), occurring worldwide as white-gray masses of orthorhombic CRYSTALS. Thick beds of anhydrite are formed where old seawater lagoons have evaporated. It is used in producing sulfuric acid and ammonium sulfate and as a drying agent.

Aniline (aminobenzene, $C_6H_5NH_2$)
A primary aromatic AMINE. It is a toxic, oily, colorless liquid, readily oxidized to various products, and a weak BASE. It is made by the reduction of nitrobenzene or by reaction of chlorobenzene with AMMONIA. Aniline is used in making synthetic dyes, rubber, pharmaceuticals, explosives and resins. (See also AROMATIC COMPOUNDS; DIAZONIUM COMPOUNDS.)

Animal behavior
The observed actions and interactions of animals, particularly interactions with other members of the same species. The categories of behavior most often studied include territorial defense, communication, group formation and cohesion, formation of hierarchies within groups, courtship, mating, rearing of the young, food-finding and migration. Animal behavior is studied both in the laboratory and in the field (ETHOLOGY). The study of animal behavior in an evolutionary and genetic context is known as SOCIOBIOLOGY.

Animals
Members of the animal KINGDOM. All are multicellular (EUKARYOTIC) organisms which lack chloroplasts for photosynthesis and therefore rely on other living organisms for their food (they are HETEROTROPHS). They are diploid (having two sets of chromosomes) and produce two different types of gametes, known as eggs and sperm. None of the above characteristics can be used to define the animal kingdom, however, as each of them is shared with certain other groups. One characteristic that is unique to the animals is the series of changes that occurs during their early embryonic development, involving a BLASTULA and a GASTRULA stage. Animals are generally thought of as being essentially motile organisms, but in fact many are sessile, including the sponges, sea anemones, corals and barnacles. (See also DEUTEROSTOME; PROTOSTOME; INVERTEBRATE; VERTEBRATE.) (◆◆ page 13, 16, 44)

Anion
A negatively charged ION that moves to the ANODE in ELECTROLYSIS; often a BASE.

Ankylosis
Fusion or stiffness of a JOINT, restricting movement. It may be caused by injury, surgery or disease. Ankylosing spondylitis is a condition more commonly found in men, in which chronic, progressive INFLAMMATION of the joints of the spine leads to stiffness and immobility.

Annealing
The slow heating and cooling of METALS and GLASS to remove stresses that have arisen in CASTING, cold working or machining. The annealed material is tougher and easier to process further. (See METALLURGY.)

Annelids (true worms)
A phylum containing worms that have segmented bodies, coelomic body cavities, a central nervous system composed of cerebral ganglia from which branches extend along the lower part of the body, and, typically, bristles called chaetae that project from the body wall. Annelids are classified into three classes, the Polychaeta, Oligochaeta and Hirudinea. The polychaetes include the ragworms, are marine, and have groups of long chaetae, called parapodia, extending in bundles from extensions of the body wall. The oligochaetes include the earthworms, which are terrestrial, and some small aquatic species. They have relatively few chaetae per segment and lack parapodia. Members of the Hirudinea, the leeches, are found mainly in fresh water. They lack chaetae and parapodia, but possess suckers at each end of the body. (◆ page 16)

Annual
Plant that completes its life-cycle in one growing season and then dies. Annuals propagate themselves only by seeds. They include such plants as marigolds, cornflowers, peas and tomatoes. Many of the garden plants referred to as annuals are not truly annual: they are perennial plants from warmer climates that cannot withstand cold winters. (See also BIENNIAL; PERENNIAL.)

Annual rings
Concentric rings each representing one year's growth, visible in cross-sections of woody plants, especially trees. Each ring is usually composed of two growth layers, a broad, large-celled layer representing spring growth and a narrow, denser

layer showing summer growth. The relative amounts of these layers are affected by the environment, a fact that forms the basis of DENDROCHRONOLOGY.

Anode
The positive ELECTRODE of a BATTERY, electric CELL or ELECTRON TUBE. ELECTRONS, which conventionally carry negative charge, enter the device at the CATHODE and leave by the anode. (➧ page 118)

Anodizing
A process for building up a corrosion-resistant or decorative oxide layer on the surface of metal (usually ALUMINUM) objects. The item to be coated is made the ANODE in a CELL containing an aqueous solution of sulfuric, chromic or oxalic acid as electrolyte. The desired oxide coating is formed when a current is passed through the cell (see ELECTROLYSIS). Further treatment can render this oxide layer waterproof, electrically insulating or brightly colored.

Anorexia nervosa
Pathological loss of appetite and loathing of body fat, with secondary MALNUTRITION and HORMONE changes. It often affects young women with diet obsession and may reflect underlying psychiatric disease. (See also BULIMIA NERVOSA.)

Anorthite
Mineral occurring in IGNEOUS ROCKS, consisting of calcium aluminum silicate ($CaAl_2Si_2O_8$); vitreous white or gray crystals. It is one of the three end-members (pure compounds) of the FELDSPAR group.

Anoxia see HYPOXIA.

Antabuse
Disulfiram (tetraethylthiuram disulfide), a drug occasionally used in the treatment of ALCOHOLISM. It prevents the breakdown of ACETALDEHYDE, a product of ETHANOL metabolism, producing unpleasant symptoms.

Antacids
Mild ALKALIS or BASES taken to neutralize excess STOMACH acidity for relief of DYSPEPSIA.

Antares
Alpha Scorpii, a DOUBLE STAR comprising a red supergiant 480 times larger than the Sun and a blue star of unknown type 3 times larger than the Sun (apparent magnitudes +1.23 and +5.5). It is 100pc from the Earth.

Antenna (aerial)
A component in an electrical circuit that radiates or receives RADIO waves. In essence a transmitting antenna is a combination of conductors that converts AC electrical ENERGY into ELECTROMAGNETIC RADIATION. The simple dipole consists of two straight conductors aligned end on and energized at the small gap that separates them. The length of the dipole determines the frequency for which this configuration is most efficient. It can be made directional by adding electrically isolated director and reflector conductors in front and behind. Other configurations include the folded dipole, the highly directional loop antenna, and the dish type used for MICROWAVE links. Receiving antennas can consist merely of a short DIELECTRIC rod or a length of wire for low-frequency signals. For VHF and microwave signals, complex antenna configurations similar to those used for transmissions must be used. (See also RADIO TELESCOPE.)

Antennae
Sensory organs found on the head of insects, myriapods, crustaceans and onychophorans. There is one pair in most groups, two in the crustaceans. Antennae are sensitive to touch and to chemicals (smell and taste). (➧ page 53)

Anther see FLOWER.

Anthocyanins
Pigments producing most red, purple and blue colors in higher plants. Their main function is to provide flowers and fruit with bright colors, to attract insects and other animals for the purposes of pollination and seed dispersal.

Anthracite see COAL.

Anthrax
A rare BACTERIAL DISEASE causing characteristic SKIN pustules and LUNG disease; it may progress to SEPTICEMIA and death. Anthrax spores, which can survive for years, may be picked up from infected animals (such as sheep or cattle), or from unsterilized bone meal. Treatment is with penicillin, and people at risk are vaccinated; the isolation of animal cases and disinfection of spore-bearing material is essential. It was the first disease in which a particular bacterium was shown (by KOCH) to be causative, and it had one of the earliest effective vaccines, developed by PASTEUR.

Anthropology
The study of humans from biological, cultural and social viewpoints. There are two main disciplines, physical anthropology and cultural anthropology, the latter embracing social anthropology. Physical anthropology is the study of humans as a biological species, their past evolution and contemporary physical characteristics. In its study of prehistoric man (paleoanthropology) it has many links with ARCHEOLOGY. In cultural anthropology, ethnography is the study of the culture of a single group, ethnology the comparative study of the cultures of two or more groups. Cultural anthropology is also concerned with cultures of the past. Social anthropology is concerned primarily with social relationships.

Antibiotics
Substances originally discovered in microorganisms, but now mainly produced synthetically, that kill or prevent growth of other microorganisms; their properties are utilized in the treatment of bacterial and fungal infection. PASTEUR noted the effect, and ALEXANDER FLEMING in 1929 first showed that the mold *Penicillium notatum* produced penicillin, a substance able to destroy certain bacteria. It was not until 1940 that FLOREY and CHAIN were able to manufacture sufficient penicillin for clinical use. The isolation of Streptomycin by WAKSMAN, of Gramicidin by DUBOS, and of the Cephalosporins were among early discoveries of antibiotics useful in human infection. Numerous varieties of antibiotics now exist and the search continues for new ones. Semi-synthetic antibiotics, in which the basic molecule is chemically modified, have increased the range of naturally occurring substances.

The mode of action of antibiotics ranges from preventing cell-wall synthesis to interference with PROTEIN and NUCLEIC ACID metabolism. Bacteria resistant to antibiotics either inherently lack susceptibility to their mode of action or have acquired resistance through PLASMIDS transferred from other, already resistant bacteria.

Among the more important antibiotics are the penicillins, Cephalosporins, tetracyclines, quinolones, Streptomycin, Gentamicin, Chloramphenicol and Rifampicin. Each group has its own particular range of target organisms. Some antibiotics can cause side effects, such as ALLERGY. Many antibiotics are effective by mouth but INJECTION may be necessary; topical application can also be used.

Antibodies and antigens
As one of the body's defense mechanisms, PROTEINS called antibodies are made by specialized white cells to counter foreign proteins known as antigens. Common antigens are found in VIRUSES, BACTERIA and their products (including TOXINS). A specific antibody is made for each antigen. Antibody reacts with antigen in the body, leading to a number of effects including enhanced PHAGOCYTOSIS by white cells, activation of complement (a substance capable of damaging cell membranes) and HISTAMINE release. Antibodies are produced faster and in greater numbers if the body has previously encountered the particular antigen. IMMUNITY to second attacks of diseases such as MEASLES and CHICKENPOX, and VACCINATION against diseases not yet contracted, are based on this principle.

Anticoagulants
Drugs that interfere with blood CLOTTING, used to treat or prevent THROMBOSIS and clot EMBOLISM. The two main types are heparin, which is injected and has an immediate but shortlived effect, and the coumarins (including Warfarin), which are taken by mouth and are longer lasting. They affect different parts of the clotting mechanism.

Antidepressants
Drugs used in the treatment of DEPRESSION. They are of three types: tricyclics, monoamine oxidase inhibitors (MAOIs) and the newer tetracyclics. They seem to work by altering the balance of NEUROTRANSMITTERS in the brain.

Antigens see ANTIBODIES AND ANTIGENS.

Antihistamines
Drugs that counteract HISTAMINE action; they are useful in HAY FEVER and HIVES and in some insect bites. They also act as SEDATIVES and may relieve MOTION SICKNESS.

Antimatter
A variety of MATTER differing from the matter that predominates in our part of the UNIVERSE in that it is composed of antiparticles rather than particles. Individual antiparticles, many of which have been found in COSMIC RAY showers or produced using particle ACCELERATORS, differ from their particle counterparts in that they are oppositely charged (as with the antiproton-PROTON pair) or in that their magnetic moment is orientated in the opposite sense with respect to their SPIN (as with the antineutrino and neutrino). In our part of the Universe antiparticles are very short-lived, being rapidly annihilated in collisions with their corresponding particles, their mass-energy reappearing as a gamma-ray PHOTON. (The reverse is also true: a high-energy GAMMA RAY sometimes spontaneously forms itself into a positron-ELECTRON pair.) However, it is by no means inconceivable that regions of the Universe exist in which all the matter is antimatter, composed of what are to us antiparticles. The first antiparticle, the positron (antielectron), was discovered by C.D. ANDERSON in 1932. (See also SUBATOMIC PARTICLES.)

Antimony (Sb)
Brittle, silvery white metal in Group VA of the PERIODIC TABLE, occurring mainly as STIBNITE, which is roasted to give antimony (III) oxide. Its ALLOTROPY resembles that of ARSENIC. Antimony, though rather unreactive, forms trivalent and pentavalent oxides, halides and oxyanions. It is used in SEMICONDUCTORS and in lead ALLOYS, chiefly BABBITT METAL, PEWTER, type metal and in lead storage batteries. Certain antimony compounds are used in the manufacture of medicines, paints, matches, explosives and of fireproofing materials. AW 121.75, mp 630.8°C, bp 1600°C, sg 6.691 (20°C). (➧ page 130)

Antiparticles see ANTIMATTER; SUBATOMIC PARTICLES.

Antiseptics (germicides)
Substances that prevent the growth or spread of microorganisms without necessarily killing them. Disinfectants prevent spread, and also kill any existing germs. They are used to avoid SEPSIS. Modern antisepsis was pioneered by Ignaz P. Semmelweis (1818-1865), JOSEPH LISTER and ROBERT KOCH, and dramatically reduced deaths from childbirth and surgery. Commonly used antiseptics and disinfectants include IODINE,

ATMOSPHERE

The Earth's atmosphere consists mainly of nitrogen (78 percent by volume) and oxygen (21 percent). Of the other constituents the most abundant are water vapor, argon and carbon dioxide. Average pressure at ground level is 1013 millibars (mb) – equivalent to the weight of a column of mercury 0·76m high, or that of a 10m column of water – and this is termed "one atmosphere". The greenhouse effect sustains the mean surface temperature at about 290K (17°C).

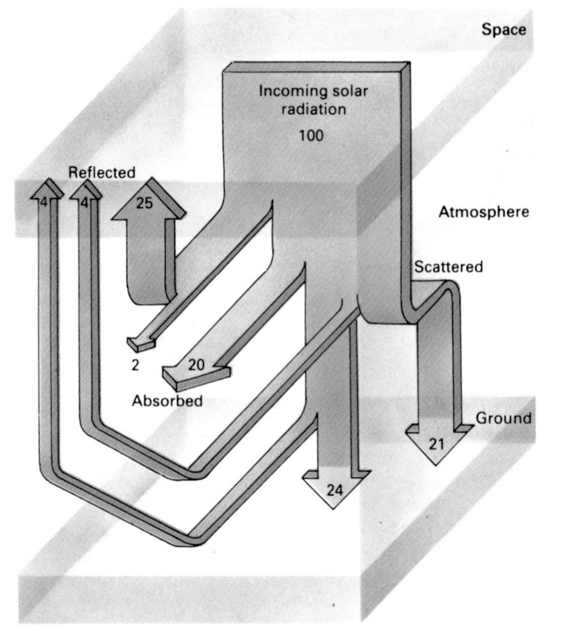

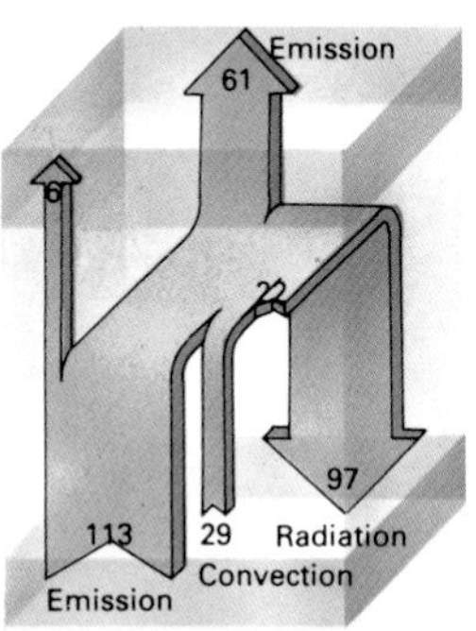

◄▲ *The atmosphere helps to maintain a balance between incoming and outgoing radiation, as shown by the distribution of incoming solar radiation (left), and of long-wave (infrared) radiation from the Earth (above). Atmospheric gains and losses equal each other, so average surface temperature is stable.*

▼ *The general circulation pattern of the atmosphere is complicated by the Earth's rotation. Rising equatorial air masses are deflected westwards and produce a spiral motion. A zone of turbulence and instability is created where temperate and polar air masses meet, resulting in an unstable climate in mid-latitudes.*

Wind circulation

Polar cell
Temperate cell
Tropical cell
NE Trades
Equatorial trough
SE Trades
Polar front

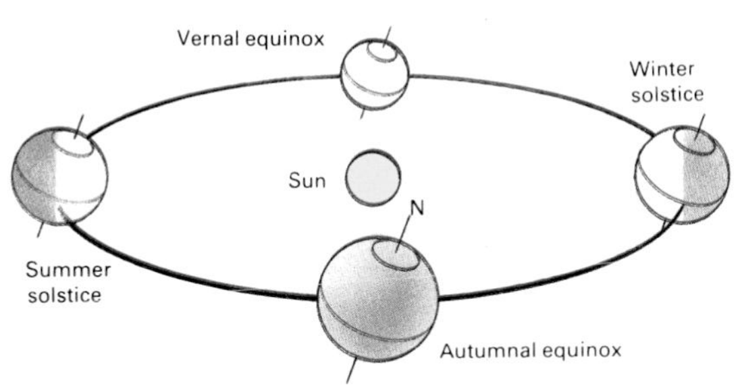

The seasons

► *As the Earth orbits the Sun, the tilt of its axis gives rise to the seasons. At vernal equinox, the Sun is above the equator; at summer solstice, above the Tropic of Cancer; at autumnal equinox, above the equator; at winter solstice, above the Tropic of Capricorn.*

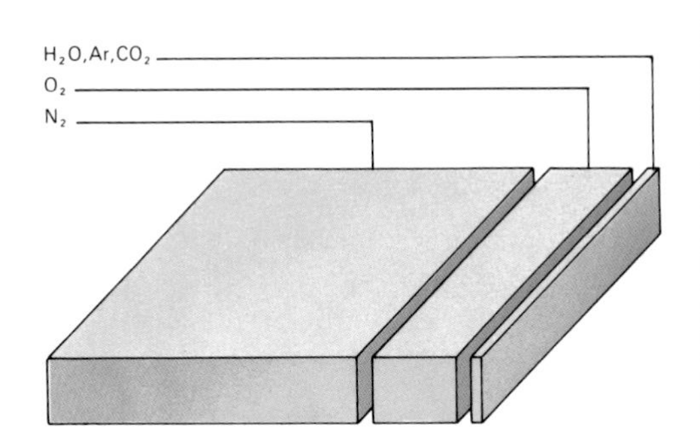

▲ *Nitrogen and oxygen together make up 97 percent by volume of the Earth's atmosphere.*

▼ *The lowest atmospheric layer, the troposphere, which contains most of the clouds and water vapor, is heated by infrared radiation from the ground; its temperature falls with increasing height. The stratosphere is heated by incoming ultraviolet light, as is the thermosphere, where the ionized gases of the ionosphere are formed.*

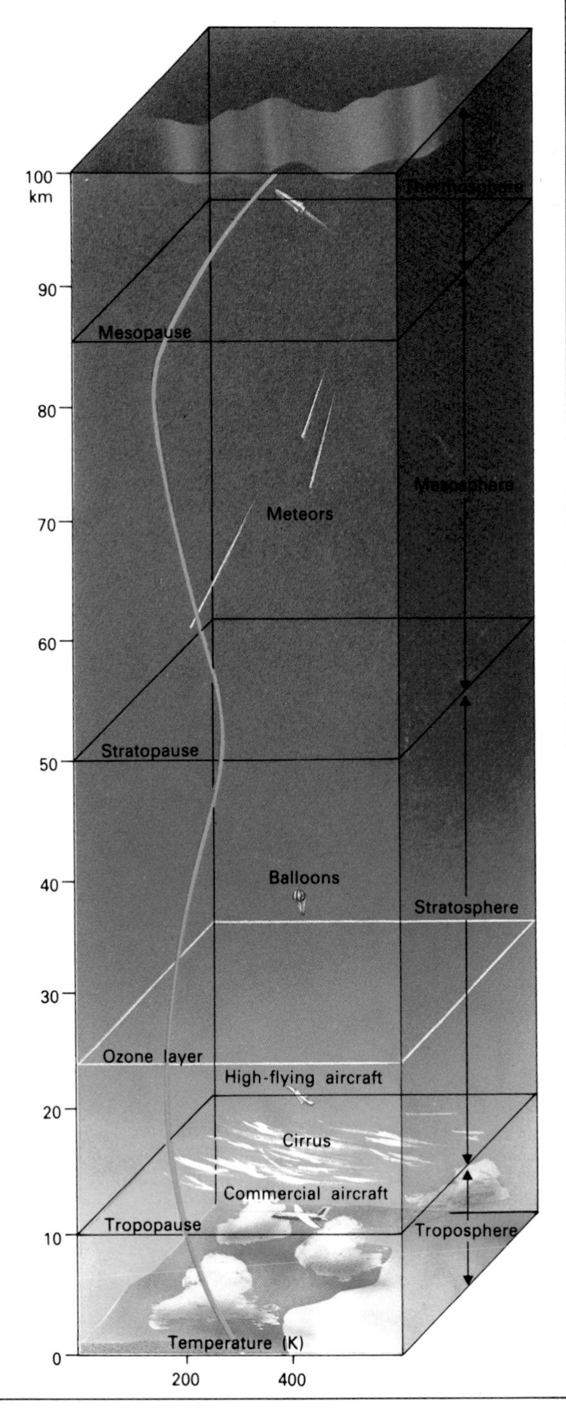

CHLORINE, hypochlorous acid, ETHANOL, PHENOLS, quaternary ammonium salts (see AMMONIA), FORMALDEHYDE, hydrogen PEROXIDE, potassium permanganate, and acriflavine (an acridine dye). Heat, ULTRAVIOLET and ionizing radiations also have antiseptic effects. (See also ASEPSIS; STERILIZATION.)

Antiserum
Serum containing antibodies directed against a particular antigen. It is administered to counteract development of the pathological features associated with the presence of the antigen in the body (eg snake venom antiserum).

Antitoxins
Antibodies produced in the body against the TOXINS of some bacteria. They are also formed after INOCULATION with toxoids, which are chemically inactivated toxins that can still confer IMMUNITY.

Antlers
Paired bony protuberances, either branched or unbranched, on the heads of male deer (stags, bucks, bulls, etc). They are shed and regrown every year, generally becoming larger and more branched each time until the deer reaches his prime, after which the size stabilizes or declines slightly. During growth the antlers are covered by a thin layer of furry skin, known as velvet, that supplies the developing bone with nutrients. Except in the female caribou or reindeer, antlers are undoubtedly a product of sexual selection, and are used primarily for fighting other males in contests over females.

Anxiety
A state of apprehension and fear, often accompanied by rapid heartbeat, palpitations, dry mouth and sweaty palms. A normal response to some situations, it can take over a person's life.

Aorta
The chief systemic ARTERY, distributing oxygenated blood from the heart to the whole body except the LUNGS via its branches. (See also BLOOD CIRCULATION; HEART.) (➧ page 28)

Apatite
The chief PHOSPHATE mineral, found in the Kola peninsula, USSR, N Africa, Montana and Florida, and mined for FERTILIZER and as the major ore of PHOSPHORUS. Its chemical composition is $Ca_5(PO_4)_3X$, where X=F (fluorapatite, the most common), Cl (chlorapatite), OH (hydroxyapatite) or a mixture of all three; it forms hexagonal crystals.

APD (amidronate sodium)
One of a group of compounds called biphosphonates, a drug that binds to the calcium orthophosphate molecules of bone, and inhibits OSTEOCLASTS, thus preventing bone loss in cases of Paget's disease and bone cancer.

Apes see PRIMATES.

Aphasia
A speech defect resulting from injury to certain areas of the brain and resulting in an inability to communicate using speech or writing; it may be partial (dysphasia) or total. Common causes are cerebral THROMBOSIS, HEMORRHAGE and brain TUMORS. (See SPEECH AND SPEECH DISORDERS.)

Aphelion see ORBIT.

Apoda see CAECILIANS.

Apogee see ORBIT.

Apollo program
US SPACE EXPLORATION program. Its prime objective was not only to land men on the Moon but also to carry out research into its nature and origins. Of 17 missions, 6 were unmanned, 2 Earth orbital, 2 lunar orbital, 1 was aborted by in-flight accident, and 6 made lunar landings.

Appeasement behavior
Action taken by one animal to allay AGGRESSION toward it by another member of its species.

Appendicitis
Inflammation of the APPENDIX, often caused by obstruction to its narrow opening followed by swelling and bacterial infection. Acute appendicitis may lead to rupture of the organ, formation of an ABSCESS or PERITONITIS. Symptoms include abdominal pain, usually in the right lower ABDOMEN, tenderness, nausea and FEVER. Early surgical removal of the appendix is essential or it may rupture, which can prove fatal.

Appendix
Narrow tubular structure opening into the cecum (see GASTROINTESTINAL TRACT) found in some vertebrates, including humans. The human appendix contains lymphoid tissue (see LYMPH) and is probably vestigial. (See APPENDICITIS.)

Appleton, Sir Edward Victor (1892-1965)
English physicist who discovered the Appleton layer (since resolved as two layers termed F_1 and F_2) of ionized gas molecules in the IONOSPHERE. His work in atmospheric physics won him the 1947 Nobel Prize for Physics and contributed to the development of RADAR. During WWII he helped to develop the ATOMIC BOMB.

Apsides, Line of
The imaginary straight line connecting the two points of an elliptical ORBIT that represent the orbiting body's greatest (higher apsis) and least (lower apsis) distances from the body around which it is revolving.

Apterygotes
A group of primitive, wingless insects. Unlike insects such as fleas, lice and ants they are not descended from winged forms, but represent early offshoots of the evolutionary line that led to the flying insects. They include the springtails or Collembola, the bristletails or Thyusanura, and two other, less well-known groups, the Diplura and the Protura, both of which live in the soil or under bark and avoid daylight. Like the more advanced flying insects, the apterygotes have only three pairs of legs, but some have vestigial limbs on their abdominal segments, indicating their descent from an ancestor with legs on every segment, as in the living ONYCHOPHORANS.

Aquamarine
Transparent pale blue or blue-green semiprecious GEM stone; a variety of BERYL.

Aqua regia
Corrosive mixture of one part NITRIC ACID and three parts HYDROCHLORIC ACID that dissolves gold and platinum. It is used in analysis of minerals and alloys, and as a cleaning agent.

Aquarius
The Water Bearer, a large but faint constellation on the ECLIPTIC; the 11th sign of the ZODIAC.

Aquifer
An underground rock formation through which GROUNDWATER can easily percolate. SANDSTONES, GRAVEL beds and jointed LIMESTONES make good aquifers.

Aquila
The Eagle, a large autumn constellation in the N Hemisphere, lying in the plane of the Milky Way. (See also ALTAIR.)

Arachnids
A group of terrestrial invertebrates that includes scorpions, spiders, harvestmen, ticks and mites, pseudoscorpions and camel-spiders. Some 60,000 species are known, most being terrestrial and living in soil, leaf litter and low vegetation. Most are carnivorous, but the mites and ticks are blood-sucking parasites. The arachnids are a subdivision of the CHELICERATES, which also includes the sea spiders and king crabs, both marine forms. (See also ARTHROPODS.) (➧ page 16)

Arachnoid membrane
Membrane covering the brain and carrying a network of blood vessels to its surface.

Arago, Dominique François Jean (1786-1853)
French physicist and mathematician whose work helped to establish the wave theory of LIGHT. He discovered the polarization of light in quartz crystals (1811), and in 1825 he was awarded by the Royal Society the Copley Medal for demonstrating the magnetic effect of a rotating copper disk.

Arber, Werner (b. 1929)
Swiss microbiologist who shared the 1978 Nobel Prize for Physiology or Medicine with NATHANS and SMITH for their contribution to MOLECULAR GENETICS.

Archaeopteryx
A species of bird known from a set of fossils found in Jurassic limestone in Germany. The animal lived about 150 million years ago and had wings and feathers much like those of today's birds, but still had teeth, hooks on its wings (the remains of the three inner digits) and a long tail. (The "tail" of modern birds is made of long feathers attached to the stump-like remains of the true tail – *Archaeopteryx* had an elongated, fleshy tail made of vertebrae, muscles and skin, with feathers all along it.) These features clearly show that *Archaeopteryx* was intermediate between modern birds and small, bipedal, fast-running dinosaurs of the theropod group. This does not necessarily mean that it was ancestral to living birds, as is often assumed – it may have been a side-branch of early bird evolution. Nevertheless, it is of immense value in confirming the dinosaur ancestry of the birds, something that T.H. Huxley proposed on purely anatomical grounds before the fossil was found.

The first *Archaeopteryx* fossil to be recognized as such was found in 1861; other fossils had been found previously but were classified as dinosaurs because the feathers were not preserved. Four more fossils were subsequently found during the 19th and 20th centuries. The widely reported suggestion that all these fossils were skillful forgeries is clearly false, given that they were found over a period of almost a hundred years in freshly excavated limestone. Furthermore, detailed examination of the fossils has revealed many features that show them, beyond any doubt, to be genuine.

Archean
The portion of the PRECAMBRIAN before about 2390 million years ago. It has always been regarded as the period in which no life existed on Earth. However, as more and more ancient indications of life are uncovered the later part of the Archean is now beginning to be called the Archeozoic – the time of ancient life. (➧ page 106)

Archebacteria
A group of bacteria that are believed to be relics of some of the earliest living cells. They show various primitive features, such as an inability to tolerate oxygen (which was not present in the Earth's original atmosphere). Analysis of their genetic material shows them to be a separate group, with a long period of evolutionary divergence from the other bacteria; the latter are sometimes referred to as the eubacteria, or "true bacteria".

The archebacteria include three main groups: the methanogenic bacteria, which produce methane from carbon dioxide; the thermophiles, which inhabit hot springs; and the extreme halophyles, which tolerate high concentrations of salt.

Archeology
The study of the past through identification and interpretation of the material remains of human cultures. A comparatively new science, involving many academic and scientific disciplines, including ANTHROPOLOGY, history, PALEOGRAPHY and philology, it makes use of numerous scientific

techniques. Excavation is a painstaking procedure, as great care must be taken not to damage any object or fragment of an object, and each of the different levels of excavation must be carefully documented and photographed. The location of suitable sites for excavation is assisted by historical accounts, topographical surveys and aerial photography. Dating is accomplished in several ways. The first method is by comparison of the relative depths of objects that are discovered. Analysis of the types of pollen in an object can provide an indication of its date. The most widespread dating technique is RADIOCARBON DATING, incorporating the corrections formulated through discoveries in DENDROCHRONOLOGY.

Archimedes (*c*287-*c*212 BC)
Greek mathematician and physicist who spent most of his life at his birthplace, Syracuse (Sicily). He founded the science of HYDROSTATICS with his enunciation of Archimedes' Principle. This states that the force acting to buoy up a body partially or totally immersed in a fluid is equal to the weight of the fluid displaced. In MECHANICS he studied the properties of the LEVER and applied his experience in the construction of military catapults and grappling irons. He is also said to have invented the Archimedes screw, a machine for raising water still used to irrigate fields in Egypt.

Arcturus
Alpha Boötis, a red giant star 11pc distant. It has a very high PROPER MOTION and an apparent magnitude of −0.04. (➧ page 228)

Argentite
A soft, dark SULFIDE ore of SILVER (Ag_2S), related to CHALCOCITE. It occurs in Norway, Czechoslovakia, South America, Mexico and Nevada.

Argol see TARTARIC ACID.

Argon (Ar)
The commonest of the NOBLE GASES, comprising about 1% of the ATMOSPHERE. It is used as an inert shield for arc welding and for the production of silicon and germanium crystals, to fill electric light bulbs and fluorescent lamps, and in argon-ion LASERS. AW 39.9, mp −189°C, bp −186°C. (➧ page 130)

Aries
The Ram, first constellation of the ZODIAC. In N skies it is a winter constellation.

Aries, First Point of
The vernal equinox, a point of intersection of the EQUATOR and the ECLIPTIC used as the zero of celestial longitude. Owing to PRECESSION, the First Point of Aries is currently in PISCES.

Aristarchus of Samos (*c*310-*c*250 BC)
Alexandrian Greek astronomer who realized that the Sun is larger than the Earth and who is reported by ARCHIMEDES to have taught that the Earth orbited a motionless Sun.

Aristotle (384-322 BC)
Greek philosopher and scientist, a pupil of PLATO, the foremost systematizer of the knowledge of the ancient world and founder of the Peripatetic School of philosophy. In science his best work concerned biology. He proposed new principles for the classification of animals; he effectively founded the science of comparative ANATOMY, and contributed greatly to EMBRYOLOGY. His physics and cosmology were less successful though none the less influential: his views in this area, in the guise of Aristotelianism, dominated the mind of science down to the Renaissance. He rejected the atomic theory and held the essence of matter to reside in the four qualities – hot, cold, wet and dry – the combination of these constituting the four Aristotelian elements – earth, air, fire and water. The heavens, kept in motion by the Unmoved Mover, are composed of a fifth "quintessential" element – ether – and circle a fixed, central and spherical Earth.

Armature
The part of an electric motor or generator that includes the principal current-bearing windings. In small motors it usually comprises several coils of wire wound on a soft iron core and mounted on the drive shaft, though on larger AC motors the armature is often the stationary component. When the current flows in the armature winding of a motor, it interacts with the magnetic field produced by the field windings, giving rise to a TORQUE between the rotor and the stator. In the generator the armature is rotated in a magnetic field giving rise to an ELECTROMOTIVE FORCE in the windings.

Armillary sphere
A model displaying the mutual dispositions of the imaginary circles of classical astronomy in which metal circles were used to represent the celestial equator (see CELESTIAL SPHERE), the ECLIPTIC, the TROPICS, the arctic and antarctic circle, the hours of the day, the HORIZON and a MERIDIAN. Derived from ancient astronomical instruments, armillary spheres became particularly popular in the 17th and 18th centuries.

Aromatic compounds
Major class of organic compounds containing one or more planar rings of atoms having special stability due to their electronic structure. This includes TORUS-shaped ORBITALS above and below the plane of the ring, known as a π-electron system, containing ($4n+2$) electrons (ie 6, 10, 14 and so on). Such systems can be represented by RESONANCE structures of alternate single and double bonds (see BOND, CHEMICAL) round the ring. Typical properties of aromatic compounds include ease of formation, tendency to react by substitution rather than addition, and modification of the properties of attached groups. The most important aromatic compounds are BENZENE and its derivatives, including PHENOLS, TOLUENE, BENZALDEHYDE, BENZOIC ACID, BENZYL ALCOHOL, ANILINE and SALICYLIC ACID. Compounds with more than one benzene ring (polycyclic) in clude NAPHTHALENE and anthracene. Non-benzenoid aromatics include cyclopentadienyl and cycloheptatrienylium ions (see FERROCENE) and azulene. Many HETEROCYCLIC COMPOUNDS are aromatic.

Arrhenius, Svante August (1859-1927)
Swedish physical chemist whose theory concerning the DISSOCIATION of SALTS in solution (see CONDUCTIVITY) earned him the 1903 Nobel Prize for Chemistry and laid the foundations for the study of ELECTROCHEMISTRY.

Arrhythmia
A disorder of the normal rhythmic beating of the heart. The commonest form, ectopic beats, is usually harmless, but more serious arrhythmias may occur in heart disease. Severe disruption of rhythm leads to uncontrollable twitching (fibrillation), which prevents the heart from beating and is therefore fatal if sustained. The ECG is used in diagnosis.

Arrow-worms
A phylum of worm-like marine animals with bodies divided into head, trunk and tail, the Chaetognatha. The trunk bears lateral fins. Larval Chaetognathans are free-swimming; adults are planktonic or tube dwellers.

Arsenic (As)
Metalloid in Group VA of the PERIODIC TABLE. Its chief ore is ARSENOPYRITE, which is roasted to give arsenic (III) oxide, or white arsenic, used as a poison. Arsenic has two main allotropes (see ALLOTROPY): yellow arsenic, As_4, resembling white PHOSPHORUS; and gray (metallic) arsenic. It burns in air and reacts with most other elements, forming trivalent and pentavalent compounds, all highly toxic. It is used as a doping agent in TRANSISTORS; gallium arsenide is used in LASERS. AW 74.9, subl 610°C, sg 1.97 (yellow), 5.73 (gray). (➧ page 130)

Arsenopyrite (mispickel)
Silvery white mineral with metallic luster, crystallizing in the monoclinic system, iron sulfarsenide (FeAsS). It is the chief ore of ARSENIC, and is found in the US, Canada, Germany, England and Scandinavia.

Artery
Blood vessel that carries BLOOD from the HEART to the tissues (see BLOOD CIRCULATION). The arteries are elastic and expand with each PULSE. In most vertebrates, the two main arteries leaving the heart are the pulmonary artery, which carries blood from the body to the LUNGS to be reoxygenated, and the AORTA, which supplies the body with oxygenated blood. Major arteries supply each limb and organ and within each they divide repeatedly into arterioles and CAPILLARIES. Fish have only one arterial system, which leads from the heart via the GILLS to the body. (See also ATHEROSCLEROSIS.) (➧ page 28)

Artesian well
A well in which water rises under hydrostatic pressure above the level of the AQUIFER in which it has been confined by overlying impervious strata. Often pumping is necessary to bring the water to the surface, but true artesian wells (named for the French province of Artois where they were first constructed) flow without assistance.

Arthritis
INFLAMMATION, with pain and swelling, of JOINTS. Bacterial infection, with PUS in the joint, and GOUT, due to deposition of crystals in SYNOVIAL FLUID, may lead to serious joint destruction.

Rheumatoid arthritis is a systemic disease manifested mainly in joints, with inflammation of synovial membranes and secondary destruction. Tendons may be disrupted and deformity can result. Arthritis also occurs in many other systemic diseases, including RHEUMATIC FEVER, SYSTEMIC LUPUS ERYTHEMATOSUS, PSORIASIS and some SEXUALLY TRANSMITTED DISEASES. Osteoarthritis, properly called osteoarthrosis, is a wear-and-tear arthritis, causing pain and limitation of movement.

Arthropods
Invertebrate animals that have a hard EXOSKELETON containing CHITIN and jointed limbs to allow movement despite this "coat of armor". The arthropods were once placed together in a single phylum, but it is now believed that they include three or four quite separate lineages that evolved independently from soft-bodied ancestors. These four principal groups are: CRUSTACEANS (crabs, shrimps, woodlice, waterfleas and copepods); INSECTS and MYRIAPODS (collectively known as the Uniramia); CHELICERATES (spiders, scorpions, mites and king crabs, and the extinct sea scorpions); and the extinct TRILOBITES.

Artificial insemination
Introduction of SPERM into the vagina by means other than copulation. The technique is widely used for breeding livestock as it produces many offspring from one selected male. Human artificial insemination may be by donor or by husband, including frozen sperm from a dead partner.

Artificial intelligence (AI)
In 1968 Marvin Minsky, cofounder of the Massachusetts Institute of Technology's AI Group, defined AI as "the science of making machines do things that would require intelligence if done by man". But AI means different things to different people: some regard it as a technique for making COMPUTERS operate more effectively; others believe that its usefulness lies in advancing knowledge of how the human BRAIN works. Minsky's

CIRCULATION AND RESPIRATION

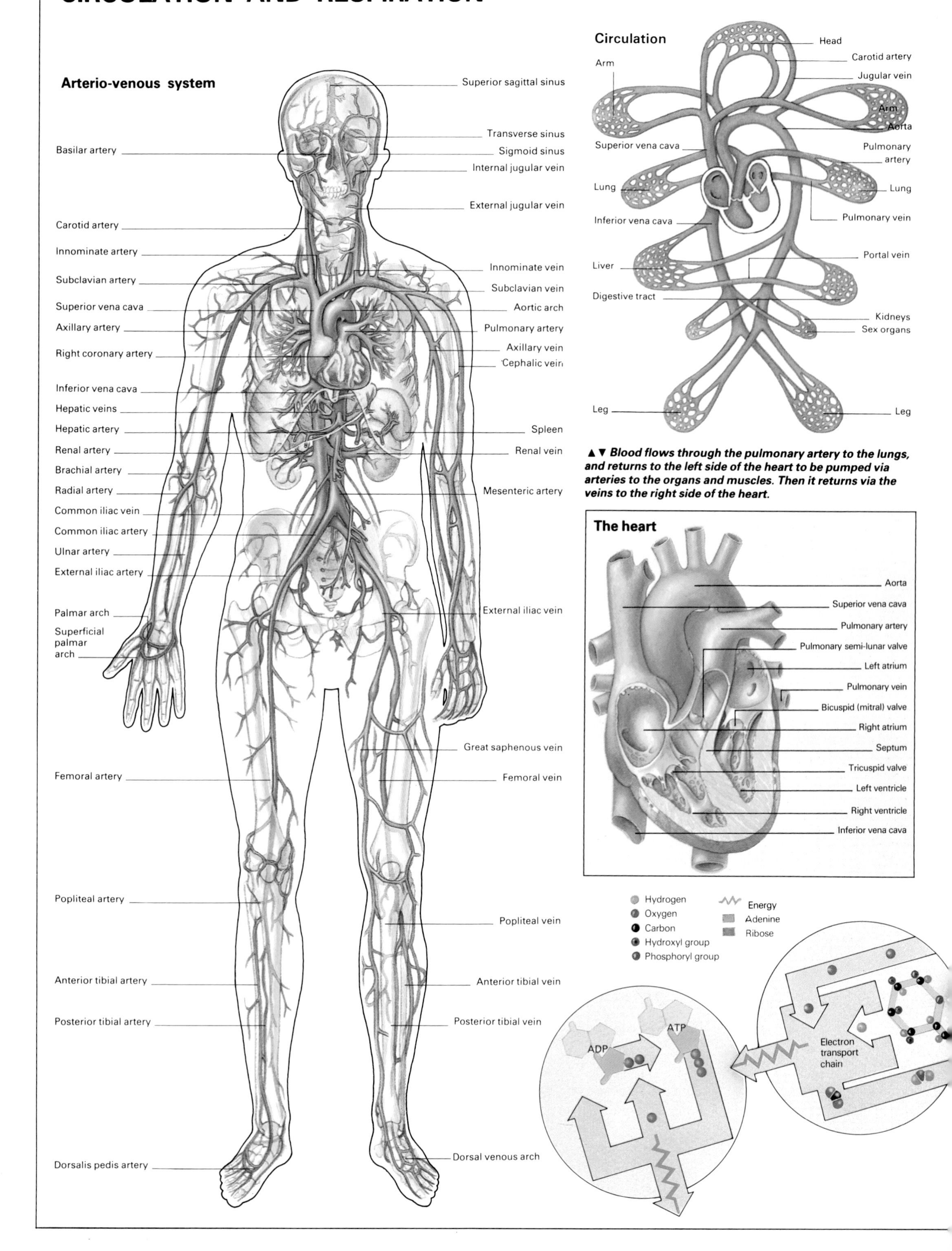

▲▼ ***Blood flows through the pulmonary artery to the lungs, and returns to the left side of the heart to be pumped via arteries to the organs and muscles. Then it returns via the veins to the right side of the heart.***

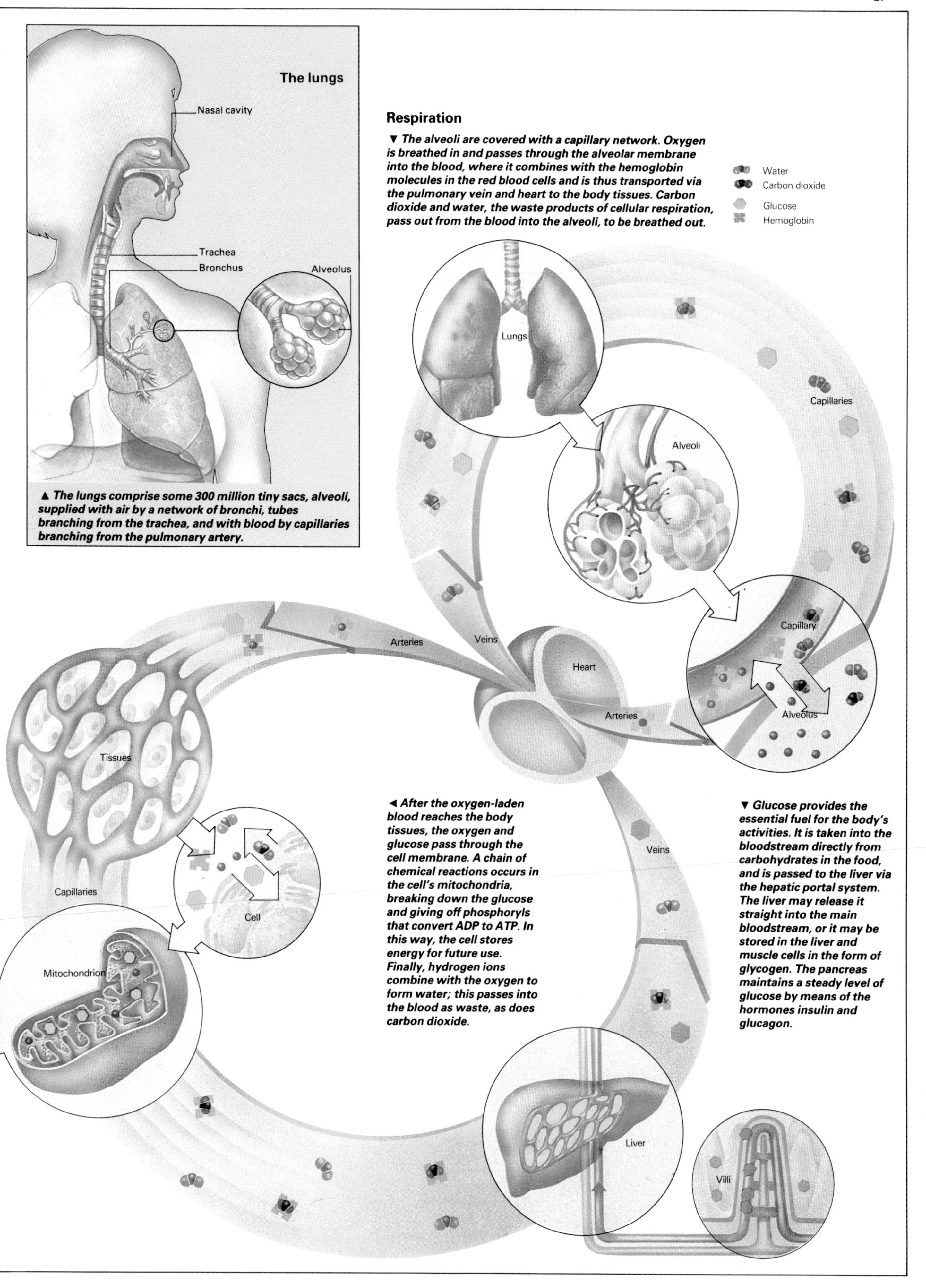

▲ ***The lungs comprise some 300 million tiny sacs, alveoli, supplied with air by a network of bronchi, tubes branching from the trachea, and with blood by capillaries branching from the pulmonary artery.***

Respiration

▼ ***The alveoli are covered with a capillary network. Oxygen is breathed in and passes through the alveolar membrane into the blood, where it combines with the hemoglobin molecules in the red blood cells and is thus transported via the pulmonary vein and heart to the body tissues. Carbon dioxide and water, the waste products of cellular respiration, pass out from the blood into the alveoli, to be breathed out.***

◄ ***After the oxygen-laden blood reaches the body tissues, the oxygen and glucose pass through the cell membrane. A chain of chemical reactions occurs in the cell's mitochondria, breaking down the glucose and giving off phosphoryls that convert ADP to ATP. In this way, the cell stores energy for future use. Finally, hydrogen ions combine with the oxygen to form water; this passes into the blood as waste, as does carbon dioxide.***

▼ ***Glucose provides the essential fuel for the body's activities. It is taken into the bloodstream directly from carbohydrates in the food, and is passed to the liver via the hepatic portal system. The liver may release it straight into the main bloodstream, or it may be stored in the liver and muscle cells in the form of glycogen. The pancreas maintains a steady level of glucose by means of the hormones insulin and glucagon.***

definition however is mostly interpreted as meaning enabling computers to understand natural (as opposed to computer) language, solving problems, learning and making logical deductions. Another AI pioneer, Joseph Feigenbaum, defined it thus in 1983, "If we can imagine an artifact that can collect, assemble, choose among, understand, perceive, and know, then we have an artificial intelligence". (See also EXPERT SYSTEM; FIFTH GENERATION; ROBOT.)

Artificial limbs see PROSTHETICS.

Artificial organs
Mechanical devices that can perform the functions of bodily organs. The heart-lung machine can maintain BLOOD CIRCULATION and oxygenation and has enabled much new cardiac SURGERY. Artificial kidneys clear waste products from the blood (see UREMIA) by DIALYSIS and may take over KIDNEY function for life.

Artificial respiration
The means of inducing RESPIRATION when it has ceased, as after DROWNING, ASPHYXIA, in COMA or respiratory PARALYSIS. It must be continued until natural breathing returns, and ensuring a clear airway via the mouth to the lungs is essential. The most common first aid methods are: "mouth-to-mouth", in which air is breathed via the mouth into the lungs and is then allowed to escape, and the less effective Holger Nielsen technique where rhythmic movements of the CHEST alternately force air out and encourage its entry. If prolonged artificial respiration is needed, mechanical pumps are used, and these may support respiration for months or even years.

Artificial selection
The method by which man has altered certain animals and plants by selecting for breeding those individuals that display desired characteristics. These include fast growth rates in cattle, heavy crop yields and disease resistance in plants. (See also DOMESTICATION.)

Artiodactyls
An order of mammals that includes the hippos, camels, deer, giraffes, antelope, cows, buffalo, bison, sheep, goats, pronghorn, pigs and peccaries. All are even-toed – that is, they have two or four toes, usually two, ie cloven hooves, a specialization that originally evolved for fast running. Many artiodactyls are RUMINANTS. (See also PERISSODACTYLS; UNGULATES.)

Asbestos
Name of various fibrous minerals, chiefly CHRYSOTILE and AMPHIBOLE. Canada and the USSR are the chief producers. It is a valuable industrial material because it is refractory, alkali- and acid-resistant and an electrical insulator. But its use can be dangerous; see ASBESTOSIS.

Asbestosis
A form of PNEUMOCONIOSIS caused by inhaling fine particles of asbestos. It may predispose to cancers such as MESOTHELIOMA.

Aschelminthes
A group of small rather heterogeneous animals previously classified in separate phyla. They include the ROTIFERS, microscopic forms found mainly in freshwater, and the NEMATODES, or roundworms, which are parasitic or free-living.

Ascomycetes
One of the two major divisions of the higher FUNGI, the other being the BASIDIOMYCETES. The fruiting bodies of the Ascomycetes include the cup fungi, morels and truffles. Within the fruiting bodies, the spores are produced in a characteristic sac known as an ascus. The fungal components of most lichens are Ascomycetes, as are many yeasts. In modern taxonomic schemes, the name Ascomycetes has been superseded by Ascomycotina, but the old name is still widely used.

Ascorbic acid (vitamin C) see VITAMIN.

Asepsis
The principle in modern SURGERY of excluding GERMS. Means include the STERILIZATION of instruments, dressings, gowns and gloves, and the use of ANTISEPTICS.

Aspartame (aspartylphenylalanine)
A synthetic dipeptide used as an artificial sweetener, particularly in soft drinks. Weight for weight it is 200 times sweeter than sugar and does not have the bitter aftertaste associated with SACCHARIN.

Asphyxia
The complex of symptoms due to inability to take oxygen into or excrete carbon dioxide from the LUNGS. The commonest causes are DROWNING, suffocation or strangling, inhalation of toxic gases, and obstruction of LARYNX, TRACHEA or BRONCHI (which can occur in severe cases of CROUP, ASTHMA and DIPHTHERIA). Early ARTIFICIAL RESPIRATION is essential.

Association
In PSYCHOLOGY, the mental linking of one item with others: eg black and white, Tom with Dick and Harry, etc. The connections are described by the primary (similarity and contiguity) and secondary (frequency, recency, vividness and primacy) laws of association.

Associationism
A psychological school that held that the sole mechanism of human learning consisted in the permanent association in the intellect of impressions that had been repeatedly presented to the senses. The "association of ideas" was the dominant theme in British PSYCHOLOGY for 200 years.

Astatine (At)
Radioactive HALOGEN, occurring naturally in minute quantities, and prepared by bombarding BISMUTH with ALPHA PARTICLES. The most stable isotope, At^{210}, has half-life 8.3h. Tracer studies show that astatine closely resembles IODINE. AW 210, mp 300°C, bp 350°C. (➧ page 130)

Asteroids
The thousands of planetoids or minor planets, ranging in diameter from a few meters to 1000km (CERES), most of whose orbits lie in the Asteroid Belt between the orbits of Mars and Jupiter. Vesta is the only asteroid visible to the naked eye, though Ceres was the first to be discovered (1801 by PIAZZI). Their total mass is estimated to be 0.001 that of the Earth. A second asteroid belt beyond the orbit of Pluto has been postulated. The "Apollo" asteroids are those that cross the orbit of the Earth; of these, Phaethon passes closest to the Sun. Chiron's orbit passes from inside that of Saturn to just beyond Uranus. (See also METEORITE; SOLAR SYSTEM.)

Asthenosphere
In the structure of the Earth, a layer of material less rigid than the layers above and below it. It lies within the upper mantle, about 175km thick, and with its upper surface at a depth of about 75km. Its presence is detected by the slowing down of earthquake waves as they pass through it. In the motions of plate tectonics, in which the surface of the Earth is divided into discrete plates that move in relation to one another, the asthenosphere acts as the lubricating layer for the motion. The rigid plates constitute the LITHOSPHERE. (➧ page 134)

Asthma
Respiratory disease marked by recurrent attacks of wheezing and acute breathlessness. It is due to abnormal bronchial sensitivity, which makes breathing out slow, thus limiting the respiration rate. It is usually associated with ALLERGY to house dust mite, pollen, FUNGI, furs and other substances which may precipitate an attack. Chest infection, exercise or emotional upset may also provoke an attack. The symptoms are caused by spasm of bronchioles (see BRONCHI) and the accumulation of thick MUCUS. CYANOSIS may occur in severe attacks.

Astigmatism
A defect of VISION in which the LENS of the EYE distorts objects horizontally or vertically, corrected using cylindrical lenses. Also, an ABERRATION of lenses having spherical surfaces.

Aston, Francis William (1877-1945)
British physicist who designed the first mass spectrograph (see MASS SPECTROSCOPY) and used it to identify and separate the ISOTOPES of the nonradioactive elements. This work earned him the Nobel Prize for Chemistry in 1922 and led to the formulation of the whole number rule for isotopic weights (see ATOMIC WEIGHT) .

Astringent
Agent used to shrink mucous membranes and to dry up secretions or wet lesions. They may act by vasoconstriction, dehydration (as with ETHANOL) or by denaturing proteins (as with TANNIN).

Astrolabe
An astronomical instrument dating from the Hellenic period, used to measure the altitude of celestial bodies and, before the introduction of the SEXTANT, as a navigational aid. It consisted of a vertical disk with an engraved scale across which was mounted a sighting rule or "alidade".

Astrology
The art and science of divining the future from the study of the heavens. Originating in ancient Mesopotamia as a means for predicting the fate of states and their rulers, the astrology that found its way into Hellenistic culture applied itself also to the destinies of individuals. Together with the desire to devise accurate CALENDARS, astrology provided a key incentive leading to the earliest systematic ASTRONOMY and was a continuing spur to the development of astronomical techniques until the 17th century. The majority of Classical and medieval astronomers, PTOLEMY and KEPLER among them, practiced astrology, often earning their livelihoods thus. Astrology exercised its greatest influence in the Greco-Roman world and again in Renaissance Europe (despite the opposition of the Church) and, although generally abandoned after the 17th century, it has continued to excite a fluctuating popular interest down to the present. The key datum in Western astrology is the position of the stars and planets, described relative to the 12 divisions of the ZODIAC, at the moment of an individual's birth.

Astronomical Unit (AU)
A unit of distance equal to the semi-major axis of the Earth's orbit (approximately the mean distance of the Earth from the Sun), used for describing distances within the Solar System. Its value is 149.6Gm. (➧ page 240)

Astronomy
The study of the heavens. At the crossroads of agriculture and religion, astronomy, the earliest of the sciences, was of great practical importance in ancient civilization. Before 2000 BC, Babylonians, Chinese and Egyptians all sowed their crops according to calendars computed from the regular motions of the Sun and Moon.

Although early Greek philosophers were more concerned with the physical nature of the heavens than with precise observation, later Greek scientists (see ARISTARCHUS; HIPPARCHUS) returned to the problems of positional astronomy. The vast achievement of Greek astronomy was epitomized in the writings of Claudius PTOLEMY. His *Almagest*, passing through Arabic translations, was eventually transmitted to medieval Europe and remained the chief authority among astronomers for over 1400 years.

Throughout this period the main purpose of positional astronomy had been to assist in the casting of accurate horoscopes, the twin sciences of astronomy and ASTROLOGY having not yet parted company. The structure of the Universe meanwhile remained the preserve of (Aristotelian) physics. The work of COPERNICUS represented an early attempt to harmonize an improved positional astronomy with a true physical theory of planetary motion. Against the judgment of antiquity that Sun, Moon and planets circled the Earth as lanterns set in a series of concentric transparent shells, in his *de Revolutionibus* (1543) Copernicus argued that the Sun lay motionless at the center of the planetary system.

Although the Copernican (or heliocentric) hypothesis proved to be a sound basis for the computation of navigators' tables (the need for which was stimulating renewed interest in astronomy), it did not become unassailably established in astronomical theory until NEWTON published his mathematical derivation of KEPLER'S LAWS in 1687. In the meantime KEPLER, working on the superb observational data of TYCHO BRAHE, had shown the orbit of Mars to be elliptical and not circular, and GALILEO had used the newly invented TELESCOPE to discover SUNSPOTS, the phases of Venus and four moons of Jupiter.

Since the 17th century the development of astronomy has followed on successive improvements in the design of telescopes. In 1781 WILLIAM HERSCHEL discovered Uranus, the first discovery of a new PLANET to be made in historical times. Measurement of the PARALLAX of a few stars in 1838 first allowed the estimation of interstellar distances. Analysis of the FRAUNHOFER LINES in the spectrum of the Sun gave scientists their first indication of the chemical composition of the STARS.

In the present century the scope of observational astronomy has extended to ELECTROMAGNETIC RADIATION emitted by celestial objects from the shortest wavelengths (GAMMA RAYS) to the largest (RADIO), taking in X-RAYS, ULTRAVIOLET, optical and INFRARED. As the Earth's atmosphere absorbs most of these radiations from space, telescopes have been launched by rockets and carried on satellites. Thus, astronomers have discovered QUASARS, PULSARS and neutron stars, and have been able to show in detail the faintest parts of distant galaxies and nebulae. In their turn, these discoveries have enabled cosmologists to develop more self-consistent models of the UNIVERSE. (See also COSMOLOGY; OBSERVATORY.) (♦ page 134, 228, 240)

Astrophotography

Photography of celestial objects, usually by using a TELESCOPE to focus an image onto a photographic plate. Of paramount importance in, for example, measurement of stellar PARALLAX, astrophotography has almost entirely replaced direct observation. The photographic plate has itself been replaced for certain applications by more sensitive electronic imaging devices.

Astrophysics

Study of the physical and chemical nature of celestial objects and events, using the data produced by, for example, SPECTROSCOPY and RADIO ASTRONOMY. By investigating and applying the laws of nature, astronomers formulate theories of the behavior and evolution of stars, galaxies and the Universe (see COSMOLOGY).

Ataxia

Impaired coordination of body movements resulting in unsteady gait, difficulty in fine movements and speech disorder. Caused by disease of the cerebellum or spinal cord, ataxia occurs with MULTIPLE SCLEROSIS, certain hereditary conditions and in the late stages of syphilis.

Atherosclerosis

Disease of arteries in which the wall becomes thickened and rigid, and blood flow is hindered. Fatty deposits (containing CHOLESTEROL) are formed in the inner lining of an ARTERY, followed by scarring and calcification. Excess saturated fats in the blood may play a role in the formation of these deposits. Narrowing or obstruction of cerebral arteries may lead to STROKE, while that of coronary arteries causes ANGINA PECTORIS and CORONARY THROMBOSIS. Reduced blood flow in limbs may cause CRAMP, ULCERS and GANGRENE.

Atlas

The uppermost VERTEBRA of the spinal column, supporting the skull, and forming a pivot joint with the axis vertebra below it, thus allowing the head to turn. (See also SKELETON.)

Atmosphere

The envelope of GAS, VAPOR and AEROSOL particles surrounding the EARTH, retained by gravity and forming a major constituent in the environment of most forms of terrestrial life, protecting it from the impact of METEORS, COSMIC RAY particles and harmful solar radiation. The composition of the atmosphere and most of its physical properties vary with ALTITUDE, certain key properties being used to divide the whole into several zones, the upper and lower boundaries of which change with LATITUDE, the time of day and the season of the year. About 75% of the total MASS of atmosphere and 90% of its water vapor and aerosols are contained in the troposphere, the lowest zone. Excluding water vapor, the air of the troposphere contains 78% NITROGEN, 20% OXYGEN, 0.9% ARGON and 0.03% CARBON dioxide, together with traces of the other NOBLE GASES; and METHANE, HYDROGEN and nitrous oxide. The water vapour content fluctuates within wide margins as water is evaporated from the OCEANS, carried in CLOUDS and precipitated upon the continents. The air flows in meandering currents, transferring ENERGY from the warm equatorial regions to the colder poles (see also GREENHOUSE EFFECT). The troposphere is thus the zone in which weather occurs (see METEOROLOGY), as well as that in which most air-dependent life exists. Apart from occasional INVERSIONS, the TEMPERATURE falls with increasing altitude through the troposphere until at the tropopause (altitude 7km at the poles, 16km on the equator) it becomes constant (about 217K), and then slowly increases again into the stratosphere (up to about 48km). The upper stratosphere contains the OZONE layer, which filters out the dangerous ULTRAVIOLET RADIATION incident from the SUN. Above the stratosphere, the mesosphere merges into the IONOSPHERE, a region containing various layers of charged particles (IONS) of immense importance in the propagation of RADIO waves, being used to reflect signals between distant ground stations. At greater altitudes still, the ionosphere passes into the exosphere, a region of rarefied HELIUM and hydrogen gases, in turn merging into the interplanetary medium. In all, the atmosphere has a mass of about 5.2×10^{18}kg, its DENSITY being about 1.23kg/m at sea level. Its WEIGHT results in its exerting an average air pressure of 101.3kPa (1013mbar) near the surface, this fluctuating greatly with the weather and falling off rapidly with height (see also PRESSURE). The other PLANETS of the SOLAR SYSTEM (with the possible exception of PLUTO), though only two of their SATELLITES, all have distinctive atmospheres, though none of these contains as much life-supporting oxygen as does that of the Earth. (♦♦ page 25, 44.)

Atmosphere (atm)

CGS UNIT of PRESSURE; see BAR.

Atmosphere refraction

The REFRACTION of light rays passing through the ATMOSPHERE because of variations in its density and temperature, which produce corresponding variations in its refractive index. Under standard conditions, rays of light passing through the atmosphere are slightly curved so that the apparent positions of celestial bodies are displaced toward the ZENITH by a small amount; ie celestial bodies appear slightly higher above the horizon than their true positions. The effect is greater close to the horizon. Unusual density variations may produce MIRAGES, shimmer and other deceptive effects. (♦ page 25)

Atoll

A typically circular CORAL reef enclosing a LAGOON with no central island. Many atolls, often supporting low arcuate islands, are found in the Pacific Ocean.

Atom

Classically one of the minute, indivisible, homogeneous material particles of which material objects are composed (see ATOMISM), and in 20th-century science the name given to a relatively stable package of MATTER, typically about 0.1nm across, and itself made up of at least two SUBATOMIC PARTICLES. Every atom consists of a tiny nucleus (containing positively charged PROTONS and electrically neutral NEUTRONS) with which is associated a number of negatively charged ELECTRONS. These, though individually much smaller than the nucleus, occupy a hierarchy of ORBITALS that represent the atom's electronic ENERGY LEVELS, and fill most of the space taken up by the atom. The number of protons in the nucleus of an atom (the atomic number, Z) defines of which chemical ELEMENT the atom is an example. In an isolated neutral atom the number of electrons equals the atomic number, but in an electrically charged ION of the same atom there is either a surfeit or a deficit of electrons. The number of neutrons in the nucleus (the neutron number, N) can vary between different atoms of the same element, the resulting species being called the ISOTOPES of the element. Most stable isotopes have slightly more neutrons than protons. Although the nucleus is very small, it contains nearly all the MASS of the atom – protons and neutrons having very similar masses, and the mass of the electron (about 0.05% of the proton mass) being almost negligible. Counting the proton mass as one, this means that the mass of the atom is roughly equal to the total number of its protons and neutrons. This number $Z+N$, is known as the mass number of the atom, A. In equations representing nuclear reactions the atomic number of an atom is often written as a subscript preceding the chemical symbol for the element and the mass number as a superscript following it. Thus an atomic nucleus with mass number 16 and containing 8 protons belongs to an atom of "oxygen-16", written ${}^{16}_{8}O$. The average of the mass numbers of the various naturally occurring isotopes of an element, weighted according to their relative abundance, gives the chemical ATOMIC WEIGHT of the element. Subatomic particles fired into atomic nuclei can cause nuclear reactions giving rise either to new isotopes of the original element or to atoms of a different element, and emitting ALPHA PARTICLES, BETA RAYS or GAMMA RAYS.

The earliest atomistic concept, regarding the atom as that which could not be subdivided, was implicit in the first modern, chemical atomic theory, that of JOHN DALTON (1808). Although it survives in the once common chemical definition of atom – the smallest fragment of a chemical element that retains the properties of that element and can take part in chemical reactions – chemists now recognize that it is the MOLECULE and not the atom that is the natural chemical unit of matter. The atomic nuclei form the vertebra of the molecules; but it is the interaction of the VALENCE electrons associated with these nuclei, rather than

CELL DIVISION

Meiosis

***1** Meiosis is the creation of sex cells that may combine to form a new human being. It differs from other forms of cell division in that the sex cells (sperm or eggs) or gametes formed by this process contain only half the normal number of chromosomes and genes.*

***2** The chromosomes form a distinct shape and group into 23 pairs. One from each pair was originally inherited from the person's father, the other from the mother.*

***3** Each chromosome shortens, thickens, and splits lengthwise to form two chromatids, held together at a central point in an X-shape. At this stage each chromosome thus carries double the usual amount of genetic material.*

***4** Paternal and maternal chromosomes line up together closely.*

Parents: yellow round YYRR; green wrinkled yyrr
Gametes: YR, yr
F_1
Gametes: YR, Yr, yR, yr; YR, Yr, yR, yr
F_2: YR YR, YR Yr, Yr YR, YR yR, Yr Yr, yR YR, YR yr, Yr yR, yR Yr, yr YR, Yr yr, yR yR, yr Yr, yR yr, yr yR, yr yr

Summary of F_2: 9 yellow round + 3 yellow wrinkled + 3 green round + 1 green wrinkled

Mendelian genetics

◀ The way in which genes interact was demonstrated by Gregor Mendel for peas, with the characteristics of yellowness (dominant gene) or greenness (recessive), and round (dominant) or wrinkled seeds (recessive). When a plant carrying both dominant genes was bred with one carrying the two recessive genes, the progeny produced round yellow seeds (F1). When two of these were crossed, their progeny (F2) showed four different possibilities in a 9:3:3:1 ratio. This effect can be explained in terms of the two pairs of characters segregating independently. The dominant gene of each pair determines whether or not the recessive characteristic is manifested, but the recessive gene remains intact and can be passed on to succeeding generations.

The inheritance of sex

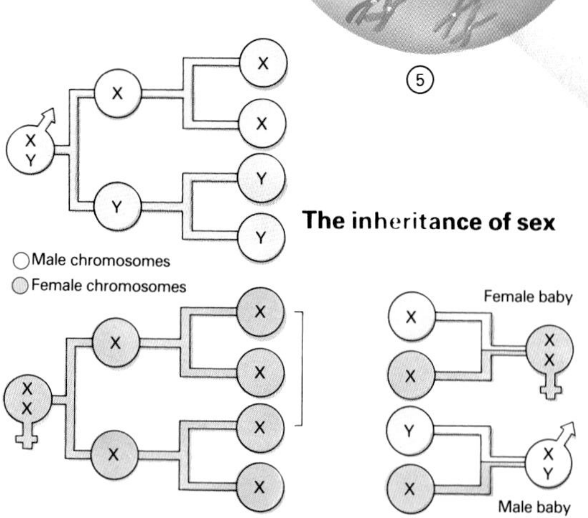

***▲** A child's sex is determined by the chromosomes inherited from its parents. In the process of sperm-creation, the father's chromosomes are split so that half his sperm carry the X, the other half the Y. If a Y sperm fertilizes an X ovum, the child will be male; if the sperm carries X, the child will be female, with two X chromosomes and no Y.*

Mitosis

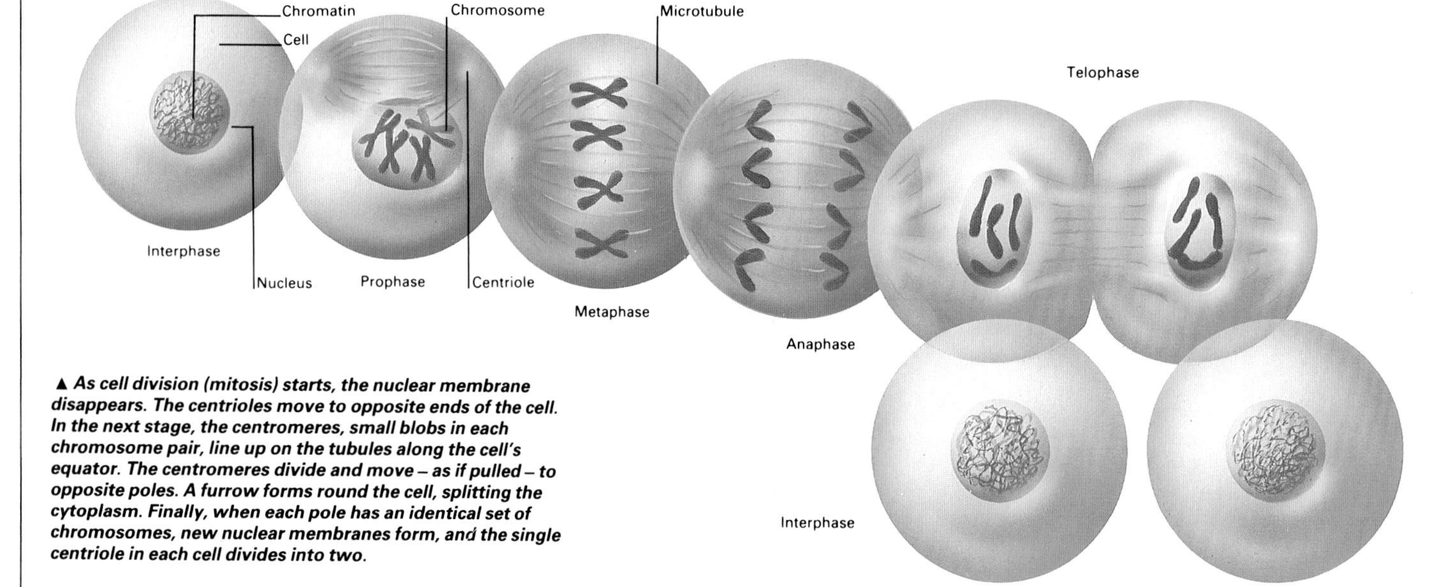

***▲** As cell division (mitosis) starts, the nuclear membrane disappears. The centrioles move to opposite ends of the cell. In the next stage, the centromeres, small blobs in each chromosome pair, line up on the tubules along the cell's equator. The centromeres divide and move – as if pulled – to opposite poles. A furrow forms round the cell, splitting the cytoplasm. Finally, when each pole has an identical set of chromosomes, new nuclear membranes form, and the single centriole in each cell divides into two.*

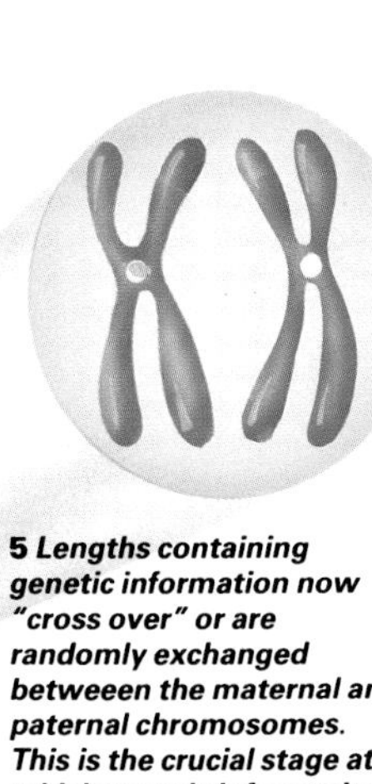

► *A computer-graphic representation of the structure of DNA.*

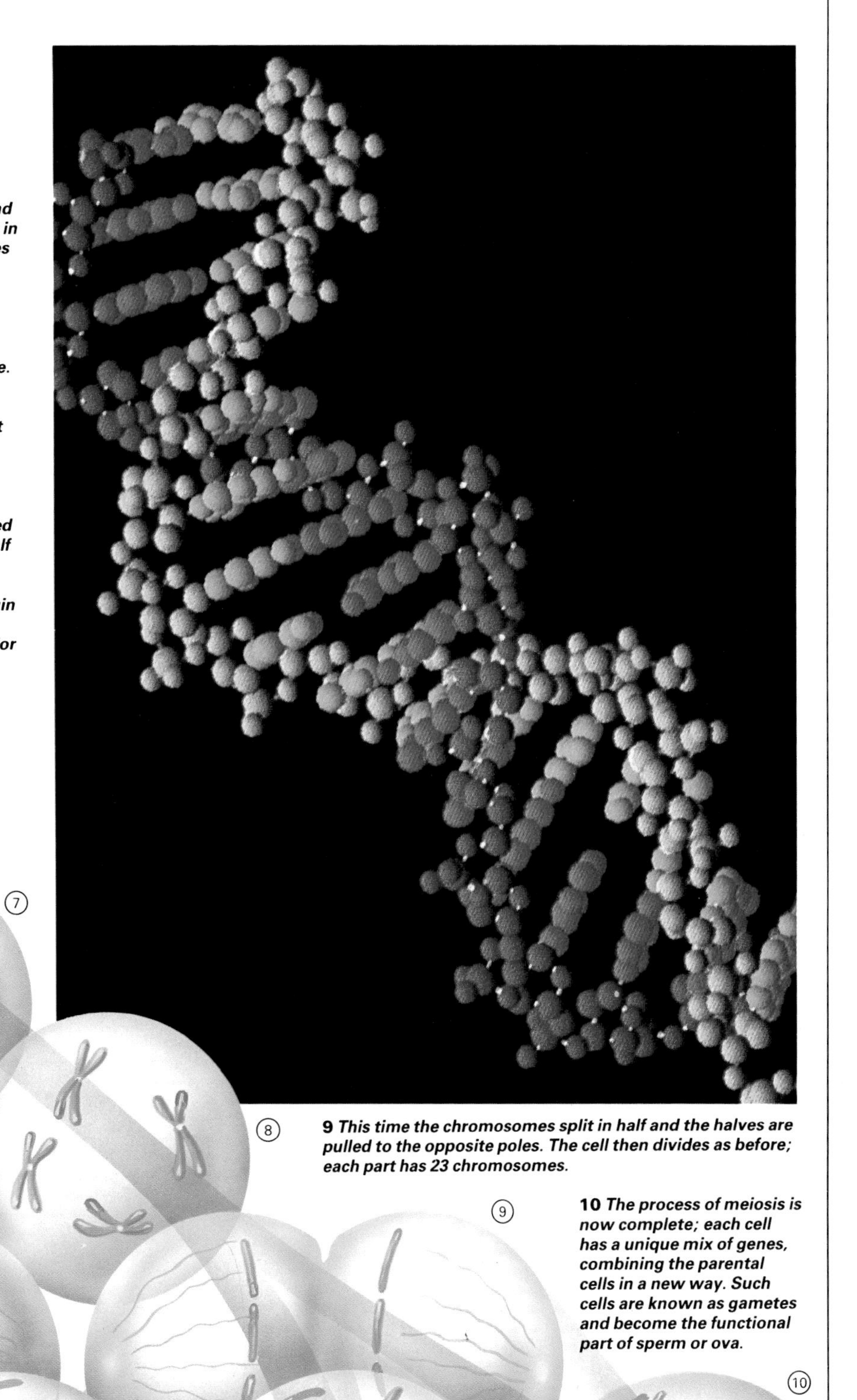

5 *Lengths containing genetic information now "cross over" or are randomly exchanged betweeen the maternal and paternal chromosomes. This is the crucial stage at which genetic information is mixed, and is the stage at which mutations may occur as some lengths may be mismatched, transferred to a new position or inverted. Many such mutations are impossible; other may develop new traits.*

6 *Poles form in the cell and the chromosomes collect in the middle. Chromosomes from each pair are then pulled to opposite sides.*

7 *The cell begins to divide. Each part now contains only 23 chromosomes, although each one is split into two chromatids.*

8 *After the cell has divided completely, with each half containing a set of 23 chromosomes, each still doubled, poles again begin to form in each and the chromosomes line up prior to the next cell division.*

9 *This time the chromosomes split in half and the halves are pulled to the opposite poles. The cell then divides as before; each part has 23 chromosomes.*

10 *The process of meiosis is now complete; each cell has a unique mix of genes, combining the parental cells in a new way. Such cells are known as gametes and become the functional part of sperm or ova.*

the properties of the individual component atoms, that is responsible for the chemical behavior of matter. (➧ page 37, 49, 130, 147)

Atomic bomb
A weapon of mass destruction deriving its energy from nuclear FISSION. The first atomic bomb was exploded at Alamogordo, New Mexico, on 16 July 1945. As in the bomb dropped over Hiroshima, Japan a few weeks later (6 August), the fissionable material was uranium-235, but when Nagasaki was destroyed by another bomb three days after that, plutonium-239 was used. Together the Hiroshima and Nagasaki bombs killed more than 100,000 people. Since the early 1950s, the power of the fission bomb (equivalent to some 20,000 tons of TNT in the case of the Hiroshima bomb) has been vastly exceeded by that of the HYDROGEN BOMB, which depends on nuclear FUSION.

Atomic clock
A device that utilizes the exceptional constancy of the FREQUENCIES associated with certain electron SPIN reversals (as in the CESIUM clock) or the inversion of AMMONIA molecules (ammonia clock) to define an accurately reproducible time scale.

Atomic energy see NUCLEAR ENERGY.

Atomic number (Z). see ATOM.

Atomic reactor see NUCLEAR REACTOR.

Atomic waste disposal see NUCLEAR ENERGY.

Atomic weight (AW)
The mean mass of the ATOMS of an ELEMENT weighted according to the relative abundance of its naturally occurring ISOTOPES and measured relative to some standard. Since 1961 this standard has been provided by the CARBON isotope C^{12} whose atomic mass is defined to be exactly 12. On this scale atomic weights for the naturally occurring elements range from 1.008 (HYDROGEN) to 238.03 (URANIUM). (➧ page 130, 147)

Atomism
The theory that all matter consists of atoms – minute indestructible particles, homogeneous in substance but varied in shape, developed in the 5th century BC.

Atom smasher see ACCELERATORS, PARTICLE.

Atopy
Allergy, or the condition of being likely to have allergic reactions.

ATP (adenosine triphosphate)
A ubiquitous and highly important chemical substance in living organisms. Its role in the cell is to store energy and make it available for reactions where it is needed; thus it is the link between CATABOLISM and ANABOLISM, and is often referred to as the "energy currency" of the cell. The energy is stored in the bond with the third phosphate group, and when this is split off, to give ADP (adenosine diphosphate), energy is released. Adenosine itself is a NUCLEOTIDE. A derivative of ATP, adenosine monophosphate (AMP) forms a cyclical molecule (cyclic AMP, or cAMP), which is an important messenger substance, mediating the effects of many hormones.

Atrophy
Wasting away of bodily TISSUES or organs because of disease, MALNUTRITION, disuse or old age. (See also MUSCULAR DYSTROPHY.)

Atropine
An ALKALOID derived from HYOSCYAMINE obtained from belladonna (deadly nightshade). It decreases the effects of the parasympathetic NERVOUS SYSTEM, and is used to dilate the pupils of the eyes, to increase heartrate, to reduce the secretion of mucus and saliva, and to relax spasm.

Aureomycin see ANTIBIOTICS; TETRACYCLINES.

Auriga
The Charioteer, winter constellation of N skies, containing CAPELLA, Alpha Aurigae, the sixth brightest star in the night sky.

Aurora (polar lights)
Striking display of lights seen in night skies near the Earth's geomagnetic poles. The *aurora borealis* (northern lights) is seen in Canada, Alaska and N Scandinavia, the *aurora australis* (southern lights) is seen in Antarctic regions. The auroras are caused by the collision of air molecules in the upper atmosphere with charged particles from the Sun that have been accelerated and "funneled" by the Earth's magnetic field. Particularly intense auroras are associated with high solar activity. Night-time AIRGLOW is termed the permanent aurora.

Auscultation
Diagnosis by listening, usually of the heart and lungs with a stethoscope.

Autism
Withdrawal from reality and relations with others. Pathological autism occurs in various psychoses, especially SCHIZOPHRENIA. Certain children who fail to establish normal communication with others or social responses are termed autistic; it may represent a juvenile form of schizophrenia.

Autoclave
A strong-walled pressure vessel suitable for heating liquid above its BOILING POINT, used in the study of high-pressure chemical reactions, for sterilizing and cooking and for impregnating wood. The industrial autoclave derives from Denis Papin's "steam digester" of 1679, the prototype for the modern pressure cooker.

Auto-immune diseases
Conditions in which the immune system (See IMMUNITY) attacks the body's own tissues as though they were foreign. Many diseases, such as diabetes and rheumatoid arthritis, may be partly auto-immune.

Autonomic nervous system see NERVOUS SYSTEM.

Autopsy
A postmortem examination, the dissection of a corpse to determine the cause of death and the nature and progress of the prior disease by the recognition of abnormal anatomy (see PATHOLOGY). In FORENSIC MEDICINE it may determine identity and the manner and time of death.

Autosuggestion see SUGGESTION.

Autotroph
An organism that can make its own food, using only inorganic compounds and solar energy (PHOTOSYNTHESIS) or chemical energy (CHEMOSYNTHESIS). Green plants, cyanobacteria and other bacteria are autotrophs; animals, fungi and most bacteria are heterotrophs, requiring also organic compounds to sustain life. Autotrophs form the primary link in the food chain.

Auxins
Plant HORMONES that promote lengthwise growth and control ABSCISSION and the plant's responses to light and gravity (see TROPISMS). Natural auxins are derivatives of indole (see HETEROCYCLIC COMPOUNDS). Synthetic auxins are used for crop control and as HERBICIDES.

Avalanche
Mass of snow, ice and mixed rubble moving down a mountainside or over a precipice, usually caused by loud noises or other shock waves acting on snow already unstable from subsurface thawing. They can be major factors in EROSION.

Aves see BIRDS.

Avicenna (980-1037)
Latin name of Abu-Ali al-Husayn ibn Sina, the greatest of the Arab scientists of the medieval period. His *Canon of Medicine* remained one of the standard medical texts in Europe until the Renaissance.

Avogadro, Count Amedeo (1776-1856)
Italian physicist who first realized that gaseous ELEMENTS might exist as MOLECULES that contain more than one ATOM, thus distinguishing molecules from atoms. In 1811 he published Avogadro's hypothesis – that equal volumes of all GASES under the same conditions of TEMPERATURE and PRESSURE contain the same number of molecules – but his work in this area was ignored by chemists for over 50 years. The Avogadro Number ($\mathcal{N}$), the number of molecules in one MOLE of substance, 6.02×10^{23}, is named for him.

Avoirdupois
The system of weights customarily used in the US for most goods except gems and drugs, for which are employed respectively TROY WEIGHTS and APOTHECARIES WEIGHTS. There are 7000 grains or 16 ounces in the pound avoirdupois.

Axelrod, Julius (b. 1912)
US biochemist who shared the 1970 Nobel Prize for Medicine or Physiology with BERNARD KATZ and ULF VON EULER for their independent contributions toward elucidating the chemistry of the transmission of nerve impulses (see NERVOUS SYSTEM). Axelrod identified a key ENZYME in this mechanism.

Axis see ATLAS; VERTEBRAE.

Axolotl
The larva of the salamander *Ambystoma mexicanum*. They usually never become adult (see PEDOMORPHOSIS), but in captivity they can be induced to develop into normal adult salamanders by feeding them with thyroid extract. They are 10-17.5mm in length.

Axon
The fiber of a nerve cell (NEURON) that conducts impulses away from the cell body. At the end of the axon is a SYNAPSE, where the nerve impulse is transmitted to another cell by means of NEUROTRANSMITTERS. The axons of motor nerves can be over a meter in length. (➧ page 126)

Azeotropic mixture
Constant-boiling-point mixture, a SOLUTION of two or more liquids that on DISTILLATION behaves as a pure liquid: its boiling-point is invariable at a given pressure, and the composition of the vapor phase is the same as that of the liquid. (See also PHASE EQUILIBRIA.)

Azide
A chemical compound, either organic or inorganic, characterized by having a $-N{=}N{=}N$ group as part of its structure. Many azides are explosive; lead azide is used as a detonator.

Azimuth
In navigation and astronomy, the angular distance measured from 0-360° along the horizon eastward from an observer's north point to the point of intersection of the horizon and a great circle (see CELESTIAL SPHERE) passing through the observer's ZENITH and a star or planet.

Azo compounds
Class of organic compounds of general formula $R{-}N{=}N{-}R'$. The most important azo compounds have AROMATIC groups for R and R′, and are made by coupling of diazonium compounds with NUCLEOPHILES such as PHENOLS or aromatic AMINES. They comprise more than half the DYES commercially available.

Azurite
Blue mineral consisting of basic COPPER carbonate ($Cu_{32}[CO_3]_2$), occurring with MALACHITE, notably in France, SW Africa and Arizona. It forms monoclinic crystals used as GEMS, and was formerly used as a pigment.

Babbitt metal
An ALLOY containing 89% TIN, 9% ANTIMONY and 2% COPPER, devised in 1839 by US inventor Isaac Babbitt (1799-1862) for lining bearings. Today the term babbitt metal is also applied to other high-tin and high-lead bearing alloys.

Bacillus
A genus of rod-shaped BACTERIA of the family Bacillaceae. The term is sometimes also used for any cylindrical bacterium.

Backbone
In animals, see VERTEBRATES; INVERTEBRATES; in humans, see SKELETON; VERTEBRAE.

Bacon, Francis; Lord Verulam, Viscount St Albans (1561-1626)
English philosopher and statesman who rose to become Lord Chancellor (1618-21) to James I, but who is chiefly remembered for the stimulus he gave to scientific research in England.

Bacteria
Unicellular microorganisms between 0.3 and 2μm in diameter. They differ from plant and animal cells in lacking a nucleus and other cell organelle: they are PROKARYOTES. Together with the CYANOBACTERIA, they form one of the major KINGDOMS of life, the Monera.

Some bacteria are SYMBIOTIC, living either in MUTUALISTIC relationships (as in RUMINANTS, for example), or as commensals (see COMMENSALISM), which coexist harmlessly with host cells. Other bacteria are PARASITES that live inside host cells or body cavities, and damage the host organism by producing toxins, which may cause tissue damage (see BACTERIAL DISEASES). Such bacteria are often described as pathogens rather than parasites. This distinction between a commensal and a parasite/pathogen is not absolute: *Escherischia coli* is a commensal in the human intestine, but may cause infection in the urinary tract.

Bacteria are surrounded by a rigid cell wall and many are incapable of movement, but certain types can swim using hairlike (single-fibril) flagella (more correctly called an UNDULIPODIUM) or by other means (see below). Bacteria vary in their food requirements: some are AUTOTROPHS and can obtain energy by PHOTOSYNTHESIS or CHEMOSYNTHESIS, whereas others are HETEROTROPHS and therefore need organic substances for nutrition. Aerobic (see AEROBE) bacteria need oxygen to survive, whereas anaerobic (see ANAEROBE) species do not and are poisoned by it. Most such anaerobes are primitive forms belonging to the ARCHEBACTERIA. Bacteria generally reproduce asexually by binary FISSION, but some species can transfer small fragments of genetic material, known as PLASMIDS, from one to another; this is not true sexual reproduction, but it parallels sexual processes in higher organisms. Certain bacteria can survive adverse conditions by forming highly resistant spores.

Bacteria are used industrially in biotechnological processes and, more traditionally, in the making of many dairy products. Some bacteria, especially the ACTINOMYCETES, produce ANTIBIOTICS, used in destroying pathogenic bacteria.

CLASSIFICATION There is no standard way of classifying bacteria. The higher bacteria are filamentous and the cells may be interdependent – they include the family Actinomycetes. The lower bacteria are subdivided according to shape: cocci (round), bacilli (cylindrical), vibrios (curved), and spirilla (spiral). Cocci live singly, in pairs (DIPLOCOCCI), in clusters (STAPHYLOCOCCI) or in chains (STREPTOCOCCI) – as a group they are of great medical importance. Spirochetes form a separate group from the above: they have a less rigid cell wall and internalized undulipodia which enable them to move by a corkscrew-like motion. The response reflects a fundamental difference in the construction of their cell walls. Bacteria are also classified in terms of their response to the GRAM STAIN: those absorbing it are termed gram-positive, those not, gram-negative. (See also BACTERIOLOGY; MICROBIOLOGY.)(♦♦ page 13, 44)

Bacterial diseases
Diseases caused by BACTERIA or their products. Many bacteria have no effect and some are beneficial, while only a small number lead to disease. This may be a result of bacterial growth, the INFLAMMATION in response to it or of TOXINS (eg TETANUS, BOTULISM and CHOLERA). Bacteria may be contracted from the environment, other animals or humans, or from other parts of a single individual. Infection of SKIN and soft tissues with STAPHYLOCOCCUS or STREPTOCOCCUS leads to BOILS, carbuncles, impetigo, cellulitis, SCARLET FEVER and ERYSIPELAS. ABSCESS represents the localization of bacteria, while SEPTICEMIA is infection circulating in the BLOOD. Sometimes a specific bacterium causes a specific disease (eg ANTHRAX, DIPHTHERIA, TYPHOID FEVER), but any bacteria in some organs cause a similar disease: in LUNGS, PNEUMONIA occurs; in the urinary tract, CYSTITIS or nephritis, and in the BRAIN coverings, MENINGITIS. Many SEXUALLY TRANSMITTED DISEASES are due to bacteria. In some diseases (eg TUBERCULOSIS, LEPROSY, RHEUMATIC FEVER), many manifestations are due to hypersensitivity (see IMMUNITY) to the bacteria. While ANTIBIOTICS have greatly reduced death and ill-health from bacteria, and VACCINATION against specific diseases (eg WHOOPING COUGH) has limited the number of cases, bacteria remain an important factor in disease.

Bacteriology
The science that deals with BACTERIA, their characteristics and their activities as related to medicine, industry and agriculture. Bacteria were discovered in 1676 by ANTON VON LEEUWENHOEK. Modern techniques of study originate from about 1870 with the use of stains and the discovery of culture methods using plates of nutrient AGAR media. Much pioneering work was done by LOUIS PASTEUR and ROBERT KOCH. (See also BACTERIAL DISEASES; NITROGEN FIXATION; SPONTANEOUS GENERATION.)

Bacteriophage (phage)
A VIRUS that attacks bacteria. They have a thin PROTEIN coat surrounding a central core of DNA (or occasionally RNA), and a small protein tail. The phage attaches itself to the bacterium and injects the DNA or RNA into the cell. This genetic material alters the metabolism of the bacterium, and several hundred phages develop inside it; eventually the cell bursts, releasing the new, mature phages.

Badlands
Arid to semiarid areas of pinnacles, ridges and gullies, usually lacking in vegetation, formed by heavy EROSION of non-uniform rock. The Big Badlands of South Dakota are a particularly notable example.

Baekeland, Leo Hendrik (1863-1944)
Belgian-born chemist who, after emigrating to the US in 1889, devised Velox photographic printing paper (selling the process to EASTMAN in 1899) and went on to discover BAKELITE, the first modern synthetic PLASTIC.

Baer, Karl Ernst von (1792-1876)
Estonian-born German embryologist who discovered the mammalian egg (see REPRODUCTION) and the NOTOCHORD of the vertebrate embryo. He is considered to have been one of the founders of comparative EMBRYOLOGY.

Baeyer, Johann Friedrich Wilhelm Adolf von (1835-1917)
German organic chemist who proposed a "strain theory" to account for the relative stabilities of ring compounds (see ALICYCLIC COMPOUNDS). He was awarded the 1905 Nobel Prize for Chemistry for his research on dyestuffs: in 1878 he had become the first to synthesize indigo.

Bailey, Liberty Hyde (1858-1954)
US horticulturalist and botanist who, as professor of horticulture at Cornell University (1888-1913), did much to put US horticulture on a scientific footing.

Bailly, Jean Sylvain (1736-1793)
French astronomer and politician. After studying the satellites of Jupiter and writing a five-volume history of astronomy (1775-87), he turned to politics, becoming mayor of Paris 1789-91; he was executed in the Terror.

Baily's Beads
Named for Francis Baily (1774-1844), the apparent fragmentation of the thin crescent of the Sun just before totality in a solar ECLIPSE, caused by sunlight shining through mountains at the edge of the lunar disk.

Baird, John Logie (1888-1946)
Scotish inventor who, by first transmitting moving pictorial images (1925), inaugurated the television era. Although the Baird system was tested when the British public television service began in 1936, it was dropped in favor of a rival that used electronic rather than mechanical scanning.

Bakelite
Synthetic RESIN discovered by LEO BAEKELAND, made by chemical reaction of FORMALDEHYDE and PHENOL. It is a thermosetting PLASTIC (see also POLYMERS). A hard, strong material, it is used as an electrical insulator, an adhesive and a paint binder.

Balance
Instrument used for measuring the WEIGHT of an object, typically by comparison with objects of known weight. The equal-arm balance, known to ancient Egyptians and Mesopotamians, consists of two identical pans hung from either end of a centrally suspended beam. When objects of equal weight are placed in each pan, the beam swings level because the MOMENTS of the gravitational FORCES acting on each object and pan about the central pivot or fulcrum are equal in magnitude and opposite in sense. Other types of beam balance involve fixed weights sliding along or hung below unequal beam arms, but the principle of the balancing of equal and opposite gravitational moments remains the same. The relatively inaccurate spring balance utilizes HOOKE's Law to determine the weight of the specimen from the extension it produces in a coiled spring, and the much finer TORSION balance utilizes the resistance of a wire to being twisted. Modern chemical microbalances can measure weights as small as 1μg.

Balance of nature
Late 19th- and early 20th-century concept of nature existing in an equilibrium maintained by interdependencies between different organisms. In fact this balance can easily be disrupted, as

many factors (eg climatic changes, POLLUTION) may cause dynamic change in large or small natural populations. (See also ECOLOGY.)

Baldness see ALOPECIA.

Baleen
A material found as fibrous plates hanging in rows from the roof of the mouth in whales; it is often known as whalebone. Its function is to strain PLANKTON, on which these whales feed, from the water. Strong and elastic, it was used for many purposes before the advent of plastic and spring steel.

Ballistics
The science concerned with the behavior of projectiles, traditionally divided into three parts.
INTERIOR BALLISTICS is concerned with the progress of the projectile before it is released from the launching device. In the case of a gun this involves determining the propellant charge, barrel design and firing mechanism needed to give the desired muzzle velocity and stabilizing spin to the projectile.
EXTERNAL BALLISTICS is concerned with the free flight of the projectile. At the beginning of the 17th century GALILEO determined that the trajectory (flight path) of a projectile should be parabolic, as indeed it would be if the effects of air resistance, the rotation and curvature of the Earth, the variation of air density and gravity with height, and the rotational INERTIA of the projectile could be ignored. The shockwaves accompanying projectiles moving faster than the speed of sound (see SUPERSONICS) are also the concern of this branch.
TERMINAL (PENETRATION) BALLISTICS deals with the behavior of projectiles on impacting at the end of their trajectory. The velocity- to-mass ratio of the impact particle is an important factor and results are of equal interest to the designers of ammunition and of armorplate.
FORENSIC BALLISTICS is a relatively recent development in the science, and now plays an important role in the investigation of gun crimes.

Baltimore, David (b. 1938)
US virologist who shared the 1975 Nobel Prize for Physiology or Medicine with R. DULBECCO and H.M. TEMIN in recognition of their work linking VIRUSES with the development of some cancerous tumors.

Bandwidth
In telecommunications, the difference (expressed in HERTZ) between the upper and lower limits of the band range of FREQUENCIES either needed or available to transmit a given signal, or which can be adequately passed by a component device.

Bang's disease see BRUCELLOSIS.

Banks, Sir Joseph (1743-1820)
British botanist and president of the ROYAL SOCIETY OF LONDON (1778-1820), the foremost British man of science of his time. He accompanied Cook as naturalist on his first expedition (1768-71) and was a key figure in the establishment of the Royal Botanic Gardens at Kew, London.

Banneker, Benjamin (1731-1806)
American mathematician and astronomer, notable as the first American Negro to gain distinction in science and the author of celebrated *Almanacs* (1791-1802).

Banting, Sir Frederick Grant (1891-1941)
Canadian physiologist who, with C.H. BEST, first isolated the hormone INSULIN from the pancreases of dogs (1922). For this he shared the 1923 Nobel Prize for Physiology or Medicine with J.J.R. MACLEOD, who provided experimental facilities.

Bar
A unit of pressure in the CGS system equal to 100kPa. The millibar (mbar or mb) – 100Pa – is commonly used in meteorology. The standard ATMOSPHERE is 1013.25mbar.

Bárány, Robert (1876-1936)
Austrian-born physiologist who received the 1914 Nobel Prize for Physiology or Medicine for research on the functioning of the organs of the inner EAR. From 1916 he continued his research in Sweden.

Barbels
Tactile organs protruding about the mouth of certain fish, notably the barbels (*Barbus*) of the Family Cyprinidae.

Barbiturates
A class of drugs acting on the central NERVOUS SYSTEM that may be SEDATIVES, anesthetics or anticonvulsants. They depress nerve cell activity, the degree of depression and thus clinical effect varying in different members of the class.

Bardeen, John (b. 1908)
US physicist who shared the 1956 Nobel Prize for Physics with SHOCKLEY and BRATTAIN for their development of the TRANSISTOR. In 1972 he became the first person to win the Physics Prize a second time, sharing the award with COOPER and SCHRIEFFER for their development of a comprehensive theory of SUPERCONDUCTIVITY.

Barite
Commonest barium mineral, consisting of barium sulfate (see BARIUM); white or yellow orthorhombic crystals. It is very dense and is mined for use in oil well drilling muds and as the chief source of barium compounds.

Barium (Ba)
Silvery white ALKALINE-EARTH METAL resembling CALCIUM; chief ores BARITE and witherite ($BaCO_3$). Barium is used to remove traces of gases from vacuum tubes; its compounds are used in making flares, fireworks, paint pigments and poisons. AW 137.3, mp 730°C, bp 1640°C, sg 3.5 (20°C). Barium sulfate ($BaSO_4$), highly insoluble and opaque to X-rays, can be safely ingested for X-ray examination of the GASTROINTESTINAL TRACT. (♦ page 130)

Barium tests
X-ray examination of the GASTROINTESTINAL TRACT using radio-opaque substances, usually barium salts. In barium swallow and meal, an emulsion is swallowed and the ESOPHAGUS, STOMACH and DUODENUM are X-rayed. A follow-through may be performed later to outline the small intestine. For barium enema, a suspension is passed into rectum and large intestine. CANCER, ULCERS, diverticulae and forms of ENTERITIS and COLITIS may be revealed.

Bark
General term for the covering of stems of woody plants, comprising the secondary phloem, cork cambium and CORK. The bark is impervious to water and protects the stem from excessive evaporation; it also protects the more delicate tissues within. (See also TREE.)

Barkla, Charles Glover (1877-1944)
British physicist who was awarded the 1917 Nobel Prize for Physics for research on the scattering of X-RAYS by GASES, in particular for his discovery of the "characteristic radiation" scattered by different elements.

Barn (b)
Unit of nuclear CROSS-SECTION (area) equal to $10^{-28}m^2$.

Barnard, Edward Emerson (1857-1923)
US astronomer who discovered the fifth satellite of JUPITER (1892). In 1916 he discovered Barnard's star, a red dwarf STAR only 6 ly from the Earth and which has the largest known stellar PROPER MOTION.

Barometer
An instrument for measuring air pressure (see ATMOSPHERE), used in WEATHER FORECASTING and for determining altitude. Most commonly encountered is the aneroid barometer, in which the effect of the air in compressing an evacuated thin cylindrical corrugated metal box is amplified mechanically and read off on a scale or, in the barograph, used to draw a trace on a slowly rotating drum, thus giving a continuous record of the barometric pressure. The aneroid instrument is that used for aircraft altimeters. The earliest barometers, as invented by TORRICELLI in 1643, consisted simply of a glass tube about 800mm long closed at one end and filled with MERCURY before being inverted over a pool of mercury. Air pressure acting on the surface of the pool held up a column of mercury about 760mm tall in the tube, a "Torricellian" vacuum appearing in the closed end of the tube. The height of the column was read as a measure of the pressure. In the Fortin barometer, devised by Jean Fortin (1750-1831) and still used for accurate scientific work, the lower mercury level can be finely adjusted and the column height is read off with the aid of a VERNIER SCALE.

Barrier reef
CORAL reef, lying roughly parallel to a shore, but separated from it by a lagoon.

Bartholin
Family of Danish physicians. Caspar Berthelsen Bartholin (1585-1629) was author of the much-used *Institutiones Anatomicae* (1611). This was enlarged by his son, Thomas Bartholin (1616-1680), noted also for his study of the human lymphatic system (1652). Erasmus Bartholin (1625- 1698), brother of Thomas, in 1669 discovered the phenomenon of DOUBLE REFRACTION in Iceland spar crystals. Caspar Bartholin (1655-1738), son of Thomas, discovered Bartholin's glands, which secrete liquid at the sides of the vagina.

Barton, Sir Derek Harold Richard (b. 1918)
British organic chemist who shared the 1969 Nobel Prize for Chemistry with ODD HASSEL for the development of CONFORMATIONAL ANALYSIS.

Baryons
In particle physics, a class of SUBATOMIC PARTICLES (comprising the nucleons and hyperons) built from three QUARKS.(♦ page 49)

Basal metabolic rate (BMR)
A measure of the rate at which an animal at rest uses energy. Human BMR is a measure of the heat output per unit time from a given area of body surface, the subject being at rest under certain standard conditions. It is usually estimated from the amounts of oxygen and carbon dioxide exchanged in a certain time. (See METABOLISM.)

Basalt
A dense IGNEOUS ROCK, mainly plagioclase FELDSPAR, fine-grained and dark gray to black in color; volcanic in origin, it is widespread as lava flows or intrusions. Basalt can assume a striking columnar structure, as exhibited in the Palisades along the Hudson River, or in Devils Postpile in California, and it can also form vast plateaus, such as the 518,000 sq km Deccan of India. Most oceanic islands of volcanic origin, such as Hawaii, are basaltic.

Base
In chemistry, the complement of an acid. Bases used to be defined as substances that react with acids to form SALTS, or as substances that give rise to hydroxyl ions (see HYDROXIDE) in aqueous solution. Some such inorganic strong bases are known as ALKALIS. In modern terms, bases are species that can accept a HYDROGEN ion from an acid, or that can donate an electron-pair to a Lewis ACID. In biochemistry, "base" often refers to the PURINES and PYRAMIDINES, which are found in NUCLEOTIDES and form the basis of the GENETIC CODE.

Basidiomycetes
One of the two major divisions of the higher

CHEMICAL BONDING

Name and formula	Lewis	Structural	Ball-and-stick	Space-filling
Fluorine (F_2)				
Carbon dioxide (CO_2)				
Water (H_2O)				
Sulfur trioxide (SO_3)				
Ammonia (NH_3)				
Methane (CH_4)				
Benzene (C_6H_6)				
Sulfur (S_8)				

Chemical formulae

The formula of a chemical compound shows how many atoms it contains, but gives little information about its structure or architecture. Formulae of ionic solids show only the simplest ratio of atoms within the structure, but do not tell how many ions surround each ion in the solid network. Similarly the formulae of covalently bound molecules give no indication of molecular geometry. Several types of molecular diagrams can be drawn up to represent the architecture of the molecules. The simplest type shows the atoms joined by straight lines. Water (H_2O) and hydrogen sulfide (H_2S) are represented in this way as linear molecules, or, more accurately, the oxygen and sulfur atoms can be displaced to indicate the bond angle between the two H-O and H-S bonds, of 105° and 92° respectively. Multiple bonds are shown using two or three lines as necessary. This type of representation can also be used to give some idea of the arrangement of the molecule in three dimensions. In two dimensions, methane (CH_4) and ammonia (NH_3) can be drawn, but it is known that in methane the hydrogens surround the carbon atom tetrahedrally, and that the ammonia molecule has trigonal pyramidal geometry.

Other ways of representing molecules include the "ball-and-stick" model, which provides a useful three-dimensional structure of the molecule; and a space-filling model, which gives a clearer concept of the overall shape, but the individual bonds may be less easy to see.

Electron dot formulae help in the understanding of covalent bonding. Dots are used to show the number of valence electrons of an atom, and the shared electrons in covalent bonds are easily represented using these "Lewis" structures.

Metallic bonding

Covalent bonding

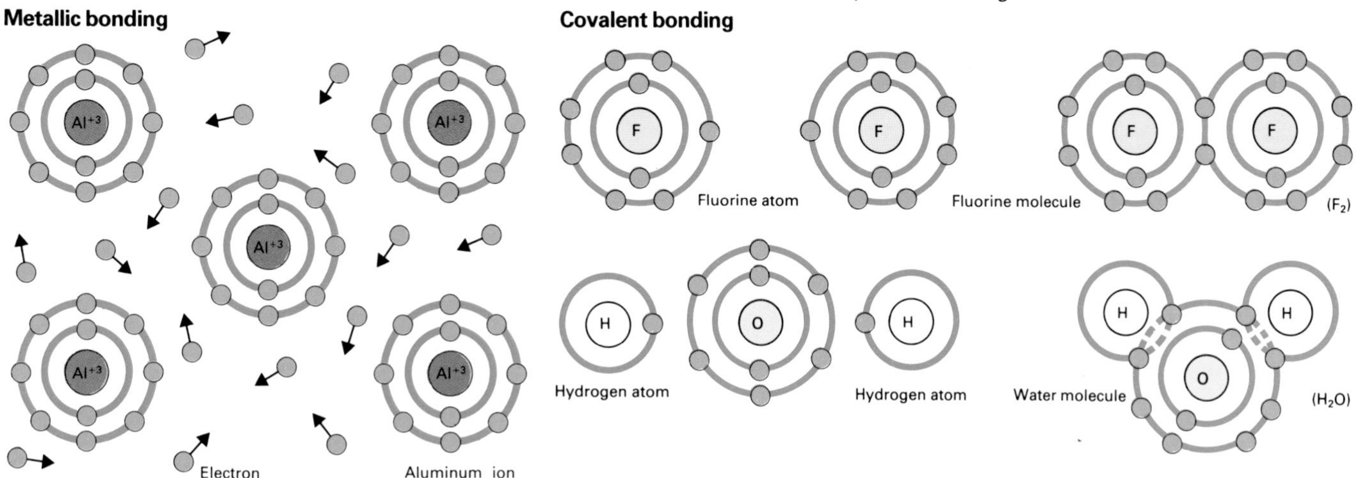

Ionic bonding

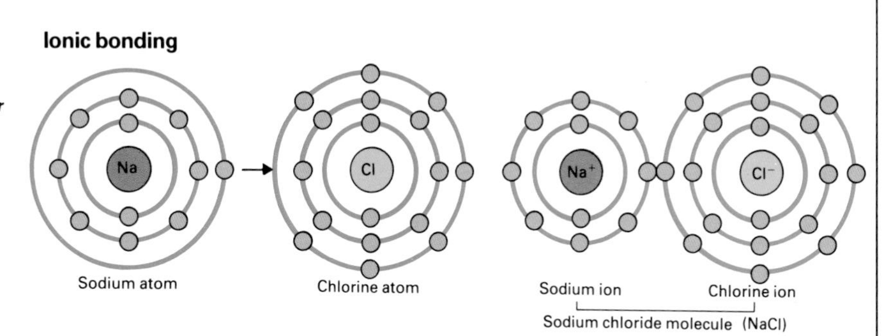

▲ In metallic bonds, the outer electrons of each atom float freely in the metal, and the ions are packed together geometrically. In covalent bonds the outer or valence electrons are shared between the atoms, or "paired" in order to fill the electron shells around the atoms. Thus in water, oxygen is two electrons short of a filled shell, and each hydrogen is one electron short of a filled shell. When the atoms come together, all three have their shells completed. In ionic bonds the outer electrons are exchanged: this gives the two ions an electric charge – one positive and one negative – and they are attracted electrostatically. Thus in common salt the sodium and chlorine ions are attracted to form sodium chloride.

FUNGI, the other being the ASCOMYCETES. The fruiting bodies of the Basidiomycetes include the familiar mushrooms and toadstools, as well as bracket fungi, stinkhorns and puffballs. Within the fruiting body the spores are produced in a characteristic structure, known as a basidium, from which the spores bud off. In modern taxonomic schemes the name Basidiomycete has been superseded by Basidiomycotina, but the old name is still widely used.

Basov, Nikolai Gennadievich (b. 1922)
Soviet physicist who shared the 1964 Nobel Prize for Physics with TOWNES and PROKHOROV for research in quantum physics which led to the development of LASERS and MASERS.

Bateson, William (1861-1926)
English biologist, known as the "father of GENETICS". In 1900 he translated MENDEL's classic HEREDITY paper into English and thereafter his work did much to promote general acceptance of the Mendelian theory.

Batholith
Large subterranean mass of IGNEOUS ROCK formed by the intrusion of MAGMA across the enclosing rock beds, and its subsequent cooling. Batholiths have an extent of over $100km^2$, and frequently form the cores of mountain ranges.

Bathyscaphe
Submersible deep-sea research vessel, invented by AUGUSTE PICCARD in the late 1940s, comprising a small, spherical, pressurized passenger cabin suspended beneath a cigar-shaped flotation hull.

Bathysphere
A hollow steel sphere suspended by cables from a surface ship and used for deep-sea research before the development of the BATHYSCAPHE. The first bathysphere was built and used by the engineer Otis Barton and the naturalist WILLIAM BEEBE in 1930.

Battani, abu-Abdullah Muhammad ibn-jabir, al- (*c*858-929)
Also known as Albategnius or Albitenius; Arab mathematician and astronomer who improved on the results of Claudius PTOLEMY by applying trigonometry to astronomical computations. He had a powerful influence on medieval European ASTRONOMY.

Battery
A device for converting internally stored chemical ENERGY into direct-current ELECTRICITY. The term is also applied to various other electricity sources, including the solar cell and the nuclear cell, but is usually taken to exclude the FUEL CELL, which requires the continuous input of a chemical fuel for its operation.

Chemical batteries consist of one or more electrochemical (voltaic) cells (comprising two ELECTRODES immersed in a conducting electrolyte) in which a chemical reaction occurs when an external circuit is completed between the electrodes. Most of the energy liberated in this reaction can be tapped if a suitable load is placed in the external circuit, impeding the flow of ELECTRONS from CATHODE to ANODE. (The conventional current, of course, flows in the opposite sense.) (➧ page 118)

Bauer, George see AGRICOLA, GEORGIUS.

Baumé, Antoine (1728-1804)
French chemist, inventor of the Baumé HYDROMETER and remembered in the Baumé hydrometer scales, used to describe the DENSITIES of liquids.

Bauxite
The main ore of ALUMINUM, consisting of hydrated aluminum oxide, usually with iron oxide impurity. It is a claylike, amorphous material formed by the weathering of silicate rocks, especially under tropical conditions. High-grade bauxite, being highly refractory, is used as a lining for furnaces. Synthetic corundum is made from it, and it is an ingredient in some quick-setting cements.

Bayliss, Sir William Maddock (1860-1924)
English physiologist who, in collaboration with Ernest Henry Starling (1866-1927) introduced the term HORMONE to describe substances that, when secreted from one part of the body, have a specific effect on another. In 1902 they discovered SECRETIN, one of the first hormones to be identified. Bayliss also introduced the saline injection for cases of surgical SHOCK during WWI.

Bayou
A minor creek or river tributary to a larger body of water, especially in Louisiana. By extension, the term is applied to muddy or sluggish bodies of water in general.

BCG (Bacillus Calmette-Guérin)
AntiTUBERCULOSIS vaccine (see VACCINATION) developed by CALMETTE and GUÉRIN. It contains a harmless strain of TB bacterium that stimulates production of anti-TB antibodies.

Beach
Stretch of sand, shingle, gravel and other material along the shore of a lake, river or sea. Caused by erosion and deposition of sediment by waves, it usually extends from the furthest point reached by waves out to a water depth of about 10m.

Beadle, George Wells (b. 1903)
US geneticist who shared part of the 1958 Nobel Prize for Physiology or Medicine with E.L. TATUM (and J. LEDERBERG) for work showing that individual GENES controlled the production of particular ENZYMES (1937-40).

Beak (bill)
General term for a rigid, projecting oral structure. All birds (for which the term "bill" is preferred) and turtles, as well as some fish, cephalopods and insects, have beaks. So too did certain dinosaurs. Beak shapes and sizes are usually highly specialized.

Beardworms
Marine worms of the phylum POGONOPHORA.

Beating
A phenomenon of importance in RADIO and ACOUSTICS resulting from the INTERFERENCE of two wave-trains of similar frequency (see WAVE MOTION) in which a new periodicity is set up in the aggregate AMPLITUDE having frequency equal to the difference of the two constituent frequencies. Beats between two musical notes of similar pitch can often be heard as an unpleasant throbbing; beating between two ultrasonic tones may result in an audible tone.

Beaufort scale
Method of measuring wind force, developed in 1806 by the British Admiral Sir Francis Beaufort. Wind strength is measured on a scale ranging from 0 to 12 (0 to 17 in Britain and the US). Internationally, the scale has now been superseded by measurement in knots.

Beaumont, William (1785-1853)
US army physician noted for his researches into the human DIGESTIVE SYSTEM. While on assignment in northern Michigan in 1822 he treated a trapper with a serious stomach wound; when the wound healed, an opening (or FISTULA) into the victim's stomach remained, through which Beaumont could extract gastric juices for analysis.

Becher, Johann Joachim (1635-1682)
German chemist, physician and economist whose conception of an active principle of combustion was developed by his pupil STAHL into the PHLOGISTON theory.

Beckmann thermometer
A mercury-in-glass THERMOMETER used in CALORIMETRY that offers an accuracy of up to $\pm 0.001K$ but that has a range of only 5K. This is achieved through its having a large bulb and fine bore. It was devised by the German organic chemist Ernest Otto Beckmann (1853-1923), who is also remembered for his discovery (1886) of the Beckmann rearrangement of ketoximes (see OXIMES) into AMIDES under acid CATALYSIS.

Becquerel
The SI unit of activity, equal to the decay of one radioactive nucleus per second. It is named in honor of the French physicist A.H. BECQUEREL, who discovered radioactivity in 1896. The becquerel is much smaller than the CURIE, the original unit of activity:

$$1BQ = 2.7 \times 10^{-11}Ci \text{ or } 2.7 \times 10^{-5}Ci.$$

Becquerel, Antoine Henri (1852-1908)
French physicist who, having discovered natural RADIOACTIVITY in a URANIUM salt in 1896, shared the 1903 Nobel Prize for Physics with PIERRE and MARIE CURIE.

Beddoes, Thomas (1760-1808)
English chemist and physician, who pioneered the inhalation of various gases in medicine. While in Bristol (from 1792) he gathered around him an important group of scientists and men of letters, including DAVY, WATT, S.T. Coleridge, Southey and Wordsworth.

Bednorz, Georg (b. 1949)
Swiss scientist awarded the 1987 Nobel Prize for Physics jointly with ALEX MÜLLER for their inspired research into high-temperature superconductivity, far extending the top range of temperature previously recorded.

Bedsores see DECUBITUS ULCERS.

Beebe, Charles William (1877-1962)
US naturalist remembered for the descents into the ocean depths he made with Otis Barton in their BATHYSPHERE. Diving off Bermuda in 1934 they reached a then-record depth of 923m.

Behavioral sciences
Those sciences dealing with human activity, individually or socially. The term, which is sometimes treated synonymously with social sciences, embraces such fields as physical and, in particular, cultural and social anthropology, psychology and sociology.

Behaviorism
School of PSYCHOLOGY based on the proposal that behavior should be studied empirically, by objective observations of reactions rather than speculatively. It had its roots in animal behavior studies, defining behavior as the actions and reactions of a living organism (including humans) in its environment; and more specifically in the work of PAVLOV in such fields as conditioned reflexes. Behaviorism developed as an effective factor in US psychology following the work of J.B. WATSON just before WWI; since then it has influenced most schools of psychological thought.

Behavior therapy
An approach to PSYCHOTHERAPY that seeks to replace inappropriate or unacceptable behavior with more appropriate behavior, eg for treating a PHOBIA.

Behring, Emil Adolf von (1854-1917)
German bacteriologist who was awarded the first Nobel Prize for Physiology or Medicine (1901) in recognition of the part he played in the development of an ANTITOXIN giving protection against DIPHTHERIA.

Békésy, Georg von (1899-1972)
Hungarian-born US physicist who was awarded the 1961 Nobel Prize for Physiology or Medicine for his development of a new theory of the physical mechanism of hearing (see EAR).

Bell, Alexander Graham (1847-1922)
Scottish-born US scientist and educator who invented the telephone (1876), founded the Bell Telephone Company, and also devised the wax-cylinder phonograph and various aids for teaching

the deaf. In later life he helped perfect the aileron for airplanes.
Bell, Sir Charles (1774-1842)
Scottish anatomist, a pioneer investigator of the working of the NERVOUS SYSTEM, whose most important discovery was to distinguish the functions of sensory and motor nerves.
Benacerraf, Baruj (b. 1920)
US pathologist who was corecipient of the 1980 Nobel Prize for Physiology or Medicine in recognition of his work on immunology.
Bends see AEROEMBOLISM.
Benign
Not threatening to life; the term is sometimes used of TUMORS that are not cancer.
Benthos
Plants and animals living on the sea bottom, as distinct from NEKTON (creatures that swim freely) and PLANKTON (creatures that drift with the current). Benthic animals include sea anemones, sea lilies, sea cucumbers, bivalve mollusks such as mussels and clams, tube-dwelling worms, and sponges. Many of them are filter-feeders, or obtain nutrients from the sediment of the ocean floor. (See also OCEAN.)
Bentonite
Fine-grained CLAY formed from volcanic ash by the hydration (formation of HYDRATES or absorption of water) of, and loss of BASES and perhaps SILICA from, tiny particles of volcanic glass present in the ash. SODIUM bentonites, which expand considerably when saturated with water, have various uses, including papermaking and the sealing of dams. CALCIUM bentonites are used to make FULLER'S EARTH.
Benzaldehyde (C_6H_5CHO)
Colorless liquid with a smell of almonds. It is an aromatic ALDEHYDE, found in bitter almonds but synthesized from TOLUENE, and is used as a flavoring and in chemical synthesis. It is oxidized to BENZOIC ACID. mp −57.1°C, bp 179°C.
Benzene (C_6H_6)
Colorless toxic liquid HYDROCARBON produced from PETROLEUM by reforming, and from coal gas and coal tar. It is the prototypical AROMATIC COMPOUND: its molecular structure, first proposed by KEKULÉ, is based on a regular planar hexagon of carbon atoms. Stable and not very reactive, benzene forms many substitution products, and also reacts with the HALOGENS to give addition products – including γ- benzene hexachloride, a powerful insecticide. It is used as a solvent, in motor fuel, and as the starting material for the manufacture of a vast variety of other aromatic compounds, especially PHENOL, STYRENE, ANILINE and maleic anhydride. mp 5.5°C, bp 80°C. (◆ page 37)
Benzoic acid (C_6H_5COOH)
White crystalline solid, an aromatic CARBOXYLIC ACID. It occurs naturally in many plants, and is made by oxidation of TOLUENE. It is mainly used as a food preservative. mp 122°C, bp 249°C. Compounds containing the benzoyl group (C_6H_5CO-) were studied by VON LIEBIG and WÖHLER.
Benzyl alcohol (phenylcarbinol, $C_6H_5CH_2OH$)
Colorless liquid, an aromatic ALCOHOL whose esters are found in flowers and used in perfumes. It is used in color-film developing and in dyeing. mp −15°C, bp 205°C.
Berg, Paul (b.1926)
US biochemist who shared the 1980 Nobel Prize for Chemistry with SANGER and GILBERT for their work on the chemistry of NUCLEIC ACIDS. In 1974 Berg was instrumental in calling a moratorium on GENETIC ENGINEERING experiments until the hazards could be evaluated.
Bergius, Friedrich (1884-1949)
German industrial chemist who, during WWI, developed a process for making GASOLINE by the high-pressure HYDROGENATION of COAL. He shared the 1931 Nobel Prize for Chemistry with KARL BOSCH for work on high- pressure reactions.
Bergstrom, Sune (b. 1916)
Swedish biochemist who shared the 1982 Nobel Prize for Physiology or Medicine with VANE and SAMUELSSON for their work on the PROSTAGLANDINS. Bergstrom's contribution was to identify the main members of the prostaglandin family.
Beriberi
Deficiency disease caused by lack of vitamin B (thiamine); it may occur in malnutrition, alcoholism or as an isolated deficiency. NEURITIS leading to sensory changes, and foot or wrist drop, palpitations, EDEMA and HEART failure are features.
Berkelium (Bk)
A TRANSURANIUM ELEMENT in the ACTINIDE series. Bk^{249} is prepared by bombarding curium-244 with neutrons. (◆ page 130)
Bernard, Claude (1813-1878)
French physiologist regarded as the father of experimental medicine. Following the work of BEAUMONT he opened artificial FISTULAS in animals to study their DIGESTIVE SYSTEMS. He demonstrated the role of the PANCREAS in digestion, discussed the presence and function of GLYCOGEN in the LIVER (1856) and in 1851 reported the existence of the vasomotor nerves.
Bernouilli's principle
Proposed by Swiss mathematician, anatomist and botanist Daniel Bernouilli (1700-1782), stating that in any small volume of space through which a fluid is flowing steadily, the total energy, comprising the pressure, potential and kinetic energies, is constant. This means that the pressure is inversely related to the velocity. This principle is applied in the design of the airfoil, the key component in making possible all heavier-than-air craft, where the faster flow of air over the longer upper surface results in reduction of pressure there and hence a lifting force acting on the airfoil (see also AERODYNAMICS.)
Berry
Botanically, a fleshy FRUIT normally with many seeds, though occasionally only one, as in the date. The tomato, melon, orange and grape are examples of berries. The name is often given to other edible, berry-like structures such as the "false fruits" of the strawberry, or the aggregate fruits of the raspberry and blackberry.
Berthelot, Pierre Eugène Marcel(l)in (1827-1907)
French chemist and statesman, who pioneered the synthesis of organic compounds not found in nature and later introduced the terms exothermic and endothermic (descriptive of chemical reactions) to THERMOCHEMISTRY. His public career was crowned in 1895 when he became foreign secretary.
Berthollet, Claude Louis, Count (1748-1822)
Savoyard-born French chemist noted for his work on CHLORINE (first using it in BLEACHING) but best remembered for his generally erroneous belief that the components in a chemical compound might be present in any of a continuous range of proportions (see COMPOSITION, CHEMICAL).
Beryl (Aluminum beryllium silicate, $Al_2[Be_3(SiO_3)_6]$)
The commonest ore of BERYLLIUM, found throughout the world mainly as hexagonal crystals in granite. EMERALD is a deep green beryl with some chromium; AQUAMARINE is a pale blue beryl. mp 1280°C.
Beryllium (Be)
A gray ALKALINE-EARTH METAL, found mainly as BERYL, prepared by reducing beryllium fluoride with magnesium. It is strong, hard and very light, and has a high melting point and high heat absorption. Combined with copper it makes a very hard alloy resistant to corrosion and fatigue. It is also used in NUCLEAR REACTOR construction to moderate neutrons, and in X-RAY tube windows. Beryllium is relatively unreactive; it forms divalent, tetracoordinate compounds which are poisonous, causing the disease berylliosis. The refractory beryllium oxide (BeO) is used in ceramics and in electronics. AW 9.01, mp 1280°C, bp 2480°C, sg 1.848 (20°C). (◆ page 130)
Berzelius, Jöns Jakob, Baron (1779-1848)
Swedish chemist who determined the ATOMIC WEIGHTS of nearly 40 elements before 1818, discovered CERIUM (1803), SELENIUM (1818) and THORIUM (1829), introduced the terms PROTEIN, ISOMERISM and CATALYSIS, and devised the modern method of writing empirical formulas (1813).
Bessel, Friedrich Wilhelm (1784-1846)
German astronomer who first observed stellar PARALLAX (1838) and set new standards of accuracy for positional astronomers. From the parallax observation, which was of 61 Cygni, he calculated the star to be about 6 ly distant, setting a new lower limit for the scale of the Universe.
Bessemer process
The first cheap, large-scale method of making STEEL from pig iron, invented in the 1850s by Sir Henry Bessemer (1813-1898). The Bessemer converter is a pivoting, pear-shaped BLAST FURNACE lined with refractory bricks. The furnace is tilted, loaded with molten pig iron, then righted. Compressed air blown through the tuyeres burns off most of the carbon and converts silicon and manganese to slag as the temperature rises. If lime is added, an afterblow removes phosphorus. The process is largely superseded by the OPEN-HEARTH PROCESS.
Best, Charles Herbert (1899-1978)
US-Canadian physiologist who assisted F.G. BANTING in the isolation of INSULIN but, to Banting's annoyance, did not share in the Nobel Prize that Banting shared with J.J.R. MACLEOD.
Beta ray
A stream of beta particles (ie ELECTRONS or POSITRONS) emitted from radioactive nuclei that are undergoing beta disintegration (see RADIOACTIVITY). Beta particles are emitted with velocities approaching that of light and can penetrate up to 1mm of lead. Positive beta rays are not emitted from any naturally occurring material.
Betelgeuse
Alpha Orionis, second brightest star in ORION. An irregularly variable red supergiant (see VARIABLE STAR) with a variable radius some 300 times that of the Sun, it is over 150pc from Earth. (◆ page 228)
Bethe, Hans Albrecht (b. 1906)
German-born US theoretical physicist who proposed the nuclear CARBON CYCLE to account for the Sun's energy output (1938). During WWII he worked on the Manhattan Project. He was awarded the 1967 Nobel Prize for Physics for his work on the source of stellar energy.
Bettelheim, Bruno (b.1903)
Austrian-born US psychologist who drew on his prewar experience as an inmate of Nazi concentration camps to describe men's behavior in extreme situations (1943). His subsequent work has mainly concerned the treatment of autistic (see AUTISM) and disturbed children.
Bicarbonate of soda (sodium bicarbonate) see SODIUM.
Bicarbonates (hydrogen carbonates)
Acid salts of carbonic acid (see CARBON) containing the ion HCO_3^-. Bicarbonates are formed by the action of carbon dioxide on carbonates in aqueous solution; this reaction is reversed on heating. Dissolved calcium and magnesium bicarbonates give rise to HARD WATER.

DIGESTION

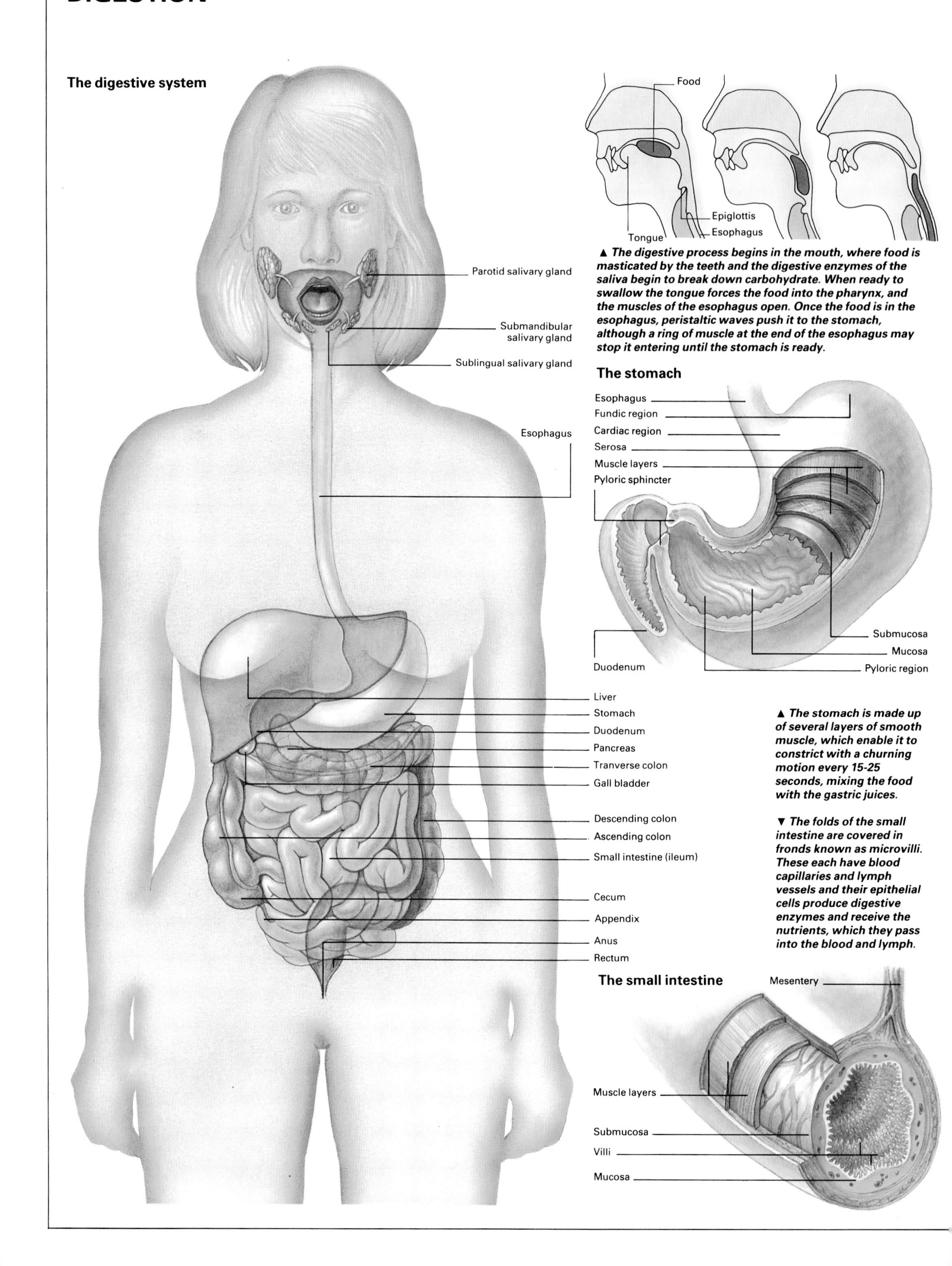

▲ The digestive process begins in the mouth, where food is masticated by the teeth and the digestive enzymes of the saliva begin to break down carbohydrate. When ready to swallow the tongue forces the food into the pharynx, and the muscles of the esophagus open. Once the food is in the esophagus, peristaltic waves push it to the stomach, although a ring of muscle at the end of the esophagus may stop it entering until the stomach is ready.

▲ The stomach is made up of several layers of smooth muscle, which enable it to constrict with a churning motion every 15-25 seconds, mixing the food with the gastric juices.

▼ The folds of the small intestine are covered in fronds known as microvilli. These each have blood capillaries and lymph vessels and their epithelial cells produce digestive enzymes and receive the nutrients, which they pass into the blood and lymph.

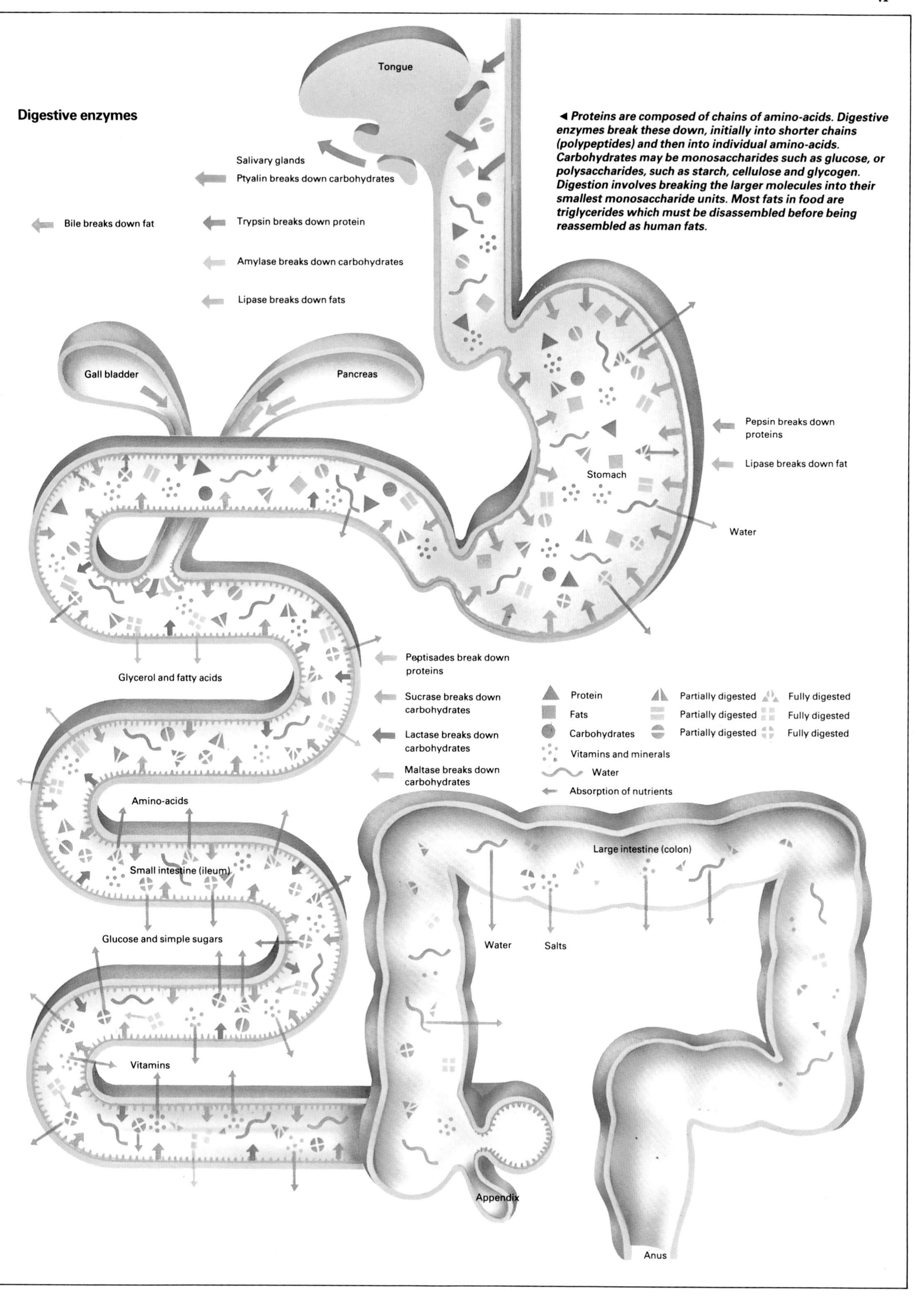

◄ Proteins are composed of chains of amino-acids. Digestive enzymes break these down, initially into shorter chains (polypeptides) and then into individual amino-acids. Carbohydrates may be monosaccharides such as glucose, or polysaccharides, such as starch, cellulose and glycogen. Digestion involves breaking the larger molecules into their smallest monosaccharide units. Most fats in food are triglycerides which must be disassembled before being reassembled as human fats.

Biceps
Either of two MUSCLES that are split in two in the upper part to form a Y-shape. Biceps brachii is the chief upper arm muscle, attached to the shoulder blade and the radius. Biceps femoris is a thigh muscle attached to the PELVIS, the FEMUR and the FIBULA. (➧ page 216)

Bichat, Marie François Xavier (1771-1802)
French anatomist and pathologist, the founder of HISTOLOGY. Although working without the MICROSCOPE, Bichat distinguished 21 types of elementary TISSUES from which the organs of the body are composed.

Biennial
Plant that completes its life cycle in two years. During the first year leaves are produced and food is stored (as in the carrot and cabbage) for use in the second year, when the plant bears flowers and fruit, and then dies. (See also ANNUAL; PERENNIAL.)

Big Bang theory see COSMOLOGY; HUBBLE; LEMAITRE.

Big dipper see GREAT BEAR.

Bilateral symmetry
In the bodies of animals, symmetry about a central axis, as seen in humans and other mammals, birds, reptiles, amphibians, fish, insects, most crustaceans and worms. By contrast, some invertebrates show RADIAL SYMMETRY.

Bile
A yellow-brown fluid secreted by the liver and containing salts derived from CHOLESTEROL. Stored and concentrated in the GALL BLADDER and released into the DUODENUM after a meal, the bile emulsifies fats and aids absorption of fat-soluble vitamins A, D, E and K. Other constituents of bile are in fact waste products. (➧➧ page 40, 233)

Bilirubin
Yellow pigment in BILE, formed from the breakdown of hemoglobin. Excess bilirubin causes jaundice.

Bill see BEAK.

Binary code
A number system using the powers of two; the system of numeration normally used in computers (a single binary digit being known as a bit). The digits used are 1 and 0; the powers of two being written as $2^0=1$; $2^1=10$; 2^2 (4 in the decimal notation) = 100; 2^3 (8 in the decimal notation) = 1000; 2^4 (16) = 10,000, etc. Thus the decimal number 25 is written 11001 (16+8+1).

Binary fission
A simple form of asexual reproduction; see FISSION.

Binary star see DOUBLE STAR.

Binet, Alfred (1857-1911)
French psychologist who pioneered methods of mental testing. He collaborated with Théodore Simon in devising the Binet-Simon tests, widely used to estimate INTELLIGENCE.

Binnig, Gerd (b. 1947)
German scientist working in Switzerland who was awarded half of the 1986 Nobel Prize for Physics jointly with HEINRICH ROHRER of Switzerland for their design of the scanning tunneling microscope.

Binocular vision
The use of two eyes, set a small distance apart in the head and aligned approximately parallel, to view a single object. Owing to PARALLAX, the images in the two eyes are slightly different, which enables the observer to perceive what is seen in three dimensions and so to judge distance, size and shape. (See also STEREOSCOPE.)

Biochemistry
The study of the substances occurring in living organisms and the reactions in which they are involved; a science on the border between biology and organic chemistry. The main constituents of living matter are water, CARBOHYDRATES, LIPIDS and PROTEINS. The total chemical activity of the organism is known as its METABOLISM.

Plants use sunlight as an energy source to produce carbohydrates from carbon dioxide and water (see PHOTOSYNTHESIS). The carbohydrates are then stored as starch, used for structural purposes, as in the CELLULOSE of plant cell walls, or oxidized through a series of reactions including the CITRIC ACID CYCLE, the energy released being stored as adenosine triphosphate (see ATP). In animals energy is obtained from other organisms, and is stored mainly as lipids. As well as forming fat deposits, lipids are components of all cell membranes.

Proteins have many functions, of which metabolic regulation is perhaps the most important. ENZYMES, which control almost all biochemical reactions, are proteins. Each enzyme is specialized for a particular reaction, and there are thousands of different enzymes in any living organism. Plants synthesize proteins using simpler nitrogenous compounds from the soil. Animals obtain proteins from food and break them down to AMINO ACIDS, from which new proteins are built up. New proteins are made according to the pattern determined by the sequence of NUCLEIC ACIDS in the DNA (or RNA).

The methods used by biochemists are similar to those of chemists, and include labeling with radioactive ISOTOPES and separation techniques such as CHROMATOGRAPHY, used to analyze very small amounts of substances. Molecular structures may be determined by X-RAY DIFFRACTION. A major landmark in biochemistry was the elucidation of the "double helix" structure of the DNA molecule by JAMES WATSON and FRANCIS CRICK in 1953. (See also BIOTECHNOLOGY; GENETIC ENGINEERING.)

Biochips
Living CELLS rely on tiny electrical signals to work. Scientists believe that by redesigning cell structures they can use living material to process data in the way that a CHIP does. That is one form of biochip; the other is the tiny conventional microchip that can be linked to organic material to control its growth and operation. Another name for this new and fast-developing science is molecular electronics.

Biogenesis
The theory that all living organisms are derived from other living organisms. It is the opposite of the theory of SPONTANEOUS GENERATION. The theory of biogenesis holds true for the Earth at the present time, but it is probable that 3000-4000 million years ago living forms could be generated chemically. (See also LIFE, ORIGIN OF.)

Biogeography
The study of the distribution of living organisms, and its relationship to climatic factors and evolutionary events. It involves looking at the present-day distribution of plant and animal species, at the means by which they are dispersed, at their climate and habitat requirements, and at the present and past barriers to migration as inferred from the geological record. CHARLES DARWIN and ALFRED RUSSELL WALLACE were among the first to study biogeography, and evidence from this source played an important part in the formulation of Darwin's theory of EVOLUTION. A major borderline between two biogeographical regions, Asia and Australasia, was first identified by Wallace, and is still known as "Wallace's line".

Biological clocks
The mechanisms which control the rhythm of various activities of plants and animals. Some activities, such as mating, migration and hibernation, have a yearly cycle; others follow the lunar month. The majority, however, have a period of roughly 24 hours, called a circadian rhythm. As well as obvious rhythms, such as the patterns of leaf movement in plants and the activity/sleep cycle in animals, many other features, such as body temperature and cell growth, oscillate daily. Although related to the day/night cycle, circadian rhythms are not directly controlled by it. Organisms in unvarying environments will continue to show 24-hr rhythms, but their "days" get gradually longer or shorter: they need an external cue to maintain exact timekeeping. This is not thought to be due to any inaccuracy in the mechanism, however. Resetting to external cues is probably important in that it allows an organism to adjust to seasonally changing daylengths.

Biological clocks are important in modern travel. After moving from one time zone to another, it takes some time for the body to adjust to the newly imposed cycle, and "jet-lag" results. Biological clocks are also important in animal NAVIGATION. Many animals, such as migrating birds or bees returning to the hive, navigate using the Sun. They can do this only if they have some means of knowing what time of day it is. Biological clocks are apparently inborn, not learned, but need to be triggered. An animal kept in the light from birth shows no circadian rhythms, but if placed in the dark for an hour or so immediately starts rhythms based on a 24-hr cycle. Once started, the cycles are almost independent of external temperature, indicating that they cannot be based on a simple rhythm of chemical reactions, which would be affected by temperature. It is still unclear exactly how the clock works, but there is evidence that membranes are involved, and a mechanism rather like an egg-timer, with substances moving gradually across a membrane, then being returned abruptly by a change in the permeability of the membrane, has been suggested.

Lunar cycles, controlled by circalunar rhythms, are found mainly in coastal organisms, which must time their activities to suit the tides. Like circadian rhythms, they are innate. In a few organisms circannual rhythms are also innate, and control features such as molting and breeding behavior in birds. Such innate yearly cycles are reinforced by information from the relative length of day and night, a mechanism known as PHOTOPERIODISM. The majority of annual rhythms are probably produced by photoperiodism alone.

Biological control
The control of pests by the introduction of natural predators, parasites or disease, or by modifying the environment so as to encourage those already present. This first took place in California in 1888, when Australian ladybird beetles were introduced to eliminate the damage to citrus trees by the cottony-cushion scale insect. Another example is the introduction of myxomatosis to combat crop damage by rabbits. The "sterile male" technique is used to control many insect pests. Large numbers of males are bred, sterilized with X-rays and released. As the females mate only once, many of them will mate with the sterile males, so the population rapidly decreases. New methods of biological control are now emerging from GENETIC ENGINEERING.

Biology
The study of living things. Broadly speaking there are two main branches of biology, the study of animals (ZOOLOGY) and the study of plants (BOTANY). The study of bacteria, protozoa and other microorganisms is described as MICROBIOLOGY. Within each of these main branches are a number of traditional divisions dealing with structure (ANATOMY, CYTOLOGY), development and function (EMBRYOLOGY, PHYSIOLOGY), classification (TAXONOMY), inheritance (GENETICS, EVOLUTION)

and interrelations of organisms with each other and with their environment (ECOLOGY). These divisions are also split into a number of specialist fields dealing with particular groups of organisms, such as MYCOLOGY, ENTOMOLOGY, HERPETOLOGY and ICHTHYOLOGY.

There are also biosciences that bridge the gap with the physical sciences of chemistry, physics and geology, eg BIOCHEMISTRY, BIOPHYSICS and PALEONTOLOGY. Similarly, there are those that relate to areas of human behavior, eg PSYCHOLOGY. Disciplines such as MEDICINE, veterinary medicine, agronomy and horticulture also have a strong basis in biology.

As the impetus to investigate the living world generally arose from a desire to improve the techniques of medicine or agriculture, most early biologists were physicians or landowners. An exception is provided by ARISTOTLE, the earliest systematist of biological knowledge and an outstanding biologist – he founded the science of comparative anatomy – but most other classical authors, such as GALEN and the members of the Hippocratic school, were primarily physicians. In the medieval period much biological knowledge became entangled in legend and allegory. The classical texts continued to be the principal sources of knowledge although new compilations, such as AVICENNA's *Canon of Medicine*, were produced by Muslim philosophers. In 16th-century Europe interest revived in descriptive natural history, the work of GESNER being notable; physicians such as PARACELSUS began to develop a chemical pharmacology, and experimental anatomy was revived in the work of VESALIUS, FABRICIUS and FALLOPIUS. The discoveries of SERVETUS, HARVEY and MALPIGHI followed. Quantitative plant physiology began with the work of VAN HELMONT and was taken to spectacular ends in the work of STEPHEN HALES. In the 17th century, microscopic investigations began with the work of HOOKE and VAN LEEUWENHOEK; GREW advanced the study of plant organs and RAY laid the foundation for LINNAEUS' classic 18th-century formulation of the classification of plants. This same era saw BUFFON devise a systematic classification of animals and VON HALLER lay the groundwork for the modern study of physiology.

The 17th century had seen controversies over the role of mechanism in biological explanation – LA METTRIE had even developed the theories of DESCARTES to embrace the mind of man; the 19th century saw similar disputes, now couched in the form of the mechanist-vitalist controversy concerning the possible chemical nature of life (see BICHAT; MAGENDIE; CLAUDE BERNARD). Evolutionary biology, foreshadowed by LAMARCK, was thoroughly established following the work of DARWIN; in anatomy, SCHWANN and others developed the cell concept; in histology, Bichat's pioneering work was continued; in physiology, organic and even physical chemists began to play a greater role, and medical theory was revolutionized by the advent of bacteriology (see PASTEUR; KOCH). The impact of MENDEL's discoveries in genetics was not felt until the early 1900s. Genetics, physiology and cell science were all advanced by the advent of BIOCHEMISTRY and the invention of the electron microscope. Possibly the high point of 20th-century biology came with the proposal of the double-helix model for DNA (see NUCLEIC ACIDS), the chemical carrier of genetic information, by FRANCIS CRICK and JAMES WATSON in 1953. This paved the way for the sequencing of the genetic material, and its direct manipulation, in GENETIC ENGINEERING and modern BIOTECHNOLOGY.

Bioluminescence
The production of nonthermal light by living organisms such as fireflies, many marine animals, and some bacteria and fungi. The effect is an example of CHEMILUMINESCENCE. In some cases its utility to the organism is not apparent, though in others its use is clear. Thus, in the firefly, the abdomen of the female glows, enabling the male to find her. Similarly, bioluminescence enables many deep-sea fish to locate each other for mating or to attract their prey. The glow in a ship's wake at night is due to luminescent microorganisms.

Biomass
In ecology, the combined weight of all living organisms within a given area or habitat (total biomass), or the combined weight of a particular species (eg oak biomass), or of organisms at a particular trophic level (eg saprotroph biomass).

Biome
An ecological region characterized by the predominant vegetation type, such as savanna. The biome is the largest biogeographical unit. (See ECOLOGY.)

Biomedical engineering
Development and application of mechanical electrical, electronic and nuclear devices in medicine. The many recent advances in biomedical engineering have occurred in four main areas: ARTIFICIAL ORGANS; new surgical techniques involving the use of LASERS, cryosurgery and ULTRASONICS; diagnosis and monitoring using thermography and computers; and PROSTHETICS.

Bionics
The science of designing artificial systems that have the desirable characteristics of living organisms. These may be simply imitations of nature, such as military vehicles with jointed legs, or, more profitably, systems that embody a principle learned from nature. Examples of the latter include RADAR, inspired by the echolocation system of bats, or the development of associative memories in COMPUTERS as in the human brain.

Biophysics
A branch of BIOLOGY in which the methods and principles of PHYSICS are applied to the study of living things. The term is less widely used now, as biophysics is an integral part of a great many modern biological studies.

Biopsy
Removal and microscopic examination of tissue from a living patient for purposes of diagnosis. The tissue is removed by needle, suction, swabbing, scraping or excision.

Biosensors see SENSORS.

Biosphere
The region inhabited by living things. It forms a thin layer around the Earth and includes the surface of the LITHOSPHERE, the HYDROSPHERE and the lower ATMOSPHERE.

Biosynthesis (anabolism)
The biochemical reactions by which living cells build up simple molecules into complex ones. These reactions require energy, which is obtained from light (see PHOTOSYNTHESIS) or from ATP, which is produced in degradation reactions. (See also METABOLISM; PROTEIN SYNTHESIS.)

Biot, Jean Baptiste (1774-1862)
French physicist who first demonstrated the extraterrestrial origin of and hence the actual existence of meteorites (see METEORS) (1803); who accompanied GAY-LUSSAC on his pioneering balloon ascent to collect data concerning the upper ATMOSPHERE (1804); and who, having shown that some organic substances show OPTICAL ACTIVITY (1815), first developed the methods of POLARIMETRY – despite his rejection of the wave theory of LIGHT.

Biotechnology
Using biological organisms, systems or processes to make or modify products. Biotechnology dates from the first fermented drinks, and for thousands of years microorganisms – mainly bacteria, yeasts and molds – have been used to make food, drugs, dyes, fertilizers, etc. As their understanding of microbes and molecular biology has grown, biotechnologists have been able to increase the output of traditional microorganisms by creating an environment in which they multiply quickly and can be used for large-scale production. Sometimes, rather than the whole microbe, they use a part of it, particularly ENZYMES that will perform some subsequent chemical conversion. Where there is no known enzyme or microbe that manufactures a substance naturally, the techniques of GENETIC ENGINEERING can be applied to create new strains. For example, foreign genes can be inserted into bacteria to endow them with novel characteristics and to induce them to synthesize particular materials, such as human growth-hormones or insulin. The biotechnological application of genetic engineering requires four main stages: isolating the required GENE; inserting the gene into the BACTERIA; inducing the bacteria to start synthesizing the product; and harvesting that product. (See also FERMENTATION.)

Biotin see VITAMINS.

Biotite
A range of iron-rich varieties of MICA, grading into PHLOGOPITE. It is a constituent of most igneous and many metamorphic rocks.

Bird banding (bird ringing)
Placing numbered metal or plastic bands on the legs of birds for identification. Birds are banded as nestlings or when trapped in nets. All relevant data is recorded at a national agency whose address is on the band. Birds found can then be reported, an important aid to studies of migration, distribution and life expectancy.

Bird migration see MIGRATION.

Birds
A class of vertebrates. They are HOMOIOTHERMS ("warm-blooded"), lay eggs and fly, though some species have lost this ability. There are about 8600 living species. To make flight possible, the skeleton is light, the body cavity is filled by a series of air-sacs, and the forelimbs are modified as wings, with large flight muscles for their operation. FEATHERS, which are not found in any other group of animals, provide the wing and tail flight-surfaces, as well as an insulating layer over the entire body. (♦ page 17)

Birdseye, Clarence (1886-1956)
US inventor and industrialist who, having observed during fur-trading expeditions to Labrador (1912-16) that many foods keep indefinitely if frozen, developed a process for the rapid commercial freezing of foodstuffs. In 1924 he organized the company later known as General Foods to market frozen produce.

Birds of prey
A general term for birds that hunt vertebrate prey, such as eagles, hawks and kestrels; it is sometimes used to include the owls as well. The term is also used taxonomically to refer to the Order Falconiformes, which comprises the eagles, buzzards, hawks, kites and Old World vultures (all members of the hawk family, Accipitridae), the falcons, kestrels and hobbies (Family Falconidae), the osprey (Family Pandionidae), the African secretary bird (Family Sagittariidae) and the New World vultures and condors (Family Cathartidae). Apart from the vultures and condors, all members of this order take live prey.

Birdsong
The pattern of notes, often musical and complex, with which birds proclaim their territory and, in some species, attract a mate. Ornithologists call all such sounds songs, though those that are harsh and unmusical are often referred to simply as the "voice".

ECOSYSTEM The Nutrient Cycles

The Earth's chemical resources are in constant use and re-use within food chains and food webs. Elements such as carbon, nitrogen and oxygen are transformed from an inorganic to an organic state and back agaain in nutrient cycles.

All life on Earth depends on oxygen. Animals, plants, fungi and bacteria use up oxygen from the air in respiration. Atmospheric oxygen is also used up when fossil fuels are burned. We depend on plant photosynthesis to replace this lost oxygen.

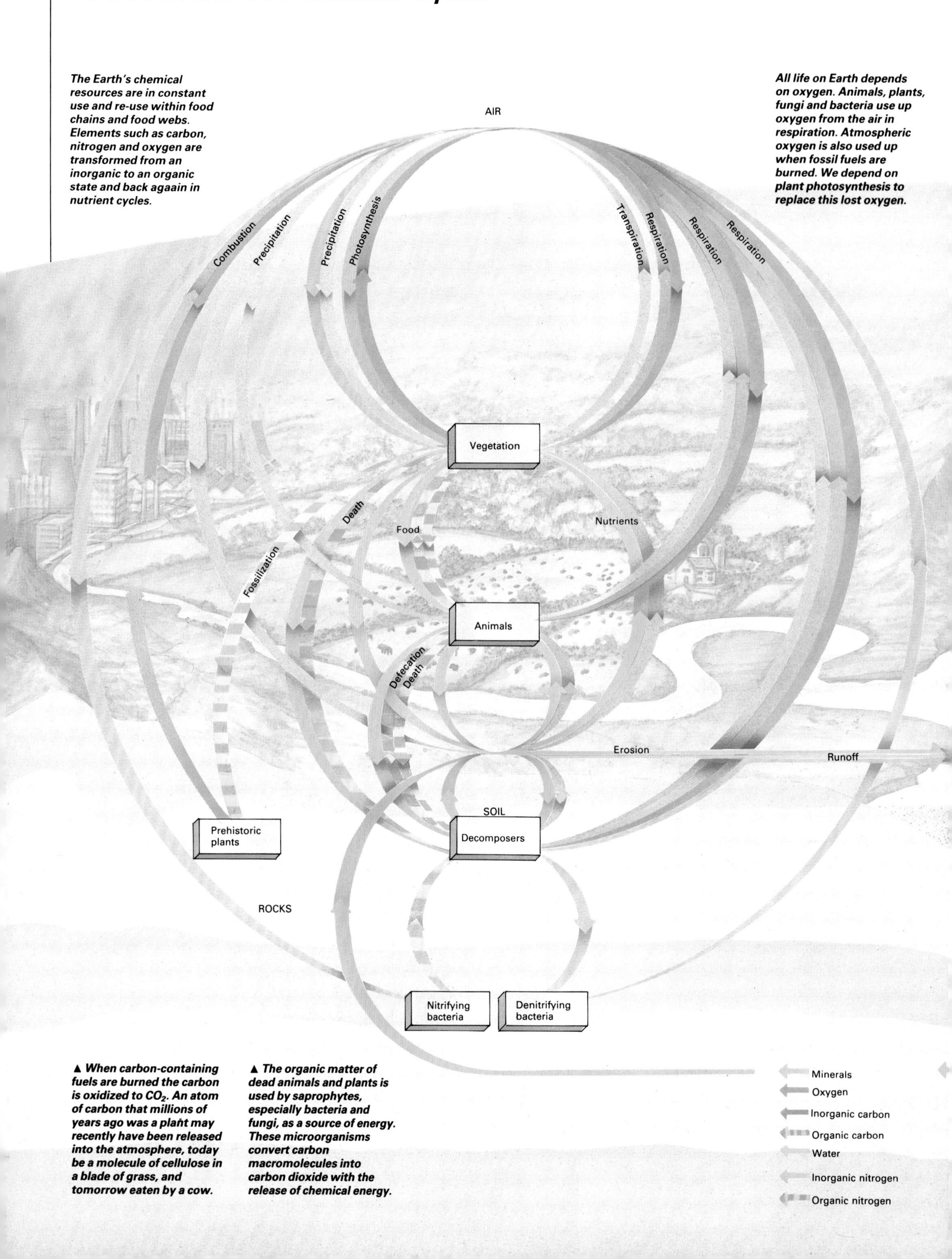

▲ When carbon-containing fuels are burned the carbon is oxidized to CO_2. An atom of carbon that millions of years ago was a plant may recently have been released into the atmosphere, today be a molecule of cellulose in a blade of grass, and tomorrow eaten by a cow.

▲ The organic matter of dead animals and plants is used by saprophytes, especially bacteria and fungi, as a source of energy. These microorganisms convert carbon macromolecules into carbon dioxide with the release of chemical energy.

Nutrient cycles exist on land and in water and the stages involved in both ecosystems are almost identical. Also, the two diverse ecosystems are closely linked. An upset of nutrient cycles in one produces a change in the other.

Limits to growth of aquatic organisms are dissolved oxygen and sunlight. In the atmosphere, oxygen is mixed with other gases and its concentration is quite constant. In aquatic ecosystems it is dissolved in water and concentrations vary considerably.

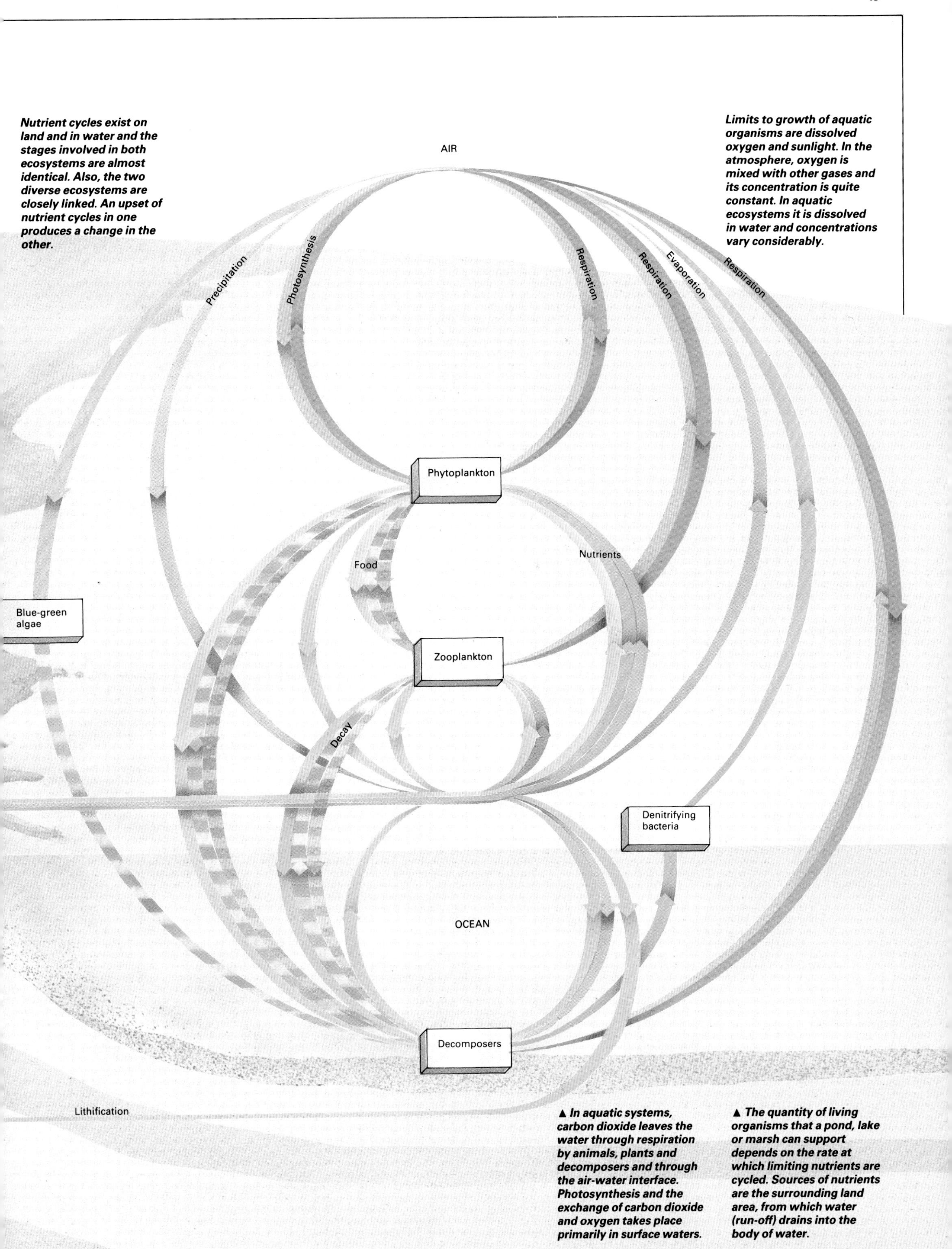

▲ In aquatic systems, carbon dioxide leaves the water through respiration by animals, plants and decomposers and through the air-water interface. Photosynthesis and the exchange of carbon dioxide and oxygen takes place primarily in surface waters.

▲ The quantity of living organisms that a pond, lake or marsh can support depends on the rate at which limiting nutrients are cycled. Sources of nutrients are the surrounding land area, from which water (run-off) drains into the body of water.

Birefringence see DOUBLE REFRACTION.

Birth

Emergence from the mother's UTERUS, or, in the case of most lower animals, from the EGG, marking the beginning of an independent life The birth process is triggered by HORMONE changes in the mother's bloodstream. Birth may be induced, if required, by oxytocin. Mild labor pains (contractions of the womb) are the first sign that a woman is about to give birth. Initially occurring about every 20 minutes, in a few hours they become stronger and occur every few minutes. This is the first stage of labor, usually lasting about 14 hours. The contractions push the baby downward, usually head first, which breaks the membranes surrounding the baby, and the AMNIOTIC FLUID escapes.

In the second stage of labor, stronger contractions push the baby through the cervix and vagina. This is the most painful part and lasts less than 2 hours. Anesthetics (see ANESTHESIA) or ANALGESICS are often given, and delivery may be aided by hand or obstetric forceps. A CESARIAN SECTION may be performed if great difficulty occurs. Some women choose "natural childbirth", in which no anesthetic is used, but pain is minimized by prior relaxation exercises.

As soon as the baby is born, its nose and mouth are cleared of fluid and breathing starts, whereupon the UMBILICAL CORD is cut and tied. In the third stage of labor the PLACENTA is expelled from the uterus and bleeding is stopped by further contractions. Birth normally occurs 38 weeks after conception. Premature births are those occurring after less than 35 weeks. Most premature babies develop normally with medical care, but if born before 28 weeks the chances of survival are poor. (See also EMBRYO; GESTATION; OBSTETRICS; PREGNANCY.)

Birth control

Prevention of unwanted births, by means of CONTRACEPTION, ABORTION, and STERILIZATION.

Birthmarks

Skin blemishes, usually congenital. There are two main types: pigmented nevuses, or moles, which are usually brown or black and may be raised or flat; and vascular nevuses, local growths of small blood vessels, such as the "strawberry mark" and the "port-wine stain".

Bismuth (Bi)

Metal in Group Va of the PERIODIC TABLE, brittle and silvery gray with a red tinge. It occurs naturally as the metal, and as the sulfide and oxide, from which it is obtained by roasting and reduction with carbon. In the US it is obtained as a byproduct of the refining of copper and lead ores. Bismuth is rather unreactive; it forms trivalent and some pentavalent compounds. Physically and chemically it is similar to LEAD and ANTIMONY. Bismuth is used in low-melting-point alloys in fire-detection safety devices; as it expands on solidification, it is used in alloys for casting dies and type metal. Bismuth (III) oxide is used in GLASS and CERAMICS; various bismuth salts are used in medicine. AW 209.0, mp 271.4°C, bp 1650°C, sg 9.747 (20°C). (▶ page 130)

Bisulfates see SULFATES.

Bit

Abbreviation for binary digit, in COMPUTER technology and information theory, the smallest conceivable unit of information, representing a choice between only two possible states: the presence or absence of a signal pulse; + or −; 0 or 1; a switch being off or on. The capacity of an information storage or handling device is measured in bits. (See also BYTE.)

Bitumen

A mixture of high molecular weight hydrocarbons, occurring naturally or as the residue left after distillation of crude oil. ASPHALT and PITCH are used as adhesives, and on roofs and road surfaces.

Bituminous coal see COAL.

Bituminous sands

Sands that contain natural BITUMEN. The largest deposit is in the Athabasca region of N Alberta, and there are substantial deposits in California and Utah. The heavy tar is extracted and synthetic crude oil produced.

Bivalves

A class of the phylum MOLLUSCA. It includes all those with a two-part hinged shell (such as the clams, oysters, cockles, mussels and scallops) and the piddocks and shipworms, whose sharp-edged, elongated shells are adapted for tunneling in wood. Most bivalves feed by taking in water, filtering food particles out of it with their enlarged gills, and then ejecting the water again. Because they process large volumes of water in this way they tend to accumulate pollutants, such as heavy metals and, where raw sewage is discharged into the sea, human pathogens. In many areas it is therefore dangerous to eat these mollusks. (See also LAMPSHELLS.)

Black, Joseph (1728-1799)

Scottish physician and chemist who investigated the properties of CARBON dioxide, discovered the phenomena of LATENT and SPECIFIC HEATS, distinguished HEAT from TEMPERATURE and pioneered the techniques used in the quantitative study of CHEMISTRY.

Black body

In theoretical physics, an object that absorbs all the ELECTROMAGNETIC RADIATION that falls on it. In practice, no object acts as a perfect black body, though a closed box admitting radiation only through a small hole is a good approximation. Black bodies are also ideal thermal radiators.

Black body radiation

The ELECTROMAGNETIC RADIATION emitted from a BLACK BODY in virtue of its thermal energy. The derivation of its properties by PLANCK in 1901 was the occasion of the proposal and first success of QUANTUM THEORY. The energy emitted from a black body is proportional to the fourth power of its (absolute) temperature (Stefan-BOLTZMANN law). The intensity SPECTRUM of radiation from a black body takes the form of a skew hump which tails off in its longer-wavelength branch. Its precise shape is described in WIEN's laws, the first of which (Wien's displacement law) states that the greatest emission occurs at a wavelength that is inversely proportional to the absolute temperature.

Black death see PLAGUE.

Blackett, Patrick Maynard Stuart, Baron (1897-1974)

British physicist who, having developed the Wilson CLOUD CHAMBER into an instrument for observing COSMIC RAYS, won the 1948 Nobel Prize for Physics for the results he obtained using it.

Blackhead see SKIN.

Black hole

A region of space into which matter has fallen and within which the gravitational field is so powerful that nothing, not even ELECTROMAGNETIC RADIATION (including light), can escape. The matter that has formed the black hole becomes infinitely compressed in a central point known as a SINGULARITY. A black hole may be formed by the total gravitational collapse of a massive star that has run out of nuclear fuel and can no longer support itself. The minimum mass of material that can collapse in this way, in the present-day Universe, is about three solar masses, but it has been suggested that very much smaller masses could have been compressed into tiny "primordial" black holes during the BIG BANG. At the other end of the scale, supermassive black holes of millions or even billions of solar masses may exist in the cores of active galaxies.

Because no electromagnetic radiation can escape from black holes, they may be detected only through their gravitational influence on other bodies or through the emission of X- and gamma rays by infalling matter, which becomes violently heated close to their boundaries (see EVENT HORIZON). (▶ page 228)

Bladder

Any hollow muscular sac; used especially of the urinary bladder (see also GALL BLADDER; SWIM BLADDER). URINE trickles continually into the bladder from the KIDNEYS. The bladder empties through the urethra, a tube that issues from its base, being normally closed by the external sphincter muscle. (▶ page 233)

Blast furnace

Furnace in which a blast of hot, high-pressure air is used to force combustion; used mainly to reduce IRON ore to pig iron, and also for lead, tin and copper. It consists of a vertical, cylindrical stack surmounting the bosh (the combustion zone) and the hearth from which the molten iron and slag are tapped off.

Modern blast furnaces are about 30m high and 10m in diameter, and can produce more than 1800 tonnes per day. Layers of iron oxide ore, coke and limestone are loaded alternately into the top of the stack. The burning coke heats the mass and produces carbon monoxide, which reduces the ore to iron; the limestone decomposes and combines with ash and impurities to form a slag, which floats on the molten iron. The hot gases from the top of the stack are burned to preheat the air blast.

Blastula

A hollow sphere composed of a single layer of cells, formed by cleavage of a fertilized OVUM; an early stage in the development of the EMBRYO. In mammals a similar cluster, the blastocyst, is formed, with an inner cell mass and a spherical envelope that develops into the PLACENTA. (See also EMBRYOLOGY.)

Bleaching

Process of whitening materials using sunlight, ULTRAVIOLET RADIATION or chemicals that reduce or oxidize DYES into a colorless form. Hydrogen PEROXIDE is used to bleach wool, silk and cotton; HYPOCHLORITES, including BLEACHING POWDER, are used for cotton; and sodium chlorite for synthetic fibers.

Bleaching powder

White powder consisting of calcium HYPOCHLORITE and basic calcium chloride, made by reacting CALCIUM hydroxide with CHLORINE. It is used for BLEACHING and as a disinfectant, but in time loses its strength.

Bleeding see HEMORRHAGE.

Blende see SPHALERITE.

Bleuler, P. Eugen (1857-1939)

Swiss psychiatrist who introduced the term SCHIZOPHRENIA (1908) as a generic term for a group of mental illnesses which he had learned to differentiate in a classic research project. He was an early supporter of SIGMUND FREUD but later criticized his dogmatism.

Blight see PLANT DISEASES.

Blindness

Severe loss or absence of vision caused by injury to the eyes, congenital defects, or diseases including CATARACT, DIABETES, GLAUCOMA, LEPROSY, TRACHOMA and VASCULAR disease. Malnutrition (especially vitamin A deficiency) may also cause blindness in children. Infant blindness can result if the mother had RUBELLA early in pregnancy. Cortical blindness is a disease of the higher perceptive centers in the BRAIN concerned with vision. Blindness due to cataract may be relieved by removal of the eye lens and the use of glasses. Prevention or early recognition and treatment of predisposing conditions is essential to save sight, as established blindness is rarely recoverable.

Blind spot
The area of the retina of each eye where the optic nerve and blood vessels enter, about 2mm in diameter. It has no light-sensitive receptors. In binocular vision the two spots do not receive corresponding images, and so are not noticeable. (♦ page 213)

Blink comparator (blink microscope)
Astronomical instrument used to detect differences between apparently similar star pictures, viewed as in a STEREOSCOPE, by rapidly obscuring each alternately. Anything that has moved flickers. The planet PLUTO was discovered this way. It can also detect variable stars and those of large proper motion.

Blister
A swelling filled with serum or blood formed between two layers of skin following burns, friction or contact with certain corrosive chemicals (vesicants). Blisters also occur in certain skin diseases.

Bloch, Felix (1905-1983)
Swiss-born US physicist who shared the 1952 Nobel Prize for Physics with E.M. PURCELL for developing a method for determining the magnetic fields of NEUTRONS in atomic nuclei (see ATOM). This was developed into the nuclear magnetic resonance (NMR) method of determining chemical structures (see SPECTROSCOPY).

Bloch, Konrad Emil (b. 1912)
German-born US biochemist who shared the 1964 Nobel Prize for Physiology or Medicine with F. LYNEN for developing an isotopic labeling technique (see ISOTOPES), which he used to elucidate the path by which CHOLESTEROL is synthesized in the body.

Bloembergen, Nicolaas (b. 1920)
Dutch-born US physicist who shared half of the 1981 Nobel Prize for Physics with SCHAWLOW for the development of laser SPECTROSCOPY, especially for their pioneering work with LASERS to probe the structure of the ATOM.

Blood
The fluid found in the CIRCULATORY SYSTEM; it transports food and oxygen around the body, and removes carbon dioxide and wastes from the tissues for excretion. Blood also transports HORMONES to various tissues and organs for chemical signaling, and digested food from the gut to the LIVER. It contains the cells of the immune system for prevention of infection, and clotting factors to help stop bleeding, carrying these to all parts of the body. Blood also plays a major part in HOMEOSTASIS, as it contains BUFFERS that keep the acidity (pH) of the body fluids constant and, by carrying heat from one part of the body to another, tends to equalize body temperature.

The adult human has about 5 liters of blood, half PLASMA and half blood cells (erythrocytes or red cells, leukocytes or white cells, and thrombocytes or platelets). The formation of blood cells (hemopoiesis) occurs in bone MARROW, but some of the cells are processed by the lymphoid tissue, particularly the thymus. Red cells (about 5 million per mm^3) are produced at a rate of over 100 million per minute and live only about 120 days. They have no nucleus, but contain a large amount of the red pigment HEMOGLOBIN, responsible for oxygen transfer from lungs to tissues and carbon dioxide transfer from tissues to lungs. (Some lower animals employ copper-based HEMOCYANINS instead of hemoglobin. Others, eg cockroaches, have no respiratory pigments.) White cells (about 6000 per mm^3) are concerned with defense against infection (see IMMUNITY). Platelets, which live for about 8 days and are much smaller than white cells and about 40 times as numerous, assist in the initial stages of blood CLOTTING, together with at least 12 plasma clotting factors and fibrinogen. (See also HEMOPHILIA; THROMBOSIS).

Blood from different individuals may differ in the type of antigen on the surface of its red cells. Consequently, in a blood TRANSFUSION, if the blood groups of the donor and recipient are incompatible a dangerous reaction occurs, involving the production of antibodies to the transfused blood, which in turn causes aggregation of the red cells of the donor in the recipient's circulation. Many blood group systems have been discovered, the first and most important being the ABO system by KARL LANDSTEINER in 1900. In this system blood is classified by whether the red cells have antigens A (blood group A), B (group B), A and B (group AB), or neither A nor B antigens (group O). Another important antigen is the Rhesus antigen (or Rh factor). People who have the Rh factor (84%) are designated Rh+, those who do not, Rh−. Rhesus antibodies are not usually produced but may develop in certain circumstances. Where Rh− women are pregnant with Rh+ babies, blood leakage from baby to mother causes production of antibodies by the mother, which do not develop fast enough to affect the baby being carried but may progressively destroy the blood of any subsequent Rh+ baby. (See also EDEMA; POLYCYTHEMIA; SEPTICEMIA; SERUM; RHESUS FACTOR.) (♦♦ page 28, 233)

Blood circulation
The movement of BLOOD from the HEART through the ARTERIES, CAPILLARIES and VEINS and back to the heart. In humans the circulatory system has two distinct parts: the pulmonary circulation, in which blood is pumped from the right ventricle to the left atrium via the blood vessels of the lungs (where the blood is oxygenated and carbon dioxide is eliminated); and the systemic circulation, in which the oxygenated blood is pumped from the left ventricle to the right atrium via the blood vessels of the body tissues (where – in the capillaries – the blood is deoxygenated and carbon dioxide is taken up). As it leaves the heart the blood is under considerable pressure. Sustained high blood pressure, or hypertension, occurs in kidney and hormone diseases and in old age, but generally its cause is unknown. It may predispose to ATHEROSCLEROSIS and heart, brain and kidney damage. Low blood pressure occurs in SHOCK, TRAUMA and ADDISON'S DISEASE. (See also CIRCULATORY SYSTEM.) (♦ page 28)

Blood poisoning see SEPTICEMIA.

Blood pressure
Blood must be pumped from the heart at enough pressure to reach all the tissues. It is usually measured in the upper arm. The pressure is high in certain conditions, eg narrowing of the arteries. Systolic pressure is the heartbeat pressure, diastolic pressure the pressure between beats.

Bloodstone
A dark green variety of CHALCEDONY containing nodules of red JASPER.

Blotting
Southern blotting is a technique for identifying sequences of DNA. Variants for identifying RNA and for protein sequences are called Northern and Western blotting.

Blubber
The layer of fat below the skin of whales and other marine mammals; it is thick and provides buoyancy, insulation and energy reserves.

Blue baby
Infant born with a HEART defect (a hole between the right and left sides of the heart) or malformation of the arteries that permits much of the BLOOD to bypass the LUNGS. The resulting lack of oxygen causes CYANOSIS.

Blue-green algae
Obsolete name for CYANOBACTERIA. (♦ page 44)

Blue vitriol (copper (II) sulfate) see COPPER.

Blumberg, Baruch Samuel (b. 1925)
US physiologist who shared the 1976 Nobel Prize for Physiology or Medicine with D.C. GAJDUSEK for research into the nature of infection. He discovered the "Australia antigen" which is associated with serum HEPATITIS.

Bode's Law
A numerical relationship which was found to hold between the radii of the ORBITS of the then-known PLANETS. It was formulated by the German astronomer Johann Elert Bode (1747-1826). Bode started from the sequence 0, 3, 6, 12, 24, 48, 96, 192,... and added 4 to each member; 4, 7, 10, 16, 28, 52, 100, 196,...; then, if the radius of the orbit of the Earth was taken to be 10, he found that MERCURY fell into place at 4, VENUS at ~7, MARS at 16, JUPITER at 52 and SATURN at ~100. The discovery of URANUS at ~196 initiated a hunt for a missing planet at 28, which was supplied by the asteroid CERES but when NEPTUNE was discovered in 1846, it failed to satisfy the "law", which immediately declined in importance.

Body language
Information conveyed to other people, often unconsciously, by postures or movements of the body. (See also NONVERBAL COMMUNICATION.)

Boerhaave, Hermann (1668-1738)
Dutch chemist and physician, renowned in his day as a leading man of science, although he made no discoveries of lasting importance. His teaching did much to establish Leiden as a medical center of international repute.

Bog
Commonly, any marsh or swamp; specifically, a low-lying area, usually formed by the action of GLACIERS, which is poorly drained, may contain shallow water, and in which organic matter accumulates. Bog waters are generally acidic.

Bohr, Niels Henrik David (1885-1962)
Danish physicist who proposed the Bohr model of the ATOM while working with SIR ERNEST RUTHERFORD in Manchester, England, in 1913. Bohr suggested that a hydrogen atom consisted of a single electron performing a circular orbit around a central proton (the nucleus), the energy of the electron being quantized (ie the electron could only carry certain well-defined quantities of energy – see QUANTUM THEORY). At one stroke this accounted both for the properties of the atom and for the nature of its characteristic radiation (a spectrum comprising several series of discrete sharp lines). In 1927 Bohr proposed the COMPLEMENTARITY PRINCIPLE to account for the apparent paradoxes which arose on comparing the wave and particle approaches to describing subatomic particles. After escaping from Copenhagen in 1943 he went to the UK and then to the USA, where he helped develop the atomic bomb, but he was always deeply concerned about the graver implications for humanity of this development. In 1922 he received the Nobel Prize for Physics in recognition of his contributions to atomic theory. His son, Aage Niels Bohr (b. 1922), shared the 1975 Nobel Prize for Physics with B. MOTTELSON and J. RAINWATER for contributions made to the physics of the atomic nucleus.

Boil
An ABSCESS in a hair FOLLICLE, usually caused by infection with STAPHYLOCOCCUS. A sty is a boil on the eyelid; a carbuncle is a group of contiguous boils, usually on the back of the neck.

Boiler
Device used to convert water into STEAM by the

action of HEAT (see also BOILING POINT), usually to drive a steam engine. A boiler requires a heat source (ie a FURNACE), a surface whereby the heat may be conveyed to the water, and enough space for steam to form. The two main types of boiler are the fire-tube, where the hot gases are passed through tubes surrounded by water; and the water-tube, where the water is passed through tubes surrounded by hot gases. Fuels include coal, oil and fuel gas; nuclear energy is also used. HERO designed boilers, but used them only in toys. Steam power proper was barely considered until the 17th century, and little used before the 18th century.

Boiling point (bp)
The temperature at which the VAPOR PRESSURE of a liquid becomes equal to the external pressure, so that boiling occurs; the temperature at which a liquid and its vapor are at equilibrium. Measurement of boiling point is important in chemical ANALYSIS and the determination of MOLECULAR WEIGHTS. (See also EVAPORATION; PHASE EQUILIBRIUM.)

Bolometer
An instrument used to measure radiant ENERGY, usually in the infrared and microwave regions of the spectrum of ELECTROMAGNETIC RADIATION. It comprises a lens or stop system which focuses the test radiation on a thermoconductive device (usually a THERMISTOR), which is set in a WHEATSTONE BRIDGE circuit with another non-illuminated reference thermistor. Sensitive bolometers are used in conjunction with spectroscopes to measure the intensities of spectral lines (see SPECTROSCOPY).

Boltzmann, Ludwig (1844-1906)
Austrian physicist who made fundamental contributions to THERMODYNAMICS, classical statistical mechanics and KINETIC THEORY. The Boltzmann constant (k), the quotient of the universal gas constant R and the AVOGADRO number ($\mathcal{N}$), is used in statistical mechanics.

Bolus
Food passing through the gut is formed into balls, called boluses.

Bond, Chemical
The links which hold ATOMS together in compounds. In the 19th century it was found that many substances, known as covalent compounds, could be represented by structural FORMULAS in which lines represented bonds. By using double and triple bonds, most organic compounds could be formulated with constant VALENCES of the constituent atoms. STEREOISOMERISM showed that the bonds must be localized in fixed directions in space. Electrovalent or ionic compounds (see ELECTROCHEMISTRY) consist of oppositely charged IONS arranged in a lattice; here the bonds are nondirectional electrostatic interactions.

The theory that atoms consist of electrons orbiting in shells around the nucleus (see PERIODIC TABLE) led to a simple explanation of both kinds of bonding: atoms combine to achieve highly stable, filled outer shells containing 2, 8 or 18 electrons, either by transfer of electrons from one atom to the other (ionic bond), or by the sharing of one electron from each atom so that both electrons orbit around both nuclei (covalent bond). In the coordinate bond, a variant of the covalent bond, both shared electrons are provided by one atom. QUANTUM THEORY has now shown that electrons occupy ORBITALS having certain shapes and energies, and that, when atoms combine, the outer atomic orbitals are mixed to form molecular orbitals. The energy difference constitutes the bond energy – the energy required to break the bond by separating the atoms. The energy and length of chemical bonds, and the angles between them, may be investigated by SPECTROSCOPY and X-RAY DIFFRACTION. (See also HYDROGEN BONDING.) (◆ page 37)

Bond, William Cranach (1789-1859)
US astronomer, first director of the Harvard Observatory and a pioneer of ASTROPHOTOGRAPHY. In 1850 he made the first DAGUERREOTYPE of a celestial object and discovered the third, "crape", ring of SATURN.

Bondi, Sir Hermann (b. 1919)
Austrian-born British cosmologist who with T. GOLD in 1948 formulated the Steady State theory of the Universe (see COSMOLOGY).

Bone
The hard tissue that forms the SKELETON of VERTEBRATES other than the cartilaginous fish (sharks and rays). Bones support the body, project its organs, act as anchors for MUSCLES and as levers for the movement of limbs. Bone consists of living cells (osteocytes) embedded in a matrix of COLLAGEN fibers with calcium salts, similar in composition to hydroxyapatite (see APATITE), deposited between them. Some carbonates are also present. All bones have a shell of compact bone in concentric layers (lamellae) around the blood vessels, which run in small channels (Haversian canals). Within this shell is porous or spongy bone, and in some bones there is a hollow cavity containing MARROW. The bone is enveloped by a fibrous membrane, the periosteum, which is sensitive to pain, unlike the bone itself, and that has a network of nerves and blood vessels that penetrate the bone surface. After primary growth has ended, bone formation (ossification) occurs where the periosteum joins the bone, where there are many bone-forming cells (OSTEOBLASTS). Ossification begins in the embryo at the end of the second month, mostly by transformation of CARTILAGE: some cartilage cells become osteoblasts and secrete collagen, together with a hormone that causes calcium salts to be deposited. Bone is constantly being reshaped by the osteoblasts and by OSTEOCLASTS, which remove calcium. This process also helps to keep calcium and phosphate levels steady in the blood. Vitamin D makes calcium available from the food to the blood, and its deficiency leads to RICKETS. The two ends of a "long" bone (the epiphyses) ossify separately from the shaft, and are attached to it by cartilaginous plates at which lengthwise growth takes place. Radical growth is controlled by the periosteum, and at the same time the core of the bone is eroded by osteoclast cells to make it hollow. Primary growth is stimulated by the pituitary and sex HOROMONES; it is completed in adolescence, when the epiphyses fuse to the shaft. Disorders of bone include OSTEOMYELITIS and CANCER. Dead bone is not readily absorbed and can be a focus of infection. In old age thinning and weakening of the bones by loss of calcium (osteoporosis) is common, particularly in women.

Bony fish
The common name of the OSTEICHTHYES, a taxonomic class that includes the subclasses ACTINOPTERYGII (ray-finned fish) and SARCOPTERYGII (lobe-finned fish). "Bony fish" is sometimes also used rather loosely to refer to the ray-finned fish alone, as they comprise one of the two major groups of fish, the other being the sharks, rays and skates, or cartilaginous fish (CHONDRICHTHYES). As cartilage precedes bone in the development of the vertebrate skeleton, it was once believed that the cartilaginous fish represented a more primitive stage than the bony fish, but this idea has now been discarded. The term "bony fish" is now going out of favor in either usage, because the primitive jawless fish or AGNATHA, which are not members of the Osteichthyes, also have bony skeletons, and it is clear that the ancestors of the cartilaginous fish were bony too. The distinction between cartilaginous fish on the one hand and bony fish on the other is misleading. (◆ page 17)

Boötes
The Herdsman, a constellation of the N Hemisphere, containing the star ARCTURUS.

Boranes
Covalent HYDRIDES of boron, of unusual molecular structure: they have hydrogen-bridge bonding, and the boron atoms form the vertices of polyhedra. Boranes are volatile, reactive and often flammable in air. They are used as reducing agents in chemical synthesis.

Borax
Mineral name for sodium tetraborate, $Na_2B_4O_7.10H_2O$, found mainly in California. (For sodium borates see SODIUM; KERNITE.)

Borazon
Cubic form of boron nitride (BN), resembling DIAMOND in structure and properties. It is made by heating hexagonal boron nitride (a white powder resembling GRAPHITE) to 1500°C at 65,000atm pressure. Borazon is as hard as diamond, and more useful industrially because it is more resistant to oxidation and heat.

Bordet, Jules Jean-Baptiste Vincent (1870-1961)
Belgian bacteriologist and immunologist awarded the 1919 Nobel Prize for Physiology or Medicine for his discovery of the substances later named "complement" and the process of complement fixation (see ANTIBODIES AND ANTIGENS). He also discovered the BACILLUS responsible for WHOOPING COUGH.

Bore (eagre)
Tidal phenomenon of rivers that widen gradually toward broad mouths, and that are subject to high tides. During spring flood, larger quantities of water from the sea than can normally flow upriver are driven into the rivermouth, resulting in a high wave that travels upriver at great speed. Perhaps the best-known bores are those of the Ganges, the Severn and the Bay of Fundy.

Boreal forest
The region of coniferous forest lying to the north of the deciduous forest vegetation zone, between about 45°N and 70°N. An area of mixed forest occurs at the lower latitudes, where the two zones meet. In the far north, the boreal forest grades into TAIGA and TUNDRA.

Borelli, Giovanni Alfonso (1608-1679)
Italian astronomer, physicist and physiologist. After making contributions to astronomy, including the proposal that COMETS travel along elliptical paths, he turned his knowledge of physics to the working of the living body and successfully explained MUSCLE action on mechanical principles.

Boric acid (boracic acid, H_3BO_3)
Colorless crystalline solid, a weak inorganic acid. It gives boric oxide (B_2O_3) when strongly heated; SODIUM borate typifies its salts. Boric acid is used as an external antiseptic, in the production of glass and as a welding flux.

Borlaug, Norman Ernest (b. 1914)
US agricultural scientist who was awarded the 1970 Nobel Peace Prize for his part in the development of improved varieties of CEREAL CROPS, important in the green revolution.

Born, Max (1882-1970)
German theoretical physicist active in the development of quantum physics, whose particular contribution was the probabilistic interpretation of the SCHRÖDINGER wave equation, thus providing a link between WAVE MECHANICS and the QUANTUM THEORY. Sharing the Nobel Prize for Physics with BOTHE in 1954, he devoted his later years to the philosophy of physics.

Bornite
Reddish brown SULFIDE mineral of iron and

ELEMENTARY PARTICLES

Subatomic particles

The atom consists of very light, negatively charged electrons, circling around a small, dense nucleus which is made up of two kinds of particles: positively charged protons and electrically neutral neutrons. However, other subatomic particles have also been found, existing for short periods in high-energy particle accelerators, or occurring naturally in cosmic rays. Study of the interactions of these high-energy particles has revealed much about the fundamental particles that make them up.

Many subatomic particles have been shown to be made up of pointlike particles called quarks: some, known as baryons, or heavy particles, have three quarks, whereas others (known as mesons) have two. Six types of quark have been identified, of widely varying mass and different charge: the combinations of quarks in the baryons and mesons give particles of increasing mass. The heavy particles are unstable, tending quickly to transmute into lighter ones – thus only the lightest baryons are found in normal matter.

Another class of elementary particle exists, known as the leptons. The electron is the lightest of these, and the most common.

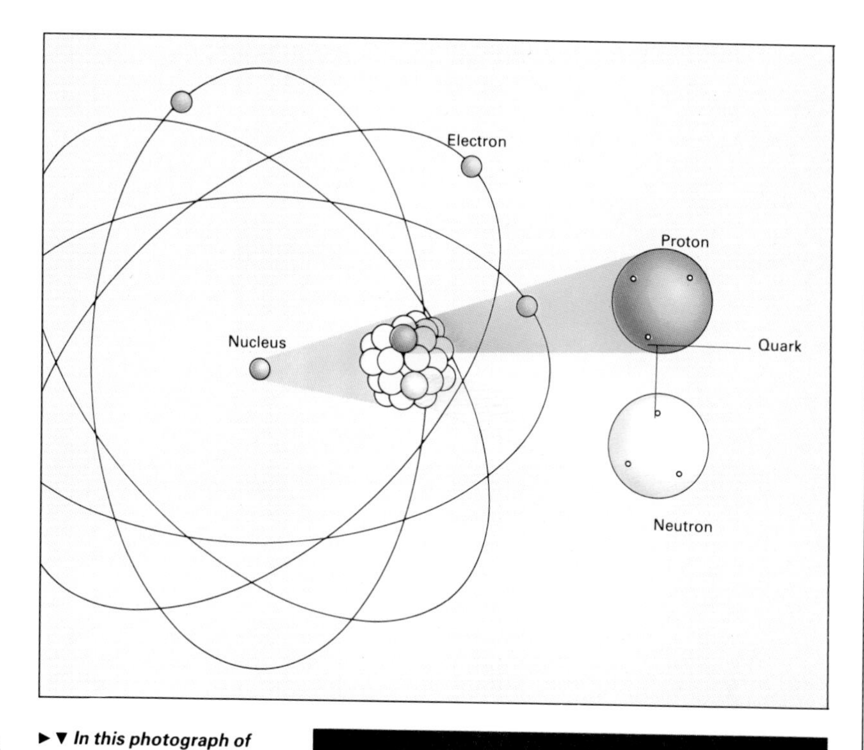

Elementary particles

Baryons		Quarks	Mass (MeV)	Lifetime (s)
Proton	p	uud	938.3	$>10^{32}$y
Neutron	n	udd	939.6	898
Lambda	Λ	uds	1115.6	2.6×10^{-10}
Sigma	Σ^+	uus	1189.4	0.8×10^{-10}
	Σ^0	uds	1192.5	5.8×10^{-20}
	Σ^-	dds	1197.3	1.5×10^{-10}
Xi	Ξ^0	uss	1314.9	2.9×10^{-10}
(cascade)	Ξ^-	dss	1321.3	1.6×10^{-10}
	Λ_c^+	udc	2282.0	2.3×10^{-13}

Quarks		Charge	Mass (MeV)
Up	u	$+2/3$	5
Down	d	$-1/3$	7
Charm	c	$+2/3$	1400
Strange	s	$-1/3$	150
Top	t	$+2/3$	?
Bottom	b	$-1/3$	1,800

Mesons		Quarks	Mass (MeV)	Lifetime (s)
Pion	$\pi^\pm$	$u\bar{d},\bar{u}d$	139.6	2.6×10^{-8}
	π^0	$(u\bar{u},d\bar{d})$	135.0	0.8×10^{-16}
Kaon	$K^\pm$	$u\bar{s},s\bar{u}$	493.7	1.2×10^{-8}
	K_S^0	$(d\bar{s},s\bar{d})$	497.7	0.9×10^{-10}
	K_L^0			5.2×10^{-8}
	$D^\pm$	$c\bar{d},\bar{c}d$	1869.4	9.2×10^{-13}
	D^0	$c\bar{u}$	1864.7	4.4×10^{-13}
	$F^\pm$	$c\bar{s},\bar{c}s$	1971	1.9×10^{-13}
	$B^\pm$	$u\bar{b},\bar{u}b$	5270.8	14×10^{-13}
	B^0	$b\bar{d}$	5274.2	

$\bar{d}$, $\bar{u}$, $\bar{s}$, $\bar{c}$, $\bar{b}$, – Antiquarks

Leptons		Charge	Mass (MeV)	Lifetimes (s)
Electron	e	−1	0.51	stable
Electron-neutrino	ν_e	0	$<0.46\times10^{-4}$	stable
Muon	μ	−1	105.6	2.197×10^{-6}
Muon-neutrino	ν_μ	0	<5.0	stable
Tau	τ	−1	1,784	3.4×10^{-13}
Tau-neutrino	ν_τ	0	<164	stable

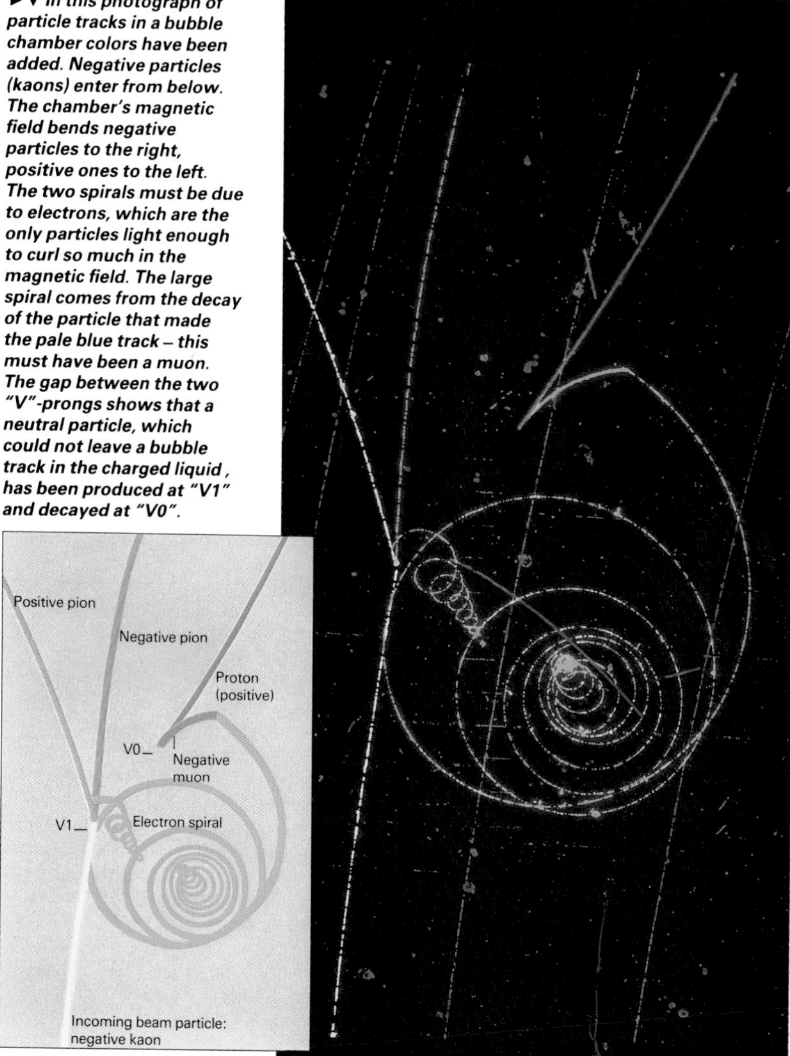

►▼ *In this photograph of particle tracks in a bubble chamber colors have been added. Negative particles (kaons) enter from below. The chamber's magnetic field bends negative particles to the right, positive ones to the left. The two spirals must be due to electrons, which are the only particles light enough to curl so much in the magnetic field. The large spiral comes from the decay of the particle that made the pale blue track – this must have been a muon. The gap between the two "V"-prongs shows that a neutral particle, which could not leave a bubble track in the charged liquid, has been produced at "V1" and decayed at "V0".*

copper (Cu_5FeS_4); a COPPER ore, called "peacock ore" because of its tarnish when fractured. Its occurrence is widespread, especially in Chile, Peru and Tasmania. It alters to CHALCOCITE.

Boron (B)
Nonmetallic element in Group IIIA of the PERIODIC TABLE, occurring as KERNITE and BORAX in California and Turkey. Boron has three black crystalline allotropes and an amorphous form, and is best prepared by reduction of the halides with hydrogen. Normally inert, it becomes reactive at high temperatures; it is trivalent. Boron is a trace element vital to plant growth; it is used to produce heat-resistant alloy steels, and boron fibers are used as a high-strength construction material. Boron-steel rods absorb neutrons in nuclear reactors. The borates are salts of BORIC ACID; the SODIUM salts are the most important. AW 10.8, mp 2130°C, bp 3700°C, sg 2.34. (See also BORANES; BORAZON.) (♦ page 130)

Bosch, Karl (1874-1940)
German industrial chemist who adapted the HABER PROCESS for AMMONIA manufacture for large-scale industrial use. In 1931 he shared the Nobel Prize for Chemistry with F. BERGIUS for their work on high- pressure synthesis.

Bose, Sir Jagadis Chandra (1858-1937)
Indian biologist who devised and used sensitive instruments for measuring the growth and response to stimuli of plants.

Bose-Einstein statistics
In QUANTUM MECHANICS, the statistical behavior of a system of indistinguishable particles with a number of discrete states, each of which may be occupied at any one time by any number of particles. SUBATOMIC PARTICLES that show this behavior are termed bosons. (See also FERMI-DIRAC STATISTICS.)

Boson see SUBATOMIC PARTICLES.

Botany
The study of plant life. Botany and ZOOLOGY are the major divisions of BIOLOGY. There are many specialized disciplines within botany, including morphology, physiology, GENETICS, ECOLOGY and TAXONOMY.

The plant morphologist studies the form and structure of plants, particularly the whole plant and its major components, while the plant anatomist concentrates upon the cellular and subcellular structure, perhaps using ELECTRON MICROSCOPY. The functioning of plants is studied by the plant physiologist and plant biochemist. An important practical branch of genetics is plant BREEDING. Other specialists study the plant in relation to its uses (economic botany), to PLANT DISEASES (plant pathology) or to the agricultural importance of plants (agricultural botany). (See also AGRONOMY; BIOCHEMISTRY; PLANT.)

The forerunners of the botanists were men who collected herbs for medical use long before philosophers turned to the scientific study of nature. However, the title of "father of botany" goes to THEOPHRASTUS, a pupil of ARISTOTLE, whose *Inquiry into Plants* sought to classify the types, parts and uses of the members of the plant kingdom. The first successful classification of the plant kingdom in which modern binomial names first appeared (1753) was that of LINNAEUS. While NEHEMIAH GREW and JOHN RAY laid the foundations for plant anatomy and physiology in the 17th and 18th centuries, and HOOKE identified the cell (1665) with the aid of the microscope, these subjects were not actively pursued until the 19th century, when ROBERT BROWN identified the nucleus and THEODOR SCHWANN proposed his comprehensive cell theory. The work of CHARLES DARWIN revolutionized the theory of classification; that of GREGOR MENDEL pointed the way to a true science of plant breeding.

Bothe, Walter Wilhelm Georg Franz (1891-1957)
German experimental physicist who devised the coincidence method for the detection of COSMIC RAY showers (1929). This work won him a share (with MAX BORN) in the 1954 Nobel Prize for Physics.

Botulism
A serious type of FOOD POISONING caused by a toxin produced by the anaerobic bacteria *Clostridium botulinum* and *C. parabotulinum*, which normally live in soil but may infect inadequately sterilized canned food. The toxin paralyzes the nervous system. Thorough cooking destroys both bacteria and toxin.

Boussingault, Jean-Baptiste Joseph Dieudonné (1802-1887)
French agricultural chemist who studied the GERMINATION of SEEDS and promoted the use of inorganic fertilizers containing NITROGEN and PHOSPHORUS compounds.

Bovet, Daniel (b. 1907)
Swiss-born Italian pharmacologist who discovered the first ANTIHISTAMINE and developed CURARE and curare-like compounds for use as muscle-relaxants during surgery. He was awarded the 1957 Nobel Prize for Physiology or Medicine.

Bovids
Members of the Bovidae, a very large family of mammals within the Order ARTIODACTYLA. It includes the many different types of antelope, the sheep and goats, and the cattle and their wild relatives – buffalo, bison, bison, yak, anoa and gaur.

Boyd-Orr, John, Baron (1880-1971)
British agricultural scientist and nutritionist who was awarded the 1949 Nobel Peace Prize for his services to the Food and Agriculture Organization (FAO) of which he was director 1945-48.

Boyle, Robert (1627-1691)
British natural philosopher, often called the father of modern CHEMISTRY for his rejection of the theories of the alchemists and his espousal of ATOMISM. He was a founder member of the ROYAL SOCIETY OF LONDON, and was noted for his pneumatic experiments.

Boyle's Law (Mariotte's Law)
An empirical relation reported by BOYLE (1662) and MARIOTTE (1676) but actually discovered by Boyle's assistant R. Townely, which states that given a fixed MASS of GAS at constant TEMPERATURE, its VOLUME is inversely proportional to its PRESSURE. Real gases deviate considerably from this law.

Brachiopods see LAMPSHELLS.

Bradley, James (1693-1762)
English astronomer who discovered the ABERRATION OF LIGHT (1728) and the Earth's nutation (see NUTATION, ASTRONOMICAL).

Bragg, Sir William Henry (1862-1942)
British physicist who shared the 1915 Nobel Prize for Physics with his son, Sir William Lawrence Bragg (1890-1971), for learning how to deduce the atomic structure of CRYSTALS from their X-RAY DIFFRACTION patterns (1912).

Brahe, Tycho (1546-1601)
Danish astronomer, the greatest exponent of naked-eye positional ASTRONOMY. KEPLER became his assistant in 1601 and was driven to postulate an elliptical orbit for MARS only because of his absolute confidence in the accuracy of Brahe's data. Brahe is also remembered for the "Tychonic system", in which the planets circled the Sun, which in turn orbited a stationary Earth, this being the principal 17th-century rival of the Copernican hypothesis.

Brain
A complex organ that, together with the SPINAL CORD, comprises the central NERVOUS SYSTEM and coordinates all nerve-cell activity. In INVERTEBRATES the brain is an enlarged GANGLION in the head region, reaching its highest development in the insects and cephalopod mollusks (where it is in the form of a ring around the esophagus). In VERTEBRATES the brain is more developed – tubular in lower vertebrates, larger, more differentiated and more rounded in higher ones. In higher mammals, including humans, the brain is dominated by the highly developed cerebral cortex. The brain is composed of many billions of interconnecting nerve cells (see NEURONS) and supporting cells (neuroglia). The BLOOD CIRCULATION, in particular the regulation of blood pressure, is designed to ensure an adequate supply of oxygen to these cells: if this supply is cut off, neurons die in only a few minutes. The brain is well protected inside the SKULL, and is surrounded, like the spinal cord, by three membranes, the meninges. Between the two inner meninges lies the CEREBROSPINAL FLUID (CSF), an aqueous solution of salts and GLUCOSE. CSF also fills the four ventricles (cavities) of the brain and the central canal of the spinal cord. If the circulation of CSF between ventricles and meninges becomes blocked, HYDROCEPHALUS results. Relief of this may involve draining CSF to the atrium of the heart.

The human brain may be divided structurally into three parts. The hindbrain consists of the medulla oblongata, which contains vital centers to control heartbeat and breathing; the pons, which, like the medulla oblongata, contains certain cranial nerve nuclei and numerous fibers passing between the higher brain centers and the spinal cord; and the cerebellum, which regulates balance, posture and coordination. The midbrain, a small but important center for REFLEXES in the brain stem, also contains nuclei of the cranial nerves and the reticular formation, a diffuse network of neurons involved in regulating arousal – SLEEP and alertness. The forebrain consists of the thalamus, which relays sensory impulses to the cortex; the hypothalamus, which controls the autonomic nervous system, food and water intake and temperature regulation, and to which the PITUITARY GLAND is closely related (see also PINEAL BODY); and the cerebrum. The cerebrum makes up two-thirds of the entire human brain and has a deeply convoluted surface; it is divided into two interconnected halves or hemispheres. The main functional zones of the cerebrum are the surface layers of gray matter, the cortex, below which is a broad white layer of nerve fiber connections, and the basal ganglia, concerned with muscle control. (Disease of the basal ganglia causes PARKINSON'S DISEASE.) Each hemisphere has a motor cortex, controlling voluntary movement, and a sensory cortex, receiving cutaneous sensation, both relating to the opposite side of the body. Other areas of cortex are concerned with language (see APHASIA; SPEECH AND SPEECH DISORDERS), memory, and perception of the special senses (sight, smell, sound); higher functions such as abstract thought may also be a cortical function (see also INTELLIGENCE; LEARNING). The brain has four interconnecting VENTRICLES, one in each hemisphere and two central ones, that contain cerebrospinal fluid (CSF) and a cavity that connects the ventricle to the subdural space. In some forms of spina bifida the ventricles are greatly enlarged and a "tap" must be inserted to drain the CSF. Diseases of the brain include infections – specifically MENINGITIS, ENCEPHALITIS, syphilis (see SEXUALLY TRANSMITTED DISEASES) and ABSCESSES; also trauma, TUMORS, STROKES, MULTIPLE SCLEROSIS, and degenerative diseases with early ATROPHY, either generalized or localized. Investigation of brain diseases includes

X-RAYS using various contrast methods, SPINAL TAP (lumbar puncture) – to study CSF abnormalities – and the use of the ELECTROENCEPHALOGRAM. (♦ page 111, 126, 213)

Brain death
Absence of electrical activity in the brain, usually taken as a measure of death even when other organs are working, possibly with the assistance of life-support machines.

Brainwashing
The manipulation of an individual's will, generally without his knowledge and against his wishes. Most commonly, it consists of a combination of isolation, personal humiliation, disorientation, systematic indoctrination and alternating punishment and reward.

Brass
An ALLOY of COPPER and ZINC, known since Roman times, and widely used in industry and for ornament and decoration. Up to 36% zinc forms α- brass, which can be worked cold; with more zinc a mixture of α- and β- brass is formed, which is less ductile but stronger. Brasses containing more than 45% zinc (white brasses) are unworkable and have few uses. Some brasses also contain other metals: lead to improve machinability, aluminum or tin for greater corrosion-resistance, and nickel, manganese or iron for higher strength.

Brassicas
Plants of the genus *Brassica*, many of which are grown as crops, eg cabbage, kale, Brussels sprouts, cauliflower, broccoli, mustard, turnip, and oil-seed rape.

Brattain, Walter Houser (b. 1902)
US physicist who shared the 1956 Nobel Prize for Physics with SHOCKLEY and BARDEEN for their development of the TRANSISTOR.

Braun, Karl Ferdinand (1850-1918)
German physicist who shared the 1909 Nobel Prize for Physics with MARCONI for his discovery that certain crystals could act as RECTIFIERS, and for his proposal that these could be used in (crystal-set) RADIOS.

Braun, Wernher Magnus Maximilian von (1912-1977)
German ROCKET engineer who designed the first self-contained missile, the V-2, which was used against the UK in 1944. In 1945 he went to America, where he led the team that put the first US artificial SATELLITE in ORBIT (1958).

Breasts (mammary glands)
The milk-secreting glands in mammals. The breasts develop alike in both sexes, with about 20 ducts being formed leading to the nipples, until puberty, when the female breasts develop in response to sex HORMONES. In PREGNANCY the breasts enlarge and milk-forming tissue grows around multiplied ducts; later milk secretion and release in response to suckling occur under the control of specific hormones. Disorders of the breast include mastitis, breast CANCER (see also MASTECTOMY) and adenosis.

Breathing see RESPIRATION.

Breccia see CONGLOMERATE; TALUS.

Breeder reactor
A NUCLEAR REACTOR that produces more nuclear fuel than it consumes, used to convert material that does not readily undergo FISSION into material that does. Commonly, nonfissile URANIUM-238 is converted into PLUTONIUM-239. (See also NUCLEAR ENERGY.)

Breeding
The selective breeding of plants and animals is the production of new varieties or strains by a program of selection spanning several generations of the organism concerned. Conscious breeding usually also involves controlled crosses between selected parents. MENDEL discovered the laws of GENETICS through breeding experiments, and his discoveries put breeding on a scientific basis. Later developments included the use of agents that increase the rate of MUTATION; chemicals that can induce POLYPLOIDY in plants following hybridization between two distinct species; and the production of F_1 hybrids from two inbred lines, which show much faster growth and higher yield because of hybrid vigor or heterosis (see HYBRIDIZATION). More recently, GENETIC ENGINEERING has greatly broadened the scope of breeding by making it possible to introduce a favorable gene directly from one organism to another.

Breeding behavior
Any behavior by which animals attract members of the opposite sex for the purposes of reproduction. Such behavior includes visual displays, calls and song and the production of scent for attraction and stimulation. (See also COURTSHIP.)

Brentano, Franz Clemens (1838-1917)
Austrian philosopher and psychologist, a Roman Catholic priest from 1864 to 1873, who founded the school of intentionalism and taught both SIGMUND FREUD and HUSSERL.

Breuer, Josef (1842-1925)
Austrian physician who pioneered the methods of PSYCHOANALYSIS and collaborated with FREUD in writing *Studies in Hysteria* (1895). He also discovered the role of the semicircular canals of the inner EAR in maintaining balance (1873).

Brewster, Sir David (1781-1868)
Scottish man of science who discovered Brewster's Law in optics and did much to popularize science in 19th-century Britain. Brewster's Law states that the maximum polarization surface occurs when the reflected and refracted rays are at right angles. (See POLARIZED LIGHT.)

Bridgman, Percy Williams (1882-1961)
US physicist who won the 1946 Nobel Prize for Physics for his investigation of substances at very high pressures. His work led to the production of synthetic DIAMONDS (1955). In the philosophy of science he championed the view that scientific terms are only meaningful if they can be given "operational definitions".

Bright's disease
A form of acute NEPHRITIS that may follow infections with certain STREPTOCOCCUS bacteria. Blood and protein are lost in the urine; there may be EDEMA and raised blood pressure. Recovery is usually complete but a few patients progress to chronic KIDNEY disease.

Brill, Abraham Arden (1874-1948)
Austrian-born US psychiatrist, the "father of American PSYCHOANALYSIS", who introduced the Freudian method to the US and translated many of FREUD's works into English.

Brine
A concentrated solution of SALT and other compounds. Natural brine occurs as seawater and as underground deposits. It is a major source of salt, BROMINE, IODINE and potassium compounds, and is also used as a food preservative and in refrigeration. (See also EVAPORITES.)

British Thermal Unit (BTU)
The quantity of ENERGY required to raise 1lb of water through 1°F. The International Steam Tables define the Btu_{IT} as $251.9958cal_{IT}$; this is equivalent to 1055.056 joules.

Brittlestars
A group of ECHINODERMS of the Class Ophiuroidea. They have long, flexible arms, clearly differentiated from the disk-shaped central body.

Broadleaved trees
Trees belonging to the flowering plants or ANGIOSPERMS. They are called broadleaved trees to contrast them with the CONIFERS and other GYMNOSPERM trees, which mostly have narrow, needle-like leaves.

Broglie, Louis Victor Pierre Raymond de see DE BROGLIE, LOUIS VICTOR PIERRE RAYMOND.

Bromine (Br)
Dark-red fuming liquid, toxic and corrosive, with a pungent odor; one of the HALOGENS, intermediate in properties between CHLORINE and IODINE. It occurs as bromides, mainly in seawater, from which it is extracted by oxidation with chlorine. Silver bromide, being light-sensitive, is used in PHOTOGRAPHY. Ethylene dibromide, the chief bromine product, is used in antiknock additives. Alkyl bromides (see ALKYL HALIDES) are used as fumigants and solvents. AW 79.9, mp −7.3°C, bp 58.9°C, sg 3.12 (20°C). (♦ page 130)

Bronchi
The tubes through which air passes from the TRACHEA to the LUNGS. The trachea divides into the two primary bronchi, one to each lung, which divide into smaller branches and finally into the narrow bronchioles that connect with the alveolar sacs. The bronchi are lined with a mucous membrane that has motile CILIA to remove dust, etc. (♦ page 28)

Bronchitis
Inflammation of BRONCHI. Acute bronchitis, often due to virus infection, is accompanied by cough and fever and is short-lived. Chronic bronchitis is a more serious, often disabling and finally fatal disease. The main cause is smoking, which irritates the lungs and causes overproduction of mucus. The cilia fail, and sputum has to be coughed up. Bronchi thus become liable to recurrent bacterial infection, sometimes progressing to PNEUMONIA. Areas of lung become non-functional, and ultimately CYANOSIS and HEART failure may result.

Bronchoscope
A tube with a light and lens system, used to examine the TRACHEA and BRONCHI, and also to perform a BIOPSY.

Brønsted, Johannes Nicolaus (1879-1947)
Danish physical chemist remembered for formulating (independently of T.M. Lowry, 1874-1936) the Brønsted-Lowry theory of ACIDS and bases.

Bronze
An ALLOY of COPPER and TIN, known since the 4th millennium BC. It is a hard, strong alloy with good corrosion-resistance (the patina formed in air is protective). Various other components are added to improve hardness or machinability, such as aluminum, iron, lead, zinc and phosphorus. Aluminum bronzes, and some others, contain no tin.

Brood parasite
An animal that relies on another species to rear its young. The best-known example is the cuckoo, which lays its egg in the nest of a reed warbler, dunnock, wagtail or some other small bird. Among insects, several species of bees and wasps lay their eggs in the nests of other bees or wasps, which unknowingly rear the alien young; these species are known as cuckoo-bees and cuckoo-wasps.

Brood parasitism always involves some sort of MIMICRY. The eggs of the cuckoo generally resemble those of its host, and in the widowbirds the young have identical throat-markings and begging behavior to that of the particular species parasitized. They also resemble the host's young in their plumage, and only grow their distinctive adult plumage on leaving the nest. Some insect brood-parasites resemble their host even as adults, to facilitate entry to the nest.

Brown dwarf
A STAR of such low mass (probably less than 0.08 of the Sun's mass) that thermonuclear reactions never get started in its core. Brown dwarfs are expected to be very cool and dim, shining mainly at INFRARED wavelengths as a result of the heat

generated by their contracting from interstellar gas clouds. Because they are so faint only a few have been discovered so far, but they may exist in large numbers. An example is the companion to the star Van Biesbroeck 8, which is about 21 light years from Earth.

Brown fat
Fat that has a rich blood supply, making it look brown. Most common in newborn babies and in animals that hibernate, it is easily converted into sugars for energy.

Brown, Herbert C. (b. 1912)
US chemist who shared the 1979 Nobel Prize for Chemistry with WITTIG for devising new ways of making organic chemicals with boron- and phosphorus-containing compounds.

Brownian motion
Frequent, random fluctuation, illustrated by the motion of particles of the dispersed phase of a fluid COLLOID; first described by ROBERT BROWN (1827) after observation of a SUSPENSION of pollen grains in water. It is a result of the bombardment of the colloidal particles by the MOLECULES of the continuous phase (see KINETIC THEORY): a chance greater number of impacts in one direction changes the direction of motion of the particle. It is believed that all molecules of FLUIDS undergo Brownian motion. Observation of Brownian motion in colloids is valuable in studying DIFFUSION.

Brown, Michael S. (b. 1941)
US scientist. Corecipient with J.L. GOLDSTEIN of the 1985 Nobel Prize for Physiology or Medicine for their discoveries concerning the regulation of cholesterol metabolism.

Brown, Robert (1773-1858)
Scottish botanist who first observed BROWNIAN MOTION (1827) and who identified and named the plant CELL nucleus (1831).

Bruce, Sir David (1855-1931)
Australian-born British microbiologist who discovered the organisms responsible for undulant fever, nagana and African SLEEPING SICKNESS. The BACILLUS causing the first of these (now known as BRUCELLOSIS) is named *Brucella* in his honor.

Brucellosis (Bang's disease)
A BACTERIAL DISEASE of cattle, goats and swine, caused by *Brucella*. It causes ABORTION, and affected animals have to be slaughtered to prevent spread.

Bruise
Bleeding into injured skin, causing discoloration until the red blood cells are broken down and removed, which takes several days.

Bruxism
The habit of tooth-grinding.

Bryophytes
Division of the PLANT kingdom that contains the most primitive of the green land plants. They are confined largely to damp habitats. The Bryophyta are normally divided into three classes: Hepaticae (liverworts), Anthoceratae (hornworts) and Musci (mosses). Bryophytes have a characteristic life cycle in which the gametophyte is dominant and the sporophyte is attached to and dependent on the gametophyte for nutrition. (See also ALTERNATION OF GENERATIONS.) (➧ page 139)

Bryozoa
Also known as moss animals, sea mats or ectoprocts, these are sessile, filter-feeding colonial animals that create flat carpets of glass-like "cells" on the surface of seaweeds, or form branched fronds. Each "cell" is a rigid calcareous box containing an individual animal, that feeds by means of a tentacular organ, a lophophore (see LOPHOPHORATES). The name ectoprocts refers to the fact that the anus is outside the lophophore; in the ECDOPROCTS, by contrast, it is inside the ring of tentacles.

Bubble chamber
Device invented by GLASER (1952) to observe the paths of SUBATOMIC PARTICLES with energies too high for a CLOUD CHAMBER to be used. A liquid (eg liquid HYDROGEN) is held under PRESSURE just below its BOILING POINT. Sudden reduction in pressure lowers this boiling point: boiling starts along the ionized trails of energetic particles. Their paths may thus be photographed as a chain of tiny bubbles, before pressure is reapplied and the cycle restarts. (➧ page 49)

Bubo
Old name for a swelling in the armpit or groin due to swollen LYMPH NODES. The plague was sometimes bubonic, sometimes pneumonic (respiratory).

Bubonic plague see PLAGUE.

Buchner, Eduard (1860-1917)
German organic chemist, awarded the 1907 Nobel Prize for Chemistry for his discovery that FERMENTATION did not require the presence of complete YEAST cells but only an extract containing the enzyme ZYMASE. This discovery inaugurated ENZYME chemistry.

Bud
A condensed shoot in which the stem is very short, the inner leaves are closely packed and the outer scale-leaves form a protective covering. In the spring, or when the conditions that kept the bud dormant are removed, the stem elongates rapidly and the leaves unfold. The term bud is also used in zoology for a point from which new growth develops.

Budding
A form of asexual REPRODUCTION in which a parent organism produces smaller daughter organisms. It is common in yeasts, hydroids and some polychaete worms.

The term is also used to describe a form of GRAFTING used particularly for roses and fruit trees. The grafted portion, or scion, is a bud that is inserted under the bark of a stock that is usually not more than one year old.

Buffer
A solution in which pH is maintained at a nearly constant value. It consists of a relatively concentrated solution of a weak ACID and its conjugate BASE, and works best if their concentrations are roughly equal, when hydrogen ion concentration equals the DISSOCIATION constant of the acid.

Buffon, Georges Louis Leclerc, Comte de (1707-1788)
French naturalist who was the first modern taxonomist of the ANIMAL KINGDOM and who led the team which produced the 44-volume *Histoire naturelle* (1749-1804).

Bulb
A short, underground storage stem composed of many fleshy scale-leaves, which are swollen with stored food, and an outer layer of protective scale-leaves. Bulbs are a means of overwintering; in the spring, flowers and foliage leaves are rapidly produced when growing conditions are suitable.

Bunion
Deformity of the joint of the big toe, made worse by ill-fitting shoes.

Bunsen, Robert Wilhelm Eberhard (1811-1899)
German chemist who, after important work on organo-arsenic compounds, went on (with G.R. KIRCHHOFF) to pioneer chemical SPECTROSCOPY, discovering the elements CESIUM (1860) and RUBIDIUM (1861). He also helped to popularize the gas burner known by his name.

Burette
Graduated glass tube with a stopcock, used in VOLUMETRIC ANALYSIS to measure the volume of a liquid, especially of one of the reagents in a TITRATION. A gas burette measures gas volume by the volume of liquid displaced.

Burnet, Sir Frank Macfarlane (1899-1986)
Australian physician and virologist who shared the 1970 Nobel Prize for Physiology or Medicine with P.B. MEDAWAR for his suggestion that the ability of organisms to form ANTIBODIES in response to foreign tissues was acquired and not inborn. Medawar followed up the suggestion and performed successful skin transplants in mice.

Burnham, Sherbourne Wesley (1838-1921)
US astronomer noted for his studies of DOUBLE STARS, reported in his General Catalog of Double Stars (1906).

Burning glass see MAGNIFYING GLASS.

Burns and scalds
Injuries caused by heat, electricity, radiation or caustic substances, in which protein denaturation causes death of tissues. (Scalds are burns caused by boiling water or steam.) Burns cause PLASMA to leak from blood vessels into the tissues, and in severe burns substantial leakage leads to SHOCK.

Burr (bur)
A seed with barbed spines that catch onto animal fur, thus promoting their dispersal.

Bursa
Fibrous sac containing SYNOVIAL FLUID which reduces friction where TENDONS move over bones. Extra bursae may develop where there is abnormal pressure or friction.

Bursitis
Inflammation of a BURSA, commonly caused by excessive wear and tear (as in housemaid's knee) or by rheumatoid ARTHRITIS, GOUT or various bacteria.

Bush, Vannevar (1890-1974)
US electrical engineer, director of the Office of Scientific Research and Development in WWII. In the 1930s he developed a "differential analyzer" – in effect the first analog COMPUTER.

Bushel (bu)
Unit of dry measure, equivalent to four pecks or 35.238 liter (US Customary System) or 36.369 liter (British Imperial System).

Butadiene (1,3-Butadiene, $CH_2{=}CHCH{=}CH_2$)
Gaseous HYDROCARBON made by dehydrogenation of BUTANE and butene. It is mostly polymerized with STYRENE to make synthetic rubber. mp −109°C, bp −4°C. (See also ALKENES; DIELS-ALDER REACTION.)

Butane (C_4H_{10})
Gaseous ALKANE found in NATURAL GAS and also made by the cracking of PETROLEUM. It is used in bottled gas and in the manufacture of high-octane GASOLINE. It has two isomers, n-butane and 2-methylpropane (isobutane).

Butenandt, Adolf Friedrich Johann (b. 1903)
German chemist who shared the 1939 Nobel Prize for Chemistry (with L. RUZICKA) for his work in isolating and determining the chemical structures of the human sex HORMONES estrone, androsterone and progesterone.

Butte
A small, flat-topped hill formed when EROSION dissects a MESA.

Buys-Ballot's Law
In meteorology, the law states that when an observer stands with his back to the wind in the N Hemisphere, there is high pressure to his right and low pressure to his left. In the S Hemisphere the reverse holds true. It was named for the Dutch meteorologist Christopher Hendrik Didericus Buys-Ballot (1817-1890) who published it in 1875. (See also FERREL'S LAW.)

Byte
In COMPUTER technology a series of BITS that operates as a unit. Thus a character may be represented by a 6-bit byte, or an instruction by an 8-bit byte.

FLOWERING PLANTS Structure and Growth

Flowering plants, or angiosperms, represent over 80 percent of all living green plants. They are characterized by having seeds that develop completely enclosed in the tissue of the parent plant. The tissue surrounding the seed develops from the ovary wall or carpel, which, like other parts of the flower, is derived from a modified leaf.

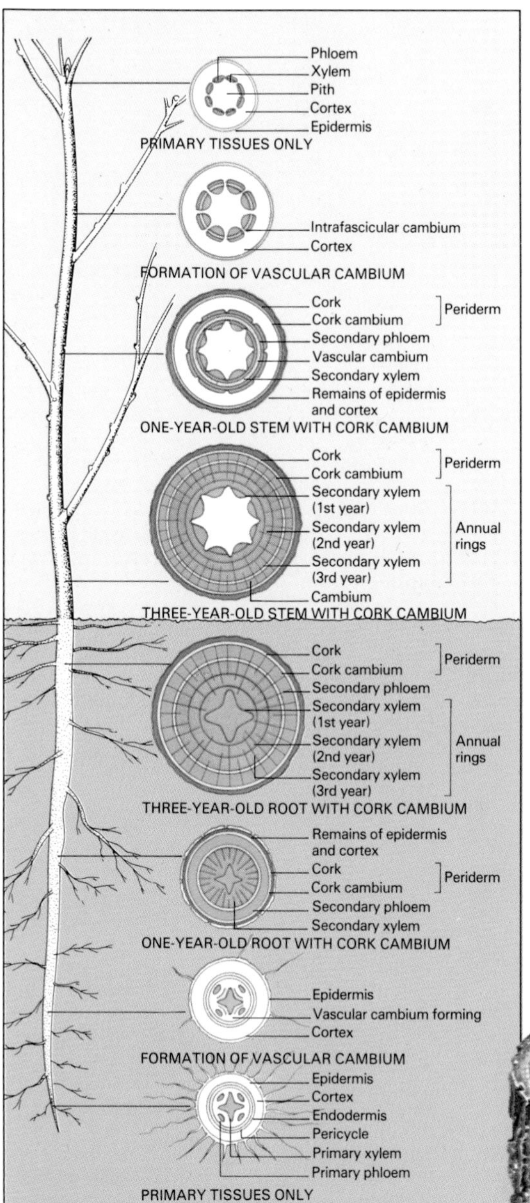

▲► Some flowering plants have highly differentiated wood tissues. The diagram above illustrates the formation of these tissues. The cross-section illustrates their mature form. The outermost layer, cork, protects and insulates. New layers of cork are produced by the cork cambium. The next layer of tissue, phloem, conducts essential food materials; as it dies it contributes tissues to the cork layer. The vascular cambium is the life-giving zone, often just one cell thick; by divisions it produces new phloem cells to the outside and new wood cells to the inside. The bulk comprises the true wood or xylem. The outermost layer, sapwood, is formed of tubular cells that conduct water from the roots to the leaves. As this function deteriorates waste products are deposited in the cells to form the solid heartwood.

Parts of a flower

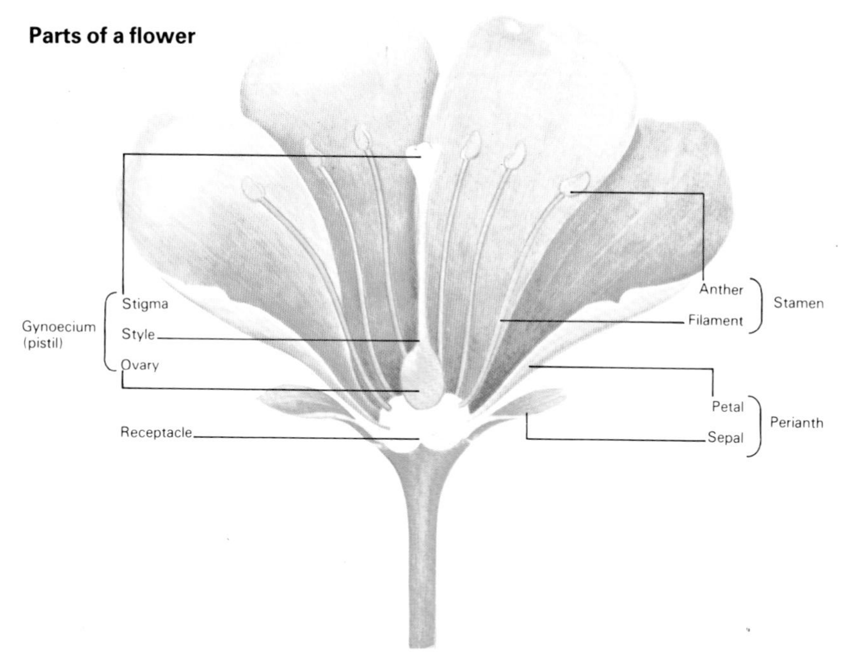

Parts of a leaf

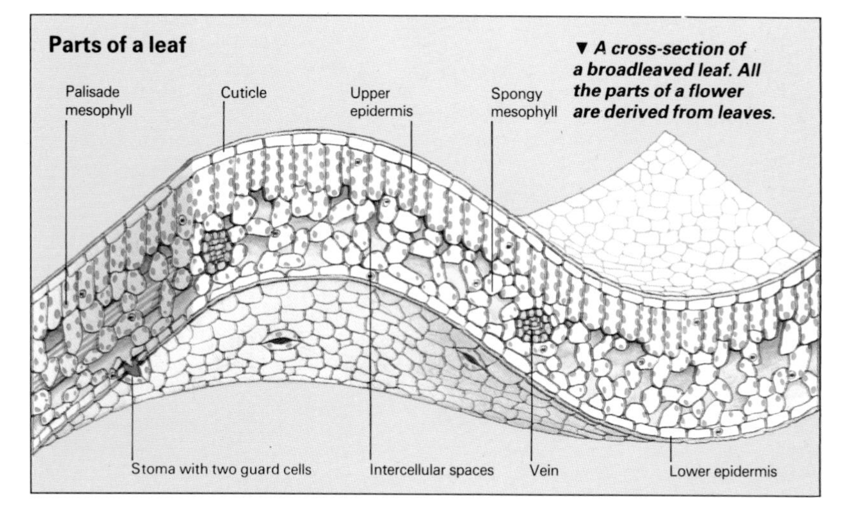

▼ A cross-section of a broadleaved leaf. All the parts of a flower are derived from leaves.

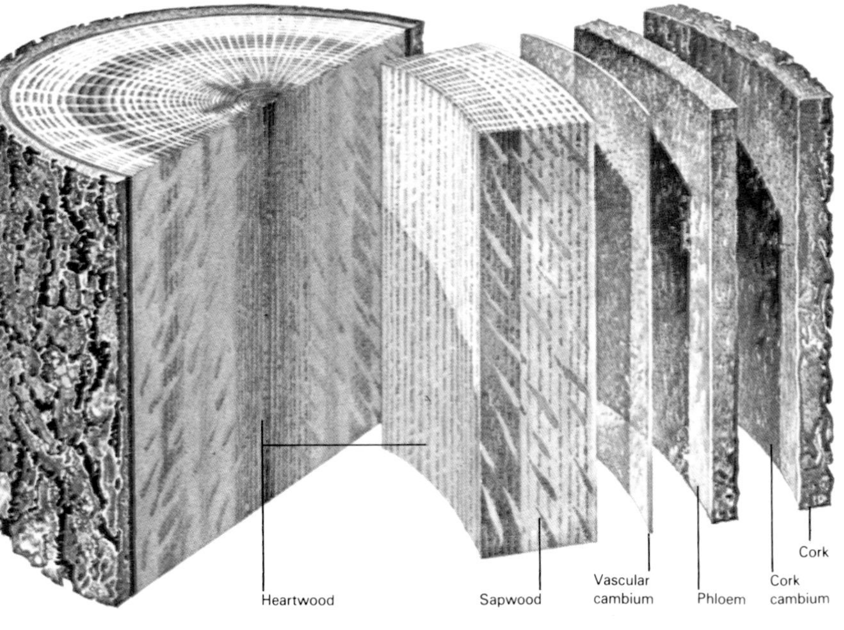

Cachexia
Wasting and weight loss due to poor health.

Cadmium (Cd)
Soft, silvery white metal in Group IIB of the PERIODIC TABLE; an anomalous TRANSITION ELEMENT. It is found as greenockite (CdS) and in ZINC ores, from which it is extracted as a by-product. Cadmium is intermediate in chemical properties between ZINC and MERCURY, forming mainly Cd^{2+} compounds. Cadmium is used for electroplating to give corrosion-resistance; in storage batteries; in low-melting ALLOYS; and as a moderator in nuclear reactors. Some compounds are used in pigments. AW 112.4, mp 321°C, bp 770°C, sg 8.65 (20°C). (➧ page 130)

Caecilians
A group of legless, burrowing amphibians, only distantly related to the other living amphibians (the frogs and toads, or Anura, and the newts and salamanders, or Urodela), and classified as a separate order, the Apoda. Many live underground in tropical rainforests and warm temperate forests, feeding on invertebrates; others are aquatic and swim like eels. Several species are ovoviviparous. Caecilians have a number of curious features, including scales in the skin, an ancestral feature that has been lost in other modern amphibians.

Caesarian section see CESARIAN SECTION.

Caesium see CESIUM.

Caffeine (trimethylxanthine, $C_8H_{10}N_4O_2$)
An ALKALOID extracted from coffee, and also found in tea, cocoa and cola. Caffeine stimulates the central nervous system and heart, and is a DIURETIC. It increases alertness, in excess causing insomnia, and is mildly addictive. mp 238°C.

Cainozoic see CENOZOIC.

Caisson disease (bends) see AEROEMBOLISM.

Calcaneum
The heel bone. In some people it can become enlarged and callused as a result of pressure from the shoe.

Calciferol (Vitamin D_2) see VITAMINS.

Calcination
The process of heating materials in air to drive off moisture, carbon dioxide or other volatile compounds, and sometimes to oxidize them. Ores are often calcined after grinding, and plaster, cement and some pigments are made by calcination.

Calcite
Mineral form of calcium carbonate (see CALCIUM) of widespread occurrence; hexagonal symmetry. LIMESTONE, MARBLE and CHALK are types of calcite. Iceland spar, a very pure calcite, is transparent and exhibits DOUBLE REFRACTION of light, so being useful in optical instruments. Lime, cement and fertilizers are manufactured from calcite.

Calcitonin
A hormone secreted by the thyroid glands, which has the opposite effects to parathyroid hormone (parathormone), which is secreted by the parathyroid glands.

Calcium (Ca)
A fairly soft, silvery white ALKALINE-EARTH METAL, the fifth most abundant element. It occurs naturally as CALCITE, GYPSUM and FLUORITE. The metal is prepared by ELECTROLYSIS of fused calcium chloride. Calcium is very reactive, reacting with water to give a surface layer of calcium hydroxide, and burning in air to give the nitride and oxide. Calcium metal is used as a reducing agent to prepare other metals, as a getter in vacuum tubes, and in alloys. AW 40.1, mp 840°C, bp 1490°C, sg 1.55 (20°C). (➧ page 130)

Calcium compounds are important constituents of animal skeletons: calcium phosphate forms the bones and teeth of vertebrates, and many seashells are made of the carbonate.

CALCIUM CARBONATE ($CaCO_3$), colorless crystalline solid, occurring naturally as calcite and aragonite, which loses carbon dioxide on heating above 900°C. It is an insoluble BASE.

CALCIUM CHLORIDE ($CaCl_2$), a colorless crystalline solid. Being very deliquescent, it is used as an industrial drying agent. mp 782°C.

CALCIUM FLUORIDE (fluorite, CaF_2), colorless phosphorescent crystalline solid, used as windows in ultraviolet and infrared SPECTROSCOPY. mp 1423°C, bp c2500°C.

CALCIUM HYDROXIDE (slaked lime, $Ca(OH)_2$), colorless crystalline solid, slightly soluble in water, prepared by hydrating calcium oxide and used in industry and agriculture as an ALKALI, in mortar and in glass manufacture.

CALCIUM OXIDE (quicklime, CaO), white crystalline powder, made by calcination of calcium carbonate minerals, which reacts violently with water to give calcium hydroxide and is used as a refractory to neutralize acid effluents and to remove sulfur dioxide from gaseous effluents. mp 2580°C, bp 2850°C.

CALCIUM SULFATE ($CaSO_4$), colorless crystalline solid, occurring naturally as gypsum and anhydrite. When the dihydrate is heated to 128°C, it loses water, forming the hemihydrate, plaster of paris. This re-forms the dihydrate as a hard mass when mixed with water, and is used to make plaster casts.

Calculus
In dentistry, the "fossilized" plaque on teeth.

Calculi (stones)
Solid concretions of calcium salts or organic compounds, formed in the KIDNEY, BLADDER or GALL BLADDER. There may be no symptoms, but they may pass down or block tubes, causing colic or obstruction in an organ.

Caldera
Extremely large crater of volcanic (see VOLCANISM) origin, caused by the inward collapse of a volcanic cone. (➧ page 142)

Calendar
A system for reckoning the passing of time. The principal problem in drawing up calendars arises from the fact that the solar day, the lunar month and the tropical year – the most important natural time units – are not simple multiples of each other. In practice a solution is found in basing the system either on the phases of the Moon (lunar calendar) or on the changing of the seasons (solar calendar). The difficulty that the days eventually get out of step with the Moon or the seasons is solved by adding in (intercalating) one or more extra days or months at regular intervals in an extended cycle of months or years.

Years are commonly numbered in Western societies from the birth of Christ – as computed by a 6th-century monk. Years since that epoch are labeled AD, years before, BC. There is no year 0, AD 1 following directly from 1 BC. Astronomers, on the other hand, figure years BC as negative numbers one less than the date BC and include a year 0 (=1 BC). The astronomers' year 10 is thus the same as 11 BC.

Calibration
A most important step in preparing a scientific instrument or experimental apparatus for use, in which it is provided with a numerical scale in accordance with an internationally agreed procedure or by comparison with a standard instrument or measure. Thus an ordinary thermometer may be calibrated simply by noting its reading first in freezing and then in boiling water (other conditions being specified), but for greater accuracy, complex calibration procedures must be used to within carefully controlled tolerances.

Californium (Cf)
A TRANSURANIUM ELEMENT in the ACTINIDE series. Numerous isotopes have been synthesized by various bombardment methods; most undergo spontaneous nuclear FISSION. (➧ page 130)

Calipers
Devices used in measuring. Simple calipers comprise a pair of metal legs, pivoted about a shared screw at one end, the other ends being turned inward (for outer dimensions) or outward (for inner dimensions). They are usually used in conjunction with a rule. Vernier calipers resemble a sliding wrench and incorporate a VERNIER SCALE.

Callus
Connective tissue initially formed around a FRACTURE, and slowly ossified as repair proceeds (see BONE).

Callus (callosity) see CORNS AND CALLUSES.

Calmette, Albert Léon Charles (1863-1933)
French bacteriologist who with C. GUÉRIN discovered the BCG (Bacillus Calmette-Guérin) vaccine, which has greatly reduced the incidence of TUBERCULOSIS.

Calms, Regions of
Areas where the sea is of mirror-like calmness and where the wind-speed is less than 1 knot. Such wind-speed is denoted 0 on the BEAUFORT SCALE. These conditions are characteristic of the HORSE LATITUDES.

Calomel (Mercury (I) chloride) see MERCURY.

Caloric theory of heat
The view, formalized by LAVOISIER toward the end of the 18th century, that heat consists of particles of a weightless, invisible fluid, caloric, found between the atoms of material substances. The theory fell from favor as physicists began to appreciate the equivalence of work and HEAT.

Calorie
The name of various units of HEAT. The calorie or gram calorie (c or cal), originally defined as the quantity of heat required to raise lg of water through 1°C at 1 atm pressure, is still widely used in chemical THERMODYNAMICS. The large calorie, kilogram calorie or kilocalorie (Cal or kcal), 1000 times as large, is the "calorie" of dietitians. The 15° calorie (defined in terms of the 1°C difference between 14.5°C and 15.5°C) is 4.184 joules; the International Steam Table calorie (cal_{IT}) of 1929, originally defined as 1/860 watt-hour, is now set equal to 4.1868J in SI UNITS.

Calorimetry
The measurement of the HEAT changes associated with chemical reactions and physical processes. This is done using various types of calorimeter. The water calorimeter is a thermally insulated metal cup of known thermal properties containing a known mass of water. When a reaction is carried out in the vessel, any heat liberated or absorbed is taken or given up by the water, the heat changes occurring being monitored by means of a thermometer dipped in the water. Heats of combustion are measured using a bomb calorimeter in which the reaction is carried out in an enclosed, pressurized chamber immersed in the water.

Other types of apparatus used to measure SPECIFIC and LATENT HEATS include ice and steam calorimeters. (See also THERMOCHEMISTRY.)

Calotype see TALBOT, WILLIAM HENRY FOX.

Calvin, Melvin (b. 1911)
US biochemist who gained the 1961 Nobel Prize for Chemistry after having led the team that unraveled the details of the chemistry of PHOTOSYNTHESIS.

Cam
A mechanical device that, on rotating, gives a regular, repetitive motion to a member held in contact with it.

Cambium
A meristematic tissue (see MERISTEM) that lies between the XYLEM and PHLOEM in the vascular tissue of plants. In woody plants a complete ring of cambium develops, which produces new xylem cells on the inside and phloem on the outside by a process known as secondary thickening. In some trees a layer of cork cambium develops that then produces a corky BARK.

Cambrian
The earliest period of the PALEOZOIC (see GEOLOGY), dated roughly 590-505 million years ago, immediately preceding the ORDOVICIAN. Cambrian rocks contain the oldest FOSSILS that can be used for dating (see PRECAMBRIAN). (➧ page 106)

Camcorder
A video camera, or camera that records moving images on magnetic tape.

Camouflage
The development of coloration and form that allows a living organism to merge with its background and remain unseen by others. Most camouflage is for the purposes of protection from predators, but there are also camouflaged predators that use their cryptic coloration to lie in wait for their prey. Thus many predatory cats are camouflaged, as are the angler fish and flower mantids.

Forms of camouflage range from simply being colored like the background through to a close physical resemblance to a leaf, twig, flower, pebble, bird-dropping, or other inanimate object. Such elaborate types of camouflage are akin to MIMICRY. Apart from modifications of coloring and body form, this kind of camouflage involves a behavioral component. For example, twig caterpillars hold their bodies stiffly at an acute angle to a plant stem and keep completely still in order to increase their resemblance to a twig.

Some camouflaged animals, including certain species of chameleon and flatfish, can alter their coloration to suit different backgrounds by expanding or contracting pigment cells in the skin.

cAMP (cyclic adenosine monophosphate) see ATP.

Camphor
A white crystalline compound distilled from the wood and young shoots of the camphor tree (*Cinnamomum camphora*). Camphor has a strong characteristic odor that repels insects.

Canadian Shield (Laurentian Shield)
The area of N America (including the eastern half of Canada and small portions of the US) that has remained more or less stable since PRECAMBRIAN times. Its surface rocks, which are IGNEOUS and METAMORPHIC, are among the oldest in the world, younger structures having disappeared through EROSION, in some areas by GLACIERS of the PLEISTOCENE.

Cancer
The Crab, a spring constellation in the N Hemisphere, the fourth sign of the ZODIAC. At the time the zodiacal system was adopted, Cancer marked the northernmost limit of the ECLIPTIC. A hazy object near the center of Cancer is a cluster of stars named Praesepe, the Beehive.

Cancer
A group of diseases in which some body cells change their nature, start to divide uncontrolably and may revert to an undifferentiated type. They form a malignant TUMOR, which enlarges and may spread to adjacent tissues; in many cases cancer cells enter the BLOOD or LYMPH systems and are carried to distant parts of the body. There they form secondary "colonies" called metastases. Such advanced cancer is often rapidly fatal, causing gross emaciation. Cancer may present in very many ways – as a lump, some change in body function, bleeding, ANEMIA or weight loss – occasionally the first symptoms being from a metastasis. Less often tumors produce substances mimicking HORMONE action or producing remote effects such as NEURITIS.

Cancers are classified according to the type of tissue in which they originate. The commonest type, carcinoma, occurs in glandular tissue, skin, or visceral linings. Sarcoma occurs in connective tissue, muscle, bone and cartilage. Glioma is a sarcoma of brain neuroglia, unusual in that it does not spread elsewhere. Lymphoma, which includes HODGKIN'S DISEASE, is a tumor of the lymphatic system (see LYMPH); LEUKEMIA can be regarded as a cancer of white blood cells or their precursors.

Although the cause of many cancers remains unknown, exposure to a number of different agents is seen to increase the individual's risk of developing the disease. These agents include X-rays, radioactivity, ultraviolet radiation and certain chemicals (some dyes, oils and tars). In a number of cases VIRUSES are suspected of being the causative agents. (See also INTERFERON.)

Candela (cd)
The photometric base unit in SI UNITS. It is defined as the luminous intensity, in the perpendicular direction, of a surface of 1/600,000sq m of a black body at the temperature of freezing platinum under a pressure of 101,325 newtons per sq m. (See BLACKBODY RADIATION; PHOTOMETRY.)

Candlepower
The ability of a light source to radiate as expressed in CANDELAS.

Canids
Members of the Canidae, a family within the Order CARNIVORA. It includes the domestic dog, the wolves, jackals, coyote, hunting dogs and foxes.

Canis Major
The Great Dog, a constellation of the S Hemisphere visible during winter in Northern skies. It contains SIRIUS, the brightest star in the night sky. Mythologically, Canis Major and CANIS MINOR were ORION's hunting dogs.

Canis Minor
The Little Dog, a constellation on the celestial equator (see CELESTIAL SPHERE) visible in Northern skies during winter. It contains the binary star PROCYON (see DOUBLE STAR).

Cannabis
General term for the habit-forming, recreational drug obtained from plants of the genus *Cannabis* (*C. sativa* and *C. indica*). The term marijuana usually refers to the leaves, seeds and flowers of the plant and hashish to the resin, though both may be considered synonymous with cannabis. The effects are variable, ranging from euphoria and a sense of well-being, to fear or paranoia. Its use is illegal almost worldwide.

Cannizzaro, Stanislao (1826-1910)
Italian chemist who discovered the Cannizzaro reaction in which BENZALDEHYDE is converted to BENZYL ALCOHOL and BENZOIC ACID (1853), but he is chiefly remembered for the republication of AVOGADRO's hypothesis at the Karlsruhe conference of 1860. This at last allowed chemists to distinguish EQUIVALENT WEIGHTS from true ATOMIC WEIGHTS.

Canopus
Alpha Carinae, the second brightest star in the sky, with an apparent magnitude of −0.72. It is 210 times larger than the Sun and 360pc from Earth. (➧ page 151)

Canyon
Steep-sided VALLEY formed through EROSION by a river of hard rock lying in horizontal strata (see STRATIGRAPHY). The best-known example is the Grand Canyon, Arizona.

Capacitance
The ratio of the electric charge (see ELECTRICITY) on a conductor to its POTENTIAL, or, for a CAPACITOR, the ratio of its charge to the potential difference between its plates. Capacitance is measured in farads (F).

Capacitor (condenser)
An electrical component used to store electric charge (see ELECTRICITY) and to provide REACTANCE in alternating-current circuits. In essence, a capacitor consists of two conducting plates separated by a thin layer of insulator. When the plates are connected to the terminals of a BATTERY, a current flows until the capacitor is "charged", having one plate positive and the other negative. The ability of a capacitor to hold charge, its capacitance, C, is the ratio of quantity of electricity on its plates, Q, to the potential difference between the plates, V. The electric energy stored in a capacitor is given by $\frac{1}{2}CV^2$. The capacitance of a capacitor depends on the area of its plates, their separation and the DIELECTRIC constant of the insulator. Small fixed capacitors are commonly made with metal-foil plates and paraffin-paper insulation; to save space the plates and paper are rolled up into a tight cylinder. Variable capacitors used in RADIO tuners consist of movable intermeshing metal vanes separated by an air gap. Electrolytic capacitors, which must be connected with the correct polarity, also find use in flash guns. (➧ page 118)

Capella
Alpha Aurigae, the sixth brightest star in the night sky. It is a DOUBLE STAR, 13.8pc distant, each component having apparent magnitude +0.85, with possibly two further dim components.

Capillaries
Minute blood vessels concerned with supplying oxygen and nutrients to and removing waste products from the tissues. In the lungs capillaries pick up oxygen from the alveoli and release carbon dioxide. These processes occur by DIFFUSION. The capillaries are supplied with blood by ARTERIES and drained by VEINS. (➧ page 28)

Capillarity
The name given to various SURFACE TENSION phenomena in which the surface of a liquid confined in a narrow-bore tube rises above or is depressed below the level it would have if it were unconfined. When the attraction between the molecules of the liquid and those of the tube exceeds the combined effects of gravity and the attractive forces within the liquid, the liquid rises in the tube until EQUILIBRIUM is restored. Capillarity is of immense importance in nature, particularly in the transport of fluids in plants and through the soil.

Capricornus
The Sea Goat, a fairly inconspicuous constellation of the S Hemisphere, lacking any bright stars, and the tenth sign of the ZODIAC. Lying between the constellations of AQUARIUS and SAGITTARIUS, Capricornus in ancient times lay at the southernmost limit of the ECLIPTIC.

Capsule
Botanically a dry, dehiscent FRUIT developed from two or more carpels, containing numerous

seeds. Seeds are shed by a number of methods, including pores at the top of capsule (in the poppy) and detachment of the apex (in the pimpernel).

Carapace
The hard outer covering on many invertebrate and a few vertebrate animals. The carapace of insects and other arthropods is made of the polysaccharide CHITIN, and forms part of the external skeleton or EXOSKELETON. In crustaceans it is hardened by deposits of calcium carbonate (chalk). The carapace of vertebrates, such as the turtle, serves a protective function, and is made of bone.

Carat
A unit of mass used for weighing precious stones. Since 1913 the internationally accepted carat has been the metric carat (CM) of 200mg. The purity of GOLD is also expressed in carats (usually spelled "karat" in the US). Here one karat is a 24th part; thus, pure gold is 24-karat; "18-karat gold" contains 75% gold and 25% other noble metals, and so on.

Carbides see CARBON.

Carbohydrates
Organic compounds that play an important part in living organisms. They have the general formula $(CH_2O)_n$, hence the name, "hydrates of carbon". The simplest carbohydrates are sugars, which may consist of five carbon atoms (pentoses) or six carbon atoms (hexoses). These sugar molecules generally exist as rings, and are known as monosaccharides; examples include glucose and fructose. When two such sugar molecules join together they produce a disaccharide such as sucrose (table sugar – one glucose and one fructose unit per molecule). A chain made up of three to six monosaccharides is known as an oligosaccharide, while long-chain polymers of sugar molecules are called polysaccharides. The polysaccharides act as energy-storage compounds, including starch in plants and glycogen in animals. Other polysaccharides are of structural importance, such as CELLULOSE, which makes up plant cell walls, and CHITIN, which is found in the cell walls of fungi and the exoskeletons of various invertebrates. A third role for polysaccharides is in glycoproteins, where relatively short chains are attached to a protein molecule, as in the immunoglobulins and various cell wall proteins; the role of these carbohydrate units is not yet well understood. Carbohydrates are formed by PHOTOSYNTHESIS and stored by plants in the form of starch; they represent an important source of food for many heterotrophic organisms. (♦♦ page 40, 233)

Carbolic acid
Old name for PHENOL, particularly when used as an ANTISEPTIC.

Carboloy
A common name for the ultrahard bearing and cutting-tool material, TUNGSTEN carbide (WC).

Carbon (C)
Nonmetal in Group IVA of the PERIODIC TABLE. It is unique among elements in that a whole branch of chemistry (ORGANIC CHEMISTRY) is devoted to it, because of the vast number of compounds it forms. The simple carbon compounds described below are usually regarded as inorganic.

Carbon occurs in nature both uncombined (DIAMOND) and as CARBONATES, carbon dioxide in the atmosphere, and PETROLEUM. It exhibits ALLOTROPY, occurring in three contrasting forms: DIAMOND, GRAPHITE and "white" carbon, a transparent allotrope discovered in 1969 by subliming graphite. So-called amorphous carbon is actually microcrystalline graphite; it occurs naturally, and is found as coke, CHARCOAL and carbon black (obtained from the incomplete burning of petroleum, and used in pigments and printer's ink, and to reinforce rubber). Amorphous carbon is widely used for ADSORPTION, because of its large surface area. A new synthetic form is carbon fiber, which is very strong and is used to reinforce plastics and to make electrically conducting fabrics. (♦♦ page 28, 37, 130, 147)

Carbon has several ISOTOPES: C^{12} (used as a standard for ATOMIC WEIGHTS) is much the most common, but C^{13} makes up 1.11% of natural carbon. C^{10}, C^{11}, C^{14}, C^{15} and C^{16} are all radioactive. C^{14} has the relatively long half-life of 5730yr, and is continuously formed in the atmosphere by COSMIC RAY bombardment; it is used in RADIOCARBON DATING.

The element (especially as diamond) is rather inert, but all forms will burn in air at a high temperature to give carbon monoxide in a poor supply of oxygen, and carbon dioxide in excess oxygen. Fluorine will attack carbon at room temperature to give carbon tetrafluoride, and strong oxidizing agents will attack graphite. Carbon will combine with many metals at high temperatures, forming carbides. Carbon shows a covalency of four, the bonds pointing toward the vertices of a tetrahedron, unless multiple bonding occurs. AW 12.011.

CARBIDES Binary compounds of carbon with a metal, prepared by heating the metal or its oxide with carbon. Ionic carbides are mainly acetylides (C_2^{2-}) which react with water to give ACETYLENE, or methanides (C^{4-}) which give METHANE. There are also metallic interstitial carbides, and the covalent boron carbide (B_4C) and silicon carbide (see CARBORUNDUM).

CARBON DIOXIDE (CO_2) A colorless, odorless gas. It is nontoxic, but can cause suffocation. The air contains 0.03% carbon dioxide, which is exhaled by animals and absorbed by plants (see RESPIRATION; PHOTOSYNTHESIS; CARBON CYCLE). Carbon dioxide is prepared in the laboratory by reacting a CARBONATE with acid; industrially it is obtained by calcining LIMESTONE, burning coke in excess air, or from FERMENTATION. At atmospheric pressure, it solidifies at −78.5°C to form "dry ice" (used for refrigeration and CLOUD seeding) which sublimes above that temperature; liquid carbon dioxide, formed under pressure, is used in fire extinguishers. Carbon dioxide is also used to make carbonated drinks. When dissolved in water an equilibrium is set up, with CARBONATE, BICARBONATE and HYDROGEN ions formed, and a low concentration of carbonic acid (H_2CO_3).

CARBON DISULFIDE (CS_2) A colorless liquid, of nauseous odor due to impurities; highly toxic and flammable. Used as a solvent and in the manufacture of rayon and cellophane. mp −111.9°C, bp 46.5°C, sg 1.261 (22°C).

CARBON MONOXIDE (CO) A colorless, odorless gas. It is produced by burning carbon or organic compounds in a restricted supply of oxygen, for example in poorly ventilated stoves, or the incomplete combustion of gasoline in automobile engines. It is manufactured as a component of WATER GAS. It reacts with the halogens and sulfur, and with many metals, to give carbonyls. Carbon monoxide is an excellent reducing agent at high temperatures, and is used for smelting metal ores (see BLAST FURNACE; IRON). It is also used for the manufacture of METHANOL and other organic compounds. It is a component of manufactured gas, but not of natural gas. Carbon monoxide is toxic because it combines with hemoglobin, the red BLOOD pigment, to form pink carboxyhemoglobin, which is stable, and will not perform the function of transporting oxygen to the tissues. mp −205°C, bp −191.5°C.

Carbonates
Salts of carbonic acid (see CARBON), containing the CO_3^{2-} ion. A solution of carbon dioxide in water reacts with a base to form a carbonate. Carbonates are decomposed by heating to give the metal oxide and carbon dioxide. They also react with acids to give carbon dioxide. Many minerals are carbonates, the most important being CALCITE, DOLOMITE and MAGNESITE. (For sodium carbonate see SODIUM.)

Carbon black see CARBON.

Carbon cycle
In ecology, the pathways by which carbon compounds are circulated in the environment. There are three main reservoirs of carbon: in the atmosphere, where it exists as carbon dioxide gas (CO_2); in plants, where it exists as complex compounds, principally CELLULOSE, LIGNIN and storage CARBOHYDRATES (starches); and in fossil fuels such as coal, oil and natural gas. The pivotal reaction of the carbon cycle is PHOTOSYNTHESIS, whereby plants and other photosynthesizers "fix" carbon dioxide and use it to make complex carbon compounds. The process of RESPIRATION, in which these compounds are broken down for food (either by plants or by those that eat them), returns the carbon to the atmosphere as carbon dioxide. In general, the rate of photosynthesis exceeds that of respiration, allowing a buildup of carbon reserves in the form of BIOMASS.

The carbon cycle today is very different from the carbon cycle 300 years ago, because of the large-scale burning of fossil fuels. Fossil fuels were originally formed by the photosynthetic activities of marine microorganisms in the Cambrian (oil) or the forests of the Carboniferous (coal) period, which "locked up" a great deal of carbon that had previously been present in the Earth's atmosphere as carbon dioxide. The burning of fossil fuels to power machinery and provide us with heat and light is increasing the concentration of carbon dioxide in the atmosphere and causing a warming of the planet, due to the GREENHOUSE EFFECT. The massive destruction of forests, particularly those of the tropics which have a very high biomass, is also affecting the carbon cycle by pumping more carbon dioxide into the atmosphere (if the trees are burned) and by depleting the Earth's overall photosynthetic capacity.

Carbon cycle (carbon-nitrogen cycle)
In physics, the chain of nuclear FUSION reactions, catalyzed by CARBON nuclei (see CATALYSIS), which is the main source of ENERGY in the hotter STARS, though of minor importance in the SUN and cooler stars where proton-proton fusion is the chief reaction. Overall, four PROTONS are converted to one ALPHA PARTICLE, with destruction of some matter and consequent evolution of energy (see RELATIVITY). Most of the gamma-radiation produced is absorbed within the star, and energy is released as heat and light. The carbon cycle was first described by HANS BETHE.

Carboniferous
The penultimate period of the PALEOZOIC era, lasting from about 360 to 286 million years ago. In the United States it is regarded as two periods – the lower MISSISSIPPIAN and the upper PENNSYLVANIAN. (See also GEOLOGY.) (♦ page 106)

Carboranes
Compounds derived from polyhedral BORANES by substitution of CARBON for BORON atoms. They are relatively unreactive and thermally stable, and form many derivatives, including highly stable organo-inorganic polymers.

Carborundum (silicon carbide, SiC)
Black, cubic crystalline solid, made by heating coke with SILICA in an electric furnace. It is almost as hard as DIAMOND (whose structure it resembles), and hence it is used as an abrasive. It is inert, refractory and a good heat conductor, so is used in making high-temperature bricks; at high

temperatures it is a SEMICONDUCTOR. subl 2700°C (with decomposition).

Carboxylic acids
A major class of organic compounds, of general formula RCOOH where R is an organic group; those acids where R is a straight-chain alkyl group (see ALKANES) are known as FATTY ACIDS, and are named systematically as alkanoic acids from the corresponding alkane. Some carboxylic acids occur free in nature, including FORMIC and ACETIC acids – the two simplest – and CITRIC, LACTIC, MALIC and TARTARIC acids. These, and others including BENZOIC, OXALIC and SALICYLIC acids, are found also as their salts and ESTERS. Many of the fatty acids, including OLEIC, PALMITIC and STEARIC acids, occur in oils and fats (TRIGLYCERIDES) as esters of glycerol (see also SOAPS AND DETERGENTS). Carboxylic acids are made by HYDROLYSIS of esters, ACID ANHYDRIDES or ACID CHLORIDES, or by oxidation of ALDEHYDES or primary ALCOHOLS. They are weak ACIDS, the exact strength depending on the ELECTRONEGATIVITY of the group R, and so are often used with their salts as BUFFERS. Many derivatives of carboxylic acids are important in nature or chemical synthesis: they include acid anhydrides, acid chlorides, AMIDES, esters, NITRILES and peroxy-acids (see PEROXIDES). (See also AMINO ACIDS.)

Carbuncle see BOIL.

Carcinogen
An agent, such as certain chemicals or viruses, that induces cancer.

Carcinoma see CANCER.

Cardiac
Relating to the HEART, as in cardiac surgery; also to the cardia or upper part of the STOMACH.

Caries
Decay or softening of hard tissues, usually TEETH, but also used for BONE, especially spinal TUBERCULOSIS. Dental caries is bacterial decay of dentine and enamel, hastened by dietary carbohydrates that ferment in bacterial PLAQUE and by poor oral hygiene. Fluoride in small quantities conveys considerable protection.

Carinates
The major group of birds, including all except the RATITES. They are so named from the Latin *carina*, meaning a keel, in reference to the keel-like breastbone to which flight muscles are attached; the ratites, being flightless, have no need of such a breastbone. Not all carinates fly, however: the group includes the penguins and various other flightless forms.

Carminative
Any drug claimed to relieve wind by stimulating the stomach and intestine.

Carnelian (cornelian)
Translucent variety of CHALCEDONY, colored red by colloidal HEMATITE: a semiprecious GEM stone widely used in Classical times for intaglio signets. (See also SARD.)

Carnivora
An order of MAMMALS that includes the raccoons, bears, wolves, dogs and foxes, badgers, otters, weasels and martens, hyenas, lions and other cats, civets, genets and mongooses, and the pandas. Most are fish- or meat-eaters and have large canine teeth, pointed cheek teeth, and claws on all the toes. However, several species include some plant material and invertebrates in their diet (eg red fox, some badgers, most civets) and the bears are truly omnivorous. The giant panda feeds only on bamboo, but has clearly evolved from carnivorous ancestors.

Carnivore
A meat-eating animal. The word is used in at least three different ways. In ecology, it may be used to mean any animal that eats another animal, as opposed to HERBIVORES, which eat plants. It can also refer exclusively to flesh-eating animals – those that prey on mammals, birds and other vertebrates – and so exclude animals that eat insects (insectivores), for example. Finally, the word is sometimes used to mean only those mammals belonging to the Order CARNIVORA, a taxonomic group that includes several omnivores and one herbivore, although the majority are "meat-eaters".

Carnivorous plant see INSECTIVOROUS PLANTS.

Carnot, Nicholas Léonard Sadi (1796-1832)
French physicist who, seeking to improve the EFFICIENCY of the steam engine, devised the Carnot cycle (1824) on the basis of which LORD KELVIN and R.J.E. CLAUSIUS formulated the second law of THERMODYNAMICS. The Carnot cycle, which postulates a heat engine working at maximum thermal efficiency, demonstrates that the efficiency of such an engine does not depend on its mode of operation but only on the TEMPERATURES at which it accepts and discards heat ENERGY.

Carnotite
Potassium uranyl vanadate with the formula ($K_2[UO_2]_2[VO_4]_2.nH_2O$), a soft, yellow radioactive mineral, a major ore of URANIUM and VANADIUM. It is found in Colorado, Wyoming, South Dakota, Pennsylvania, Siberia, Zaire and Australia.

Carotenoids
A group of yellow, orange and red pigments found in almost all animals and plants, and responsible for the color of carrots, tomatoes, oranges, lemons, shrimps, lobsters, and many flowers. In leaf CHLOROPLASTS the carotenoid colors are masked by CHLOROPHYLL until this is lost just before leaf-fall, resulting in a spectacular display of carotenoid colors. The color is due to a lone conjugated double-bond system (see ALKENES; RESONANCE) formed by condensation of ISOPRENE units. There are two main types of carotenoids, the oxygen-containing xanthophylls and the hydrocarbon carotenes. VITAMIN A is formed from certain carotenes.

Carpal
Relating to the hand. Carpal tunnel syndrome is tingling and numbing of the hand resulting from fibrous thickening of the wrist bones, which press on a nerve.

Carpel see FLOWER.

Carrel, Alexis (1873-1944)
French surgeon who won the 1912 Nobel Prize for Physiology or Medicine for developing a technique for suturing (sewing together) blood vessels, thus paving the way for organ TRANSPLANTS and blood TRANSFUSION.

Carrier
Person who carries the agents responsible for infectious DISEASES and is able to infect others, while often remaining well.

Cartilage
Tough, flexible connective tissue found in all vertebrates, consisting of cartilage cells (chondrocytes) in a matrix of COLLAGEN fibers and a rubbery protein gel. The skeleton of the vertebrate embryo is formed wholly of cartilage, but in the majority of groups much of this is converted to BONE during growth. Only in the cartilaginous fish (sharks and rays) does the cartilage remain unmineralized. There are three main types of cartilage: hyaline, which is translucent and glossy and is found in the joints, nose, trachea and bronchi; elastic, found in the external ear, Eustachian tube and larynx; and fibrocartilage, which attaches tendons to bone and forms the disks between the vertebrae.

Cartilaginous fish see CHONDRICHTHYES.

Cartography see MAP.

Casein
The chief milk PROTEIN, found there as its calcium salt. It is precipitated from skim milk with acid or rennet, washed and dried. Highly nutritious, it is used in the food industry. In alkaline solution casein forms a COLLOID used as a glue, a binder for paint pigments and paper coatings, and to dress leather. Casein is also used to make PLASTICS.

Cassini, Giovanni Domenico (1625-1712)
Alternatively known as Jean Dominique Cassini. An Italian-French astronomer who discovered four satellites of SATURN, the Cassini division in the ring of Saturn, and estimated the scale of the Solar System from the PARALLAX of MARS (1672).

Cassiopeia
In astronomy, a Northern circumpolar constellation (see CIRCUMPOLAR STARS) whose five principal stars form a prominent "W".

Cassiterite
A mineral consisting of stannic oxide (SnO_2) with iron impurity; the chief ore of TIN. It is usually brown to black, with submetallic luster; it forms prismatic crystals (tetragonal), but is usually massive granular. sg 6.8-7.1.

Castor
Alpha Geminorum, second brightest star of GEMINI. About 14.4pc from Earth, it has at least six components: a binary, each component of which is itself a binary, is orbited at distance by a third binary (see DOUBLE STAR).

Castration
Removal of the testes of a male animal, usually by surgery but also by constricting their blood supply. It is employed for sterilization, to produce docility, to cause the animal to gain more flesh of better quality, and to prevent secondary sexual characteristics from developing. Spaying is the removal of the OVARIES of a female animal.

Castration complex
The nexus of fears (see COMPLEX) concerned with possible or threatened loss of the generative organs, especially the penis (see REPRODUCTION). The term is used analogously to describe fears of loss of sexuality or the capacity for erotic pleasure in males or females. (See OEDIPUS COMPLEX.)

CAT (computer assisted tomography) see TOMOGRAPHY

Catabolism
Biochemical processes whereby larger molecules are broken down into smaller ones, either to yield food energy (as in GLYCOLYSIS, the CITRIC ACID CYCLE and the ELECTRON TRANSPORT CHAIN), or for the purposes of EXCRETION. Together with ANABOLISM (in which larger molecules are built up from smaller ones), it makes up the processes of METABOLISM. (See also ATP.)

Catalepsy
Rare nervous disorder characterized by episodes of rigid immobility.

Catalysis
The changing of the rate of a chemical reaction by the addition of a small amount of a substance which is unchanged at the end of the reaction. Such a substance is called a catalyst, though this term is usually reserved for those that speed up reactions; additives that slow down reactions are called inhibitors. Catalysts are specific for particular reactions. In a reversible reaction, the forward and back reactions are catalyzed equally, and the EQUILIBRIUM position is not altered. Catalysis is either homogeneous (the catalyst and reactants being in the same phase, usually gas or liquid), in which case the catalyst usually forms a reactive intermediate that then breaks down; or heterogeneous, in which ADSORPTION of the reactants occurs on the catalytic surface. Heterogeneous catalysis is often blocked by impurities called poisons. Catalysts are widely used in industry, as in the CONTACT PROCESS, the HYDROGENATION of oils, and the cracking of PETROLEUM. All living organisms are dependent on the complex catalysts

called ENZYMES that regulate biochemical reactions.

Cataplexy
Episodic muscular weakness, often precipitated by emotion, its severity relating to depth of emotion. Associated with NARCOLEPSY.

Cataract
Disease of the EYE lens, regardless of cause. The normally clear lens becomes opaque, and light transmission and perception are reduced. Congenital cataracts occur especially in children born to mothers who have had RUBELLA in early pregnancy, and in a number of inherited disorders. Certain disturbances of METABOLISM or HORMONE production can also cause cataracts. Eye trauma and INFLAMMATION are other causes. Some degree of cataract formation is common in old age, and it is among the commonest causes of BLINDNESS in developing countries.

Catarrh
Vague term usually referring to excess MUCUS or discharge of PUS from mucous membranes of nose or sinuses, but sometimes referring to sputum.

Catastrophism
In geology, the early 19th-century theory that major changes in the geological structure of the Earth occurred only during short periods of violent upheaval (catastrophes), which were separated by long periods of comparative stability. It fell from prominence after LYELL's enunciation of the rival doctrine of UNIFORMITARIANISM.

Catatonia
A form of SCHIZOPHRENIA in which the individual oscillates between excitement and stupor.

Catechol (pyrocatechol)
The ortho-isomer of dihydroxybenzene, used as an antioxidant, an antiseptic and a photographic chemical. MW 110.11, mp 105°C, bp 245°C.

Catharsis
In PSYCHOANALYSIS (where it is generally termed abreaction), the bringing into the open of a previously repressed memory or emotion, in the hope of relieving unconscious emotional stress. In medicine, the term is used for the artificial induction of vomiting (see EMETIC).

Catheter
Hollow tube passed into body organs for investigation or treatment. Urinary catheters are used for relief of BLADDER outflow obstruction and sometimes for loss of nervous control of bladder; they also allow measurement of bladder function and special X-ray techniques. Cardiac catheters are passed through arteries or veins into chambers of the heart to study its functioning and anatomy.

Cathode
A negatively charged ELECTRODE, found particularly in ELECTRON TUBES, CATHODE RAY TUBES, and electrochemical CELLS, and used in combination with its positive counterpart, an ANODE, as a source of ELECTRONS or to produce an electric field. (➧ page 118)

Cathode rays
ELECTRONS emitted by a CATHODE when heated. First studied by Julius Plücker (1801-1868) in 1858, they can be drawn off in vacuo by the attraction of an ANODE to form a beam which causes fluorescence (see LUMINESCENCE) in appropriately coated screens or X-RAY emission from metal targets.

Cathode ray tube
The principal component of OSCILLOSCOPES and television sets. It consists of an evacuated glass tube containing at one end a heated CATHODE and an ANODE, and widened at the other end to form a flat screen, the inside of which is coated with a fluorescent material. Electrons emitted from the cathode are accelerated toward the anode, and pass through a hole in its center to form a fine beam which causes a bright spot where it strikes the screen. Because of the electric charge carried by the electrons, the beam can be deflected by transverse electric or magnetic fields produced by electrodes or coils between the anode and screen: one such set allows horizontal deflection, and another vertical. The number of electrons reaching the screen can be controlled by the voltage applied to a third electrode, commonly in the form of a wire grid, placed between the cathode and anode so as to divert to itself a proportion of the electrons emitted by the former. It is thus possible to move the spot about the screen and vary its brightness by the application of appropriately timed electrical signals, and sustained images may be produced by causing the spot to traverse the same pattern many times a second. In the oscilloscope, the form of a given electrical signal, or any physical effect capable of conversion into one, is investigated by allowing it to control the vertical deflection while the horizontal deflection is scanned steadily from left to right, while in television sets half-tone pictures can be built up by varying the spot brightness while the spot scans out the entire screen in a series of close horizontal lines.

Cation
A positively charged ION, normally of a metal, which moves to the CATHODE in ELECTROLYSIS.

Catkin
A type of INFLORESCENCE found on some trees, usually consisting of a spike of unisexual flowers that have no petals or sepals. In most catkins POLLINATION is by the wind. Typical examples are hazels, birches and poplars.

CAT scan
Short for computer aided tomography, an X-ray technique for producing cross-sectional images of a living body.

Cat's eye
Any of several GEM stones which, when cut to form a convex surface, show a thin line of light like a cat's eye (chatoyancy). The commonest type is CHALCEDONY; the rarest is cymophane, a green variety of CHRYSOBERYL, found in Sri Lanka. The effect is due to minute parallel inclusions or cavities.

Caustic soda (sodium hydroxide) see SODIUM; ALKALI.

Cauterization
Application of heat or caustic substances. Used for minor skin or mucous-membrane lesions (especially of the nose and cervix) to remove abnormal tissue and encourage normal healing.

Cave
Any chamber formed naturally in rock and, usually, open to the surface via a passage. Caves are found most often in LIMESTONE, where rainwater, rendered slightly acid by dissolved carbon dioxide from the atmosphere, drains through joints in the stone, slowly dissolving it. Enlargement is caused by further passage of water and by fragments of rock that fall from the roof and are dragged along by the water. Such caves often form in connected series; they may display STALACTITES AND STALAGMITES and their collapse may form a GORGE. Caves are also formed by selective EROSION by the sea of cliff bases. Very occasionally they occur in LAVA, either where lava has solidified over a mass of ice that has later melted, or where the surface of a mass of lava has solidified, molten lava beneath bursting through and flowing on.

Cavendish, Henry (1731-1810)
English chemist and physicist who showed HYDROGEN (inflammable air) to be a distinct GAS, water to be a compound and not an elementary substance and the composition of the ATMOSPHERE to be constant. He also used a torsion BALANCE to measure the DENSITY of the Earth.

Cavitation
The formation of bubbles of vapor in a liquid, strictly applicable only to formation of bubbles through reduction in PRESSURE without corresponding TEMPERATURE change. Cavitation reduces efficiency in, for example, TURBINES and PROPELLERS; it rapidly erodes (see EROSION) the moving surfaces, and produces unwanted VIBRATIONS.

Cayley, Sir George (1773-1857)
British inventor who pioneered the science of AERODYNAMICS. He built the first man-carrying glider (1853) and formulated the design principles later used in airplane construction, though he recognized that in his day there was no propulsion unit that was sufficiently powerful and yet light enough to power an airplane.

Celestial sphere
In ancient times, the sphere to which it was believed all the stars were attached. In modern times, an imaginary sphere of indefinite but very large radius upon which, for purposes of angular computation, celestial bodies are considered to be situated. The celestial poles are defined as those points on the sphere vertically above the terrestrial poles, and the celestial equator by the projection of the terrestrial EQUATOR onto the sphere. (See also ECLIPTIC.) Astronomical coordinate systems are based on these great circles (circles whose centers are also the center of the sphere) and, in some cases, on the observer's celestial HORIZON. In the most frequently used, the equatorial system, terrestrial latitude corresponds to declination – a star directly overhead in New York City will have a declination of +41° (S Hemisphere declinations are preceded by a minus sign), New York City having a latitude of 41°N and terrestrial longitude to right ascension, which is measured eastward from the First Point of ARIES. Right ascension is measured in hours, one hour corresponding to 15° of longitude.

Celiac disease
A disease of the small intestine (see GASTROINTESTINAL TRACT), among the commonest causes of food malabsorption. ALLERGY to part of gluten, a component of wheat, causes severe loss of absorptive surface. In children, failure to thrive and DIARRHEA are common signs, while in adults weight loss, ANEMIA, diarrhea, TETANY and vitamin deficiency may bring it to attention. Complete exclusion of dietary gluten leads to full recovery.

Cell
The basic unit of living matter from which almost all organisms are built. The major exceptions are the FUNGI and SLIME MOLDS, whose hyphae are not divided into true cells. The bacteria, protozoa and other microorganisms are considered as consisting of a single cell. There are two basic types of cell: prokaryotic and eukaryotic. The former is found only in the bacteria and cyanobacteria, the latter in all other organisms.

Eukaryotic cells can be divided into three parts: an outer MEMBRANE, a nucleus, which is usually near the center of the cell, and an extranuclear region called the CYTOPLASM.

The plasma membrane that surrounds all cells is thin and flexible. It allows substances to diffuse in and out, and is also able to select some substances and exclude others. The membrane plays a vital part in deciding what enters a cell. Plant cell membranes are surrounded by a thick, rigid, CELLULOSE cell wall, and a similar cell wall surrounds prokaryotic cells, and the hypha of a fungus.

Around the nucleus of the eukaryotic cell is the nuclear envelope, a double membrane that contains tiny pores to allow molecules to pass between the cytoplasm and the nucleus. Within the cytoplasm is the much-folded endoplasmic reticulum, which is a continuation of the nuclear membrane. The endoplasmic reticulum has a major role in BIOSYNTHESIS by organizing key ENZYMES, and is always associated with the RIBOSOMES where PROTEIN SYNTHESIS takes place. All these processes are controlled by the CHROMOSOMES, which are sited in the nucleus and are complex structures made of DNA and proteins.

The cytoplasm includes many organelles. Among the most important are the rod-shaped MITOCHONDRIA, which contain the enzymes necessary for the release of energy from food by the process of RESPIRATION. Other organelles include the Golgi bodies, which are involved in the synthesis of membranes, and the lysosomes, which contain enzymes involved in the autolysis and controlled destruction of tissues. The cytoplasm of green plants also contains CHLOROPLASTS, where PHOTOSYNTHESIS occurs. The jelly-like material in which these organelles are embedded is known as the cytosol. It contains protein fibers and MICROTUBULES that make up the skeleton of the cell or CYTOSKELETON, helping to transport materials around the cell and to maintain its shape.

New cells are formed by a process of division called MITOSIS. Gamete (reproductive) cells are formed by MEIOSIS, which is a division that halves the number of chromosomes. Cells differentiate in a multicellular organism to perform specific functions, producing cells as different as a nerve cell and a muscle cell. Cells of similar types are grouped into TISSUES.

Prokaryotic cells differ from those of eukaryotes in not having the genetic material, in the form of a loose circular filament of DNA, separated from the cytoplasm by a membrane. Prokaryotic cells also lack cell organelles, apart from ribosomes, though some of the more highly developed forms have complex infoldings of the cell membrane, which fulfil the same functions as mitochondria or chloroplasts. There is no cytoskeleton in prokaryotic cells, and they are much smaller than eukaryotic cells. (◆ page 13, 32)

Cell, Electrochemical
Device for interconverting chemical and electrical ENERGY. For power cells, which transform chemical into electrical energy, see BATTERY. In electrolytic cells the reverse process occurs: by applying an ELECTROMOTIVE FORCE across two electrodes in an electrolyte, a chemical reaction is effected (see ELECTROLYSIS). (◆ page 118)

Cell fusion
Artificially joining cells to combine the desirable characteristics of different cells into one cell (see GENETIC ENGINEERING).

Cellulose
The main constituent of the cell walls of higher plants and most algae. Cotton is 90% cellulose. Cellulose is a CARBOHYDRATE with a similar structure to starch. In its pure form it is a white solid that absorbs water until completely saturated, but dissolves in only a few solvents, notably strong alkalis and some acids. It can be broken down by heat and by some bacteria, but it passes through the human digestive tract unchanged and is helpful only in stimulating movement of the intestines. Mammals that can derive nourishment from cellulose, such as cattle, horses and rabbits, rely on bacteria living in the gut to digest it for them. Industrially, it is used in manufacturing textile fibers, cellophane, celluloid, and the cellulose PLASTICS, notably NITROCELLULOSE (used also in explosives). (◆ page 13, 40)

Celsius, Anders (1701-1744)
Swedish astronomer, chiefly remembered for his proposal (1742) of a centigrade TEMPERATURE scale which had 100° for the freezing point and 0° for the boiling point of water. The modern centigrade temperature scale is known as the Celsius scale in his honor.

Cenozoic (Cainozoic)
The era of geological time containing the TERTIARY and QUATERNARY periods. (◆ page 106)

Centaurus
The Centaur, a constellation in the S Hemisphere. (See ALPHA CENTAURI; PROXIMA CENTAURI.)

Center of gravity
The point about which gravitational FORCES on an object exert no net turning effect, and at which the mass of the object can for many purposes be regarded as concentrated. A freely suspended object hangs with its center of gravity vertically below the point of suspension, and an object will balance, though it may be unstable, if supported at a point vertically below the center of gravity. In free flight, an object spins about its center of gravity, which moves steadily in a straight line; the application of forces causes the center of gravity to accelerate in the direction of the net force, and the rate of spin to change according to the resultant turning effect.

Centigrade degree (C°)
A unit of TEMPERATURE difference equal to 1 kelvin (K), originally defined by dividing the interval between the boiling and freezing points of water into 100 equal divisions. It is used as the basis of the Celsius temperature scale (see CELSIUS, ANDERS).

Centimeter (cm)
A unit of length equal to 0.01 meter.

Central dogma
An important tenet of biology, formulated by FRANCIS CRICK, which holds that information can only pass out from the genetic material to the rest of the body, and not vice versa. The idea had its origins in the theories of AUGUST WEISMANN, who believed that the body could be divided into immortal "germ line" cells, carrying the hereditary material, and transient "somatic" cells. Weismann maintained that the soma could not influence the information content of the germ cells, thus refuting theories about the inheritance of ACQUIRED CHARACTERISTICS. Weismann was wrong in his details but right in principle, as became clear when the details of DNA and PROTEIN SYNTHESIS emerged in the 1950s: proteins are made using messenger RNA (copied from DNA) as a template, but there is no means by which RNA or DNA can by synthesized using proteins as a template. It was on this basis that Crick proposed the central dogma. More recently an enzyme known as reverse transcriptase has been discovered (in RETROVIRUSES), which can produce DNA from an RNA template. It has been suggested that this enzyme might, under certain circumstances, allow acquired characteristics to be inherited, though in a very limited way.

Central processing unit
The part of a computer that draws data from the memory, processes them in the arithmetic and logic unit and outputs or stores the results, following the instructions of the operator or program.

Centrifugal force see CENTRIPETAL FORCE.

Centrifuge
A machine for separating mixtures of solid particles and immiscible liquids of different densities and for extracting liquids from wet solids by rotating them in a container at high speed. The ultracentrifuge, invented by T. SVEDBERG, uses very high speeds to measure (optically) sedimentation rates of macromolecular solutes and so determine molecular weights.

Centriole
An organelle found in almost all animal cells. Centrioles are located in the middle of the microtubule organizing centers – two centers from which the MICROTUBULES of the cell radiate, including the spindle that forms during cell division (MITOSIS and MEIOSIS). The centrioles are themselves made up of 27 short, parallel microtubules, grouped in nine sets of three and forming a tubular bundle. There is a distinct similarity in the structure and role of the centrioles and that of the basal bodies found at the root of flagella and cilia. Plant cells lack centrioles, though they do have microtubule-organizing centers.

Centripetal force
The FORCE applied to a body to maintain it moving in a circular path. To maintain a body of MASS m, traveling with instantaneous VELOCITY v, in a circular path of radius r, a centripetal force F, acting toward the center of the circle, given by $F=mv^2/r$ must be applied to it. If a body is resting in a rotating frame, it experiences a centrifugal force, apparently acting away from the center of rotation, numerically equal to the external centripetal force. Because, from the point of view of an observer external to the rotating frame, the centrifugal force has no real existence, it is often termed a fictitious force.

Cephalization
The tendency in the evolution of animals for sense organs and associated nervous tissue (see BRAIN; GANGLION) to be concentrated at one end of the body – the end that "faces" the environment. Cephalization has led to the development of a distinct head in most animals.

Cephalochordates
A subphylum of the CHORDATES that lack a vertebral column (spine) and are therefore distinguished from the VERTEBRATES. The best-known member is AMPHIOXUS, which is capable of short bursts of swimming activity but spends most of its life partially buried in sand. Amphioxus resembles the larvae of the TUNICATES (Urochordata), and it may have evolved from a tunicate ancestor by a process of NEOTENY – becoming sexually mature without adopting the sessile, adult form. A parallel evolutionary path could have led to the ancestral vertebrate. The fact that vertebrate embryos go through a stage that has a NOTOCHORD, like that of cephalochordates, supports this idea. (◆ page 17)

Cephalopods
A class of the phylum MOLLUSCA. It includes the octopuses, squid and cuttlefish, the most intelligent and fast-moving of the invertebrates.

Cephalothorax
A fused head and thorax; typical of ARACHNIDS.

Cepheid variables
Stars whose brightness varies regularly with a period of 1-60 days. They are pulsating stars that fluctuate in size and temperature during each cycle. The length of their cycle is directly proportional to their absolute magnitude, making them useful "mileposts" for computing large astronomical distances. (See VARIABLE STAR.)

Ceramics
Materials produced by treating non-metallic inorganic materials (originally CLAY) at high temperatures. Modern ceramics include such diverse products as porcelain and china, furnace bricks, electric insulators, ferrite magnets (see SPINEL), rocket nosecones and abrasives. In general, ceramics are hard, chemically inert under most conditions, and can withstand high temperatures in industrial applications. Many are refractory metal OXIDES.

Cereal crops
Annual plants of the grass family including wheat, rice, corn, barley, sorghum, millet, oats and rye. Their grain forms the staple diet for most of the world. Though lacking in calcium and vitamin A, they have more CARBOHYDRATE than any other

food, as well as PROTEIN and other VITAMINS. Cereal crops are relatively easy to cultivate and can cope with a wide range of climates. About 711 million hectares of the world's arable land are sown with cereal crops each year. The US leads in production of corn, oats and sorghum.

Cerebellum
The lobe at the rear of the BRAIN that controls movement and balance. (♦ page 126)

Cerebral cortex
The outer layer of the forebrain (see BRAIN) of mammals, and seat of thinking. In humans it is large, and convoluted like a walnut. (♦ page 126)

Cerebral palsy
A diverse group of conditions caused by brain damage around the time of birth, resulting in a variable degree of nonprogressive physical and mental handicap. While abnormalities of muscle control are the most obvious, loss of sensation and some degree of deafness are common accompaniments. Speech and intellectual development can also be impaired but may be entirely normal. SPASTIC PARALYSIS of both legs with mild arm weakness (diplegia), or of one half of the body (hemiplegia), are common forms. A number of cases have abnormal movements (athetosis) or ATAXIA. Common causes include birth trauma, ANOXIA, prematurity, Rhesus incompatibility (see BLOOD) and cerebral HEMORRHAGE.

Cerebrospinal fluid
Watery fluid circulating in the chambers (ventricles) of the BRAIN and between layers of the meninges covering the brain and SPINAL CORD. It is a filtrate of BLOOD and is normally clear, containing salts, glucose and some protein. It may be sampled and analyzed by SPINAL TAP to diagnose diseases such as multiple sclerosis.

Cerebrum see BRAIN.

Cerenkov, Pavel Alekseyevich see CHERENKOV, PAVEL ALEKSEYEVICH.

Cerenkov radiation see CHERENKOV RADIATION.

Ceres
Largest of the ASTEROIDS (1000km in diameter) and the one first discovered (by PIAZZI, 1801). Its orbit was first computed by K.F. GAUSS and found to satisfy BODE'S LAW. Its "year" is 1681 days and its maximum apparent magnitude is +7.

Cerium (Ce)
Most abundant of the RARE EARTHS; one of the LANTHANUM SERIES of elements. AW 140.1, mp 800°C, bp 3000°C, sg 6.657 (125°C). (♦ page 130)

Cermets (ceramels)
Composite materials made from mixed metals and ceramics. The transition elements are most often used. Powdered and compacted with an oxide, carbide or boride, etc., they are heated to just below their melting point, when bonding occurs. Cermets combine the hardness and strength of metals with a high resistance to corrosion, wear and heat. This makes them invaluable in jet engines, cutting tools, brake linings and nuclear reactors.

Cerussite
White mineral form of lead (II) carbonate ($PbCO_3$), a major ore of LEAD, found in Spain, SW Africa, Australia and Colorado. It is formed by weathering of GALENA. White lead, used as a paint pigment, is a basic lead (II) carbonate ($PbCO_3+Pb(OH)_2$).

Cesarian section
Birth of a child from the uterus by abdominal operation. Cesarian section may be necessary if the baby is too large to pass through the PELVIS, if it shows delay or signs of ANOXIA during labor, or in cases where maternal disease does not allow normal labor. It may be performed effectively before labor has started.

Cesium (Cs)
A very soft, silvery white ALKALI METAL, found mainly in the mineral pollucite, and made by reduction of cesium chloride. It is highly reactive, similar to potassium, and burns spontaneously in moist air. When exposed to light it emits electrons, and has applications in television cameras. It is used in vacuum tubes to remove oxygen and water; and a recent application is in ion-propulsion rocket engines. AW 132.9, mp 28.4°C, bp 686°C, sg 1.873 (20°C). (♦ page 130)

Cetaceans
An order of MAMMALS that includes the whales, dolphins and porpoises. Cetaceans are aquatic but, being mammals, they breathe air.

CGS units
A metric system of units based on the CENTIMETER (length), GRAM (mass) and SECOND (time), generally used among scientists until superseded by SI UNITS.

Chadwick, Sir James (1891-1974)
English physicist who was awarded the 1935 Nobel Prize for Physics for his discovery of the NEUTRON (1932).

Chagas' disease
PARASITIC DISEASE found only on the American continent, caused by *Trypanosoma cruzi* and carried by insects. In the acute form there is swelling around the eye, fever, malaise, enlargement of lymph nodes, liver and spleen, and edema. Most cases recover fully. The chronic form causes disease of the heart and of the gastrointestinal tract.

Chain reaction see NUCLEAR ENERGY.

Chain, Sir Ernst Boris (1906-1979)
German-born UK biochemist who helped develop PENICILLIN for clinical use. For this he shared with FLOREY and FLEMING the 1945 Nobel Prize for Physiology or Medicine.

Chalcedony
Mineral consisting of fibrous microcrystalline QUARTZ. Common in gravels, it occurs in several forms, including some semiprecious GEM stones. Notable are BLOODSTONE, CAT'S EYE, ONYX, CHRYSOPRASE, JASPER, CARNELIAN, SARD and AGATE.

Chalcocite
Dark gray or black mineral, copper (I) sulfide (Cu_2S). A major COPPER ore, it is found (often with BORNITE) in the Ural Mountains, Africa, South America, Alaska, Connecticut and the southern US.

Chalcopyrite ($CuFeS_2$)
The most important COPPER ore, very similar to PYRITE, with which it is often associated. It is found in the US, Canada, Australia, W Europe and South America.

Chalk
Soft, white rock formed mainly of fine-grained, porous LIMESTONE and consisting of calcareous remains of minute marine animals. Chalk is widely used in lime and cement manufacture and as a fertilizer. It is also used in cosmetics, crayons and oil paints; school chalk is today usually made from chemically produced calcium carbonate.

Chamberlain, Owen (b. 1920)
US physicist who shared the 1959 Nobel Prize for Physics with E. SEGRÉ in recognition of their discovery of the antiproton (see ANTIMATTER).

Chandrasekhar, Subrahmanyan (b. 1910)
Pakistan-born US astrophysicist who was co-recipient of the 1983 Nobel Prize for Chemistry with FOWLER for his theoretical studies of the physical processes important to the structure and evolution of STARS. He is also known for the Chandrasekhar limit, which is the maximum mass that an old star can have without further collapsing to form a BLACK HOLE.

Change of life see MENSTRUATION.

Charcoal
Form of amorphous CARBON produced when wood, peat, bones, cellulose or other carbonaceous substances are heated with little or no air present. A highly porous residue of microcrystalline GRAPHITE remains.

Charcot, Jean Martin (1825-1893)
French physician and founder of modern NEUROLOGY, whose many researches advanced knowledge of HYSTERIA, MULTIPLE SCLEROSIS, locomotor ATAXIA, ASTHMA and aging. FREUD was one of his many pupils.

Charge, Electric see ELECTRICITY.

Charge coupled device
A SEMICONDUCTOR device used in imaging systems, such as telescopes, to produce a video signal. A PHOTON striking one of an array of capacitors generates a charge which is amplified.

Charles, Jacques Alexandre César (1746-1823)
French physicist who, with Nicholas Robert, made the first ascent in a hydrogen balloon in 1783. In about 1787 he discovered Charles' Law which, stated in modern terms, records that for an ideal GAS at constant PRESSURE, its VOLUME is directly proportional to its absolute TEMPERATURE.

Chelate
Chemical complex formed from a polydentate LIGAND and a metal ION, thus making a ring. They are more stable than the corresponding unidentate complexes, and are used to sequester metal ions (see HARD WATER), as well as in chemical ANALYSIS and for separating metals. Some biochemical substances, including CHLOROPHYLL and HEMOGLOBIN, are chelates.

Chelicerates
A group of invertebrates once considered part of the phylum ARTHROPODA, but probably meriting a phylum to itself. It includes the spiders, ticks, mites, harvestmen, scorpions and sun spiders (Class Arachnida), the sea spiders (Class Pycnogonida), the horseshoe crabs (Class Merostomata) and their relatives, the extinct sea scorpions or eurypterids. These disparate animals are classed together on the basis of having a pair of pincer-like mouthparts known as chelicerae. (♦ page 16)

Chemiluminescence
LUMINESCENCE caused by a chemical reaction, usually oxidation, as of phosphorus in air. The molecules are excited to a high ENERGY LEVEL, and emit light as they return to the ground state. This process in living organisms is called BIOLUMINESCENCE.

Chemistry
The science of the nature, composition and properties of material substances, and their transformations and interconversions. In modern terms, chemistry deals with ELEMENTS and compounds, with the ATOMS and MOLECULES of which they are composed, and with the reactions between them. It is thus basic to natural phenomena and modern technology alike. Chemistry may be divided into five major parts: ORGANIC CHEMISTRY, the study of carbon compounds (which form an idiosyncratic group); INORGANIC CHEMISTRY, dealing with all the elements except carbon, and their compounds; chemical ANALYSIS, the determination of what a sample contains and how much of each constituent is present; BIOCHEMISTRY, the study of the complex organic compounds in biological systems; and PHYSICAL CHEMISTRY, which underlies all the other branches, encompassing the study of the physical properties of substances and the theoretical tools for investigating them. Related sciences include GEOCHEMISTRY and METALLURGY. Practical chemistry originated with the art of the metallurgists and artisans of the ancient Middle East. Their products included not only refined and alloyed metals but also dyes and glasses, and their methods were and remain shrouded in professional secrecy. Their chemical theory, expressed in terms of the prevailing

theology, involved notions such as the opposition of contraries and the mediation of a mediating third. Classical Greek science generally expressed itself in the theoretical rather than the practical, as the conflicting physical theories of THALES and ANAXAGORAS, ANAXIMENES and ARISTOTLE demonstrate. An important concept, that matter exists as atoms – tiny individual material particles – emerged about this time (see ATOMISM) though it did not become dominant for another 2000 years. During the Hellenistic age a new practical chemistry arose in the study of ALCHEMY. These early alchemists sought to apply Aristotelian physical theory to their practical experiments. Alchemy was the dominant guise of chemical science throughout the medieval period. Like the other sciences, it passed through Arab hands after the collapse of the Roman world, though great practical advances were made during this time with the discovery of alcohol DISTILLATION and methods for preparing NITRIC and SULFURIC ACIDS. Chemical theory, however, remained primitive and practitioners guarded their secret recipes by employing obscure and even mystical phraseology.

The 16th century saw new clarity brought to the description of metallurgical processes in the writings of GEORGIUS AGRICOLA and the foundation by PARACELSUS of the new practical science of IATROCHEMISTRY with its emphasis on chemical medicines. JAN BAPTIST VAN HELMONT, the greatest of his successors, began to use quantitative experiments. In the 17th century mechanist atomism enjoyed a revival with ROBERT BOYLE leading a campaign to banish obscurantism from chemical description. The 18th century saw the rise and fall of the PHLOGISTON theory of combustion, promoted by STAHL and adopted by all the great chemists of the age: BLACK, SCHEELE and PRIESTLEY (all of whom found their greatest successes in the study of GASES). The phlogiston theory fell before the oxygen theory of LAVOISIER and his associated binomial nomenclature, and the new century (the 19th) saw the proposal of DALTON's atomic theory, AVOGADRO's hypothesis (neglected for 50 years until revived by CANNIZZARO) and the foundation of ELECTROCHEMISTRY which, in the hands of DAVY, rapidly yielded two new elements, SODIUM and POTASSIUM. During the 19th century chemistry gradually assumed its present form, the most notable innovations being the periodic table of MENDELEYEV, the BENZENE ring-structure of KEKULÉ, the systematic chemical THERMODYNAMICS of GIBBS and BUNSEN's chemical SPECTROSCOPY. In the opening years of the present century the new atomic theory revolutionized chemical theory and the interpretation of the elements was deciphered. Since then successive improvements in experimental techniques (eg CHROMATOGRAPHY; isotopic labeling; MICROCHEMISTRY) and the introduction of new instruments (infrared, nuclear magnetic resonance and mass spectroscopes) have led to continuing advances in chemical theory. These developments have also had a considerable impact on industrial chemistry and biochemistry. Perhaps the most significant recent change in the chemist's outlook has been that his interest has moved away from the nature of chemical substance toward questions of molecular structure, the energetics of chemical processes and reaction mechanisms. (♦ page 130)

Chemosynthesis
The production of food using the energy stored in inorganic molecules. The ability to chemosynthesize is found only in certain bacteria. Although such bacteria represent a second major group of AUTOTROPHS, their worldwide contribution as primary producers is minuscule compared to that of photosynthetic organisms. However, there are certain habitats in which they are important, as in the deep sea around hydrothermal vents, where chemosynthetic sulfur bacteria utilize the inorganic compounds spewed out in the superheated water from the vent. A whole community of other organisms is supported by these bacteria, which ultimately derive their energy from the heat of the Earth's core.

Chemotherapy
The use of chemical substances to treat disease. More specifically, the term refers to the use of agents for treating malignant disease. The drug must interfere with the growth of bacterial, parasitic or TUMOR cells, without significantly affecting host cells, and is especially successful in LEUKEMIA and lymphoma; in carcinoma it is usually reserved for disseminated tumor. The term is also less commonly used to refer to treatment with a synthetic antimicrobial agent.

Cherenkov (or Cerenkov), Pavel Alekseyevich (b. 1904)
Russian physicist who first observed CHERENKOV RADIATION (1934). He shared the 1958 Nobel Prize for Physics with FRANK and TAMM, who correctly interpreted this phenomenon in 1937.

Cherenkov radiation
ELECTROMAGNETIC RADIATION emitted when a high-energy particle passes through a dense medium at a velocity greater than the velocity of light in that medium. It was first detected in 1934 by P.A. CHERENKOV, and may be seen when radioactive materials are stored under water.

Chert see FLINT.

Chickenpox (varicella)
A VIRUS disease due to *Varicella zoster*, affecting mainly children, usually in EPIDEMICS. It is contracted from other cases or from cases of SHINGLES and is contagious. It causes malaise, fever and a characteristic vesicular rash, mainly on trunk and face, that appears in crops. Infrequently it becomes hemorrhagic or lung involvement occurs. It is rarely serious in the absence of underlying disease. Chickenpox and shingles are caused by the same virus.

Chilblain
Itchy or painful red swelling of extremities, particularly toes and fingers, in predisposed subjects. A tendency to cold feet and exposure to extremes of temperature appear to be causative factors.

Childbirth see BIRTH.

Chile saltpeter
An impure form of sodium nitrate (see SODIUM) occurring in large quantities in Chile.

Chimera
A type of fish, also known as a rat-tail, ratfish or rabbitfish. The last name refers to its huge eyes and somewhat rabbit-like mouth, the others to its long, tapering rat-like tail and its generally odd appearance. There are about 25 species, and they are placed in the Subclass Holocephali of the Class Chondrichthyes, the CARTILAGINOUS FISH. They are not closely related to the other cartilaginous fish, the sharks and rays. Many live in the deep ocean and all are found in cold water. They are poor swimmers and tend to keep to the bottom. Their teeth are fused into a horny beak, an adaptation to a diet of mollusks and other shelled animals.

China clay see KAOLIN.

Chinaware see CERAMICS.

Chinook
FOEHN wind blowing eastward from the Rocky Mountains of western N America, mainly in winter. The term is also applied to the warm, wet wind coming off the Pacific before it passes over the Rockies.

Chip
A tiny INTEGRATED CIRCUIT on a chip of SILICON about 1cm square. A chip can accommodate more than 100,000 TRANSISTORS, using LSI or VLSI technology (see ELECTRONICS). When memory and logic components are built in, it is known as a MICROPROCESSOR. (See also COMPUTER; ELECTRONICS.)

Chirality
The property of asymmetric molecules that are mirror images of each other. ENANTIOMERS, as they are also called, have identical physical properties, except for the direction in which their solutions rotate plane-polarized light. Complex organic molecules may contain several asymmetric or chiral centers, in each of which a tetrahedrally bonded carbon atom is attached to four different atoms or molecular groupings.

Chiropractic
A health discipline based on a theory that disease results from misalignment of VERTEBRAE. Manipulation, massage, dietary and general advice are the principal methods used.

Chitin
A complex, insoluble polysaccharide (see CARBOHYDRATES) found in the EXOSKELETON of insects, myriapods, crustaceans, arachnids and other invertebrates formerly grouped together as ARTHROPODS. Chitin also makes up the cell walls of fungi, except in the oomycetes. Traces of it are found in the cuticle of annelid worms, and a similar compound makes up bacterial cell walls.

Chitons
Coat-of-mail shells, a small group of mollusks with jointed shells that mostly feed on algae and other organisms growing on rocks. If in danger, the chiton can roll up into a ball and is then completely protected by its shell. The general body plan of chitons is very simple, and is thought to resemble that of the hypothetical molluskan ancestor. Some authorities have suggested that the chiton's shell indicates a segmented ancestry for the mollusks, but others disagree. (See also NEOPILINA.)

Chloral (trichloroacetaldehyde, CCl_3CHO)
Colorless, oily liquid made by reacting CHLORINE and ETHANOL or ACETALDEHYDE. It is used chiefly in the manufacture of DDT. MW 147.4, mp −57.5°C, bp 98°C.

Chloral hydrate ($CCl_3CH(OH)_2$) is a colorless crystalline solid made by reacting chloral and water, and is used as a SEDATIVE, since it is a DEPRESSANT of the central nervous system. It is toxic in excess, and especially when mixed with alcohol to make "Mickey Finns" or "knockout drops". Habitual use produces DRUG ADDICTION and gastritis. mp 57°C.

Chloramphenicol see ANTIBIOTICS.

Chlorine (Cl)
Greenish yellow gas with a pungent odor, a typical member of the HALOGENS, occurring naturally as chlorides (see HALIDES) in seawater and minerals. It is made by electrolysis of SALT solution, and is used in large quantities as a bleach, as a disinfectant for drinking water and swimming pools, and in the manufacture of plastics, solvents and other compounds. Being toxic and corrosive, chlorine and its compound PHOSGENE have been used as poison gases. Chlorine reacts with most organic compounds, replacing hydrogen atoms (see ALKYL HALIDES) and adding to double and triple bonds. AW 35.5, mp −101°C, bp −34°C. (♦ page 130)

CHLORIDES The commonest chlorine compounds, are typical HALIDES. Other chlorine compounds include a series of oxides, unstable and highly oxidizing, and a series of oxyanions – HYPOCHLORITE, chlorite, chlorate and perchlorate – with the corresponding OXYACIDS, all powerful oxidizing agents. Calcium hypochlorite (see BLEACHING POWDER) and sodium chlorite are used as bleaches; chlorates are used as weedkillers and to make matches and fireworks; perchlorates are used as explosives and rocket fuels. (See also HYDROGEN CHLORIDE.)

Chlorite

Group of green CLAY minerals, related to the MICAS, and consisting of alternate single layers of BIOTITE, containing ferrous iron (Fe^{2+}) and brucite [$(Mg,Al)_6(OH)_{12}$]. They are widespread alteration products.

Chlorofluorocarbons

Organic compounds of exceptional stability that are used as propellants in some aerosol cans, as refrigerants, and in various other products, including the insulating boxes supplied by fast-food outlets. It is their stability that is valued in these applications, but this property also makes them hazardous environmentally because they do not break down when released into the atmosphere. In common with certain other gases they help to trap heat and thus contribute to the GREENHOUSE EFFECT, but the major cause for concern is their impact on the OZONE layer.

Chlorophyll

Green pigment found in plant CHLOROPLASTS. It absorbs light and converts it into chemical energy, thus playing a basic part in PHOTOSYNTHESIS. Photosynthetic bacteria also contain chlorophyll, though of a slightly different type. Chlorophylls are CHELATE compounds in which a magnesium ion is surrounded by a PORPHYRIN system.

Chlorophyta

A division of the ALGAE, also known as the green algae. (➧ page 139)

Chloroplast

An organelle containing CHLOROPHYLL, found in plant cells. Higher plant cells contain up to 50 chloroplasts, while most algal cells have only one. Variously shaped, they are of the order of 5μm in diameter, and are surrounded by a semipermeable membrane. They contain ordered stacks of membranes within which all PHOTOSYNTHESIS reactions take place. Chloroplasts are a type of PLASTID. They are thought to be derived from CYANOBACTERIA that invaded larger cells, or were ingested by them, and then took up a symbiotic relationship with the host cell (see ENDOSYMBIOSIS theory). (➧ page 13)

Chloroprene (2-chlorobuta-1, 3-diene)

An unsaturated chlorine containing hydrocarbon used in the manufacture of the synthetic rubber polychloroprene (neoprene). Polychloroprene is more resistant than natural rubber to attack by oxygen, and it is also more resistant to burning. MW 88.4, bp 59.4°C.

Choke (inductor)

Device used in electric CIRCUITS to oppose changes in the magnitude or direction of current flow (see ELECTRICITY); ie it is a coil of high self-inductance (see INDUCTANCE). The term is also used for the VALVE regulating the air supply to the carburetor of an INTERNAL-COMBUSTION ENGINE.

Cholecystectomy

Operation for removal of the GALL BLADDER.

Cholera

A BACTERIAL DISEASE caused by *Vibrio cholerae*. It is endemic in many parts of the East and EPIDEMICS occur elsewhere. Abdominal pain and diarrhea, which rapidly becomes severe and watery, are main features, with rapidly developing dehydration and shock. Without rapid and adequate fluid replacement, death rapidly ensues. VACCINATION gives limited protection for six months.

Cholesterol ($C_{27}H_{46}O$)

A STEROL found in all cell membranes, especially in the NERVOUS SYSTEM, where it is an important constituent of MYELIN. Cholesterol is a precursor in the synthesis of adrenal and sex HORMONES. It is broken down by the liver to give BILE SALTS. Large amounts are synthesized in the liver, intestines and skin, and cholesterol in the diet supplements this. If there is excess in the bloodstream abnormal deposition of cholesterol in the arteries results, and this is associated with ATHEROSCLEROSIS. Doctors therefore advise avoiding high-cholesterol foods such as animal fats, butter and cream. Cholesterol is a major constituent of gallstones (see CALCULI).

Chondrichthyes (cartilaginous fish)

A class of the vertebrates that includes the sharks, dogfish, rays and skates (elasmobranchii), and the CHIMERAS or rat-tails. They are characterized by having cartilaginous (see CARTILAGE) skeletons, and scales that are tooth-like. They also lack a SWIMBLADDER. Today, they are marine, but freshwater forms are known from a fossil record that extends back some 300 million years. (➧ page 17)

Chondrostei

A group of mainly fossil fish of the ACTINOPTERYGII, represented today by the sturgeons, paddlefish and, in some classifications, the bichirs.

Chondrule

Small, roughly spherical granules of OLIVINE, PYROXENE, GLASS or other MINERALS found in members of that class of stony meteorites (see METEOR) known as chondrites.

Chordates

A phylum that includes TUNICATES and CEPHALOCHORDATES as well as the better-known VERTEBRATE groups (subphylum Vertebrata): the fish, amphibians, reptiles, birds and mammals. The latter is divided into the AGNATHA, or jawless fish, and the Gnathostomata, or jawed vertebrates. All chordates are characterized by having, at some stage in their lives, a rod of stiffening tissue that runs the length of the body and is known as a NOTOCHORD. This persists in cephalochordates such as AMPHIOXUS, but is replaced by a bony or cartilaginous spinal column in the vertebrates. (➧ page 17)

Chorea

Abnormal, nonrepetitive involuntary movements of the limbs, body and face. It may start with clumsiness, but later uncontrolable and bizarre movements occur. It is a disease of basal ganglia. Sydenham's chorea, or Saint Vitus' dance, is a childhood illness associated with STREPTOCOCCUS infection and RHEUMATIC FEVER; recovery is usually full. Huntingdon's chorea is a rare hereditary disease, usually coming on in middle age and associated with progressive dementia.

Chromatic aberration see ABERRATION, OPTICAL.

Chromatography

A versatile technique of chemical separation and ANALYSIS, capable of dealing with many-component mixtures, and large or small amounts. The sample is injected into the moving phase, a gas or liquid stream which flows over the stationary phase, a porous solid or a solid support coated with a liquid. The various components of the sample are adsorbed (see ADSORPTION) by the stationary phase at different rates, and separation occurs. Each component has a characteristic velocity relative to that of the solvent, and so can be identified. In liquid-solid chromatography the solid is packed into a tube, the sample is added at the top, and a liquid eluant is allowed to flow through; the different fractions of effluent are collected. A variation of this method is ion-exchange chromatography, in which the solid is an ION-EXCHANGE resin from which the ions in the sample are displaced at various rates by the acid eluant. Other related techniques are paper chromatography (with an adsorbent paper stationary phase) and thin-layer chromatography (using a layer of solid adsorbent on a glass plate). The other main type of chromatography – the most sensitive and reliable – is gas-liquid chromatography (glc), in which a small vaporized sample is injected into a stream of inert eluant gas (usually nitrogen) flowing through a column containing nonvolatile liquid adsorbed on a powdered solid. The components are detected by such means as measuring the change in thermal conductivity of the effluent gas.

Chromite

A hard, black SPINEL mineral, an iron (II) chromium (III) mixed OXIDE ($FeCr_2O_4$); the chief ore of CHROMIUM.

Chromium (Cr)

Silvery white, hard metal in Group VIB of the PERIODIC TABLE; a TRANSITION ELEMENT. It is widespread, the most important ore being CHROMITE. This is reduced to a ferrochromium alloy by carbon or silicon; pure chromium is produced by reducing chromium (III) oxide with aluminum. It is used to make hard and corrosion-resistant ALLOYS and for chromium ELECTROPLATING. Chromium is unreactive. It forms compounds in oxidation states +2 and +3 (basic) and +6 (acidic). Chromium (III) oxide is used as a green pigment, and lead chromate (VI) as a yellow pigment. Other compounds are used for tanning leather and as mordants in dyeing. (See also ALUM.) AW 52.0, mp 1860°C, bp 2600°C, sg 7.20 (20°C). (➧ page 130)

Chromosomes

Thread-like bodies found in the nucleus of EUKARYOTIC cells; they carry the hereditary material, DNA. They also contain various proteins, and have a complex internal structure that may play a part in controlling gene expression. A given species always has the same number of chromosomes (the chromosome complement), though it may have one set (HAPLOID) or two sets (DIPLOID). More than two sets is known as the POLYPLOID condition. Not only the number but also the length and shape of the chromosomes is characteristic for each species; these features are described as the karyotype. Information about evolutionary pathways can often be revealed by analysis of the karyotype of related species, as speciation is often marked by chromosome breakage or duplication, or rearrangements of genetic material between chromosomes. In birds, mammals and some other animals one chromosome pair consists of two dissimilar chromosomes and is involved in sex determination (see SEX CHROMOSOMES). The simple loop of DNA found in prokaryotes is sometimes referred to as the "bacterial chromosome". (➧ page 32)

Chromosphere see SUN.

Chronic

Of diseases, likely to last for a long time, often throughout life. (See also ACUTE.)

Chronometer

An extremely accurate clock, especially one used in connection with celestial NAVIGATION at sea (see also CELESTIAL SPHERE). It differs from the normal clock in that it has a fusee, by means of which the power transmission of the mainspring is regulated so that it remains approximately uniform at all times; and a balance made of metals of different coefficients of expansion to minimize the effects of temperature changes. The device is maintained in gimbals to reduce the effects of rolling and pitching. A chronometer's accuracy is checked daily and its error noted; the daily change in error is termed the daily rate. Chronometers are always set to GREENWICH MEAN TIME. The first chronometer was invented by JOHN HARRISON (1735). (See also ATOMIC CLOCK.)

Chrysalis

Another name for the PUPA stage in the life-cycle of insects, usually used for butterflies and moths.

Chrysoberyl
Mixed OXIDE mineral of BERYLLIUM and ALUMINUM ($BeAl_2O_4$), forming hard crystals in the orthorhombic system. ALEXANDRITE and CAT'S EYE are GEM varieties.
Chrysoprase
A GEM variety of CHALCEDONY, colored apple green by colloidal nickel silicate. It is found in Silesia and California.
Chrysotile
Mineral forming the chief variety of ASBESTOS, a fibrous SERPENTINE.
Chylomicron
A minute fat droplet suspended in blood, which transports it within the body.
Cilia
Hair-like outgrowths from the surfaces of cells, which provide the means of locomotion for some unicellular organisms such as the ciliate PROTOZOA; and for some small invertebrates such as comb-jellies (CTENOPHORES). In higher animals, cilia provide means for moving fluids over the surfaces of cells, as in the respiratory tract of mammals. Cilia are made up of protein MICROTUBULES that generate movement in the cilium by sliding over each other, utilizing energy in the form of ATP. The same basic mechanism occurs in FLAGELLA; these are longer than cilia, have a more whip-like movement, and usually occur singly or in pairs, whereas cilia are present in large numbers. (◆ page 213)
Ciliates
A major group of PROTOZOA, constituting the Class Ciliata. They are among the most complex single cells, with a macronucleus and a micronucleus, and in some species a distinctive gullet for ingesting food. Thousands of CILIA are found on the cell surface, providing movement by their synchronized beating in motile species, creating water currents for feeding in sessile ones. One well-known species is *Paramecium*. (◆ page 13)
Cinnabar
Red SULFIDE mineral consisting of mercury (II) sulfide (HgS), the chief ore of MERCURY, found as massive deposits or hexagonal crystals. It is used as the pigment vermilion. A black, cubic form of mercury (II) sulfide, metacinnabar, also occurs.
Circadian rhythm
An internal rhythm of activity and rest, or other physiological changes; it is shown by most living organisms, and lasts for about 24 hours (hence *circa diem* – "about a day"). Experiments show that the rhythm persists even when all natural light, and any other cues that might indicate the day-night cycle, are excluded. However, in the absence of external cues the cycle begins to "drift", with the times of waking and sleeping getting a little later or a little earlier each day. In natural conditions, the cycle can be reset each day so that it is always synchronized with external conditions. The "drift" in the clock is an adaptation to changing daylengths and allows the cycle to adjust to different seasons of the year. A circadian rhythm is just one type of BIOLOGICAL CLOCK.
Circuit, Electric
Assemblage of electrical CONDUCTORS (usually wires) and components through which current from a power source such as a BATTERY or GENERATOR flows (see ELECTRICITY). Components may be connected one after another (in series) or side by side (in parallel). If current may flow between two points their connection is a closed circuit; if not, an open circuit; and if RESISTANCE between them is virtually zero, a short circuit: a switch when off is open, when on, a short circuit. Short circuits between the terminals of the power source are dangerous. (See also ELECTRONICS; KIRCHHOFF'S LAWS.) (◆ page 118)
Circulatory system
A system of internal transport found in the bodies of all vertebrates and most invertebrates. It is responsible for transporting oxygen and nutrients to the tissues of the body, and for removing waste products. At its simplest it consists of a set of tubes, through which the circulatory fluid is pushed as a side-effect of normal body movements. In larger invertebrates and in vertebrates a pumping organ, or heart, is required to push the circulatory fluid around. Some invertebrates, such as the mollusks, have an "open" circulatory system in which the afferent and efferent vessels are not directly joined; instead the blood coming from the heart empties into a body cavity, which is then drained by vessels leading to the heart. Other invertebrates have a "closed" system like that of vertebrates, in which the arteries (leading from the heart) and the veins (leading to the heart) are linked by fine vessels known as capillaries where exchange of nutrients and gases with the body's tissues occurs. In animals with an open system, the circulatory fluid is sometimes referred to as hemolymph, the term BLOOD being reserved for that of animals with a closed system. (For details of the circulatory system of humans, see BLOOD CIRCULATION.) (◆ page 16-17, 28)
Circumcision
Removal of the foreskin, the loose skin at the end of the penis.
Circumpolar stars
The stars that can be seen every night of the year from any particular latitude, and that appear to circle the celestial pole (see CELESTIAL SPHERE).
Cirque (corrie or cwm)
Steep-sided hollow formed by glacial EROSION, usually occupied by a lake where the GLACIER has retreated, or by NÉVÉ where the glacier is still present.
Cirrhosis
Chronic disease of the LIVER with disorganization of normal structure and replacement by fibrous scars and regenerating nodules. It is the end result of many liver diseases, all of which cause liver-cell death; most common are those associated with alcoholism and following some cases of hepatitis, while certain poisons and hereditary diseases are rare causes. All liver functions are impaired, but symptoms often do not occur until early liver failure develops with EDEMA, ascites, JAUNDICE, COMA, emaciation, or gastrointestinal tract HEMORRHAGE; blood clotting is often abnormal and plasma proteins are low. The liver damage is not reversible.
Citral (geranial)
A terpene aldehyde found in lemon grass oil and used widely as a perfume and flavoring agent, on account of its strong lemon scent. MW 152.24, bp 229°C.
Citric acid ($C_6H_8O_7$)
Colorless crystalline solid, a CARBOXYLIC ACID widespread in plant and animal tissue, especially in citrus fruits. It is made commercially by FERMENTATION of crude sugar with the fungus *Aspergillus niger*, and is used in the food, pharmaceutical and textile industries, and for cleaning metals. It is vital in cell metabolism (see CITRIC ACID CYCLE). MW 192.1, mp 153°C.
Citric acid cycle (Krebs cycle, tricarboxylic acid cycle)
A vital cycle of chemical reactions forming a stage in the oxidation of food in cells (see RESPIRATION) and thus producing energy. The previous stage of oxidation (GLYCOLYSIS) produces pyruvate (a three-carbon molecule), which is then converted to the acetyl derivative of coenzyme A with the loss of one carbon atom as carbon dioxide. The two-carbon acetyl group then adds to oxaloacetic acid to produce citric acid. In a sequence of enzyme-catalyzed oxidation reactions, this is converted back to oxaloacetic acid with the loss of two more carbon atoms as carbon dixoide. The net result of the cycle is the total oxidation of the three-carbon pyruvate to carbon dioxide and water. The oxidizing agents are NADH and $fADH_2$, coenzymes that accept hydrogen atoms and form high-energy bonds with them. They are then reoxidized by cytochromes in the ELECTRON TRANSPORT CHAIN, the energy produced being stored as ATP. Except in bacteria and cyanobacteria (PROKARYOTES), the enzymes needed in the cycle are located in the MITOCHONDRIA.
Civil engineering
That branch of engineering concerned with the design, construction and maintenance of stationary structures such as buildings, bridges, highways, dams, etc. The term "civil engineer" was first used *c*1750 by John Smeaton, who built the Eddystone Lighthouse. Nowadays, civil engineering incorporates modern technological advances in the structures required by industrial society. The branches of the field include: surveying, concerned with the selection of sites; hydraulic and sanitary engineering dealing with public water supply and sewage disposal, etc.; transportation engineering, which deals with highways, airports and so on; structural engineering, which is concerned with the actual planning and construction of permanent installations: and environmental, town or city planning, which improves existing urban environments and plans the development of new areas.
Class
In TAXONOMY, a group of related orders. Related classes are grouped together in a phylum. Examples of classes include Class Aves (birds) and Class Mammalia, both part of the Subphylum Vertebrata, Phylum Chordata.
Clathrate
A chemical complex in which one type of molecule encloses another molecule or atom. Hydrates of the rare gases, in which groups of water molecules cluster around a rare gas atom (eg argon), are examples of crystalline clathrates.
Claude, Albert (1899-1983)
Belgian-US biologist who shared the 1974 Nobel Prize for Physiology or Medicine with G.E. PALADE and C. DE DUVE for demonstrating the usefulness of the ELECTRON MICROSCOPE and the CENTRIFUGE in biological studies. He pioneered the study of the internal structure of CELLS.
Claudication
Literally, limping. Caused when blood-flow is insufficient to bring enough oxygen to leg tissues.
Clausius, Rudolf Julius Emanuel (1822-1888)
German theoretical physicist who first stated the second law of THERMODYNAMICS (1850) and proposed the term ENTROPY (1865). He also contributed to KINETIC THEORY and the theory of ELECTROLYSIS.
Claustrophobia see PHOBIA.
Clavicle (collarbone)
A bone that forms part of the shoulder girdle in many vertebrates. Humans have two collarbones running between the breastbone and the shoulder blade. In birds the two collarbones are joined to form the furcula or wishbone.
Claws
Sharp, pointed extensions of the digits in many tetrapod vertebrates. They are formed from the protein KERATIN. In the ungulate mammals the claws are modified to form hooves, and in the primates they have become flatter and are known as nails.
Clay
Any SOIL material with a particle size of less than 2-4μm in diameter, ie finer grained than SILT or SAND; an earthy material that becomes plastic

when wet, including mud (which is used in oil drilling). Clays are used as catalysts (see CATALYSIS) in PETROLEUM refining, for making molds for casting and, when molded and fired, for ceramics, pottery and porcelain, bricks and tiles. They are also used in making cement and rubber, and as ion-exchange agents for softening hard water. Clay rocks, including mudstones and shales, are microcrystalline rocks composed mainly of clay-size particles. Their mineralogical composition is very variable, but they usually contain a high proportion of clay minerals, hydrated aluminum and magnesium SILICATES including BENTONITE, CHLORITE, diaspore (hydrated ALUMINUM oxide), illite (hydrated MICA – see also GLAUCONITE), KAOLINITE and MEERSCHAUM.

Cleavage
The tendency of a mineral to split along a definite plane parallel to an actual or possible CRYSTAL face: eg GALENA, whose crystals are cubic, cleaves along three mutually perpendicular planes (parallel to 100, 010, 001). Such cleavage is useful in identifying minerals. ROCK cleavage generally takes place between roughly parallel beds whose resistance under deformation to internal SHEARING differs.

Cleft palate
Common developmental deformity of the PALATE in which the two halves are not joined in the midline; it is often associated with HARELIP. It can be familial or follow disease in early pregnancy, but usually appears spontaneously. It causes a characteristic nasal quality in the cry and voice. Plastic surgery can close the defect and allow more normal development of the voice and teeth.

Climacteric
Another name for the MENOPAUSE.

Climate
The sum of the weather conditions prevalent in an area over a period of time. Weather conditions include temperature, rainfall, sunshine, wind, humidity and cloudiness. Climates may be classified into groups. The system most used today is that of Vladimir Köppen, with five categories (A, B, C, D, E), broadly defined as follows: A Equatorial and tropical rainy climates; B Arid climates; C Warmer forested (temperate) climates; D Colder forested (temperate) climates; and E Treeless polar climates.

These categories correspond to a great extent to zoning by LATITUDE; this is because the closer an area is to the EQUATOR, the more directly it receives sunlight and the less ATMOSPHERE there is through which that sunlight must pass. Other factors are the rotation of the Earth on its axis (diurnal differences), and the revolution of the Earth about the Sun (seasonal differences).

PALEOCLIMATOLOGY The study of climates of the past has shown that there have been considerable long-term climatic changes in many areas: this is seen as strong evidence for CONTINENTAL DRIFT (see also PLATE TECTONICS). Other theories include variation in the solar radiation (see SUN) and change in the EARTH's axial tilt. Man's influence has caused localized, short-term climatic changes. (See CLOUDS; METEOROLOGY; RAIN; TROPICS; WEATHER FORECASTING AND CONTROL; WIND.)

Climax vegetation
The type of vegetation that develops on a site if SUCCESSION is allowed to proceed to its natural conclusion. In most parts of the world the climax vegetation is forest of some type, but dry conditions result in scrub, grassland or desert vegetation, while in subpolar regions the extreme cold excludes most trees and produces tundra as the climax vegetation.

Clone
A population of cells or individuals derived from a single ancestor. All have the same genetic make-up. Cloning is a vital tool in the development of GENETIC ENGINEERING.

Clotting
The formation of semisolid deposits in a liquid by coagulation, often by the denaturing of previously soluble ALBUMIN. Thus clotted cream is made by slowly heating milk so that the thick cream rises; the curdling of skim milk to make cheese is also an example of clotting. Clotting of BLOOD is a complex process set in motion when it comes into contact with tissues outside its ruptured vessel. These contain a factor, thromboplastin, which activates a sequence of changes in the PLASMA clotting factors (12 enzymes). Alternatively, many surfaces, such as glass and fabrics, activate a similar sequence of changes. In either case, factor II (prothrombin, formed in the liver), with calcium ions and a platelet factor, is converted to thrombin. This converts factor 1 (fibrinogen) to fibrin, a tough, insoluble polymerized protein which forms a network of fibers around the platelets (see BLOOD) that have stuck to the edge of the wound and to each other. The network entangles the blood cells and contracts, squeezing out the serum and leaving a solid clot. (See also ANTICOAGULANTS; EMBOLISM; HEMOPHILIA; HEMORRHAGE; THROMBOSIS.)

Cloud chamber
Device, invented by C.T.R. WILSON (1911), used to observe the paths of subatomic particles. In simplest form, it comprises a chamber containing saturated VAPOR (see SATURATION) and some liquid, one wall of the chamber (the window) being transparent, another retractable. Sudden retraction of this wall lowers the temperature, and the gas becomes supersaturated (and thus metastable). Passage of SUBATOMIC PARTICLES through the gas leaves charged IONS that serve as seeds for CONDENSATION of the gas into droplets. These fog trails (condensation trails) may be photographed through the window. (See also BUBBLE CHAMBER.)

Clouds
Visible collections of water droplets or ice particles that, because they fall so slowly, may be regarded as suspended in the ATMOSPHERE. Clouds whose lower surfaces touch the ground are usually called FOG. The water droplets are very small, indeed of colloidal size (see COLLOID; AEROSOL); they must coagulate or grow before falling as RAIN or SNOW. This process may be assisted by cloud seeding (to relieve the effects of drought); supercooled clouds are seeded with particles of (usually) dry ice (ie solid carbon dioxide) to encourage CONDENSATION of the droplets, ideally causing RAIN or SNOW.

CLOUD FORMATION Clouds are formed when air containing water vapor cools in the presence of suitable condensation nuclei (eg dust particles). This may occur through CONVECTION, when warm, moist air from near the Earth's surface penetrates upward into regions of lower PRESSURE; here they expand, thus cooling past the dew point (see CONDENSATION), the temperature at which SATURATION of water in air is reached. The vapor condenses out to form cloud droplets. They may also form when warm air flows over a mountain; the air travels in a vertical wave (see WAVE MOTION), parts of which may be higher than its condensation level, resulting in a stationary cloud at the crest of the wave. Other modes of formation occur. (See METEOROLOGY.)

TYPES OF CLOUDS There are three main cloud types. Cumulus (heap) clouds, formed by convection, are often mountain- or cauliflower-shaped and are found from about 600m up as far as the tropopause, even temporarily into the stratosphere (see ATMOSPHERE). Cirrus (hair) clouds are composed almost entirely of ice crystals. They appear feathery, and are found at altitudes above about 6000m. Stratus (layer) clouds are low-lying, found between ground level and about 1500m. Other types of cloud include cirrostratus, cirrocumulus, altocumulus, altostratus, cumulonimbus, stratocumulus and nimbostratus. (♦ page 25)

Clubfoot
Deformity of the FOOT, with an abnormal relationship of the foot to the ankle; most commonly the root is turned in and down. Abnormalities of fetal posture and ligamentous or muscle development, including CEREBRAL PALSY and SPINA BIFIDA, may be causative. Correction includes gentle manipulation, PHYSIOTHERAPY, plaster splints and sometimes SURGERY.

Clubmosses
Primitive vascular plants of the Order Lycopodiales, related to the ferns (see PTERIDOPHYTES). They have creeping stems that branch dichotomously, and small leaves arranged spirally. They are found mainly in the tropics, but some occur in temperate climates. (♦ page 139)

Cluster see STAR CLUSTER.

Cnidarians
A phylum of mainly marine animals once included in the COELENTERATE phylum. They include jellyfish, sea anemones, corals and colonial forms, such as the Portuguese man o'war. Cnidarians have bodies composed of just two layers of cells separated by a jelly-like layer known as the mesoglea. The main forms are the polyp (as in sea anemones and hydroids), which is cylindrical and sessile with the tentacles uppermost, and the medusa (as in jellyfish), which is bell- or saucer-shaped with pendulous tentacles, and usually free-swimming. In some species there is a two-stage life-cycle, in which the polyp and medusoid forms alternate. (♦ page 16)

CNS
The central nervous system, consisting of the brain and spinal cord.

Coal
Hard, black mineral burned as a FUEL. With its byproducts coke and coal tar it is vital to many modern industries.

Coal is the compressed remains of tropical and subtropical plants, especially those of the CARBONIFEROUS and PERMIAN periods. Changes in the world climatic pattern explain why coal occurs in all continents, even Antarctica. Coal formation began when plant debris accumulated in swamps, partially decomposing and forming PEAT layers. A rise in sea level or land subsidence buried these layers below marine sediments, whose weight compressed the peat, transforming it under high-temperature conditions to coal; the greater the pressure, the harder the coal.

Coals are analyzed in two main ways: the "ultimate analysis" determines the total percentages of the elements present (carbon, hydrogen, oxygen, sulfur and nitrogen); the "proximate analysis" gives an empirical estimate of the amounts of moisture, ash, volatile materials and fixed carbon. Coals are classified, or ranked, according to their fixed-carbon content, which increases progressively as they are formed. In ascending rank, the main types are: lignite or brown coal, which weathers quickly, may ignite spontaneously, and has a low calorific value (see FUEL), but is used in Germany and Australia; subbituminous coal, mainly used in generating stations; bituminous coal, the commonest type, used in generating stations and the home, and often converted into coke; and anthracite, a lustrous coal that burns slowly and well, and is the preferred domestic fuel.

Coat-of-mail shells
Another name for the CHITONS.

Cobalt (Co)
Silvery white, hard, ferromagnetic metal in Group VIII of the PERIODIC TABLE; a TRANSITION ELEMENT. It occurs in nature largely as sulfides and arsenides, and in nickel and copper ores. An ALLOY of cobalt, aluminum, nickel and iron ("Alnico") is used for magnets; other cobalt alloys, being very hard, are used for cutting tools. Cobalt is used as the matrix for tungsten carbide in drill bits. Chemically it resembles IRON and NICKEL: its characteristic oxidation states are +2 and +3. Cobalt compounds are useful colorants (notably the artists' pigment cobalt blue). Cobalt catalysts facilitate HYDROGENATION and other industrial processes. The RADIOISOTOPE cobalt-60 is used in RADIOTHERAPY. AW 58.9, mp 1495°C, bp 2900°C, sg 8.9 (20°C). (➧ page 130)

Cobalt bomb
Device used in CANCER treatment as a source of gamma rays (see RADIOACTIVITY; RADIOTHERAPY). It uses Co^{60}, a RADIOISOTOPE of cobalt. Because of its long half-life Co^{60} is used widely as a radioactive tracer; and it gives its name to a theoretical nuclear weapon (see ATOMIC BOMB), also called the cobalt bomb, whose fallout might remain deadly for years.

Coca
A shrub, *Erythroxylon coca*, whose leaves contain various ALKALOIDS, especially COCAINE. Native to the Andes, it is widely cultivated elsewhere. The leaves have been chewed by South American Indians for centuries to quell hunger and to dull pain. Cocaine-free coca extracts are used in making cola drinks.

Cocaine
An ALKALOID from the coca leaf, the first local anesthetic agent and model for those currently used; it is occasionally used for surface ANESTHESIA. It is a drug of abuse, taken for its euphoriant effect by chewing the leaf, as snuff or by intravenous INJECTION. Although physical dependence does not occur, its abuse may lead to acute psychosis. It also mimics the actions of the sympathetic NERVOUS SYSTEM. (See also DRUGS.)

Cocci
Spherical bacteria that often stick together in necklace-like chains.

Coccyx (tailbone)
Bone at the base of the spine, consisting of five fused VERTEBRAE. (➧ page 216)

Cochlea
Part of the inner EAR, concerned with the mechanism of hearing. It is a spiral structure containing fluid and specialized membranes on which receptor nerve cells lie. Sound is conducted to the fluid by the bones of the middle ear. (➧ page 213)

Cockcroft, Sir John Douglas (1897-1967)
English physicist who first "split the atom". With E.T.S WALTON he built a PARTICLE ACCELERATOR and in 1932 initiated the first manmade nuclear reaction by bombarding LITHIUM atoms with PROTONS, producing ALPHA PARTICLES. For this work Cockcroft and Walton received the 1951 Nobel Prize for Physics. In 1946 Cockcroft became the first director of the UK's atomic research laboratory at Harwell, and in 1959 the first Master of Churchill College, Cambridge.

Cocoon
The protective capsule enclosing the eggs or the PUPA of many animals. In silk moths the cocoon enclosing the pupa is formed by secretions of silk.

Codeine
A mild NARCOTIC, ANALGESIC and COUGH suppressant related to MORPHINE. It reduces bowel activity, causing CONSTIPATION, and is used to cure DIARRHEA.

Cod liver oil
An oil rich in vitamins A and D, a convenient way of enriching diets which contain inadequate amounts, particularly used for children who might be susceptible to RICKETS. Both vitamins are dangerous in excess.

Codon
A set of three adjacent nucleotides in the molecules of DNA and RNA (NUCLEIC ACIDS); each codon, or triplet, codes for a single amino acid in a PROTEIN molecule, or indicates where the message stops. There are only four different nucleotides in DNA and RNA, not enough to code individually for the 20 amino acids that make up proteins. So the nucleotides are grouped in threes, giving 64 different codons ($4 \times 4 \times 4 = 64$). Of these, 61 code for amino acids – several codons usually coding for the same amino acid – and 3 signal for termination of the protein chain. (See also GENETIC CODE; PROTEIN SYNTHESIS.)

Coelacanth
Latimeria chalumnae, a member of the fossil fish group, the CROSSOPTERYGII, which was thought to be extinct until discovered off the east coast of Africa in 1938.

Coelenterata
An obsolete phylum of simple, invertebrate animals, now divided into two groups, the CNIDARIANS and the CTENOPHORES.

Coeliac disease see CELIAC DISEASE.

Coelom
A body cavity found in most larger invertebrates and all vertebrates – the so-called coelomate animals. The main organs of the body are contained within the coelom.

Coevolution
The process whereby two species, or two groups of organisms, evolve together, interacting in such a way that they influence the course of each other's evolutionary pathway. The interaction may be to their mutual advantage, as in the coevolution of flowering plants and pollinating insects (see POLLINATION), or it may be a predator-prey relationship, as between the grazing ungulates of the African plains and the big cats that pursue them – a coevolution that has produced speed, agility, stealth and sharp senses in both groups. Coevolution can also take place between a parasite (or brood parasite) and its host, and between organisms that are competing for the same resource.

Cohen, Stanley (b. 1922)
US scientist who received the 1986 Nobel Prize for Physiology or Medicine together with RITA LEVI-MONTALCINI for their discoveries of growth factors.

Cohesion
The tendency of different parts of a substance to hold together. This is due to forces acting between its MOLECULES: a molecule will repel one close to it but attract one that is farther away; somewhere between these there is a position where WORK must be done either to separate the molecules or to push them together. This situation results both in cohesion and in ADHESION. Cohesion is strongest in a SOLID, less strong in a LIQUID, and least strong in a GAS.

Cohn, Ferdinand Julius (1828-1898)
German botanist renowned as one of the founders of BACTERIOLOGY. He showed that BACTERIA could be classified in fixed species and discovered that some of these formed endospores which could survive adverse physical conditions. He was also the first to recognize the value of KOCH's work on the ANTHRAX bacillus.

Colchicine ($CH_{22}H_{25}NO_6$)
Poisonous ALKALOID found in the autumn crocus or meadow saffron (*Colchicum autumnale*). It is used in plant breeding to stimulate the formation of POLYPLOIDS.

Cold-blooded animals (poikilotherms)
A rather inaccurate term often used for POIKILOTHERMS or ectotherms. Such animals are as warm as, or warmer than, "warm-blooded" animals (HOMOIOTHERMS) when the external temperature is high, but they cannot generate their own heat metabolically in cold conditions.

Cold, Common (coryza)
A mild illness of the nose and throat caused by various types of VIRUS. General malaise and RHINITIS, initially watery but later thick and tenacious, are characteristic; sneezing, cough, sore throat and headache are also common, but significant fever is unusual. Spread is from person to person.

Cold sore
Vesicular SKIN lesion of lips or nose caused by *Herpes simplex* VIRUS. Often associated with periods of general ill-health or infections such as the COMMON COLD. The virus, which is often picked up in early life, persists in the skin between attacks. Recurrences may be reduced by special drugs, applied during an attack.

Colectomy
Removal of the colon. As the colon removes water from the feces, people who have undergone colectomy produce watery feces.

Coleoptera
An order of insects, the beetles. They are characterized by having the front pair of wings modified into hardened wing-cases or elytra.

Colic
Intermittent severe abdominal pain interspersed with pain-free intervals. It is due to irritation or obstruction of hollow viscera, in particular the GASTROINTESTINAL TRACT, ureter (see KIDNEY) and GALL BLADDER or bile ducts (see LIVER).

Coliform
A word used for a group of bacteria often found in the intestine. The presence of coliforms in food, water, etc. is a sign of contamination with feces.

Colitis
INFLAMMATION of the colon (see GASTROINTESTINAL TRACT). Infection with VIRUSES, BACTERIA or PARASITES may cause colitis, often with ENTERITIS. Inflammatory colitis can occur without bacterial infection in the chronic diseases ulcerative colitis and Crohn's disease. Impaired blood supply may also cause colitis. Symptoms include COLIC and DIARRHEA (with slime or blood). Severe colitis can cause serious dehydration or SHOCK.

Collagen
A tough, fibrous protein occurring as a major component of CARTILAGE, BONE, LIGAMENTS and TENDONS. Animal hide is chiefly collagen, converted by tanning into leather. When collagen is boiled it yields gelatin and glue.

Collarbone see CLAVICLE.

Collective unconscious
Term used, especially by JUNG, for those parts of the UNCONSCIOUS derived from multiple, rather than individual, experience.

Collembolans (springtails)
A group of primitive flightless insects; see APTERYGOTES.

Colloid (colloidal solution)
A system in which two (or more) substances are uniformly mixed so that one is extremely finely dispersed throughout the other. A colloid may be viewed intuitively as a halfway stage between a SUSPENSION and a SOLUTION, the size of the dispersed particles being larger than simple MOLECULES, smaller than can be viewed through an optical MICROSCOPE (more precisely, they have at least one diameter in the range 1μm-1nm). Typical examples of colloids include fog and butter. Colloids may be classified in two ways: one by the natures of the particles (dispersed phase) and medium (continuous phase); the other by, as it were, the degree of permanency of the colloid.

In the latter case, one may define a lyophilic colloid as one that forms spontaneously when the two phases are placed in contact; and a lyophobic colloid as one that can be formed only with some difficulty and maintained for a moderate period of time only under special conditions. Colloids have interesting properties, perhaps the most notable of which is light DISPERSION: it is due to colloidal particles in the atmosphere that the sky is blue in the daytime and the sunset red. Moreover, the property of ADSORPTION of molecules and IONS at the interface between particles and continuous phase plays a major part in water purification. (See also AEROSOL; BROWNIAN MOTION; DIALYSIS; ELECTROPHORESIS; OSMOSIS; TYNDALL; ULTRAMICROSCOPE.)

Colon see GASTROINTESTINAL TRACT.

Color

The way the brain interprets the wavelength distribution of the light entering the eye. The phenomenon of color has two aspects: the physical or optical, concerned with the nature of the light, and the physiological or visual, dealing with how the eye sees color. (➧ page 115)

The light entering the eye is either emitted by or reflected from the objects we see. Hot objects emit light with wavelengths occupying a broad continuous band of the electromagnetic SPECTRUM, the position of the band depending on the temperature of the object – the hotter the object, the shorter the wavelengths emitted (see BLACKBODY RADIATION). (We tend to think of the spectrum of visible light in terms of the band of colors revealed by NEWTON when he split up a beam of sunlight using a PRISM; in these terms, the shorter the wavelength, the bluer the light.) Other objects emitting light do so either at particular wavelengths or in narrow bands of the spectrum (see SPECTROSCOPY). Where these emission bands fall within the visible spectrum, such objects appear colored, otherwise black. Objects reflecting diffuse light appear colored in virtue of combined SCATTERING and ABSORPTION effects, though the nature of the light source is also important.

The EYE can see colors only when the light is relatively bright; the rods used in poor light see only in black and white. The cones used in color VISION are of three kinds, responding to light from the red, green or blue portions of the visible spectrum. The brain adds together the responses of the different sets of cones and produces the sensation of color. The three colors to which the cones of the eye respond are known as the three primary colors of light. By mixing different proportions of these three colors, any other color can be simulated, equal intensities of all three producing white light. This is known as the production of color by addition, the effect being used in color television tubes where PHOSPHORS glowing red, green and blue are employed. Color pigments, working by transmission or reflection, produce colors by subtraction, abstracting light from white and displaying only the remainder. Again a suitable combination of a set of three pigments – cyan (blue-green), magenta (blue-red) and yellow (the "complementary" colors of the three primaries) – can simulate most other colors, a dense mixture of all three producing black. This effect is used in color photography, but in color printing an additional black pigment is commonly used.

Most colors are not found in the spectrum. These nonspectral colors can be regarded as intermediates between the spectral colors and black and white. Many schemes have been proposed for the classification and standardization of colors. The most widely used is that of Albert Henry Munsell which describes colors in terms of their hue (basic color), saturation (intensity or density), and lightness or brightness (the degree of whiteness or blackness).

Color blindness

More properly, color deficiency, an inability to discriminate between certain COLORS, an inherited trait. It is a disorder of the RETINA. The commonest form is red-green color blindness usually found in men (about 8%), the other types being rare.

Colorimetry

The techniques for measuring and describing the phenomena of COLOR. Since the experience of color is largely subjective and similar experiences can occur with quite different sets of physical phenomena, many different instruments (colorimeters) and criteria are employed. Of two basic approaches, one involves the visual comparison of colors (using color comparators), while the other employs the instruments and techniques of PHOTOMETRY and SPECTROSCOPY.

Colostomy

The operation of connecting the colon to the outside of the abdomen, usually after removal of the rectum. A form of STOMA.

Colostrum

The first thin, watery fluid secreted by the MAMMARY GLAND after birth in mammals. Although low in fat, it is rich in protein, including ANTIBODIES, which are important in giving temporary passive IMMUNITY to the offspring.

Columbite

Lustrous black OXIDE mineral of iron, manganese and niobium [(Fe,Mn)Nb_2O_6], the chief ore of NIOBIUM. TANTALUM replaces niobium in all proportions up to 100% (tantalite). Columbite crystallizes in the orthorhombic system.

Columbium

Former name of NIOBIUM, especially in the US.

Coma

State of unconsciousness in which a person cannot be roused by sensory stimulation and is unaware of his surroundings. Body functions continue but may be impaired, depending on the cause. These include poisoning, head injury, DIABETES, and brain diseases, including STROKES and CONVULSIONS. Severe malfunction of LUNGS, LIVER or KIDNEYS may also lead to coma.

Coma (in optics) see LENS.

Comb jellies see CTENOPHORES.

Combustion (burning)

The rapid OXIDATION of FUEL in which heat and usually light are produced. In slow combustion (eg a glowing charcoal fire) the reaction may be heterogeneous, the solid fuel reacting directly with gaseous oxygen; more commonly, the fuel is first volatilized, and combustion occurs in the gas phase (a flame is such a combustion zone, its luminance being due to excited particles, molecules and ions). In the 17th and 18th centuries combustion was explained by the PHLOGISTON theory, until LAVOISIER showed it to be due to combination with oxygen in the air. In fact the oxidizing agent need not be oxygen: it may be another oxidizing gas such as nitric oxide or fluorine, or oxygen-containing solids or liquids such as nitric acid (used in rocket fuels). If the fuel and oxidant are premixed, as in a bunsen burner, the combustion is more efficient, and little or no SOOT is produced. Very rapid combustion occurs in an explosion (see EXPLOSIVES), when more heat is liberated than can be dissipated, or when a branched chain reaction occurs (see FREE RADICALS). Each combustion reaction has its own ignition temperature below which it cannot take place, eg *c*400°C for coal. Spontaneous combustion occurs if slow oxidation in large piles of such materials as coal or oily rags raises the temperature to the ignition point (See also INTERNAL-COMBUSTION ENGINE.)

Comet

A nebulous body that orbits the Sun. In general, comets can be seen only when they are comparatively close to the Sun, though the time between their first appearance and their final disappearance may be as much as years. As they approach the Sun, a few comets develop tails (some comets develop more than one tail) of lengths of the order 1-100Gm, though at least one tail 300Gm in length – more than twice the distance from the Earth to the Sun – has been recorded. The tails of comets are always pointed away from the Sun, so as the comet recedes into space its tail precedes it. There are two principal types of tail, the dust tail and the ion (or plasma) tail. The tiny dust particles that make up the dust tail are blown from the head of the comet by radiation pressure, while the ionized gas of the plasma tail is driven out by the SOLAR WIND.

The head of the comet contains the nucleus. Nuclei may be as little as 100m or possibly as much as 100km in radius, and are thought to be composed primarily of frozen gases and ice mixed with smaller quantities of meteoritic material. The nucleus of HALLEY'S COMET measures about 16×8×8km. Most of the mass of a comet is contained within the nucleus, though this may be less than 0.000,001 that of the Earth. Surrounding the nucleus is the bright coma, possibly as much as 100Mm in radius, which is composed of gases and small particles erupting from the nucleus.

Cometary orbits are usually very eccentric ellipses, with some perihelions (see ORBIT) closer to the Sun than that of MERCURY, aphelions as much as 100,000AU from the Sun. The orbits of some comets take the form of hyperbolas, and it is thought that these have their origins altogether outside the SOLAR SYSTEM, that they are interstellar travelers.

In Greco-Roman times it was generally believed that comets were phenomena restricted to the upper atmosphere of the Earth. In the late 15th and 16th centuries it was shown by Michael Mästlin (1550-1631) and BRAHE that comets were far more distant than the Moon. NEWTON interpreted the orbits of comets as parabolas, deducing that each comet was appearing for the first time. It was not until the 17th century that HALLEY showed that some comets returned periodically.

Commensalism

A relationship between two organisms of different species in which one organism – the commensal – benefits from the partnership while the other neither gains nor loses. The commensal often gets shelter, transport or food from its host. (See also SYMBIOSIS.)

Communicable diseases see DISEASE.

Compact disk see OPTICAL DISK.

Compass

Device for determining direction parallel to the Earth's surface. Most compasses make use of the Earth's magnetic field (see GEOMAGNETISM); if a bar magnet (see MAGNETISM) is pivoted at its center so that it is free to rotate horizontally, it will seek to align itself with the horizontal component in its locality of the Earth's magnetic field. A simple compass consists of a magnet so arranged and a compass card marked with the four cardinal points and graduated in degrees. In ship compasses, to compensate for rolling, the card is attached to the magnet and floated or suspended in a liquid, usually alcohol. Aircraft compasses often incorporate a GYROSCOPE to keep the compass horizontal. The two main errors in all magnetic compasses are variation (the angle between lines of geographic longitude and the local horizontal component of the Earth's magnetic field) and deviation (local, artificial magnetic effects, such as nearby electrical equipment). Both vary with the siting of the compass, and may, with more or less difficulty, be compensated for. (See also NAVIGATION.) A radio compass, used in aircraft, is an automatic radio DIRECTION FINDER, calibrated with respect to the station to which it is tuned.

Competitive inhibition see ENZYMES.

Compiler see COMPUTER.

Complementarity principle
A philosophical thesis proposed in 1927 by NIELS BOHR, seeking to make intelligible the then newly developed WAVE MECHANICS. Bohr recognized that the hardware used in any subatomic physics experiment materially affected the results obtained, and inferred from this that atomic systems could only be described and understood in terms of a series of complementary partial views.

Complex
A nexus of ideas and feelings from both the CONSCIOUS and UNCONSCIOUS that has an effect, favorable or adverse, on the actions or emotional state of the individual. ADLER used the term in connection with the superiority and inferiority complexes, FREUD in connection with the CASTRATION COMPLEX and OEDIPUS COMPLEX.

Composition, Chemical
The proportion by weight of each ELEMENT present in a chemical compound. The law of definite proportions, discovered by J.L. PROUST, states that pure compounds have a fixed and invariable composition. A few compounds, termed non-stoichiometric, disobey this law: they have lattice vacancies or extra atoms, and the composition varies within a certain range depending on the formation conditions. The law of multiple proportions, discovered by DALTON, states that, if two elements A and B form more than one compound, the various weights of B which combine with a given weight of A are in small whole-number ratios. (See also BOND, CHEMICAL; EQUIVALENT WEIGHT). (◆ page 37)

Compost see MANURE.

Compound
A chemical substance in which the molecules are composed of atoms of more than one element. (◆ page 37)

Compound eye
A type of eye found in INSECTS and CRUSTACEANS, so called because it consists of a number of organs called ommatidia. Each ommatidium perceives part of the visual field so that the animal's view of the world is composed of a mosaic of images. Predatory insects such as dragonflies have compound eyes made up of thousands of ommatidia.

Compton, Arthur Holly (1892-1962)
US physicist who discovered the Compton effect (1923), thus providing evidence that X-RAYS could act as particles as predicted in QUANTUM THEORY. Compton found that when monochromatic X-rays were scattered by light elements, some of the scattered radiation was of longer wavelength, ie of lower ENERGY than the incident. Compton showed that this could be explained in terms of the collision between an X-ray PHOTON and an ELECTRON in the target. For this work he shared the 1927 Nobel Prize for Physics with C.T.R. WILSON.

Compulsion
An irresistible UNCONSCIOUS force which makes an individual perform conscious (see CONSCIOUSNESS) thoughts or actions that he would not normally perform. The force may also come from outside, ie from someone whose character dominates the individual (see also BRAINWASHING; OBSESSIONAL NEUROSIS).

Computer
Any device that performs calculations. The term is usually limited to those electronic devices that are given a program to follow, data to store or to calculate with, and means with which to present results or other (stored) information.

PROGRAMMING A computer program consists essentially of a set of instructions that tells the computer which operations to perform, in what order to perform them, and the order in which subsequent data will be presented to it; for ease of use, the computer may already have subprograms built into its memory, so that, on receiving an instruction such as Log X, it will automatically go through the program necessary to find the logarithm of that piece of data supplied to it as X. Every model of computer has a different machine language or code; that is, the way in which it should ideally be programmed; however, this language is usually difficult and cumbersome for an operator to use. Thus a special program known as a compiler is retained by the computer, enabling it to translate computer languages such as Algol, Basic and Fortran, which are easily learned and used, into its own machine code. Programmers also make extensive use of algorithms to save programming and operating time.

INPUT Programs and data are fed into computers as electrical pulses from a terminal keyboard, a magnetic medium (disk or tape), a sensor (measuring, say, temperature or rate of flow of a liquid), optical readers, or a modem. Mostly these generate digital signals (either "on" or "off"), but in the case of a sensor the signal is analog which an ADC (analog-to-digital converter) changes to a binary signal.

STORAGE Machine languages in computers generally take the form of a BINARY CODE so that the two characters 0 and 1 may be easily represented by + and −. Computers have an internal read-only memory (ROM), which cannot be erased or altered, containing permanent data and instructions for starting and operating the computer. The mediums for auxiliary data storage may take the form of magnetic tapes, disks, bubbles or drums, optical disks or, originally, paper tape and card.

With all magnetic storage, particular items of information are stored at specific sites, called addresses, so that instructions for their retrieval may be given to the computer. Magnetic tapes are used much as they are in a tape recorder: a magnetic head "writes" by magnetizing the surface in one direction (north to south) to represent a binary 0, and the other direction (south to north) for a 1. The reading head then senses changes in the magnetic polarization. Seven or nine parallel tracks run the length of the tape, and a row of bits, one in each track across the tape, is a binary code for one number, character or symbol. Up to 6250 magnetic changes per inch can be recorded on tape, representing 6250 bits. The data on a tape can be accessed only in sequence.

Magnetic disks are the favoured medium for storage; here the tracks are circular, and provide random access memory (RAM) – that is, the computer can read the required data directly, without, as on tape, reading all the data between the last information accessed and the information sought. It provides much faster access, the maximum time for the read/write head to locate any specific area being that taken for one revolution (approximately 300 revolutions per second). On small computers, floppy disks are the norm. Made of a flexible Mylar and coated with a magnetic material, a 5¼in disk can store more than 1 megabyte. Hard disks, including Winchester disks, are made of a rigid base, eg ceramics or aluminum, and can store 300 megabytes or more. Hard disks are usually non-removable, and although more expensive than floppies they provide faster access. The newer laser disk technologies offer capacities in excess of 500 megabytes.

In magnetic bubble memory, data are encoded in the form of magnetically charged bubbles onto a thin film of magnetic silicate. A sensor detects the positive or negative condition of the bubbles, and transmits an electronic pulse as each bubble passes its read head.

Optical disks offer mass storage (2 million bytes per disk), and can store not only data but also pictures and sound, or a combination of these. They have the disadvantage, however, that they can only be "written" once, and the data cannot be altered or erased.

CENTRAL PROCESSING UNIT At the heart of a computer is the central processing unit (CPU). The ALU (arithmetic logic unit) performs the arithmetic and logic operations on the data. This is almost exclusively done by addition, and performed using binary arithmetic. More complicated procedures are performed algorithmically, suitable subprograms being built into the computer. Again the characters 0 and 1 are represented by + and −, where this may refer to a closed or open switch, a direction of magnetic flux (see MAGNETISM), etc. The control unit receives instructions, interprets the addresses and operational parts, and executes the instructions by sending signals to the appropriate units in the computer, including the ALU, memory, and input and output devices. A built-in clock in the CPU ensures that all operations are properly synchronized.

OUTPUT Before being fed out, the information must be converted from machine code back into the programmer's computer language, numerical data being translated from the binary into the DECIMAL SYSTEM. The information is then transmitted to a printer, visual display unit, speech synthesizer, electromechanical device (which controls, say, a washing machine or heating system), or to disk, tape, etc. for storage.

TYPES OF COMPUTER At one time all computers were mainframes, but with the introduction of microcomputers the term is today reserved for large machines which can be accessed by several users simultaneously on separate VDUs, and which operate at up to 100 times the speed of smaller computers. They are used where large volumes of data are processed, eg large companies, universities and government offices. A microcomputer is a small computer which uses one or more MICROPROCESSORS as its CPU. Microcomputers are used in businesses, the home and in schools. They may be linked together to form a NETWORK which may equal the mainframe in its computing power and accessibility.

Engineers classify the development of computers into four distinct stages. The first generation, which started in about 1951, consisted of physically large computers that incorporated vacuum tube circuitry and stored programs. Auxiliary storage was mostly on magnetic tape. The second generation lasted from the late 1950s until the early 1960s, and used solid-state transistor circuits, which required less power and produced less heat. Auxiliary storage was on disks as well as tape. The third generation, which ended in the early 1970s, incorporated integrated circuits. We are now in the fourth generation, with small low-cost microcomputers using microprocessors and memory chips. The next stage is the FIFTH GENERATION. (See also ELECTRONICS.)

Comte, Auguste (1798-1857)
French philosopher, the founder of positivism and a pioneer of scientific sociology. His thinking was essentially evolutionary; he recognized a progression in the development of the sciences, starting from mathematics and progressing through astronomy, physics, chemistry and biology towards the ultimate goal of sociology. He saw this progression reflected in man's mental development. This had proceeded from a theological stage to a metaphysical one. Comte now

sought to help inaugurate the final scientific or positivistic era.

Conchoid fracture
A form of ROCK and MINERAL fracture producing a curved, ribbed surface resembling the shell of certain mollusks.

Concretion
Irregularly shaped nodule of rock embedded in sedimentary rock of a different composition, in which it grew by deposition on a nucleus. Examples include FLINT in chalk.

Concussion
Brief loss of consciousness following head injury, characterized by partial AMNESIA for events preceding and following the trauma. Permanent brain damage is found only in cases of repeated concussion, as in boxers who develop the punch-drunk syndrome.

Condensation
Passage of substance from gaseous to liquid or solid state: CLOUDS are a result of condensation of water vapor in the ATMOSPHERE (see also RAIN). Warm air can hold more water vapor than cool air; if a body of air is cooled it will reach a temperature (the DEW point) where the water vapor it holds is at SATURATION level. Further decrease in temperature without change in pressure will initiate water condensation. Such condensation is greatly facilitated by the presence of condensation nuclei ("seeds"), small particles (eg of smoke) about which condensation may begin. Condensation trails behind high-flying jet aircraft result primarily from water vapor produced by the engines increasing the local concentration (see also CLOUD CHAMBER; GAS; VAPOR). Condensation is important in all processes using steam; and in DISTILLATION, where the liquid is collected, and condensed by removal of its LATENT HEAT of vaporization in an apparatus called a condenser. In chemistry a condensation reaction is one in which two or more MOLECULES link together with elimination of a relatively small molecule, eg water.

Condenser see CAPACITOR; CONDENSATION.

Conditioned reflex see REFLEX.

Conditioning
Term used to describe two quite different LEARNING processes. In the first, a human or animal response is generated by a stimulus that does not normally generate such a response (see conditioned REFLEX; PAVLOV). In the second, animals (and by extension humans) are trained to perform certain actions to gain rewards or escape punishment (see LEARNING).

Conductance
A measure of the ability of a body to conduct ELECTRICITY. Measured in siemens, it is the RECIPROCAL of RESISTANCE. (See also CONDUCTIVITY.)

Conduction, Heat
Passage of heat through a body without large-scale movements of matter within the body (see CONVECTION). Mechanisms involved include the transfer of vibrational ENERGY from one MOLECULE to the next through the substance (dominant in poor conductors), and energy transfer by ELECTRONS (in good electrical conductors) and PHONONS (in crystalline solids). In general, solids, especially metals, are good conductors, liquids and gases poor. (See also RADIATION.)

Conductivity (specific conductance)
The CONDUCTANCE of a 1-meter cube of a substance, measured between opposite faces. Measured in siemens per meter, conductivity is the RECIPROCAL of resistivity (see RESISTANCE), and expresses the substance's ability to conduct electricity. The equivalent conductivity Λ of an electrolytic solution is the conductivity of a solution divided by its concentration in gram-equivalents (see EQUIVALENT WEIGHT) per cubic meter, and is usually measured with an electrolytic CELL in a WHEATSTONE BRIDGE. The degree of ionic DISSOCIATION (α) was found by ARRHENIUS to be given by

$$\alpha = \frac{L}{L_0}$$

where Λ_0 is the value of Λ extrapolated to zero concentration; this equation has since been shown to apply only to weak electrolytes (see ELECTROLYSIS).

Conductors, Electric
Substances, usually metals, whose high CONDUCTIVITY makes them useful for carrying electric current (see ELECTRICITY). They are most often used in the form of WIRES or CABLES. The best conductor is SILVER, but for reasons of economy COPPER is most often used. (See also SEMICONDUCTORS; SUPERCONDUCTIVITY.) (➧ page 118)

Cone
In plants, the typical reproductive body of the CONIFERS, such as pines and firs. Cones are also seen in cycads and some clubmosses, and in one angiosperm tree, the alder, where the cones are actually short, woody catkins. Despite their different origins, alder cones serve the same purpose as other cones – retaining the seed during adverse conditions, then opening to release it when conditions are best for dispersal and growth. Certain light-sensitive cells in the retina of the vertebrate eye are also known as cones. They are able to detect different colors, unlike the other visual cells, known as rods. (See VISION.)

Conformational analysis
The study of molecular conformations, ie the spatial arrangements of the atoms that can be interconverted merely by rotation about single bonds (see BOND, CHEMICAL). In general, different conformations are transient and nonisolable, unlike different configurations (see STEREOCHEMISTRY). But if large substituents, such as bromine or phenyl groups, are attached to adjacent atoms, steric hindrance or interference may cause the eclipsed conformation to be less stable than the staggered. This may affect reaction rates. Conformational effects are most important in ALICYCLIC COMPOUNDS.

Congenital
Present from birth, rather than acquired later in life.

Conglomerate (pudding stone)
In geology, a consolidated GRAVEL consisting of rock fragments, rounded by transportation, cemented in a sedimentary matrix. If the fragments are angular it is known as breccia. The fragments vary in size from boulders to small pebbles, grading into SANDSTONE. Poorly sorted conglomerates, having fragments of many different rocks, include tillite (of glacial origin) and graywacke. (See also AGGLOMERATE.)

Conifers
An order of trees, Order Coniferales, belonging to the GYMNOSPERM group. The members of the order, namely the pines, firs, spruces, hemlocks, Douglas firs, cedars, larches, cypresses, thujas, junipers and redwoods, are all trees or large shrubs, and all bear CONES. The YEWS and their allies are included with the conifers by some authorities but not by others. Cycads also bear cones, as did the extinct tree-like clubmosses, but they are not conifers. (➧ page 139)

Conjugation
A form of sexual reproduction occurring in some single-celled and simple multicellular organisms, notably the ciliate protozoa and some unicellular and filamentous algae, in which two individual cells fuse and the CHROMOSOMES of one pass into the other. The term is also used for the process of genetic transfer in BACTERIA whereby one bacterium passes a PLASMID (a small, discrete, partially autonomous piece of DNA) to another bacterium along a protein tubule known as a pilus.

Conjunction
This occurs when the Earth, the Sun and a planet are in a straight line (as projected onto the plane of the Solar System). A planet on the far side of the Sun is in superior conjunction, a planet between Earth and Sun in inferior conjunction (see also OPPOSITION). Planets may be in conjunction with each other (see ASTROLOGY).

Conjunctivitis
INFLAMMATION of the conjunctiva, the fine layer of skin covering the EYE and inner eyelids. It is a common but usually harmless condition caused by ALLERGY (as part of HAY FEVER), foreign bodies, or infection with VIRUSES or BACTERIA. It causes irritation, watering and sticky discharge, but does not affect vision.

Connective tissue
Tissue that joins together, or gives support to, the other organs of the body. It consists principally of an extracellular matrix: a gel of polysaccharides (see CARBOHYDRATE) containing microscopic protein fibers, often collagen or elastin. There are relatively few cells compared to other tissues. Connective tissues include cartilage, bones, tendons, adipose tissue (fat), the elastic supportive tissues around the lungs, and the framework of organs such as the liver and spleen. The blood is sometimes also classed as a connective tissue.

Conscious
In psychoanalysis, the structure of the mind in which logical, conscious thought takes place. Because of confusion over the roles of conscious, CONSCIOUSNESS, UNCONSCIOUS and UNCONSCIOUSNESS (eg in dreams), the conscious is now often referred to as the EGO.

Consciousness
In psychology and psychoanalysis, the human capacity for self-awareness as contrasted with its lack in other animals. According to FREUD it differs from UNCONSCIOUSNESS in that it applies and abides by rules, and in that it recognizes distinctions of space and time.

Conservation
Care of any delicate, fragile entity; used particularly of the natural environment, or biosphere. Conservation measures include: setting aside nature reserves, where little or no human activity is allowed; curbing POLLUTION that threatens living organisms or alters the environment; controlling activities that lead to soil erosion, or to irreversible changes in the global environment, such as the removal of tropical rainforest; preventing the hunting of endangered animals and the gathering of endangered plants. With the rapidly growing pressure on the environment that has characterized the late 20th century, the need for conservation has become more urgent and the techniques of conservation better developed, though the level of implementation of these techniques is still inadequate. In severely damaged environments conservationists now aim to preserve islands of virgin habitat linked by "corridors" of relatively undisturbed vegetation along which migration can take place, so that large animals have an adequate home range and smaller ones can recolonize areas where they have died out after a period of local adverse conditions. In many areas, such as marshes and grassland, conservation involves actively halting the natural process of SUCCESSION, which would otherwise replace the present vegetation with another type. This is justified on the basis that, as a result of human activity, new habitats of the original type are not being created elsewhere, as they would be under natural conditions. Thus the fauna of the area

being conserved – migratory birds overwintering on mudflats in an estuary, for example – has no alternative to the existing area.

Constellation
A group of stars forming a pattern in the sky, though otherwise unconnected. In ancient times the patterns were interpreted as pictures, usually of mythical characters. The ECLIPTIC passes through twelve constellations, known as the zodiacal constellations (see ZODIAC).

Constipation
A decrease in the frequency of bowel actions from the norm for an individual; also increased hardness of stool, in lay terminology. Often precipitated by inactivity or a change in diet, it is occasionally due to GASTROINTESTINAL TRACT disease.

Contact process
Major process for manufacturing SULFURIC ACID. SULFUR is burned in air to give sulfur dioxide, which is then oxidized to sulfur trioxide with a vanadium pentoxide or platinum catalyst. Arsenic, which poisons the catalyst, must first be removed. The sulfur trioxide is absorbed in concentrated sulfuric acid to yield oleum, which is diluted with water to the concentration required.

Continent
One of the seven major divisions of land on Earth: Africa, Antarctica, Asia, Australia, Europe, North America, South America. These continents have evolved over the last 200 million years from a single landmass, PANGEA. (See also CONTINENTAL DRIFT; CONTINENTAL SHELF; GONDWANA; LAURASIA; PLATE TECTONICS.) (➧ page 142-3)

Continental drift
Theory first rigorously formulated by WEGENER, later amplified by DU TOIT, to explain a number of geological and paleontological phenomena. It suggests that originally the land on Earth comprised a single, vast CONTINENT, PANGEA, which broke up. Continental drift is now recognized to be a consequence of the theory of PLATE TECTONICS. (➧ page 142-3)

Continental shelf
The portion of a landmass that is submerged in the OCEAN to a depth of less than 200m, resulting in a rim of shallow water surrounding the landmass. The outer edge of the shelf slopes towards the ocean bottom (see ABYSSAL PLAINS), and is called the continental slope. (➧ page 142)

Continuous creation see COSMOLOGY.

Contraception
The avoidance of conception, and thus of PREGNANCY. Many different methods exist, none of which is absolutely certain. In the rhythm method, sexual intercourse is restricted to the days immediately before and after MENSTRUATION, when fertilization is unlikely. The condom is a rubber sheath, fitting over the penis, into which ejaculation occurs; the diaphragm is a complementary device that is inserted into the vagina before intercourse. Both are more effective with spermicide creams. Intrauterine devices (IUDs) are plastic or copper devices that are inserted into the uterus and interfere with IMPLANTATION. They are convenient but may lead to infection or to increased blood loss and pain at menstruation. Oral contraceptives ("the Pill") are sex hormones of the ESTROGEN and PROGESTERONE type that, if taken regularly through the menstrual cycle, inhibit the release of eggs from the ovary or make the cervical mucus impenetrable to sperm. While they are the most reliable form of contraception, they are not free from adverse effects. When the Pill is stopped, periods and ovulation may not return for some time, and this can cause difficulty in assessing fetal maturity if pregnancy follows without an intervening period. While the more effective forms of contraception carry a slightly greater risk, this must be set against the risks of pregnancy. New forms of contraceptive include the sponge, which contains spermicide and is inserted into the vagina. Research is taking place into a male contraceptive, and a female contraceptive that is a hybrid between a diaphragm and a condom.

Contusion
Another name for a bruise.

Convection
Passage of heat through a fluid by means of large scale movements of material within the body of the fluid (see CONDUCTION). If, for example, a liquid is heated from below, parts close to the heat source expand and, because their DENSITY is thus reduced, rise through the liquid; near the top, they cool and begin to sink. This process continues until HEAT is uniformly distributed throughout the liquid. Convection in the ATMOSPHERE is responsible for many climatic effects (see METEOROLOGY). (See also RADIATION.)

Convergent evolution
The process by which unrelated animals or plants occupying a similar NICHE come to resemble one another. Convergence is due to the similar effects of NATURAL SELECTION on the organisms living under the same conditions. An example of convergence is the resemblance between dolphins and large fish.

Convulsions
Abnormal involuntary movements, also known as fits or seizures. They are usually rhythmic and associated with disturbance of consciousness and EPILEPSY.

Coombs' test
A test that uses antiserum to detect the presence of ANTIGENS on the surface of red blood cells.

Cooper, Leon N. (b. 1930)
US scientist awarded with J. BARDEEN and J.R. SCHRIEFFER the 1972 Nobel Prize for Physics for their jointly developed theory of superconductivity, usually called the BCS theory.

Coordinates see CELESTIAL SPHERE.

Copepods
A group of tiny, bullet-shaped crustaceans, some with extremely long antennae that are used in locomotion. They are very abundant in the marine plankton, and are also found in freshwater environments. Most copepods can either seize prey or live as filter feeders, depending on the sort of food available; some species are parasitic.

Copernicus, Nicolas (Niklas Koppernigk, 1473-1543)
Polish astronomer who displaced the Earth from the center of man's conceptual universe and made it orbit a stationary Sun. Belonging to a wealthy German family, he spent several years in Italy mastering all that was known of mathematics, medicine, theology and astronomy before returning to Poland where he eventually settled into life as lay canon at Frauenburg. His dissatisfaction with the Earth-centered (geocentric) cosmology of PTOLEMY was made known to a few friends in the manuscript *Commentariolus* (1514), but it was only on the insistence of Pope Clement VII that he expanded this into the *De revolutionibus orbium coelestium* (*On the revolutions of the heavenly spheres*) which, when published in 1543, announced the Sun-centered (heliocentric) theory to the world. Always the theoretician rather than a practical observer, Copernicus' main dissatisfaction with Ptolemy was philosophical. He sought to replace the equant, EPICYCLE and deferent of Ptolemaic theory with pure circular motions, but in adopting a moving Earth theory he was forced to reject the whole of the scholastic physics (without providing an alternative – this had to await the work of GALILEO) and postulate a much greater scale for the Universe. Although the heliocentric hypothesis was not immediately accepted by the majority of scientists, its proposal did begin the period of scientific reawakening known as the Copernican Revolution.

Copolymer
A synthetic polymer made from two or three different monomers, rather than a single monomer. Many synthetic plastics, fibers and rubbers are copolymers.

Copper (Cu)
Soft, red metal in Group IB of the PERIODIC TABLE; a TRANSITION ELEMENT. Copper has been used since *c*6500 BC. It occurs naturally as the metal and as the ores CUPRITE, CHALCOPYRITE, antlerite, CHALCOCITE, BORNITE, AZURITE and MALACHITE. The metal is produced by roasting the concentrated ores and smelting, and is then refined by electrolysis. Copper is strong, tough, and highly malleable and ductile. It is an excellent conductor of heat and electricity, and most copper produced is used in the electrical industry. It is also a major component of many ALLOYS, including brass, bronze, German silver, cupronickel and beryllium copper (very strong and fatigue resistant). Many copper alloys are called bronzes, though they need not necessarily contain tin: copper+tin+phosphorus is phosphor bronze, and copper+aluminum is aluminum bronze. Copper is a vital trace element: in humans it catalyzes the formation of HEMOGLOBIN; in mollusks and crustaceans it is the basic constituent of HEMOCYANIN. Chemically, copper is unreactive, dissolving only in oxidizing acids. It forms cuprous compounds (oxidation state +1), and the more common cupric salts (oxidation state +2), used as fungicides and insecticides, in pigments, as mordants for dyeing, as catalysts, for copper plating, and in electric cells. Copper (II) sulfate ($CuSO_4.5H_2O$) or blue vitriol is a blue crystalline solid occurring naturally as chalcanthite; it is used as above. AW 63.5, mp 1084°C, bp 2590°C, sg 8.96 (20°C). (➧ page 130)

Copperas (iron (II) sulfate) see IRON.

Copper pyrites see CHALCOPYRITE.

Cordillera
An extended fold mountain system, often composed of a number of parallel ranges. In some parts of the world they only appear as chains of islands.

Core
In the structure of the Earth, the innermost portion. It is divided into two parts. The inner core is a sphere about 1600km in diameter and is solid. The outer core is 1820km thick, bringing its outer surface to 2885km from the Earth's surface, and is thought to be liquid. The whole structure is estimated to have a mass of 1.9×10^{24}kg and an average density of about 11,000kg/m^3. The core probably consists of iron and nickel with a small proportion of lighter elements, possibly carbon, sulfur, silicon, oxygen or hydrogen. (➧ page 134)

Cori, Carl Ferdinand (1896-1984)
Czech-born US biochemist who shared the 1947 Nobel Prize for Medicine or Physiology with his wife, Gerty Theresa Radnitz Cori (1896-1957), for their joint elucidation of the processes by which GLYCOGEN is broken down and reformed in the body. (B.A. HOUSSAY also shared in the 1947 Prize for Physiology.)

Coriolis effect
A FORCE that, like centrifugal force (see CENTRIPETAL FORCE), apparently acts on moving objects when observed in a frame of reference that is itself rotating. Because of the rotation of the observer, a freely moving object does not appear to move steadily in a straight line as usual, but rather as if, besides an outward centrifugal force, a "Coriolis force" acts on it, perpendicular to its motion, with a strength proportional to its MASS, its VELOCITY,

and the rate of rotation of the frame of reference. The effect, first described in 1835 by Gaspard de Coriolis (1792-1843), accounts for the familiar circulation of air flow around CYCLONES, and numerous other phenomena in METEOROLOGY, COSMIC RADIATION, OCEANOGRAPHY and BALLISTICS.

Cork
Protective, waterproof layer of dead cells found as the outer layer of stems and roots in older woody plants. The cells have thick walls impregnated with suberin, a waxy material. The cork oak, *Quercus suber*, of S Europe and North Africa produces a profuse amount of cork, which is harvested commercially every 3 to 4 years. (See also CAMBIUM.) (page 53)

Corm
A short, stout, underground stem. It is an organ of VEGETATIVE REPRODUCTION consisting of a stem base swollen with food material and bearing buds in the remains of the leaves of the previous year's growth. The crocus and gladiolus are examples.

Cormack, Allan Macleod (b. 1924)
US physicist who received the 1979 Nobel Prize for Physiology or Medicine with HOUNSFIELD for his mathematical contribution to the development of computer-assisted TOMOGRAPHY.

Cornea
The transparent part of the outer EYE, partly responsible for much of the focusing power of the eye. Made of specialized cells and connective tissue, after damage or infection it may be replaced by a graft from a cadaver. (page 213)

Cornelian see CARNELIAN.

Cornforth, Sir John Warcup (b. 1917)
Australian-born British biochemist who shared the 1975 Nobel Prize for Chemistry with V. PRELOG for his work on the STEREOCHEMISTRY of enzyme-catalyzed reactions.

Corns and calluses
Localized thickenings of the outer layer of the skin, produced by continual pressure or friction. The soles of the feet form some spontaneous calluses; they form on the hands in response to wear, but cannot form on many parts of the body. Calluses project above the skin and are rarely troublesome; corns are smaller, and are forced into the deep, sensitive layers of skin, causing pain or discomfort.

Corolla see FLOWER.

Corona
The outer atmosphere of the SUN or other STAR. The term is also used for the halo seen around a celestial body, due to DIFFRACTION of its light by water droplets in thin CLOUDS of the Earth's ATMOSPHERE. Around high voltage terminals a faint glow appears due to the ionization (see ION) of the local air. The result of this ionization is an electrical discharge known as corona discharge, the glow being called a corona.

Coronary thrombosis
The primary cause of myocardial infarction or heart attack, the commonest causes of serious illness and death in Western countries. The coronary ARTERIES, which supply the HEART with oxygen and nutrients, may become diseased, which reduces blood flow. Significant narrowing may lead to superimposed THROMBOSIS which causes sudden complete obstruction, and results in death or damage to a substantial area of heart tissue. Severe persistent pain in the center of the CHEST is common, and it may lead to shock or lung congestion. Predisposing factors, including obesity, smoking, high blood pressure, excess blood FATS (including CHOLESTEROL) and DIABETES must be recognized and treated.

Corpuscle see BLOOD.

Corpus luteum
In humans, the gland that develops from the GRAAFIAN FOLLICLE after it has discharged the egg. If the egg is fertilized, the corpus luteum secretes estrogen and progesterone to maintain the pregnancy. In the absence of conception it degenerates, and MENSTRUATION takes place. (page 111, 150)

Corrosion
The insidious destruction of metals and alloys by chemical reaction (mainly OXIDATION) with the environment. In moist air most metals form a surface layer of oxide, which, if it is coherent, may slow down or prevent further corrosion. Tarnishing is the formation of such a discolored layer, mainly on copper or silver. (Rust – hydrated iron (III) oxide, FeO(OH) – offers little protection, so iron corrodes rapidly.) Industrial AIR POLLUTION greatly speeds up corrosion, oxidizing and acidic gases (especially sulfur dioxide) being the worst culprits. Corrosion is usually an electrochemical process (see ELECTROCHEMISTRY): small cells are set up in the corroding metal, the potential difference being due to the different metals present or to different concentrations of oxygen or electrolyte; corrosion takes place at the anode. It also occurs preferentially at grain boundaries and where the metal is stressed. Prevention methods include bonderizing; a protective layer of paint, varnish or electroplate; or the use of a "sacrificial anode" composed of zinc or aluminum that is in electrical contact with the metal, and is preferentially corroded. Galvanizing – coating iron objects with zinc – works on the same principle.

Cortex
In plant stems and roots, the layer of mostly unspecialized packing cells between the EPIDERMIS and the PHLOEM. It is used to store food and other substances including resins, oils and tannins; in stems it may contain CHLOROPLASTS; in roots it transports water and ions inwards. The term is also used for the outer layer of the BRAIN, KIDNEYS or ADRENAL GLANDS.

Cortisol (hydrocortisone)
The principal naturally occurring STEROID HORMONE, secreted by the ADRENAL GLAND.

Cortisone
A naturally occurring STEROID, not secreted by the ADRENAL GLAND and not normally found in the BLOOD. Synthetically produced cortisone is used to treat a number of conditions including ADDISON'S DISEASE.

Corundum (α-alumina)
Rhombohedral form of aluminum oxide (see ALUMINUM), occurring worldwide (see also EMERY); transparent varieties include RUBY and SAPPHIRE. Artificial corundum, or β-alumina, is made by calcining BAUXITE. Corundum is the hardest natural substance after diamond (Mohs hardness 9), and is used as an abrasive and for bearings.

Coryza
Another name for the COMMON COLD.

Cosmic egg see COSMOLOGY; LEMAITRE.

Cosmic rays
Charged particles, mainly the nuclei of HYDROGEN and other ATOMS, that isotropically bombard the Earth's upper ATMOSPHERE at VELOCITIES close to that of light. These primary cosmic rays interact with molecules of the upper atmosphere to produce what are termed secondary cosmic rays, which are considerably less energetic: they are SUBATOMIC PARTICLES that change rapidly into other types of particles. Initially, secondary cosmic rays, which pass frequently and harmlessly through our bodies, were detected by use of the GEIGER COUNTER and the cloud chamber, though now many other kinds of detector are used in studies of cosmic rays. It is thought that cosmic rays are produced by SUPERNOVAS, though some may be of extragalactic origin. (page 49)

Cosmogony
The science or pseudoscience of the origins of the UNIVERSE.

Cosmology
The study of the structure and evolution of the Universe. Ancient and medieval cosmologies were many, varied and imaginative, usually oriented around a stationary, flat Earth at the center of the Universe, surrounded by crystal spheres carrying the Moon, Sun, planets and stars, though ARISTARCHUS understood that the Earth was spherical and circled the Sun. With increasing sophistication of observational techniques and equipment, more realistic views of the Universe emerged (see BRAHE; COPERNICUS; GALILEO; KEPLER; NEWTON). Modern cosmological theories take into account EINSTEIN's Theory of RELATIVITY and the recession of galaxies shown by the RED SHIFT in their spectra (see SPECTRUM).

The most important of the evolutionary theories is the Big Bang theory resulting from HUBBLE's observations of the galaxies. This theory proposes that the Universe began by exploding from a hot, superdense state some 10 to 20 billion years ago. As the expanding mixture of matter and radiation cooled and diluted, it eventually condensed to form galaxies and stars. An alternative model, the Steady State theory of BONDI, HOYLE and GOLD, put forward in 1948, proposed that the Universe had existed and will exist forever in its current form. This required the continuous creation of new matter so that the average density and appearance of the Universe remain the same at all times.

In 1965 PENZIAS and WILSON discovered that the Universe possesses an inherent radio "background noise", and it was suggested by R. Dicke that this was the relic of the RADIATION produced by the Big Bang. Further researches have indicated the probability that space is filled with uniform BLACKBODY thermal radiation corresponding to a temperature of around 3K. This supports an evolutionary theory, and the Steady State theory was largely abandoned.

The Big Bang theory also accounts neatly for the observed amounts of helium, deuterium and lithium in the Universe by proposing that these elements were formed by fusion reactions everywhere in the expanding fireball of matter and radiation during the first few minutes of the history of the Universe. However, the Big Bang theory leaves several unanswered questions, the most important of which is why the Universe should be so uniform – eg matter is distributed evenly through the Universe, and background radiation has the same strength from all directions in space. A possible solution to this and other questions has been a theoretical development called inflation. The inflationary Universe describes a phase of exponential expansion inflating a tiny seed of the Universe by a factor of 10^{50} or more during the first instant of the Big Bang. Just as a surface of a balloon expanded by a similar amount would look like the surface of a flat plane, so this rapid expansion of the Universe would smooth out the inhomogeneities. The inflationary Universe is an active area of research, and new developments are likely.

Considerable contemporary debate centers on whether the Universe is "open" or "closed". If the Universe is open it will expand forever; if it is closed it will eventually cease to expand and thereafter will collapse until it reaches a superdense state again. The oscillating Universe theory suggests that this collapse may trigger a new Big Bang, and that the Universe expands and contracts in a cyclical fashion. Whether or not the Universe continues to expand forever depends on its mean density. If the mean density of the

Universe exceeds a value known as the critical density (equivalent to about three hydrogen atoms per cubic meter of space), then gravity will eventually halt the expansion and the Universe will be closed. However, if the mean density is less than this value, the Universe is open. The inflationary hypothesis requires that the mean density be almost precisely equal to the critical density. At present it is not certain whether the Universe is open or closed. (See also ASTRONOMY; BLACK HOLE; PULSAR; QUASAR.)

Cot death
An unexplainable death in its sleep of a seemingly well baby.

Cottrell, Frederick Gardner (1877-1948)
US chemist who invented the electrostatic precipitator used to reduce AIR POLLUTION from factory flues (1910), and founded the nonprofit Research Corporation to plow back the profits of invention into further scientific research (1912).

Cotyledon
A seed leaf, that is, a leaf forming part of the embryo in seeds. In MONOCOTYLEDONS there is only one cotyledon and in DICOTYLEDONS there are two. They contain food reserves used in the early stages of GERMINATION. In some plants they stay below ground (hypogeal germination), but in others they are brought above the ground, turn green and carry out PHOTOSYNTHESIS (epigeal germination) for a time. They bear no resemblance to normal foliage leaves. (See also SEED.)

Cough
Sudden explosive release of air from the LUNGS, which clears respiratory passages of obstruction and excess MUCUS or PUS. Persistent cough implies disease.

Coulomb (C)
The SI UNIT of electric charge, defined as the quantity of ELECTRICITY transferred by 1 ampere in one second. Named after C.A. DE COULOMB.

Coulomb, Charles Augustin de (1736-1806)
French physicist noted for his researches into FRICTION, TORSION, ELECTRICITY and MAGNETISM. Using a torsion BALANCE, he established Coulomb's Law of electrostatic forces (1785). This states that the force between two charges is proportional to their magnitudes and inversely proportional to the square of their separation. He also showed that the charge on a charged conductor lies solely on its surface.

Counterglow see ZODIACAL LIGHT.

Cournand, André Frédéric (b. 1895)
French-born US physiologist who shared the 1956 Nobel Prize for Physiology or Medicine with D.W. RICHARDS and W. FORSSMANN for the development of heart catheterization (see CATHETER) and other work which has led to a much fuller understanding of heart and lung diseases.

Courtship
In animals, a sequence of ritualized behavior patterns that lead to mating. Their purpose is to prepare each of the partners for mating, so that their hormonal states are synchronized, and, in territorial species, to overcome the mutual antagonism that individuals feel for one another so that the close physical contact of mating can occur. Courtship rituals also allow an animal to ensure that its mate is of the correct species, and to assess its mate for health and general suitability. For example, the courtship of many birds involves the male bringing food to the female – this gives her the chance to assess the male's abilities as a provider for their young. In species where the parents stay together after mating and cooperate in rearing the young, courtship rituals do not necessarily end with mating; elements of the courtship ritual may be regularly repeated as a means of reinforcing the pair bond. Courtship rituals are one of the ways in which SEXUAL SELECTION operates.

Covalent bond see BOND, CHEMICAL.

Cowpox
A disease of cattle, caused by a VIRUS related to the smallpox virus. The IMMUNITY to SMALLPOX conferred on humans by a previous infection with cowpox (contracted by milking infected cattle) led Benjamin Jesty and later Edward Jenner to experiment with VACCINATION against smallpox.

CPU (central processing unit) see COMPUTER.

Crab nebula
M1, a bright NEBULA in the constellation TAURUS, the remnants of the SUPERNOVA of 1054. It is about 2000pc from the Earth and is associated with a PULSAR.

Cracking
Process by which heavy HYDROCARBON molecules in PETROLEUM are broken down into lighter molecules and isomerized by means of high temperatures or catalysis. It yields branched chain ALKANES and ALKENES of high OCTANE rating for GASOLINE, and also simple gaseous alkenes for chemical synthesis.

Cram, Donald (b. 1919)
US scientist awarded the 1987 Nobel Prize for Chemistry together with CHARLES PEDERSEN and JEAN-MARIE LEHN for their contributions to the study of molecular recognition ("host-guest" chemistry), which have also opened up new fields of research in medicine, biology and the science of materials.

Cramp
The painful contraction of muscle – often in the legs. The cause is usually unknown. It may be brought on by exercise or lack of salt; it also occurs in muscles with inadequate blood supply.

Crepitus
The creaking sound of infected lungs heard through the stethoscope, or of damaged bones moving against each other.

Crepuscular
Active at dusk and dawn.

Cresol (methylphenol, $CH_3C_6H_4OH$)
A member of the PHENOLS; there are three isomers, synthesized from TOLUENE. Distillation of creosote or PETROLEUM yields a mixture of cresols and xylenols ("cresylic acid") used as a disinfectant and in the manufacture of resins and tricresyl phosphate (used to make gasoline additives and plasticizers).

Cretaceous
Final period of the MESOZOIC era, about 144 to 65 million years ago, lying after the JURASSIC and before the TERTIARY (see GEOLOGY). (♦ page 106)

Cretinism
Congenital disease caused by lack of THYROID HORMONE in late fetal life and early infancy, which interferes with normal development, including that of the brain. It may be due to congenital inability to secrete the hormone or, in certain areas of the world, to lack of dietary iodine (which is needed for hormone formation). The typical appearance, with coarse skin, puffy face, large tongue and slow responses, usually enables early diagnosis.

Crick, Francis Harry Compton (b. 1916)
English biochemist who, with J.D. WATSON, proposed the double-helix model of DNA. For this, one of the most spectacular advances in 20th-century science, they shared the 1962 Nobel Prize for Physiology or Medicine with M.H.F. WILKINS, who had provided them with the X-ray data on which they had based their proposal. Crick's subsequent work has been concerned with deciphering the functions of the individual CODONS in the genetic code.

Crinoids
A class of ECHINODERMS that includes the sea lilies and feather stars. They have cup-shaped bodies and long tentacles. Sea lilies are sedentary and are attached to the sea bed by a stalk. Feather stars are mobile and have no stalk. The crinoids included many extinct forms.

Critical mass
The MASS of a radioactive material needed to achieve a self-sustaining FISSION process. This requires that the NEUTRONS emitted by a given nuclear fission are more likely to encounter further fissionable nuclei than to escape. It is of the order of several kilograms for URANIUM-235.

Crohn's disease
Inflammation, thickening and ulceration of the ileum (see GASTROINTESTINAL TRACT) named after US physician B.B. Crohn (b. 1894).

Cronin, James (b. 1931)
US physicist who shared the 1980 Nobel Prize for Physics with FITCH for their discovery of the CP violation which showed that nature is not as symmetrical as previous theories had shown, and led to a greater understanding of SUBATOMIC PARTICLES.

Cronstedt, Axel Fredrik, Baron (1722-1765)
Swedish chemist who discovered the element NICKEL, introduced the use of the blowpipe into chemical ANALYSIS and published one of the first chemical classifications of minerals.

Crookes, Sir William (1832-1919)
English physicist who discovered the element THALLIUM (1861), invented the RADIOMETER (1875) and pioneered the study of CATHODE RAYS. By 1876 he had devised the Crookes tube, a glass tube containing two ELECTRODES and pumped out to a very low gas PRESSURE. By applying a high voltage across the electrodes and varying the pressure, he was able to produce and study cathode rays and various glow discharges.

Crop
In some birds, a pouch-like enlargement of the esophagus in which food is stored. Some invertebrates also have a food-storage cavity known as a crop.

Cross-eye see STRABISMUS.

Crossing over
Exchange of sections of their length between the members of a homologous pair of CHROMOSOMES. This results from the breakage and rejoining of the chromosomes as they lie alongside each other during MEIOSIS. It can alter the LINKAGE groups and is an important source of genetic VARIATION. (See also RECOMBINATION.) (♦ page 32)

Crossopterygii
A subclass of the SARCOPTERYGII, or lobe-finned fish, mainly containing fossil fish; these are thought to include the ancestors of the AMPHIBIANS. The group was presumed to be extinct until a member, the COELACANTH, was discovered living off the east coast of Africa.

Cross-section, Nuclear
A measure of the strength of the interaction between two atomic or SUBATOMIC PARTICLES. A beam of one kind of particle is directed at a particle of the other kind, and the number of particles scattered or absorbed is measured as an equivalent area (often in BARNS) presented to the beam by the target particle within which all incident particles are affected.

Croup
A condition common in infancy due to VIRUS infection of LARYNX and TRACHEA and causing characteristic stridor, or spasm of larynx when the child breathes in. Often a mild and short illness, it occasionally causes so much difficulty in breathing that OXYGEN is needed.

Crucible
A container used for chemical reactions at high temperatures, for CALCINATION, or for melting metals. Crucibles can be made of earthenware,

porcelain, quartz or graphite; or of metals such as iron, platinum, nickel or silver. Metal crucibles may be damaged by alloy formation if metals are melted in them. Industrial crucibles are generally large and lined with refractory materials.

Crucifers
Members of the Family Cruciferae, which includes such plants as the wallflower, shepherd's purse and cuckoo flower, as well as many crops – cabbage and other BRASSICAS, watercress and radish.

Crust
In the structure of the Earth, the outermost solid layer. There are two types of crust. Oceanic crust consists largely of silica and magnesia (geophysicists call this material SIMA). It is constantly being formed at ocean ridges and destroyed in ocean trenches. The lighter continental crust is embedded in the oceanic crust and stands proud of it, forming the continents. It is made up predominantly of silica and alumina (the SIAL of geophysicists) and, unlike oceanic crust, is not continually created and destroyed. Crust and the topmost layer of mantle form the moving plates of PLATE TECTONICS. (♦ page 134, 142-3)

Crustaceans
A group of invertebrates with jointed legs and a hard EXOSKELETON made up of CHITIN and proteins, impregnated with calcium carbonate (chalk). The crustaceans include crabs, shrimps, woodlice or sowbugs, barnacles and copopeds. Most crustaceans are aquatic, breathing by gills, and have two pairs of antennae. (See also ARTHROPODS.) (♦ page 16)

Cryogenics
The branch of physics dealing with the behavior of matter at very low temperatures, and with the production of those temperatures. Early cryogenics relied heavily on the Joule Thomson effect (named for JAMES JOULE and William Thomson, later LORD KELVIN) by which temperature falls when a gas is permitted to expand without an external energy source. Using this, JAMES DEWAR liquefied HYDROGEN in 1895 (though not in quantity until 1898), and H. KAMERLINGH-ONNES liquefied HELIUM in 1908 at 4.2K (see also ABSOLUTE ZERO). Several cooling processes are used today. Down to about 4K the substance is placed in contact with liquefied gases which are permitted to evaporate, so removing HEAT energy (see LATENT HEAT). The lowest temperature that can be reached thus is around 0.3K. Further temperature decrease may be obtained by paramagnetic cooling (adiabatic demagnetization). Here a paramagnetic material (see PARAMAGNETISM) is placed in contact with the substance and with liquid helium, and subjected to a strong magnetic field (see MAGNETISM), the heat so generated being removed by the helium. Then, away from the helium, the magnetic field is reduced to zero. By this means, temperatures of the order of 10^{-2}-10^{-3}K have been achieved (though, because of heat leak, such temperatures are always unstable). A more complex process, nuclear adiabatic demagnetization, has been used to attain temperatures as low as 2×10^{-7}K. Near absolute zero, substances can display strange properties. Liquid helium II has no VISCOSITY (see SUPERFLUIDITY) and can flow up the sides of its container. Some elements display SUPERCONDUCTIVITY: an electric current started in them will continue indefinitely.

Cryolite (Na_3AlF_6)
A mineral found in Greenland; white monoclinic crystals. Molten cryolite is used to dissolve bauxite in the electrolytic production of aluminum.

Cryptic coloration
A coloration that aids the concealment of animals. In many the coloration matches the animal's natural background, but in others it serves to break up the outline of the animal, making it less noticeable. (See also CAMOUFLAGE.)

Cryptogam
A name given by early botanists to ALGAE, mosses, LIVERWORTS and FERNS, which do not have prominent organs of reproduction, compared to the cones of the GYMNOSPERMS or the flowers of the ANGIOSPERMS. The term is no longer used.

Cryptosporidiosis
A food-borne diarrheal disease of mammals and humans, caused by a protozoan called *Cryptosporidium*. This multiplies in the cells of the intestine; it has an incubation period of 6 or 9 days, representing 2 or 3 complete life-cycles. It is most commonly carried in sausagemeat or untreated milk. It is usually mild and self-limiting, but can be severe and even fatal in people with immune-deficiency conditions such as AIDS.

Cryptozoic
In GEOLOGY, an obsolete term for the PRECAMBRIAN.

Crystals
Homogenous solid objects with naturally formed plane faces. The order in their external appearance reflects the regularity of their internal structure, this internal regularity being the keynote of the crystalline state. Although external regularity is most obvious in natural crystals and those grown in the laboratory, most other inorganic solid substances (with the notable exceptions of PLASTICS and GLASS) also exist in the crystalline state, though the crystals of which they are composed are often microscopic in size. True crystals must be distinguished from cut gemstones which, although often internally crystalline, exhibit faces chosen according to the whim of the lapidary rather than developed in the course of any natural growth process. The study of crystals and the crystalline state is the province of crystallography. Crystals are classified according to the symmetry elements that they display. An alternative method of reaching the same classification uses the crystallographic axes used in describing the crystal's faces. This gives the 32 crystal classes, which can be grouped into the seven traditional crystal systems (six if trigonal and hexagonal are counted together). Crystals are allotted to their proper class by considering their external appearance, the symmetry of any etch marks made on their surfaces, and their optical and electrical properties (see DOUBLE REFRACTIONS; PIEZOELECTRICITY). Although such observations do enable the crystallographer to determine the type of "unit cell" that, when repeated in space, gives the overall LATTICE structure of a given crystal, they do not allow him to determine the actual dispositions of the constituent atoms or ions. Such patterns can only be determined using X-RAY DIFFRACTION techniques. When a crystal is composed solely of particles of a single species and the attractive forces between molecules are not directionally localized, as in the crystals of many pure metals, the atoms tend to take up one of two structures both of which allow a maximum degree of close-packing. These are known as hexagonal close-packed (hcp) and face-centered cubic (fcc), and can be looked upon as different ways of stacking planes of particles in which each is surrounded by six neighbors. Where there is a degree of directionality in the bonding between particles (as in diamond) or more than one particle species is involved (as in common salt), the crystals exhibit more complex structures, single substances often adopting more than one structure under different conditions (POLYMORPHISM). Although much crystallography assumes that crystals perfectly exhibit their supposed structure, real crystals, of course, contain minor defects such as grain boundaries and DISLOCATIONS. Many of the most important properties and uses of crystals depend on these defects (see SEMICONDUCTOR).

Ctenophores (comb jellies)
Exclusively marine, gelatinous animals, characterized by the presence of eight rows of comb plates that bear CILIA used for swimming. Ctenophores were once classified with the CNIDARIANS as COELENTERATES, but are now considered to represent a separate phylum. The simplest type of comb jelly is almost spherical, and is often called a sea gooseberry. (♦ page 16)

Cuesta (escarpment)
A ridge or hill with one side gently sloping, the other a steep cliff (scarp), formed by selective EROSION in areas where the rock strata are tilted.

Culture
A controlled growth of living cells in an artificial medium. The cells may be microorganisms isolated and studied in a pure culture, or they may be cells from animal or plant tissue. The culture medium usually contains water, GELATIN or AGAR, salts and various nutrients.

Cumulus see CLOUDS.

Cuprite
Red mineral consisting of copper(I) oxide (Cu_2O); a major ore of COPPER, found in Europe, Australia, Arizona, South America and SW Africa. It is formed by oxidation of copper sulfide ores.

Curare
Arrow poison used by South American Indian hunters, extracted from various plants, chiefly of the genera *Strychnos* and *Chondodendron*, killing by respiratory paralysis. Curare is a mixture of ALKALOIDS, the chief being *d*-tubocurarine. By competing with ACETYLCHOLINE it blocks nerve impulse transmission to muscles, producing relaxation and paralysis. It has revolutionized modern surgery by producing complete relaxation without a dangerous degree of ANESTHESIA being required. (See also TETANUS.)

Curie (Ci)
A unit of RADIOACTIVITY now defined as the quantity of a radioactive material in which 3.7×10^{10} disintegrations per second are occurring.

Curie, Marie (1867-1934)
Born Marja Sklodowska, Polish-born French physicist who, with her French-born husband Pierre Curie (1859-1906), was an early investigator of RADIOACTIVITY, discovering the radioactive elements POLONIUM and RADIUM in the mineral PITCHBLENDE (1898). For this the Curies shared the 1903 Nobel Prize for Physics with A.H. BECQUEREL. After the death of Pierre, Marie went on to investigate the chemistry and medical applications of radium and was awarded the 1911 Nobel Prize for Chemistry in recognition of her isolation of the pure metal. She died of LEUKEMIA, no doubt contracted in the course of her work with radioactive materials. Pierre Curie is also noted for the discovery, with his brother Jacques, of PIEZOELECTRICITY (1880) and for his investigation of the effect of TEMPERATURE on magnetic properties. In particular he discovered the Curie point, the temperature above which ferromagnetic materials display only PARAMAGNETISM (1895). The Curies' elder daughter, IRENE JOLIOT-CURIE, was also a noted physicist.

Curium (Cm)
A TRANSURANIUM ELEMENT in the ACTINIDE series. It is prepared by bombarding plutonium-239 with ALPHA PARTICLES or americium-241 with neutrons. Cm^{244} is used as a compact power source for space uses, the heat of nuclear decay being converted to electricity. AW 247, mp 1340°C, sg 13.51. (♦ page 130)

Current, Electric see ELECTRICITY.
Current, Ocean see OCEANS.
Cusa, Nicholas of see NICHOLAS OF CUSA.
Cushing, Harvey Williams (1869-1939)
US surgeon who pioneered many modern neurosurgical techniques and investigated the functions of the PITUITARY GLAND. In 1932 he described Cushing's Syndrome, a rare disease caused by an excess of adrenal cortex hormones, and showing itself in obesity, high blood pressure and other symptoms.
Cuticle
Any non-cellular layer forming the outer covering of an organism's body; it is secreted by the EPIDERMIS. Plants have a waxy cuticle that prevents water loss. It is broken only by the STOMATA, which allow gas exchange to occur. Most invertebrates have a cuticle, often made of CHITIN or protein, or both. In some, this cuticle is rigid and acts as an EXOSKELETON. (◆ page 53)
Cuvier, Georges Léopold Chrétien Frédéric Dagobert, Baron (1769-1832)
French comparative anatomist and the founder of PALEONTOLOGY. By applying his theory of the "correlation of parts" he was able to reconstruct the forms of many fossil creatures, explaining their creation and subsequent extinction according to the doctrine of CATASTROPHISM. A tireless laborer in the service of French Protestant education, Cuvier was perhaps the most renowned and respected French scientist in the early 19th century.
Cyanamide process
Rarely used process for NITROGEN FIXATION. Calcium carbide (CaC_2) is made by heating CALCIUM carbonate with coke. Nitrogen is passed through the finely divided calcium carbide at 1000°C, giving calcium cyanamide ($CaCN_2$), which is then decomposed by steam to give calcium carbonate, which can be recycled, and AMMONIA.
Cyanides
Compounds containing the CN group. Organic cyanides are called NITRILES. Inorganic cyanides are salts of hydrocyanic acid (HCN), a volatile weak ACID; both are highly toxic. Sodium cyanide is made by the Castner process: ammonia is passed through a mixture of carbon and fused sodium. The cyanide ion (CN^-) is a PSEUDOHALOGEN, and forms many complexes. Cyanides are used in the extraction of GOLD and SILVER, ELECTROPLATING and casehardening.
Cyanobacteria
A group of unicellular or colonial organisms that belong to the PROKARYOTES. They are found in fresh and salt water, and in symbiotic relationships, as in LICHENS, for example. All are photosynthetic, and they were once mistakenly classified with the algae, and known as "blue-green algae" (some have phycobilin pigments that give them a bluish tinge). The cyanobacteria have quite complex cells for prokaryotes, with complex infoldings of the cell membrane on which the enzymes involved in photosynthesis are located. They are capable of nitrogen fixation; in filamentous forms this occurs in specialized cells known as heterocysts, which are formed only when nitrogenous compounds are in short supply.

Cyanobacteria have played an important part in the evolution of life on Earth as they were responsible for generating the oxygen in the early atmosphere, oxygen being a byproduct of photosynthesis. The appearance of oxygen resulted in the extinction of many ancient bacteria (the ANAEROBES) but allowed much larger and more active life-forms to evolve. Cyanobacteria may also have given rise to the algae and higher plants by developing a way of life inside other, larger cells (see ENDOSYMBIOSIS THEORY). (◆ page 13)
Cyanosis
The bluish tinge of skin and mucous membranes seen when there is too little OXYGEN bound to HEMOGLOBIN in the BLOOD. If generalized it may be due to inadequate oxygen reaching the blood or failure of the BLOOD CIRCULATION. If it is only at the extremities, it indicates slow blood flow due to VASOCONSTRICTION.
Cybernetics
A branch of information theory that compares the communication and control systems built into mechanical and other manmade devices with those present in biological organisms. For example, fruitful comparisons may be made between data processing in computers and various brain functions; the fundamental theories of cybernetics may be applied with equal validity to both.
Cycads
A group of seed-bearing plants (SPOROPHYTES) classified as GYMNOSPERMS. They look rather like palm trees, with short, stubby trunks, but are in fact unrelated to them. Their seeds are borne in massive cones. One primitive feature that the cycads have retained is motile sperm; these sperm swim to the egg cells in a layer of moisture within the cone. (◆ page 139)
Cyclamates
The sodium or calcium salts of N-cyclohexylsulfamic acid ($C_6H_{11}NHSO_3H$), white crystalline solids widely used as potent SWEETENING AGENTS until they were found to cause CANCER in rats. They were banned in the US in 1969.
Cyclohexane (C_6H_{12})
An alicyclic compound made up of a ring of six carbon atoms. It is the saturated analog of benzene, from which it can be obtained by hydrogenation. It is is used widely as an industrial organic solvent. MW 84.16, mp 6.55°C, bp 80.74°C.
Cyclone
An atmospheric disturbance in regions of low pressure (see ATMOSPHERE; METEOROLOGY) characterized by a roughly circular ground plan, a center towards which ground WINDS move, and at which there is an upward air movement, usually spiraling. Above the center, in the upper TROPOSPHERE, there is a general outward movement. The direction of spiraling is counterclockwise in the N Hemisphere, clockwise in the S Hemisphere, owing to the CORIOLIS EFFECT. Anticyclones, which occur in regions of high pressure, are characterized by an opposite direction of spiral. (See also HURRICANE; TORNADO.)
Cyclonite see RDX.
Cyclotron
A type of charged particle ACCELERATOR in which the particles travel a spiral path in a strong magnetic field, thus repeatedly traversing the same electric field regions and achieving energies much greater than those attainable with a linear accelerator.
Cygnus
The Swan, a large, approximately cruciform constellation of the N Hemisphere containing Deneb (absolute magnitude −7) and the fine binary Albireo.
Cygnus X-1
An X-ray emitting binary system (see DOUBLE STAR) that consists of a blue supergiant STAR, HDE 226868, and an invisible companion (itself estimated to have a mass 9-11 times that of the SUN), which is widely believed to be a BLACK HOLE. The X-ray emission is believed to come from a disk of high-temperature gas, drawn from the visible star and circulating around the black hole.
Cyst
A fluid- or fat-filled sac, lined by fibrous connective tissue or surface EPITHELIUM. It may form in an enlarged normal cavity (eg sebaceous cyst), it may arise in an embryonic remnant (eg branchial cyst), or it may occur as part of a disease process. Some benign or malignant TUMORS take the form of cysts. They may present as swellings or may cause pain (eg some ovarian cysts). Multiple cysts in KIDNEY and LIVER occur in indented polycystic diseases; here kidney failure may develop.
Cystic fibrosis
An inherited disease with onset in infancy or childhood, in which glandular secretions are abnormally viscous. Blockage of ducts is common. Chronic LUNG disease, with thick sputum and liability to infection, is typical, as is malabsorption, with pale bulky feces and MALNUTRITION. Sweat contains excessive salt (the basis for a diagnostic test) and heat exhaustion may result. Although long-term outlook in this disease has recently improved, there is still a substantial mortality before adult life.
Cystitis
INFLAMMATION of the BLADDER, usually due to infection. A common condition in women, sometimes precipitated by intercourse. It occasionally leads to kidney disease or upper urinary tract infection. Burning pain and increased frequency of urination are usual symptoms. Recurrent cystitis may suggest an underlying disorder of the bladder.
Cytochromes
A group of proteins, found in all AEROBIC ORGANISMS, that transfer energy within cells. They contain a CHELATE complex of an iron ion within a PORPHYRIN ring, as in HEMOGLOBIN. Cytochromes undergo a series of OXIDATION AND REDUCTION reactions, transferring electrons from a substrate (NADH or $FADH_2$) to each other and finally to oxygen (an electron-acceptor), in the ELECTRON TRANSPORT CHAIN. In cellular RESPIRATION the substrates are the hydrogen-acceptors in the CITRIC ACID CYCLE, and the cytochromes are therefore harvesting some of the energy released by that cycle. The energy released by transfer from one cytochrome to another is stored as ATP.
Cytokinins
Plant hormones that stimulate cell enlargement, bud formation and seed germination. They also slow down the aging process, for example in the leaves of deciduous trees.
Cytology
The branch of BIOLOGY dealing with the study of CELLS, their structure, function, biochemistry, etc. (See also HISTOLOGY.)
Cytoplasm
The contents of the cell, outside the nucleus, in EUKARYOTIC cells. Strictly speaking it means the entire extranuclear material, including organelles such as the mitochondria. However, it is often used to mean only the protein gel in which those organelles are embedded; the correct name for this is the CYTOSOL. (◆◆ page 13, 32, 204)
Cytosine see NUCLEIC ACIDS; NUCLEOTIDES.
Cytoskeleton
The protein fibers and MICROTUBULES that are found within the cytosol of eukaryotic cells. They give structure and shape to the cell, control the movement of important molecules around it, and are involved in cell movement. (◆ page 13)
Cytosol
That part of the CYTOPLASM other than the cell organelles and membranous structures. It consists of a protein gel containing the CYTOSKELETON, and many enzymes that are involved in cell metabolism. The cytosol has a definite structure, which differs from one region of the cell to another.
Cytotoxic
Harmful to living cells. Cytotoxic drugs prevent cell division and are used in cancer treatment.

Daguerre, Louis Jacques Mandé (1789-1851)
French theatrical designer and former partner of NIEPCE who in the late 1830s developed the DAGUERREOTYPE process, the first practical means of producing a permanent photographic image.

Daguerreotype
The first practical photographic process, invented by DAGUERRE in 1837 and widely used in portraiture until the mid-1850s. A brass plate coated with silver was sensitized by exposure to iodine vapor and exposed to light in a camera for several minutes. A weak positive image produced by mercury vapor was fixed with a solution of salt. Hypo soon replaced salt as the fixing agent and after 1840 gold (III) chloride was used to intensify the image. (See PHOTOGRAPHY.)

Dale, Sir Henry Hallet (1875-1968)
British biologist who discovered ACETYLCHOLINE and described its properties together with those of HISTAMINE. In 1936 he shared the Nobel prize for Physiology or Medicine with OTTO LOEWI, having identified the chemical that Loewi had found to be secreted by certain nerve endings as acetylcholine and demonstrated its role in transmitting nerve impulses.

D'Alembert, Jean Le Rond see ALEMBERT, JEAN LE ROND D'.

D'Alembert's principle
The observation that NEWTON's third law of motion (that to every action there is an equal and opposite reaction) applies not only to systems in static EQUILIBRIUM but also to those in which at least one component is free to move. Thus when a FORCE F is applied to an unconstrained body of MASS m endowing it with ACCELERATION a, there is an equal inertial reaction – ma.

Dalén, Nils Gustaf (1869-1937)
Swedish engineer who was awarded the 1912 Nobel Prize for Physics in recognition of his invention of the automatic "sun valve". This device, which rapidly went into worldwide use, allowed the gas lights in unmanned buoys and lightships to be controlled by the amount of natural light.

Dalton (unit)
The unit of mass equal to 1/12th of the mass of the carbon isotope ^{12}O; also called the atomic mass unit (amu).

Dalton, John (1766-1844)
English Quaker scientist renowned as the originator of the modern chemical atomic theory. First attracted to the problems of GAS chemistry through an interest in meteorology, Dalton discovered his Law of Partial Pressures in 1801. This states that the PRESSURE exerted by a mixture of gases equals the sum of the partial pressures of the components and holds only for ideal gases. (The partial pressure of a gas is the pressure it would exert if it alone filled the volume.) Dalton believed that the particles or ATOMS of different ELEMENTS were distinguished from one another by their weights, and, taking his cue from the laws of definite and multiple proportions (see COMPOSITION, CHEMICAL), he compiled and published in 1803 the first table of comparative ATOMIC WEIGHTS. This inaugurated the new quantitative atomic theory. Dalton also gave the first scientific description of COLOR BLINDNESS. The red-green type from which he suffered is still known as Daltonism.

Dam, Carl Peter Henrik (1895-1976)
Danish biochemist who discovered and investigated the physiological properties of VITAMIN K. For this he shared the 1943 Nobel Prize for Physiology or Medicine with E.A. DOISY, who isolated the vitamin and determined its structure.

Danidell cell see BATTERY.

Darwin, Charles Robert (1809-1882)
English naturalist who first formulated the theory of EVOLUTION by NATURAL SELECTION. Between 1831 and 1836 the young Darwin sailed round the world as the naturalist on board HMS *Beagle*. In the course of this he made many geological observations favorable to LYELL's uniformitarian geology, devised a theory to account for the structure of coral islands and was impressed by the facts of the geographical distribution of plants and animals. He became convinced that species were not fixed categories, as was commonly supposed, but were capable of variation, though it was not until he read MALTHUS' *Essay on the Principle of Population* that he discovered a mechanism whereby ecologically favored varieties might form the basis for new distinct species. Darwin published nothing for 20 years until, on learning of A.R. WALLACE's independent discovery of the same theory, he collaborated with the younger man in a short Linnean-Society paper. The next year (1859) the theory was set before a wider public in *The Origin of Species*. The rest of his life was spent in further research in defense of his theory, though he always avoided entering the popular controversies surrounding his work and left it to others to debate the supposed consequences of "Darwinism".

Darwin, Erasmus (1731-1802)
English physician, philosopher, scientist and poet who proposed a theory of EVOLUTION by the inheritance of ACQUIRED CHARACTERISTICS in his *Zoonomia* of 1794-96. Perhaps the foremost physician of his day, he was a leading member of the LUNAR SOCIETY OF BIRMINGHAM. CHARLES DARWIN was his grandson.

Database (databank)
A collection of information and data, which can be indexed by keywords. A subscriber can access individual items, often by telephonic links, and manipulate them by COMPUTER.

Date Line, International
An imaginary line on the Earth's surface, with local deviations, along longitude 180° from Greenwich. As the Earth rotates, each day first begins and ends on the line. A traveler going east over the line sets his calendar back one day, and one going west adds one day.

Dating see DENDROCHRONOLOGY; RADIOCARBON DATING.

Dausset, Jean Baptiste Gabriel (b. 1916)
Frenchman who was co-recipient of the 1980 Nobel Prize for Physiology or Medicine for his research on the genetic basis of IMMUNITY, particularly the discovery of the human histocompatability system.

Davis, William Morris (1850-1934)
US geographer and leading geomorphologist of the late 19th century. Davis is best remembered for his erosion-cycle concept, characterizing the successive stages in the history of a newly uplifted landmass as its youth, its maturity and its old age (see EROSION).

Davisson, Clinton Joseph (1881-1958)
US physicist who shared the 1937 Nobel Prize for Physics with G.P. THOMPSON for his demonstration of the DIFFRACTION of ELECTRONS. This confirmed DE BROGLIE's view that matter might act in a wavelike manner in appropriate circumstances.

Davy, Sir Humphry (1778-1829)
English chemist who pioneered the study of ELECTROCHEMISTRY. Electrolytic methods yielded him the elements SODIUM, POTASSIUM, MAGNESIUM, CALCIUM, STRONTIUM and BARIUM (1807-08). He also recognized the elemental nature of and named CHLORINE (1810). His early work on nitrous oxide (see NITROGEN) was done at Bristol under T. BEDDOES, but most of the rest of his career centered on the ROYAL INSTITUTION where he was assisted by his protégé, M. FARADAY, from 1813. A major practical achievement was the invention of a miner's safety lamp, known as the Davy Lamp, in 1815-16.

Day
Term referring either to a full period of 24 hours (the civil day) or to the (usually shorter and varying) period between sunrise and sunset when a given point on the Earth's surface is bathed in light rather than darkness (the natural day). Astronomers distinguish the sidereal day from the solar day, depending on whether the reference location on the Earth's surface is taken to return to the same position relative to the stars or to the Sun. The civil day is the mean solar day, some 168 seconds longer than the sidereal day, ie the sideral day is equivalent to 23 hours 56 minutes 04.1 seconds of mean time. The sidereal day itself is subdivided into 24 hours of sidereal time. In most modern states the day is deemed to run from midnight to midnight, though in Jewish tradition the day is taken to begin at sunset.

DDT (dichlorodiphenyltrichloroethane)
A synthetic contact INSECTICIDE, widely used until the 1970s, which kills a wide variety of insects, including mosquitoes, lice and flies, by interfering with their nervous systems. Its use, in quantities as great as 100,000 tonnes yearly, almost eliminated many insect-borne diseases, including TYPHUS, YELLOW FEVER and PLAGUE. Being chemically stable and physically inert, it persists in the environment for many years. Its concentration in the course of natural food chains (see ECOLOGY) led to the buildup of dangerous accumulations in some fish and birds. This prompted the US to restrict the use of DDT in 1972. In any case the development of insect strains resistant to DDT was already reducing its effectiveness as an insecticide. Although DDT was first made in 1874, its insecticidal properties were only discovered in 1939 (by P.H. MÜLLER).

Deaf mute see DUMBNESS.

Deafness
Failure of hearing may have many causes. Conductive deafness is due to disease of the outer or middle EAR, while perceptive deafness is due to disease of the inner ear or nerves of hearing. Common physical causes of conductive deafness are obstruction with wax or foreign bodies and injury to the eardrum. Middle ear disease is another important cause: in acute otitis, common in children, the ears are painful, with deafness, fever and discharge; in secretory otitis or glue ear, also in children, deafness and discomfort result from poor drainage; chronic otitis, in any age group, leads to a deaf discharging ear, with drum perforation. Otosclerosis is a common disease of middle age in which fusion or ANALYSIS of the small bones of the ear causes deafness. Perceptive deafness may follow infections in pregnancy (eg RUBELLA) or be hereditary. Acute virus infection and trauma to the inner ear (eg blast injuries or chronic occupational noise exposure) are important causes. Damage to the ear blood supply or the

auditory nerves by drugs, TUMORS or MULTIPLE SCLEROSIS may lead to perceptive deafness, as may the later stages of MENIERE'S DISEASE. Deafness of old age, or presbycusis, is of gradual onset, mainly as a result of loss of nerve cells.

Death

The complete and irreversible cessation of LIFE in an organism or part of an organism. In humans death is conventionally accepted as the time when the heart ceases to beat, there is no breathing and when the brain shows no evidence of function. Ophthalmoscopic examination of the eye shows that columns of BLOOD in small vessels are interrupted and static. As it is now possible to resuscitate and maintain heart function and to take over breathing mechanically, it is not uncommon for the brain to have suffered irreversible death but for "life" to be maintained artificially. The concept of "brain death" has been introduced, in which no spontaneous breathing, no movement and no specific REFLEXES are seen on two occasions. When this state is reached, artificial life support systems can be reasonably discontinued. The ELECTROENCEPHALOGRAPH has been used to diagnose brain death but is now considered unreliable.

After death, ENZYMES are released that begin the process of autolysis or spontaneous self-destruction of the cells. This paves the way for decomposition, which involves BACTERIA and other SAPROTROPHS. In the hours following death, changes occur in muscle which cause rigidity or RIGOR MORTIS.

Death can also affect just part of an organism, when it is often known as necrosis. In animals this may occur following loss of blood supply, for example, and involves loss of cell organization, autolysis and GANGRENE. The part may separate or be absorbed, but if it becomes infected this is liable to spread to living tissue. Cells also die as part of the normal turnover of a structure after POISONING or infection (eg in the LIVER), from compression (eg by TUMOR), or as part of a degenerative disease. In FUNGI, and to a lesser extent in plants, death of part of an organism is a normal feature of growth. (◆ page 44)

De Broglie, Louis-Victor Pierre Raymond, Prince (1892-1987)

French physicist who was awarded the 1929 Nobel Prize for Physics for his suggestion that acknowledged particles should display wave properties under appropriate conditions in the same way that ELECTROMAGNETIC RADIATION sometimes behaved as if composed of particles.

Debye, Peter Joseph Wilhelm (1884-1966)

Dutch-born German-US physical chemist chiefly remembered for the Debye-Hückel theory of ionic solutions (1923). He was awarded the 1936 Nobel Prize for Chemistry.

Decapods

Members of the CRUSTACEAN Order Decapoda, which includes the shrimps, crabs and lobsters. They are so named because the first three pairs of limbs on the thorax are modified as feeding appendages, leaving five pairs of thoracic legs (hence "ten-legs") for walking; in fact the first pair of these is often modified as claws, leaving only four pairs of walking legs.

Decay, Radioactive see RADIOACTIVITY.

Decibel (dB)

Unit used to express power ratio, equal to one-tenth of a BEL, widely used by acoustics and telecommunications engineers. It is commonly employed in describing NOISE levels relative to the threshold of hearing. Doubling the noise level adds 3 to the decibel rating. (◆ page 225)

Deciduous trees

Trees that shed their leaves at a certain season, when the environmental conditions are becoming unsuitable for active growth and just before the tree enters a period of dormancy. In temperate regions leaf loss occurs during the winter, but in warmer climates with seasonal rainfall it occurs during the dry season. (See also EVERGREENS; BROADLEAVED TREES.)

Decimal system

A number system using the powers of ten; our everyday system of numeration. The digits used are 0, 1, 2, 3, 4, 5, 6, 7, 8, 9; the powers of 10 being written $10^0=1$, $10^1=10$, $10^2=100$, $10^3=1000$, etc. To each of these powers is assigned a place value in a particular number; thus $4\times10^3)+(0\times10^2)+(9\times10^1)+(2\times10^0)$ is written 4092. Similarly, fractions may be expressed by setting their denominators equal to powers of 10:

$$3/4=75/100=7/10+5/100=(7\times10^{-1})+(5\times10^{-2})$$

which is written as 0.75. Not all numbers can be expressed in terms of the decimal system: one example is the fraction 1/3, which is written 0.3333..., the row of dots indicating that the 3 is to be repeated an infinite number of times. Fractions like 0.3333... are termed repeating decimals. Approximation is often useful when dealing with decimal fractions.

Declination

The angular distance of a celestial body from the celestial equator (see CELESTIAL SPHERE) along the MERIDIAN through the body. Bodies north of the equator have positive declinations, those south, negative. Together with right ascension, declination defines the position of a body in the sky.

Declination, Magnetic see EARTH.

Decomposition, Chemical

A reaction in which a chemical compound is split up into its elements or simpler compounds. Heat, or light of a suitable wavelength, will decompose many compounds, and some decompose spontaneously. Ionic compounds may be decomposed by ELECTROLYSIS. Double decomposition is a reaction of the type

$$AC+BD\rightarrow AD+BC$$

in which radicals are exchanged.

Decompression sickness (bends) see AEROEMBOLISM.

Decubitus ulcer

Also known as a bedsore or pressure sore, a skin ulcer that commonly occurs during prolonged bed rest, eg during a long illness when the patient is immobile. They can be prevented by encouraging patients to move as much as they are able, and by turning unconscious or paralyzed patients.

De Duve, Christian René (b. 1917)

British born Belgian-US biochemist who shared the 1974 Nobel Prize for Physiology or Medicine with A. CLAUDE and G. PALADE for their pioneer studies of CELL anatomy. De Duve's particular contribution was the discovery of LYSOSOMES.

Deep-sea animals see ABYSSAL FAUNA.

Defense mechanism

Any measure or measures taken by the individual's UNCONSCIOUS to protect his EGO from unpleasant or humiliating situations: eg SUBLIMATION, REPRESSION or projection. The more embracing term, defense reaction, describes the behavior resulting from such measures and also from analogous conscious measures.

Deferent see EPICYCLE.

Defibrillator

An apparatus for stopping heart FIBRILLATION by applying electric currents to the chest wall or directly to the heart.

Deficiency diseases see DISEASE; VITAMINS.

De Forest, Lee (1873-1961)

US inventor of the TRIODE (1906), an electron tube with three electrodes (cathode, anode and grid) which could operate as a signal amplifier as well as a rectifier. The triode was crucial to the development of RADIO.

Deforestation

The process of cutting down forests. It is often considered necessary for the provision of farmland or living areas. However, it is of major concern to environmentalists for a number of reasons. Removal of trees and their roots causes soil to be broken up and washed away; the removal of forests can cause the extinction of the individual plants and animals dependent on them. It is thought that large-scale deforestation may reduce the rate at which carbon dioxide can be reconverted back into oxygen, and so give rise to a GREENHOUSE EFFECT.

Degaussing

Removing the magnetization (see MAGNETISM) of a permanent magnet or other object by withdrawing it slowly from a strong oscillating magnetic field, so that it is repeatedly magnetized in opposite directions but to progressively smaller degrees.

Degenerate

A term sometimes applied to parasitic organisms that lack features found in their free-living relatives. The term is in some ways misleading, as such parasites usually possess other traits that suit them for their specialized way of life.

Degenerating diseases see DISEASE.

Dehydration

The removal of WATER from a substance (drying). To remove the elements of water, as in the dehydration of ALCOHOLS to ETHERS, requires a powerful dehydrating agent such as concentrated SULFURIC ACID. Generally, however, water is present as such, as a HYDRATE or merely absorbed, in which case milder methods suffice, such as equilibration in a desiccator with SILICA GEL or deliquescent compounds (see DELIQUESCENCE). Dry air may be passed over solids in heated drums, causing EVAPORATION. Gases are dried (for air conditioning, or before liquefying them) by compression and refrigeration. Foods are preserved by drying; this is done for convenience and compactness. Milk and eggs are dried by spraying into hot air. Modern freeze drying – sublimation of ice from frozen foods under vacuum – retains texture and flavor. In humans dehydration of the body occurs through DIARRHEA, VOMITING, CHOLERA or merely lack of water to replace PERSPIRATION.

Deimos

The outer moon of MARS, circling the planet in 30.3h at a distance of 23,500km. Its diameter is about 8km. (◆ page 221)

Delbrück, Max (1906-1981)

German-born US biologist whose discovery of a method for detecting and measuring the rate of mutations in BACTERIA opened up the study of bacterial GENETICS.

Deliquescence

The absorption of atmospheric moisture by a solid until it dissolves to form a saturated solution. If it merely forms a crystalline HYDRATE it is termed hygroscopic. The phenomenon depends on the relative HUMIDITY: sugar, for example, deliquesces above 85% humidity. (See also EFFLORESCENCE.)

Delirium

Altered state of consciousness in which a person is restless, excitable, hallucinating and only partly aware of his surroundings. It is seen in high FEVER, POISONING, drug withdrawal, disorders of METABOLISM and organ failure.

Delirium tremens

Specific delirium due to acute alcohol withdrawal in ALCOHOLISM. It occurs within days of abstinence and is often precipitated by injury, surgery or imprisonment. The sufferer becomes restless, disorientated, extremely anxious and tremulous; fever and profuse sweating are usual. Characteristically, hallucinations of insects or animals cause abject terror.

Delphinus
The Dolphin, a small summer constellation in N skies. Four of its stars form a diamond sometimes known as Job's Coffin.

Delta
A flat alluvial plain at a rivermouth or at the confluence of two rivers, formed of fertile mud deposited by the slow-moving water. Typically, the stream divides and subdivides until a fan-shaped plain covered by a complex of channels results. The form of a delta depends on the rate of SEDIMENTATION and of EROSION by the sea. (See also ESTUARY.)

Delta rays
Short ELECTRON tracks surrounding the track left by a fast charged particle in CLOUD CHAMBERS, BUBBLE CHAMBERS, or photographic emulsions. The primary particle dislodges electrons from the ATOMS of the medium concerned, and the faster of these leave short tracks of their own before being brought to rest.

Dementia
A permanent loss of the ability to reason, accompanied by generalized disorder of mental processes. It is caused by brain disease or injury.

Demyelinating diseases
Diseases such as MULTIPLE SCLEROSIS, in which the myelin sheaths that insulate most nerves are destroyed and nerve conduction is impaired.

Denatured alcohol see ETHANOL.

Dendrite
A branched, treelike CRYSTAL form, common in ice (especially frost) and certain minerals, and chiefly important in metals, which often consist of dendrites embedded in a matrix of the same or (for alloys) different composition. In anatomy, a dendrite is a tapering receptor branch of a NEURON.

Dendrochronology
The dating of past events by the study of tree-rings found in timber from archeological sites. A hollow tube is inserted into the trunk of an ancient living tree growing in the same area and a core from bark to center removed. The ANNUAL RINGS are counted, examined and compared with rings from the timber to be dated. Because the width of the rings varies from year to year, distinctive patterns can be recognized for particular centuries or decades, and the timber dated by comparison with these. Using dead trees of overlapping age, the technique can be extended even further back in time. Through such studies important corrections have been made to RADIOCARBON DATING.

Dengue fever (breakbone fever)
A VIRUS infection carried by mosquitoes, with FEVER, headache, prostration and characteristically severe muscle and joint pains. There is also a variable skin rash through the roughly week-long illness. It is a disease of warm climates, and may occur in EPIDEMICS.

Density
The ratio of MASS to volume for a given material or object. Substances that are light for their size have a low density, and vice versa. Objects whose density is less than that of water will float in water, while a hot air balloon will rise when its average density becomes less than that of air. The term is also applied to properties other than mass; eg charge density refers to the ratio of electric charge (see ELECTRICITY) to volume. (➧ page 134)

Dentistry
The study, management and treatment of diseases and conditions affecting the mouth, jaws, teeth and their supporting tissues.

Dentition see TEETH.

Deoxyribonucleic acid see DNA; NUCLEIC ACIDS.

Deoxyribose ($C_5H_{10}O_4$)
A pentose monosaccharide that forms part of the NUCLEOTIDE units from which DNA is made.

Depression
A common psychiatric disease with pathologically depressed mood and characteristic disturbances of bodily function and sleep. It can be divided into two types: neurotic or exogenous depression, in which a precipitating external cause can be identified; and psychotic or endogenous depression, when no external cause can be traced.

Depression
In METEOROLOGY, a midlatitude CYCLONE.

Dermatitis
SKIN conditions in which inflammation occurs. These include ECZEMA, contact dermatitis (see ALLERGY) and seborrheic dermatitis (dandruff). Acute dermatitis leads to redness, swelling, blistering and crusting, while chronic forms usually show scaling or thickening of skin.

Dermatology
Subspeciality of MEDICINE concerned with the diagnosis and treatment of SKIN DISEASES.

Dermatophyte
Any parasitic fungus, such as ringworm, that lives on the skin.

Descartes, René (Renatus Cartesius, 1596-1650)
Mathematician, physicist and the foremost of French philosophers. After being educated in his native France and spending time in military service (1618-19) and traveling, Descartes spent most of his creative life in Holland (1625-49). He entered the service of Queen Christiana of Sweden shortly before his death. He formulated a new methodology of science on the deductive logic of mathematical reasoning, first publishing his conclusions in his *Discourse on Method* (1637). The occult qualities of late scholastic science were to be done away with; only ideas which were clear and distinct were to be employed. To discover what ideas could be used to form a certain basis for a unified *a priori* science, he introduced the method of universal doubt. The first certitude he discovered was his famous *cogito ergo sum* (I think therefore I am) and on the basis of this, the existence of other bodies and of God, he worked out his philosophy. In science, Descartes, denying the possibility of a vacuum, explained everything in terms of motion in a plenum of particles whose sole property was extension. This yielded his celebrated but ultimately unsuccessful vortex theory of the Solar System, and statements of the principle of inertia and the laws of ordinary REFRACTION. In mathematics, he founded the study of analytical geometry. In biology, his views were mechanistic; he regarded animals as complex machines.

Deserts
Areas where life has extreme difficulty in surviving. Deserts cover about one third of the Earth's land area. There are two types.

COLD DESERTS In cold deserts, water is unavailable during most of the year as it is trapped in the form of ice. Cold deserts include the Antarctic polar icecap, the barren wastes of Greenland, and much of the TUNDRA. (See also GLACIER.) Eskimos, Lapps and Samoyeds are among the ethnic groups inhabiting such areas in the N Hemisphere. Their animal neighbors include seals and the polar bear.

HOT DESERTS These typically lie between latitudes 20° and 30° N and S, though they exist also farther from the equator in the centers of continental landmasses. They can be described as areas where water precipitation from the ATMOSPHERE is greatly exceeded by surface EVAPORATION and plant TRANSPIRATION. The best-known, and largest, is the Sahara. GROUNDWATER exists but is normally far below the surface; here and there it is accessible as SPRINGS or WELLS (see ARTESIAN WELL; OASIS). In recent years IRRIGATION has enabled reclamation of much desert land. Landscapes generally result from the surface's extreme vulnerability to EROSION (see also SOIL EROSION). Features include arroyos, BUTTES, DUNES, MESAS and wadis. The influence of humans may assist peripheral areas to become susceptible to erosion, and thus temporarily advance the desert's boundaries. (See also DUST BOWL; SANDSTORM.) Plants may survive by being able to store water, like the cactus; by having tiny leaves to reduce evaporation loss, like the paloverde; or by having extensive root systems to trap maximum moisture, like the mesquite. Animals may be nomadic, or spend the daylight hours underground.

De Sitter, Willem see SITTER, WILLEM DE.

Desquamation
Shedding of the outer skin. In reptiles the entire skin is periodically molted like a glove. In hairless animals like humans it happens all the time and is imperceptible, but becomes rapid and noticeable after sunburn or during some skin diseases.

Detergents see SOAPS AND DETERGENTS.

Deuterium ("heavy hydrogen", D or $_1H^2$)
An ISOTOPE of HYDROGEN discovered in 1931, whose nucleus (the deuteron) has one PROTON and one NEUTRON (see also TRITIUM). It forms 0.014% of the hydrogen in naturally occurring hydrogen compounds, such as water, and is chemically very like ordinary hydrogen, except that it reacts more slowly. It is obtained as HEAVY WATER by the fractional electrolysis of WATER. Deuterium is the major fuel for nuclear FUSION (see HYDROGEN BOMB) and is used in tracer studies. (See also NUCLEAR REACTORS.) AW 2.0, mp −254.6°C, bp −249.7°C.

Deuteromycetis (fungi imperfecti) see FUNGI.

Devonian
The fourth period of the PALEOZOIC, which lasted from about 408 to 360 million years ago. (See GEOLOGY.) (➧ page 106)

De Vries, Hugo see VRIES, HUGO DE.

Dew
Water droplets produced on clear, calm nights by CONDENSATION of water vapor in the air. It is deposited on surfaces freely exposed to the sky that have cooled by RADIATION of heat. Dew forms when the air TEMPERATURE falls to the dew point. This, the temperature at which the air becomes saturated with water vapor, rises rapidly with the HUMIDITY. Dew is an important source of moisture for desert plants.

Dewar, Sir James (1842-1923)
British chemist and physicist who proposed various structures (including the "Dewar structures") for BENZENE, invented the vacuum bottle (Dewar bottle – 1892) and pioneered the techniques of low-temperature physics. He developed methods for liquefying gases and discovered the magnetic properties of liquid OXYGEN and OZONE.

Dextrin
A polysaccharide (see CARBOHYDRATES) obtained from STARCH by heating or partial HYDROLYSIS with acids, diastase or the bacterium *Bacillus macerans*. Soluble in water, it is used as an adhesive and a size for paper and textiles.

Dextrose see GLUCOSE.

Diabase (dolerite)
Dark IGNEOUS ROCK intermediate in grain size between BASALT and GABBRO. Consisting of plagioclase FELDSPAR and PYROXENE, it is a widespread intrusive rock, quarried for crushed and monumental stone (known as "black granite").

Diabetes mellitus
A common disease, affecting between 0.5 and 1% of the population, characterized by the absence or inadequate secretion of INSULIN, the principal

hormone controlling BLOOD sugar. The exact cause is unknown, though both hereditary and environmental factors are believed to play a part. Though it may start at any time, two main groups are recognized: juvenile (beginning in childhood, adolescence or early adult life) – due to inability to secrete insulin; and late onset (late middle life or old age) – associated with obesity and with a relative lack of insulin. High blood sugar may lead to coma, often with keto-ACIDOSIS, excessive thirst and high urine output, weight loss, ill-health and liability to infections. The disease may be detected by urine or blood tests. It causes disease of small blood vessels as well as premature ATHEROSCLEROSIS, RETINA disease, CATARACTS, KIDNEY disease and NEURITIS. Poor blood supply, neuritis and infection may lead to chronic leg ULCERS.

Dialysis
Process of selective DIFFUSION of ions and molecules through a semipermeable membrane which retains COLLOID particles and macromolecules. It is accelerated by applying an electric field (see ELECTROPHORESIS). It is used for desalination. A therapeutic form of dialysis is used to mimic the functions of the kidney in people with chronic renal failure. (See also OSMOSIS.)

Diamagnetism
Very weak magnetization (see MAGNETISM) of a material in a direction opposing the magnetizing field, due to ELECTRON orbital distortion. It is a property of all materials, though sometimes masked by stronger PARAMAGNETISM or ferromagnetism.

Diamond
Allotrope of CARBON (see ALLOTROPY), forming colorless cubic crystals. Diamond is the hardest known substance, with a Mohs hardness of 10, which varies slightly with the orientation of the crystal. Thus diamonds can be cut only by other diamonds. They do not conduct electricity, but conduct heat extremely well. Diamond burns when heated in air to 900°C; in an inert atmosphere it reverts to GRAPHITE slowly at 1000°C, rapidly at 1700°C. Diamonds occur naturally in dikes and pipes of KIMBERLITE, notably in South Africa (Orange Free State and Transvaal), Tanzania, and in the US at Murfreesboro, Arkansas. They are also mined from secondary (alluvial) deposits, especially in Brazil, Zaire, Sierra Leone and India. The diamonds are separated by mechanical panning, and those of GEM quality are cleaved (or sawn), cut and polished. Inferior, or industrial, diamonds are used for cutting, drilling and grinding. Synthetic industrial diamonds are made by subjecting graphite to very high temperatures and pressures, sometimes with fused metals as solvent.

Diapause
In insects, a period of "suspended animation", when development is arrested, growth ceases and the metabolism is reduced to a minimal level. It is an established part of the species' life cycle and occurs at a particular point, often the egg or pupal stage. Diapause frequently coincides with the winter months and is analogous to the HIBERNATION of homoiothermic ("warm-blooded") animals. External conditions, such as daylength, low temperatures or changes in food quality at the end of the summer, induce diapause. A period of low temperatures is often needed to break the state of dormancy. These responses are mediated by hormones produced by the brain.

Diaphragm
Thin muscular and fibrous structure dividing the contents of the CHEST from those of the ABDOMEN. It is involved to a varying degree in quiet breathing, and also contracts during deep breathing and straining. Through it pass the ESOPHAGUS, AORTA and inferior VENA CAVA.

Diarrhea
Loose and/or frequent bowel motions. A common effect of FOOD POISONING, GASTROINTESTINAL TRACT infection (eg DYSENTERY, CHOLERA) or INFLAMMATION (eg ABSCESS, COLITIS, gastroenteritis), drugs and systemic diseases. Benign or malignant TUMORS of the colon and rectum may also cause diarrhea.

Diastole see BLOOD PRESSURE.

Diastrophism
The large-scale deformation of the crust of the Earth to produce such features as continents, oceans, mountains and rift valleys. (See also EARTH; FAULT; FOLD; PLATE TECTONICS.)

Diathermy
The use of electrically generated heat, particularly in surgery. Two ELECTRODES are connected to the patient: a localized point electrode is used to cause local tissue destruction, while a larger surface electrode, which dissipates the electrical energy over a wider area and thus avoids damage, is used to complete the circuit. Diathermy allows small blood vessels to be occluded and is often used to incise the GASTROINTESTINAL TRACT, when it gives a bacteria free, nonbleeding edge. It has also been used to remove hairs, but the FOLLICLES may be scarred.

Diatoms
Single-celled photosynthetic fresh and salt water algae, important as food to many small animals. They are a major constituent of the marine plankton. Over millions of years their silicaceous skeletons have formed deposits on the sea bed, one in California being over 300m thick. These deposits are excavated as fuller's or diatomaceous earth, and the fine silica grains used in metal polishes. Diatoms are variously classified with the ALGAE or with the PROTISTS.

Dick, George Frederick (1881-1967) and **Gladys Henry Dick** (1881-1963)
US physicians (husband and wife) who discovered the organism responsible for causing SCARLET FEVER (1923) and developed a means (the Dick test) for estimating an individual's susceptibility to the disease.

Dicotyledons (dicots)
Flowering plants or ANGIOSPERMS that produce seeds with two seed leaves (COTYLEDONS) and thus differ from MONOCOTYLEDONS, or monocots, which produce only one. They are also characterized by having netveined leaves and a ring of bundles in the stem vascular system. Their flowering parts are in fours or fives, or multiples of these, rather than threes, as in monocots. (➧ page 139)

Dielectric
An electrical insulator in which the application of an electric field (see ELECTRICITY) causes polarization which in turn produces a field opposed to the original field, thus reducing the resultant field strength by a factor known as the dielectric constant for the material concerned. Their properties are exploited in CAPACITORS, and to reduce dangerously strong fields. They also have optical applications, since refractive index (see REFRACTION) is the square root of dielectric constant.

Diels, Otto Paul Hermann (1876-1954)
German organic chemist who discovered carbon suboxide (C_3O_2) in 1906 and who with KURT ALDER discovered the DIELS-ALDER REACTION (1928). For this latter work he and Alder shared the 1950 Nobel Prize for Chemistry.

Diels-Alder reaction (diene synthesis)
Reaction discovered by OTTO DIELS and KURT ALDER, important in making insecticides and fungicides. A conjugated diene (see ALKENES) such as 1, 3-BUTADIENE or ISOPRENE, adds readily to a "dienophile" containing a double or triple bond activated by an adjacent NUCLEOPHILE group.

Dietetic foods
Special foods for conditions in which normal diet leads to disease or ill-health, or where dietary manipulation modifies disease. Examples include a GLUTEN-free diet for CELIAC DISEASE; polyunsaturated fat diet for excess blood fat and ATHEROSCLEROSIS; milk avoidance in lactose intolerance; low CARBOHYDRATE in DIABETES; low phenylalanine foods for PHENYLKETONURIA, and low PROTEIN diet for KIDNEY or LIVER failure.

Diffraction
The property by which a WAVE MOTION (such as ELECTROMAGNETIC RADIATION, SOUND or water waves) deviates from the straight line expected geometrically and thus gives rise to INTERFERENCE effects at the edges of the shadows cast by opaque objects, where the wave-trains that have reached each point by different routes interfere with each other. Opaque objects thus never cast completely sharp shadows, though such effects only become apparent when the dimensions of the obstruction are of the same order as the wavelength of the wave motion concerned. It is diffraction effects that place the ultimate limit on the resolving power of optical instruments, RADIO TELESCOPES and the like. Diffraction is set to work in the diffraction grating. Here, light passed through a series of very accurately ruled slits or reflected from a series of narrow parallel mirrors produces a series of spectrums by the interference of the light from the different slits or mirrors. Gratings are ruled with from 70 lines/mm (for infrared work) to 1800 lines/mm (for ultraviolet work).

Diffusion
The gradual mixing of different substances placed in mutual contact due to the random thermal motion of their constituent particles. Most rapid with gases and liquids, it does also occur with solids. Diffusion rates increase with increasing TEMPERATURE; the rates at which gases diffuse through a porous membrane vary as the inverse of the square root of their MOLECULAR WEIGHT. Gaseous diffusion is used to separate fissile URANIUM-235 from nonfissile uranium-238, the gas used being uranium hexafluoride (UF_6).

Digestion
The breakdown of food so that it can be absorbed. In some unicellular organisms this is by PHAGOCYTOSIS followed by enzyme breakdown of large molecules within the cell. In larger animals it occurs outside the cells but within a digestive system, through the liberation of ENZYMES. Among fungi, some bacteria, houseflies, etc. digestion is achieved by secreting enzymes onto the food, and then sucking up or otherwise absorbing the liquefied nutrients.

In mammals the digestive system consists of the mouth, esophagus, stomach and intestines (see GASTROINTESTINAL TRACT); these are the principal absorbing surfaces, and also secrete enzymes. The related organs, the liver and pancreas, are also considered part of the digestive system; they secrete enzymes, bile salts, etc. into the tract via ducts.

Different enzymes act best at different pH, and gastric juice and BILE respectively regulate the acidity of the STOMACH and alkalinity of the small intestine. PROTEINS are broken down by pepsin in the stomach and by trypsin, chymotrypsin and peptidases in the small intestine. CARBOHYDRATES are broken down by specialized enzymes, mainly in the small intestine. FATS are broken down physically by stomach movement, enzymatically by lipases, and emulsified by bile salts. Food is mixed and propelled by PERISTALSIS, while nerves and locally regulated HORMONES, including gastrin and secretin, control both secretion and motility. Absorption of most substances occurs in the small intestine through a specialized, high-surface-area mucous membrane; some molecules pass through unchanged but most in altered form. Absorption may be either by an active transport system involving chemical or physical interaction in the gut wall, or simply by a passive DIFFUSION process. Some VITAMINS and trace metals have specialized transport systems. Most absorbed food passes via the blood system to the liver, where much of it is metabolized and toxic substances are removed. Some absorbed fat is passed into the LYMPH. BACTERIA colonize most of the small intestine and are important in certain digestive processes. Malabsorption occurs when any part of the digestive system becomes defective. Pancreas and liver disease, obstruction to bile ducts, alteration of the bacterial population and inflammation of the intestine are common causes. (⧫⧫ page 40, 111)

Digestive system
That part of the body that breaks down or modifies dietary intake into a form that is absorbable and usable. (⧫⧫ page 16-17, 40, 111)

Dike (dyke)
A tabular body of IGNEOUS ROCK which, unlike a SILL, cuts across the beds of surrounding rock. Dikes commonly occur in swarms, which may be parallel or radial.

Dilation and curettage
An operation in which the neck of the UTERUS is dilated (stretched) and curetted (scraped). It is obsolete as a form of abortion but is useful for a variety of menstrual disorders.

Dimensions
As applied to a physical quantity, an indication of the role it plays in equations. The dimensions of a mechanical quantity, in terms of mass [M], length [L] and time [T], can be deduced from the units in which it is expressed. Thus VELOCITY, measured in m/s, has dimensions [L]/ [T]. Dimensions are purely conventional, having no real physical significance. This is clear in the case of electromagnetic quantities where dimensions vary according to the units system used. Dimensions nevertheless find use in dimensional analysis.

Dinoflagellates
Microscopic single-celled organisms, occurring in vast numbers in fresh and salt water, and forming an important part of the marine PLANKTON. Many dinoflagellates contain CHLOROPHYLL, others do not; some authorities classify them with the ALGAE, others with the PROTISTS. They are propelled by two whip-like hairs (FLAGELLA) and each one is covered by a layer of cellulose; they are important as food for many animals.

Dinosaurs
Extinct REPTILES that flourished for 160 million years from the late TRIASSIC to the end of the CRETACEOUS. They ranged in size from small forms no larger than a domestic chicken to giants such as *Diplodocus*, which was 27m long and weighed about 30 tonnes. Early in their history two distinct dinosaur groups evolved: the SAURISCHIA and the ORNITHISCHIA. At the end of the Cretaceous period, about 65 million years ago, dinosaurs disappeared. The reasons for this sudden extinction are not certain and are hotly debated by paleontologists. (⧫ page 106)

Diode
Originally an electron tube having two electrodes (cathode and anode) used as a rectifier, but now extended to include semiconductor devices performing a similar function (see ELECTRONICS).

Dioecious plants
Those in which the male and female organs are borne on separate plants. Examples are willow, hemp and asparagus. (See also MONOECIOUS PLANTS.)

Diopter (reciprocal meter)
A unit used to express the focal power of optical lenses. The power of a converging LENS in diopters is a positive number equal to the reciprocal of its focal length in meters. Diverging lenses have negative powers.

Diphtheria
BACTERIAL DISEASE, now uncommon, causing FEVER, malaise and sore throat, with a characteristic "pseudomembrane" on throat or PHARYNX; also, the LYMPH nodes may enlarge. The LARYNX, if involved, leads to a hoarse voice, breathlessness and coarse wheezing; this may progress to respiratory obstruction requiring surgical intervention. The bacteria produce TOXINS that can damage nerves and HEART muscle; cardiac failure and abnormal rhythm, or PARALYSIS of palate and of eye movement, and peripheral NEURITIS may follow.

Diplococcus
Spherical or ovoid nonmotile BACTERIA, so named because they occur in pairs. Diseases caused by diplococci include gonorrhea, whooping cough and some forms of pneumonia and meningitis.

Diploid see MEIOSIS.

Dip needle (inclinometer)
An instrument for measuring magnetic dip (see EARTH), consisting of a magnetic needle mounted free to pivot in the vertical plane within a graduated circle. For use, the instrument is carefully leveled and oriented into the magnetic meridian.

Dipnoi (lungfish)
A subclass of the SARCOPTERYGII that contains a large number of fossil forms and three living freshwater groups, found in Africa, Australia and South America respectively. As their name suggests, the lungfish have lungs, organs found in early fish but that evolved into SWIMBLADDERS in the majority of the ACTINOPTERYGII. The African and South American lungfish are unusual in their ability to estivate in cocoons of mud during droughts.

Dipole (in radio) see ANTENNA.

Dipole moment
An electric dipole is a pair of equal and opposite electric charges a short distance apart (see ELECTRICITY). All ordinary manifestations of MAGNETISM are the result of magnetic dipoles, whether these arise in the context of permanent magnets or electromagnets. In either case the dipole moment is a vector quantity descriptive of the dipole. Atomic nuclei and asymmetric molecules often exhibit dipole or other multipole properties.

Dipsomania
A form of ALCOHOLISM characterized by bouts of drinking alternating with abstention.

Diptera
The order of insects that includes the flies, mosquitoes and gnats. They are characterized by having only one pair of wings (hence "diptera" – two wings). The hindwings are much reduced in size and are known as halteres. They are adapted for use in flight control, beating in phase with the wings.

Dirac, Paul Adrien Maurice (1902-1984)
English theoretical physicist who shared the 1933 Nobel Prize for Physics with P. SCHRÖDINGER for their contributions to WAVE MECHANICS. Dirac's theory (1928) took account of RELATIVITY and led him to postulate the existence of the positive ELECTRON or positron, later discovered by S.C. ANDERSON. Dirac was also the co-discoverer of FERMI-DIRAC STATISTICS.

Direction finder
A device used to locate the direction of an incoming RADIO signal. Usually a loop ANTENNA is rotated until maximum reception strength is achieved, giving the line of the transmission. If this is repeated from a different position, the transmitting station may be located. The radiocompass used in air and sea NAVIGATION is a direction finder: position is determined by finding the directions of two or more transmitters.

Dislocation
A CRYSTAL defect in which the normal crystal lattice is distorted (eg by the interposition of an extra half plane of atoms – an edge dislocation). The type and number of dislocations in a crystal help to determine its electrical and mechanical properties.

Dispersion
In optics, the separation of a mixture of LIGHT radiations according to COLOR (ie wavelength). This can occur in REFRACTION (when it is responsible for the RAINBOW and the production of a SPECTRUM with a PRISM), in DIFFRACTION (as is applied in the grating spectroscope), or in SCATTERING (giving rise, eg to the blue color of the sky). The dispersive power of an optical medium is a measure of the extent to which its refractive index varies with wavelength.

Displacement activity
Term used by ethologists for actions performed by an animal in stressful situations that appear to be irrelevant to prevailing conditions. An example is grooming in a situation where fighting would normally be expected. A possible explanation of such activity is that it results from the effects of conflicting impulses.

Dissociation
In chemistry, the reversible decomposition of a compound, often effected by heat. Ionic dissociation is the dissociation of an ionic compound – a weak electrolyte (see ELECTROLYSIS) – into IONS when dissolved in water or other ionizing SOLVENTS. It accounts in part for the phenomena of electrolytic CONDUCTIVITY.

Dissonance
In acoustics, the simultaneous sounding of two or more tones which, because of BEATING, seem unpleasant to the human ear. Which musical intervals are considered dissonant is largely a matter of cultural tradition.

Distillation
Process in which substances are vaporized and then condensed by cooling, probably first invented by the Alexandrian School and used in ALCHEMY, the still and the alembic being employed. It may be used to separate a volatile liquid from nonvolatile solids, as in the production of pure water from seawater, or from less volatile liquids, as in the distillation of liquid air to give oxygen, nitrogen and the noble gases. If the boiling points of the components differ greatly, simple distillation can be used: on gentle heating, the components distill over in order (the most volatile first) and the pure fractions are collected in different flasks. Mixtures of liquids of similar boiling points require fractionation for efficient separation. This technique employs multiple still heads and fractionating columns in which some of the vapor is condensed and returned to the still, equilibrating as it does so with the rising vapor. In effect, the mixture is redistilled several times; the number of theoretical simple distillations, or theoretical plates, represents the separating efficiency of the column. The theory of distillation is an aspect of PHASE EQUILIBRIA studies. For ideal solutions, obeying RAOULT's Law, the vapor always contains a higher proportion than the liquid of the more volatile component; if this is not the case, an AZEOTROPIC MIXTURE may be formed. When two immiscible liquids are distilled, they come over in the proportion of their VAPOR PRESSURES at a temperature below the boiling point of either. This is utilized in steam distillation, in which superheated steam is passed into the still and comes over together with the volatile liquid. It is useful when normal distillation would require a temperature high enough to cause

decomposition, as is vacuum distillation, in which the pressure reduction lowers the boiling points. A further refinement is molecular distillation, in which unstable molecules travel directly in high vacuum to the condenser.

Disulfiram see ANTABUSE.

Diuretics
Drugs that increase urine production by the KIDNEY. Alcohol and CAFFEINE are mild diuretics. Thiazides and other diuretics are commonly used in treatment of HEART failure, EDEMA, high blood pressure, liver and kidney disease.

Diurnal
Active during the day, as opposed to the night (nocturnal) or at dawn (crepuscular).

Divide (water parting or watershed)
A region of high ground that lies between and determines the flow of two unconnected drainage systems. In North America, the Continental Divide is formed by the Rocky Mountains.

Dixon, Joseph (1799-1869)
US manufacturer, the inventor of the high-temperature GRAPHITE crucible and other graphite products.

DNA (deoxyribonucleic acid)
A long-chain molecule found in the CHROMOSOMES that carries the genetic information (except in RNA VIRUSES). DNA is a NUCLEIC ACID, with not one but two sugar-phosphate chains twisted around each other to form a double helix. Linking the chains rather like the rungs of a ladder are the bases, each interlocking with its appropriate opposite number, for a base can be partnered by another base of only one other kind. The molecule replicates by gradually splitting down the middle, while the bases of each half bond with the appropriate bases of free nucleotide units to form a pair of identical DNA molecules. The replication process is controlled by enzymes.

The elucidation of DNA structure, one of the greatest advances of 20th-century biology, is chiefly associated with the Nobel prizewinners WATSON, CRICK and WILKINS. DNA controls living processes by coding for all the PROTEINS that the cell contains by means of the GENETIC CODE. (See also BIOTECHNOLOGY and GENETIC ENGINEERING.) (◆ page 13, 32)

DNA fingerprinting see RESTRICTION MAPPING.

Dobzhansky, Theodosius (1900-1975)
Russian-born US biologist, famed for his study of the fruit fly, DROSOPHILA, which demonstrated that a wide genetic range could exist in even a comparatively well-defined species. Indeed the greater the "genetic load" of unusual genes in a species, the better equipped it is to survive in changed circumstances. (See also EVOLUTION; HEREDITY.)

Doctor, Medical see MEDICINE.

Dog star see SIRIUS.

Doisy, Edward Adelbert (1893-1986)
US biochemist who first crystallized the female sex HORMONE estrone (1929). He shared the 1944 Nobel Prize for Physiology or Medicine with H. DAM after isolating, determining the structure of, and synthesizing VITAMIN K (1936).

Doldrums
Regions of low wind, calms and strong upward air movement around the EQUATOR, produced by the convergence of the SE and NE TRADE WINDS. Sailing ships were often becalmed there.

Dolerite British term for DIABASE.

Dolomite
A common mineral, calcium magnesium carbonate ($CaMg(CO_3)_2$), with white or colorless rhombohedral crystals, often found associated with limestone and marble and in magnesium-rich METAMORPHIC ROCKS. It is used as an ornamental and building stone, and is a source of magnesium.

Domagk, Gerhard (1895-1964)
German pharmacologist who discovered the antibacterial action of the dye Prontosil Red. This led to the discovery of other SULFA DRUGS. In recognition of this Domagk was offered the 1939 Nobel Prize for Physiology or Medicine, though he was not allowed to accept it at the time.

Domains, Magnetic see MAGNETISM.

Domestication
The process by which some animals have been adapted to the uses of humans. Domestication usually involves genetic changes in the organisms concerned. (See also ARTIFICIAL SELECTION.)

Dominance
In genetics, the ability of one ALLELE of a gene to mask the effects of another (recessive) allele. The effect is seen when a diploid organism is heterozygous for a particular gene – that is, it has two different alleles on its two homologous chromosomes. In most cases, only one of those alleles will be expressed in the PHENOTYPE, and this is described as the dominant allele. Where both are expressed, the condition is referred to as codominance; where one is expressed but its effect weakened, this is incomplete dominance. The more usual situation, where the heterozygote has exactly the same phenotype as a homozygote for the dominant allele, is known as complete dominance.

Dominance can arise in several different ways, but it often occurs because the protein coded for by the recessive allele is either defective or is not produced at all. In the case of enzymes (the major class of proteins in the body), a 50% reduction in the enzyme production level generally has no detectable physiological effect, so the normal enzyme, produced by the dominant allele, masks the deficiency of the recessive allele. (◆ page 32)

Dominance hierarchy
The organization of animals within groups according to social status. This order is usually maintained by mutual threatening or warning signals that make fighting unnecessary. The dominant animals have priority over other members of the group in feeding, sex relations, and other activities. Such hierarchies are most often seen in higher animals such as birds and mammals.

Donati, Giovanni Battista (1826-1873)
Italian astronomer who first studied the spectra of COMETS (1864). He discovered several comets including "Donati's Comet" of 1858.

Doppler effect
The change observed in the wavelength of a sonic, electromagnetic or other wave (see WAVE MOTION) because of relative motion between the wave source and an observer. As a wave source approaches an observer, each pulse of the wave is closer behind the previous one than it would be were the source at rest relative to the observer. This is perceived as an increase in frequency, the pitch of a sound source seeming higher, the color of a light source bluer. When a sound source achieves the speed of sound, a sonic boom results. As a wave source recedes from an observer, each pulse is emitted farther away from him than it would otherwise be. There is hence a drop in pitch or a reddening in COLOR (see LIGHT; SPECTRUM). The Doppler effect, named for Christian Johann Doppler (1803-1853) who first described it in 1842, is of paramount importance in astronomy. Observations of stellar spectra can determine the rates at which stars are moving towards or away from us, while observed red shifts in the spectra of distant galaxies are generally interpreted as an indication that the Universe as a whole is expanding (but see RED SHIFT).

Dorado
The Swordfish or Goldfish, a S Hemisphere constellation containing the Large Magellanic Cloud (see MAGELLANIC CLOUDS).

Dormancy
A resting state that occurs in living organisms when growth stops and the internal METABOLISM is slowed down. In animals, dormancy during the winter is termed HIBERNATION, while dormancy in the summer or dry season, such as in the lungfish and earthworm, is termed aestivation. In plants, dormancy is manifest in the lack of growth of perennial plants such as grasses during adverse conditions, the loss of foliage by deciduous trees in winter, and the production of dormant buds and underground perennating organs such as BULBS, CORMS and TUBERS. In some bacteria, fungi and algae thick-walled dormant spores are produced that survive adverse conditions.

Dosimeter
A device worn by persons working in situations where they are exposed to ionizing radiations (see RADIOACTIVITY), which measures the dose of radiation to which they have been exposed.

Double blind
A method for assessing the value of a drug or treatment; neither the patient nor the person giving the treatment knows what it contains.

Double refraction (birefringence)
The property of certain CRYSTALS to split a ray of unpolarized light into two rays plane-polarized at right-angles to each other (see POLARIZED LIGHT). One of these, the ordinary ray, is refracted according to the ordinary laws of REFRACTION, but the other, the extraordinary ray, is refracted with a refractive index that depends on the direction from which the original ray was incident. Double refraction in Iceland spar is used in the NICOL PRISM to produce plane-polarized light.

Double star (binary star)
A pair of stars revolving around a common center of gravity. Less frequently the term "double star" is applied to two stars that merely appear close together in the sky, though in reality at quite different distances from the Earth (optical pairs), or to two stars whose motions are linked but that do not orbit each other (physical pair). About 50% of all stars are members of either binary or multiple star systems, in which there are more than two components. It is thought that the components of binary and multiple star systems are formed simultaneously. Visual binaries are those that can be seen telescopically to be double. There are comparatively few visual binaries, as the distances between components are small relative to interstellar distances, but examples are CAPELLA, PROCYON, SIRIUS and ALPHA CENTAURI. Spectroscopic binaries, though they cannot be seen telescopically as doubles, can be detected by RED and BLUE SHIFTS in their spectra, their orbit making each component alternately approach and recede from us. Eclipsing binaries are those whose components, due to the orientation of their orbit, periodically mutually eclipse each other as seen from the Earth.

Dow, Herbert Henry (1866-1930)
Canadian-born US chemist and industrialist who developed an electrolytic method for extracting BROMINE from certain natural BRINES. He founded the Dow Chemical Company to exploit this process in 1897.

Down's syndrome (mongolism)
A relatively common (1 in 1000 births) congenital disorder caused by a specific abnormality in a patient's chromosomes. It causes characteristic facial appearance, hand shape and skin patterns; floppiness in the baby; MENTAL RETARDATION; and delayed growth. Congenital diseases of the HEART and GASTROINTESTINAL TRACT are also common. Down's syndrome is more common in babies born to mothers aged over 40 years.

Draco
The Dragon, a large N Hemisphere constellation. Alpha Draconis was the POLE STAR *c*3000 BC (see PRECESSION). Gamma Draconis is the second magnitude star Eltanin. Draco also contains the planetary NEBULA NGC 6543.

Drainage
The runoff of water from an area, either naturally or, in agriculture, under artificial control. In nature, drainage generally takes the form of a pattern of streams, which feed rivers and lakes and flow, usually, to the sea (see HYDROLOGIC CYCLE). An area all of whose rainwater drains into a particular body of water is called a watershed or catchment basin. (See also GROUNDWATER.) Systems of artificial drainage depend on the nature of the soil as well as local topography. Two main systems are used: surface and subsurface. Surface systems usually comprise a pattern of ditches; subsurface systems a pattern of conduits and tunnels. These usually lead to a natural stream.

Draper, John William (1811-1882)
English-born US chemist who investigated the chemical action of light, first photographed the Moon (1840) and obtained a photograph of the solar spectrum (1844). His son, Henry Draper (1837-1882), US physician and amateur astronomer, obtained the first photograph of a nonsolar stellar spectrum (1872) and the first photograph of a NEBULA (the Orion) in 1880.

Dreams
Fantasies, usually visual, experienced during sleep and in certain other situations. About 25% of an adult's sleeping time is characterized by rapid eye movements (REM) and brain waves that, registered on the ELECTROENCEPHALOGRAPH, resemble those of a person awake (EEG). This REM-EEG state occurs in a number of short periods during sleep, each lasting a number of minutes, the first coming some 90min after sleep starts and the remainder occurring at intervals of roughly 90min. It would appear that it is during these periods that dreams take place, as people woken during a REM-EEG period will report and recall visual dreams in some 80% of cases; people woken at other times report dreams only about 40% of the time, and of far less visual vividness. Observation of similar states in animals suggests that at least all mammals experience dreams. Dreams can also occur, though in a limited way, while falling asleep (hypnogocic) or waking (hypnopompic); the origin and nature of these dreams is not known.

Drift
The material left behind on the retreat of a GLACIER. The unstratified material deposited directly on land is called TILL; fluvio-glacial drift, well stratified, is that transported by melted waters of the glacier. Drift may be composed of particles from fine sand to huge boulders, and may be up to 100m deep.

Dropsy see EDEMA.

Drosophila (fruit fly)
A member of the INSECTA that has been used extensively in the study of GENETICS. It has the advantages of being small, of breeding extremely rapidly, and of exhibiting giant chromosomes in its salivary glands.

Drought
Temporary, often disastrous climatic condition of extreme dryness, when an area's natural water supplies are insufficient for plant, specifically crop, needs. It occurs when loss of water from the soil by evaporation or otherwise greatly exceeds the water precipitation from the atmosphere; this may result from high winds, low humidity and heat. Areas with a well-defined dry season suffer seasonal drought. (See also DESERT; DUST BOWL.)

Drug addiction
An uncontrollable craving for a particular drug, usually a NARCOTIC, which develops into a physical or sometimes merely psychological dependence on it. Generally the individual acquires greater tolerance for the drug, and therefore requires larger and larger doses, to the point where he may take doses that would be fatal to the nonaddict. Should the supply be cut off he will suffer withdrawal symptoms that are psychologically grueling and often physically debilitating. Many drugs, such as cannabis and tobacco, are not addictive in the strictest sense but more correctly habit-forming. Others, such as the OPIUM derivatives, particularly HEROIN, MORPHINE and COCAINE, are extremely addictive. With others, such as LSD (and most other HALLUCINOGENIC DRUGS) and the AMPHETAMINES, the situation is unclear. An inability to abstain from regular self-dosage with a drug is described as a drug habit.

Drugs
Chemical agents that affect biological systems. In general they are taken to treat or prevent disease, but certain drugs, such as the OPIUM NARCOTICS, AMPHETAMINES, BARBITURATES and cannabis, are taken for their psychological effects and are termed recreational drugs, or drugs of addiction or abuse (see DRUG ADDICTION). Many drugs are the same as or similar to chemicals occurring naturally in the body and are used either to replace the natural substance (eg THYROID HORMONE) when deficient, or to induce effects that occur with abnormal concentrations, as with STEROIDS or oral CONTRACEPTIVES. Other agents are known to interfere with a specific mechanism. Many other drugs are obtained from nature, like ANTIBIOTICS from some fungi or bacteria and digitalis from plants. In addition, there are a number of entirely synthetic drugs (eg barbiturates), some of which are based on active parts of naturally occurring drugs.

In devising drugs for treating common conditions, an especially desirable factor is that the drug should be capable of being taken by mouth; that is, that it should be able to pass into the body unchanged in spite of being exposed to stomach acidity and the enzymes of the digestive system. In many cases this is possible, but there are some important exceptions, as with insulin, which has to be given by injection. This method may also be necessary if vomiting or gastrointestinal tract disease prevent normal absorption. In most cases the level of the drug in the blood or tissues determines its effectiveness. Factors affecting this include: the route of administration; the rate of distribution in the body; the degree of binding to PLASMA PROTEINS or fat; the rates of breakdown (eg by the LIVER) and excretion (eg by the KIDNEYS); the effect of disease on the organs concerned with excretion, and interactions with other drugs taken at the same time. There is also an individual variation in drug responsiveness which is also apparent with adverse effects. These arise because drugs acting on one system commonly act on others. Side effects may be nonspecific (nausea, diarrhea, malaise or skin rashes); allergic (HIVES, ANAPHYLAXIS), or specific to a drug (abnormal heart rhythm with digitalis). Drugs may cross the PLACENTA to reach the FETUS during pregnancy, interfering with its development and perhaps causing deformity as happens with THALIDOMIDE.

Drugs may be used for symptomatic relief or to control a disease. This can be accomplished by killing the infecting agents; by preventing specific infections; by restoring normal control over muscle (anti-Parkinsonian agents) or mind (antidepressants); by replacing a lost function or supplying a deficiency (eg vitamin B_{12} in ANEMIA); by suppressing inflammatory responses (steroids, aspirin); by improving the functioning of an organ (digitalis); by protecting a diseased organ by altering the function of a normal one (eg diuretics for heart failure), or by toxic actions on CANCER cells (cancer CHEMOTHERAPY). The scientific study of drugs is the province of PHARMACOLOGY.

Drumlins
Elongate hillocks, formed of TILL, found usually in swarms in lowland areas formerly covered by GLACIERS. They usually taper away from a steep slope that faced the oncoming ice.

Drupe
Botanically, a fleshy FRUIT comprising an outer skin (exocarp), a fleshy pulp (mesocarp) and an inner woody stone (endocarp) enclosing a single seed. Examples are found in the cherry, peach and plum.

Dry cell see BATTERY.

Dry ice
Solid carbon dioxide, used as a refrigerant for transporting perishables. It is made by compressing carbon dioxide gas to about $7MN/m^2$ at $-57°C$, when it liquefies; it is then expanded adiabatically to atmospheric pressure and cools, solid carbon dioxide "snow" separating. This is compressed into blocks. subl$-78.5°C$.

Dry rot
An invasive form of wood decay found in houses, resulting in loss of strength of timbers. The causal agents are the basidiomycete fungi *Merulius lacrymans* (in Britain) and *Poria incrassata* (in North America). Preventative measures include ensuring that conditions ideal for growth of the fungus are avoided, ie relative HUMIDITY is kept low, no free moisture is allowed and wood surfaces are covered. Treatment for the disease includes painting the wood surface with CREOSOTE and using FUNGICIDES to kill the fungus.

Dubos, René Jules (b. 1901)
French-born US microbiologist who discovered tyrothricin (1939), the first ANTIBIOTIC to be used clinically.

Ductility
The property of metals, alloys and some other substances to be drawn out or extruded (see EXTRUSION) without rupture or loss of strength. GOLD is the most ductile metal at normal temperatures. (See MALLEABILITY; MATERIALS, STRENGTH OF.)

Ductless glands see ENDOCRINE GLANDS.

Dulbecco, Renato (b. 1914)
Italian-born Anglo-American physiologist who shared the 1975 Nobel Prize for Physiology or Medicine with D. BALTIMORE and H. TEMIN for their work on cancer-causing VIRUSES.

Dulong, Pierre Louis (1785-1838)
French chemist who with A.T. PETIT discovered Dulong and Petit's Law. This states that the SPECIFIC HEATS of elements are inversely proportional to their ATOMIC WEIGHTS, and is equivalent to the observation that the atomic heats of most elements are about 26.4J/K.mol.

Dumas, Jean Baptiste André (1800-1884)
French organic chemist who discovered that CHLORINE could substitute for HYDROGEN in HYDROCARBONS and was thus led to propose his "law of substitution" (1834) which revolutionized the theory of organic chemistry. From 1868 he was secretary of the Academy of Sciences.

Dumbness
Failure of speech development, usually associated with congenital DEAFNESS, is the most common cause in childhood. APHASIA and hysterical mutism are the usual adult causes. If comprehension is intact, writing and sign language are alternative forms of communication, but in aphasia language

is usually globally impaired. (See SPEECH AND SPEECH DISORDERS.)

Dune
Hillock of sand built up by a prevailing wind, found mostly in DESERT areas. They have several forms, the commonest being barchans, crescent shaped with the horns pointing downwind, and transverse, elongate dunes at right angles to the wind direction.

Dunning, John Ray (1907-1975)
US physicist who first measured the ENERGY released in nuclear FISSION (1939) and the next year demonstrated the fission of URANIUM-235. He later helped to develop the gas-diffusion method of ISOTOPE separation used in making the first two ATOMIC BOMBS.

Duodenum
The first part of the small intestine, approx 30cm long, leading from the STOMACH to the jejunum (see GASTROINTESTINAL TRACT). The BILE and pancreatic ducts end in it and its injury may result in a FISTULA. Peptic ULCERS are common in the duodenum. (◆ page 40)

Dupuytren's contracture
Contracture of the fibrous sinews of the palm so that the fingers cannot be straightened, first described by the French surgeon Guillaume Dupuytren (1775-1835).

Duralumin
Aluminum-based ALLOY typically containing 4% copper, 1% magnesium, 0.7% manganese and 0.5% silicon. After heat treatment and aging it is hard and strong as steel, and, being light, is used in aircraft construction.

Dust
Fine particles, usually inorganic, that may be easily picked up by the wind and remain suspended in the air for long periods. It may be produced by volcanic action (dust from the Krakatoa explosion of 1883 circled the Earth several times, some taking years to settle), by wind EROSION (see DUST BOWL; SANDSTORM), by the breaking up of meteoroids (see METEOR) in the atmosphere, by salt spray from the oceans, and by industrial processes and auto exhausts. POLLEN is an example of an organic dust. Many dusts are serious health risks, especially the radioactive dust present in nuclear FALLOUT. (See also AIR POLLUTION; INTERSTELLAR MATTER.)

Dust bowl
Area of some 400,000km^2 in the S Great Plains region of the US that, during the 1930s, the Depression years, suffered violent dust storms as a result of accelerated soil EROSION. Grassland was plowed up in the 1910s and 1920s to plant wheat: a severe drought bared the fields, and high winds blew the topsoil (see SOIL) into huge dunes (see SANDSTORM). Despite rehabilitation programs, farmers plowed up grassland again in the 1940s and 1950s, and a repetition of the tragedy was averted only by the action of Congress.

Du Vigneaud, Vincent (1901-1978)
US biochemist who was awarded the 1955 Nobel Prize for Chemistry in recognition of his synthesis of the hormone OXYTOCIN (1954). He had earlier worked out the structure of the VITAMIN biotin (1942).

Dwarfism
Small stature may be a family characteristic or associated with disease of CARTILAGE or BONE development. Various HORMONE disorders may also result in dwarfism. The condition may also arise from spine or limb deformity (eg SCOLIOSIS), MALNUTRITION, RICKETS, chronic infection or visceral disease.

Dwarf star see STAR.

Dyes and dyeing
Dyes are colored substances which impart their color to textiles to which they are applied and for which they have a chemical affinity. They differ from pigments in being used in solution in an aqueous medium. Dyeing was practiced in the Fertile Crescent and China by 3000 BC, using natural dyes obtained from plants and shellfish. These were virtually superseded by synthetic dyes – more varied in color and applicability – after the accidental synthesis of mauve by SIR WILLIAM PERKIN (1856). The raw materials are aromatic hydrocarbons obtained from coal tar and petroleum. These are modified by introducing chemical groups called chromophores which cause absorption of visible light (see also COLOR). Other groups, auxochromes, such as amino or hydroxyl, are necessary for substantivity – ie affinity for the material to be dyed. This fixing to the fabric fibers is by hydrogen bonding, adsorption, ionic bonding or covalent bonding in the case of "reactive dyes" (see BOND, CHEMICAL). If there is no natural affinity, the dye may be fixed by using a MORDANT before or with dyeing. Vat dyes are made soluble by reduction in the presence of alkali, and after dyeing the original color is re-formed by acidification and oxidation; indigo and anthraquinone dyes are examples. Dyes are also used as biological stains (see MICROSCOPE), indicators and in photography. (See also AZO COMPOUNDS; FLUORESCEIN.)

Dyke see DIKE.

Dynamics
The branch of mechanics concerned with the actions of forces on bodies, with particular respect to the motions produced. (See MECHANICS.)

Dynamite
High EXPLOSIVE invented by ALFRED NOBEL, consisting of NITROGLYCERIN absorbed in an inert material such as kieselguhr or wood pulp. Unlike nitroglycerin itself, it can be handled safely, not exploding without a detonator. In modern dynamite sodium nitrate replaces about half the nitroglycerin. Gelatin dynamite or gelignite, contains also some nitrocellulose.

Dynamo
Also known as a generator, a machine for converting mechanical power into electrical power, normally by rotating a coil in a magnetic field (see GENERATOR, ELECTRIC). (◆ page 118)

Dyne (dyn)
Unit of FORCE in the CGS UNITS system, equal to 0.00001 newtons in SI UNITS.

Dysentery
An infection of the GASTROINTESTINAL TRACT causing severe diarrhea with blood and mucus. It is usually a short-lived illness but may cause dehydration. The organism may be carried in feces in the absence of symptoms. Bacillary dysentery is caused by bacteria of the genus *Shigella* and is associated with poor hygiene. Amebic dysentery is a chronic disease, usually seen in warm climates; constitutional symptoms occur and the disease may resemble noninfective COLITIS.

Dyslexia
Difficulty with reading, often a developmental problem possibly associated with suppressed left-handedness and spatial difficulty; it requires special training. It may also be acquired as a result of birth injury, failure of learning, visual disorders or as part of APHASIA (see SPEECH AND SPEECH DISORDERS).

Dysmenorrhea
Painful MENSTRUATION, most common in young women.

Dyspareunia
Painful sexual intercourse. In men it is usually due to infection, but in women may also be due to anatomical problems or to previous psychological or sexual trauma.

Dyspepsia (indigestion)
A vague term usually describing an abnormal visceral sensation in the upper abdomen or lower chest, often of a burning quality. Relief by ANTACIDS or milk is usual. HEARTBURN from esophagitis and pain of peptic (gastric or duodenal) ULCERS are usual causes.

Dysphasia see APHASIA; SPEECH AND SPEECH DISORDERS.

Dyspnoea
Difficulty in breathing. It can be caused by a variety of respiratory diseases, or by nervous tension.

Dysprosium (Dy)
One of the LANTHANUM SERIES of elements. AW 162.5, mp 1410°C, bp 2600°C, sg 8.550 (25°C). (◆ page 130)

Ear
Any sense organ concerned with hearing. In higher mammals it also plays a part in balance. It may be divided into the outer ear, extending from the cartilaginous external ear-flap, or pinna, to the tympanic membrane or ear drum and the inner ear embedded in the SKULL bones, consisting of cochlea and labyrinth. Between these two is the middle ear, an air-filled space containing small bones or ossicles, which communicates with the PHARYNX via the Eustachian tube. This allows the middle ear to be at the same pressure as the outer, and permits secretions to drain away. Three ossicles (malleus, incus, stapes) form a bony (ossicular) chain articulating between the ear drum and part of the cochlea; tiny muscles are attached to the drum and ossicles and can affect the intensity of sound transmission. The cochlea is a spiral structure containing fluid and specialized membranes on which hearing receptors are situated. The labyrinth consists of three semicircular canals, the utricle and the saccule, all of which contain fluid and receptor cells. Nerve fibers pass from the cochlea and labyrinth to form the eighth cranial nerve, which transmits the sound information obtained by the ear to the brain.

In hearing, sound waves are funneled into the ear by the pinna, and cause vibration of the ear drum. The drum and ossicular chain, which transmits vibration to the cochlea, effect some amplification. The cochlea translates those vibrations into nerve impulses, while the labyrinth is mainly concerned with balance.

Disease of the ear usually causes DEAFNESS or persistent ringing in the ears (TINNITUS). Peripheral disorders of balance include VERTIGO and ATAXIA, which may be accompanied by nausea or vomiting. MENIERE'S DISEASE is an episodic disease affecting both systems. (➧ page 213, 225)

Earth
The largest of the inner planets of the SOLAR SYSTEM, the third planet from the Sun and, so far as is known, the sole home of life in the Solar System. To an astronomer on Mars, several things would be striking about our planet. Most of all, he would notice the relative size of our MOON: there are larger moons in the Solar System, but none so large compared with its planet – indeed, some astronomers regard the Earth as one component of a "double planet", the other being the Moon. Our Martian astronomer would also notice that the Earth shows phases, just as the Moon and Venus do when viewed from Earth. And, if he were a radio astronomer, he would detect a barrage of radio "noise" from our planet – clear evidence of the presence of intelligent life.

The Earth is rather larger than Venus. It is slightly oblate (flattened at the poles), the equatorial diameter being about 12,756.4km, the polar diameter about 12,713.6km. It rotates on its axis in 23h 56min 04.1s (one sidereal day), though this is increasing by roughly 0.00001s annually due to tidal effects (see TIDES); and revolves about the Sun in 365d 6h 9min 9.5s (one sidereal year: see SIDEREAL TIME). Two other types of year are defined: the tropical year, the interval between alternate EQUINOXES (365d 5h 48min 46s); and the anomalistic year, the interval between moments of perihelion (see ORBIT), 365d 6h 13min 53s. The Earth's equator is angled about 23.5° to the ECLIPTIC, the plane of its orbit. The direction of the Earth's axis is slowly changing owing to PRECESSION. The planet has a mass of about 5.98×10^{21} tonnes, a volume of about $1.08\times10^{21}m^3$, and a mean DENSITY of about 5.52 tonnes/m^3.

Like other planetary bodies, the Earth has a magnetic field (see MAGNETISM). The magnetic poles do not coincide with the axial poles (see NORTH POLE; SOUTH POLE), and moreover they "wander". At or near the Earth's surface, magnetic declination (or variation) is the angle between true N and compass N (lines joining points of equal variation are isogonic lines); and magnetic dip (or inclination) the vertical angle between the MAGNETIC FIELD and the horizontal at a particular point. Isomagnetic lines can be drawn between points of equal intensity of the field. There is also evidence to suggest that the direction of the field reverses from time to time. These changes are of primary interest to the paleomagnetist (see PALEOMAGNETISM). The Earth is surrounded by radiation belts, probably the result of charged particles from the Sun being trapped by the Earth's magnetic field (see VAN ALLEN RADIATION BELTS; AURORA).

There are three main zones of the Earth: the ATMOSPHERE; the HYDROSPHERE (the world's waters); and the LITHOSPHERE, the solid body of the world. The atmosphere shields us from much of the harmful radiation of the Sun, and protects us from excesses of heat and cold. Water covers much of the Earth's surface (over 70%) in both liquid and solid (ice) forms (see GLACIER; OCEANS). There are permanent polar icecaps. The Earth's solid body can be divided into three regions. The core (diameter about 7000km), at a temperature of about 3000K, is at least partly liquid, though the central region (the inner core) is probably solid. Probably mainly of NICKEL and IRON, the core's density ranges between about 9.5 and perhaps over 15 tonnes/m^3. The mantle (outer diameter about 12,686km), probably mainly of OLIVINE, has a density of about 5.7 tonnes/m^3 toward the core, 3.3 tonnes/m^3 toward the crust, the outermost layer of the Earth and the one to which all human activity is confined. It is some 35km thick (much less beneath the oceans) and composed of three types of rocks: IGNEOUS ROCKS, SEDIMENTARY ROCKS and METAMORPHIC ROCKS. FOSSILS in the strata of sedimentary rocks give us a geological time scale (see GEOLOGY). The Earth formed about 4550 million years ago; life appeared probably about 3500 million years ago, and humans about 2 million years ago. Life has thus been present for about 77% of the Earth's history, man for about 0.05%, and civilization for less than 0.0001%.

It is now known that the Earth's configuration of continents and oceans has changed radically – through geological time – the map as it were, has changed. Originally, this was attributed to continents drifting, and the process was called CONTINENTAL DRIFT (see also ALFRED WEGENER). However, although this term is still used descriptively, the changes are now realized to be a manifestation of the theory of PLATE TECTONICS, and so a result of the processes responsible also for earthquakes, mountain building and many other phenomena. (➧➧ page 25, 134, 221)

Earthquake
A fracture or implosion beneath the surface of the Earth, and the shock waves that travel away from the point where the fracture has occurred. The immediate area where the fracture takes place is the focus or hypocenter, the point immediately above it on the Earth's surface is the epicenter, and the shock waves emanating from the fracture are called seismic waves.

Earthquakes occur to relieve a stress that has built up within the crust or mantle of the EARTH; fracture results when the stress exceeds the strength of the rock. The reasons for the stress buildup are to be found in the theory of PLATE TECTONICS. If a map is drawn of the world's earthquake activity, it can immediately be seen that earthquakes are confined to discrete belts. These belts signify the borders of contiguous plates; shallow earthquakes aregenerally associated with MID-OCEAN RIDGES, where creation of new material occurs, deep ones with regions where one plate is being forced under another.

Seismic waves are of two main types. Body waves travel from the hypocenter, and again are of two types: P (compressional) waves, where the motion of particles of the Earth is in the direction of propagation of the wave; and S (shear) waves, where the particle motion is at right angles to this direction. Surface waves travel from the epicenter, and are largely confined to the Earth's surface; Love waves are at right angles to the direction of propagation; Rayleigh waves have a more complicated, backward elliptical movement in the direction of propagation.

The experienced intensity of an earthquake depends mainly on the distance from the source. Local intensities are gauged in terms of the Mercalli Intensity Scale, which runs from I (Detectable only by SEISMOGRAPH) through to XII (Catastrophic). Comparison of intensities in different areas enables the source of an earthquake to be located. The actual magnitude of the event is gauged according to the RICHTER SCALE. The study of seismic phenomena is called seismology. (See also FAULT; TSUNAMI.) (➧ page 142-3)

Ebbinghaus, Hermann (1850-1909)
German psychologist who developed experimental techniques for the study of rote LEARNING and memory. In later life he devised means of intelligence testing and researched into color vision.

Ecchymosis
Bruises due to disease, eg leaking of blood vessels, rather than injury.

Eccles, Sir John Carew (b. 1903)
Australian physiologist who shared the 1963 Nobel Prize for Physiology or Medicine with ALAN HODGKIN and ANDREW HUXLEY. Using their findings, he had been able to establish the chemical bases of the electrical changes during transmission of nervous impulses across the SYNAPSES (see also NERVOUS SYSTEM).

Ecdysis
The periodic shedding of the EXOSKELETON in insects, crustaceans, arachnids and other ARTHROPODS to allow for growth. Because the exoskeleton is rigid, a steady increase in size is prohibited. The animal therefore grows until it has filled its old exoskeleton, then seeks out a safe, secluded place in which to molt. The old cuticle splits open and is shed, often being eaten for the nutrients it contains. Meanwhile the body expands rapidly, by uptake of water or air, before the new exoskeleton has had time to harden.

Echinoderms
A phylum of about 5000 marine animals that includes the Crinoidea (sea lilies and feather stars), Asteroidea (starfish), Ophuroidea (brittlestars), Echinoidea (sea urchins, heart urchins and sand dollars) and the Holothuroidea (sea cucumbers). All echinoderms have a skeleton

composed of plates of CALCITE, radial (usually five-fold) symmetry and a water-vascular system by which they extend their tentacles and move about. Fossil echinoderms include crinoids and blastoids; their history goes back to the Lower Cambrian, 590 million years ago. (♦ page 17)

Echiuroids
A phylum of ovoid or bulbous, worm-like marine animals of uncertain affinities.They have a muscular proboscis, a single pair of chaetae and are unsegmented. Echiuroids inhabit rock crevices or U-shaped burrows excavated in mud or sand.

Echo
A wave signal reflected back to its point of origin from a distant object, or, in the case of RADIO signals, a signal coming to a receiver from the transmitter by an indirect route. Echoes of the first type can be used to detect and find the position of reflecting objects (echolocation). High-frequency SOUND echolocation is used both by bats for navigation and to detect prey and by man in marine SONAR. RADAR, too, is similar in principle, though this uses UHF radio and MICROWAVE radiation rather than sound energy. The range of a reflecting object can easily be estimated for ordinary sound echoes: since sound travels about 340m/s through the air at sea level, an object will be distant about 170m for each second that passes before an echo returns from it.

Echolocation
In animals, the ability to issue sounds and interpret their echoes from the surroundings, a process similar to RADAR that is useful for nocturnal animals, those living in caves, and those hunting in the ocean's depths where little light penetrates. At its most sophisticated this sensory ability can be as accurate as vision in pinpointing prey items, avoiding obstacles and building up a picture of the surroundings. It was first discovered in insect-eating bats (Microchiroptera), which emit in the ultrasonic range (inaudible to humans) and are accomplished enough to detect fine wires stretched across their flightpath. The larger, fruit-eating bats of the Suborder Megachiroptera fly mainly at dusk, have large eyes and use echolocation as a secondary sense. These bats issue audible sounds, which give far less sensitive echolocation than ultrasounds; they are thought to have evolved this ability independently of the Microchiroptera. Certain cave-dwelling birds, the oilbird and the cave swiftlets, also practice echolocation, both of them using audible clicks. Dolphins and other toothed whales have ultrasonic echolocation.

Eclampsia
A rare, sometimes fatal, condition of late pregnancy, of unknown cause, consisting of fits. It is preceded by pre-eclampsia, formerly known as toxemia of pregnancy, a relatively common condition characterized by edema (swelling), high blood pressure, and albumin in the urine.

Eclipse
The partial or total obscurement of one celestial body by another; also the passage of the Moon through the Earth's shadow. The components of a binary star (see DOUBLE STAR) may eclipse each other as seen from the Earth, in which case the star is termed an eclipsing binary. The Moon frequently eclipses stars or planets, and this is known as OCCULTATION.

A lunar eclipse occurs when the Moon passes through the umbra of the Earth's shadow. This happens usually not more than twice a year, as the Moon's orbit around the Earth is tilted with respect to the ECLIPTIC. The eclipsed Moon is blood-red in color because some of the Sun's light is refracted by the Earth's atmosphere into the umbra. A partial lunar eclipse occurs when only part of the umbra falls on the Moon.

In a solar eclipse the Moon passes between the Sun and the Earth. A total eclipse occurs when the observer is within the umbra of the Moon's shadow: the disk of the Sun is covered by that of the Moon, and the solar corona (see SUN) becomes clearly visible. The maximum possible duration of a total eclipse is about 7½ min. Should the observer be outside the umbra but within the penumbra, or should the Earth pass through only the penumbra, a partial eclipse will occur.

An annular eclipse is seen when the Moon is at its farthest from the Earth, its disk not being large enough totally to obscure that of the Sun. The Moon's disk is seen surrounded by a brilliant ring of light.

Ecliptic
The great circle traced out on the CELESTIAL SPHERE by the apparent motion of the Sun during the year, corresponding to the motion of the Earth around the Sun. The ecliptic passes through 12 constellations, known as the constellations of the ZODIAC.

Ecology
The study of living organisms in relation to their ENVIRONMENT. The whole Earth can be considered as a large ecological unit: the term BIOSPHERE is used to describe the atmosphere, Earth surface, oceans and ocean floors within which living organisms exist. However, it is usual to divide the biosphere into a large number of ecological sub- units or ecosystems. Typical examples of ecosystems are a pond, a deciduous forest or a desert. The overall climate and topography within an area are major factors determining the type of ecosystem that develops. Within any ecosystem minor variations give rise to smaller communities within which animals and plants occupy their own particular NICHES.

The most important factor for any organism is its source of energy or food. Thus, within any ecosystem complex patterns of feeding relationships are built up. These are represented by the ecologist as food chains or food webs. Plants and photosynthetic bacteria (AUTOTROPHS) are the primary source of food energy; they derive it from PHOTOSYNTHESIS, utilizing light, water and carbon dioxide. Herbivores then obtain their food by eating plants. In their turn, herbivores are preyed upon by carnivores, who may also be the source of food for other carnivores. Animal and plant waste is decomposed by SAPROTROPHS such as BACTERIA and FUNGI, within the habitat, and this returns the raw materials to the air, soil or water. The number of links within a food chain are normally three or four, with five, six and seven less frequently. The main reason for the limited length of food chains is that the major part of the energy stored within a plant or animal is wasted at each stage in the chain. Thus if it were possible for a carnivore to occupy, say, position 20 within a food chain, the area of vegetation required to supply the energy needed for the complete chain would be the size of a continent.

Within any ecosystem raw materials such as nitrogen, carbon, oxygen and hydrogen (in water) are continually being recycled via a number of processes including the NITROGEN CYCLE, CARBON CYCLE, photosynthesis, and respiration.

Most natural ecosystems are in a state of equilibrium or balance so that few changes occur in the natural flora and fauna. However, when changes occur in the environment, either major climatic changes or alterations made by man, an imbalance results and the ecosystem changes. The natural sequence of change that leads to a new period of equilibrium is called a SUCCESSION and may take any length of time, from a few years for the establishment of a new species to several centuries for the change from grassland to forest. Over millions of years largely stable ecosystems have evolved. Natural changes, such as adaptations to changing of climate, tend to be gradual. However, humans often cause much more sudden changes – the introduction of a disease to a hitherto uninfected area, the cutting down of forests, drainage of marshes or the pollution of rivers, seas, soils or the atmosphere. The effects of this type of change upon an ecosystem can be rapid, catastrophic and irreversible. All forms of life depend entirely upon the sensitive balance within the environment, and any change with worldwide effects could have devastating consequences. (See also CONSERVATION; GREENHOUSE EFFECT.) (♦ page 44-5)

Ecosystem see ECOLOGY.

Ectoderm
In animals, the outer layer of the body. It is derived from the outer layer of cells in the blastula (see EMBRYOLOGY), and gives rise to the skin and nervous system, including the brain. (See also ENDODERM; MESODERM.)

Ectoplasm
The outer layer of CYTOPLASM in a cell. It is more solid than the inner layer of ENDOPLASM, and is involved in cell movement.

Ectoprocts
Another name for the BRYOZOA or moss animals.

Ectotherm
A "cold-blooded" animal or POIKILOTHERM.

Eczema
Form of DERMATITIS, usually with redness and scaling. It runs in families, often being worst in childhood, and is associated with HAY FEVER and ASTHMA.

Eddington, Sir Arthur Stanley (1882-1944)
English astronomer and astrophysicist who pioneered the theoretical study of the interior of STARS and who, through his *Mathematical Theory of Relativity* (1923), did much to introduce the English-speaking world to the revolutionary new theories of EINSTEIN.

Edelman, Gerald Maurice (b. 1929)
US biochemist who shared with RODNEY PORTER the 1972 Nobel Prize for Physiology or Medicine for his researches into the chemical structures of antibodies (see ANTIBODIES AND ANTIGENS).

Edema
The accumulation of excessive watery fluid outside the cells of the body, causing swelling of a part. Some edema is seen locally in INFLAMMATION. The commonest type is gravitational edema, where fluid swelling is in the most gravity-dependent parts, typically the feet. HEART or LIVER failure, MALNUTRITION and KIDNEY disease are common causes, while disease of VEINS or LYMPH vessels in the legs also leads to edema. Serious edema may form in the LUNGS in heart failure and in the BRAIN in some disorders of METABOLISM, trauma, TUMORS and infections.

Edentates
An order of MAMMALS with only three living families: the anteaters, armadillos and sloths. The anteaters are completely toothless (edentate) but the sloths and armadillos have primitive molars. Extinct edentates include the giant ground sloth, *Megatherium*.

Edgerton, Harold Eugene (b. 1903)
US electrical engineer known for his development of rapid-flash STROBOSCOPES and application of them to high speed PHOTOGRAPHY.

Edison, Thomas Alva (1847-1931)
Prolific US inventor, with over 1000 patents issued to his name. His first successful invention, an improved stock-ticker (1869), earned him the capital to set up as a manufacturer of telegraphic apparatus. He then devised the diplex method of telegraphy which allowed one wire to carry four messages at once. Moving to a new "invention

factory" (the first large-scale industrial research laboratory) at Menlo Park, New Jersey, in 1876, he devised the carbon transmitter and a new receiver which made A.G Bell's telephone commercially practical. His tin-foil phonograph followed in 1877, and in the next year he started to work toward devising a practical incandescent lightbulb. By 1879 he had produced the carbon-filament bulb and electric lighting became a reality, though it was not until 1882 that his first public generating station was supplying power to 85 customers in New York. Moving his laboratories to West Orange, New Jersey, in 1887 he set about devising a motion-picture system (ready by 1889) though he failed to exploit its entertainment potential.

In all his career he made only one important scientific discovery, the Edison effect – the ability of ELECTRICITY to flow from a hot filament in a vacuum lamp to another enclosed wire but not the reverse (1883) – and, because he saw no use for it, he failed to pursue the matter. His success was probably more due to perseverance than any special insight; as he himself said: "Genius is one percent inspiration and ninety-nine percent perspiration."

EEG (electroencephalogram)
A recording of the electrical activity of the brain. It is useful for diagnosing neurological disorders and for determining BRAIN DEATH.

Efficiency
In THERMODYNAMICS and the theory of MACHINES, the ratio of the useful WORK derived from a machine to the ENERGY put into it. The mechanical efficiency of a machine is always less than 100%, some energy being lost as HEAT in FRICTION. When the machine is a heat engine, its theoretical thermal efficiency can be found from the second law of thermodynamics but actual values are often rather lower. A typical gasoline engine may have a thermal efficiency of only 25%, a steam engine 10%.

Efflorescence
In chemistry, spontaneous loss of water from a crystalline HYDRATE, which crumbles on its surface to an anhydrous powder. SODIUM carbonate and sulfate are common examples. Like its converse, DELIQUESCENCE, it depends on the relative HUMIDITY, occurring if the partial VAPOR PRESSURE of water at the solid exceeds that of the air.

Egg
The female GAMETE, germ cell or OVUM. Popularly, the term is used to describe those animal eggs that are deposited by the female and develop outside her body, such as the eggs of snails, amphibians, reptiles and birds; those that develop internally, as in mammals, are also known as eggs. The egg is a single cell that develops into a new individual after FERTILIZATION by a single sperm cell, or male gamete. In animals it is formed in a primary sex organ or GONAD called the ovary. In fish, reptiles and birds there is a food store of yolk enclosed within its outer membrane. In ANGIOSPERMS the female reproductive organs form part of the FLOWER; the egg cell is found within the ovules, which upon fertilization develop into the embryo and SEED. Unlike sperm, eggs are all immotile; they are generally much larger than the sperm. (⧫⧫ page 32, 204)

Ego
That part of the mind that develops from a person's experience of the outside world. According to FREUD the ego is said to reconcile the demands of the ID, SUPEREGO and reality.

Ehrlich, Paul (1854-1915)
German bacteriologist and immunologist, the founder of CHEMOTHERAPY and an early pioneer of HEMATOLOGY. His discoveries include: a method of staining (1882), and hence identifying, the TUBERCULOSIS bacillus (see also ROBERT KOCH); the reasons for immunity in terms of the chemistry of ANTIBODIES AND ANTIGENS, for which he was awarded (with METCHNIKOFF) the 1908 Nobel Prize for Physiology or Medicine; and the use of the drug Salvarsan to cure syphilis (see SEXUALLY TRANSMITTED DISEASES), the first DRUG to be used in treating the root cause of a disease (1911).

Eidetic image
An exceptionally vivid mental image which the individual "sees" either projected onto a suitable background (eg a wall) or with eyes closed in a darkened room. The image may be a memory (eidetic memory) or a fantasy, and may be either voluntary or spontaneous. Eidetic imagery is most common among children.

Eiffel, Alexandre Gustave (1832-1923)
French engineer best known for his design and construction of the Eiffel Tower, Paris (1887-89), from which he carried out experiments in AERODYNAMICS. In 1912 he founded the first aerodynamics laboratory.

Eigen, Manfred (b. 1927)
German physicist awarded, with R. NORRISH and G. PORTER, the 1967 Nobel Prize for Chemistry for studies of extremely fast chemical reactions.

Eijkman, Christiaan (1858-1930)
Dutch pathologist. Following a trip to Indonesia (1886) to investigate BERIBERI he was able to show that the disease resulted from a dietary deficiency. This led to the discovery of VITAMINS. For his work he was awarded (with SIR F.G. HOPKINS) the 1929 Nobel Prize for Physiology or Medicine.

Einstein, Albert (1879-1955)
German-born Swiss-American theoretical physicist, the author of the theory of RELATIVITY. In 1905 Einstein published several papers of major significance. In one he applied PLANCK'S QUANTUM THEORY to the explanation of photoelectric emission. For this he was awarded the 1921 Nobel Prize for Physics. In a second he demonstrated that it was indeed molecular action which was responsible for BROWNIAN MOTION. In a third he published the special theory of relativity, with its postulate of a constant VELOCITY for LIGHT (c) and its consequence, the equivalence of MASS (m) and ENERGY (E), summed up in the famous equation $E=mc^2$. In 1915 he went on to publish the general theory of relativity. This came with various testable predictions, all of which were spectacularly confirmed within a few years.

Einstein was on a visit to the US when Hitler came to power in Germany and, being a Jew, decided not to return to his native land. The rest of his life was spent in a fruitless search for a "unified field theory" which could combine QUANTUM MECHANICS with GRAVITATION theory. After 1945 he also worked hard against the proliferation of nuclear weapons, although he had himself, in 1939, alerted President F.D. Roosevelt to the danger that Germany might be the first to develop an ATOMIC BOMB, and had thus played an important part in the setting up of the Manhattan Project. (⧫ page 151)

Einsteinium (Es)
A TRANSURANIUM ELEMENT in the ACTINIDE series, first found in the debris from the first HYDROGEN BOMB, and now prepared by bombardment of lighter actinides. (⧫ page 130)

Einthoven, Willem (1860-1927)
Dutch physiologist awarded the 1924 Nobel Prize for Physiology or Medicine for his invention of, and investigation of heart action with, the ELECTROCARDIOGRAPH. In 1903 he devised the string galvanometer, a single fine wire placed under tension in a MAGNETIC FIELD. Current passed through the wire causes a deflection which can be measured, for greater accuracy, by microscope. This GALVANOMETER was sensitive enough for him to use it to record the electrical activity of the HEART.

Elasmobranchs
A collective name for the sharks, rays, mantas and dogfish – that is, all members of the CHONDRICHTHYES (cartilaginous fish) apart from the aberrant CHIMERAS or rat-tails.

Elasticity
The ability of a body to resist tension, torsion, shearing or compression and to recover its original shape and size when the stress is removed. All substances are elastic to some extent, but if the stress exceeds a certain value (the elastic limit), which is soon reached for brittle and plastic materials, permanent deformation occurs. Below the elastic limit, bodies obey Hooke's Law (see MATERIALS, STRENGTH OF).

Elastomer
Any POLYMER suitable for use as a rubber.

Electrical engineering
Branch of technology dealing with the practical applications of ELECTRICITY and ELECTRONICS, and thus concerned with generation of electric power, design and construction of electrical and electronic components, and the use of these components in integrated, functional systems.

Electric arc
A high-current electric discharge between two ELECTRODES. The current is carried by the gas PLASMA maintained by the discharge. Sodium and neon lights exemplify large, relatively cool arcs, while arc-welding uses a small, very hot arc between a slowly consumed electrode and the workpiece. LIGHTNING is a naturally occurring arc.

Electric current see ELECTRICITY.

Electric field
What is said to exist where stationary electric charges (see ELECTRICITY) experience a force: the field is defined with the strength and direction of the force on a unit charge. Electric fields are produced by (other) electric charges, and by changing magnetic fields as in a dynamo (see ELECTROMAGNETISM). (⧫ page 118)

Electric furnace
Any of various industrial furnaces heated electrically. In an arc furnace, batches of steel are melted by the heat of electric arcs struck between graphite electrodes and the charge. In induction furnaces, used mainly for remelting batches of steel, a refractory crucible is surrounded by a large water-cooled copper coil. When this is connected to a high-frequency AC supply, eddy currents are induced in the batch which heat it. Often a lower-frequency field is also applied to help stir the charge. Chamber furnaces are simply large electric ovens, heated from the walls.

Electric generator see GENERATOR, ELECTRIC.

Electricity
The phenomena of charged particles at rest and in motion. Electricity is used to provide a highly versatile form of ENERGY, electrical devices being used in heating, lighting, machinery, telephony and ELECTRONICS. Electric charge is an inherent property of matter, ELECTRONS carrying a negative charge of 1.602×10^{-19} coulomb, and atomic nuclei normally carrying a similar positive charge for each electron in the ATOM. When the balance is disturbed (in the case of glass, by rubbing it, for example), a net charge is left on an object, and the study of such isolated charges is called electrostatics. Like charges repel and unlike charges attract each other with a FORCE proportional to the two charges and inversely proportional to the square of their separation (the inverse square law): electrostatic repulsion, for example, will be familiar in newly combed hair. The force is normally interpreted in terms of an ELECTRIC FIELD produced by one charge with which the other charge

interacts. The fields produced by any number of charges may be superposed independently, and the result conveniently represented graphically by field lines, beginning at positive and ending at negative charges, showing by their direction that of the field, and by their density its strength. Pairs of equal but opposite charges separated by a small distance are called dipoles, the product of charge and separation being called the DIPOLE MOMENT. These experience a TORQUE in an electric field tending to align them with the field, but no net force unless the field is nonuniform. The amount of work done in moving a unit charge from one point to another against the electric field is called the electric potential difference or voltage between the points, and is measured in VOLTS (volts (V)=joules/coulomb). The ratio of a charge added to a body to the voltage produced is called the CAPACITANCE of the body; for most practical purposes, the Earth provides a reference potential with an infinite capacitance.

In some materials, known as electric CONDUCTORS, there are charges free to move about – for example, valence electrons in metals, or IONS in salt solutions – and in these, the presence of an electric field produces a steady flow of charge with positive charges flowing in the direction of the field (negative charges moving the opposite way); such a flow constitutes an electric current, measured in AMPERES (amperes (A)=coulombs/second). The field implies a voltage between the ends of the conductor, which is normally proportional to the current (OHM'S LAW), the ratio being called the RESISTANCE of the conductor, and measured in OHMS (ohms (Ω)=volts/amps); it normally rises with TEMPERATURE. Materials with high resistance to currents are classed as insulators. (See also SEMICONDUCTOR, DIELECTRIC.) The energy acquired by the charges in falling through the field is dissipated as HEAT – and LIGHT, if a sufficient temperature is reached – the total POWER output being the product of current and voltage. Thus a 1 kW fire supplied at 110 V draws a current of about 9 A, and the hot element has a resistance of about 12 Ω.

Electric sources such as BATTERIES or GENERATORS convert chemical, mechanical, or other energy into electrical energy (see ELECTROMOTIVE FORCE), and will pump charge through conductors much as a water pump circulates water in a radiator heating system (see CIRCUIT, ELECTRIC). Batteries create a constant voltage, and so produce a steady or direct current (DC); many generators on the other hand provide a voltage which changes in sign many times a second, and so produce an alternating current (AC) in which the charges move to and fro instead of continuously in one direction. This system has advantages in generation, transmission and application, and is now used almost universally for domestic and industrial purposes.

An electric current is found to produce a MAGNETIC FIELD circulating around it, to experience a force in an externally generated magnetic field, and to be itself generated by a changing magnetic field; for more details of these properties on which most electrical machinery depends, see ELECTROMAGNETISM.

Static electricity was known to the Greeks; an inverse square law was hinted at by J. PRIESTLEY in 1767 and later confirmed by H. CAVENDISH and C.A. DE COULOMB. G.S. OHM formulated his law of conduction in 1826, though its essentials were known before then. The common nature of all the "types of electricity" then known was shown in 1826 by M. FARADAY, who also originated the concept of electric field lines. (➧ page 118)

Electric meter
A single-phase AC watt-hour meter used to measure the electricity consumption of domestic consumers. It is a type of simple electric motor with a disk free to rotate in the magnetic field set up by two sets of coils, one in series with (the current coil), and the other in parallel to (the potential coil), the applied load. The rate of rotation is proportional to the power being used and so, by counting the rotations mechanically, the total energy consumption can be measured.

Electric motor see MOTOR, ELECTRIC.

Electric power see ELECTRICITY; POWER.

Electrocardiograph
Instrument for recording the electrical activity of the HEART, producing its results in the form of multiple tracing called an electrocardiogram (ECG). These are conventionally recorded with twelve combinations of electrodes on the limbs and chest wall. The electrical impulses in the conducting tissue and muscle of the heart pass through the body fluids, while the position of the electrodes determines the way in which the heart is "looked at" in electrical terms. ECGs allow coronary thrombosis, abnormal heart rhythm, disorders of the heart muscle and pericardium to be detected, as well as diseases of the metabolism that affect the heart.

Electrochemical series (electromotive series)
A sequence of elements (chiefly metals) listed in order of their standard redox potentials – ie the potential developed by an electrode of the element immersed in a molar solution (see MOLECULAR WEIGHT) of one of its salts (see ELECTROCHEMISTRY; OXIDATION AND REDUCTION). Metals high in the series are generally more reactive than those lower down, and displace them from aqueous solutions of their salts. (See also ELECTRONEGATIVITY; IONIZATION POTENTIAL.)

Electrochemistry
Branch of PHYSICAL CHEMISTRY dealing with the interconversion of electrical and chemical energy (see CELL, ELECTROCHEMICAL). Many chemical species are electrically charged IONS (see BOND, CHEMICAL), and a large class of reactions – OXIDATION AND REDUCTION – consists of electron-transfer reactions between ions and other species. If the two half-reactions (oxidation, reduction) are made to occur at different ELECTRODES, the electron-transfer occurs by the passing of a current through an external circuit between them (see BATTERY; FUEL CELL). The ELECTROMOTIVE FORCE driving the current is the sum of the electrode potentials (in volts) of the half-reactions, which represent the free energy (see THERMODYNAMICS) produced by them. Conversely, if an emf is applied across the electrodes of a cell, it causes a chemical reaction if it is greater than the sum of the potentials of the half-reactions (see ELECTROLYSIS). Such potentials depend both on the nature of the reaction and on the concentrations of the reactants. Cells arising through concentration differences are one cause of CORROSION. (See also CONDUCTIVITY; ELECTROMETALLURGY; ELECTROPHORESIS.)

Electrocution
The usually fatal effect of passing a high-energy electric current through the body. ELECTRICITY passing through the body fluid, which acts as a resistor, causes BURNS at sites of connection and along the electrical pathway. CONVULSIONS and rhythm disturbance in the heart are usual; the latter are the immediate cause of death.

Electrode
A component in an electric CIRCUIT at which current is transferred between ordinary metal conductors and a gas or electrolyte. A positive electrode is an ANODE and a negative one a CATHODE.

Electrodynamics
The study of the interaction of charged particles with ELECTROMAGNETIC FIELDS, as summarized in the definitions of ELECTRIC and MAGNETIC FIELDS through the FORCES experienced by charged particles and described by the MAXWELL equations.

Electroencephalograph
Instrument for recording the BRAIN's electrical activity using several small electrodes on the scalp. Its results are produced in the form of a multiple tracing called an electroencephalogram (EEG). It is useful for diagnosing neurological disorders and for determining BRAIN DEATH.

Electroluminescence
The emission of LIGHT from some PHOSPHORS (particularly ZnS) and SEMICONDUCTORS under the influence of an applied ELECTRIC FIELD. Electrophotoluminescent screens (in which the effect is PHOTON-activated) are used medically to intensify X-ray pictures. (See also LUMINESCENCE.)

Electrolysis
Production of a chemical reaction by passing a direct current through an electrolyte – ie a compound that contains ions when molten or in solution. (See ELECTROCHEMISTRY.) The CATIONS move toward the CATHODE and the ANIONS toward the ANODE, thus carrying the current. At each electrode the ions are discharged according to FARADAY's laws: (1) the quantity of a substance produced is proportional to the amount of electricity passed; (2) the relative quantities of different substances produced are proportional to their EQUIVALENT WEIGHTS. Thus one gram-equivalent of any substance is produced by the same amount of electricity, known as a faraday (96,500 coulombs). Electrolysis is used to extract electropositive metals from their ores (see ELECTROCHEMICAL SERIES), and to refine less electropositive metals, to produce SODIUM hydroxide, CHLORINE, HYDROGEN, OXYGEN and many others; and in ELECTROMETALLURGY. (➧ page 118)

Electrolyte
When salts are dissolved they break down into charged particles called electrolytes. The principal blood electrolytes are sodium, potassium, calcium, chloride and bicarbonate. (➧ page 118)

Electromagnetic field
Any combination of ELECTRIC and MAGNETIC FIELDS, the two being closely related. (➧ page 118)

Electromagnetic force
The force between electrically charged bodies. It is one of the four fundamental forces recognized by physicists. At the quantum level it is transmitted by PHOTONS, and is well described by the theory of quantum electrodynamics. (See also QUANTUM THEORY, SUBATOMIC PARTICLES, UNIFIED FIELD THEORY.)

Electromagnetic radiation (radiant energy)
The form in which ENERGY is transmitted through space or matter using a varying ELECTROMAGNETIC FIELD. Classically, radiant energy is regarded as a wave MOTION. In the mid-19th century MAXWELL showed that an oscillating (vibrating) electric charge would be surrounded by varying electric and magnetic fields. Energy would be lost from the oscillating charge in the form of transverse waves in these fields, the waves in the electric field being at right angles both to those in the magnetic field and to the direction in which the waves are traveling (propagated). Moreover, the VELOCITY of the waves would depend only on the properties of the medium through which they passed; in a vacuum its value is a fundamental constant of physics – the electromagnetic constant, c=299,792.5km/s. At the beginning of the 20th century PLANCK proposed that certain properties of radiant energy were best explained by regarding it as transporting energy in discrete amounts or packets called quanta. EINSTEIN later proposed the name PHOTON for the electromagnetic quantum. The energy of each photon is

proportional to the frequency of the associated radiation (see QUANTUM THEORY). The different kinds of electromagnetic radiation are classified according to the energy of the photons involved, the range of energies being known as the electromagnetic SPECTRUM. (Looked at in other ways, this spectrum arranges the radiations according to wavelength or frequency.) In order of decreasing energy the principal kinds are GAMMA RAYS, X-RAYS, ULTRAVIOLET RADIATION, LIGHT, INFRARED RADIATION, MICROWAVES and RADIO waves. In general, the higher the energies involved, the better the properties of the radiation are described in terms of particles (photons) rather than waves. Radiant energy is emitted from objects when they are heated (see BLACKBODY RADIATION) or otherwise energetically excited (see LUMINESCENCE; SPECTROSCOPY). It is used to channel and distribute both energy and information. (See also ABSORPTION.) (♦♦ page 21, 118)

Electromagnetic spectrum see SPECTRUM.

Electromagnetism

The study of ELECTRIC and MAGNETIC FIELDS, and their interaction with electric charges and currents. The two fields are in fact different manifestations of the same physical field, and are interconverted according to the speed of the observer. Apart from the effects noted under ELECTRICITY and MAGNETISM, the following are found:

1. Moving charges (and hence currents) in magnetic fields experience a FORCE, perpendicular to the field and the current, and proportional to their product. This is the basis of all electric MOTORS, and was first applied for the purpose by M. FARADAY in 1821.

2. A change in the number of magnetic field lines passing through a circuit "induces" an electric field in the circuit, proportional to the rate of the change. The basis of most GENERATORS, this was also established by M. Faraday, in 1831.

3. An effect analogous to the above, but with magnetic and electric fields interchanged, and usually much smaller. This was hypothesized by J.C. MAXWELL, who in 1862 deduced from it the possibility of self-sustaining electromagnetic waves traveling at a speed which coincided with that of LIGHT, thereby identifying the nature of visible light, and predicting other waves such as the RADIO waves found experimentally by H. HERTZ shortly afterwards. (♦ page 118)

Electrometallurgy

Branch of METALLURGY that uses electricity. It includes the use of the ELECTRIC FURNACE, and also the use of ELECTROLYSIS for extracting and refining metals, electroplating, anodizing and electroforming.

Electromotive force (emf)

Loosely, the voltage produced by a BATTERY, GENERATOR or other source of ELECTRICITY, but more precisely, the product of the current it produces in a circuit and the total circuit RESISTANCE, including that of the source itself. The actual voltage across the source is usually somewhat lower.

Electron

A stable SUBATOMIC PARTICLE, with rest MASS 9.1091×10^{-31}kg (roughly 1/1836 the mass of a HYDROGEN atom) and a negative charge of 1.6021×10^{-19}C, the charges of other particles being positive or negative integral multiples of this. Electrons are one of the basic constituents of ordinary MATTER, commonly occupying the ORBITALS surrounding positively charged atomic nuclei. The chemical properties of ATOMS and MOLECULES are largely determined by the behavior of the electrons in their highest-energy orbitals. Both CATHODE RAYS and BETA RAYS are streams of free electrons passing through a gas or vacuum. The unidirectional motion of electrons in a solid conductor constitutes an electric current. Solid conductors differ from nonconductors in that in the former some electrons are free to move about while in the latter all are permanently associated with particular nuclei. Free electrons in a gas or vacuum can usually be treated as classical particles, though their wave properties become important when they interact with or are associated with atomic nuclei. The anti-electron, with identical mass but an equivalent positive charge, is known as a positron (see ANTIMATTER). (♦ page 21, 37, 49)

Electronegativity

The relative power of an atom in a molecule to attract electrons. A concept variously defined and estimated since its proposal by L. PAULING, it depends on the atom's VALENCE state (see BOND, CHEMICAL) and is useful only as a qualitative guide to the polarity of bonds and molecules. Electropositive metals are generally high in the ELECTROCHEMICAL SERIES. (See also NUCLEOPHILES.)

Electron gun

That part of an ELECTRON TUBE which produces, accelerates, focuses and deflects a beam of ELECTRONS. A CATHODE emits electrons which pass through a grid to the steering ANODES.

Electronics

An applied science dealing with the development and behavior of ELECTRON TUBES, SEMICONDUCTORS and other devices in which the motion of electrons is controlled; it covers the behavior of electrons in gases, vacuums, conductors and semiconductors. Its theoretical basis lies in the principles of ELECTROMAGNETISM and solid-state physics discovered in the late 19th and early 20th centuries. Electronics began to grow in the 1920s with the development of RADIO. During WWII, the US and UK concentrated resources on the invention of RADAR and pulse transmission methods and by 1945 they had enormous industrial capacity for producing electronic equipment. The invention of the TRANSISTOR in 1948 as a small, cheap replacement for vacuum tubes led to the rapid development of COMPUTERS, transistor radios, etc. Now, with the widespread use of integrated circuits, electronics plays a vital role in communications (telephone networks, information storage, etc.) and industry. All electronic circuits contain both active and passive components and transducers (eg MICROPHONES) which change ENERGY from one form to another. Sensors of light, temperature, etc., may also be present. Passive components are normally conductors and are characterized by their properties of RESISTANCE (R), CAPACITANCE (C) and INDUCTANCE (L). One of these usually predominates, depending on the function required. Active components are electron tubes or semiconductors; they contain a source of power and control electron flow. The former may be general purpose tubes (DIODES, TRIODES, etc., the name depending on the number of ELECTRODES) which rectify, amplify or switch electric signals. Image tubes (in television receivers) convert an electric input into a light signal; photoelectric tubes (in television cameras) do the reverse. Semiconductor diodes and transistors, which are basically sandwiches made of two different types of semiconductor, now usually perform the general functions once done by tubes, being smaller, more robust and generating less heat. These few basic components can build up an enormous range of circuits with different functions. Common types include power supply (converting AC to pulsing DC and then smoothing out the pulsations); switching and timing (the logic circuits in computers are in this category); amplifiers, which increase the amplitude or power of a signal, and oscillators, used in radio and television transmitters and which generate AC signals. Demands for increased cheapness and reliability of circuits have led to the development of microelectronics. In printed circuits, printed connections replace individual wiring on a flat board to which about two components per cm^3 are soldered. Integrated circuits assemble thousands of components and interconnections in a single structure, formed directly by evaporation or other techniques as films about 0.03mm thick on a substrate. In monolithic circuits, components are produced in a tiny chip of semiconductor by selective diffusion. The first integrated circuits used SSI (small-scale integration, with fewer than 20 transistors), followed by MSI (medium-scale integration, up to 100 transistors) and LSI (large-scale integration, over 10,000 transistors). LSI made the construction of the MICROPROCESSOR possible. VLSI (very large-scale integration) is the next stage, with over 20,000 transistors on each integrated circuit. (See also CHIPS.)

Electron microscope

A microscope using a beam of ELECTRONS rather than LIGHT to study objects too small for conventional MICROSCOPES. First constructed by Max Knoll and Ernst Ruska around 1930, the instrument now consists typically of an evacuated column of magnetic lenses with a 20-1000kV electron gun at the top and a fluorescent screen or photographic plate at the bottom; it can thus be thought of as a kind of CATHODE RAY TUBE. The various lenses allow the operator to see details almost at the atomic level (0.3 nm) at up to a million times magnification (though many specimens deteriorate under the electron bombardment at these limits), and to obtain diffraction patterns from very small areas. In the scanning electron microscope, the beam is focused to a point and scanned over the specimen area while a synchronized television screen displays the transmitted or scattered electron intensity. Electron microscopes are used for structural, defect and composition studies in a wide range of biological and inorganic materials. (♦ page 123)

Electron spin resonance see MAGNETIC RESONANCE DETECTION.

Electron transport chain

In organisms that use oxygen for respiration (AEROBES), the second part of the respiration process, which harvests the energy generated by the first part of the process, the CITRIC ACID CYCLE. The electron transport chain receives electrons from the electron-carriers NADH and fADH, these having been furnished with electrons by the citric acid cycle. It passes these electrons along from one carrier molecule to another, losing energy at each stage. At certain points the energy is harvested as ATP molecules, which can be used to supply energy elsewhere in the cell, as required. In the final step of the electron transport chain the electrons are passed to oxygen, which immediately reacts with a proton (hydrogen ion) to form water. The carrier molecules that make up the electron transport chain are cytochromes and other pigments.

Electron tube (valve)

An evacuated glass or metal tube which may contain gas at low pressure, through which ELECTRONS flow between two or more ELECTRODES. A heated filament, the cathode, emits electrons which are attracted to the positively charged anode. By varying the voltage on intermediate (grid) electrodes, the electron flow can be regulated and the tube made to work as an AMPLIFIER. (See also DIODE; CATHODE-RAY TUBE; RECTIFIER; TRIODE.)

Electronvolt (eV)

A unit of ENERGY used in atomic and high-energy physics, defined as the kinetic energy acquired by an ELECTRON in passing through a POTENTIAL difference of 1 volt in a vacuum. The electronvolt is about $1.602,19\times10^{-19}$ joules.

Electrophiles see NUCLEOPHILES.

Electrophoresis
The DIFFUSION of charged particles through FLUIDS or GELS under the influence of an ELECTRIC FIELD. Particles with different sizes and charges diffuse at different rates, so the effect can be used to separate and identify large MOLECULES such as PROTEINS.

Electroplating
Process of depositing a thin metal coating on base-metal objects, to improve their appearance or corrosion resistance. The object is made the CATHODE of a cell containing a salt of the metal to be deposited, which is made the ANODE; on ELECTROLYSIS the metal dissolves from the anode and deposits on the object. Chromium, nickel, copper, silver and gold are commonly used. In electropolishing, the reverse process, the object is made the anode; preferential solution of irregularities gives a high polish. (See also ANODIZING.)

Electrostatic generator
An instrument producing a very high direct voltage, particularly that developed by Robert J. van de Graaff (1901-1967). An insulating belt is driven around a pair of rollers some distance apart, a small electric discharge at one end producing positively charged IONS which are carried by the belt to the other end and lifted off by a small metal comb, the charge accumulating on a polished metal sphere surrounding this end. Such generators can develop up to 20MV with a 15m sphere and are used as particle ACCELERATORS.

Electrostatics
The study of static ELECTRICITY.

Electrum
Pale yellow ALLOY of GOLD with up to 40% SILVER, occurring naturally as gold ore, and used for ornament. The term has been used for AMBER.

Element, Chemical
Simple substance composed of ATOMS of the same atomic number, and so incapable of chemical degradation or resolution. They are generally mixtures of different ISOTOPES. Of the 107 known elements, about 90 occur naturally, and the rest have been synthesized (see TRANSURANIUM ELEMENTS). The elements are classified by physical properties as METALS, METALLOIDS and NONMETALS, and by chemical properties and atomic structure according to the PERIODIC TABLE. Most elements exhibit ALLOTROPY, and many are molecular (eg oxygen, O_2). The elements have all been built up in STARS from HYDROGEN by complex sequences of nuclear reactions, eg the CARBON CYCLE. (➧ page 130, 147)

Elementary particles see SUBATOMIC PARTICLES.

Elephantiasis
Condition in which there is massive swelling and enlargement of the SKIN and subcutaneous tissue, due to the obstructed flow of LYMPH. This may be a congenital disease, due to trauma, CANCER, or infection with FILARIASIS, TUBERCULOSIS and some SEXUALLY TRANSMITTED DISEASES. Recurrent secondary bacterial infections are common and chronic skin ULCERS may form.

Ellis, Henry Havelock (1859-1939)
British writer chiefly remembered for his studies of human sexual behavior and psychology. His major work was *Studies in the Psychology of Sex* (1897-1928).

Embolism
Blockage of an artery by an air bubble or blood clot. Small quantities of air entering arteries usually disperse, but large ones prevent blood flow. Clots of blood frequently enter the circulation via veins, and occasionally lodge in a tissue, especially the lungs.

Embryo
In seed plants (sporophytes) the part of the seed that will develop into a new plant, excluding the COTYLEDONS. In animals the stage of the development from a fertilized EGG, through the differentiation of the major organs, to the emergence of a functional organism, whether a LARVA, NYMPH or adult. In humans the fertilized egg divides repeatedly, forming a small ball of cells that fixes by IMPLANTATION to the wall of the uterus (womb); differentiation into PLACENTA and three primitive layers (ENDODERM, MESODERM and ECTODERM) follows. These layers then undergo further division into distinct organ precursors, and each of these develops by a process of migration, differentiation and differential growth. Much of development depends on the formation of cavities, either by splitting of layers or by enfolding. The heart develops early at the front, probably splitting into a simple tube before being divided into separate chambers. The nervous system develops as an infolding of ectoderm, which then becomes separated from the surface. The overall control of these processes is not yet fully understood. Infection (especially RUBELLA) in the mother, or the taking of certain drugs during pregnancy may lead to abnormal development and so to congenital defects. By convention, the human embryo is described as a fetus after three months' gestation. (➧ page 204)

Embryology
Study of the development of EMBRYOS of animals, including humans.

Emerald
Valuable green GEM stone, a variety of BERYL. The best emeralds are mined in Colombia, Brazil and the USSR. Since 1935 it has been possible to make synthetic emeralds.

Emery
An impure CORUNDUM containing MAGNETITE and other minerals, occurring on the island of Naxos and in Asia Minor. It is used as an abrasive (Mohs hardness 8), and as a non-skid material in floors and stairs.

Emotion
In psychology, a term that is only loosely defined. Generally, an emotion is a sensation that causes physiological changes (as in pulse rate, breathing) as well as psychological changes (as disturbance) which result in, usually, compulsive adaptations in the individual's behavior. Some psychologists differentiate types of emotion: eg into primary (eg fear), complex (eg envy) and sentiment (eg love, hate); but such schemata are controversial.

Emphysema
Progressive condition in which the air spaces of the LUNGS become enlarged due to destruction of their walls. Often associated with chronic BRONCHITIS, it is usually a result of smoking but may be a congenital or occupational disease. It causes breathlessness, especially on exertion, and a cough producing sputum. Subcutaneous emphysema refers to air in the subcutaneous tissues, usually as the result of surgery or trauma.

Emulsion
A COLLOID in which both phases are initially liquid.

Enantiomer
One of a pair of molecules that are non-superimposable mirror images of each other. An enantiomeric pair of molecules differ from each other in only one characteristic, the direction in which their solutions rotate plane-polarized light.

Encephalitis
Inflammation affecting the substance of the brain. It is a rare complication of certain common diseases (eg mumps, *Herpes simplex* infections) and a specific manifestation of other less common ones. Typically an acute illness with HEADACHE and FEVER, it may produce evidence of patchy lesions in the brain tissue, such as personality change, EPILEPSY, localized weakness or rigidity. It may progress to impairment of consciousness and COMA.

Encke, Johann Franz (1791-1865)
German astronomer who discovered one of the divisions in the rings of SATURN and, using observations of a transit of Venus, established a good value for the distance of the Sun. He also discovered Encke's Comet, whose period is only 3.3yr (see COMET).

Enders, John Franklin (1897-1986)
US microbiologist who shared the 1954 Nobel Prize for Physiology or Medicine with F.C. ROBBINS and T.H. WELLER for their cultivation of POLIOMYELITIS virus in non-nerve tissues, so opening the gate for the development of polio vaccines.

Endocarditis
Inflammation of the endocardium, the membrane that lines the heart and its valves.

Endocrine glands
Ductless glands in the body that secrete HORMONES directly into the BLOOD stream. They include the PITUITARY GLAND, THYROID and PARATHYROID GLANDS, ADRENAL GLANDS and part of the PANCREAS, TESTES and OVARIES. Each secretes a number of hormones that affect body function, development, mineral balance and METABOLISM. They are under complex control mechanisms, including FEEDBACK from their metabolic function and from other hormones. The pituitary gland, which is itself regulated by the HYPOTHALAMUS, has a regulatory effect on the thyroid, adrenals and gonads.

Endoderm
In animals, the inner layer of the body, usually derived from the inner layer of cells in the BLASTULA (see EMBRYOLOGY). The endoderm gives rise to the digestive tract in lower animals, and to the inner lining of the digestive tract in higher animals. (See also ECTODERM; MESODERM.)

Endometriosis
The cells that line the UTERUS from the endometrium. These can spread to other tissues, such as the surface of the ovaries, causing premenstrual pain.

Endoplasm
In cells, the inner part of the CYTOPLASM. It is more liquid than the outer layer or ECTOPLASM.

Endoplasmic reticulum
A structure found in EUKARYOTIC cells that appears as a complex, folded mass of membranes under the microscope. It has been demonstrated that it is in fact a membranous bag, with many separate, flattened compartments (*cisternae*) connected by membranous tubules. The interior of the bag, known as the ER lumen, is sealed off from the cytosol of the cell, and the movement of molecules into and out of it is controlled by the ER membrane. The ER lumen provides the cell with a means of separating newly synthesized molecules that belong in the cytosol from those that do not, the latter being stored in the ER lumen and then carried to their target organelles in transport vesicles; these vesicles are formed from the ER membrane itself by a budding-off process. The endoplasmic reticulum is also important in organizing certain biosynthetic reactions, and there are various enzymes associated with it; almost all the proteins and lipids needed for the synthesis of cell membranes are produced by the ER. Areas of ER concerned with protein synthesis also have a great many ribosomes attached – these areas are known as rough ER; those without ribosomes are smooth ER. The latter is concerned with lipid synthesis. (➧ page 13)

Endoprocts (entoprocts)
A phylum of colonial animals mostly found in shallow marine water and rarely exceeding a length of 2mm. They feed by filtering food particles from the water. Members of the best-known genus, *Pedicellina*, are composed of a branching stem, or stolon, on which are borne zooids – cup-shaped structures with tentacles arising from the upper rim. (See also BRYOZOA (ectoprocts).)

Endopterygotes
One of the two major subclasses of flying INSECTS, the other being the EXOPTERYGOTES. It includes the most advanced of the insect groups – the lice, bugs, lacewings, butterflies and moths, flies, fleas, ants, bees, wasps and beetles. The eggs hatch into larvae that are quite unlike the parent and must undergo METAMORPHOSIS during a pupal stage to achieve adulthood.

Endorphins
Substances that relieve pain. Produced in the body, they were originally known as endogenous morphines. Enkephalins, similar to endorphins, are produced in the brain only.

Endoskeleton
Any SKELETON that is enclosed within an animal's body. The VERTEBRATES (fish, amphibians, reptiles, birds and mammals) are the prime examples of animals with an endoskeleton. Those sponges with strengthening spicules in their bodies are also said to possess an endoskeleton, and the internal hydrostatic skeletons of the annelid worms and similar invertebrates could be described as endoskeletons. The hard, spiny, calcareous "skin" of the starfish, sea urchins and their relatives (ECHINODERMS) is also considered an endoskeleton as it is covered by a thin layer of living tissue. (See also EXOSKELETON.)

Endosperm
In the seeds of ANGIOSPERMS, a food-storage tissue that is additional to, or replaces, the COTYLEDONS. The endosperm has a strange mode of development, being formed when a diploid cell (from the female parent) is fertilized by a second male gamete from the pollen grain at the same time as the ovule itself is fertilized. Thus the new cell is triploid (having three sets of chromosomes). It divides rapidly to give a food-rich mass of triploid cells.

Endosymbiosis theory
The idea that certain organelles in the cells of EUKARYOTES originated as smaller bacterial cells that invaded, or were ingested by, larger cells, and took up a symbiotic relationship with them. Certain parts of the theory are now widely accepted but others remain controversial. Thus the chloroplasts of plant cells are probably derived from CYANOBACTERIA that originally provided a photosynthetic capacity to their hosts. The mitochondria may once have been aerobic bacteria that conferred powers of oxygen respiration on larger, anaerobic cells. An endosymbiotic origin for both these types of organelles is now supported by a great deal of evidence, including that of structural and biochemical similarities, and of analogous symbiotic relationships arising more recently. Less widely accepted is the idea that the flagella and cilia of higher cells, together with their basal bodies, the apparently related CENTRIOLES, and the protein microtubules of the spindle (see MITOSIS), are all derived from SPIROCHAETE BACTERIA.

Endotherm
A "warm-blooded" animal or HOMOIOTHERM.

Endothermic reaction
A chemical reaction that absorbs heat.

Energy
To the economist, a synonym for fuel; to the scientist, one of the fundamental modes of existence, equivalent to and interconvertible with MATTER. The MASS-energy equivalence is expressed in the Einstein equation, $E=mc^2$, where E is the energy equivalent to the mass m, c being the electromagnetic constant (speed of LIGHT). Since c is so large, a tiny mass is equivalent to a vast amount of energy. However, this energy can only be realized in nuclear reactions and so, although the conversion of mass may provide energy for the STARS, this process does not figure much in physical processes on Earth (except in nuclear power installations). The law of the conservation of mass-energy states that the total amount of mass-energy in the UNIVERSE or in an isolated system forming part of the Universe cannot change. In an isolated system in which there are no nuclear reactions, this means that the total quantities both of mass and of energy are constant. Energy then is generally conserved.

Energy exists in a number of equivalent forms. The commonest of these is HEAT – the motion of the MOLECULES of matter. Ultimately all other forms of energy tend to convert into thermal motion. Another form of energy is the motion of ELECTRONS, ELECTRICITY. Moving electrons give rise to electromagnetic fields and these too contain energy. A pure form of electromagnetic energy is ELECTROMAGNETIC RADIATION (radiant energy) such as light. According to the QUANTUM THEORY, the energy of electromagnetic radiation is "quantized", referable to discrete units called PHOTONS, the energy E carried by a photon of radiation of FREQUENCY v being given by $E=hv$, where h is the PLANCK CONSTANT. When macroscopic bodies move, they too have energy in virtue of their motion; this is their kinetic energy and is given by $\frac{1}{2}mv^2$ where m is the mass and v the velocity of motion. To change the velocity of a moving body, or to set it in motion, a FORCE must be applied to it and work must be done. This work is equivalent to the change in the kinetic energy of the body and gave physicists one of their earliest definitions of energy: the ability to do work. When work is done against a restraining force, potential energy is stored in the system, ready to be released again. The restraining force may be electromagnetic, torsional, electrostatic, tensional or of any other type. On Earth when an object of mass m is raised up to height h, its gravitational potential energy is given by mgh, where g is the acceleration due to gravity. If the object is let go, it falls and will strike the ground with velocity v, its potential energy being converted into kinetic energy $\frac{1}{2}mv^2$. SOUND energy is kinetic energy of the vibration of air. Chemical energy is the energy released from a chemical system in the course of a reaction. Although all forms of energy are equivalent, not all interconversion processes go with 100% EFFICIENCY (the energy deficit always appears as heat – see THERMODYNAMICS). The SI UNIT of energy is the joule.

Energy level
A stationary state of a physical system characterized by its having or being able to have a fixed quantity of ENERGY. QUANTUM MECHANICS assumes that physical systems can exist only in a well-defined set of energy levels. The emission of ELECTROMAGNETIC RADIATION is associated with transitions of electronic and molecular systems between energy levels (see SPECTROSCOPY). (◆ page 21)

Engine
A device for converting stored ENERGY into useful WORK. Most engines in use today are heat engines which convert HEAT into work, though the EFFICIENCY of this process, being governed according to the second law of THERMODYNAMICS, is often very low. Heat engines are commonly classified according to the fuel they use (as in gasoline engine); by whether they burn their fuel internally or externally (see INTERNAL-COMBUSTION ENGINE), or by their mode of action (whether they are reciprocating, rotary or reactive). (See GAS TURBINE; JET PROPULSION; TURBINE.)

Engineering
Essentially, the managing of ENGINES. The term also describes the application of the sciences (including mathematics), and the development and uses of technology, to the service of humans. Its branches include: aeronautical engineering; chemical engineering; the application of chemistry in the conversion of raw materials into desired products (see CHEMISTRY); CIVIL ENGINEERING; ELECTRICAL ENGINEERING; HUMAN ENGINEERING; marine engineering, the design and construction of structures and processes for naval purposes; mechanical engineering, the design and use of machines, with its tool, mechanical drawing; and nuclear engineering (see NUCLEAR ENERGY).

Enteritis
Inflammation of the small intestine (see GASTROINTESTINAL TRACT) causing abdominal COLIC and DIARRHEA. It may result from VIRUS infection, certain BACTERIAL DISEASES or FOOD POISONING, which are in general self-limited and mild. The noninfective inflammatory condition known as Crohn's disease causes a chronic relapsing regional enteritis, which may result in weight loss, ANEMIA, abdominal mass or VITAMIN deficiency, as well as colic and diarrhea.

Enthalpy (heat content) see THERMODYNAMICS.

Entomology
The study of insects.

Entropy
The name of a quantity in THERMODYNAMICS, statistical mechanics and information theory variously representing the degree of disorder in a physical system, the extent to which the ENERGY in a system is available for doing WORK, the distribution of the energy of a system between different modes, or the uncertainty in a given item of knowledge. In thermodynamics ABSOLUTE entropies cannot be determined, only changes in entropy. The infinitesimal entropy change δS when a quantity of HEAT δQ is transferred at absolute TEMPERATURE T is defined as $\delta S=\delta Q/T$. One way of stating the second law of thermodynamics is to say that in any change in an isolated system, the entropy (S) increases: $\Delta S \geqslant 0$. This increase in entropy represents the energy that is no longer available for doing work in that system.

Environment
In ecology, the surroundings in which animals and plants live, including both the physical factors and other organisms. Organisms are affected by many different physical features in their environment, such as temperature, water, gases, light and pressure, and by biotic factors such as food resources, competition with other species, predators and disease.

More generally, the word is used to refer to the entire biosphere, particularly when discussing CONSERVATION. In genetics it has a slightly different meaning again: the sum total of the effects of external factors on an organism, as opposed to the effects of heredity; see PHENOTYPE.

Enzymes
Molecules that act as catalysts (see CATALYSIS) for the chemical reactions upon which LIFE depends. They are all PROTEINS, and are generally specific for just one reaction or for a group of related reactions, such as the degradation of a particular type of molecule. Enzymes are responsible for the production of all the organic materials present in living CELLS, for providing the mechanisms for energy production and utilization, for the digestion of food, the replication of DNA, and for maintaining the intracellular environment within

fine limits. They are frequently organized within subcellular organelles, such as MITOCHONDRIA or CHLOROPLASTS, which catalyze a whole sequence of chemical events in a manner analogous to a production line. Enzymes are themselves synthesized by other enzymes on templates derived from DNA, the hereditary material. An average cell contains about 3000 different enzymes. In order to function correctly, many enzymes require the assistance of metal IONS known as cofactors, or accessory substances known as coenzymes. Cells also contain special activators and inhibitors, which switch particular enzymes on and off as required. In some cases a substance closely related to the substrate (the substance on which the enzyme acts) will act as an inhibitor by competing for the enzyme and preventing the normal action on the substrate; this is termed competitive inhibition. Alternatively, the product of a reaction may inhibit the action of the enzyme so that no more is produced until its level has dropped to a particular threshold; this is known as FEEDBACK control. Enzymes either synthesize or break down chemical compounds, or transform them from one type to another. These differing actions form the basis of the classification of enzymes into oxidoreductases, transferases, hydrolases, lyases, isomerases and synthetases. Enzymes normally work inside living cells but some (eg digestive enzymes) can work outside the cell. Enzymes are used commercially in "biological" washing powders, food processing, brewing and BIOTECHNOLOGY. (◆◆ page 13, 40, 204)

Eocene
The second epoch of the TERTIARY period, lasting from about 55 million to about 38 million years ago. (See also GEOLOGY.)

Ephedrine
A drug related to ADRENALINE. It may act as a central nervous system STIMULANT and is used to dilate the BRONCHI in ASTHMA, to dilate the pupils, and occasionally for its effects on the BLADDER.

Epicanthic fold
Fold of skin at the inner corner of the eye, found in Asiatic people, Eskimos and American Indians.

Epicenter
The point on the Earth's surface directly above the focus of an EARTHQUAKE. (◆ page 142)

Epicycle
A circle whose center lies on the circumference of a larger circle. In geocentric cosmologies, such as that of PTOLEMY, the planets were thought to move around epicycles, the centers of which lay on larger deferent circles centered on the Earth.

Epidemic
The occurrence of a disease in a geographically localized population over a limited period of time; it usually refers to an INFECTIOUS DISEASE, such as influenza, that spreads from case to case or by carriers. Epidemics arise from importation of infection, from changes in the microbe responsible, environmental changes favoring infectious organisms or following altered host susceptibility. A pandemic is an epidemic of international proportions. Infectious disease is said to be endemic in an area if cases are continually occurring there.

Epidermis
The outermost cellular layer of an animal or plant. In plants, insects and some other invertebrates the epidermis is covered by a non-cellular layer, the CUTICLE, which is waterproof. (◆ page 213)

Epidural
Literally, under the dura, one of the membranes surrounding the brain and spinal cord. An epidural injection of anesthetic is often made in the lower spinal cord to block the pain of childbirth.

Epigastrium
The upper part of the abdomen, between the ribs and the waist.

Epiglottis
A structure made of CARTILAGE and covered with mucous membrane, situated in the PHARYNX in front of the glottis or upper windpipe, from which it tends to divert food. It becomes swollen and inflamed in the childhood condition of acute epiglottitis and may cause respiratory obstruction.

Epilepsy
A chronic disease of the brain, characterized by susceptibility to CONVULSIONS or other transient disorders of NERVOUS SYSTEM function and due to abnormal electrical activity within the cerebral cortex. There are many types, of which four are common. *Grand mal* convulsions involve rhythmic jerking and rigidity of the limbs, associated with loss of consciousness, urinary incontinence, transient cessation of breathing and sometimes CYANOSIS, foaming at the mouth and tongue biting. *Petit mal* is largely a disorder of children in which very brief episodes of absence or vacancy occur, when the child is unaware of the surroundings, and is associated with a characteristic ELECTROENCEPHALOGRAPH disturbance. In focal epilepsy the nature of the attack will depend on the precise area of the brain involved. Fits or convulsions may therefore be specific to quite a small part of the body. Temporal lobe or psychomotor epilepsy is often characterized by abnormal visceral sensations, unusual smells, visual distortion or memory disorder, and may or may not be followed by unconsciousness. *Status epilepticus* is when attacks of any sort occur repetitively without consciousness being regained in between; it requires emergency treatment.

Epilepsy may be either primary (idiopathic) due to an inborn tendency, often appearing in early life, or it may be symptomatic of brain disorders such as those following trauma or brain SURGERY, ENCEPHALITIS, cerebral ABSCESS, TUMOR, or vascular disease.

Epinephrine see ADRENALINE.

Epiphyses
The ends of the long bones of the body (eg the FEMUR or HUMERUS), which are growth centers during growth and form the articulating surfaces at JOINTS. Each epiphysis remains partly cartilaginous during growth. The CARTILAGE is eventually converted into mature BONE, which is then incorporated into the main shaft. Bone-growth disorders such as RICKETS and some HORMONE disorders affect the epiphyses.

Epiphyte
A plant that grows on another but obtains no nourishment from it. Various lichens, mosses, ferns, bromeliads and orchids are epiphytes, particularly on trees. Epiphytes thrive in warm, wet climates but are also known in temperate zones. Many plants that normally grow on the ground, or on rocks or walls, can become epiphytes if the opportunity arises. (See also COMMENSALISM; PARASITE.)

Episiotomy
A cut in the perineum, the tissue at the back of the vaginal opening, made to assist the passage of a baby when the tissues might otherwise be jaggedly torn.

Epithelioma
CANCER of the skin EPITHELIUM.

Epithelium
Surface tissue covering an organ or structure. Examples include skin and the mucous membranes of the lungs, gut and urinary tract. A protective layer specialized for water resistance or absorption, it usually shows a high cell-turnover rate.

Epoxy resin
Class of thermosetting POLYMERS of the polyether type, formed from epichlorhydrin and a dihydric ALCOHOL or PHENOL (eg bisphenol-A), and cross-linked by a curing agent. Inert, strong, adhesive, and good insulators, they are mixed with fillers and plasticizers and used for construction, coating and bonding.

Epsom salts (epsomite)
The mineral MAGNESIUM sulfate heptahydrate, found at Epsom, England, and elsewhere. It has been used as a laxative.

Equator
An imaginary line drawn about the Earth such that all points on it are equidistant from the N and S poles (see NORTH POLE; SOUTH POLE). All points on it have a latitude of 0°. (See also CELESTIAL SPHERE; LATITUDE AND LONGITUDE.)

Equids
Members of the Family Equidae, which includes the horses, zebras, onagers and asses. They belong to the Order PERISSODACTYLA.

Equilibrium
A state in which a mechanical, electrical, thermodynamic or other system will remain if undisturbed. In "stable equilibrium" (as with a well-sprung automobile body) the system returns to its original position if disturbed; the position of stable equilibrium thus determines the rest position of the system. In "unstable equilibrium" (as with a tall pole balanced on one end) it moves farther away. Stable and unstable equilibria correspond to configurations with minimal and maximal ENERGY respectively. Systems in thermodynamic equilibrium, eg the contents and air in an unheated room, have the same TEMPERATURE throughout.

Equilibrium, Chemical
The EQUILIBRIUM condition of a reversible reaction, in which the concentrations of the reactants and products have no tendency to change. In this case, the free energy of the system (see THERMODYNAMICS) is a minimum, and it may be shown that for the general reaction $aA+bB \rightleftharpoons cC+dD$ there is an equilibrium constant K given approximately by

$$K=[C]^c[D]^d/A^a[B]^b]$$

where [A] is the concentration of A, and so on. Chemical KINETICS yields the same equation, since at equilibrium the rates of the forward and back reactions are equal, and so K is also given by the ratio of the rate constants of these two reactions ($K=k_1/k_{-1}$).

Equinoxes
(1) The two times each year when day and night are of equal length. The spring or vernal equinox occurs in March, the autumnal equinox in September. (◆ page 25)

(2) The two intersections of the ECLIPTIC and equator (see CELESTIAL SPHERE). The vernal equinox is in PISCES (see also ARIES, First Point of), the autumnal between VIRGO and LEO.

Equivalent weight
The weight of an element or compound which combines with or displaces the equivalent weight of any other element or compound; for an element, it equals the atomic weight (see ATOM) divided by the VALENCE. This presupposes the law of equivalent proportions, which states that the ratio of the weights of two elements A and B which combine with the same weight of an element C, is the same as the ratio of the weights of A and B which combine with each other, or a small integral multiple of it. Since an element may have more than one valence, and a compound may react in more than one way, they may have more than one equivalent weight. The normality of a solution is its concentration in gram-equivalent weights per liter; a solution whose normality is 1 is called normal. (See also ELECTROLYSIS; MOLECULAR WEIGHT.)

Erasistratus (*c*304-*c*250BC)
Greek physician of the Alexandrian School who is credited with the foundation of PHYSIOLOGY as a separate discipline.

Eratosthenes of Cyrene (276-196 BC)
The Librarian of the Alexandrian Library, remembered for his remarkably accurate determination of the circumference of the Earth and for his map of the then known world.

Erbium (Er)
One of the LANTHANUM SERIES of elements. AW 167.3, mp 1520°C, bp 2600°C, sg 9.066 (25°C). (♦ page 130)

Erg
The unit of WORK in the CGS system (see CGS UNITS), equal to 10^{-7} joules in SI UNITS.

Ergot
A disease of grasses and sedges caused by fungal species of the genus *Claviceps*. The name is also give to the black or brown masses of dormant mycelia (sclerotia) formed in the flower heads of the host plant. These ergots contain toxic ALKALOIDS, which can cause serious poisoning and hallucinations (ergotism or St Anthony's fire).

Eridanus
The River; large, long CONSTELLATION of the S celestial hemisphere. Of particular interest is Epsilon Eridani: this, at 3.31 pc, is the closest star to us to resemble our Sun. (♦ page 228)

Erikson, Eric Homburger (b. 1902)
German-born US psychoanalyst who defined eight stages, each characterized by a specific psychological conflict, in the development of the ego from infancy to old age. He also introduced the concept of the identity crisis.

Erlanger, Joseph (1874-1965)
US physiologist who shared with HERBERT GASSER the 1964 Nobel Prize for Physiology or Medicine for their discovery that different nerve fibers have different functions, carrying different types of impulses and at different speeds. (See NERVOUS SYSTEM.)

Eros
An ASTEROID measuring roughly 35×16×8km discovered in 1898 by G. Witt. Eros' eccentric orbit brings it close to Earth every seven years, sometimes within 22 million km. Its orbital period is 643 days.

Erosion
The wearing away of the Earth's surface by natural agents. Running water constitutes the most effective eroding agent, the process being accelerated by the transportation of particles eroded or weathered farther upstream: it is these that are primarily responsible for further erosion. GROUNDWATER may cause erosion by dissolving certain minerals in the rock (see also KARST). OCEAN WAVES and especially the debris that they carry may substantially erode coastlines. GLACIERS are extremely important eroding agents, eroded material becoming embedded in the ice and acting as further abrasives. Many common landscape features are the results of glacial erosion (eg DRUMLINS, FJORDS). Rocks exposed to the atmosphere undergo weathering: mechanical weathering usually results from temperature changes (eg in exfoliation, the cracking off of thin sheets of rock due to extreme daily temperature variation); chemical weathering results from chemical changes brought about by, for example, substances dissolved in RAIN water. Wind erosion may be important in dry, sandy areas. (See also DAVIS, WILLIAM; SOIL EROSION.) (♦♦ page 44, 208)

Erysipelas
An infection of the skin and underlying tissues with the bacterium *Streptococcus pyogenes*. The face and scalp are most commonly affected, and become red and swollen. The patient will have a high FEVER.

Erythema
Redness of the SKIN due to increased CAPILLARY BLOOD flow; it occurs in INFLAMMATION including DERMATITIS, and numerous rashes.

Erythrocyte
Red BLOOD cell, a discoid cell without a nucleus, which contains HEMOGLOBIN and is responsible for oxygen transport in the body.

Esaki, Leo (b. 1925)
Japanese-born US physicist awarded, with I. GIAEVER and B. JOSEPHSON, the 1973 Nobel Prize for Physics for his work on tunneling (see WAVE MECHANICS).

Escape velocity
The minimum velocity that a less massive body must achieve in order to escape from the gravitational attraction of a more massive body; sometimes known as the parabolic velocity. The Earth's escape velocity is 11.2km/s. Less massive planets have smaller escape velocities (Mars: 5.1 km/s), more massive planets greater escape velocities (Jupiter: 61km/s).

Escherischia coli
A bacterium that normally lives as a commensal in the human intestine. Different human populations have slightly different strains of the bacterium, and "holiday diarrhea" is often caused by exposure to an unfamiliar strain. This bacterium has been intensively studied by geneticists, and is the organism most often used as a host in GENETIC ENGINEERING because it is so well understood.

Escarpment see SCARP.

Esker
Serpentine ridge of glacial DRIFT, up to several kilometers long, formed from deposits at the mouth of a subglacial stream as the GLACIER retreated.

Esophagus
The thin tube leading from the PHARYNX to the STOMACH. Food passes down it as a bolus by gravity and PERISTALSIS. Its diseases include reflex esophagitis (HEARTBURN) caused by hiatus HERNIA, ULCER, stricture and CANCER. (♦ page 40)

Esters
Organic compounds formed by CONDENSATION of an ACID (organic or inorganic) with an ALCOHOL, water being eliminated. This reaction, esterification, is acid-catalyzed; its reverse, HYDROLYSIS, is acid- or base-catalyzed; an EQUILIBRIUM is set up in aqueous solution. Many esters occur naturally – those of low molecular weight have fruity odors and are used in flavorings, perfumes and as solvents; those of higher molecular weight are FATS and WAXES.

Estivation see AESTIVATION; DORMANCY.

Estradiol
An estrogenic steroid (female sex hormone).

Estrogens
Female sex HORMONES concerned with the development of secondary sexual characteristics and maturation of reproductive organs. They are under the control of pituitary gland GONADOTROPHINS and their amount varies before and after MENSTRUATION and in PREGNANCY. After the menopause their production decreases. Many pills used for CONTRACEPTION contain estrogen, as do some preparations given to menopausal women. (♦ page 111)

Estuary
The typically funnel-shaped part of a river near its mouth where fresh- and seawater mix, and which is affected by TIDES. At ebb tide both tide and river current assist in the EROSION of the estuary. Estuaries may also form by local subsidence of the coast. (See also FIRTH.) Many estuaries provide important harbors.

Ethane (C_2H_6)
An ALKANE occurring in natural gas and formed in petroleum CRACKING. It is catalytically dehydrogenated to produce ETHYLENE. MW 30.1, mp −183°C, bp −88.6°C.

Ethanol (ethyl alcohol, C_2H_5OH)
Also known as grain alcohol, the best-known ALCOHOL; a colorless, inflammable, volatile, toxic liquid, the active constituent of alcoholic beverages. Of immense industrial importance, ethanol is used as a solvent, in antifreeze, as an ANTISEPTIC and in much chemical synthesis. Its production is controlled by law, and it is heavily taxed unless made unfit for drinking by adulteration (denatured alcohol); see METHANOL. Most industrial ethanol is the AZEOTROPIC MIXTURE containing 5% water. It is made by FERMENTATION of sugars or by catalytic hydration of ETHYLENE. MW 46.1 mp −114°C, bp 78°C.

Ether
A hypothetical medium postulated by late 19th-century physicists in order to explain how LIGHT could be propagated as a wave motion through otherwise empty space. Light was thus thought of as a mechanical WAVE MOTION in the ether. The whole theory was discredited following the failure of the MICHELSON-MORLEY EXPERIMENT to detect any motion of the Earth relative to the supposed stationary ether.

Ethers
Organic compounds of general formula R-O-R′, where R and R are HYDROCARBON radicals. They are prepared by catalytic dehydration of ALCOHOLS, catalytic hydration of ALKENES, or by reacting an ALKYL HALIDE with a sodium alkoxide (see ALCOHOLS). Ethers are chemically fairly inert, though they are split by hydrogen HALIDES, and form explosive PEROXIDES on standing in air. Diethyl ether $(C_2H_5)_2O$, the most important, was widely used as an anesthetic (see ANESTHESIA) and an industrial solvent. It is a volatile, inflammable liquid, immiscible with water. MW 74.1, mp −116°C, bp 35°C.

Ethology
The study of the behavior of animals in the wild rather than in the laboratory. Its founders were KONRAD LORENZ and NIKO TINBERGEN, whose ideas on instinctive behavior supplied the stimulus for renewed study of animals in the wild.

Ethyl compounds
Organic compounds containing the ethyl group, C_2H_5 (see ALKANES). The most important are: ETHANOL, diethyl ETHER, ethyl ACETATE, LEAD tetraethyl, and the ethyl halides (see ALKYL HALIDES).

Ethylene (ethene, C_2H_4)
The simplest ALKENE, made by CRACKING of ETHANE and PROPANE or NAPHTHA. It is a colorless gas, used to ripen fruit and to make ETHANOL, diethyl ETHER, ethyl chloride, ACETALDEHYDE, ACETIC ACID, ethylene dibromide, vinyl acetate, ethylene dichloride, POLYETHYLENE and STYRENE. MW 28.05, mp 169°C, bp 103°C. It is oxidized by air over a silver catalyst to ethylene oxide, a reactive gas used as a fumigant and to make plastics and emulsifiers, and hydrated to ETHYLENE GLYCOL.

Ethylene glycol ($HOCH_2CH_2OH$)
Colorless syrupy liquid, an ALCOHOL, made from ETHYLENE. It is used to make polyester POLYMERS, and, mixed with water, as an antifreeze and de-icer. MW 62.1, mp −12°C, bp 198°C.

Etiology
Study of the origins and causes of disease.

Eugenics
The advocation of scientifically directed selection in order to improve the genetic endowment of human populations. Eugenic control was first suggested in the 1880s by SIR FRANCIS GALTON, who suggested that those with "good" traits should be encouraged to have children while those with "bad" traits should be discouraged or even

forbidden to have families. These ideas were put into practice by the Nazi regime in Germany

Euglena
A genus of the PROTOZOA. Some of its species have chlorophyll and are capable of PHOTOSYNTHESIS, but unlike plants are not able to make all their own food, needing certain molecules preformed. Euglena species all have a semi-rigid cell wall, and move about by means of a flagellum.

Euglenophyta (euglenoids) see ALGAE.

Eukaryote
An organism with large cells containing a NUCLEUS and a variety of membranous subcellular organelles in the extranuclear area or CYTOPLASM. These organelles include the MITOCHONDRIA (where respiration occurs), CHLOROPLASTS (where photosynthesis occurs – plant and algal cells only), ENDOPLASMIC RETICULUM, GOLGI APPARATUS and LYSOSOMES (animal cells only). The nucleus contains the genetic material DNA, which is arranged in several discrete linear bodies, known as CHROMOSOMES. These have a complex structure involving proteins as well as DNA. A prokaryote cell, by contrast, has its DNA in a single loop, with no associated proteins; there are no membranous organelles (though there may be infoldings of the cell membrane) and no nucleus. Another feature of eukaryotic cells, which is lacking in prokaryotes, is a protein CYTOSKELETON. Eukaryotes also have complex cilia and flagella, made up of parallel strands of protein capable of moving in relation to each other. Prokaryotes do have a flagellum-like organ but it is much simpler in structure, consisting of a single protein molecule that can be moved only at the base. It has been suggested that the bacterial flagellum should be renamed a "unipodium" to emphasize the difference between it and the eukaryote flagellum.

The eukaryotes include all living organisms other than the bacteria and cyanobacteria. It is believed that they evolved from symbiotic associations between different types of prokaryotic cells (see ENDOSYMBIOSIS THEORY). (◀ page 13)

Euler, Ulf Svante von (1905-1983)
Swedish physiologist, President of the Nobel Foundation from 1965, who shared with JULIUS AXELROD and SIR BERNARD KATZ the 1970 Nobel Prize for Physiology or Medicine for their independent work on the chemistry of the transmission of nerve impulses (see NERVOUS SYSTEM).

Euler-Chelpin, Hans Karl August Simon von (1873-1964)
German-born Swedish biochemist who shared with ARTHUR HARDEN the 1929 Nobel Prize for Chemistry for their work on ENZYME action in the FERMENTATION of sugar.

Eumycotina see FUNGI.

Europium (Eu)
One of the LANTHANUM SERIES of elements. AW 152.0, mp 820°C, bp 1450°C, sg 5.243 (25°C).

Eusocial animal
One that is "truly social". This is usually defined as follows: living in a communal nest that contains at least two overlapping generations, caring for the young of other individuals in the group, and having some castes that never reproduce. Most eusocial species are insects (see SOCIAL INSECTS), but there is also a eusocial mammal, the naked mole-rat.

Eustachian tube
A narrow passage that in mammals connects the space behind the eardrum with the back of the throat. It is anatomically comparable with one of the gill passages of fish. (▶ page 209)

Eustachio, Bartolomeo (1524-1574)
Italian anatomist. Aiming at first to vindicate GALEN against VESALIUS and others, he brought a new skill and accuracy to dissection. The Eustachian tube of the EAR is named for him. (See also ANATOMY.)

Euthanasia
The practice of hastening or causing the death of a person suffering from incurable disease. While frequently advocated by various groups, its practical and legal implications are so contentious that it is illegal in most countries.

Eutheria (placental mammals)
A subclass of MAMMALS that includes all species except those classified in the MONOTREMATA and METATHERIA (marsupials). The embryo of placental mammals is nourished in the uterus, attached to a highly organized PLACENTA, until a comparatively late stage in its development.

Eutrophication
The increasing concentration of plant nutrients and FERTILIZERS in lakes and estuaries, partly by natural drainage and partly by POLLUTION. It leads to excessive growth of algae and aquatic plants, with oxygen depletion of the deep water, causing various undesirable effects.

Evaporation
The escape of molecules from the surface of a liquid into the vapor state. Only those molecules with above-average ENERGY are able to overcome the cohesive forces holding the liquid together and escape from the surface. Eventually all the molecules left in the liquid have below-average energy; its temperature is now lower. In an enclosed space, the pressure of the vapor above the surface eventually reaches a maximum, the saturated vapor pressure (SVP). This varies according to the substance concerned and, like the rate of evaporation, increases with temperature, equalling atmospheric pressure at the liquid's BOILING POINT. (◀ page 44)

Evaporites
Sedimentary deposits of salts that have been precipitated from solution owing to evaporation from a body of water (see EVAPORATION; SOLUTION). Evaporite deposits have the least soluble salts at the bottom (CALCIUM salts), followed by the very soluble halite (common SALT) and MAGNESIUM and POTASSIUM salts. Most important commercially are GYPSUM ($CaSO_4.2H_2O$), ANHYDRITE ($CaSO_4$) and halite (NaCl). (See also SALT DOME; SEDIMENTATION.)

Event horizon
The boundary of a BLACK HOLE beyond which an outside observer can detect nothing.

Evergreen
A plant that retains its leaves all the year around, though they are continually being shed and replaced. Examples are the holly, laurel and pine. (See also DECIDUOUS TREES.)

Evolution
The process by which living organisms have changed since the origin of life. The formulation of the theory of evolution by NATURAL SELECTION is credited to CHARLES DARWIN. His observations while sailing around the world on HMS *Beagle*, taken together with elements from MALTHUS' population theory and his own meticulous studies in biogeography, classification, natural history, paleontology and animal breeding, led him to the concept of natural selection. The theory also later occurred independently to A.R. WALLACE. A different, and now discredited, theory of evolution based solely on the presumed inheritance of ACQUIRED CHARACTERISTICS had earlier been proposed by LAMARCK. Darwin proposed the mechanism of natural selection on the basis of three observations: that animals and plants produced far more offspring than were required to maintain the size of their population; that the size of any natural population remained more or less stable over long periods; and that the members of any one generation exhibited variation. From the first two he argued that in any generation there was a high mortality rate, and from the third that, under certain circumstances, some of the variants had a greater chance of survival than did others. The surviving variants were, by definition, those most suited to the prevailing environmental conditions. Any change in the environment, in either physical conditions or other organisms, led to adjustment in the population such that certain new variants were favored and gradually became predominant.

The missing link in Darwin's theory was the mechanism by which heritable variation occurs. Unknown to him, a contemporary, GREGOR MENDEL, had demonstrated the principles of GENETICS and had deduced that the heritable characteristics were controlled by discrete particles. We now call these particles GENES, and know them to be carried on the CHROMOSOMES. Darwin's variants were caused by RECOMBINATION and MUTATION of the genes. The fact that heredity is particulate allows mutations to be inherited without any dilution of their effects. Dawin himself had seen heredity as a blending process, and was aware that this meant dilution of any novel traits, making it difficult for natural selection to produce lasting effects. Thus Mendel's findings complemented Darwin's theories; and in the early 20th century they were combined in a general evolutionary theory known as neo-Darwinism.

Today, the evidence for evolution is overwhelming and comes from many branches of biology. For instance, the comparative anatomy of the arm of a man, the foreleg of a horse, the wing of a bat and the flipper of a seal reveals that these superficially different organs have a very similar internal structure, indicating a common ancestor. Vestigial organs such as the appendix of man and the wing of the ostrich are now of no use, but in related species such as herbivores and flying birds they are clearly of vital importance. Evidently these individuals have progressively evolved in different ways from a common ancestor. The hierarchical classification of plants and animals into species, genus, family, etc. (see TAXONOMY) is a direct reflection of the natural pattern that would be expected if evolution from common ancestors has occurred. Fossils provide convincing evidence of evolution. Thus, the anatomy of birds indicates descent from small, bipedal theropod dinosaurs. The fossil ARCHAEOPTERYX, a flying animal with some therapod features but with feathers, wings and other bird-like characteristics, confirms that this is correct. (See also LIFE, ORIGIN OF.)

Exclusion principle
First stated by W. PAULI in 1926, it says that no two electrons in an atom can have identical QUANTUM status.

Excretion
The removal of the waste products of metabolism, either by storing them in insoluble forms or by removing them from the body. Excretory organs are also responsible for maintaining the correct balance of body fluids. In vertebrates the excretory organs are the KIDNEYS: blood flows through these and water and waste products are removed as URINE. Other forms of excretory organs include the myriapods, the contractile vacuoles of Protozoa and the nephridia of annelids. In plants, excretion usually takes the form of producing insoluble salts of waste products. Excretion does not include the evacuation of undigested food material from the intestines; it refers only to the removal of waste substances from within the body cells. (▶ page 233)

Exocrine gland
A gland, such as the salivary or Bartholin's glands, that discharges its products into the place

where they will act, unlike endocrine glands, which secrete their hormones into the blood stream, which carries them to the target tissues.

Exopterygotes
One of the two major subclasses of flying INSECTS. It includes the dragonflies, mayflies, cockroaches, mantids, termites, earwigs, stoneflies, stick insects (walkingsticks), grasshoppers and crickets. Rather than having a distinct larval stage and undergoing metamorphosis in a pupa, as in the ENDOPTERYGOTES, these insects hatch from the egg as miniature adults, or nymphs, lacking only wings and reproductive organs. These develop progressively as the nymph grows and molts. The name of the group refers to the fact that the wings develop externally, rather than within a pupa. The exopterygotes are generally considered to be more primitive than the endopterygotes; they evolved earlier, and yet account for only 12% of the living insect species.

Exoskeleton
A skeleton that lies on the surface of an animal's body. In this position it not only performs the mechanical functions common to any SKELETON but, in addition, affords protection. Exoskeletons are particularly well developed in arthropods such as crabs, lobsters and insects. See also ARTHROPODA.

Exosphere
The outermost zone of the Earth's ATMOSPHERE (altitude greater than 500km), where terrestrial GRAVITATION is too weak an effect to prevent the escape of uncharged particles. (◀ page 25)

Exothermic reaction
A chemical reaction that absorbs heat.

Expanding Universe see COSMOLOGY; UNIVERSE.

Expansion
The increase in volume of a body as a result of changing conditions, normally increasing TEMPERATURE or decreasing PRESSURE, the latter being more important for GASES. Contraction is the reverse process. In most solids and liquids, increasing temperature increases the random thermal motion of their atoms, which tend to move apart, ie expansion occurs. The amount of expansion is usually expressed as a coefficient of expansion – the fractional change in length or volume per unit temperature change – and is specific for a given material. Water is unusual in that it expands on cooling from 4°C to 0°C. This means that ice floats on water at 0°C and rivers freeze from the surface downward.

Experiment see SCIENTIFIC METHOD.

Expert system
A computer program that solves problems in the way a human does. The computer is provided with a set of rules from a human expert, and instructed how to draw conclusions from it. Also known as intelligent knowledge based systems, or IKBS; a key area in ARTIFICIAL INTELLIGENCE.

Explosives
Substances capable of very rapid COMBUSTION (or any other exothermic reaction – see THERMOCHEMISTRY) to produce hot gases whose rapid expansion is accompanied by a high-velocity shock wave, shattering nearby objects. The detonation travels 1000 times faster than a flame. The earliest known explosive was gunpowder, invented in China in the 10th century AD, and in the West by Roger Bacon (1242). Explosives are classified as primary explosives, which explode at once on ignition, and are used as detonators; and high explosives, which if ignited at first merely burn, but explode if detonated by a primary explosion. The division is not rigid. Military high explosives are usually mixtures of organic nitrates, TNT, RDX, PICRIC ACID and PETN, which are self-oxidizing. Commercial blasting explosives are less powerful mixtures of combustible and explosive substances; they include DYNAMITE (containing NITROGLYCERIN, ammonium nitrate and sometimes NITROCELLULOSE), ammonals (ammonium nitrate + aluminum) and Sprengel explosives (an oxidizing agent mixed with a liquid fuel such as nitrobenzene just before use). Obsolete explosives include the dangerous chlorates and perchlorates, and the uneconomical liquid oxygen explosives (LOX). Explosives which do not ignite firedamp are termed "permissible", and may be used in coal mines. Propellants for guns and rockets are like explosives, but burn fast rather than detonating.

Extinction
The total disappearance of any living species or the disappearance of a species from a given area. Extinction may be the result of a number of factors, including physical changes in the environment and competition from other species. Today, however, almost all extinctions are the direct or indirect result of human activity.

Extraction
Method of separating a desired substance from a SOLUTION containing one or more other substances. The solution is shaken with a solvent immiscible with it, into which the solute required (but not the others) is largely transferred, according to the distribution law that its concentrations in the two phases are in a constant ratio. The process is repeated if necessary. Continuous extraction is used if the distribution ratio is small. Metal ions are extracted as a CHELATE by adding a ligand to the solvent.

Extraversion see INTROVERSION AND EXTRAVERSION.

Extrusion
A way of producing metal and plastic components of constant cross-section (eg tubes, sheets) by forcing the material through a die. In cold extrusion a billet of metal is surrounded by liquid lubricant in a suitable chamber; hydrostatic PRESSURE is increased, forcing the metal through orifices at one end of the chamber. In hot extrusion, used for plastics, the material is melted throughout before being forced through a die and rapidly cooled.

Eye
Any sense organ concerned with VISION. The simplest types of eye cannot focus on an object as they lack a lens, and their main function is to distinguish light from dark. Such eyes are found in many lower animals, but the faster-moving invertebrates, such as the insects and cephalopod mollusks (squid, octopuses, etc.) have more highly developed eyes, consisting of a lens, or lenses, that can focus light onto the light-receptive cell, and provide a sharp image of the world. This sort of eye is also seen in the vertebrates, where the eye is roughly spherical in shape, has a tough fibrous capsule with the transparent cornea in front, and is moved by specialized eye muscles. In humans the exposed surface of the eye is kept moist with tears from lacrymal glands. Most of the eye contains vitreous humor, a substance with the consistency of jelly that fills the space between the lens and the retina, while in front of the lens there is watery or aqueous humor. The colored iris or aperture surrounds a hole known as the pupil. The focal length of the lens can be varied by specialized ciliary muscles. The retina is a layer containing receptor cells (rods and cones) which are sensitive to light. The optic nerve leads back from the retina to the BRAIN. Rods and cones contain the visual pigment RHODOPSIN, which is bleached by light and thus sets off the nerve cell reaction. The eye of cephalopod mollusks is remarkably similar to the vertebrate eye, whereas insects and crustaceans have a COMPOUND EYE, made up of large numbers of separate visual units. (▶ page 213)

Fabre, Jean Henri (1823-1915)
French entomologist who used direct observations of insects in their natural environments in his pioneering researches into insect instinct and behavior.
Fabricius, David (1564-1617)
German Protestant minister and astronomer who discovered the first VARIABLE STAR (1596). His son, Johannes Fabricius (1587-1615), was one of the first to observe SUNSPOTS and hence report the SUN's rotation.
Fabricius Ab Aquapendente, Hieronymus (Girolamo Fabrizi, 1537-1619)
Italian physician, the pupil and successor of FALLOPIUS at Padua and teacher of WILLIAM HARVEY. Fabricius made a detailed study of the valves in VEINS and pioneered the modern study of EMBRYOLOGY.
Face
Front part of the head, bordered by hairline and chin, and consisting of the EYES, NOSE, mouth, EARS, forehead, cheeks and jaw. It is particularly concerned with sensibility and communication: it bears the special sense organs of vision, hearing, SMELL, TASTE and balance, as well as especially sensitive SKIN. VOICE and facial expression (fine facial movements) are mediated via the face. Neck mobility allows these organs to be directed quickly and easily toward an object without turning the body.
Fahrenheit, Gabriel Daniel (1686-1736)
German-born Dutch instrument maker who introduced the mercury-in-glass THERMOMETER and discovered the variation of boiling points with atmospheric pressure, but who is best remembered for his Fahrenheit temperature scale. This has 180 divisions (degrees) between the freezing point of water (32°F) and its boiling point (212°F). Although still commonly used in the US, elsewhere the Fahrenheit scale has been superseded by the CELSIUS scale.
Fainting (syncope)
Transient loss of consciousness associated with an abrupt fall in blood pressure. In the upright position, head and brain are dependent on a certain blood pressure to maintain BLOOD CIRCULATION through them; if the pressure falls for any reason, inadequate flow causes consciousness to recede, often with the sense of things becoming more distant. The body goes limp and falls, so that, unless artificially supported, the effect of gravity on brain flow is lost and consciousness is rapidly regained. Fainting may result from sudden emotional shock in susceptible individuals, HEMORRHAGE, ANEMIA or occur with transient rhythm disorders of the HEART.
Falling star see METEOR.
Fall line
A line along which a number of nearby rivers have WATERFALLS, marking the progress of the rivers from hard to softer rock. As this marks the farthest inland point navigable from the sea, and because the falls can supply hydroelectric power, many important industrial centers have sprung up along fall lines.
Fallopian tube
Narrow tube leading from the surface of each ovary within the female pelvis to the uterus. Its abdominal end has fine finger-like projections (fimbria), which waft eggs into the tube after ovulation. Fertilization may occur in the tube, and if followed by IMPLANTATION there, the PREGNANCY is ectopic and ABORTION, which may be life- threatening, is inevitable. In STERILIZATION, the tubes are sealed by being cut or clamped. The operation is effectively irreversible. (➧ page 204)
Fallopius, Gabriel (Fallopia, 1523-1562)
Italian anatomist, a supporter of VESALIUS. He carried out important work on many anatomical structures; and is best known for his descriptions of the FALLOPIAN TUBES, whose function he discovered.
Fallout, Radioactive
Deposition of radioactive particles from the ATMOSPHERE on the Earth's surface. Three types of fallout follow the atmospheric explosion of a nuclear weapon. Large particles are deposited as intense but short-lived local fallout within about 250km of the explosion; this dust causes radiation burns. Within a week, smaller particles from the TROPOSPHERE are found around the latitude of the explosion. Long-lived RADIOISOTOPES such as strontium-90, carried to the STRATOSPHERE by the explosion, are eventually deposited worldwide.
False-color photography
The processing of special photographic emulsions to display information using unnatural colors, used in terrestrial infrared PHOTOGRAPHY (eg vegetation studies) and astronomy. False-color ultraviolet photographs of the Sun can be obtained using the SPECTROHELIOGRAPH.
Family
In the classification of living things, a group of related genera (see GENUS). Related families are grouped into ORDERS. (See TAXONOMY.)
Farad (F)
The SI UNIT of CAPACITANCE, defined as the capacitance of a CAPACITOR for which a one coulomb charge raises its potential by one volt. The capacitance of most practical capacitors is measured in micro- or picofarads.
Faraday, Michael (1791-1867)
English chemist and physicist, the pupil and successor of H. DAVY at the ROYAL INSTITUTION, who discovered BENZENE (1824), first demonstrated electromagnetic INDUCTION (see also HENRY, JOSEPH) and invented the dynamo (1831 – see GENERATOR, ELECTRIC), and who, with his concept of magnetic lines of force, laid the foundations of classical field theory later built upon by J. CLERK MAXWELL. In the course of many years of researches, he also discovered the laws of ELECTROLYSIS which bear his name and, in showing that polarization of plane-polarized light was rotated in a strong magnetic field, demonstrated the existence of a connection between LIGHT and MAGNETISM.
Fathom
A unit of length used to describe depth at sea. One fathom equals six feet (1.8288m).
Fats
Another name for LIPIDS. The word is also used in a more narrow sense to mean TRIGYLCERIDES only, ESTERS of CARBOXYLIC ACIDS with GLYCEROL which are produced by animals and plants and form natural storage material. Fats are insoluble in water and occur naturally as either liquids or solids; those liquid at 20°C are normally termed oils and are generally found in plants and fishes. Oils generally contain esters of OLEIC ACID, which can be converted to esters of the solid STEARIC ACID by HYDROGENATION in the presence of finely divided nickel. This process is basic to the manufacture of margarine. Fats are the most concentrated sources of energy in the human diet, giving over twice the energy of STARCHES. Diets containing high levels of animal fats have been implicated as causative factors in heart disease, and replacement of animal fat by plant oils (eg peanut oil, sunflower oil) has been suggested. Fats, particularly of fish and plant origin, represent important items of commerce. Of major importance are soybean oil (9-10 million tonnes), sunflower, palm, peanut, cottonseed, rapeseed and coconut oil (2.5-4 million tonnes each) and olive and fish oil (over 1 million tonnes each). Major producers include the US (soybean oil), the USSR (sunflower oil and cottonseed oil) and India (peanut oil). (➧ page 130, 233)
Fatty acids
Straight-chain CARBOXYLIC ACIDS, of general formula RCOH, where R is a chain of CH_2 units, numbering 15 in palmitic acid and 17 in oleic acid and stearic acid, these being the three most common fatty acids in living organisms. In long-chain fatty acids such as these, the R group is hydrophobic (repels water), unlike the polar carboxylic group. This gives the molecule a dual nature, which is also seen in derivatives of fatty acids such as TRIGLYCERIDES, PHOSPHOLIPIDS and DETERGENTS. The properties of biological MEMBRANES are dependent on this duality, as are those of detergents.

Where the bonds between the carbon atoms in the molecule are all single, the fatty acid is said to be "saturated" with hydrogen, because it contains the maximum number of hydrogen atoms. Where there are one or more double bonds in the chain the fatty acid is said to be unsaturated, because a double bond between carbon atoms reduces the number of hydrogen atoms in the chain. A double bond also puts a kink in the otherwise linear chain. Whereas saturated fatty acids can lie alongside one another in a neat, stable arrangement, unsaturated fatty acids cannot because of these kinks in the carbon chains. This gives unsaturated fatty acids and their compounds (triglycerides etc.) a lower melting point than saturated fatty acids, so they are liquid at room temperature.

In the human diet, a high intake of saturated fatty acid compounds has been shown to increase the likelihood of heart disease.
Fauces
The back of the throat, containing the tonsils.
Fault
A fracture in the Earth's crust on either side of which there has been relative movement (see EARTH). Faults seldom occur along a single PLANE: usually a vast number of roughly parallel faults take place in a belt (fault-zone) a few hundred meters across. The side of a fault on which the strata have moved relatively downward is the downthrow side; the other the upthrow side. The difference in vertical height between the sides is the throw, the lateral displacement the heave. The angle between the horizontal and the fault plane is the dip: where this is roughly 90°, the fault is a normal fault. (See also DIASTROPHISM; HORST; RIFT VALLEY.) (➧ page 142, 208)
Feather
Outgrowths of the skin of birds, composed principally of the protein KERATIN. Feathers function to insulate and waterproof the body and to provide flight surfaces. They also have colors that are important as camouflage and in displays. There are several different types of feather, the main ones being contour feathers, flight feathers and down for insulation.
Feces
Ejected waste from the intestine, consisting of undigested cellulose, bacteria and water, colored

by bile and smelling of the products of bacterial action. (➧ page 233)

Fechner, Gustav Theodor (1801-1887)
German physicist and founder of experimental psychology, usually remembered for his reformulation of Ernst Heinrich Weber's (1795-1878) conclusions concerning the increase in a stimulus needed to make someone aware that there was a difference. Fechner's Law (or the Weber-Fechner Law) states that the perceived intensity of a sensation increases with the logarithm of its stimulus.

Feedback
The use of the output of a system to control its performance. Many examples of feedback systems can be found in biology, particularly in biochemistry and physiology. Thus the end product of a particular reaction may inhibit the enzyme controlling that reaction or shut down production of the enzyme. This occurs when the product begins to accumulate in the cell, preventing any further increase. Also used as a regulatory device in technological systems.

Feldspar
Widely distributed mineral group. Potash feldspars, principally ORTHOCLASE and MICROCLINE, are potassium aluminum silicates where sodium or barium may partly replace potassium. Plagioclase feldspars form a series derived from sodium aluminum silicate (ALBITE) with calcium replacing sodium in all proportions up to 100% (ANORTHITE). Feldspar is used in making glass and glazes.

Felids
Members of the Family Felidae, which includes all the cats, from tigers and lions down to small hunting (and domestic) cats. Felids belong to the Order CARNIVORA.

Femur
In vertebrates the proximal BONE of the hind leg, the thigh bone. In the human skeleton it is the largest and longest bone. (➧ page 216)

Feral animals
Animals that have escaped from domestication or captivity and returned to the wild, such as the dingo of Australia.

Ferenczi, Sandor (1873-1933)
Hungarian psychoanalyst, and an early colleague of FREUD, best known for his experiments in PSYCHOTHERAPY, in course of which he broke away from Freud's classic psychoanalytic theory. (See PSYCHOANALYSIS.)

Fermentation
The breakdown of GLUCOSE and other sugars by bacteria or yeasts in the absence of oxygen. LOUIS PASTEUR first demonstrated that fermentation is a biochemical process, each type being caused by one species (see also BUCHNER, EDUARD). In all species, however, the first step is GLYCOLYSIS, the degradation of glucose by ENZYME catalysis to pyruvic acid. (In aerobic respiration, this then enters the CITRIC ACID CYCLE.) In fermentation pyruvic acid may be reduced to LACTIC ACID, as with the yogurt- and cheese-making bacteria, or degraded to carbon dioxide and ETHANOL, as with the yeasts. Considerable energy is released during glycolysis; some is stored as the high-energy compound ATP, and the rest is given off as heat. Fermentation by yeast has been used for centuries in brewing and making bread and wine. Special fermentations involving other species have been used industrially for the manufacture of acetone, butanol, glycerol, citric acid, glutamic acid (monosodium glutamate) and many other compounds. Modern BIOTECHNOLOGY relies on fermentation methods to mass-produce materials.

Fermi, Enrico (1901-1954)
Italian atomic physicist who was awarded the 1938 Nobel Prize for Physics. His first important contribution was his examination of the properties of a hypothetical gas whose particles obeyed Pauli's EXCLUSION PRINCIPLE; the laws he derived can be applied to the ELECTRONS in a metal, and explain many of the properties of metals (see FERMI-DIRAC STATISTICS). Later he showed that NEUTRON bombardment of most elements produced their RADIOISOTOPES.

Fermi-Dirac statistics
In QUANTUM MECHANICS, the statistical behavior of a system of indistinguishable particles with a number of discrete states, each of which may be occupied at any one time by a single particle only. SUBATOMIC PARTICLES that show this behavior are termed fermions. (See also BOSE-EINSTEIN STATISTICS.)

Fermium (Fm)
A TRANSURANIUM ELEMENT in the ACTINIDE series, found in the debris from the first HYDROGEN BOMB, and now prepared by bombardment of lighter actinides. (➧ page 130)

Ferrel's Law
The proposition that moving air masses tend to be deflected to the right in the Northern hemisphere and to the left in the Southern, first proposed by US meteorologist William Ferrel (1817-1891). (See also BUYS-BALLOT'S LAW.)

Ferrocene (dicyclopentadienyl iron, $(C_5H_5)_2Fe$)
An orange crystalline solid; a typical, stable SANDWICH COMPOUND – the first to be prepared (1951). The molecular structure is a pentagonal ANTIPRISM. mp 174°C, subl 100°C.

Ferromagnetic materials see MAGNETISM.

Ferromagnetism see MAGNETISM.

Ferrous sulfate (iron (II) sulfate) see IRON.

Fertilization
The union of two GAMETES, or male and female sex cells, to produce a cell from which a new individual ZYGOTE develops. In DIPLOID organisms the sex cells are produced by reduction division or MEIOSIS, so they contain half the normal number of CHROMOSOMES, and fertilization therefore produces a cell with the normal number of chromosomes again. In HAPLOID organisms there is reduction division following fertilization, to halve the number of chromosomes present. Fertilization may take place outside the organism's body or inside the female. (➧ page 204)

Fertilizers
Materials added to the soil to provide elements needed for plant nutrition, and so improve crop yield. The elements needed in large quantities are nitrogen, in the form of nitrate ions, phosphorus, as phosphate ions, and potassium. Sulfur, calcium and magnesium are also needed, but are usually adequately supplied in the soil or incidentally in other fertilizers. Small amounts of TRACE ELEMENTS are also needed, and these are usually supplied in fertilizers. Natural or "organic" fertilizers such as manure, compost and seaweed generally contain all these elements, plus fibrous matter that improves the humus content of the soil. This is lacking in artificial fertilizers produced by chemical processes. Unless used carefully, these artificial fertilizers can create environmental problems: they are easily leached from the soil by rain, to pollute lakes (causing EUTROPHICATION) and water supplies (see NITRATES).

Fetus
The human embryo from three months of age to birth. During fetal life organ development is consolidated and specialization extended so that function may be sufficiently mature at birth; some organs start to function before birth in preparation for independent existence. During the fetal period most increase in size occurs, both in the fetus and in the placenta and womb. The fetus lies in a sac of AMNIOTIC FLUID which protects it and allows it to move about. Blood circulation in the fetus is adapted to the placenta as the source of oxygen and nutrients and site for waste excretion, but alternative channels are developed so that within moments of birth they may take over. Should the fetus be delivered prematurely, immaturity of the LUNGS may cause respiratory distress, and that of the LIVER may result in JAUNDICE. (➧ page 204)

Fever
Raising of body TEMPERATURE above normal (37°C or 98.6°F in humans), usually caused by DISEASE. Substances that cause fever are called pyrogens and are derived from cell products; they alter the set level of temperature-regulating centers in the HYPOTHALAMUS. Fever may be continuous, intermittent or remittent, the distinction helping to determine the cause.

Feynman, Richard Phillips (1918-1988)
US physicist awarded with SCHWINGER and TOMONAGA the 1965 Nobel Prize for Physics for their independent work on quantum electrodynamics (see QUANTUM MECHANICS). With GELL-MANN he proposed the quark as a fundamental SUBATOMIC PARTICLE.

Fiber
A thin thread of natural or artificial material. Animal fibers include wool, from the fluffy coat of the sheep, and silk, the fiber secreted by the silkworm LARVA to form its cocoon. Vegetable fibers include cotton, flax, hemp, jute and sisal: they are mostly composed of LIGNIN, though CELLULOSE is also important. Mineral fibers are generally loosely termed ASBESTOS. These fibrous mineral SILICATES are mined in South Africa, Canada and elsewhere. Man-made fibers are of two types: regenerated fibers, extracted from natural substances (eg rayon is cellulose extracted from wood pulp); and synthetic fibers. Most paper is made from wood fiber.

Fiberglass
GLASS drawn or blown into extremely fine fibers that retain the tensile strength of glass while yet being flexible. The most used form is fused QUARTZ, which when molten can be easily drawn and which is resistant to chemical attack. Most often, the molten glass is forced through tiny orifices in a platinum plate, on the far side of which the fine fibers are united (though not twisted) and wound onto a suitable spindle. Fiberglass mats (glass wool) are formed from shorter fibers at random directions bonded together with a thermosetting RESIN; they may be pressed into predetermined shapes. Fiberglass is used in INSULATION, automobile bodies, etc.

Fibiger, Johannes Andreas Grib (1867-1928)
Danish pathologist awarded the 1926 Nobel Prize for Physiology or Medicine for his discovery of a technique of inducing CANCER in rats.

Fibrin see CLOTTING.

Fibroid
A harmless growth of muscle tissue in the UTERUS, sometimes causing MENORRHAGIA or, because of its size, abdominal distension. Several may be present simultaneously.

Fibrositis
Inflammation and pain of the fibrous tissue of muscle sheaths.

Fibula
The smaller of the two bones in the lower leg, apposed to the tibia, and part of the ankle joint. (➧ page 216)

Field-emission microscope
A lower resolution relative of the FIELD-ION MICROSCOPE, in which the image is produced by ELECTRONS emitted by the tip itself when negatively charged.

Field-ion microscope
An instrument producing striking images of the arrangement of individual ATOMS in materials drawn out into, or evaporated on to, a fine tip,

typically 40nm in radius. Invented by Erwin Wilhelm Müller (b. 1911) in 1936, the microscope is lensless, the image being produced on a fluorescent screen by IONS created in a low-pressure gas by the intense ELECTRIC FIELD at the tip when it is positively charged to a few kilovolts.

Fifth generation
The name that Japan gave to advanced COMPUTERS that will respond to visual, audible and touch stimuli as well as to keyboards and conventional SENSORS. Japan's program, administered by ICOT (Institute for New Generation Computer Technology), has been copied by most other developed nations in one form or another. In general, the aims for fifth-generation computers include: advanced CHIPS that are far faster and need less power than today's; the use of ARTIFICIAL INTELLIGENCE and EXPERT SYSTEMS; input and output devices that mimic human abilities; and advanced SOFTWARE to link the other capabilities. (See also ROBOTS.)

Filament
In flowers, the stalk of the stamen that bears the anther at its apex. Also, the thread-like row of cells that constitutes the body in certain ALGAE such as *Spirogyra*. (◀ page 53)

Filariasis
A group of PARASITIC DISEASES of warm climates, transmitted by mosquitoes, causing FEVER, LYMPH node enlargement, ABSCESSES, epididymal inflammation and signs of ALLERGY; ELEPHANTIASIS may result.

Filter feeders (suspension feeders)
Animals that live by filtering tiny particles of food from water. They are found in a great many different groups, including the sponges, the cnidarians (corals, some anemones and some jellyfish), annelid worms (fan worms and peacock worms), mollusks (bivalves or lamellibranchs), crustaceans (barnacles), echinoderms (sea lilies, feather stars and some sea cucumbers), tunicates, phoronid worms, moss animals and lampshells. Many different body organs are adapted to the task of filtering: in bivalves and tunicates it is the gills, in barnacles the highly modified legs, in cnidarians and annelids specialized tentacles, and in the phoronids, moss animals and lampshells a tentacular organ known as a lophophore.

Filtration
Separation of solid particles from a liquid or gas by passing it through a porous mesh on which the solid collects. Commonly used are filter cloths, filter paper, wire mesh, sintered glass, or – where a large volume of water is to be filtered – sand beds. Filters have many varied uses, for example in cigarettes, vacuum cleaners, gasoline and diesel engines, coffee-making, air-conditioning units, water purification systems, and chemical preparation and analysis. A magnetic filter is used to remove iron or steel particles from oils, etc. in machine tools and some engines.

Finlay, Carlos Juan (1833-1915)
Cuban physician who first proposed (1881) that YELLOW FEVER is transmitted by the mosquito. Despite his considerable research, this was unproved until 1900.

Finsen, Niels Ryberg (1860-1904)
Danish physician awarded the 1903 Nobel Prize for Physiology or Medicine for his discovery and use of the curative properties of certain wavelengths of LIGHT.

Fire see COMBUSTION.

Fireball
A particularly bright METEOR, brighter than MAGNITUDE −5. The term is also applied to LIGHTNING of globular form.

Firn see NÉVÉ.

Firth
Term used primarily in Scotland to describe usually a long narrow ESTUARY, often a FJORD, and sometimes a STRAIT. The best known are those of the rivers Forth, Clyde and Tay; the Solway Firth, into which several rivers drain; the Moray Firth, between the NE tip of Loch Ness and the North Sea; and the Pentland Firth, the strait between NE Scotland and the Orkney Islands.

Fischer, Emil Hermann (1852-1919)
German organic chemist and pioneer of BIOCHEMISTRY, awarded the 1902 Nobel Prize for Chemistry for his work on the structures of SUGARS and PURINES, simple members of both of which families he synthesized. He also synthesized PEPTIDES and studied ENZYME action in the breaking down of PROTEINS.

Fischer, Ernst Otto (b. 1918)
German chemist awarded with G. WILKINSON the 1973 Nobel Prize for Chemistry for their work on ORGANOMETALLIC COMPOUNDS.

Fischer, Hans (1881-1945)
German organic chemist awarded the 1930 Nobel Prize for Chemistry for his elucidation of the structure of, and synthesis of, hemin – a substance closely related to heme (see HEMOGLOBIN). He later worked on the analysis of the CHLOROPHYLLS.

Fischer, Hermann Emil (1852-1919)
German scientist awarded the 1902 Nobel Prize for Chemistry for his work on sugar and purine synthesis.

Fish
A general name for four distinct groups of aquatic vertebrates: the jawless fish or AGNATHA, the cartilaginous fish or CHONDRICHTHYES, the ray-finned fish or ACTINOPTERYGII, and the lobe-finned fish or SARCOPTERYGII. The Actinopterygii and Sarcopterygii together make up the so-called BONY FISH or Osteichthyes, but the Agnatha also have bone.

Fission
The division of unicellular, or sometimes multicellular, organisms to produce genetically identical offspring. Binary fission is the most common form, and results in the production of two equal-sized daughters. It is seen in bacteria, protozoa, unicellular algae and simple animals such as sea anemones (see CNIDARIANS). Fission is one method of asexual REPRODUCTION (See also BUDDING.)

Fission, Nuclear
The splitting of the nucleus of a heavy ATOM into two or more lighter nuclei with the release of a large amount of ENERGY. Fission power is used in NUCLEAR REACTORS and the ATOMIC BOMB.

Fistula
An abnormal passage connecting two parts of the body, eg the skin and the intestine, caused by infection or injury.

Fitch, Val L. (b. 1923)
US scientist who shared the 1980 Nobel Prize for Physics with CRONIN for their discovery of an effect called CP violation which showed that not all particle interactions are symmetrical (see SUBATOMIC PARTICLES).

Fixation
An ambivalent (see AMBIVALENCE) attachment to an object or habit typical of an earlier stage of psychological development. Specific examples are regression to infantile behavior during stress, and COMPULSION toward objects reminiscent in some way of the individual's childhood.

Fjord
Narrow, deep sea inlet, steep-sided and bounded by mountains, formed by past glacial EROSION of a stream or river valley. Usually there is a shallow rock threshold, probably of terminal MORAINE, at the seaward entrance.

Flagellata (mastigophora) see PROTOZOA.

Flame see COMBUSTION.

Flame test
Preliminary test in qualitative chemical analysis. A small amount of the sample is introduced into a nonluminous flame; certain metal ions, on excitation, impart a characteristic color to the flame, eg yellow for sodium, red for strontium, green for copper. For flame photometry see SPECTROPHOTOMETRY.

Flamsteed, John (1646-1719)
The first Astronomer Royal, appointed 1675 with the foundation of the GREENWICH OBSERVATORY. In course of compiling his major star catalog, *Historia Coelestis Britannica*, he was pestered for his data by NEWTON: Newton prevailed, and a muddled version, edited by HALLEY, was published in 1712. Flamsteed's own edition appeared posthumously (1725).

Flare, Solar
A temporary brilliance in the SUN's chromosphere associated with a sunspot or group of sunspots. Large flares may last for as long as an hour or more, small ones only a few minutes. A flare results in the emission of electromagnetic radiation ranging from GAMMA RAYS and X-RAYS to RADIO WAVES, and of atomic particles – ELECTRONS, PROTONS and ALPHA PARTICLES. Flares are believed to be produced by the release of magnetic energy stored in the magnetic fields of complex sunspot groups or active regions.

Flash photolysis
Technique for investigating very fast chemical reactions (see KINETICS, CHEMICAL). A very intense flash of light of very short duration (from a LASER or flash lamp) is passed through a reaction mixture, usually gaseous. Instant dissociation occurs, producing FREE RADICALS, whose subsequent fast reactions are followed by automatic spectroscopy. (See also PHOTOCHEMISTRY.)

Flash point
The lowest temperature at which the VAPOR above a volatile LIQUID forms a combustible mixture with the OXYGEN in the air. At the flash point the TEMPERATURE is too low for sustained COMBUSTION to occur, but when a pilot flame is introduced to the mixture a brief flash occurs.

Flatworms see PLATYHELMINTHES.

Fleming, Sir Alexander (1881-1955)
British bacteriologist, discoverer of lysozome (1922) and penicillin (1928). Lysozome is an ENZYME present in many body tissues and lethal to certain bacteria; its discovery prepared the way for that of ANTIBIOTICS. His discovery of PENICILLIN was largely accidental; and it was developed as a therapeutic later, by HOWARD FLOREY and ERNST CHAIN. All three received the 1945 Nobel Prize for Physiology or Medicine for their work.

Flexner, Simon (1863-1946)
US medical scientist who did important work in bacteriology and pathology, especially on the nature of POLIOMYELITIS, and contributed toward the development of a serum treatment for spinal MENINGITIS.

Flight
Sustained travel through the air. The only animals capable of true flight are the insects, bats and most birds. The extinct PTEROSAURS could also fly. Most insects have two pairs of membranous wings that flap together. In flies and beetles one pair is modified for other purposes. In the more advanced flying insects the muscles distort the thorax, indirectly forcing the wings up and down – at rates of up to 1000 beats per second in midges. Other insects have wings that are operated directly by the muscles. In birds power for flying comes from the breast muscles, which control both the up and down strokes of the wing. Wing shape depends on the kind of flight. The only true flying mammals are the bats, whose wings are composed of skin supported by the elongated finger bones.

Flightless birds
Members of the Class Aves that have lost the power of flight. The principal group is the RATITES, which includes the ostrich, emu, rhea, kiwi, cassowaries, and various extinct forms. Other flightless birds, belonging to the CARINATES, are the penguins, the flightless cormorant of the Galapagos, the extinct great auk of the north Atlantic and the kakapo, a flightless nocturnal parrot found in New Zealand. Flightless forms have often evolved on islands where there are no mammalian predators, and these species are particularly vulnerable to human hunters and feral animals such as cats. Many such island birds have become extinct, and many more are endangered. Notable extinctions of the past include the elephant bird, the moas and the dodo.

Flint (chert)
Sedimentary rock composed of microcrystalline QUARTZ and CHALCEDONY. It is found as nodules in LIMESTONE and CHALK, and as layered beds, and was formed mainly by alteration of marine sediments of silicaceous organisms, and by replacement, preserving many FOSSIL outlines. A hard rock, flint may be chipped to form a sharp cutting edge, and was used by Stone Age peoples for their characteristic tools.

Floods and flood control
River floods are one of mankind's worst enemies. They are often caused by unusually rapid thawing of the winter snows: the river, unable to hold the increased volume of water, bursts its banks. Heavy rainfall may have a similar effect. Coastal flooding may result from an exceptionally high TIDE combined with onshore winds, or from a TSUNAMI.

River floods can be forestalled by artificially deepening and broadening river channels or by the construction of suitably positioned dams. Artificial levees may also be built (in nature, levees occur as a result of sediment deposited while the river is in flood; they take the form of built-up banks). Vegetation planted on uplands helps to reduce surface runoff (see DRAINAGE).

Flood control can create new problems to replace the old. In Egypt the Aswan Dam has halted the once regular flooding of the Nile, robbing farmlands of a rich annual deposit of silt. But flood control made possible the civilization of ancient Mesopotamia and plays a vital part in modern water conservation. (See also RIVERS AND LAKES.)

Florey, Howard Walter, Baron (1898-1968)
Australian-born British pathologist who worked with E.B. CHAIN and others to extract penicillin from *Penicillium notatum* mold for use as a therapeutic drug (1938-44). He shared with Chain and ALEXANDER FLEMING the 1945 Nobel Prize for Physiology or Medicine.

Flory, Paul J. (1910-1985)
US scientist awarded the 1974 Nobel Prize for Chemistry for his fundamental achievements, both theoretical and experimental, in the physical chemistry of the macromolecules.

Flotation, Froth
A process used to recover valuable minerals from low-grade areas. The pulverized ore is mixed with water and flotation reagents. When air is pumped into the mixture the mineral particles that preferentially adhere to the air bubbles rise to the surface in a froth that can then be skimmed off.

Flower
The reproductive organ of a flowering plant or ANGIOSPERM. There is a great variety of floral structure, but the basic organs are the same. Each flower is borne on an expanded stalk or receptacle that bears the floral organs. The sepals are the first of these organs and are normally green and leaflike. Above the sepals there is a ring of petals, which are normally colored and vary greatly in shape. The ring of sepals is termed the calyx and the ring of petals the corolla. Collectively the calyx and corolla are called the perianth. Above the perianth are the reproductive organs, comprising the male organs, the stamens (collectively known as the androecium) and female organs, the carpels (the gynoecium). Each stamen consists of a slender stalk, or filament, which is capped by the pollen-producing anther. Each carpel has a swollen base, the ovary, which contains the ovules that later form the seed. Each carpel is connected by a slender style to the stigma, which usually resembles a pinhead and which receives the pollen. Together, the style and stigma are sometimes termed the pistil.

In many plants, the flowers are grouped together to form an INFLORESCENCE. The immense number of variations of flower form are adaptations that aid either insect or wind POLLINATION. (◀ page 53)

Flowering plants see ANGIOSPERMS.

Flu see INFLUENZA.

Fluid
A substance which flows (undergoes a continuous change of shape) when subjected to a tangential or shearing FORCE. LIQUIDS and GASES are fluids, both taking the shape of their container. But while liquids are virtually incompressible and have a fixed volume, gases expand to fill whatever space is available to them.

Fluid mechanics
The study of moving and static FLUIDS, dealing with the FORCES exerted on a fluid to hold it at rest and the relationships with its boundaries that cause it to move. The scope of the subject is wide, ranging from HYDRAULICS, concerning the applications of fluid flow in pipes and channels, to aeronautics, the study of airflow relating to the design of airplanes and ROCKETS. Any fluid process, such as flow around an obstacle or in a pipe, can be described mathematically by a specific equation that relates the forces acting, the dimensions of the system and its properties such as TEMPERATURE, PRESSURE and DENSITY. Newton's laws of motion and VISCOSITY, the first and second laws of THERMODYNAMICS and the laws of conservation of MASS, ENERGY and MOMENTUM are applied as appropriate. Much use is also made of experimental evidence from models, wind tunnels, etc., to determine the process equation. Many types of flow occur: in laminar flow in a closed pipe, distinct layers of fluid slide over each other, their velocity decreasing to zero at the pipe wall; in turbulent flow the fluid is mixed by eddies and vortices and a statistical treatment is needed. (See also ARCHIMEDES; PASCAL'S LAW; REYNOLDS NUMBER.)

Flukes
Members of the PLATYHELMINTHES or flatworms. Flukes are parasitic and are responsible for diseases in a number of animals. (◀ page 16)

Fluorescence see LUMINESCENCE.

Fluoridation
Addition of small quantities of fluorides (see FLUORINE) to public water supplies, bringing the concentration to 1 ppm, as in some natural water. It greatly reduces the incidence of tooth decay by strengthening the teeth.

Fluorine (F)
The lightest of the HALOGENS, occurring naturally as FLUORITE, CRYOLITE and fluorapatite (see APATITE). A pale yellow toxic gas, fluorine is made by electrolysis of potassium fluoride in liquid HYDROGEN FLUORIDE. It is the most reactive, electronegative and oxidizing of all elements, reacting with almost all other elements to give fluorides (see HALIDES) of the highest possible oxidation state. It displaces other nonmetals from their compounds. Most nonmetal fluorides are highly reactive, but sulfur hexafluoride (used as an electrical insulator) and carbon tetrafluoride are inert (see STEREOCHEMISTRY). Fluorine is used in rocket propulsion, in URANIUM production, and to make FLUOROCARBONS. AW 19.0, mp −219.6°C, bp 188°C. (◀▶ page 37, 130)

Fluorite
A common mineral composed of calcium fluoride (see CALCIUM), also called fluorspar. It forms cubic crystals with a wide color range. Fluorite is used as a flux in the iron and steel industries, and as a FLUORINE ore.

Fluorocarbons
HYDROCARBONS in which hydrogen atoms are replaced (wholly or in part) by fluorine. Because of the stability of the carbon-fluorine bond, they are inert and heat-resistant. Thus they can be used in artificial joints in the body, and where hydrocarbons would be decomposed by heat, as in spacecraft heatshields, the coating of nonstick pans or as lubricants. Liquid fluorocarbons are used as refrigerants. (See also TEFLON; FREON.)

Fluoroscope
Device used in medical diagnosis and engineering quality control which allows the direct observation of an X-RAY beam that is being passed through an object under examination. It contains a fluorescent screen which converts the X-ray image into visible light (see LUMINESCENCE) and, often, an image intensifier.

FM (frequency modulation) see RADIO.

Focal length see LENS, OPTICAL.

Focus (optics) see LENS OPTICAL.

Focus (of earthquake) see EARTHQUAKE.

Foehn (föhn)
Dry, warm wind coming down the leeward slopes of mountains due to air having lost its moisture while ascending the windward slope, then warming on its descent. (See also CHINOOK.)

Foetus see FETUS.

Fog
In essence, a CLOUD touching or near to the Earth's surface. A fog is a SUSPENSION of tiny water (sometimes ice) particles in the air. Fogs are a result of the air's HUMIDITY being high enough for CONDENSATION to occur around suitable nuclei, and are found most often near coasts and large inland bodies of water. In industrial areas, fog and smoke may mix to give smog. Persistent advection fogs occur when warm, moist air moves over cold land or water.

Fold
A buckling in rock strata. Folds convex upward are called anticlines; those convex downward, synclines. They may be tiny or up to hundreds of kilometers across. Folds result from horizontal pressures in the EARTH's crust. The upper portions of anticlines have often been eroded away. (See also MOUNTAIN.) (▶ page 208)

Folic acid see VITAMINS.

Follicle
In mammals, the pit in which the base of each hair sits; it is supplied with an oily secretion by a sebaceous gland. Also used of the cluster of cells that surrounds the developing eggs in the ovaries of vertebrates (▶ page 204). In botany, a type of dry FRUIT.

Food chain
In ecology, a sequence of organisms, each of which provides food for the next one in the chain. The starting point of any food chain is an AUTOTROPH, which acts as the primary producer. Feeding on the primary producer is (usually) a herbivorous animal or protist, which in turn is eaten by a carnivorous animal or protist. Such organisms are known as HETEROTROPHS, as they cannot manufacture complex carbon compounds for themselves. The waste matter and dead bodies

of all these organisms are fed upon by organisms such as bacteria and fungi, which are known as SAPROTROPHS. The different points in a food chain are known as the trophic levels. The number of trophic levels is limited to about six, because any animal at a higher trophic level would require an inordinately large foraging area to supply it with sufficient food. Food chains are only a very simplified depiction of energy flow in an environment; food webs give a somewhat more accurate view, but they too are oversimplified, as the activities of parasites, scavengers, filter-feeders and others that greatly complicate the picture are usually omitted.

Food poisoning
General term for a range of conditions resulting from ingestion of unwholesome food, usually resulting in COLIC, VOMITING, DIARRHEA and general malaise. While a number of virus, contaminant, irritant and allergic factors may play a part in some cases, three specific types are common: those due to STAPHYLOCOCCUS, *Clostridium welchii* and SALMONELLA bacteria. Inadequate cooking, allowing cooked food to stand for long periods in warm conditions, and contamination of cooked food with bacteria from humans or uncooked food are the usual causes. *Staphylococci* may be introduced from a BOIL or from the nose of a food handler; they produce a toxin if allowed to grow in cooked food. Sudden vomiting and abdominal pain occur 2-6 hours after eating. *Clostridium welchii* poisoning causes colic and diarrhea, 10-12 hours after ingestion of contaminated meat. *Salmonella* enteritis causes colic, diarrhea, vomiting and often fever, starting 12-24 hours after eating; poultry and human carriers are the usual sources. BOTULISM is a rare, but often fatal form of food poisoning, caused by toxins produced by the bacterium *Clostridium botulinum*.

Fool's gold see PYRITE.

Foot
Weight-bearing structure of vertebrate legs, and, in humans, of the lower limbs only. Also, the single muscular organ used for locomotion in gastropod mollusks (slugs and snails).

Foot (ft)
Unit of measurement of length, corresponding to 12 inches or 0.3048m.

Foot and mouth disease
A virus infection of cattle and pigs, sometimes affecting domestic animals and humans. Vesicles of the skin and mucous membranes and fever are usual. It is highly contagious, and epidemics require the strict limitation of stock movement and the slaughter of affected animals.

Foot-pound (weight) or **foot-pound-force** (ft lbf)
A unit of WORK, that is done when a MASS of one pound is lifted through one foot against gravity, equal to 1.35582 joules.

Foraminifera
A group of PROTOZOA, classified with the amebae in the Class Sarcodina. They have sculpted shells of various characteristic shapes, with tiny pores through which slender pseudopodia (extensions of the cell) can protrude; this is a modified form of PHAGOCYTOSIS in which the pseudopodia absorb microscopic particles of food that adhere to them. Foraminiferans are important constituents of both the plankton and the sea-floor community.

Force
In mechanics, the physical quantity which, when it acts on a body, either causes it to change its state of motion (ie imparts to it an ACCELERATION), or tends to deform it (ie induces in it an elastic strain – see MATERIALS, STRENGTH OF). Dynamical forces are governed by NEWTON's laws of motion, from the second of which it follows that a given force acting on a body produces in it an acceleration proportional to the force, inversely proportional to the body's mass and occurring in the direction of the force. Forces are thus VECTOR quantities with direction as well as magnitude. They may be manipulated graphically like other vectors, the sum of two forces being known as their resultant. The SI UNIT of force is the newton, a force of one newton being that which will produce an acceleration of $1m/sec^2$ in a mass of 1 kilogram.

Forebrain
In EMBRYOLOGY, the division of the BRAIN that develops into the cerebral cortex, the basal ganglia, THALAMUS, HYPOTHALAMUS, olfactory bulbs, RETINA and optic nerves. (➧ page 126)

Foreign body
Medically, any object that is, but should not be, present in the body. Examples include splinters, grit in wounds, and objects that are pushed in deliberately – eg small children may push peanuts into their nostrils.

Forensic medicine
The branch of medicine concerned with legal aspects of death, disease or injury. Forensic medical experts are commonly required to examine corpses found in possibly criminal circumstances. They may be asked to elucidate probable cause and approximate time of death, to investigate the possibility of poisoning, trauma or suicide, to analyze links with possible murder weapons and to help to identify decayed or mutilated bodies.

Forests
Extensive tracts of land whose vegetation is composed mainly of trees. They are of considerable importance to humans: they have provided fuel and building material since prehistory; their fruits have served as food; and, particularly today, they play a valuable part in countering SOIL EROSION (see also CONSERVATION).

CONIFEROUS FORESTS have as characteristic trees the conifers pine, spruce and fir – and are found in temperate regions as far north as the edges of the TUNDRA.

DECIDUOUS FORESTS, found in temperate zones, are characterized by DECIDUOUS TREES, eg oak, elm, beech and birch.

RAIN FORESTS are found in those equatorial regions where there is no dry season. In terms of numbers of species they are the richest type of environment on Earth, with millions of animals and plants that have not yet been identified. Their present rate of destruction is a matter of concern, particularly for its impact on the CARBON CYCLE and its enhancement of the GREENHOUSE EFFECT.

MONSOON FORESTS are found in monsoon regions, and resemble rain forests except that they are more open. Their trees have adapted to the marked dry season.

THORN FORESTS occur in tropical and subtropical areas where rainfall is insufficient to maintain larger trees. They are characterized by shrubs and small trees, eg mesquite, acacia. (See also TREES; WOOD.)

Formaldehyde (HCHO)
Colorless, acrid, toxic gas; the simplest ALDEHYDE, more reactive than the others. It is made by catalytic air oxidation of METHANOL vapor or of NATURAL GAS. Formaldehyde gas is unstable, and is usually stored as its aqueous solution, formalin, used as a disinfectant and preservative for biological specimens; on keeping, formalin deposits a polymer, paraformaldehyde, which regenerates formaldehyde on heating. Formaldehyde is condensed with UREA and PHENOLS to make PLASTICS, with AMMONIA to give hexamethylenetetramine (a urinary ANTISEPTIC also used to make RDX), and with ACETALDEHYDE to give pentaerithrytol and hence PETN. It is also used in tanning and textile manufacture. MW 30.0, mp −92°C, bp −19°C.

Formic acid (HCOOH)
Colorless, acrid liquid; the simplest CARBOXYLIC ACID, a stronger ACID than the others. It occurs in the stings of ants and nettles. It is made (via sodium formate) by heating sodium hydroxide with carbon monoxide under pressure, and used in tanning, as a latex coagulant and to reduce dyes. MW 46.0, mp 8.4°C, bp 100.6°C.

Formula, Chemical
A symbolic representation of the composition of a MOLECULE. The empirical formula shows merely the proportions of the atoms in the molecule, as found by chemical analysis, eg water H_2O, acetic acid CH_2O. (The subscripts indicate the number of each atom if more than one.) The molecular formula shows the actual number of atoms in the molecule, eg water H_2O, acetic acid $C_2H_4O_2$. The atomic symbols are sometimes grouped to give some idea of the molecular structure, eg acetic acid CH_3COOH. This is done unambiguously by the structural formula, which shows the chemical BONDS and so distinguishes between ISOMERS. The space formula shows the arrangement of the atoms and bonds in three-dimensional space, and so distinguishes between STEREOISOMERS; it may be drawn in perspective or represented conventionally. Loosely associated compounds, such as LIGAND complexes, are often shown with a dot, eg copper (II) sulfate pentahydrate, $CuSO_4.5H_2O$. Special symbols are sometimes used for common groups and ligands, eg ethyl Et, phenyl Ph, ethylenediamine en.

Forssman, Werner (1904-1979)
German surgeon awarded, with A. COURNAND and D.W. RICHARDS, the 1956 Nobel Prize for Physiology or Medicine for his discovery of the technique of introducing medicine directly into the heart via a CATHETER.

Fossil fuel
Any fuel that is obtained from the Earth and has its origin in organic matter. Peat, lignite, coal and anthracite are the solid fossil fuels, derived from plant material that has had its carbon content increased by chemical breakdown after being buried and fossilized. Petroleum is the liquid fossil fuel, derived from the substance of marine organisms and distilled and concentrated within the rocks of the Earth. Natural gas is the gaseous fossil fuel, usually consisting of hydrocarbon gases derived from deposits of coal or oil.

Fossils
The remains, traces or impressions of living organisms that inhabited the Earth during past ages. Traces may include footprints, burrows or preserved droppings.

The bodies of organisms may be fossilized or petrified (literally, turned to stone) in one of two ways. In permineralization, the pore spaces of the hard parts are infilled by certain minerals (eg SILICA, PYRITE, CALCITE) that infiltrate from the local groundwater. The resulting fossil is thus a mixture of mineral and organic matter. In many other cases, mineralization (or replacement) occurs, where even the hard parts of the organism are dissolved away but the form is retained by deposited minerals. Where this has happened very gradually, even microscopic detail may be preserved, but generally only the outward form remains. The soft parts of the body are rarely preserved, except by very fine, silty sediments. Only in exceptional circumstances is the organism preserved in its entirety, eg mammoths in the Siberian permafrost, insects in amber, or human and animal remains in peat bogs. (See also PALEONTOLOGY.)

Foucault, Jean Bernard Léon (1819-1868)
French physicist who demonstrated the rotation of the Earth with the FOUCAULT PENDULUM. He

also invented the GYROSCOPE and made the first reasonably accurate determination of the velocity of LIGHT.

Foucault pendulum
A PENDULUM comprising an iron ball at the end of a long steel wire which, on being set swinging, maintains its direction of swing while the Earth rotates beneath it. When one was demonstrated by J.B.L. FOUCAULT in 1851, it provided the first direct evidence for the rotation of the Earth.

Fourth dimension see SPACE-TIME.

Fovea
A tiny area in the RETINA with few nonreceptor cells and a concentration of cones, making it the most sensitive part of the eye for both acuity and color vision. (➧ page 213)

Fowler, William (b. 1911)
US physicist who shared the 1983 Nobel Prize for Physics with S. CHANDRASEKHAR for their independent work on the birth and death of STARS.

Fractionation see DISTILLATION.

Fractures
Mechanical breaks or cracks in BONE caused by trauma or underlying DISEASE. Most follow sudden bending, twisting or shearing forces, but prolonged stress (eg long marches) may lead to smaller fractures. Fractures may be compound, in which bone damage is associated with SKIN damage, with consequent liability to infection; or simple, in which the overlying skin is intact. Comminuted fractures are those in which bone is broken into many fragments. Greenstick fractures are partial fractures where bone is bent, not broken, and occur in children. Severe pain, deformity, loss of function, abnormal mobility of a bone and HEMORRHAGE, causing swelling and possibly shock, are important features; damage to nerves, arteries and underlying viscera (eg lung, spleen, liver and brain) are serious complications. Pathological fractures occur when lack of mineral content, tumors etc. weaken the structure of bone, allowing fracture with trivial or no apparent injury.

Francium (Fr)
A radioactive ALKALI METAL, resembling CESIUM, which has been obtained only in trace quantities. Its most stable isotope, Fr^{223}, has a half-life of 21 minutes. (➧ page 130)

Franck, James (1882-1964)
German physicist who shared with G. HERTZ the 1925 Nobel Prize for Physics for their experiments showing the internal structure of the atom to be quantized (see QUANTUM THEORY). With E.V. Condon, he was responsible for the Franck-Condon principle, which assumes that the nuclei in vibrating molecules do not have time to move during electronic transitions.

Frank, Ilya Mikhailovich (b. 1908)
Russian physicist who, with TAMM, provided an explanation for CHERENKOV RADIATION (1937), first observed by CHERENKOV (1934). For their work all three shared the 1958 Nobel Prize for Physics.

Franklin, Benjamin (1706-1790)
American printer, statesman and scientist who first achieved fame and wealth as publisher of *Poor Richard's Almanack* (1732). His famous kite experiment (1752) demonstrated the electrical nature of lightning. His inventions included the lightning conductor, the efficient Pennsylvanian fireplace (Franklin stove) and bifocal lenses. His most significant theoretical contribution was the single-fluid theory of electricity.

Frasch process
Process for extracting SULFUR from sulfur-bearing CALCITE deposits, invented by Herman Frasch (1851-1914) in 1891. Three concentric pipes are lowered down a bore-hole to the ore. Superheated water at 165°C is pumped down the outer pipe and compressed hot air is blown down the inner pipe. This forces a frothy mixture of molten sulfur and water up the middle pipe. Very pure sulfur is produced.

Fraunhofer, Joseph von (1787-1826)
German optician who mapped the dark lines (FRAUNHOFER LINES) in the solar spectrum and reinvented the DIFFRACTION grating.

Fraunhofer lines
Dark lines that appear in the SPECTRUM of the SUN. They are due to the ABSORPTION of the radiation of particular FREQUENCIES (and thus ENERGIES) by ATOMS in the outer layers of the solar ATMOSPHERE. Analysis of the solar spectrum thus leads to the identification of these atoms. The lines were first accurately mapped by J. VON FRAUNHOFER. The more prominent are denoted by letters: A and B are due to terrestrial OXYGEN; C to HYDROGEN; D to SODIUM; E to IRON, and so on.

Freckle
Small brown pigmented spot in the SKIN of exposed areas, numerous in some individuals; they are brought out by sunlight and are harmless.

Free energy see THERMODYNAMICS.

Free radicals
Molecules or atoms which have one unpaired electron (see ORBITAL), and hence an unused VALENCE. Most are very reactive and shortlived, but if the odd electron can be delocalized by RESONANCE (eg in triphenylmethyl) they may be stable. Free radicals can be studied by SPECTROSCOPY, chiefly electron-spin resonance. They are produced by heat, irradiation, FLASH PHOTOLYSIS and ELECTROLYSIS, and are important in forming POLYMERS and in explosive chain reactions.

Freezing point
The TEMPERATURE at which a liquid begins to solidify – not always well-defined or equal to the melting point (see FUSION). It usually rises with PRESSURE, solids being slightly more dense than liquids, though water is a notable exception; it is lowered by solutes in the liquid, the amount providing an accurate means of determining MOLECULAR WEIGHTS. The solid separating from a solution usually has a different composition from the liquid, and repeated freezing can be used to separate substances. Pure substances can often be "supercooled" below their freezing point for limited periods, as the formation of the solid CRYSTAL requires enucleation by rough surfaces in contact with or particles suspended in the liquid.

Freon
Trade name for a group of volatile CARBON compounds (derivatives of METHANE and ETHANE) containing fluorine and chlorine or bromine. They are nonflammable and nontoxic, and are used as refrigerants and aerosol propellants, though the latter use has declined considerably because of fears that they may damage the ozone layer.

Frequency
The rate at which a periodic WAVE MOTION executes complete cycles of its variation. Frequencies are measured in hertz (Hz), ie in cycles per second. Musical sounds typically have frequencies in the range 36-20,000Hz; alternating-current ELECTRICITY supply is at 60Hz in the US, but usually 50Hz in the rest of the world. According to QUANTUM THEORY, the frequency (v) of ELECTROMAGNETIC RADIATION provides a measure of the ENERGY (E) of its quanta (PHOTONS): $E=hv$, where h is the PLANCK CONSTANT. Again, the wavelength (λ), frequency (f) and velocity (v) of a harmonic wave are related by the equation $v=f\lambda$. (➧ page 118, 225)

Frequency modulation (FM) see RADIO.

Fresnel, Augustin Jean (1788-1827)
French physicist who evolved the transverse-wave theory of light through his work on optical interference. He worked also on reflection, refraction, diffraction and polarization, and developed a compound lens system still used for many lighthouses.

Freud, Anna (1895-1982)
Austrian-born British pioneer of child psychoanalysis. Her book *The Ego and Mechanisms of Defense* (1936) is a major contribution to the field. After escaping with her father SIGMUND FREUD from Nazi-occupied Austria (1938), she established an influential child-therapy clinic in London.

Freud, Sigmund (1856-1939)
Austrian neurologist and psychiatrist, founder and author of almost all the basic concepts of PSYCHOANALYSIS. He graduated as a medical student from the University of Vienna in 1881; and for some months in 1885 he studied under J.M. CHARCOT. Charcot's interest in HYSTERIA converted Freud to the cause of psychiatry. Dissatisfied with HYPNOSIS and electrotherapy as analytic techniques, he evolved the psychoanalytic method, founded on DREAM analysis and FREE ASSOCIATION. Because of his belief that sexual impulses lay at the heart of NEUROSES, he was for a decade reviled professionally, but by 1905 disciples such as A. ADLER and C.G. JUNG were gathering around him; both were later to break away. For some thirty years he worked to establish the truth of his theories, and these years were especially fruitful. Fleeing Nazi antisemitism, he left Vienna in 1938 for London, where he died in 1939.

Friction
Resistance to motion arising at the boundary between two touching surfaces when it is attempted to slide one over the other. As the FORCE applied to start motion increases from zero, the equal force of "static friction" opposes it, reaching a maximum "limiting friction" just before sliding begins. Once motion has started, the "sliding friction" is less than the limiting. Friction increases with the load pressing the surfaces together, but is nearly independent of the area in contact. For a given pair of surfaces, limiting friction divided by load is a dimensionless constant known as the coefficient of friction. LUBRICATION is used to overcome friction in the bearings of machines.

Friedel-Crafts reaction
Class of substitution reactions, catalyzed by acidic metal halides (eg aluminum chloride); an ALKYL HALIDE or an ACID CHLORIDE reacts with an AROMATIC COMPOUND, the alkyl or acyl group replacing a hydrogen atom on the aromatic ring: the carbonium ion formed attacks the ring as an electrophile (see NUCLEOPHILES). (See also ALKYLATION.)

Frisch, Karl von (1886-1982)
Austrian zoologist best known for his studies of bee behavior, perception and communication, discovering the "Dance of the Bees". With TINBERGEN and LORENZ he was awarded the 1973 Nobel Prize for Physiology or Medicine for his work.

Frisch, Otto Robert see MEITNER, LISE.

Fromm, Erich (b. 1906)
German-born US psychoanalyst who has combined many of the ideas of Freud and Marx in his analysis of human relationships and development in the context of social structures and in his suggested solutions to problems such as alienation.

Front
A boundary between two air masses, one cold and dense, the other warm and less dense. Fronts are regions of uncertain weather: cloud and rain, variable HUMIDITY and WIND direction, and low

air pressure (see ATMOSPHERE). A warm front is where warm air advances, displacing cold; the reverse happens in a cold front. An occluded front occurs often toward the end of a system of STORMS: a mass of warm air is surrounded and forced upward by cold air. In the N Hemisphere the polar front separates the W-moving cold air of polar regions from the E-moving warm air flowing up from the TROPICS. Warm air flows into "bays" in the front, then cold air in from the rear, resulting in the rotating air system known as a depression. (See also METEOROLOGY.)

Frontal lobe
The foremost of the four main lobes of cerebral cortex in the BRAIN of man. It is concerned with personality and behavior, and with the emotional aspects of perception. (♦ page 126)

Frost
Frozen atmospheric moisture formed on objects when the temperature is below 0°C, the freezing point of water (see FREEZING POINT). Hoarfrost forms in roughly the same way as DEW but, owing to the low temperature, the water VAPOR sublimes (see SUBLIMATION) from gaseous to solid state to form ice crystals on the surface. The delicate patterns often seen on windows are hoarfrost. Glazed frost usually forms when RAIN falls on an object below freezing: it can be seen, for example, on telegraph wires. Rime occurs when super-cooled (see SUPERCOOLING AND SUPERHEATING) water droplets contact a surface that is also below 0°C; it may result from FOG or drizzle. The first frost of the year signifies the end of the growing season. (See also ICE; SNOW.)

Frostbite
Damage occurring in SKIN and adjacent tissues caused by freezing. (The numbness caused by cold allows considerable damage without pain.) DEATH of tissues follows and they separate off. Fingers and toes are particularly susceptible.

Frozen foods see BIRDSEYE, CLARENCE.

Fructose (fruit sugar, $C_6H_{12}O_6$)
A naturally occurring simple SUGAR (mono-saccharide) found in a wide variety of fruits and in honey. It is the sweetest of the simple sugars.

Fruit
Botanically, the structure that develops from the ovary and accessory parts (the carpel) of a FLOWER after FERTILIZATION. Fruits may be simple (derived from a single ovary, eg plum), aggregate (formed by ovaries joined together in a single flower, eg raspberry) or multiple (formed from the many flowers of an INFLORESCENCE, eg fig). Fruits may be fleshy or dry. Fleshy fruits include the BERRY and the DRUPE. Dry fruits may be dehiscent (splitting open to disperse the seeds) as in the LEGUME (or pod), FOLLICLE, CAPSULE and silique, or they may be indehiscent (not splitting open) as in the ACHENE, GRAIN, SAMARA and NUT. As well as the true fruits, formed from the carpels only, there are false fruits, in which other parts of the flower are involved; for example, in apples the "core" is the true fruit, the fleshy pulp outside it being derived from the receptacle (see POME).

The main function of the fruit is to protect the seeds and disperse them when ripe.

Fruitfly see DROSOPHILA.

Fuel
A substance that may be burned (see COMBUSTION) to produce heat, light or power. Traditional fuels include dried dung, animal and vegetable oil, wood, peat and coal, supplemented by the manufactured fuels charcoal, coal gas, coke and water gas. In this century petroleum and natural gas have come into widespread use. The term "fuel" has also been extended to include chemical and nuclear fuels (see FUEL CELL; NUCLEAR ENERGY), though these are not burned. Specialized high-energy fuels such as HYDRAZINE are used in rocket engines. The chief property of a fuel is its calorific value – the amount of heat produced by complete combustion of a unit mass or volume of fuel. Also of major importance is the proportion of incombustibles – ash and moisture – and of sulfur and other compounds liable to cause air pollution. (See also ACID RAIN; RENEWABLE ENERGY.)

Fuel cell
A direct-current power source similar to a battery but differing in that a chemical fuel must be supplied while the cell is in use. Various chemical reactions are utilized in different types of cell. The most common is the hydro-oxygen cell in which hydrogen reacts with hydroxyl (HYDROXIDE) ions in the electrolyte to form water at the ANODE, and oxygen reacts with water to form hydroxyl ions at the CATHODE, the overall reaction resulting in the formation of water and a flow of ELECTRONS through an external circuit. The cell is divided into three compartments by the electrodes, the two outer ones containing the hydrogen fuel and the oxygen, and the central one the aqueous electrolyte. The electrodes are porous so that the gases can penetrate to meet the electrolyte. Platinum and nickel are typical catalysts. Fuel cells are much more efficient converters of chemical energy than heat engines.

Fugue state see AMNESIA.

Fukui, Kenichi (b. 1918)
Japanese chemist who received the 1981 Nobel Prize for Chemistry jointly with HOFFMAN for predicting the behavior of electrons in chemical reactions by applying theories of QUANTUM MECHANICS.

Fuller, Richard Buckminster (1895-1984)
US inventor, philosopher, author, mathematician and perhaps the 20th century's most original and prolific thinker. He is best known for his concept "Spaceship Earth" and for inventing the geodesic dome.

Fuller's earth
Natural CLAY material of variable composition, once used to clean wool and cloth (fulling); now used for decolorizing OIL of all sorts by selective chemical ADSORPTION, and also as a pesticide carrier and in drilling muds.

Fulminate (mercury (II) cyanate) see MERCURY.

Fumarole
Found in volcanic regions, a hole in the ground that emits steam, carbon dioxide and other vapors typical of volcanic eruption. (See GEYSER; VOLCANISM; VOLCANO.)

Functional group
A group of connected atoms whose presence in a molecule gives rise to characteristic chemical properties and infrared absorptions; eg hydroxyl $-OH$, carboxyl$=C=O$. The properties of a molecule are roughly the sum of those of its constituent functional groups, though their interactions are also significant.

Fungal diseases (mycoses)
DISEASES caused by FUNGI, which apart from common skin and nail ailments such as athlete's foot and ringworm, usually develop in people with disorders of IMMUNITY or DIABETES and those on certain DRUGS (STEROIDS, immunosuppressives, ANTIBIOTICS). THRUSH is common in the mouth and vagina but rarely causes systemic disease. Specific fungal diseases occur in some areas of the world. In addition, numerous fungi in the environment lead to forms of ALLERGY and lung disease.

Fungi
Members of the Kingdom Fungi, such as the mushrooms, brackets, puffballs, truffles, molds and yeasts. They are EUKARYOTES, which reproduce by forming spores (either sexually or asexually) and which lack chlorophyll. They have a characteristic form of growth, in which slender tubes of protoplasm known as hyphae make up the entire body of the organism (the mycelium). These hyphae usually form a loose, spreading network in the soil – or within the tissues of their host, if parasitic – but the hyphae can also mesh together to form a compact, solid mass, as in the fruiting bodies ("toadstools", etc.) of higher fungi. These fruiting organs are not the main body of the fungus, however – in effect, fungi have no central mass or "body" as plants and animals do. They are constantly flowing into new hyphae with the old hyphae dying off behind them.

The hyphae have rigid walls made of the polysaccharides CHITIN, mannan or glucan. This is true of all but the oomycetes and some other small, aberrant groups with cellulose cell-walls, which are probably not fungi at all but degenerate algae. Internally, the hyphae may be undivided (lower fungi) or divided by septa (higher fungi). The septum is incomplete even in the higher fungi, and protoplasm can flow from one division to another. Fungal nuclei are very small and there may be one, two or several per division. Fungi reproduce sexually, but by fusion of hyphae rather than by the production of gametes. The nuclei do not necessarily unite after fusion, and nuclei from different parents may coexist in the hyphae for some time. They only fuse immediately before spore formation, the zygote undergoing MEIOSIS to produce the haploid spores. The spores generate new hyphae directly, with no embryo stage. Most fungi also lack a flagellum at all stages of their life cycle, and some taxonomists use this as a definitive feature of the fungi, relegating those minor groups of lower fungi that do have a flagellum to the Kingdom PROTOCTISTA.

All fungi are HETEROTROPHS, either feeding on the remains of other organisms (saprotrophs) or extracting nutrients from living tissues (parasites). The higher fungi, which generally form large, conspicuous fruiting bodies, can be divided into two groups, the Basidiomycotina (formerly Basidiomycetes) and the Acomycotina (formerly Ascomycetes). Some members of these groups are small plant parasites, however, such as the rusts, smuts and mildews. One group of the Ascomycotina, the yeasts, exist as single cells. They may be either saprophytic (living on fallen fruits, for example) or parasitic, and some, such as the Dutch elm disease fungus, can alternate between a yeast form and a hyphal form. Some of the Ascomycotina have formed a symbiotic relationship with algae or cyanobacteria that live and photosynthesize within their hyphae; these are known as lichens.

The lower fungi include the bread molds, water molds and many other lesser known groups. Some are debilitating parasites, particularly of fish and insects, while many live in the soil, breaking down waste matter. (♦ page 13)

Fungicide
Substance used to kill FUNGI and so to control fungal diseases in humans, domestic animals and plants.

Funk, Casimir (1884-1967)
Polish-born biochemist. Following the work of C. EIJKMAN and F.G. HOPKINS, he proposed that BERIBERI, RICKETS, SCURVY and PELLAGRA arose from lack of certain trace dietary constituents, which he christened vitamines, or "life-amines". (See also AMINES; VITAMINS.)

Funny bone
Part of the ulna (one of the two bones comprising the forearm) at the elbow, over which passes the ulnar nerve. If it is struck sharply, it causes transient unpleasant tingling and numbness in the area served by the nerve. (♦ page 126, 216)

Fur
In biology, the insulating covering found in mammals, also known as HAIRS.

Furfural ($C_4H_3O.CHO$)
Colorless liquid, an ALDEHYDE similar to BENZALDEHYDE and a HETEROCYCLIC COMPOUND derived from furan. It is made by digesting corncobs, oat and rice hulls, etc. with acid, and is used as a solvent and an intermediate in making PLASTICS and other compounds.

Furnace
A construction in which heat can be generated, controlled and used. The heat may be produced by burning a fuel such as coal, oil or gas; by electricity; by concentrating the heat of the Sun; or by atomic energy (see FISSION, NUCLEAR). Simple furnaces are often used in the home to heat water; but much larger ones are used in industry, particularly in the heat treatment of metals (see METALLURGY). These are usually lined with firebricks (see also REFRACTORY), which may also be water-cooled. (See also BLAST FURNACE; ELECTRIC FURNACE.)

Fusion
The process of melting, or passing from the solid state to liquid state. This is accompanied by an absorption of LATENT HEAT and usually occurs at a well-defined TEMPERATURE, the melting point, which rises slightly with PRESSURE and chemical purity. Many amorphous solids like GLASS have no melting point, and simply reduce their VISCOSITY over a wide temperature range.

Fusion, Nuclear
A nuclear reaction in which the nuclei of light ATOMS combine to produce heavier, more stable nuclei, releasing a large quantity of ENERGY. Fusion reactions are the energy source of the SUN and the HYDROGEN BOMB. If they could be controlled and made self-sustaining man would have a safe and inexhaustible energy source. The aim of nuclear research is to produce in the laboratory conditions under which nuclei of DEUTERIUM (which exists in great abundance in ordinary water) and TRITIUM can "fuse" to produce a heavier nucleus, helium in the case of DT fusion. Only small amounts of fuel would be needed and the products are not radioactive. But if they are to fuse, the light, positively charged nuclei must collide with sufficient energy to overcome their electrostatic repulsion. This can be done by using a particle ACCELERATOR, but to get a net energy release the material must be heated to very high temperatures (around 10^9K) when it becomes a PLASMA. However, plasmas are difficult to contain. The most advanced approach to containment is magnetic confinement, which makes use of the electromagnetic properties of the plasma. Two sets of magnets create forces which compress the plasma so that when it is heated, the nuclei within the plasma fuse, and release NEUTRONS. The heat of neutrons can then be converted to heat using lithium. Two experimental tokamaks are in operation – the Joint European Torus (JET) in Culham, England, and the Tokamak Fusion Test Reactor (TFTR) at the Princeton Plasma Physics Laboratory, USA. Another technique using magnetic confinement is the reversed-field pinch (RFP), which makes more efficient use of a weaker field to control costs. The Reversed Field Experiment (RFC) at the Center for the Study of Ionizing Gases in Italy is designed to use RFP.

Scientists are also examining other routes to fusion, including inertial confinement fusion (ICF), in which fusion is set off in a series of small controlled "explosions". If a solid lump of DT mixture, below its melting temperature of 14K, is heated very suddenly the lump will be "confined" for as long as it takes to blow apart.

Several years of experimentation lie ahead before scientists will know if fusion will release more energy than goes into heating and containing the plasma. If successful, the technical problems of building a practical power station will follow. (See also NUCLEAR ENERGY.)

G
The Universal Gravitational Constant. (See GRAVITATION.)

g
Acceleration due to gravity. (See ACCELERATION.)

Gabbro
Plutonic IGNEOUS ROCK composed of coarse-grained plagioclase FELDSPAR with PYROXENE and OLIVINE. Often rhythmically banded, it arises from fractional crystallization of MAGMA.

Gabor, Dennis (1906-1979)
Hungarian-born UK physicist who invented HOLOGRAPHY, for which he was awarded the 1971 Nobel Prize for Physics. He had developed the basic technique in the late 1940s, but practical applications had to wait for the invention of the LASER (1960) by C.H. TOWNES.

Gadolinium (Gd)
Element of the LANTHANUM SERIES. AW 157.3, mp 1330°C, bp 2900°C, sg 7.9004 (25°C). (◆ page 130)

Gajdusek, Daniel Carleton (b. 1923)
US physiologist who shared the 1976 Nobel Prize for Physiology or Medicine with S.B. BLUMBERG. Gajdusek identified the rare neurodegenerative disease, kuru, as a "slow virus" infection.

Galactic clusters
Clusters of stars lying in or near the galactic plane, each of which contains a few hundred stars. Because of their irregular shape they are also termed open clusters. The best-known galactic cluster in N skies is the PLEIADES. (◆ page 240)

Galaxy
The largest individual conglomeration of matter, containing stars, gases, dust and planets. Galaxies are classified according to their shapes, the three basic types being elliptical, spiral and irregular. Elliptical galaxies contain predominantly old stars and little or no gases. A typical spiral galaxy consists of a central nucleus, where stars are relatively densely grouped together, surrounded by a disk of gases and younger stars concentrated into a pattern of spiral arms. Spiral galaxies have arms that emerge directly from the nucleus, while barred spirals have arms that emerge from a bar of stars and interstellar matter that straddles the nucleus. Irregular galaxies have no discernible structure. Irregulars usually contain the highest proportion of gases, and ellipticals the least. It seems likely that whether a galaxy becomes elliptical, spiral or irregular depends on the amount of rotation present in the gas cloud from which it formed, and on the rate of star formation within it. In the remote future, when the available gas is used up, star formation in galaxies will cease. It is thought that a significant fraction of the matter in a galaxy will eventually collapse into a BLACK HOLE.

The masses of galaxies range from about 10E6 solar masses to several times 10E12 solar masses. Recent observations of the rotation rates of spiral galaxies indicate that they are immersed within extensive haloes of dark matter amounting to up to ten times as much mass as their visible stars and gas clouds. Our own galaxy, the Milky Way system, is a spiral with a diameter (measured across the visible disk) of about 30,000 parsecs, and contains about 10E11 stars.

Galaxies emit RADIATION of all kinds, but in differing proportions. While an ordinary spiral like the Milky Way emits mainly at optical wavelengths, radio galaxies emit most strongly at radio wavelengths. Active galaxies are those that show signs of violent activity. They are usually characterized by highly luminous compact nuclei and by relatively rapid variability. Many have jets of matter emanating from their cores, and extensive radio-emitting clouds extending far beyond their visible confines. Radio galaxies, Seyfert galaxies (spiral galaxies with compact active nuclei named for the US astronomer Carl Seyfert (1911-1960)) and – according to most astronomers – QUASARS are examples of active galaxies.

Galaxies are generally located within groups or clusters, containing from a few to a few thousand members. Superclusters are even larger, loose groupings of galaxies and clusters that contain several thousand galaxies and span regions of space up to 100 megaparsecs in diameter. The Milky Way galaxy is part of the Local Group, which comprises about thirty member systems. The nearest large galaxy is the ANDROMEDA galaxy, M31. Similarly spiral, though rather larger, it has two satellite galaxies, which are elliptical in form. It was thought to be a NEBULA within our own galaxy, but in 1924 HUBBLE showed it to be a separate galaxy. (◆ page 240)

Galen of Pergamum (*c* AD 130-*c* 200)
Greek physician at the court of the Emperor Marcus Aurelius. His writings drew together the best of Classical medicine and provided the form in which the science was transmitted through the medieval period to the Renaissance. He himself contributed many original and careful observations in anatomy and physiology.

Galena
Gray mineral consisting of lead (II) sulfide (PbS), forming cubic crystals; the main ore of LEAD. Deposits occur in Germany, the US, Britain and Australia.

Galileo Galilei (1564-1642)
Italian mathematical physicist who discovered the laws of falling bodies and the parabolic motion of projectiles. The first to turn the newly invented TELESCOPE to the heavens, he was among the earliest observers of SUNSPOTS and the phases of Venus. A talented publicist, he helped to popularize the pursuit of science. However, his quarrelsome nature led him into an unfortunate controversy with the Church. His most significant contribution to science was his provision of an alternative to the Aristotelian dynamics. The motion of the Earth thus became a conceptual possibility and scientists at last had a genuine criterion for choosing between the Copernican and Tychonic hypotheses in ASTRONOMY.

Gall, Franz Joseph (1758-1828)
German-born Viennese physician who was one of the earliest proponents of the theory of cerebral localization (that different areas of the BRAIN control different functions) and who founded the pseudoscience of PHRENOLOGY.

Gall bladder
Small sac containing BILE, arising from the bile duct which leads from the liver to the duodenum. It lies beneath the liver and serves to concentrate and store bile. When food, especially fatty food, reaches the STOMACH, local HORMONES cause gall bladder contraction and bile enters the GASTRO-INTESTINAL TRACT. In some people the concentration of bile leads to the formation of gallstones, usually containing CHOLESTEROL. These stones may cause no symptoms, or they may obstruct the gall bladder causing biliary COLIC or INFLAMMATION (cholecystitis). They may pass into the bile duct and cause biliary obstruction with JAUNDICE or, less often, pancreatitis. (◆◆ page 40, 233)

Gallium (Ga)
Bluish-white metal in Group IIIA of the PERIODIC TABLE, resembling ALUMINUM; found as a trace element in SPHALERITE, PYRITE, BAUXITE and germanite. Gallium forms trivalent salts and a few monovalent compounds. It contracts on melting, and is liquid over a greater temperature range than any other element. Its uses include doping SEMICONDUCTORS and producing TRANSISTORS. AW 69.7, mp 29.8°C, bp 2200°C, sg 5.904 (29.6°C), 6.095 (29.8°C). (◆ page 130)

Gallium arsenide (GaAs)
A rival material to SILICON for making CHIPS. GaAs chips process data much faster using less power than silicon chips. Sometimes known as "three-five material" (gallium has a valency of 3 and arsenic of 5) or a binary SEMICONDUCTOR (being a compound of two materials).

Gallon (gal)
Unit of liquid measure, equivalent to 4 quarts or 3.7854 liter (US Customary System) or 4.546 liter (British Imperial System).

Galls
Growth abnormalities that occur in plants, caused by insects, mites, nematodes, fungi or bacteria. Those produced by insects often serve to protect their larval stages. By definition, galls are self-limiting growths requiring the continued presence of the inciting organism for full development. They thereby differ from TUMORS, where abnormal growth continues in the absence of the inciting organism. However, in crown gall disease of crucifers (eg cabbage), the bacterium *Agrobacterium tumefaciens* is required for the initiation of the gall, but subsequent growth becomes tumorous and does not require the presence of the bacterium.

Gallstones see GALL BLADDER.

Galton, Sir Francis (1822-1911)
British scientist, the founder of EUGENICS and biostatistics (the application of statistical methods to animal populations); the coiner of the term "anticyclone" and one of the first to realize their meteorological significance; and the developer of one of the first fingerprint systems for identification.

Galvani, Luigi (1737-1798)
Italian anatomist who discovered "animal electricity" (*c*1786). The many varying accounts of this discovery at least agree that it resulted from the chance observation of the twitching of frog legs under electrical influence. A controversy with VOLTA over the nature of animal electricity was cut short by Galvani's death.

Galvanizing see CORROSION.

Galvanometer
An instrument used for detecting and measuring very small electric currents. Most modern instruments are of the moving-coil type in which a coil of fine wire wrapped around an aluminum former is suspended by conducting ribbons about a soft iron core between the poles of a permanent magnet. When an electric current flows through the coil, a magnetic field is set up which interacts with that of the permanent magnet, producing a torque. This turns the coil until it is fully resisted by the suspension, the displacement produced being proportional to the current. The result is read from a scale onto which a light beam is reflected from a mirror carried on the suspension ribbons. If all electromagnetic and mechanical damping can be eliminated from such an instrument, it can also be used as a ballistic galvanometer to measure small charges and capacitances. (See also AMMETER.)

Gamete (germ cell)
A cell formed for sexual REPRODUCTION, capable of uniting with another gamete to form a ZYGOTE, from which a new individual will develop; this process is termed FERTILIZATION. Each gamete contains one set of chromosomes and is said to be HAPLOID. Thus when gametes unite, the resultant cell is DIPLOID, having two sets of CHROMOSOMES. The gametes of some primitive organisms are identical cells (isogamy), and both are capable of swimming in water, but in most species the gametes are of two distinct types (anisogamy), and only the male gamete (SPERM) is mobile while the much larger female gamete (EGG) is static. In higher PLANTS the male gametes are produced by the male cones in conifers, or by the anthers in flowering plants; the female gametes (ovules) are produced by the ovary within the female cone, or the carpel of a FLOWER. In animals gametes are produced by the GONADS – the testes in the male and ovaries in the female. (◀ page 32)

Gametophyte
A phase in the life-cycle of a plant, representing the haploid generation. (See ALTERNATION OF GENERATIONS.)

Gamma globulin
The fraction of blood protein containing IMMUNOGLOBULINS.

Gamma-ray astronomy
The study of GAMMA RAYS emitted by celestial objects. Because the wavelength of gamma rays is so short, astronomers use SCINTILLATION COUNTERS or SPARK CHAMBERS to obtain directional observation. Of particular interest are gamma-ray bursts – intense blasts lasting from a tenth of a second to tens of seconds – which occur several times a year from sources widely distributed over the sky. Twenty or so gamma-ray stars have been identified in the Milky Way that other techniques have not yet identified.

Gamma rays
High-energy PHOTONS of wavelength shorter than 0.01nm. They are emitted from atomic nuclei during radioactive decay (see RADIOACTIVITY). Usually their emission follows the ejection of an electron (BETA RAY) from the nucleus. The most penetrating of the radioactive emissions, gamma rays find use in engineering quality control as a source for exposing RADIOGRAPHS, and in RADIATION THERAPY. (▶ page 118, 147)

Gamow, George (1904-1968)
Russian-born US physicist and popular science writer, best known for his work in nuclear physics, especially related to the evolution of STARS; and for his support of the Big Bang theory of COSMOLOGY. In GENETICS, his work paved the way for the subsequent discovery of the role of DNA.

Ganglion
A clump of neurons (nerve cells) involved in coordination and control of responses in animals. They are present in all but the very simplest nervous systems, and increase in size as behavior and sensory systems become more complex. In segmented animals such as insects there is usually a ganglion in each segment, with a larger ganglion at the front end known as the cerebral ganglion. In several animal groups there has been a steady evolution toward greater size of, and overall control by, this cerebral ganglion, which in higher organisms is known as a brain. (▶ page 126)

In humans the term ganglion is also used for a benign lump, often at the wrist, found close to the tendons and containing a jelly-like fluid.

Gangrene
DEATH of tissue following disruption of blood supply, often after obstruction of ARTERIES by trauma, THROMBOSIS or EMBOLISM. Dry gangrene is seen when arterial block is followed by slow drying, blackening and finally separation of dead tissue from healthy. Wet gangrene occurs when the dead tissue is infected with bacteria. Gas gangrene involves infection with gas-forming organisms (*Clostridium welchii*) and its spread is particularly rapid.

Garnet
Common SILICATE mineral group of general formula $M^{II}_3M^{III}_2(SiO_4)_3$, having six end-members. It is found in metamorphic and some igneous rocks, often as rhombododecahedral crystals. It is hard, and used as an ABRASIVE. The color is variable; GEM varieties are red, green or transparent. The garnet is popularly the birthstone for January.

Gas
One of the three states (solid, liquid, gas) into which nearly all matter above the atomic level can be classified. Gases are characterized by a low DENSITY and VISCOSITY; a high compressibility; optical transparency; a complete lack of rigidity, and a readiness to fill whatever volume is available to them and to form molecularly homogeneous mixtures with other gases. Air and STEAM are familiar examples. At sufficiently high temperatures, all materials vaporize, though many undergo chemical changes first. Gases, particularly steam and CARBON dioxide, are common products of COMBUSTION, while several available naturally or from PETROLEUM or COAL (eg HYDROGEN, METHANE) are used as fuels themselves. The great bulk of the Universe is gaseous, in the form of interstellar hydrogen. Gases will often dissolve in liquids, the solubility rising with PRESSURE and falling with TEMPERATURE; a little dissolved carbon dioxide is responsible for the bubbles in soda.

In contrast to solids and liquids, the MOLECULES of a gas are far apart compared with their own size, and move freely and randomly at a wide range of speeds of the order of 100m/s. For a given temperature and pressure, equal volumes of gas contain the same number of molecules ($2.7\times10^{25}m^{-3}$ at room temperature and atmospheric pressure). The impacts of the molecules on the walls of the container are responsible for the pressure exerted by gases, which is much larger than is often appreciated: the atmosphere exerts on everything a pressure many times larger than a person's weight and without it we would quite simply boil and burst.

For a given mass of an ideal gas (ie one in which the molecules are of negligible size and exert no forces on each other), the product of the pressure (P) and the volume (V) is proportional to the absolute temperature (T):

$$PV=RT \text{ (the general gas law).}$$

The constant of proportionality (R) is known as the universal gas constant and has the value 8.314 joules/kelvin-mole. The general gas law and most of the other properties of gases can be explained in terms of the KINETIC THEORY without reference to the internal structure of the molecules. Real gases deviate from this ideal behavior at high pressures because of the actual presence of small intermolecular forces. (▶ page 130)

Gas chromatography
A widely used technique for separating mixtures of similar molecules. The mixture is entrained in an inert carrier gas and passed through a column packed with a material which adsorbs the mixture and then desorbs its components at different rates, so that they emerge from the column as separate, pure fractions.

Gas exchange
The process whereby ANAEROBIC organisms obtain oxygen from the environment and excrete carbon dioxide. In plants, gas exchange occurs principally through openings in the CUTICLE known as STOMATA, which can be closed to reduce water losses. As in microorganisms, fungi and small animals, or those with very flattened bodies, gas exchange in plants takes place by simple diffusion. Larger animals need special means of obtaining oxygen, and most also have a circulatory system for transporting it round the body (the same system also serves to collect and remove carbon dioxide.) A few animals, such as the annelid worms, have a circulatory system for transporting gases, but no specialized oxygen-collecting apparatus; they rely on diffusion through the skin.

Specialized gas exchange organs are of several types. Most aquatic organisms breathe by means of GILLS, which are feathery or comb-like structures richly supplied with blood vessels. Gills have evolved independently in several different groups of animals, including the fish, mollusks and crustaceans. Gills work well in an aquatic environment because the water supports them. On land a different type of system is required, and several groups have evolved a large, air-filled cavity with an invaginated inner surface to increase the absorptive area. The lungs of tetrapod vertebrates and lungfish are examples of such a system, as are the lung-books of spiders and the lung-like mantle cavity of land snails and slugs (derived from the normal mollusk mantle cavity by loss of the gill and corrugation of the surface). The insects show a fundamentally different solution to breathing on land. They have a system of tubes, known as tracheae, which extend from openings at the body surface (spiracles) and ramify through the body, branching into tubes of smaller and smaller diameter that eventually reach every cell, supplying them with oxygen direct. Although insects have a circulatory system it plays only a minor part in transporting gases. Some aquatic insects (eg certain dragonfly nymphs) have secondarily acquired gills, but these link up with the tracheal system, not with the circulatory system. A fourth type of gas exchange system is shown by the echinoderms (starfishes, sea urchins, etc.), which obtain oxygen via their tube feet, these also being used in most species for locomotion and the manipulation of food. It is believed that the forerunners of the tube feet were originally used solely for gas exchange, and evolved their other function later. Similarly, in many FILTER-FEEDERS the gills have developed a new function: sieving food particles from water. (See also HEMOGLOBIN; RESPIRATION.)

Gasoline (petrol)
A mixture of volatile HYDROCARBONS having 4 to 12 carbon atoms per molecule, used as a FUEL for INTERNAL-COMBUSTION ENGINES, and as a solvent. Although gasoline can be derived from oil, coal and tar, or synthesized from carbon monoxide and hydrogen, almost all is produced from PETROLEUM by refining, CRACKING and ALKYLATION, the fractions being blended to produce fuels with desired characteristics.

Gasoline engine see INTERNAL-COMBUSTION ENGINE.

Gassendi (or Gassend), **Pierre** (1592-1655)
French philosopher important for his role in tipping the balance away from the old and toward the new science. A friend and ally of KEPLER and GALILEO, he attacked the prevalent Aristotelianism and supported ATOMISM. He also made a number of important astronomical observations, and named the aurora borealis (see AURORA).

Gasser, Herbert Spencer (1888-1963)
US physiologist who shared with J. ERLANGER the 1964 Nobel Prize for Physiology or Medicine for their investigations into the functions of nerve fibers.

Gastrectomy
Partial gastrectomy is removal of part of the stomach, usually in cases of cancer or intractable ulcer.
Gastric juice see DIGESTIVE SYSTEM.
Gastrin
Group of HORMONES, produced by part of the STOMACH, that stimulate acid secretion by the stomach, PANCREAS secretion, and possibly GALL BLADDER contraction. The secretion of gastrin is stimulated by food in the stomach and by the vagus nerve.
Gastritis
Irritation of the stomach, causing vomiting. (See also GASTROENTERITIS.)
Gastroenteritis
A group of conditions usually caused by viral or bacterial infection of the stomach and intestines, causing DIARRHEA, VOMITING and abdominal COLIC. While these are mostly mild illnesses, in young infants and debilitated or elderly adults dehydration may develop rapidly and death may result. (See also ENTERITIS; FOOD POISONING.)
Gastrointestinal tract (alimentary canal)
Part of the digestive system of animals, through which food passes. In humans it starts at the PHARYNX, passing into the ESOPHAGUS and STOMACH. From this arises the small intestine, consisting of the DUODENUM and the great length of the jejunum and ileum. This leads into the large bowel, consisting of the cecum (from which the APPENDIX arises), colon and rectum. The parts from the stomach to the colon lie suspended on a MESENTERY, through which they receive their blood supply, and lie in loops within the peritoneal cavity of the ABDOMEN. In hoofed animals that eat mainly grass and grain the stomach is divided into two or three parts called rumens, and the food is returned to the mouth to be re-chewed. In rodents, whose diet is similar, food passes through the digestive tract twice as these animals eat the soft FECES they produce at night. Some hoofed animals, eg lamas, can vomit forcefully at will to repel predators. Movement of food in the tract occurs largely by PERISTALSIS, but is controlled at key points by SPHINCTERS. There are many gastrointestinal tract diseases. In GASTROENTERITIS, ENTERITIS and COLITIS, gut segments become inflamed. Peptic ULCER affects both the duodenum and stomach, while CANCER of the esophagus, stomach, colon and rectum are relatively common. Disease of the small intestine tends to cause malabsorption. Methods of investigating the tract include BARIUM TESTS, and endoscopy, in which viewing tubes are passed in via the mouth or anus. (◆◆ page 40, 233)
Gastropoda
A class of the phylum MOLLUSCA; it includes the snails and slugs.
Gas turbine
A heat engine in which hot gas, generated by burning a fuel or by heat exchange from a nuclear reactor, drives a turbine and so supplies power. Straightforward and reliable, they were developed in the late 1930s, and are now used to power aircraft, ships and locomotives, to generate electricity and to drive compressors in pipelines. The fuel used may be fuel gas, gasoline, kerosine or even powdered coal. Some gas turbines are external-combustion engines, the working gas being heated in a heat exchanger and passed round the system in a closed cycle. Most, however, are INTERNAL-COMBUSTION ENGINES working on an open cycle: in the combustors fuel is injected into compressed air and ignited; the hot exhaust gases drive the turbines and are vented to the atmosphere, heat exchangers transferring some of their heat to the air from the compressors. (See also JET PROPULSION.)
Gauss (Gs)
Unit of magnetic flux density in CGS UNITS, equalling one maxwell per square centimeter.
Gaussan system see CGS UNITS.
Gay-Lussac, Joseph Louis (1778-1850)
French chemist and physicist best known for Gay-Lussac's Law (1808), which states that, when gases combine to give a gaseous product, the ratio of the volumes of the reacting gases to that of the product is a simple, integral one. AVOGADRO's hypothesis is based on this and on DALTON's law of multiple proportions (see COMPOSITION, CHEMICAL). He also showed, independently of CHARLES, that all gases increase in volume by the same fraction for the same increase in temperature, 1/273.2 for 1C°; and (once with BIOT) made two balloon ascents to investigate atmospheric composition and the intensity of the EARTH's magnetic field at altitude. His many contributions to inorganic chemistry include the identification of CYANOGEN.
Gegenschein (counterglow) see ZODIACAL LIGHT.
Geiger counter (Geiger-Müller tube)
An instrument for detecting the presence of and measuring radiation such as ALPHA PARTICLES, BETA-, GAMMA- and X-RAYS. It can count individual particles at rates up to about 10,000/s and is used widely in medicine and in prospecting for radioactive ores. A fine wire anode runs along the axis of a metal cylinder which has sealed insulating ends, contains a mixture of argon or neon and methane at low pressure, and acts as the cathode, the potential between them being about 1kV. Particles entering through a thin window cause ionization in the gas; electrons build up around the anode and a momentary drop in the interelectrode potential occurs, which appears as a voltage pulse in an associated counting circuit. The methane quenches the ionization, leaving the counter ready to detect further incoming particles.
Geikie, Sir Archibald (1835-1924)
Scottish geologist, Director General of the Geological Survey of Great Britain (1882-1901) and President of the Royal Society (1908-12), best known for various books including *A Textbook of Geology* (1882) and *The Ancient Volcanoes of Great Britain* (1897).
Geissler tube
A forerunner of the modern ELECTRON TUBE, invented in 1858 by Heinrich Geissler (1814-1879). It is a glass tube containing a gas at low pressure which glows with a characteristic COLOR when a high voltage is applied to the metal electrodes at the ends of the tube. Modified forms are used as spectroscopic light sources and in neon or argon signs.
Gel
A COLLOID in which the two phases combine to produce a solid or semisolid material, eg a jelly. (See also SOL.)
Gelignite see DYNAMITE.
Gell-Mann, Murray (b. 1929)
US physicist awarded the 1969 Nobel Prize for Physics for his work on the classification of SUBATOMIC PARTICLES (notably K-mesons and hyperons) and their interactions. With FEYNMAN he has proposed the quark as a basic component of all subatomic particles.
Gemini
The Twins, a constellation on the ECLIPTIC named after its two brightest stars, Castor and Pollux. The third sign of the ZODIAC, Gemini gives its name to the Geminid METEOR shower.
Gemini missions
US space program designed to develop docking and rendezvous procedures, a vital preparation for the Apollo Project (see SPACE EXPLORATION). Gemini 1 was launched 8 April 1964; 3 (the first manned) on 23 March 1965: 12 (the last) on 11 November 1966. From Gemini 4 E.H. White II became the second man to float free in space.
Gems
Stones prized for their beauty, and durable enough to be used in jewelry and for ornament. A few – AMBER, CORAL, PEARL and JET – have organic origin, but most are well-crystallized MINERALS. Gems are usually found in IGNEOUS ROCKS (mainly pegmatite dikes) and in contact METAMORPHIC zones. The chief gems have HARDNESS of 8 or more on the Mohs scale, and are relatively resistant to CLEAVAGE and fracture, though some are fragile. They are identified and characterized by their SPECIFIC GRAVITY (which also determines the size of a stone with a given weight in CARATS) and optical properties, especially refractive index (see REFRACTION). Gems of high refractive index show great brilliancy (also dependent on transparency and polish) and prismatic DISPERSION ("fire"). Other attractive optical effects include chatoyancy (see CAT'S EYE), dichroism (see DOUBLE REFRACTION), opalescence and asterism – a starshaped gleam caused by regular intrusions in the crystal lattice. Since earliest times gems have been engraved in intaglio and cameo. Somewhat later cutting and polishing were developed, the cabochon (rounded) cut being used. Not until the late Middle Ages was faceting developed, now the commonest cutting style, its chief forms being the brilliant cut and the step cut. Some gems are dyed, impregnated, heated or irradiated to improve their color. Synthetic gems are made by flame-fusion or by crystallization from a melt or aqueous solution.
Gene
In classical genetics, a discrete unit of hereditary information that is inherited in a Mendelian fashion (see GENETICS) and determines an observable characteristic. In molecular genetics, a length of DNA that codes for a single POLYPEPTIDE chain. Different versions of the same gene are known as ALLELES. (◆ page 32)
Gene pool
The total amount of information present at any time in the GENES of the reproductive members of a biological population. The frequency of any particular gene in the gene pool changes because of NATURAL SELECTION, MUTATION and GENETIC DRIFT. This change forms the basis of evolution.
Generator, Electric (dynamo)
A device converting mechanical ENERGY into electrical energy. In traditional forms, electric fields are induced by changing the magnetic field lines through a circuit (see ELECTROMAGNETISM). All generators can be, and sometimes are, run in reverse as electric motors.

The simplest generator consists of a permanent magnet (the rotor) spun inside a coil of wire (the stator); the magnetic field is thus reversed twice each revolution, and an AC voltage is generated at the frequency of rotation (see also MAGNETO). In practical designs, the rotor is usually an electromagnet driven by a direct current obtained by rectification of a part of the voltage generated, and passed to the rotor through a pair of carbon brush/slip ring contacts. The use of three sets of stator coils 120° apart allows generation of a three-phase supply. (See also ARMATURE.)

Simple DC generators consist of a coil rotating in the field of a permanent magnet: the voltage induced in the coil alternates at the frequency of rotation, but it is collected through a commutator – a slip ring broken into two semicircular parts, to each of which one end of the coil is connected, so that the connection between the coil and the brushes is reversed twice each revolution – resulting in a rapidly pulsating direct voltage. A steadier voltage can be achieved through the use of multiple coil/commutator arrangements, and except in very small generators, the permanent magnet is again replaced by an electromagnet driven by part of the generated voltage.

For large-scale generation, the mechanical power is usually derived from fossil-fuel-fired steam turbines, or from dam-fed water turbines, and the process is only moderately efficient. The magnetohydrodynamic generator, currently under development, avoids this step, and has no moving parts either. A hot conducting fluid (treated coal gas, or reactor-heated liquid metal) passes through the field of an electromagnet, so that the charges are forced in opposite directions producing a DC voltage. In another device, the electrogasdynamic generator, the voltage is produced by using a high speed gas stream to pump charge from an electric discharge, against the electric field, to a collector. (See also ELECTROSTATIC GENERATOR.)

Generators originated with the discovery of induction by M. FARADAY in 1831; the considerable advantages of electromagnets over permanent magnets were first exploited by E.W. VON SIEMENS in 1866. (♦ page 118)

Gene therapy
The technology of curing genetic diseases by altering the genetic constitution of the fetus. It has been used experimentally in animals, but its use in humans poses ethical problems.

Genetic code
The means by which the genetic information is translated into the molecules that make up living organisms. It is based on the order of different nitrogenous bases (PURINES and PYRIMIDINES) along the DNA molecule. There are four of these bases in DNA: adenine, thymine, cytosine and guanine. Adenine always pairs with thymine and cytosine with guanine, in what are known as "base pairs". This constant pairing allows DNA to replicate itself and thus produce copies of the genetic material. In the same way, a molecule of messenger RNA (mRNA) can be formed alongside part of the DNA molecule, and it is this that acts as the template for protein synthesis. Reading from the beginning of the mRNA (indicated by a "start" code), the bases are taken in groups of three. Each group of three is known as a "codon", and as there are four different bases there are 64 different codons ($4\times4\times4=64$). There are only 20 amino acids making up proteins, so several of the codons denote the same amino acid. The code is translated, and the protein molecule formed, in a complex process involving the RIBOSOMES, transfer RNA molecules, and several different enzymes (see PROTEIN SYNTHESIS).

The elucidation of the genetic code became possible following JAMES WATSON and FRANCIS CRICK's discovery of the structure of DNA in 1953. Since then, the genetic code has been studied in many different plants, animals, fungi and microorganisms, and shown to be virtually the same in all living organisms. As the code is essentially arbitrary – there is no functional reason why a certain codon should denote one amino acid rather than another – this is a compelling argument for the single origin of all present life forms. The minor variations of the code in a few protozoans are clearly modifications of the original code. (See also PROTEIN; MUTATION.) (♦ page 32)

Genetic drift
A process of evolutionary change that is not produced by any form of selection but by random events. It can occur in very small populations, where a gene can be lost from the population simply by chance, because the gametes carrying that gene happen not to achieve fertilization. Similarly, a gene can become more common by chance.

Genetic engineering
The modification of the genetic information of living organisms by direct manipulation of their DNA (rather than by the more indirect method of BREEDING). The most common approach is to take a gene from one organism and introduce it into another. For example, in plants a gene from cowpea that codes for an insect-deterring chemical has been successfully transferred to tobacco. The method of transfer is a complex process, usually involving bacteria and mobile genetic elements known as PLASMIDS that can carry the desired gene into the recipient's genetic material. With plants the work is done using small cultures of plant material in the laboratory, and whole plants are then regenerated from the successfully engineered cells. Genetic engineering can also be used to produce desirable chemical compounds such as insulin, by inserting the gene for insulin into the genome of a bacterium and then culturing the bacterium on a large scale, a form of BIOTECHNOLOGY.

Controversy over genetic engineering centers on the possible risks of releasing genetically engineered organisms, particularly bacteria, into the environment. To guard against such risks, a gene that incapacitates the released bacterium under certain conditions, and thus prevents it from surviving in the wild, is introduced into its genome. Distinctive genetic markers, such as one that produces fluorescence under ultraviolet light, can also be introduced so that the released bacteria can be distinguished from their wild relatives. Preliminary experiments in which only the marker and the incapacitating gene are introduced, and the bacterium is then released, ensure that the incapacitating gene works before the fully engineered bacterium is released.

Genetics
The study of inheritance in living organisms. Modern genetics was founded by the Austrian monk GREGOR MENDEL, who carried out breeding experiments with various plants and showed that characteristics were inherited as discrete particles – that is, their effects were not diluted over the generations, though they might be masked in one generation only to reappear in the next. This was named particulate inheritance in contrast with the idea of blending inheritance that had previously been popular. Mendel's hereditary particles were later named GENES and shown to be carried on the CHROMOSOMES.

Intensive work in classical genetics during the early part of the 20th century showed that linkage between genes was related to their closeness on the same chromosome, and this phenomenon was used to locate genes on chromosomes by "linkage mapping". The genomes of certain organisms, notably the fruit fly, *Drosophila*, and the garden pea, were studied intensively, and a great deal of basic information built up. This laid the ground for later studies in molecular genetics, which established that each gene produced a single polypeptide chain, and showed the genes to be made of DNA. The elucidation of the structure of DNA in 1953 provided a link between molecular genetics and biochemistry, and opened up vast new areas of research. Today, it is possible directly to alter the genetic material of certain simple organisms, one of the tech niques that forms part of GENETIC ENGINEERING. (See also DOMINANCE; RECESSIVENESS; HETEROZYGOUS; HOMOZYGOUS; DIPLOID; HAPLOID; MUTATION; SEX DISCRIMINATION; SEX LINKAGE.) (♦ page 32)

Genome
The full genetic complement of a given species.

Genotype
The total genetic makeup of a particular organism, consisting of all the GENES received from both parents. (See also PHENOTYPE.)

Genus
In TAXONOMY, a group of related species. Related genera are grouped together in families. The concept of a genus was introduced by LINNAEUS, who recognized that there were many groups of similar species in both the animal and plant kingdoms. In the binomial system of scientific nomenclature, which he devised, the first name (always written with a capital letter) indicates the genus and the second the species. For example, the birches belong to the genus *Betula*; the name of the silver birch is *Betula pendula*, while the downy birch is *Betula pubescens*.

Geochemistry
The study of the CHEMISTRY of the EARTH (and other planets). Chemical characterization of the Earth as a whole relates to theories of planetary formation. Classical geochemistry analyzes rocks and MINERALS. The study of PHASE EQUILIBRIA has thrown much light on the postulated processes of rock formation. (See also GEOLOGY.)

Geode
A hollow mineral formation found in certain rocks. Typically, it is almost filled by inward-growing crystal "spikes", usually of QUARTZ. Geodes range between 20mm and 1m across.

Geodesy
The branch of geophysics concerned with the determination and explanation of the precise shape and size of the EARTH. The first recorded measurement of the Earth's circumference that approximates to the correct value was that of ERATOSTHENES in the 3rd century BC. Modern geodesists use not only the techniques of SURVEYING but also information received from the observations of artificial SATELLITES.

Geoffroy de Saint-Hilaire, Etienne (1772-1844)
French naturalist who argued, against G. CUVIER, for "unity of plan", the notion that all animals contain the same anatomical elements, some being developed at the expense of others in particular species. He acknowledged that there had been development in previously existing species.

Geography
The group of sciences concerned with the surface of the Earth, including the distribution of life upon it, its physical structures, etc. Geography relies on surveying and mapping, and modern cartography (mapmaking) has rapidly adapted to the new needs of geography as it advances and develops. (See MAP; SURVEYING.)

BIOGEOGRAPHY is concerned with the distribution of life, both plant and animal (including man), about our world. It is thus intimately related to BIOLOGY and ECOLOGY.

ECONOMIC GEOGRAPHY describes and seeks to explain the patterns of the world's commerce in terms of production, trade and transportation, and consumption. It relates closely to economics.

MATHEMATICAL GEOGRAPHY deals with the size, shape and motions of the EARTH, and is thus linked with ASTRONOMY (see also GEODESY).

PHYSICAL GEOGRAPHY deals with the physical structures of the Earth, also including climatology and OCEANOGRAPHY, and is akin to physical GEOLOGY.

POLITICAL GEOGRAPHY is concerned with the world as nationally divided.

REGIONAL GEOGRAPHY studies the world in terms of regions separated by physical rather than national boundaries.

HISTORICAL GEOGRAPHY deals with the geography of the past: PALEOGEOGRAPHY at one level, exploration or past political change or settlement at another.

APPLIED GEOGRAPHY embraces the applications of all these branches to the solution of socioeconomic problems. Its subdivisions include urban geography and social geography; and it contributes to sociology. (See also ETHNOLOGY; HYDROLOGY; METEOROLOGY.)

DEVELOPMENT OF GEOGRAPHY Geography had its origins in the Greeks' attempts to understand the world in which they found themselves. Once it was realized that the Earth was round, the next step was to estimate its size. This was achieved in the 3rd century BC by ERATOSTHENES. The classical achievement in geography, like that in astronomy, was summed up by Claudius PTOLEMY. His world map was used for centuries. Geographical knowledge next leapt forward in the age of exploration that opened with the voyages of Dias and Columbus. The 17th century saw continuing discovery and greatly improved methods of survey. The earliest modern geographical treatises, including that of Varenius, also appeared in this era. The 19th century brought with it the works of F.H.A. VON HUMBOLDT and Karl Ritter, the former stressing physical and systematic geography, the latter the human and historical aspects of the science. Encompassing so many different studies, geography since the mid-19th century has become a battleground for the different schools of geographers. While some have encouraged a regional approach, others have preferred to develop a landscape concept. Others have stressed the study of the physical environment while others still have concentrated on political and economic factors. Perhaps the most recent group to come to the fore favors the collection of precise numerical data. With this they try to build mathematical models of geographical phenomena.

Geoid

Geodesic model of the Earth, the shape the Earth would have to have for the pull of GRAVITY to be constant for all points, taken at sea level over the oceans and at corresponding level on land. The result is an oblate spheroid (see ELLIPSOID; OBLATENESS) with irregularities due to differing local densities. (See also EARTH; GEODESY.)

Geology

The group of sciences concerned with the study of the Earth, including its structure, long-term history, composition and origins.

PHYSICAL GEOLOGY deals with the structure and composition of the Earth and the forces of change affecting them. The sciences that make up physical geology thus include GEODESY, GEOMORPHOLOGY, GEOPHYSICS and seismology (see EARTHQUAKE). Much of modern physical geology is based on the theory of PLATE TECTONICS.

HISTORICAL GEOLOGY deals with the Earth in past ages, and with the EVOLUTION of life upon it. It embraces such sciences as PALEOCLIMATOLOGY, PALEOMAGNETISM, PALEONTOLOGY and STRATIGRAPHY; and relies heavily on dating (see CHRONOLOGY), events being related to the geological time-scale, whose derivation is primarily stratigraphical, to a lesser extent paleontological.

ECONOMIC GEOLOGY lies between these two, and borrows from both. Concerned with the location and exploitation of the Earth's natural resources (see ORE), it is generally taken to include also the disciplines of crystallography, mineralogy and petrology (see CRYSTAL; MINERALS; ROCKS). Its practical manifestations are PROSPECTING and MINING.

GEOLOGY OF OTHER PLANETS Except with the Moon, it is not yet possible to examine the rocks of other planets, but telescope and spectroscopic examinations have revealed much, as have those of unmanned probes. VOLCANISM is known on the Moon and Mars (one volcano is some 600km across), and "moonquakes" have been detected.

DEVELOPMENT OF GEOLOGY Most early geological knowledge came from the experience of mining engineers, some of the earliest geological treatises coming from the pen of GEORGIUS AGRICOLA. The interest of the 16th century in FOSSILS was also reflected in the writings of K. VON GESNER. In the 17th century the biblical timescale of about 6000 years from the Creation to the present largely constrained the many speculative "Theories of the Earth" that were issued. The century's most notable geological observations were made by N. STENO. The late 18th century saw the celebrated controversy between A. G. WERNER's "Neptunists" and J. HUTTON's "Plutonists" about the origin of the rocks. The first decades of the 19th century, however, witnessed the decline of speculative geology as field observations became ever more detailed. William Smith (1769-1839), the "father of stratigraphy", showed how the succession of fossils could be used to index the stratigraphic column, and he and others produced impressive geological maps. C. LYELL's classic *Principles of Geology* (1830-33) restated the Huttonian principle of UNIFORMITARIANISM and provided the groundwork for much of the later development of the science. L. AGASSIZ pointed to the importance of glacial action in the recent history of the Earth (1840), while mining engineering continued to contribute to the pool of geological data. The most significant recent development in the Earth sciences has been the acceptance of the theory of PLATE TECTONICS, foreshadowed in A. WEGENER's 1912 theory of CONTINENTAL DRIFT.

Geomagnetism

The magnetic field of the EARTH; and the study of it, both as it is in the present and as it was in the past (see PALEOMAGNETISM). (See also GEOPHYSICS; MAGNETISM.)

Geomorphology

The surface features of the EARTH and the study of them, especially their origins and the processes acting on them. (See GEOLOGY.)

Geophysics

The physics of the EARTH, including studies of the ATMOSPHERE, EARTHQUAKES and VOLCANISM as well as GEODESY, GEOMAGNETISM, HYDROLOGY and OCEANOGRAPHY.

Geostationary orbit

If a satellite is in orbit 35,880km above the Earth's equator, it revolves around the Earth at the same speed as the Earth rotates. The satellite's longitude thus appears stationary.

Geothermal energy

An "alternative" energy to fossil fuels, sometimes classified as renewable energy. Deep below the Earth's surface are pockets of hot water or areas of hot (but dry) rocks. Both are known as geothermal energy. To use the former, engineers drill into AQUIFERS and pump the water to the surface for use, normally to heat a large number of dwellings. Hot dry rocks are tapped by fracturing the subterranean rock with explosives, pumping down water from the surface to be heated and then pumping it back to the surface to extract energy. Hot aquifers are in limited (and mostly experimental) use in most developed countries; hot rocks are still experimental. Natural GEYSERS used to drive turbine generators are also classified as geothermal energy. Such geysers are rare, mostly being found in the US, New Zealand and Japan.

Geotropism see TROPISMS.

Geriatrics

The branch of medicine specializing in the care of the elderly. Although concerned with the same diseases as the rest of medicine, the different susceptibility of the aged and a tendency for multiple pathology make its scope different. In particular the psychological problems of old age differ markedly from those encountered in the rest of the population.

Germ

A lay term for a microorganism capable of causing disease, including VIRUSES, BACTERIA, PARASITES and PROTOZOA.

Germanium (Ge)

Silvery gray metalloid in Group IVA of the PERIODIC TABLE; brittle crystalline solid whose structure resembles that of DIAMOND. It occurs naturally in sulfide ores of SILVER, COPPER and ZINC, and in COAL, and is extracted as a by-product of processing them. Its chemical properties are intermediate between those of SILICON and TIN; it reacts with the halogens, oxidizes in air at 600°C, and is attacked by concentrated oxidizing acids and by fused alkalis. Germanium is a SEMICONDUCTOR, and is used in electronic devices, especially TRANSISTORS; it is also used in alloys and for lenses and windows for INFRARED RADIATION. AW 72.6, mp 959°C, bp 2830°C, sg 5.323 (25°C).

Germanium forms covalent tetra- and divalent compounds. Germanium (IV) Oxide (GeO_2) is used in high-refractive-index GLASS. mp 1086°C. Germanium (IV) Chloride ($GeCl_4$), is a colorless liquid intermediate in the extraction of germanium and the preparation of most of its compounds. mp −50°C, bp 84°C. Germanes are a series of volatile hydrides resembling silanes (see SILICON). (♦ page 130)

German measles see RUBELLA.

German silver (nickel silver)

An ALLOY composed of copper, nickel and zinc. It resembles silver, and is used for cheap jewelry, cutlery etc, and as the base for silver-plated ware.

Germ cell see GAMETE.

Germicides see ANTISEPTICS.

Germination

The resumption of growth of a plant embryo contained in the SEED. Conditions required for germination include an adequate water supply, sufficient oxygen and a favorable temperature. Rapid uptake of water followed by increased rate of respiration are often the first signs of germination. During germination, stored food reserves are rapidly used up to provide the energy and raw materials required for the new growth. The embryonic root and shoot that break through the seed coat are termed the radicle and plumule respectively. (See also COTYLEDONS.) The term germination is sometimes also used to describe the initial growth of a SPORE.

Germ plasm

A special type of PROTOPLASM, supposed to be present in the reproductive cells or gametes of plants and animals, that AUGUST WEISMANN suggested passed on unchanged from generation to generation. Although the idea of germ plasm was mistaken, the basic concept – that the hereditary material is unaffected by changes in the organism that carries it – is correct; see CENTRAL DOGMA.

Gerontology

The study of the nature of old age in animals and humans.

Gesner, Konrad von (1516-1558)

Swiss naturalist whose major work, *Historia Animalium* (4 vols, 1551-58), an encyclopedic study of many varieties of animals, is considered the foundation stone of modern zoology.

Gestalt psychology

A school of psychology concerned with the tendency of the human mind to organize PERCEPTIONS into "wholes"; for example, to hear a symphony rather than a series of separate notes of different tones. Gestalt psychology, whose main proponents were WERTHEIMER, KOFFKA and KÖHLER, maintained that this was due to the mind's ability to complete patterns from the available stimuli. The school emerged as a reaction against such schools as BEHAVIORISM.

GEOLOGICAL TIME SCALE

One of the chief legacies of 19th century geology is the geological time scale; the division of geological time into named intervals separated from each other by major changes in rock type, obvious breaks in the succession, and abrupt changes in fossil groups. The coarsest division is the eon, then come eras, periods and epochs.

Archean Eon
The very ancient eon. Between the formation of the Earth and 2·5 billion years ago. Some geologists place the oldest limit at the time of the oldest rock, 3·8 billion years, and class everything before this as the Hadean Eon.

Proterozoic Eon
The eon of first life. 2·5 billion to 590 million years ago. It is a time when life is known to have existed but left no clear fossils. The 2·5 billion year boundary has been set by the radiometric dating of a number of igneous and metamorphic events. However, it is now evident that life existed for some time before that.

Phanerozoic Eon
The eon of obvious life. 590 million years ago to the present day. This is the time from which good fossils are known, due to the evolution of hard shells and skeletons at the beginning of the eon.

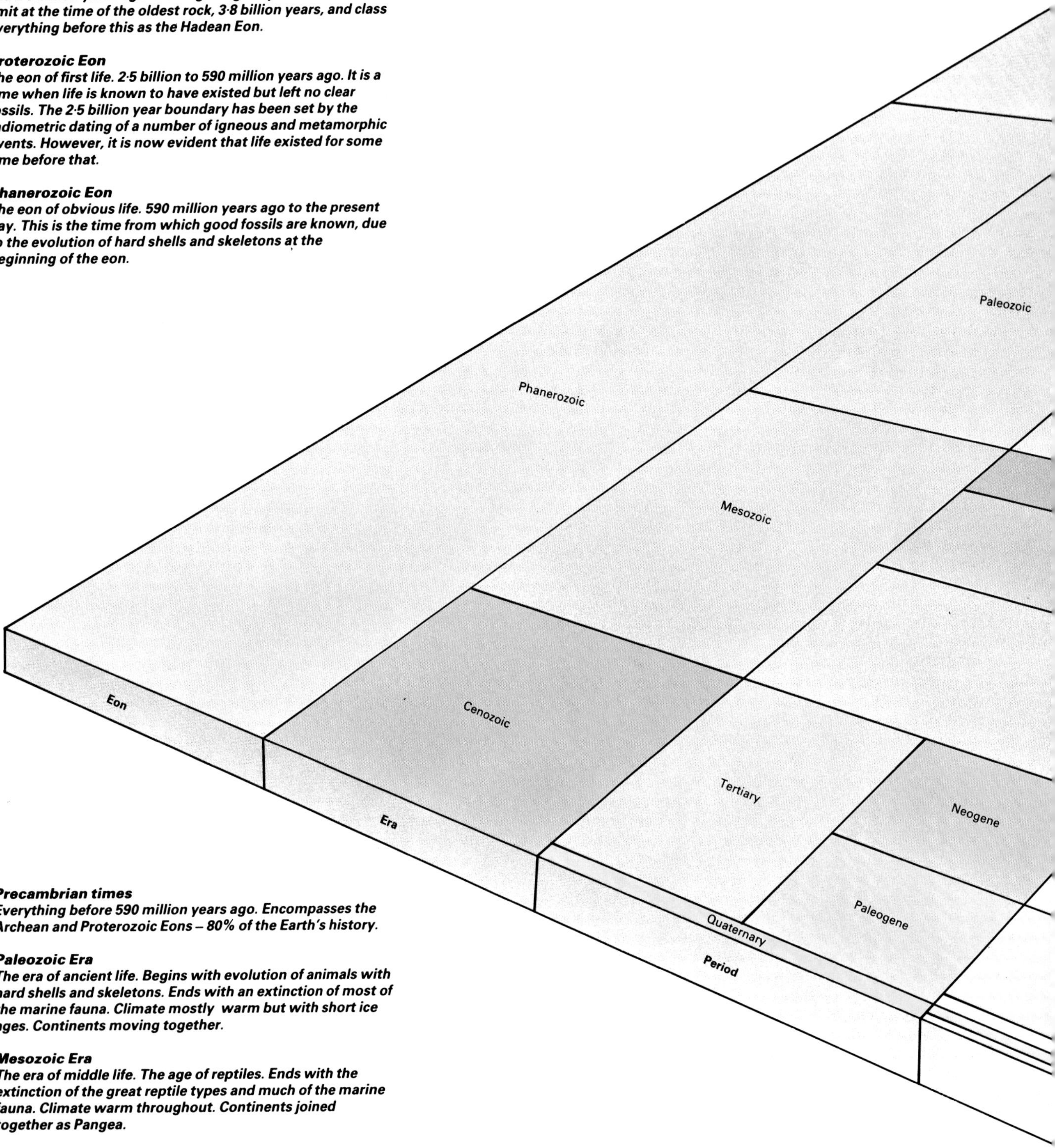

Paleozoic
Phanerozoic
Mesozoic
Eon
Cenozoic
Tertiary
Neogene
Era
Paleogene
Quaternary
Period

Precambrian times
Everything before 590 million years ago. Encompasses the Archean and Proterozoic Eons – 80% of the Earth's history.

Paleozoic Era
The era of ancient life. Begins with evolution of animals with hard shells and skeletons. Ends with an extinction of most of the marine fauna. Climate mostly warm but with short ice ages. Continents moving together.

Mesozoic Era
The era of middle life. The age of reptiles. Ends with the extinction of the great reptile types and much of the marine fauna. Climate warm throughout. Continents joined together as Pangea.

Cenozoic Era
The era of recent life. The age of mammals and of Man. Climate deteriorating towards recent ice age. Continents moving apart.

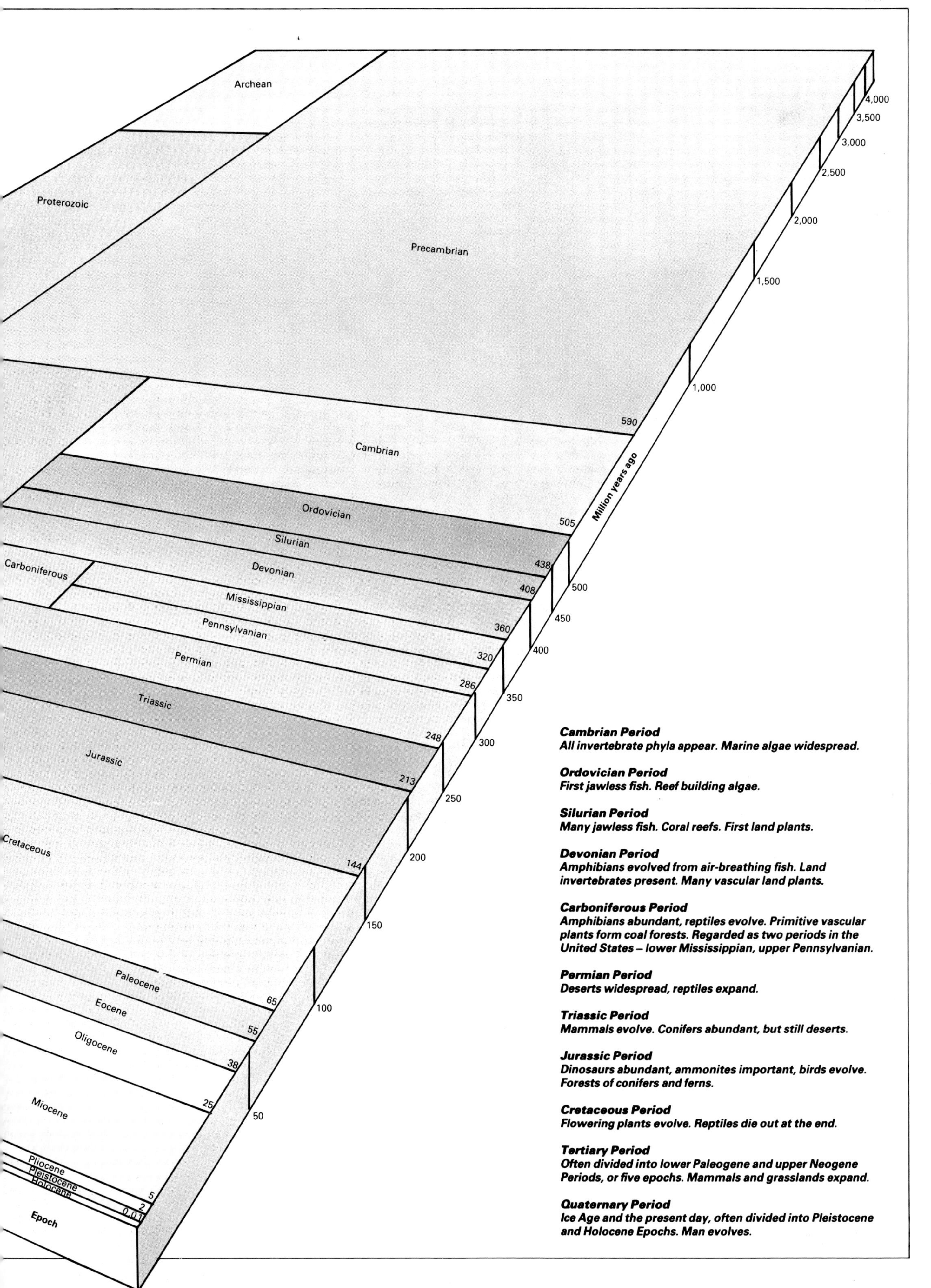

Cambrian Period
All invertebrate phyla appear. Marine algae widespread.

Ordovician Period
First jawless fish. Reef building algae.

Silurian Period
Many jawless fish. Coral reefs. First land plants.

Devonian Period
Amphibians evolved from air-breathing fish. Land invertebrates present. Many vascular land plants.

Carboniferous Period
Amphibians abundant, reptiles evolve. Primitive vascular plants form coal forests. Regarded as two periods in the United States – lower Mississippian, upper Pennsylvanian.

Permian Period
Deserts widespread, reptiles expand.

Triassic Period
Mammals evolve. Conifers abundant, but still deserts.

Jurassic Period
Dinosaurs abundant, ammonites important, birds evolve. Forests of conifers and ferns.

Cretaceous Period
Flowering plants evolve. Reptiles die out at the end.

Tertiary Period
Often divided into lower Paleogene and upper Neogene Periods, or five epochs. Mammals and grasslands expand.

Quaternary Period
Ice Age and the present day, often divided into Pleistocene and Holocene Epochs. Man evolves.

Gestation
The development of young mammals in the mother's uterus from fertilization to birth. With some exceptions, the gestation period is proportional to the adult size of the animal: for the human young, for example, the gestation period is about 270 days, but for the elephant it is closer to two years. (See also EMBRYO; IMPLANTATION; PREGNANCY.)

Geyser
A hot spring, found in currently or recently volcanic regions (see VOLCANISM), that intermittently jets steam and superheated water into the air. It consists essentially of a tube leading down to a heat source. GROUNDWATER accumulates in the tube, that near the bottom being kept from boiling by the PRESSURE of the cooler layers above. When the critical temperature is reached, bubbles rise and heat the upper layers, which expand and well out of the orifice. This reduces the pressure enough for substantial STEAM formation below, with subsequent eruption. The process then begins again. The famous Old Faithful in Yellowstone National Park, Wyoming used to erupt every 66½ minutes, but it has recently become less reliable. (See also FUMAROLE; HOT SPRINGS; SPRING.)

Giaever, Ivar (b. 1929)
Norwegian-born US physicist awarded, with L. ESAKI and B. JOSEPHSON, the 1973 Nobel Prize for Physics for his work on tunneling (see WAVE MECHANICS).

Giant star see STAR.

Giauque, William Francis (1895-1982)
US chemist who discovered the isotopes of OXYGEN (1929). He has also contributed to the science of CRYOGENICS by inventing and applying the process of adiabatic demagnetization, for which he was awarded the 1949 Nobel Prize for Chemistry.

Gibberellins
A group of plant HORMONES mainly found in the seeds, young leaves and roots of green plants. They were originally isolated from a Japanese fungus, *Gibberella fujikuroi*, which lengthens the stems of rice plants. Gibberellic acid, found in green plants, is involved in the bolting of plants like carrots.

Gibbs, Josiah Willard (1839-1903)
US physicist best known for his pioneering work in chemical THERMODYNAMICS. In *On the Equilibrium of Heterogeneous Substances* (2 vols, 1876 and 1878) he states Gibbs' Phase Rule (see PHASE EQUILIBRIA). In the course of his research on the electromagnetic theory of LIGHT, he made fundamental contributions to the art of VECTOR ANALYSIS.

Gigantism see ACROMEGALY.

Gilbert (Gb)
Unit of magnetomotive force in CGS UNITS, equaling 10/4π ampere-turns.

Gilbert, Walter (b. 1932)
US molecular biologist who shared the 1980 Nobel Prize for Chemistry for determining the NUCLEOTIDE distributions of substances, including a particular kind of RNA found in yeast.

Gilbert, William (1544-1603)
English scientist, the father of the science of MAGNETISM. Regarding the Earth as a giant magnet, he investigated its field in terms of dip and variation (see EARTH), and explored many other magnetic and electrostatic phenomena. The GILBERT is named for him.

Gill (Gi)
Name of various units of volume, usually equalling one-fourth of a pint. The US gill is 0.1183 liters; the Imperial gill is 0.1421 liters.

Gills
The respiratory organs of aquatic animals. They take in oxygen from the water and give off carbon dioxide waste. They are thin-walled so that gases pass easily through, and usually take the form of thin flat plates or finely divided feathery filaments, both structures with a very large surface area. Some invertebrates, crabs and lobsters for instance, have gills protected by an EXOSKELETON and maintain an adequate oxygen supply by pumping water over them. The gills of fish are protected by a bony operculum, and movements of the throat provide a water current over them. (See GAS EXCHANGE.)

Gilsonite
Natural asphaltic BITUMEN found in veins near the Colorado/Utah border, in the Uinta Basin. It is a lustrous black solid, used in paints and coating and insulating materials, and, more recently, converted to coke, gasoline and gas.

Gingiva
The soft tissue of the gums.

Gingivitis
Inflammation of the gums caused by bacterial infection.

Gingko (maidenhair)
A unique tree, relic of a group that flourished 200 million years ago. It belongs with the CONIFERS in the GYMNOSPERM group, but does not share their needle-like leaves or cones. Its leaves are fan-shaped with radiating veins, and the seed is enclosed by a fleshy coat that resembles the fruit of angiosperm trees.

Giorgi system see MKSA UNITS.

Gizzard
Part of the alimentary canal in a variety of animals; it is specialized for the mechanical breakdown of hard food. Situated before the main digestive region of the gut, it has very muscular walls. Fragmentation of the food may be by chitinous "teeth" in the inner wall, or by stones and grit swallowed expressly for this purpose, as in grain-eating birds such as pigeons.

Glacier
A large mass of ice that can survive for several years. In most cases, glaciers are heavy enough to flow downhill under their own weight. There are three recognized types of glacier: ice sheets and caps; mountain or valley glaciers; and piedmont glaciers.

Glaciers form wherever conditions are such that annual PRECIPITATION of snow, sleet and hail is greater than the amount that can be lost through evaporation or otherwise (see ABLATION). The occurrence of a glacier thus depends much on LATITUDE and also on local topography: there are several glaciers on the equator. Glaciers account for about 75% of the world's fresh water, and of this the Antarctic ice sheet accounts for about 85%. Mountain glaciers usually result from snow that has accumulated in CIRQUES coalescing to form glaciers; and piedmont glaciers occur when such a glacier spreads out of its valley into a contiguous lowland area. (See also DRIFT; DRUMLIN; EROSION; ESKER; FJORD; ICE; ICE AGES; ICEBERG; MORAINE; NÉVÉ; TILL.) (➧ page 208)

Glands
Structures in animals and plants that secrete special liquids. In plants they may discharge their secretions to the outside of the plant (via glandular hairs), or into special secretory canals. External secretions include NECTAR and insect attractants. In animals glands are divided into ENDOCRINE GLANDS, which secrete HORMONES into the blood stream, and exocrine glands, which usually secrete materials via ducts into internal organs or onto body surfaces. In humans, SKIN contains two types of exocrine gland: sweat glands, which secrete watery fluid (PERSPIRATION), and sebaceous glands, which secrete sebum. Lacrimal glands secrete tears; mammary glands secrete milk. Salivary glands secrete SALIVA, which partially digests food and facilitates swallowing. In the GASTROINTESTINAL TRACT mucus-secreting exocrine glands are numerous, particularly in the STOMACH and colon, where solid food or feces need lubrication. Other stomach glands secrete hydrochloric acid and pepsin as part of the digestive process. In the small intestine juices containing ENZYMES are similarly secreted by minute glandular specializations of the epithelium. The part of the PANCREAS secreting enzyme-rich juice into the DUODENUM may be regarded as an exocrine gland. Lymph nodes are sometimes incorrectly called glands, and the term "swollen glands" refers to lymph nodes swollen as a result of infection. (➧ page 111)

Glandular fever see MONONUCLEOSIS.

Glaser, Donald Arthur (b. 1926)
US physicist awarded the 1960 Nobel Prize for Physics for his invention of the BUBBLE CHAMBER (1952).

Glashow, Sheldon Lee (b. 1932)
A US physicist who shared the 1979 Nobel Prize for Physics with SALAM and WEINBERG for their theory which united the WEAK NUCLEAR FORCE with the ELECTROMAGNETIC FORCE.

Glass
Material formed by the rapid cooling of certain molten liquids so that they fail to crystallize (see CRYSTAL) but retain an amorphous structure. Glasses are in fact supercooled LIQUIDS which, however, have such high viscosity that they behave like solids for all practical purposes. Some glasses may spontaneously crystallize or devitrify. Few materials form glasses, and almost all that are found naturally or used commercially are based on SILICA and the SILICATES. Natural glass is formed by rapid cooling of MAGMA, producing chiefly OBSIDIAN, or rarely by complete thermal metamorphism (see also TEKTITES). Most modern glass products are made of soda-lime glass. Although silica itself can form a glass, it is too viscous and its melting point is too high for most purposes. Adding soda lowers the melting point, but the resultant sodium silicate is water-soluble (see WATER GLASS), so lime is added as a stabilizer, together with other metal oxides as needed for decolorizing, etc. The usual proportions are 70% SiO_2, 15% Na_2O, 10% CaO.

Glauber, Johann Rudolf (1604-1668)
German chemist who prepared a wide variety of organic and inorganic compounds to make use of their (often nonexistent) medicinal powers. In preparing hydrochloric acid from sulfuric acid and common salt, he found a residue which he claimed as a cure-all. In fact it was sodium sulfate, a mild laxative, still often called Glauber's salt.

Glaucoma
Raised fluid pressure in the EYE, leading in chronic cases to a progressive and painful deterioration of VISION. It arises from a variety of causes, often involving blocked fluid drainage.

Glauconite
A CLAY mineral of the illite type, a hydrated MICA containing considerable iron and magnesium. Formed by alteration of BIOTITE in a shallow, reducing marine environment, it commonly occurs as small green pellets in sedimentary deposits. Greensand is a mixture of glauconite with quartz SAND. (See also MARL.)

Glioma
TUMOR of glial cells, the supporting cells of the BRAIN. They never metastasize (see CANCER), but produce signs of focal damage to the brain such as weakness, visual disturbance, personality change or EPILEPSY, and often a characteristic type of HEADACHE.

Globe
A representation of the Earth as a small sphere,

mounted on an axis so that it may revolve. The oldest extant globe is from Nuremberg, 1492. Celestial and lunar globes are also made.

Globular clusters
Apparently ellipsoidal, densely packed clusters of up to a million stars orbiting a GALAXY. The MILKY WAY and the ANDROMEDA galaxy each has about 200 such clusters. They contain high proportions of cool red stars and RR Lyrae VARIABLE STARS. Study of the latter enables the distances of the clusters to be calculated. (➧ page 240)

Globulins
Large protein molecules found in blood and other tissues. ANTIBODIES are globulins.

Glomerulonephritis
A form of NEPHRITIS in which the glomeruli of the KIDNEY are inflamed.

Glomerulus see KIDNEYS

Glucose ($C_6H_{12}O_6$)
Dextrose or "grape sugar", a naturally occurring simple SUGAR (monosaccharide) found in honey and sweet fruits. It is central to the metabolism of animals and circulates in their blood, providing their cells with energy. It is broken down to pyruvate by GLYCOLYSIS, and thus enters the CITRIC ACID CYCLE and ELECTRON TRANSPORT CHAIN; through these reactions, energy is released in the form of ATP. Many animals store a long-chain compound, GLYCOGEN, which is made up of glucose units and acts as an energy reserve. Other sugars (CARBOHYDRATES) are converted to glucose by digestion before use. (➧➧ page 28, 40, 233)

Gluten
A mixture of two proteins (gliadin and glutenin) found in wheat, rye, barley and oats, but principally in bread wheat. In the rising of BREAD gluten forms an elastic network that traps the carbon dioxide, giving a desirable structure. The proportion of gluten in wheat flour varies from 8% to 15%, and determines the suitability of the flour for different uses. Some people react adversely to gluten and must eliminate it from their diet.

Glycerol (glycerin)
A colorless, viscous liquid with a sweet taste; a trihydric ALCOHOL. It forms esters with FATTY ACIDS, to give TRIGLYCERIDES and other complex LIPIDS. (➧ page 40)

Glycogen
A soluble polysaccharide (see CARBOHYDRATE) consisting of long, branching chains of glucose units. It is produced by most animals and some bacteria; in vertebrates it is stored in muscle and in the liver where it forms a readily available reserve of GLUCOSE. (➧ page 40)

Glycol (diol)
A class of dihydric ALCOHOLS having two hydroxyl groups per molecule, the lower members being viscous liquids that readily absorb water from the atmosphere. ETHYLENE GLYCOL is the most important, being used in antifreeze and engine coolants and in the manufacture of various plastics. Propylene glycol has similar uses, but being nontoxic is also used in foods, drugs and cosmetics.

Glycolysis
An enzyme-mediated cellular process by which GLUCOSE (a six-carbon molecule) is broken down to pyruvate (a three-carbon molecule; two are produced per molecule of glucose). In all, 10 different steps are involved. The process does not require the presence of OXYGEN. In anaerobic organisms it is the sole source of energy; see FERMENTATION. In aerobic organisms the pyruvate is further broken down, to carbon dioxide (CO_2) and water (H_2O) by the CITRIC ACID CYCLE and the ELECTRON TRANSPORT CHAIN.

Gneiss
Broad class of coarse-grained METAMORPHIC ROCKS with a banded, foliated structure and poor CLEAVAGE (see also SCHIST). Their composition is variable, but often approximates to that of GRANITE

Goddard, Robert Hutchings (1882-1945)
US pioneer of rocketry. In 1926 he launched the first liquid-fuel rocket. Some years later, with a Guggenheim Foundation grant, he set up a station in New Mexico, there developing many of the basic ideas of modern rocketry: among over 200 patents was that for a multistage rocket. He died before he received US Government recognition.

Goeppert-Mayer, Marie (b. 1906)
German-born US scientist awarded half of the 1963 Nobel Prize for Physics together with J.H.D. JENSEN for their discoveries concerning nuclear shell structure.

Goethite
Brown OXIDE mineral of composition FeO(OH); a major IRON ore of widespread occurrence, formed in bogs or by weathering of other iron minerals. Goethite is very similar to LIMONITE, but is crystalline (in the orthorhombic system). X-RAY DIFFRACTION analysis has shown that much supposed limonite is in fact goethite.

Goiter
Enlargement of the THYROID gland in the neck, causing swelling below the LARYNX. It may represent the smooth swelling of an overactive gland in thyrotoxicosis or, more often, the enlargement caused by multiple CYSTS and nodules without functional change. Endemic goiter is enlargement associated with IODINE deficiency, occurring in certain areas where the element is lacking in the soil and water. Rarely, goiter is due to CANCER of the thyroid.

Gold (Au)
Yellow NOBLE METAL in Group IB of the PERIODIC TABLE; a TRANSITION ELEMENT. Gold has been known and valued from earliest times and used for jewelry, ornaments and coinage. It occurs as the metal and as tellurides, usually in veins of QUARTZ and PYRITE; the chief producing countries are South Africa, the USSR, Canada and the US. The metal is extracted with CYANIDE or by forming an AMALGAM, and is refined by electrolysis. The main use of gold is as a currency reserve, a store of value. Like SILVER, it is used for its high electrical conductivity in printed circuits and electrical contacts, and also for filling or repairing teeth. It is very malleable and ductile, and may be beaten into gold leaf or welded in a thin layer to another metal (rolled gold). For most uses pure gold is too soft, and is alloyed with other noble metals, the proportion of gold being measured in CARATS. Gold is not oxidized in air, nor dissolved by alkalis or pure acids, though it dissolves in AQUA REGIA or cyanide solution because of LIGAND complex formation, and reacts with the HALOGENS. It forms trivalent and monovalent salts. Gold (III) chloride is used as a toner in photography. AW 197.0, mp 1063°C, bp 2966°C, sg 19.32 (20°C). (➧ page 130)

Gold, Thomas (b. 1920)
Austrian-born US cosmologist who, with HERMANN BONDI, proposed (1948) the Steady State model of the Universe (see COSMOLOGY).

Goldstein, Joseph L. (b. 1940)
US scientist who jointly received with M.S. BROWN the 1985 Nobel Prize for Physiology or Medicine for their discoveries concerning the regulation of cholesterol metabolism.

Golgi, Camillo (1843-1926)
Italian histologist who developed a staining technique (1873) with which he was able to explore the NERVOUS SYSTEM in great detail. He shared with RAMÓN Y CAJAL the 1906 Nobel Prize for Physiology or Medicine.

Gonadotrophins
HORMONES secreted by the PITUITARY GLAND and PLACENTA, which stimulate the production of sex hormones by the gonads: ESTROGEN and PROGESTERONE in females and ANDROGENS in males. They control the maturation and release of EGGS from the ovaries and the development of SPERM in the testes. Gonadotrophin secretion is controlled by the HYPOTHALAMUS.

Gonads
The reproductive organs of animals, which produce the GAMETES. The female gonad is the OVARY, producing EGGS, and the male gonad the testis (see TESTES), producing spermatozoa or sperm. (See also REPRODUCTION.)

Gondwana
Hypothetical S Hemisphere supercontinent formed after the split of PANGEA (see also CONTINENTAL DRIFT; LAURASIA). Stratigraphic and FOSSIL evidence suggest it comprised what are now Antarctica, Africa, Australia, India, South America and other, smaller, units.

Goniometer
Instrument for measuring angles, especially those between crystal faces. The simplest form is the contact goniometer, a protractor whose base is laid against one face, a movable arm being turned until it contacts the adjacent face. The more accurate reflecting goniometer mounts the crystal axially on a graduated circle, or more usually two graduated circles, horizontal and vertical, rotatable independently. The crystal is rotated until each face in turn reflects a collimated light beam into a fixed telescope, and so the direction of the normal to each face is determined. (The term may also be used for the DIRECTION FINDER.)

Gonorrhea see SEXUALLY TRANSMITTED DISEASES.

Goodyear, Charles (1806-1860)
US inventor of the process of VULCANIZATION (patented 1844). In 1839 he bought the patents of Nathaniel Manley Hayward (1808-1865), who had had some success by treating RUBBER with SULFUR. Working on this, Goodyear accidentally dropped a rubber/sulfur mixture onto a hot stove, so discovering vulcanization.

Gorgas, William Crawford (1854-1920)
US Army sanitarian. After WALTER REED's commission had proved (1900) CARLOS FINLAY's theory that YELLOW FEVER is transmitted by the MOSQUITO, Gorgas conducted in Havana a massive control program; he repeated this in Panama (1904-1913), facilitating the digging of the Panama Canal.

Gout
A DISEASE of PURINE metabolism characterized by increased uric acid in the BLOOD and episodes of ARTHRITIS due to uric acid crystal deposition in SYNOVIAL FLUID and the resulting INFLAMMATION. Deposition of urate in CARTILAGE and subcutaneous tissue (called tophi) and in the KIDNEYS and urinary tract (causing stones and renal failure) are other important effects. The arthritis is typically of sudden onset with severe pain, often affecting the great toe first and large JOINTS in general. It is not a symptom of excess of alcohol.

Graben see RIFT VALLEY.

Graft, Surgical see PLASTIC SURGERY; TRANSPLANTS.

Grafting
The technique of propagating plants by attaching the stem or bud of one plant (called the scion) to the stem or roots of another (the stock or rootstock). Only closely related plants can be grafted. Roses and fruit trees are often grafted so that good flowering or fruiting varieties have the benefit of strong roots.

Graham, Thomas (1805-1869)
Scottish chemist who formulated Graham's Law: the DIFFUSION rate of a gas is proportional to the inverse of the square ROOT of its density. While working further on diffusion and OSMOSIS he discovered the colloidal state (see COLLOID).

Grain (caryopsis)
Botanically, a dry one-seeded FRUIT. The grains of grasses are typical. Those of cereal crops contain a high percentage of starch.
Grain (gr)
The fundamental Anglo-American unit of weight, shared by the avoirdupois, troy and apothecaries' systems. The Imperial grain is defined as exactly equal to 0.06479891 grams.
Gram (g)
The fundamental unit of mass in the CGS version of the METRIC SYSTEM. It approximates to the mass of a cubic centimeter of water.
Gramineae
A family of flowering plants (ANGIOSPERMS) that includes the grasses, bamboos and reeds. The flowers are pollinated by wind and therefore relatively inconspicuous. The Gramineae are MONOCOTYLEDONS.
Gram's stain
A stain for BACTERIA, discovered in 1884 by the Danish bacteriologist Christian Gram (1853-1938), that divides them into two groups, gram-positive and gram-negative. As differences in the cell walls determine not only the staining difference but also behavior and antibiotic sensitivity, the stain has considerable medical value.
Grand mal see EPILEPSY.
Granit, Ragnar Arthur (b. 1906)
Finnish-born Swedish physiologist who shared the 1967 Nobel Prize for Physiology or Medicine with H.K. HARTLINE and G. WALD. Granit demonstrated that individual nerves in the EYE could distinguish light of different colors.
Granite
Coarse- to medium-grained plutonic IGNEOUS ROCK, composed of FELDSPAR (orthoclase and microcline predominating over plagioclase) and QUARTZ, often containing BIOTITE and AMPHIBOLE. It is the type of the family of granitic rocks, plutonic rocks rich in feldspar and quartz, of which the CONTINENTS are principally made. Most granite was formed by crystallization of MAGMA, though some is METAMORPHIC, and some was formed by replacement ("granitization"). It occurs as DIKES and SILLS, large masses, and enormous BATHOLITHS. A hard, weather-resistant rock, usually pink or gray, granite is used for building, paving and road curbs.
Graphite
Allotrope of CARBON (see ALLOTROPY), forming soft, black, metallic crystals, in which the atoms are arranged in layers of hexagons that easily slide over each other. It is found naturally in GNEISS and SCHIST, and synthesized from COKE. Graphite is a good conductor of HEAT and ELECTRICITY. It is used for ELECTRODES, nuclear-reactor moderators and lubricants. subl 3660°C, sg 2.25 (20°C).
Grassland
The areas of the Earth whose predominant type of vegetation consists of grasses, rainfall being generally insufficient to support higher plant forms. There are three main types. Savanna, or tropical grassland, has coarse grasses growing lm to 4m high, occasional clumps of trees and some shrubs; it is found in parts of Africa and South America. Prairie has tall, deep-rooted grasses and is found in Middle and North America, Argentina (where it is called pampas), the Ukraine, South Africa (savanna and veldt) and N Australia (downland). Steppes have short grasses and are found mainly in Central Asia. Grasslands are of great economic importance as they not only provide food for domestic animals but also often make excellent cropland for cultivation.
Gravel
In geology, a collection of rock particles whose diameter ranges from 2mm to 4mm. In general terms, gravel particles may be as large as pebbles. Gravel is used commercially in the making of concrete.
Graves, Robert James (1796-1853)
Irish physician remembered for his work on exophthalmic GOITER (Graves' disease).
Gravimeter
An instrument for detecting small variations in the Earth's gravitational field, frequently used in mineral and oil prospecting. Variations in the gravitational FORCE on a weight suspended from a spring cause it to stretch or be deflected in a way which is then measured.
Gravimetric analysis
Method of quantitative chemical ANALYSIS in which the substance to be estimated is converted to a substance which can be separated pure and entire, and which is then weighed. Commonly a highly insoluble precipitate is formed, filtered off, washed and dried (see DEHYDRATION). The weight of substance sought is calculated from the weight and composition of the precipitate.
Gravitation
One of the fundamental forces of nature, the force of attraction existing between all MATTER. It is much weaker than the nuclear or electromagnetic forces and plays no part in the internal structure of matter. Its importance lies in its long range and in its involving all masses. It plays a vital role in the behavior of the UNIVERSE: the gravitational attraction of the SUN keeps the PLANETS in their orbits, and gravitation holds the matter in a STAR together. NEWTON's law of universal gravitation states that the attractive FORCE F between two bodies of MASSES M_1 and M_2 separated by distance d is $F=GM_1M_2/d^2$ where G is the Universal Gravitational Constant (6.670×10^{-11} Nm^2kg^{-2}). The force of gravity on the Earth is a special case of the attraction between masses and causes bodies to fall toward the center of the Earth with a uniform ACCELERATION $g=GM/R^2$ where R and M are the radius and mass of the Earth. Assuming, with Newton, that the inertial mass of a body (that which is operative in the laws of motion) is identical with its gravitational mass, application of the second law of motion gives the WEIGHT of a body of mass m, the force with which the Earth attracts that body, as mg. Bodies on the Earth and Moon thus have the same mass but different weights. Again, the gravitational force on a body is proportional to its mass but is independent of the type of material it is. Newton's theory explains most of the observed motions of the planets and the TIDES and is still sufficiently accurate for most applications. The Newtonian analysis of gravitation remained unchallenged until, in the early 20th century, EINSTEIN introduced radically new concepts in his theory of general RELATIVITY. According to this, mass deforms the geometrical properties of the space around it. Einstein reaffirmed Newton's assumption regarding the equivalence of gravitational and inertial mass, proposing that it was impossible to distinguish experimentally between an accelerated coordinate system and a local gravitational field. From this he predicted that LIGHT would be found to be deflected toward massive bodies by their gravitational fields and this effect indeed was observed for starlight passing close to the Sun. It was also predicted that accelerated matter should emit gravitational waves with the velocity of light but the existence of these has not as yet been demonstrated. (♦ page 147)
Gravitational field theory
The QUANTUM THEORY that treats gravity as a field, as opposed to a force that acts instantaneously at a distance. The hypothetical SUBATOMIC PARTICLE associated with gravity is the graviton. (See also UNIFIED FIELD THEORY.)
Graviton see GRAVITATIONAL FIELD THEORY.
Gravity see GRAVITATION.
Gray (Gy)
The SI unit of absorbed dose, that is, a measure of the energy deposited in a material by ionizing radiation. It replaces the older unit, the RAD. The gray is defined as the deposition of 1J of energy per kilogram, and 1Gy=100 rad.
Gray, Asa (1810-1888)
The foremost of 19th-century US botanists. Being a prominent Protestant layman, his advocacy of the Darwinian thesis carried special force. However, he never accepted the materialist interpretation of the evolutionary mechanism and taught that NATURAL SELECTION was indeed consistent with a divine teleology.
Gray matter
The parts of the brain that are rich in nerve-cell bodies, as opposed to white matter, which is mainly nerve fibers sheathed by MYELIN. The cerebral cortex, basal ganglia, nuclei of the brain stem and the center of the SPINAL CORD are major areas of gray matter.
Graywacke see CONGLOMERATE.
Great Bear
Ursa Major, a large N Hemisphere constellation containing the seven bright stars known as the Plow or Big Dipper. Two of these, the Pointers, form roughly a straight line with POLARIS and are thus of navigational importance. Five stars of the Plow are, with SIRIUS, members of a widely separated GALACTIC CLUSTER.
Great circle routes
Routes of prime importance in air travel as they describe the shortest distances between two points on the Earth's surface. Great circles are circles on the surface of a sphere whose centers coincide with the center of the sphere (see SPHERICAL GEOMETRY): the EQUATOR is an example (see also LATITUDE AND LONGITUDE). Great circle navigation is aided by use of MAPS drawn to a gnomonic projection, where the center of perspective is the center of the Earth. On such maps great circles appear as straight lines.
Greenhouse effect
The warming of the Earth due to increasing amounts of carbon dioxide in the atmosphere. Carbon dioxide traps solar radiation reflected from the Earth's surface (reflected radiation having longer wavelengths than incident radiation) and thus increases the overall temperature of the atmosphere in much the same way as the glass of a greenhouse. The effect was predicted several years before it was actually observed. The rise in carbon dioxide has been caused by the large-scale burning of FOSSIL FUELS, which began with the Industrial Revolution and has increased dramatically during the 20th century. The destruction of large areas of forest, particularly the highly productive rainforest of the tropics, has exacerbated the sitation as the forest "fixes" carbon dioxide during photosynthesis and locks it up in the forest biomass. The steady increase in temperatures forecast for the next century will accelerate the pace of desertification in vulnerable areas such as the Sahel. (♦ page 25)
Greensand see GLAUCONITE.
Green vitriol (iron (II) sulfate) see IRON.
Greenwich Mean Time (GMT)
Local mean time along the Greenwich meridian, used as a TIME standard throughout the world.
Greenwich Observatory, Royal
Observatory established in 1675 at Greenwich, England, by Charles II to correct the astronomical tables used by sailors and otherwise to advance the art of NAVIGATION. Its many famous directors, the astronomers royal, have included J. FLAMSTEED (the first), E. HALLEY, and Sir George Airy. The original Greenwich building, now known as Flamsteed House and run as an astronomical

HORMONES The Body's Chemical Messengers

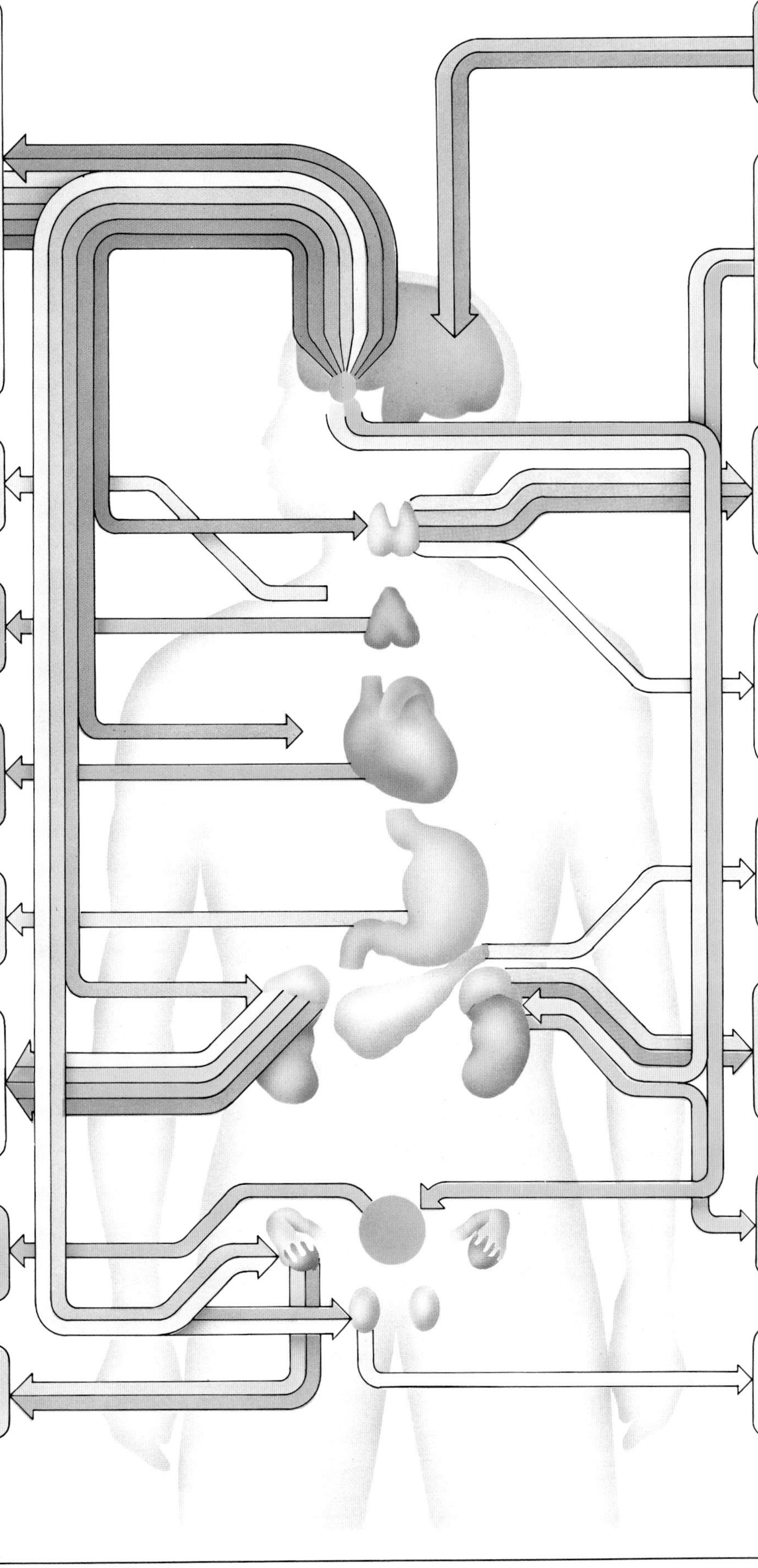

The anterior pituitary makes several hormones. Growth hormone (GH) stimulates growth, maintains cell size, and breaks down fat cells, converting fat into energy and raising blood glucose by making the liver manufacture peptides to promote growth. Melanocyte-stimulating hormone (MSH) darkens skin. Adrenocorticotropic hormone (ACTH) controls hormone secretion from the adrenal cortex. Prolactin makes the breasts produce milk. Other hormones of the anterior pituitary are thyroid-stimulating hormone (TSH), luteinizing hormone (LH) and follicle-stimulating hormone (FSH).

The brain's neurotransmitters may act as hormones elsewhere in the body; the pineal gland secretes melatonin, which seems to inhibit sexual activity.

The posterior pituitary secretes vasopressin (also known as antidiuretic hormone - ADH), which raises blood pressure by making the kidneys absorb more water and making arterioles constrict. Oxytocin stimulates the smooth muscles of the uterus during pregnancy, produces contractions at birth.

The skin manufactures Vitamin D, which acts as a hormone in controlling calcium levels in the blood, in association with PTH.

The thyroid produces calcitonin when blood calcium levels are high, to deposit calcium and phosphate in bone. Thyroxine and triiodothyronine increase heart rate, blood flow, blood pressure, gut contractions and anxiety.

The thymus produces several hormones that stimulate the growth of B-lymphocytes.

The parathyroid makes parathyroid hormone (PTH), which releases calcium from bone into blood when blood calcium levels are low. It also makes the kidneys excrete less calcium and magnesium, and increases their absorption in the gut when Vitamin D is short.

The heart atria make peptides that inhibit the production of angiotensin and ACTH, thus helping to control blood pressure.

The pancreas releases insulin and glucagon, which together regulate blood glucose levels. The pancreas also makes growth hormone inhibiting factor (GHIF) which controls the production of insulin and glucagon.

The stomach releases several hormones that stimulate or inhibit the release of gastric juices and the activity of the digestive system.

The adrenal medulla releases adrenaline and noradrenaline, in response to stress or low blood glucose. They produce the fight-or-flight response, raising blood pressure, respiration rate, muscle efficiency and metabolic rates.

The adrenal cortex is stimulated by ACTH and produces aldosterone which makes the kidneys reabsorb salt and water; glucocorticoids, which promote metabolism and provide resistance to stress; androgens; and estrogens.

The kidneys make erythropoetin, which stimulates red blood cell manufacture, and renin which stimulates the production of aldosterone in the adrenal cortex.

The placenta produces several hormones related with pregnancy, including estrogen, progesterone and relaxin.

The testes are stimulated by LH to produce testosterone which controls secondary sexual characteristics, and inhibin, which inhibits the secretion of FSH, cutting sperm production.

The ovaries produce estrogen and progesterone which regulate the menstrual cycle and pregnancy, controlled by FSH and LH.

museum, was designed by CHRISTOPHER WREN. The Observatory is now sited at Herstmonceux, Sussex, to where it was moved in 1948-1957. The Observatory itself is thus no longer sited on the Greenwich meridian, the international zero of longitude.

Gregorian calendar see CALENDAR.

Grew, Nehemiah (1641-1712)
English plant anatomist and physician who introduced the term "comparative" anatomy (1676); a principal founder of his science. His main work, *The Anatomy of Plants*, appeared in 1682.

Grignard, François Auguste Victor (1871-1935)
French chemist who shared with SABATIER the 1912 Nobel Prize for Chemistry for his discovery of GRIGNARD REAGENTS, complex compounds used in many organic syntheses.

Grignard reagents
ORGANOMETALLIC COMPOUNDS of great importance in laboratory and industrial chemical synthesis. Made by reacting ALKYL (or aryl) HALIDES with MAGNESIUM in an ETHER solution, they are commonly represented as RMgX where R is an alkyl (see ALKANES) or aryl (see AROMATIC COMPOUNDS) group. They are powerful NUCLEOPHILES, and react with many compounds, introducing the group R.

Grippe see INFLUENZA.

Ground, Electrical
An electrical connexion between apparatus and the Earth or an equivalent conducting body at zero POTENTIAL. Electricity supply systems are grounded to avoid overvoltage and to improve performance. Metal cases, frames etc. of electrical equipment are grounded to minimize risk of ELECTRIC SHOCKS in case a fault should make the exposed part "live".

Groundwater
Water accumulated beneath the Earth's surface in the pores of rocks, spaces, cracks, etc. It may be meteoric, rainwater having soaked down from above, or juvenile, where water has risen from beneath. Permeable, water-bearing rocks are AQUIFERS; rocks with pores small enough to inhibit the flow of water through them are aquicludes. Buildup of groundwater pressure beneath an aquiclude makes possible the construction of an ARTESIAN WELL. The uppermost level of groundwater saturation is the water table. (See also PERMAFROST; SPRING; WELL.)

Growth
The increase in the size of an organism, reflecting an increase either in the number of its CELLS, or in its protoplasmic material, or both. Cell number and protoplasmic content do not always increase together; cell division can occur without any increase in PROTOPLASM, giving a larger number of smaller cells. Alternatively, protoplasm can be synthesized with no cell division so that the cells become larger. Any increase in protoplasm requires the synthesis of cell components such as mitochondria, cell membrane and thousands of enzymes.

Growth curves, which plot time against growth (such as the number of cells in a bacterial culture or the weight of a plant seedling), all have a characteristic S-shape. This curve is divided into three parts: the lag phase, during which cells prepare for growth; the exponential phase, when actual growth occurs; and the stationary phase, when growth ceases. The S-type growth pattern can be readily seen in unicellular organisms. In multicellular organisms the relationships of the different types of cells complicate the pattern, but the overall growth curve is still S-shaped.

Growth factors
In cell- and tissue-culture, special proteins needed for cell growth and division.

Guano
The excrement of sea birds. There were once very thick deposits of it on certain islands off the west coast of S America. These were once mined for use as a fertilizer until artificial fertilizers were introduced.

Guericke, Otto von (1602-1686)
German physicist credited with inventing the vacuum pump. His best-known experiment was with the Magdeburg hemispheres (1654): he evacuated a hollow sphere composed of two halves placed together, and showed that two 8-horse teams were insufficient to separate the halves. (See PUMPS.) He is also credited with inventing the first electric GENERATOR.

Guérin, Camille (1872-1961)
French bacteriologist who, with CALMETTE, developed the BCG VACCINE, which is used to counter TUBERCULOSIS.

Guillaume, Charles Edouard (1861-1938)
Swiss-born French physicist best known for discovering INVAR, an iron-nickel alloy which expands and contracts only very slightly with temperature change. For his work on ferronickels he was awarded the 1920 Nobel Prize for Physics.

Guillemin, Roger Charles Louis (b. 1924)
French-born UK physiologist who shared the 1977 Nobel Prize for Physiology or Medicine with R.S. YALOW and A.V. SCHALLY for his work identifying compounds stimulating peptide HORMONE release from the brain.

Gulf Stream
Warm ocean current flowing N, then NE, off the E coast of the US. Its weaker, more diffuse continuation is the E-flowing North Atlantic Drift, which is responsible for warming the climates of W Europe. The current, often taken to include also the Caribbean Current, is fed by the N Equatorial Current, and can be viewed as the western part of the great clockwise water circulation pattern of the N Atlantic. (See also OCEAN CURRENTS.)

Gullstrand, Allvar (1862-1930)
Swedish ophthalmologist awarded the 1911 Nobel Prize for Physiology or Medicine for his work on the REFRACTION of light in the EYE. VON HELMHOLTZ had shown that the lens's surface curvature altered for focusing: Gullstrand showed that the internal components of the lens also adjust, accounting for about a third of the accommodation.

Gum
Sticky substances containing complex polysaccharides (see CARBOHYDRATES) exuded from some trees, shrubs, seeds and seaweeds. Gum arabic is produced by the African tree *Acacia senegal*. AGAR is a dried mucilaginous gum extracted from seaweeds. The original chewing gum came from a South American tree, but has now been replaced by synthetic gum.

Guncotton
A NITROCELLULOSE with a high nitrate content; an EXPLOSIVE.

Gutta percha
A naturally occurring polymer of isoprene found in the latex secreted by some plants. It is not elastic like natural rubber (which is also a polymer of isoprene), because its double bonds are in a different geometric configuration, and it is of little use commercially.

Guyot
Submarine table-mountain, found especially in the Pacific. It is thought that guyots originate as volcanic islands associated with ocean ridges. Wave EROSION reduces the island, and SEA-FLOOR SPREADING causes the guyot to be further submerged. (See also OCEANS; VOLCANISM.)

Gymnosperms
One of the two classes of seed-bearing plants (SPOROPHYTES), the other being flowering plants or ANGIOSPERMS. Gymnosperms are characterized by having naked ovules (not contained within a ovary as in flowering plants) and, therefore, naked seeds (not enclosed by a FRUIT). These seeds are formed on open scales produed in CONES. There are separate male and female cones, and pollination is by wind only. All gymnosperms are perennial plants and most are evergreen. There are several orders, the main ones being the Cycadales, the cycads or sago palms; the Coniferales or CONIFERS, including pine, larch, fir and redwood; and the Ginkgoales, containing only one species, the maidenhair or GINKGO. (➧ page 139)

Gynecology
Branch of medicine and surgery specializing in diseases of women, specifically disorders of the reproductive tract; often linked with OBSTETRICS. CONTRACEPTION, ABORTION, STERILIZATION, infertility and abnormalities of MENSTRUATION are the commonest problems. The early recognition and treatment of CANCER of the cervix after smear tests have become important.

Gynecomastia
The visible appearance of breasts in a man. All men have some breast tissue, but it develops only with hormonal disorders, obesity, and as a result of certain drugs.

Gynoecium see FLOWER.

Gypsum
Mineral consisting of calcium sulfate dihydrate. It occurs worldwide as monoclinic crystals of various colors (SELENITE), or as fibrous or massive forms (ALABASTER). Gypsum is used in building and cement.

Gyroscope
A heavy spinning disk mounted so that its axis is free to adopt any orientation. Its special properties depend on the principle of the conservation of angular MOMENTUM. Although the scientific gyroscope was devised by FOUCAULT only in the mid-19th century, the child's traditional spinning top demonstrates the gyroscope principle. The fact that it will stay upright as long as it is spinning fast enough demonstrates the property of gyroscopic inertia: the direction of the spin axis resists change. This means that a gyroscope mounted universally, in double gimbals, will maintain the same orientation in space however its support is turned, a property applied in many navigational devices. If a FORCE tends to alter the direction of the spin axis (eg the weight of a top tilting sideways), a gyroscope will turn about an axis at right-angles to the force for as long as it applied; this movement is known as precession. Instrument gyroscopes usually consist of a wheel having most of its mass concentrated at its rim to ensure a large moment of inertia and which is kept spinning in frictionless bearings by an electric motor. Once the wheel is set spinning its response to applied TORQUES can be monitored or used in control servomechanisms.

Haber, Fritz (1868-1934)
German chemist awarded the 1918 Nobel Prize for Chemistry for synthesizing AMMONIA from the elements nitrogen and hydrogen. (See also HABER PROCESS.)
Haber process
Industrial synthesis of AMMONIA invented by HABER and developed by BOSCH. NITROGEN (from the atmosphere) is mixed with HYDROGEN (from natural gas or water gas) and heated to about 500°C under 200-1000atm pressure, with a catalyst of finely divided iron containing aluminum oxide and potassium oxide. The ammonia formed is frozen out, and the unreacted gases recycled. (See also NITROGEN FIXATION.)
Habitat
An area with certain physical characteristics that support a particular community of animals and plants. In general, a habitat can be defined in physical terms, eg rocky seashore or sandy desert, but as far as any one animal or plant species is concerned, the habitat cannot be defined without reference to the other animals and plants in the community. (See also ECOLOGY; ENVIRONMENT.)
Hadrons
A class of SUBATOMIC PARTICLES including the BARYONS and the mesons. These particles are all built from quarks, and are influenced by the strong nuclear force. PROTONS and NEUTRONS are relatively stable, and are the hadronic part of bulk matter. Other hadrons (eg pi-mesons) produced by collision processes in COSMIC RAYS or in ACCELERATORS are short-lived.
Haeckel, Ernst Heinrich (1834-1919)
German biologist best remembered for his vociferous support of DARWIN's theory of EVOLUTION, and for his own theory that ontogeny (the development of an individual organism) recapitulates phylogeny (its evolutionary stages), a theory now discarded.
Haemoglobin see HEMOGLOBIN.
Haemophilia see HEMOPHILIA.
Haemorrhage see HEMORRHAGE.
Hafnium (Hf)
Hard TRANSITION ELEMENT in Group IVB of the PERIODIC TABLE, found in ores of ZIRCONIUM, which it closely resembles. It is used in nuclear reactor control rods. AW 178.5, mp 2230°C, bp 5300°C, sg 13.31 (20°C). (➧ page 130)
Hagfish
One of the two types of jawless fish, or AGNATHA.
Hahn, Otto (1879-1968)
German chemist awarded the 1944 Nobel Prize for Chemistry for his work on nuclear fission. With LISE MEITNER he discovered the new element protactinium (1918); later they bombarded uranium with neutrons, treating the uranium with ordinary barium. Meitner showed that the residue was radioactive barium formed by the splitting (fission) of the uranium nucleus.
Hahnemann, Christian Friedrich Samuel (1755-1843)
German physician, the founder of HOMEOPATHY, which has as its basis the fact that "like cures like" – a disease can be cured by treatment with a remedy that in a healthy person would induce the same symptoms.
Hahnium (Ha)
A TRANSURANIUM ELEMENT in Group VB of the PERIODIC TABLE, atomic number 105. The priority of its discovery is disputed between the USSR and the US, as in the case of RUTHERFORDIUM (also known as Kurchatovium); American scientists claimed in 1970 to have synthesized hahnium-260. It is now given the name unnilpentium (see UNNILQUADIUM). (➧ page 130)
Hail
Precipitation of pellets of ice, often associated with thunderstorms. Hailstones have diameters of 2-250mm, 2-5mm being most common. To form they require a strong updraft raising them to colder regions. Often this happens several times, the hailstone collecting a new layer of ice each time it rises, until it is too heavy to support and falls to the ground. Larger hailstones may have alternate layers of clear and white ice because of different rates of freezing. (See also ICE; RAIN; SNOW.)
Hair
Nonliving filamentous structure made of KERATIN and pigment, formed in the skin hair FOLLICLES. Hair, or fur, is a characteristic of the living mammals and the extinct pterodactyls. Hair-like processes are also seen in some insects, such as large moths and some bees. (➧ page 213)
Haldane, John Burdon Sanderson (1892-1964)
British geneticist whose work, with that of Sir Ronald Aylmer Fisher (1890-1962) and SEWALL WRIGHT, provided a basis for the mathematical study of population GENETICS.
Haldane, John Scott (1860-1936)
British physiologist best known for his researches into industrial respiratory diseases (especially in miners) caused by poor ventilation. He also contributed to a technique for dealing with the "bends" (see AEROEMBOLISM).
Hale, George Ellery (1868-1938)
US astronomer who discovered the magnetic fields of SUNSPOTS, and who invented at the same time as Henri Alexandre Deslandres (1853-1948) the SPECTROHELIOGRAPH (*c*1892). His name is commemorated in the HALE OBSERVATORIES.
Hale Observatories
Formerly the Mt Wilson and Palomar Observatories, renamed (1970) for G.E. HALE and since 1948 operated jointly by the Carnegie Institution and the California Institute of Technology. At Mt Wilson (California) are two reflecting telescopes and two solar towers; at Palomar Mountain (California) a 200-in reflector, until 1973 the largest in the world, and two Schmidt telescopes.
Hales, Stephen (1677-1761)
English plant physiologist and chemist who, in accordance with the Newtonian quantitative paradigm, devised experiments to measure blood pressure in animals, and, realizing the importance of careful weighing and measuring in chemical experiments, applied these principles in his investigations of the life of plants. His *Vegetable Staticks* was published in 1727.
Half-life
The time taken for the activity of a radioactive sample to decrease to half its original value, half the nuclei originally present having changed spontaneously into a different nuclear type by emission of particles and energy. After two half-lives, the RADIOACTIVITY will be a quarter of its original value, and so on. Depending on the type of nucleus and method of decay, half-lives range from less than a second to over 10^{10} years. The half-life concept can also be applied to other systems undergoing random decay, eg certain biological populations. (➧ page 147)
Halftone
Reproduction of a photograph or other picture containing a range of continuous tones, by using dots of various sizes but uniform tone. The dots are small enough to blend in the observer's vision to give the effect of the original. The picture is photographed through a screen on which a fine rectangular grid has been scribed (2 to 6 lines/mm); the dots arise by DIFFRACTION. From the screened negative is made a halftone plate used for printing by all processes.
Halides
Binary compounds of the HALOGENS with oxidation number −1. Metal halides are mostly ionic salts (X^-) and usually very soluble. Nonmetal halides, and a few metal halides such as tin (IV) chloride, are volatile covalent compounds, highly reactive, often violently hydrolyzed by water, and used as halogenating agents. Halide ions form stable LIGAND complexes. Their reducing power increases down the group. Halide minerals include SALT, SYLVITE, FLUORITE, APATITE and CRYOLITE. (See also ALKYL HALIDES; HYDROGEN CHLORIDE; HYDROGEN FLUORIDE.)
Halite (rock salt) see SALT.
Halitosis (bad breath)
A condition associated with either local conditions (eg bad oral hygiene, throat infection) or more generalized ones (eg liver failure).
Hall, James (1811-1898)
US geologist and paleontologist, the father of American STRATIGRAPHY. His major work was on the paleontology of the SILURIAN and DEVONIAN of New York State, *The Paleontology of New York* (13 vols, 1847-94).
Hall effect
The POTENTIAL difference that develops across a METAL or SEMICONDUCTOR placed in a transverse magnetic field when an electric current flows in it. This voltage is at right angles to both the current and magnetic field directions, and arises from the deflection of moving charge carriers (ELECTRONS or holes) by the magnetic field.
Haller, Albrecht von (1708-1777)
Swiss biologist, best known for his work on human anatomy and physiology and also a poet. A pupil of BOERHAAVE and much influenced by him, he is credited with being the founder of experimental ANATOMY. In physiology he investigated RESPIRATION, the BLOOD CIRCULATION, the NERVOUS SYSTEM and the irritability and sensibility of different types of body tissue; in all cases relying on experiment.
Halley, Edmund (1656-1742)
English astronomer. In 1677 he made the first full observation of a transit of Mercury; and in 1676-79 prepared a major catalog of the S Hemisphere stars. He persuaded NEWTON to publish the *Principia*, which he financed. In 1720 he succeeded FLAMSTEED as Astronomer Royal. He is best known for his prediction that the comet of 1680 would return in 1758 (see HALLEY'S COMET), based on his conviction that COMETS follow elliptical paths about the Sun.
Halley's comet
The first periodic comet to be identified (by HALLEY, late 17th century) and the brightest of all recurring comets. It has a period of about 76 years. Records of every appearance of the comet since 240 BC, except that of 163 BC, are extant; and it is featured on the Bayeux tapestry. The comet's reappearance in 1986 provided the first opportunity to investigate a comet at close quarters. Five spacecraft successfully investigated the comet – the two Japanese probes, Sakigake and Suisei, from long range; the two Soviet probes, Vega 1 and Vega 2, from less than 10,000km; and the European probe, Giotto, which passed by the nucleus at a range of 605km.

Water ice was proved to be the major constituent of the nucleus, which measured about 15×8×8km. The nucleus was seen to be covered by an extremely dark crust, probably made of compacted dust and perhaps only a centimeter thick, through which jets of gas and dust erupted. Features as small as 100m across were resolved. The dust, gases, plasma and magnetic fields associated with the comet were investigated in detail.

Hallucination
An experience similar to normal PERCEPTION but with the difference that sensory stimulus is either absent or too minor to explain the experience satisfactorily. Certain abnormal mental conditions (see MENTAL ILLNESS) produce hallucinations, as does taking hallucinogenic drugs. They may also result from exhaustion or FEVER; or may be experienced while falling asleep (hypnogogic) or waking (hypnopompic), and also by individuals under HYPNOSIS. A negative hallucination is lack of perception despite adequate stimulus. Mass hallucination is hallucination shared by the members of a group; it may particularly result from mass hypnosis. (See also EIDETIC IMAGE; ILLUSION.)

Hallucinogenic drugs
DRUGS that cause hallucinations, usually visual, together with personality and behavior changes. The last may arise as a result of therapy, but more usually follow deliberate exposure to certain drugs for their psychological effects. Lysergic acid diethylamide (LSD), HEROIN, MORPHINE and other OPIUM NARCOTICS, MESCALINE and PSILOCYBIN are commonly hallucinogenic and cannabis sometimes so. The type of hallucination is not predictable and many are unpleasant. Recurrent hallucinations may follow use of these drugs; another danger is that altered behavior may inadvertently cause injury or death. Although psychosis may be a result of their use, it may be that recourse to drugs represents rather an early symptom of SCHIZOPHRENIA.

Hallux
Another name for the big toe. *Hallux valgus* is a BUNION.

Halo
A luminous ring or series of arcs sometimes seen around the Sun or Moon, the result of REFRACTION or REFLECTION or both of their light by crystals of ICE in high, thin clouds. Commonest is the 22° halo, of angular radius 22° and centered on the Sun or Moon. (See also CORONA.)

Halogens
Highly reactive nonmetals in Group VIIA of the PERIODIC TABLE, comprising FLUORINE, CHLORINE, BROMINE, IODINE and ASTATINE; the general symbol X is often used. The elements have molecular formula X_2. They show a regular gradation of physical and chemical properties: with increasing atomic number, they become less volatile, darker in color, less reactive (in particular, less strongly oxidizing), and less electronegative (see ELECTRONEGATIVITY). The typical compounds of the halogens are the HALIDES, with oxidation number −1; compounds with positive oxidation numbers (usually 1, 3, 5 and 7) are also formed, with increasing stability down the group. The halogens react vigorously with most other elements and always occur combined in nature. (See also INTERHALOGENS; PSEUDOHALOGENS; ALKYL HALIDES; ACID CHLORIDES.) (➧ page 130)

Halophytes
Plants that are tolerant to saline conditions, found in the seashore and salt marshes, such as glasswort and eel-grasses (*Zostera*).

Handedness
The side of the body, and in particular to the hand, that is most used in motor tasks. Most people are right-handed and few are truly ambidextrous (either-handed). In the brain the paths for sensory and motor information are crossed, so the right side of the body is controlled by the left cerebral hemisphere and vice versa. The left hemisphere is usually dominant and also contains centers for speech and calculation. The nondominant side deals with aspects of visual and spatial relationships; other functions are represented on both sides. In some left-handed people the right hemisphere is dominant. Suppression of left-handedness may lead to speech disorder.

Hansen's disease see LEPROSY.

Haploid
Term applied to any cell or organism that contains a single set of CHROMOSOMES. The number of chromosomes present is termed the haploid number. Cells with two sets of chromosomes are said to be diploid, those with three triploid, those with four, tetraploid, etc. (See also MEIOSIS; GAMETES.)

Harden, Sir Arthur (1865-1940)
British biochemist awarded with EULER-CHELPIN the 1929 Nobel Prize for Chemistry for their work on ENZYME action in the FERMENTATION of sugars. Harden demonstrated the existence of coenzymes, and Euler-Chelpin determined the structure of the first one to be found, cozymase.

Hardening of the arteries see ATHEROSCLEROSIS.

Hardness
The resistance of a substance to scratching, or to indentation under a blow or steady load. Resistance to scratching is measured on the Mohs' scale, named for Friedrich Mohs (1773-1839), who chose 10 MINERALS as reference points, from TALC (hardness 1) to DIAMOND (10). The modified Mohs' scale is now usually used, with 5 further mineral reference points. Resistance to indentation is measured by, among others, the Brinell, Rockwell and Vickers scales. (See also MATERIALS, STRENGTH OF.)

Hard water
Water containing CALCIUM and MAGNESIUM ions and hence forming scum with soap and depositing scale in boilers, pipes and kettles.

Hardwood
The wood of an ANGIOSPERM or BROADLEAVED tree, such as oak, maple, teak or mahogany. Such wood is in general harder than that of CONIFERS and other GYMNOSPERM trees, which are known as SOFTWOODS. However, the softest wood of all, balsa, is in fact a hardwood, and some softwoods, such as pitch pine, are extremely hard. The main difference between hardwoods and softwoods is in the water-conducting elements of the xylem. In hardwoods these are known as vessels, and are joined by perforated end-plates to give a continuous water column. In softwoods, where they are known as tracheids, they are much narrower, and are connected by lateral performations only, so the water column is interrupted.

Harelip
A congenital disease with a cleft defect in the upper lip due to impaired facial development in the embryo, and often associated with CLEFT PALATE. It may be corrected by plastic surgery.

Harmattan
Wind blowing W or SW from the S Sahara. In winter it transports vast quantities of dust over the Atlantic Ocean. In summer it interacts with NE-blowing MONSOON winds, often causing TORNADOES.

Harmonic motion see SIMPLE HARMONIC MOTION.

Harmonics
Vibrations at FREQUENCIES which are whole multiples of that of a fundamental vibration: the ascending notes C, , C, G, C′ , E′ , G′ comprise a fundamental with its first five higher harmonics. Apart from their musical consonance, they are important because any periodically repeated signal – a vowel sound, for example – can be produced by superposing the harmonics of the fundamental frequency, each with the appropriate intensity and time lag. (➧ page 225)

Harrison, John (1693-1776)
British inventor of the marine CHRONOMETER (1735). A prize of £20,000 had been offered for a device to allow the accurate determination of longitude at sea: this he won with his No. 4 Marine Chronometer of 1759.

Hartley, David (1705-1757)
English physician and a founder of the school of psychology known as ASSOCIATIONISM. In his *Observations on Man* (1749), he taught that sensations were communicated to the brain via vibrations in nerve particles and that the repetition of sensations gave rise to the association of ideas in the mind.

Hartline, Haldan Keffer (1903-1983)
US physiologist awarded with R. GRANIT and G. WALD the 1967 Nobel Prize for Physiology or Medicine for his work on the functioning of the nerve cells of the retina (see EYE; VISION).

Harvest moon
The full moon occurring nearest to the autumnal equinox (about 23 September) in the N Hemisphere. For several nights the full moon rises at closely similar times (about sunset), and may be bright enough for harvesting to continue into the night. In the S Hemisphere this occurs around the spring equinox (see EQUINOXES).

Harvey, William (1578-1657)
British physician who discovered the circulation of the blood. He showed that the HEART acts as a pump and that the blood circulates endlessly about the body; that there are valves in the heart and VEINS so that blood can flow in one direction only; and that the necessary pressure comes only from the lower left-hand side of the heart. His discoveries demolished the theories of GALEN that blood was consumed at the body's periphery and that the left and right sides of the heart were connected by pores. He also made important studies of the development of the EMBRYO.

Hashish see CANNABIS.

Hassel, Odd (1897-1981)
Norwegian chemist awarded with D. BARTON the 1969 Nobel Prize for Chemistry for their work on CONFORMATIONAL ANALYSIS.

Hauptman, Herbert A. (b. 1917)
US scientist, jointly awarded with JEROME KARLE the 1985 Nobel Prize for Chemistry for their development of direct methods for the determination of crystal structures.

Haversian canals see BONE.

Haworth, Sir Walter Norman (1883-1950)
British chemist awarded with KARRER the 1937 Nobel Prize for Chemistry for his work on the structures of CARBOHYDRATES and VITAMIN C.

Hay fever
Common allergic disease causing RHINITIS and CONJUNCTIVITIS on exposure to allergen. The prototype is ALLERGY to grasses, but pollens of many trees, weeds and grasses (eg ragweed, timothy grass) may provoke seasonal hay fever in sensitized individuals. Susceptibility is often associated with ASTHMA, ECZEMA and aspirin sensitivity in the individual or his family.

Haynes, Elwood (1857-1925)
US inventor who built one of the first US AUTOMOBILES (1894) and discovered several ALLOYS, including STAINLESS STEEL (patented 1919).

Headache
The common symptom of an ache or pain affecting the head or neck, with many possible causes including FEVER, emotional tension (with spasm of neck muscles) or nasal SINUS infection.

LIGHT

Primary and secondary colors

◀ *Mixing colored lights is an additive process. Red, green and blue together stimulate all three types of color responsive cell in the eye, and the result is seen as white (far left). These primary colors of light can be mixed in pairs to give secondary colors. With a paint, the process of producing a color is subtractive – the pigment absorbs light at certain wavelengths (colors); the observed color results from removing the wavelengths from white light.*

Refraction

▼ *Light rays "bend" – change direction – when they pass from one material to another. This is refraction; it is related to the difference in the velocity of light in differing materials.*

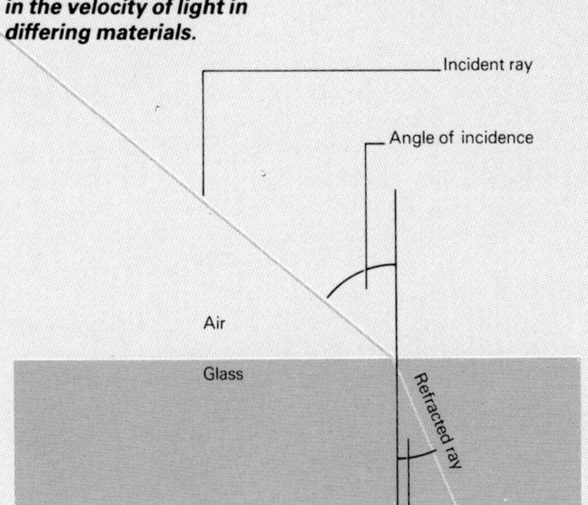

▼▶ *Refraction is used in lenses, which can converge light rays parallel to the axis of the lens together at a single focal point, f. A convex lens gives an enlarged, inverted image of an object at a point at double the focal length. This is a "real" image: it can be seen on a screen but not by eye.*

Parallel light
Convex lens
F
Focal length

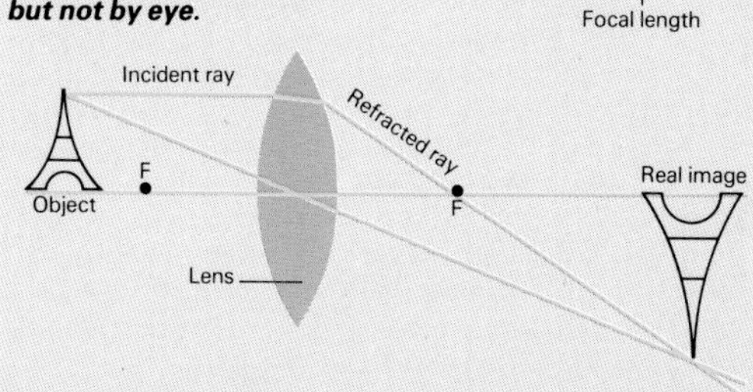

▼▶ *Diverging lenses can make parallel light rays diverge as if from a single point. A concave lens produces an upright, diminished "virtual" image, which cannot be formed on a screen but which can be seen through the lens, between the object and the lens.*

Concave lens
F
Focal length

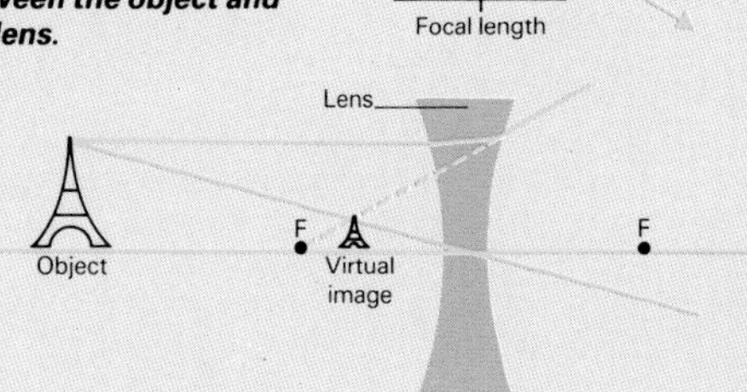

Spherical aberration

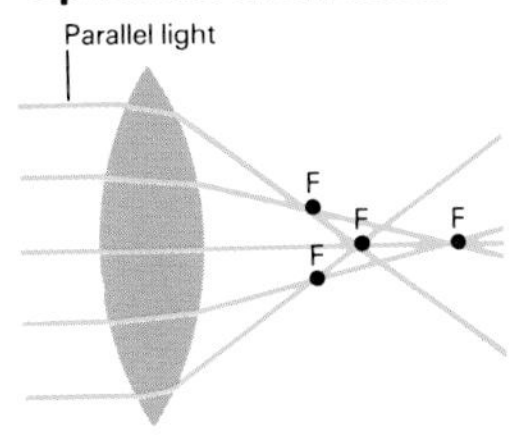

▲ *A wide beam of light is not all brought to a focus at the same point, because the angle of incidence varies towards the edge of the lens. This blurs the image – an effect known as spherical aberration.*

Chromatic aberration

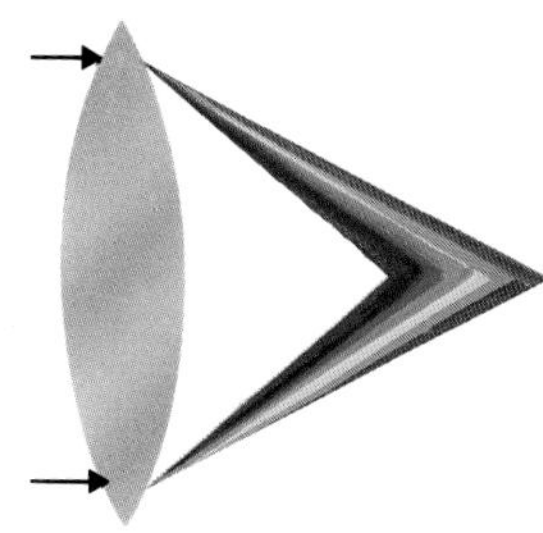

▲ *Because the refractive index of a material varies with wavelength (color), a simple lens does not have a unique focus. Thus the lens forms a series of colored images of slightly different size, and the observed image appears to have a colored fringe. This effect – chromatic aberration – often occurs in inexpensive telescopes and binoculars.*

Reflection

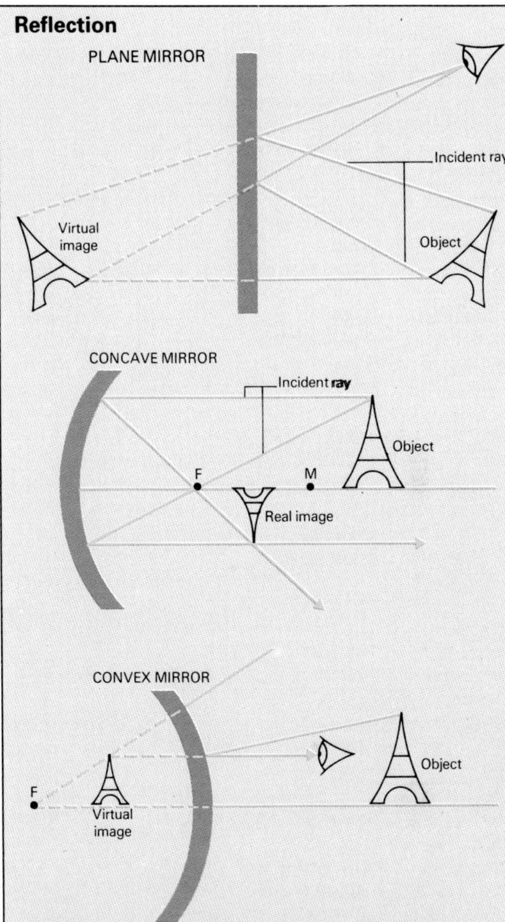

▲ *Light rays are reflected in such a way that the angle of reflection equals the angle of incidence. A plane mirror produces a virtual image by reflecting light so that it appears to the eye to come from behind the mirror. A convex mirror produces a virtual image, but diminished in size. The size of image produced by a concave mirror depends on the position of the object and the focal length of the mirror – the point to which it converges light parallel to the axis. Here an inverted, reduced real image is formed.*

Polarization

Unpolarized light
Polarizing filter
Polarized light

◀ *Ordinary, unpolarized light consists of vibrations in all directions at right angles (transverse) to the direction of travel, but these can be resolved into two directions at right angles. A polarizing filter transmits light only in one of these directions ("plane" polarization). This effect is used in polarizing sunglasses, which transmit only the light plane polarized in one direction.*

Migraine, due to abnormal reactivity of blood vessels, is typified by zig-zag or flashing visual sensations or tingling in part of the body, followed by an often one-sided severe throbbing headache. This may be accompanied by nausea, vomiting and sensitivity to light. There is often a family history. Meningeal inflammation, as in MENINGITIS and subarachnoid HEMORRHAGE, may also cause severe headache. The headache of raised intracranial pressure is often worse on waking and on coughing, and may be a symptom of brain TUMOR, ABSCESS or HYDROCEPHALUS.

Hearing see EAR.

Heart

A muscular organ found in all vertebrate and many invertebrate animals, concerned with pumping the BLOOD, and thus maintaining the blood circulation. The vertebrate heart shows a progressive development, from the simple two-chambered heart of fish to the four-chambered heart of mammals and birds. It is controlled by the autonomic nervous system. Invertebrate hearts may consist of little more than thickened sections of blood vessel that contract regularly; there may be many of these simple hearts. (See also CIRCULATORY SYSTEM.)

In humans, disorders of the heart include death or injury to areas of heart muscle following coronary thrombosis. This may lead to defects in pumping and heart failure, rhythm disorder, ANEURYSM or, rarely, cardiac rupture. Rhythm disturbance may also result from abnormal conducting or pacemaker tissue. Heart failure, in which inadequate pumping leads to imbalance between the two parts of the circulation or the failure of both, may be due to coronary thrombosis, cardiac muscle disease or fluid overload. (See also MUSCLE.) (◀ page 28, 111)

Heart attack see CORONARY THROMBOSIS.

Heartburn

Burning sensation of "indigestion" localized centrally in the upper abdomen or lower chest. It is frequently worse after large meals or on lying flat, especially with hiatus HERNIA. Acid STOMACH contents irritate the esophageal lining and may lead to ULCER.

Heart-lung machine see ARTIFICIAL ORGANS.

Heart murmur

Abnormal sound heard on listening to the chest over the HEART with a STETHOSCOPE. Murmurs arise in disease of the heart valves, with narrowing stenosis or leakage (incompetence). Holes between chambers, valve roughening and high flow rates also cause murmurs.

Heat

The form of ENERGY that passes from one body to another owing to a TEMPERATURE difference between them; one of the basic functions in THERMODYNAMICS. The energy residing in a hot body is also loosely called heat, but is better termed internal energy, since it takes several different forms. Despite an earlier view by some philosophers that heat was a form of agitation, in the 18th century the CALORIC THEORY OF HEAT predominated, until disproved by the experiments of Sir Humphry Davy and COUNT RUMFORD (1798) showing that mechanical WORK could be converted to heat. JAMES JOULE confirmed this by many ingenious experiments and found a consistent value for the mechanical equivalent of heat (the ratio of work done to heat produced). In the mid-18th century JOSEPH BLACK first clearly distinguished heat from temperature, a conceptual advance which allowed heat to be measured (see CALORIMETRY) in terms of the temperature rise of a known mass of water, the unit being the CALORIE (or the BRITISH THERMAL UNIT). In SI UNITS heat is measured, as a form of energy, in JOULES. A given mass m of any substance shows a characteristic temperature rise θ when an amount of heat Q is supplied: $Q=ms\theta$ where s is the SPECIFIC HEAT of the substance. If the substance changes its state, however, by melting, freezing, boiling or condensing, LATENT HEAT is absorbed or produced without any temperature change, the internal energy being changed by altering the molecular interrelations, not merely their degree of motion. Heat is commonly produced as required for space heating, or to power ENGINES, by conversion of chemical energy (burning fuel – see COMBUSTION), electrical energy or nuclear energy. There are three processes by which heat flows from a hotter to a cooler body: CONDUCTION and CONVECTION, in which molecular motion is transferred, and radiation, in which INFRARED RADIATION is emitted and propagates through space. (See also BLACKBODY RADIATION.) Heat transfer may be hindered by means of thermal INSULATION. Newton's law of cooling states that the rate of loss of heat by a body in a draft (forced convection) is proportional to its temperature difference from its surroundings. (See also BOLOMETER; PYROMETER; REFRIGERATION; THERMOCHEMISTRY.)

Heat death

Speculative theory of the final state of the UNIVERSE. If the Universe is an isolated system (see THERMODYNAMICS) then its ENTROPY must tend to a maximum, at which all energy is degraded to uniform heat, everything is wholly disordered and no change is possible. In the oscillating Universe model (see COSMOLOGY) it is supposed that the entropy inequality is reversed at the Universe's greatest extent, so that during contraction entropy tends to a minimum.

Heat pump

A device for transferring heat from a cold region to a hotter region by doing work (as required by the second law of THERMODYNAMICS). The working fluid or refrigerant is a condensible gas such as ammonia or freon. A motor-driven compressor compresses the gas adiabatically (raising its temperature) and delivers it to a condenser coil or "radiator" in the space to be heated. As it loses heat it liquefies and passes through an expansion valve into the evaporator, a low-pressure region where the liquid evaporates, taking heat from its relatively cool environment. The gas then returns to the compressor to complete the cycle. Heat pumps are used in domestic refrigerators (see REFRIGERATION) and also in air conditioning systems for space heating in winter and (by reversing the pump) cooling in summer.

Heat rash see PRICKLY HEAT.

Heat shield

Device to prevent overheating of space capsules on re-entry to the Earth's atmosphere. The craft is coated with a layer of ablative material (see ABLATION), often a plastic impregnated with quartz fibers. Friction with the air heats and vaporizes the outer regions. Some 80% of the heat is reradiated at the gas/liquid boundary.

Heatstroke

A condition consisting of confusion and lethargy from exposure to heat. Although there is a surprising absence of thirst, the condition is resolved after drinking copiously.

Heaviside, Oliver (1850-1925)

British physicist and electrical engineer best known for his work in telegraphy, in the course of which he developed operational calculus, a new mathematical system for dealing with changing wave-shapes. In 1902, shortly after KENNELLY, he proposed that a layer of the atmosphere was responsible for reflecting RADIO waves back to Earth. This, the E layer of the IONOSPHERE, was found by APPLETON and others (1924), and is often called the Kennelly-Heaviside Layer, or Heaviside Layer.

Heavy hydrogen see DEUTERIUM; TRITIUM.

Heavy water (deuterium oxide, D_2O)

Occurs as 0.014% of ordinary WATER, which it closely resembles. It is used as a moderator in nuclear reactors and as a source of deuterium and its compounds. It is toxic in high concentrations. Water containing TRITIUM or heavy isotopes of oxygen (O^{17} and O^{18}) is also called heavy water. mp 3.8°C. bp 101.4°C.

Hectare (Ha)

Widely used metric unit of area, equal to 10,000m^2 or 100 are. One hectare equals 2.471 acres.

Heimlich's maneuver

A remedy for choking in which another person gives the patient a vigorous bear-hug from behind, pressing upward on the upper abdomen into the rib cage.

Heisenberg, Werner Karl (1901-1976)

German mathematical physicist generally regarded as the father of QUANTUM MECHANICS, born out of his rejection of any kind of model of the atom and use of mathematical matrices to elucidate its properties. His famous UNCERTAINTY PRINCIPLE (1927) overturned traditional physics, its implications affecting areas of science far beyond the bounds of atomic physics. (See also SCHRÖDINGER.)

Heliostat

An instrument used to observe the Sun. A TELESCOPE is used in conjunction with a large flat mirror which rotates so that the image of the Sun appears stationary.

Heliozoans

A group of PROTOZOA that are related to, and share many features with, the RADIOLARIANS. Heliozoans, however, are found in freshwater rather than marine habitats.

Helium (He)

One of the NOBLE GASES, lighter than all other elements except hydrogen. It is a major constituent of the SUN and other STARS. The main source of helium is natural gas in Texas, Oklahoma and Kansas. ALPHA PARTICLES are helium nuclei. Helium is lighter than air and nonflammable, so is used in balloons and airships. It is also used in breathing mixtures for deep-sea divers, as a pressurizer for the fuel tanks of liquid-fueled rockets, in helium-neon LASERS, and to form an inert atmosphere for welding. Liquid helium He^4 has two forms. Helium I, stable from 2.19K to 4.22K, is a normal liquid, used as a refrigerant (see CRYOGENICS; SUPERCONDUCTIVITY). Below 2.18K it becomes helium II, which is a superfluid with no VISCOSITY, the ability to flow as a film over the side of a vessel in which it is placed, and other strange properties explained by QUANTUM THEORY. He^3 does not form a superfluid. Solid helium can be produced only at pressures above 25atm. AW 4.0, mp 3.5K, bp 4.22K. (▶ page 130, 147)

Helmholtz, Hermann Ludwig Ferdinand von (1821-1894)

German physiologist and physicist. During his physiological studies he formulated the law of conservation of ENERGY (1847), one of the first to do so. He was the first to measure the speed of nerve impulses (see NERVOUS SYSTEM), and invented the OPHTHALMOSCOPE (both 1850). He also made important contributions to the study of ELECTRICITY and geometry.

Helmont, Jan Baptista van (1577-1644)

Flemish chemist and physician, regarded as the father of biochemistry. He was the first to discover that there were airlike substances distinct from air, and first used the name "gas" for them.

Hematemesis

Vomiting of blood: a sign of bleeding from the esophagus or stomach.

Hematite
Hard, red OXIDE mineral, consisting of iron (III) oxide (α-Fe_2O_3), the chief IRON ore; also used in paints (ocher) and polishes (rouge). Hematite occurs worldwide, mainly in sedimentary rocks, though it is also formed by weathering of other iron minerals. In the US large deposits are found around the Great Lakes. It crystallizes in the rhombohedral system, with corundum structure.
Hematology
Branch of MEDICINE concerned with BLOOD DISEASES (eg ANEMIA, LEUKEMIA, CLOTTING disorder).
Hematoma see BRUISE.
Hematuria
The presence of blood in the urine. It is a symptom of kidney or bladder disease.
Hemichordates
A minor phylum of the CHORDATES. They include two types of animals: worm-like burrowing forms, the acorn worms, and sedentary forms, the pterobranchs. All are marine. The burrowing forms have soft bodies divided into proboscis, collar and trunk. The proboscis and collar are distensible. The sedentary forms are tube-dwellers, and often colonial.
Hemimorphite
Colorless or white mineral, formerly known as CALAMINE; a hydrated zinc silicate, formula $Zn_4Si_2O_7[OH]_2.H_2O$, of widespread occurrence, an ore of ZINC formed by alteration of other zinc minerals. It crystallizes in the orthorhombic system and exhibits PIEZOELECTRICITY.
Hemiplegia
Paralysis of one side of the body, also known as hemiparesis.
Hemiptera
An order of insects, also known as bugs or "true bugs", that are characterized by their beak-like mouthparts. These are adapted for piercing and sucking up nutritious juices, whether from a plant, as in the shield-bugs and aphids, or from an animal, as in the bed-bugs and assassin bugs.
Hemocyanin
A blue, copper-containing PROTEIN found in the BLOOD of mollusks and crustaceans. Its function is to carry oxygen to the tissues.
Hemoglobin
An oxygen-carrying protein found in the BLOOD of all vertebrates, and in a few invertebrates. It contains heme, an iron-containing molecule, which has a high affinity for oxygen. In the LUNG capillaries hemoglobin is exposed to a high oxygen concentration and oxygen is taken up. The blood then passes via the HEART into the systemic circulation. In the tissues the oxygen concentration is low, so oxygen is released from the hemoglobin before it returns to the lungs. Carbon monoxide has an even higher affinity for hemoglobin than oxygen, and thus causes death by displacing oxygen from hemoglobin. Shortage of hemoglobin produces ANEMIA. (♦ page 28)
Hemolysis
Breakdown of red blood cells. It can be caused by ANTIBODIES or poisons and is a symptom of many diseases, including jaundice or defective HEMOGLOBIN.
Hemophilia
Inherited disorder of CLOTTING in males, carried by females, who do not suffer from the disease. It consists of inability to form adequate amounts of a clotting factor (VIII) essential for the correct formation of clots. Prolonged bleeding from wounds or tooth extractions, HEMORRHAGE into JOINTS and MUSCLES with severe pain, are important symptoms. Bleeding can be stopped by giving PLASMA concentrates rich in factor VIII and, if necessary, BLOOD TRANSFUSION. Similar diseases of both sexes are Christmas disease (due to lack of factor IX) and von Willebrand's disease (factor VIII deficiency with additional CAPILLARY defect).
Hemoptysis
The presence of blood in the sputum; the act of coughing up blood. It is a sign of TUBERCULOSIS and lung cancer.
Hemorrhage
Acute loss of BLOOD from any site. Trauma to major ARTERIES, VEINS or the HEART may lead to massive hemorrhage. GASTROINTESTINAL TRACT hemorrhage is usually accompanied by loss of altered blood in vomit or feces; ULCERS and CANCER of the bowel are important causes. Antepartum hemorrhage is blood loss from the uterus in late PREGNANCY and may rapidly threaten life of both mother and baby; postpartum hemorrhage is excessive blood loss after birth resulting from inadequate womb contraction or a retained PLACENTA. STROKE due to brain hemorrhage may damage vital structures and cause COMA. FRACTURES may cause sizeable hemorrhage into soft tissues.
Hemorrhoids (piles)
Enlarged VEINS at the junction of the rectum and anus, which may bleed or come down through the anal canal, usually on defecation, and which are made worse by CONSTIPATION and straining. Bleeding from the rectum may be a sign of bowel CANCER and this may need to be ruled out before bleeding is attributed to hemorrhoids.
Hench, Philip Showalter (1896-1965)
US physician who shared with KENDALL and REICHSTEIN the 1950 Nobel Prize for Physiology or Medicine for his use of cortisone (see STEROIDS) to treat rheumatoid ARTHRITIS.
Henry (H)
The SI unit of inductance, the inductance of a circuit in which current changing at a rate of one ampere per second induces an ELECTROMOTIVE FORCE of one volt.
Henry, Joseph (1797-1878)
US physicist best known for his electromagnetic studies. His discoveries include INDUCTION and self-induction; though in both cases FARADAY published first. He also devised a much improved ELECTROMAGNET by insulating the wire rather than the core; invented one of the first ELECTRIC MOTORS; helped MORSE and WHEATSTONE devise their telegraphs; and found SUNSPOTS to be cooler than the surrounding photosphere. The HENRY is named for him.
Henry, William (1774-1836)
British chemist and physician who formulated Henry's Law, that, at a given temperature, the mass of a gas dissolved by a particular solvent is proportional to the pressure on it of the gas.
Heparin see ANTICOAGULANTS.
Hepaticae see LIVERWORTS.
Hepatitis
Inflammation of the LIVER, usually due to virus infection, causing nausea, loss of appetite, fever, malaise, JAUNDICE and abdominal pain; liver failure may result. It can occur as part of a systemic disease (eg YELLOW FEVER, MONONUCLEOSIS). In two forms infection is restricted to the liver: hepatitis A (infectious hepatitis) is an EPIDEMIC form, transmitted by feces and is of short INCUBATION; it is rarely serious or prolonged. Hepatitis B (serum hepatitis) is transmitted by bodily fluids, mainly BLOOD. It develops more slowly but may be more severe, causing death. It is common among drug addicts; carriers may be detected by blood tests, and immunization of those at risk may be helpful. Amebiasis and some DRUGS can also cause hepatitis; it is also transmitted sexually. (♦ page 233)
Hepatoma
A tumor, benign or malignant, of the liver.
Herb
In botany, any plant with soft aerial stems and leaves that die back at the end of the growing season to leave no persistent parts above ground. In everyday terms, herbs are plants used medicinally and to flavor food.
Herbarium
A collection of dried and preserved plant specimens systematically arranged and classified. Herbaria are valuable means of plant classification as they are built up over many years with plants from different sources, and therefore show the limits of variation within species.
Herbicides (weedkillers)
Chemical compounds used to kill weeds (plants growing where they are not wanted). They are mostly complex organic compounds that are also toxic to humans and to wildlife. Contact herbicides, including paraquat, kill only the parts of the plants on which they are sprayed. The use of weedkillers should be combined with good agricultural management, mechanical destruction of weeds and biological control.
Herbivore
An animal that eats plants or plant products. Also used more broadly, in ecology, to include any animal that feeds on primary producers (see FOOD CHAIN), whether plants, unicellular algae, cyanobacteria or other photosynthetic bacteria. Occasionally used in a much narrower sense to mean animals that feed on grass and other herbaceous plants, as distinct from those that feed on the leaves of trees (foliovores), fruit (frugivores) or seeds (graminovores).
Hercules
A large N Hemisphere constellation containing a superb GLOBULAR CLUSTER, M13, of about 500,000 stars, and which is just visible to the naked eye. (♦ page 240)
Heredity
The process whereby progeny inherit their physical form from their parents. (See GENETICS.) (♦ page 32)
Hermaphrodite
Any organism in which the functions of both sexes are combined. Such an individual may function in only one sexual role at a time, or each of a pair of partners may fertilize the other during reciprocal copulation. Hermaphrodism is very common among plants (which are usually referred to as being bisexual), occurs fairly widely in invertebrates (eg snails, earthworms), and is rare among vertebrates, occurring only in a few fish.
Hernia
Protrusion of abdominal contents through the abdominal wall in the inguinal or femoral part of the groin, or through the DIAPHRAGM (hiatus hernia). Hernia may occur through a congenital defect or through an area of muscle weakness. Bowel and omentum are commonly found in hernial sacs, and if there is a tight constriction at the neck of the sac (a "strangulated" hernia), the bowel may be obstructed or suffer GANGRENE. In hiatus hernia, part of the stomach lies in the chest.
Hero of Alexandria (Heron, *c*AD20-*c*62)
Greek scientist best known for inventing the aeolipile, a steam-powered engine that used the principle of jet propulsion. Other works deal with MENSURATION, optics and MECHANICS.
Heroin
OPIUM alkaloid with narcotic ANALGESIC and euphoriant properties, a valuable DRUG in severe pain of short duration (eg CORONARY THROMBOSIS) and in terminal malignant disease. It is sometimes abused (see DRUG ADDICTION), taken intravenously for its psychological effects and later because of physical addiction; it can also be smoked. SEPTICEMIA and hepatitis may follow unsterile injections, and early death is common.

MAGNETISM AND ELECTRICITY

Magnetic fields

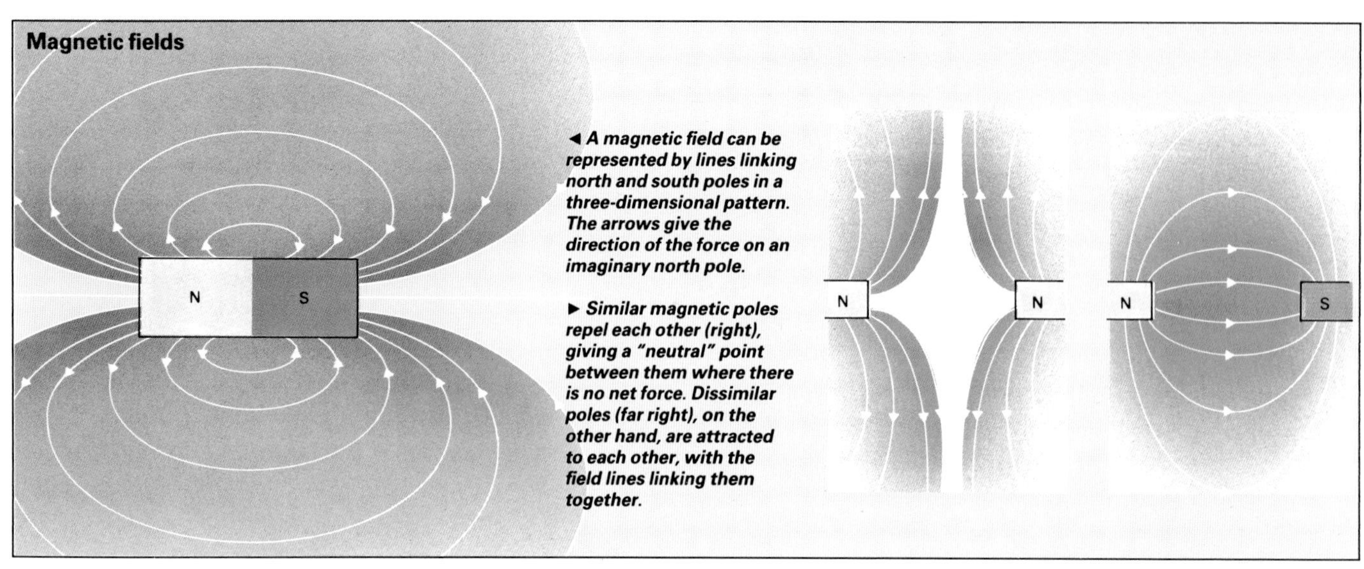

◄ *A magnetic field can be represented by lines linking north and south poles in a three-dimensional pattern. The arrows give the direction of the force on an imaginary north pole.*

► *Similar magnetic poles repel each other (right), giving a "neutral" point between them where there is no net force. Dissimilar poles (far right), on the other hand, are attracted to each other, with the field lines linking them together.*

Electric fields

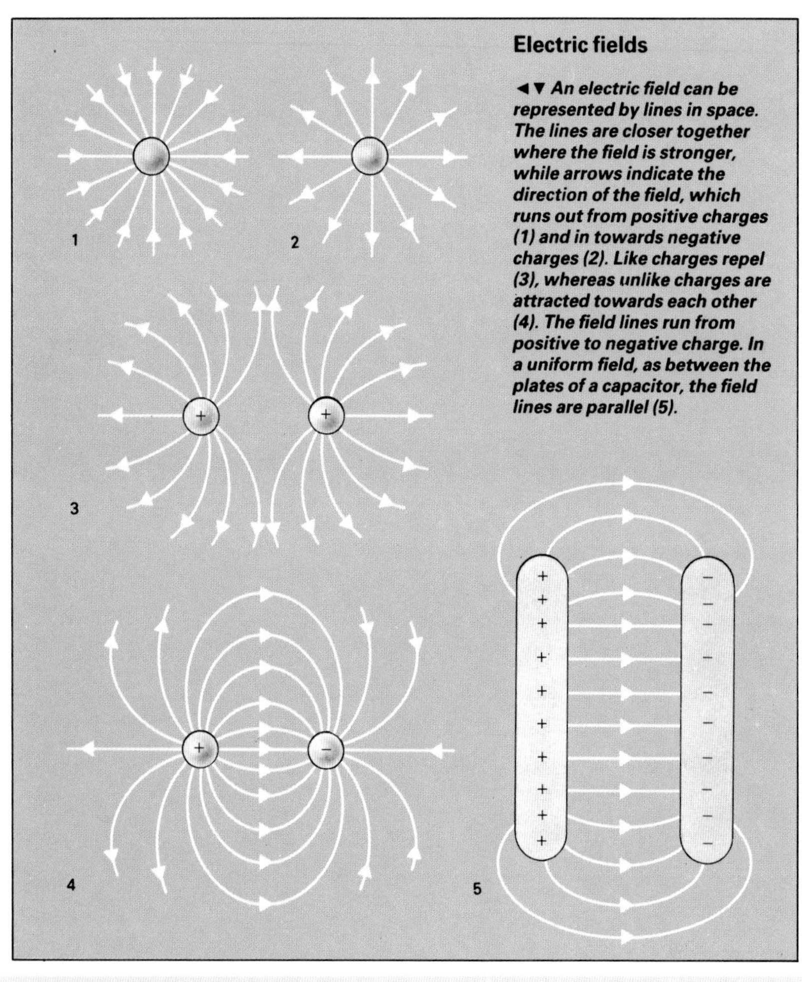

◄▼ *An electric field can be represented by lines in space. The lines are closer together where the field is stronger, while arrows indicate the direction of the field, which runs out from positive charges (1) and in towards negative charges (2). Like charges repel (3), whereas unlike charges are attracted towards each other (4). The field lines run from positive to negative charge. In a uniform field, as between the plates of a capacitor, the field lines are parallel (5).*

Current flow

N
S
Magnetic field

◄▼ *When a wire is moved in a magnetic field, a current flows if the wire is part of a circuit, because the electrons in the wire experience a force. This is the principle that underlies the operation of the generator: the wire is turned mechanically in a magnetic field, and a current is therefore generated within it.*

AC generator

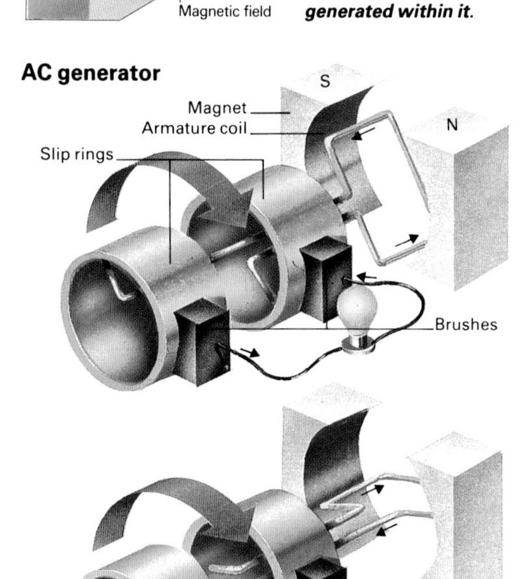

The electromagnetic spectrum

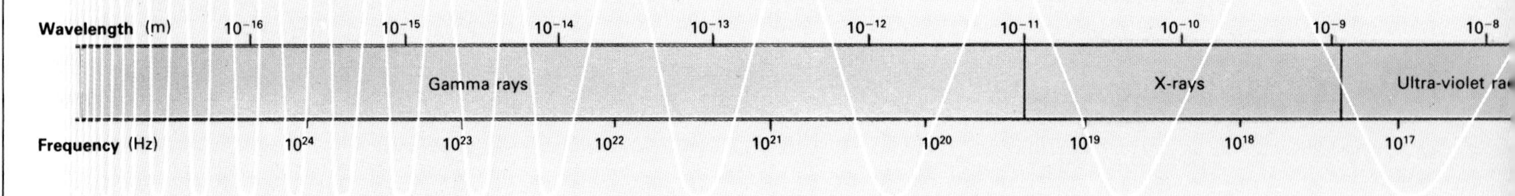

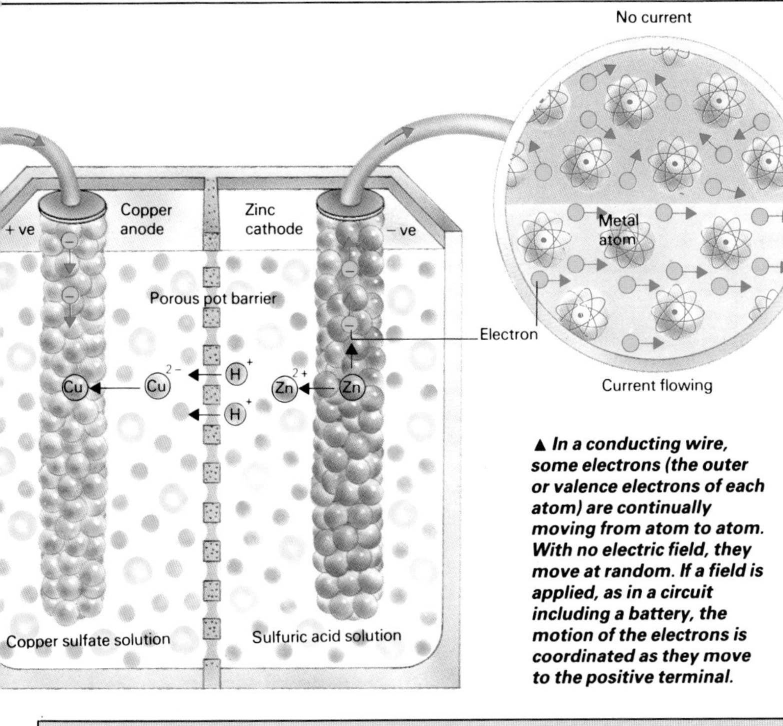

▲ *In a conducting wire, some electrons (the outer or valence electrons of each atom) are continually moving from atom to atom. With no electric field, they move at random. If a field is applied, as in a circuit including a battery, the motion of the electrons is coordinated as they move to the positive terminal.*

How a battery works

The basic operation of a battery or electrical cell is to store chemical energy and to convert it into electrical energy as required. It utilizes the natural random movement in a liquid of free ions, atoms with either too few or too many electrons.

When the terminals of a battery are connected via a circuit, electrons flow from the cathode to the anode, and an electric current flows in the opposite direction, from positive terminal to negative.

In the battery shown here, the electrolyte is sulfuric acid, the molecules of which divide into hydrogen and sulfate ions when dissolved. If the anode and cathode are connected via a circuit, a current flows as the atoms in the zinc cathode tend to give up electrons, thus becoming positive ions and joining with the sulfate ions. The hydrogen ions are attracted to the anode, where they attract electrons from the copper to form hydrogen molecules. The result is a net flow of electrons from cathode to anode. The flow continues until the cathode has been eaten away completely, or a film of non-conducting hydrogen bubbles on the anode causes internal resistance to build up.

This type of cell is known as a primary cell: the chemical changes cannot be reversed. In a secondary cell, the changes are reversible.

Electromagnetism

▼▶ *Electromagnetic waves consist of mutually perpendicular electric and magnetic fields that oscillate at the same frequency and propagate in the same direction at a velocity of about 300,000 km/s – the speed of light – in a vacuum. They are transverse, unlike the longitudinal sound waves: the electric and magnetic fields oscillate at right angles to the direction of propagation. The waves arise as a result of the relationship between electric and magnetic fields, as changes in one give rise to changes in the other. Electromagnetic waves have a huge range of possible wavelengths and frequencies, giving rise to an electromagnetic spectrum that covers more than 19 decades in wavelength and frequency. The spectrum stretches from gamma rays at the shortest wavelengths through X-rays and visible light (which covers only a small part of the spectrum) to microwaves and radiowaves, which may have a wavelength of a kilometer or more. Variations in the amplitude or frequency can carry radio or video signals.*

Electromagnetic waves

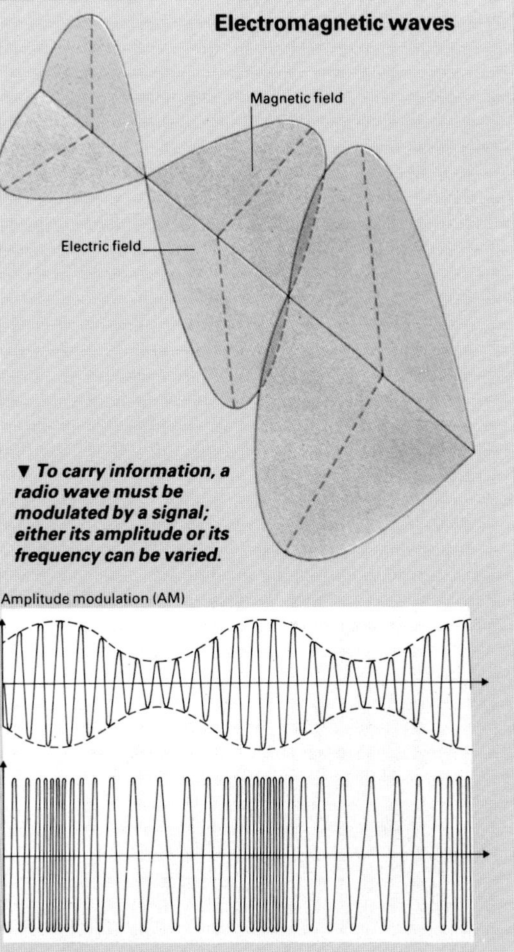

▼ *To carry information, a radio wave must be modulated by a signal; either its amplitude or its frequency can be varied.*

▼ *A resistor occurs when the flow of electrons is impeded by a constriction in the conducting wire, or by the presence of impurity atoms which the electrons strike and so lose energy.*

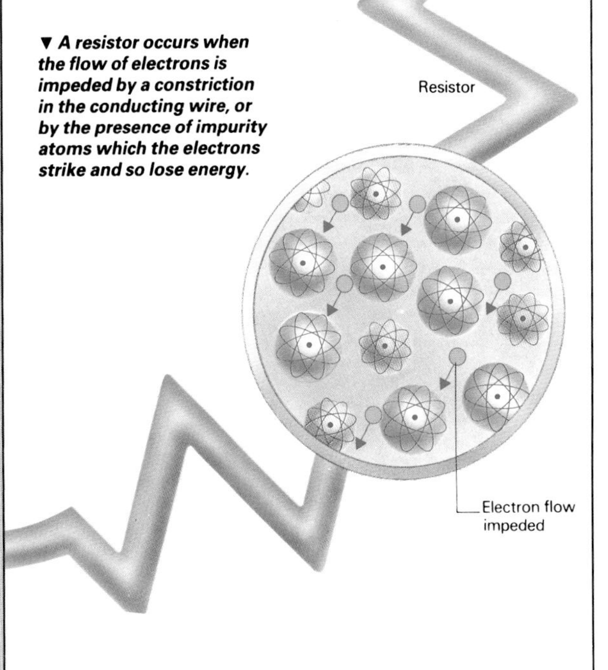

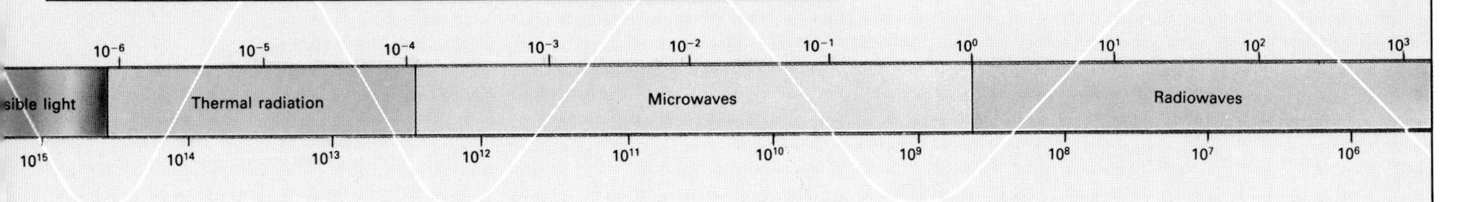

Herophilus (*c*320 BC)
Alexandrian physician regarded as the father of scientific ANATOMY, and one of the first dissectors. He distinguished nerves from tendons and partially recognized their role. His work survives only through GALEN's writings.
Herpes see COLD SORE; SHINGLES.
Herpetology
The study of reptiles and amphibians.
Herschbach, Dudley (b. 1932)
US scientist who received, together with YUAN T. LEE and JOHN C. POLANYI, the 1986 Nobel Prize for Chemistry for their work on the dynamics of chemical elementary processes.
Herschel
Family of British astronomers of German origin. Sir Frederick William Herschel (1738-1822) pioneered the building and use of reflecting TELESCOPES, discovered URANUS (1781), showed the Sun's motion in space (1783), found that some DOUBLE STARS were in relative orbital motion (1793), and studied NEBULAE. His sister Caroline Lucretia (1756-1848) assisted him and herself discovered eight COMETS. His son Sir John Frederick William Herschel (1792-1871) was the first to use SODIUM thiosulfate (hypo) as a photographic fixer, and made many contributions to ASTRONOMY, especially that of the S Hemisphere.
Hershey, Alfred Day (b. 1808)
US biologist who shared with DELBRÜCK and LURIA the 1969 Nobel Prize for Physiology or Medicine for their various researches on BACTERIOPHAGES.
Hertz (Hz)
SI unit of FREQUENCY, equal to 1 cycle per second; much used in RADIO technology. (➧ page 225)
Hertz, Gustav Ludwig (1887-1975)
German physicist who shared with J. FRANCK the 1925 Nobel Prize for Physics for their experiments showing the internal structure of the atom to be quantized, and so the value of the QUANTUM THEORY.
Hertz, Heinrich Rudolph (1857-1894)
German physicist who first broadcast and received RADIO waves (*c*1886). He showed also that they could be reflected and refracted (see REFLECTION; REFRACTION) much as light, and that they traveled at the same velocity though their wavelength was much longer (see ELECTROMAGNETIC RADIATION). In doing so he showed that light (and radiant heat) are, like radio waves, of electromagnetic nature.
Hertzsprung, Ejnar (1873-1967)
Danish astronomer who showed that there was a relation between a STAR's brightness and color: the resulting Hertzsprung-Russell Diagram (named also for HENRY RUSSELL) is important throughout astronomy and cosmology. He also conceived and defined absolute MAGNITUDE; and his work on CEPHEID VARIABLES provided a way to measure intergalactic distances. (➧ page 228)
Herzberg, Gerhard (b. 1904)
German-born Canadian spectroscopist, awarded the 1971 Nobel Prize for Chemistry for work on the electronic structure and geometry of molecules. In particular he pioneered the study of the spectra of FREE RADICALS.
Hess, Victor Franz (1883-1964)
Austrian-born US physicist who shared with C.D. ANDERSON the 1936 Nobel Prize for Physics for his discovery of COSMIC RAYS.
Hess, Walter Rudolf (1881-1973)
Swiss physiologist awarded with MONIZ the 1949 Nobel Prize for Physiology or Medicine for his determination of the control exerted by parts of the BRAIN over the functioning of internal organs.
Heterocyclic compounds
Major class of organic compounds in which the atoms are linked to form one or more rings, at least one of which includes one or more atoms other than carbon – commonly nitrogen, oxygen or sulfur. Many such compounds are of great biochemical or industrial importance. As with ALICYCLIC COMPOUNDS, saturated heterocyclic compounds resemble their ALIPHATIC analogs except for the effect of strain in small rings. There is also a large class of heterocyclic AROMATIC COMPOUNDS, highly distinctive in properties.
Heterotroph
An organism that is unable to elaborate complex carbon compounds for itself, in the way that AUTOTROPHS can, using only carbon dioxide and an energy source. Heterotrophs are dependent on other organisms to supply them with organic compounds, and all ultimately rely on autotrophs for their food, whether they consume them direct, live in symbiotic relationships with them (as in LICHENS), or feed on other heterotrophs. All animals and fungi are heterotrophs, along with the majority of protists and a great many bacteria.
Hevesey, George Charles de (or Hevesy, Georg von, 1885-1966)
Hungarian-born chemist awarded the 1943 Nobel Prize for Chemistry for his work on radioactive tracers (see RADIOISOTOPES). He was also the codiscoverer of the element HAFNIUM.
Hewish, Antony (b. 1924)
Radio astronomer, corecipient with RYLE of the 1974 Nobel Prize for Physics, Professor of Radio Astronomy at Cambridge since 1971. He headed the team which discovered the first PULSAR (1967).
Hexadecimal code
A number system using the powers of 16; the system of numeration used in computer programming. The digits used are 1, 2, 3, 4, 5, 6, 7, 8, 9, a, b, c, d, e, f (such that f represents the decimal number 15). The powers of 16 are written as $16^0=1$; $16^1=10$; 16^2 (256 in the decimal notation) $=100$, etc. Thus the decimal figure 126 is written as 7e in hex ($7\times16+14$).
Heymans, Corneille Jean François (1892-1968)
Belgian physiologist awarded the 1938 Nobel Prize for Physiology or Medicine for discovering sensory organs, close by the carotid artery and aorta, that play a part in regulating RESPIRATION.
Heyrovsky, Jaroslav (1896-1967)
Czech physical chemist awarded the 1959 Nobel Chemistry Prize for inventing the POLAROGRAPH.
Hiatus hernia
A common condition, usually asymptomatic, in which the SPHINCTER MUSCLE at the base of the ESOPHAGUS permits the reflux of stomach contents into the esophagus, causing heartburn and sometimes esophageal ulcers.
Hibernation
A state of lowered metabolism entered into by some HOMOIOTHERMIC ("warm-blooded") animals during the winter months. In hibernating animals internal preparations, such as laying down a store of fat, begin several weeks in advance. Falling temperatures trigger the onset of hibernation, and pulse rate and breathing drop to a minimum. With metabolism reduced, the animal can live until spring on food stored in its body. Winter food supplies would not be sufficient to maintain the animal in a fully active state. (See also DIAPAUSE; DORMANCY.)
Hiccup (hiccough)
A brief involuntary contraction of the DIAPHRAGM that may follow dietary or alcoholic excess and rapid eating. Rarely, it may be a symptom of UREMIA, mineral disorders or brain-stem disease.
High blood pressure see BLOOD CIRCULATION.
Hill, Sir Archibald Vivian (1886-1977)
British physiologist who shared with MEYERHOF the 1922 Nobel Prize for Physiology or Medicine for their independent work on the biochemistry of MUSCLE contraction and relaxation.
Hinshelwood, Sir Cyril Norman (1897-1967)
British physical chemist awarded with SEMENOV the 1956 Nobel Prize for Chemistry for his work on reaction rates and mechanisms (see KINETICS, CHEMICAL), especially in the reaction of hydrogen and oxygen to form WATER.
Hipparchus (*c*190-*c*120 BC)
Greek scientist, the father of systematic ASTRONOMY, who compiled the first star catalog and ascribed stars MAGNITUDES, made a good estimate of the distance and size of the Moon, probably first discovered PRECESSION, invented many astronomical instruments, worked on both plane and spherical trigonometry, and suggested ways of determining LATITUDE AND LONGITUDE.
Hippocrates (c460-c370 BC)
Greek physician generally regarded as the father of medicine, and the probable author of at least some of the Hippocratic Collection, some 60 or 70 books on all aspects of ancient MEDICINE. The authors probably formed a school centered on Hippocrates during his lifetime and continuing after his death.
Histamine
An AMINE formed from the AMINO ACID histidine. It is released into the body during allergic reactions; it also stimulates secretion of digestive juices in the stomach, dilates blood vessels, and makes smooth muscles contract.
Histology
The science of microscopic study of tissues, either after death or from live patients in the form of a BIOPSY. Tissue is fixed by chemicals that prevent degradation, and then stained with dyes that have particular affinity for different structures.
Histoplasmosis
FUNGAL disease prevalent in parts of North America and Africa, and carried by poultry, birds and bats. It may cause acute respiratory infection, a rapidly developing disseminated form, or a chronic type with FEVER, debility and specific organ involvement (especially of the LUNGS and resembling TUBERCULOSIS).
Hives (urticaria)
An itchy SKIN condition characterized by the formation of weals with surrounding ERYTHEMA. It is usually provoked by ALLERGY to food (eg shellfish, nuts, fruits), pollens, FUNGI, DRUGS (eg PENICILLIN) or parasites (SCABIES, worms). However, it may be symptomatic of infection, systemic disease or emotional disorder.
Hodgkin, Sir Alan Lloyd (b. 1914)
British physiologist awarded with A.F. HUXLEY and J. ECCLES the 1963 Nobel Prize for Physiology or Medicine for his work with Huxley on the chemical basis of nerve impulse transmission (see NERVOUS SYSTEM).
Hodgkin, Dorothy Mary, née **Crowfoot** (b.1910)
British chemist awarded the 1964 Nobel Prize for Chemistry for her work in determining the structure of VITAMIN B.
Hodgkin's disease
The most important type of LYMPHOMA or malignant proliferation of LYMPH tissue. Usually occurring in young adults, it may present with lymph node enlargement, weight loss, fever or malaise; the spleen, liver, lungs and brain may be involved.
Hoff, Jacobus Henricus van't see VAN'T HOFF, JACOBUS HENRICUS.
Hoffman, Roald (b. 1937)
Polish-born US chemist who shared the 1981 Nobel Prize for Chemistry with FUKUI for applying theories of QUANTUM MECHANICS to predict the behavior of electrons in chemical reactions.
Hofmann, August Wilhelm von (1818-1892)
German organic chemist. While he was teaching in London, W.H. PERRIN, his pupil, prepared

(1856) the first synthetic dye (see DYES AND DYEING). Hofmann returned to Germany and synthesized many new dyes, and for some 50 years thereafter Germany had the world's largest dye industry. He also discovered the Hofmann degradation process (see AMIDES).

Hofstadter, Robert (b. 1915)
US physicist who shared with MÖSSBAUER the 1961 Nobel Prize for Physics for his discoveries of the structure of protons and neutrons.

Hole see SEMICONDUCTOR.

Holley, Robert William (b. 1922)
US biochemist who shared with NIRENBERG and KHORANA the 1968 Nobel Prize for Physiology or Medicine for his work in establishing for the first time the nucleotide structure of a nucleic acid.

Holmium (Ho)
Element of the LANTHANUM SERIES of elements. AW 164.9, mp 1470°C, bp 2300°C, sg 8.795 (25°C). (▶ page 130)

Holocene
The later epoch of the QUATERNARY period representing the elapse from the end of the last ICE AGE up to and including the present; ie about the last 10,000 years. (See also GEOLOGY.) (◀ page 106)

Holography
A system of recording light or other waves on a photographic plate or other medium in such a way as to allow a three-dimensional reconstruction of the scene giving rise to the waves, in which the observer can actually see round objects by moving his head. The apparently unintelligible plate, or hologram, records the interference pattern between waves reflected by the scene and a direct reference wave at an angle to it; it is viewed by illuminating it from behind and looking through rather than at it. The high spatial coherence needed prevented exploitation of the technique, originated in 1948 by D. GABOR, until the advent of LASERS. There are two kinds of hologram: reflection, in which the 3D image seems to be behind the plate; and projection, in which the image stands out like a mirage. Holography is increasingly used in measurement and inspection (eg for irradiated nuclear fuel rods either out of or in a working reactor, and for security coding on bank cards) and for decorative or artistic artifacts. Color holograms are possible, and three-dimensional television may ultimately be feasible.

Holostei
A group of mainly fossil fish of the ACTINOPTERYGII, represented today by the garpikes (*Lepisosteus*) and bowfin (*Amia*).

Homeopathy
System of treatment founded in the early 19th century by C.F.S. HAHNEMANN, based on the theory that DISEASE is cured by DRUGS whose effects mimic it and whose efficacy is increased by the use of extremely small doses, achieved by specially prepared multiple dilutions.

Homeostasis
The self-regulating mechanisms whereby biological systems attempt to maintain a stable internal condition in the face of changes in the external environment, first described by the 19th-century French physiologist CLAUDE BERNARD. Homeostasis is partly achieved through FEEDBACK control, and hormones often play a vital part.

Homeotherm see HOMOIOTHERM.

Homoiotherm (or homeotherm)
An animal that can maintain a steady body temperature, regardless of the temperature of its surroundings, by metabolic means. Of living animals only the birds and mammals are usually described as homoiotherms, though some larger flying insects, some sharks, and certain other animals are also capable of raising their body temperature metabolically to a limited extent. In mammals and birds this is achieved by the respiration of food molecules in such a way that heat rather than ATP is produced. This raises the food requirement considerably, and they must eat approximately ten times as much as a reptile of similar size. To conserve this energy-expensive heat, homoiotherms have an insulating layer of feathers, fur and/or fat. When external temperatures are high they lose heat by a variety of methods, principally sweating or panting, both of which rely on the evaporation of water for their cooling effect. A few mammals, such as the sloths and the naked mole-rat, have abandoned homoiothermy as an adaptation to a poor diet or lack of oxygen; it can be temporarily suspended by many, as in HIBERNATION and the nightly torpor of hummingbirds and others. Among extinct animals the pterodactyls (flying reptiles) were homoiotherms, and it has been suggested that the dinosaurs also were, though this is now considered unlikely.

Homologous series
Sequence of chemical compounds which differ one from the next by a simple structural unit, and so can be given a general FORMULA. The members of a series are chemically similar, having the same FUNCTIONAL GROUPS, and their physical properties change regularly with molecular weight. Organic homologous series, such as the ALKANES, are built up by adding the methylene (CH_2) group.

Homolog
In biology, a structure or organ that has the same evolutionary origin as an apparently different structure in another species. For instance, there is little apparent similarity between a horse's leg and the flipper of a whale, but both have evolved from the limb of a mammal ancestor, and they have a similar embryonic development, so they are said to be homologous. (See EVOLUTION.)

The term homologous is also used of chromosomes. In a diploid cell, where there are two sets of chromosomes (one from each parent), the equivalent chromosomes from mother and father pair up before MEIOSIS, and form "homologous pairs".

Homo sapiens see PREHISTORIC MAN; RACE.

Homosexuality
Mutual sexual attraction between members of the same sex; in women it is also termed lesbianism.

Hooke, Robert (1635-1703)
English experimental scientist whose proposal of an inverse-square law of gravitational attraction (1679) prompted NEWTON into composing the *Principia*. From 1655 Hooke was assistant to R. BOYLE but he entered into his most creative period in 1662 when he became the ROYAL SOCIETY OF LONDON's first curator of experiments. He invented the compound MICROSCOPE, the universal joint and many other useful devices. His microscopic researches were published in the beautifully illustrated *Micrographia* (1665), a work which also introduced the term "cell" to biology. He is best remembered for his enunciation in 1678 of Hooke's Law. This states that the deformation occurring in an elastic body under stress is proportional to the applied stress (see MATERIALS, STRENGTH OF).

Hopkins, Sir Frederick Gowland (1861-1947)
British biochemist who shared with EIJKMAN the 1929 Nobel Prize for Physiology or Medicine for his work showing the necessity of certain dietary elements, now identified and known as VITAMINS, for the maintenance of health.

Horizon
The apparent line where the sky meets the land or sea. At sea, its distance varies in proportion to the square root of the height of the observer's eyes above sea level: if this is, say, 2m the horizon will be about 5.57km distant. The celestial horizon is the great circle on the CELESTIAL SPHERE at 90° from the ZENITH (the point immediately above the observer).

Hormones
Chemical messengers produced in living organisms that affect growth, differentiation, metabolism, digestive function, mineral and fluid balance. They are produced in relatively small amounts and usually act at a distance from their site of origin on specific target organs. In vertebrates and some invertebrates (eg insects) hormones are secreted by ENDOCRINE GLANDS, or analogous structures, into the bloodstream, which carries them to their point of action. The rate of secretion of hormones is affected by numerous factors, including FEEDBACK from their metabolic effects and the action of controlling hormones. The latter usually originate in the PITUITARY GLAND and those controlling the pituitary in the HYPOTHALAMUS, a part of the brain. Important hormones include INSULIN, THYROID hormone, EPINEPHRINE, GLUCAGON, GONADOTROPHINS, ESTROGEN, PROGESTERONE, ANDROGENS, pituitary growth hormone, VASOPRESSIN, thyroid stimulating hormone, adrenocorticotrophic hormone, GASTRIN and SECRETIN. (See also PHEROMONES.)

Plants also have hormone-like substances, including AUXIN, GIBBERELLINS, CYTOKININS and ABSCISIC ACID, but these differ from animal hormones in that they are not produced by a specific organ, nor directed at a specific target. Plants also have a gaseous hormone, ETHYLENE, which promotes the ripening of fruit. (◀ page 111)

Hornblende
Group of common rock-forming minerals; dark monoclinic AMPHIBOLES of general composition $Ca_2Na(Mg, Fe, Al)_5Si_6(Si, Al)_2O_{22}(OH)_2$.

Horney, Karen (1885-1952)
German-born US psychoanalyst best known for her concentration on the importance of environment in character development, so rejecting many of the basic principles of FREUD's classic psychoanalytic theory, especially his stress on the LIBIDO as the root of personality and behavior (see PSYCHOANALYSIS).

Horns
Hard, keratinous or bony structures borne on the forehead of many UNGULATES or, in the case of rhinoceroses, on the nose. They show a variety of forms. Horns are unbranched permanent structures, unlike the antlers of deer, which are cast and regrown annually. Horns may be used for defense or they may be solely a product of SEXUAL SELECTION. In some species horns are borne only by the males; in others both sexes have them but those of the male are larger.

Hornworts
A small group of BRYOPHYTES (mosses and lichens) belonging to the Class Anthocerotae. They have a very simple thallus, the cells of which contain a single chloroplast and a pyrenoid, features only occurring elsewhere in the ALGAE. They are believed to be related to the ancestral bryophyte, which probably evolved from a green alga. (▶ page 139)

Horse latitudes
Two belts, characterized by low winds, about 30°N and S of the equator. Sailing ships carrying horses to America were often becalmed here: many horses died from lack of fresh water and were cast overboard. (See also TRADE WINDS.)

Horsepower (hp)
Unit of power introduced by JAMES WATT, equivalent to 745.70 watt. Brake horsepower (bhp) is power output measured by applying a brake (usually a Prony brake) to the driving shaft. The metric horsepower, or cheval-vapeur (CV), originally the power required to raise a 75kg weight through one meter in one second, is 735.5 watt.

Horsetails
A group of plants belonging to the PTERIDOPHYTES. They have slender, whip-like leaves growing in discrete whorls at intervals up the thick, sturdy stem. Separate stems appear carrying the reproductive structures, which are borne in fleshy "cones" at their tip. (➧ page 139)
Horst
An area that has been thrust upward between two roughly parallel FAULTS.
Host
An animal, plant or other living organism that supports a parasite. Also, the partner in COMMENSALISM who does not benefit, or the tree on which an EPIPHYTE grows.
Hot springs (thermal springs)
Springs supplied by underground water heated usually by vapor from the MAGMA, and most common in recently active volcanic regions (see VOLCANISM). The water contains dissolved minerals that may form terraces around the outlet. (See also GEYSER.)
Hounsfield, Sir Godfrey Newbold (b. 1919)
British engineer who was corecipient of the 1979 Nobel Prize for Physiology or Medicine for his contribution to the development of computer-assisted TOMOGRAPHY.
Hour (h)
Unit of time equal to 60 minutes or 3600 seconds. In one hour, the Earth rotates on its axis through 15°. The time of day at any point on Earth is expressed as the number of hours and minutes that have elapsed since midnight for the time zone in which the point is situated, the time zones being fixed intervals behind or ahead of Greenwich Mean Time.
Houssay, Bernardo Alberto (1887-1971)
Argentinian physiologist awarded (with C. and G. CORI) the 1947 Nobel Prize for Physiology or Medicine for his discovery that certain HORMONES produced by the PITUITARY GLAND were responsible for regulating the blood's sugar and insulin content.
Hoyle, Sir Fred (b. 1915)
British cosmologist best known for formulating with T. GOLD and H. BONDI the Steady State theory (see COSMOLOGY), and for his important contributions to theories of stellar evolution, especially concerning the successive formation of the elements by nuclear FUSION in STARS. He is also well known as a science fiction writer and for popular books such as *Frontiers of Astronomy* (1955).
Hubble, Edwin Powell (1889-1953)
US astronomer who first showed (1923) that certain NEBULAE are in fact GALAXIES outside the MILKY WAY. By examining the RED SHIFTS in their spectra, he showed that they are receding at rates proportional to their distances (see HUBBLE'S CONSTANT).
Hubble's constant
Ratio between the distance of a GALAXY and the rate at which it is receding from us. HUBBLE first calculated this as about 500km/s per Mpc (megaparsec); however, he used incorrect estimates of the distances of the galaxies, and current revised estimates of its value are in the range 50-100km/s per Mpc.
Hubble space telescope
Due to be launched no earlier than 1988 by NASA to send back images of the remote regions of the Universe. Orbiting 600km above the Earth, the space telescope will operate in the visible range of light, infrared and ultraviolet. Free of the atmosphere, its resolution will be 10 times sharper than Earth-based telescopes. The telescope is named for the American astronomer EDWIN HUBBLE.
Hubel, David Hunter (b. 1926)
US neurobiologist who shared half of the 1981 Nobel Prize for Physiology or Medicine with WIESEL for their research on recording from single CELLS which revealed the distribution and structure of nerve cells in the BRAIN.
Huggins, Charles Brenton (b. 1901)
Canadian-born US surgeon awarded (with F.P. ROUS) the 1966 Nobel Prize for Physiology or Medicine for his discovery that TUMORS of the male PROSTATE GLAND could be controlled by injection of female sex HORMONES, the first demonstration that CANCER might be controlled by chemical agents.
Humboldt, Fredrich Heinrich Alexander, Baron von (1769-1859)
German naturalist. With the botanist Aimé Jacques Alexandre Bonpland (1773-1858) he traveled for five years through much of South America (1799-1804), collecting plant, animal and rock specimens and making geomagnetic and meteorologic observations. Humboldt published their data in 30 volumes over the next 23 years. In his most important work, *Kosmos* (1845-62), he sought to show a fundamental unity of all natural phenomena.
Humerus
Bone of the upper arm, linked to the scapula and clavicle at the shoulder and to the radius and ulna at the elbow. (➧ page 216)
Humidity
The amount of water vapor in the air, measured as mass of water per unit volume or mass of air, or as a percentage of the maximum amount the air would support without condensation, or indirectly via the DEW point. Saturation of the air occurs when the water vapor pressure reaches the VAPOR PRESSURE of liquid water at the temperature concerned; this rises rapidly with temperature. The physiologically tolerable humidity level falls rapidly with temperature, as humidity inhibits cooling by evaporation of sweat.
Humus
The organic component of soil, decomposed plant and animal material. Dark brown to black, it is intimately mixed with the inorganic soil particles. It is of great agricultural importance, mainly because of its good carbon, nitrogen, phosphorus and sulfur content.
Hunchback
Another name for KYPHOSIS.
Hundredweight (cwt)
Name of two units of weight. In the US the short hundredweight of 100lb is used. The Imperial (long) hundredweight is 112lb.
Hunger see THIRST AND HUNGER.
Hunter, John (1728-1793)
British anatomist and biologist who made many contributions to SURGERY, ANATOMY and PHYSIOLOGY. He is often regarded as the father of scientific surgery.
Hurricane
A tropical cyclone, usually of great intensity. High-speed winds spiral in toward a low-pressure core of warm, calm air (the eye): winds of over 300km/h have been measured. The direction of spiral is clockwise in the S Hemisphere, and counterclockwise in the N (see CORIOLIS EFFECT). Hurricanes form (usually between latitudes 5° and 25°) when there is an existing convergence of air near sea level toward a center. The air ascends, losing moisture as precipitation as it does so. If this happens rapidly enough, the upper air is warmed by the water's LATENT HEAT of vaporization. This reduces the surface pressure, so accelerating air convergence. Hurricanes of the N Pacific are often called typhoons. (See also CYCLONE; WIND.)
Hutton, James (1726-1797)
Scottish geologist who proposed, in his *Theory of the Earth* (1795), that the Earth's natural features result from continual processes, occurring now at the same rate as they have in the past (see UNIFORMITARIANISM). These views were little regarded until LYELL's work some decades later. (See also CATASTROPHISM.)
Huxley
Distinguished British family. Thomas Henry Huxley (1825-1895) is best known for his support of DARWIN's theory of EVOLUTION, without which acceptance of the theory might have been long delayed. Most of his own contributions to paleontology and zoology (especially taxonomy), botany, geology and anthropology were related to this. He also coined the word "agnostic". His son Leonard Huxley (1860-1933), a distinguished man of literature, wrote *The Life and Letters of Thomas Henry Huxley* (1900). Of his children, three earned fame. Sir Julian Sorell Huxley (1887-1975) is best known as a biologist and ecologist. His early interests were in development and growth, genetics and embryology. Later he made important studies of bird behavior, studied evolution and wrote many popular scientific books. Aldous Leonard Huxley (1894-1963) was one of the 20th century's foremost novelists. Sir Andrew Fielding Huxley (b. 1917) shared the 1963 Nobel Prize for Physiology or Medicine with SIR A.L. HODGKIN and SIR J. ECCLES for his work with Hodgkin on the chemical basis of nerve impulse transmission (see NERVOUS SYSTEM).
Huygens, Christiaan (1629-1695)
Dutch scientist who formulated a wave theory of LIGHT, first applied the PENDULUM to the regulation of clocks, and discovered the surface markings of MARS and that SATURN has rings. In his optical studies he stated Huygens' Principle, that all points on a wave front may at any instant be considered as sources of secondary waves that, taken together, represent the wave front at any later instant.
Hybridization
The crossing of individuals belonging to two distinct species. Mules, for example, are the result of hybridization between a horse and an ass. Hybrid offspring are usually sterile, especially in animals. Hybrids between two distinct genera are also possible, though rare.
Hydatid disease
A disease consisting of cysts of larvae of the tapeworm *Echinococcus*. A single animal or human can be infected with many cysts, which often become large.
Hydra
The Water Monster, a large S Hemisphere constellation containing the bright star Alphard and a cluster of galaxies over 30Mpc distant.
Hydramnios
An excess of AMNIOTIC FLUID.
Hydrate
A compound (usually ionic) containing a definite proportion of WATER, known as water of crystallization, which may be bound as a LIGAND to the cation, to the anion, or to both. Other hydrates, with more or less water, may be formed under different conditions. Water may be lost from a hydrate spontaneously (EFFLORESCENCE) or by heating (see also DEHYDRATION), the compound becoming anhydrous. Common hydrates include COPPER (II) sulfate, $CuSO_4.5H_2O$, SODIUM carbonate, $Na_2CO_3.10H_2O$, and the ALUMS. (See also DELIQUESCENCE.)
Hydraulics
Application of the properties of liquids (particularly water), at rest and in motion, to engineering problems. Since any machine or structure that uses, controls or conserves a liquid makes use of the principles of hydraulics, the scope of this subject is very wide. It includes methods of water supply for consumption, irrigation or navigation and the design of associated dams, canals and

MICROSCOPES

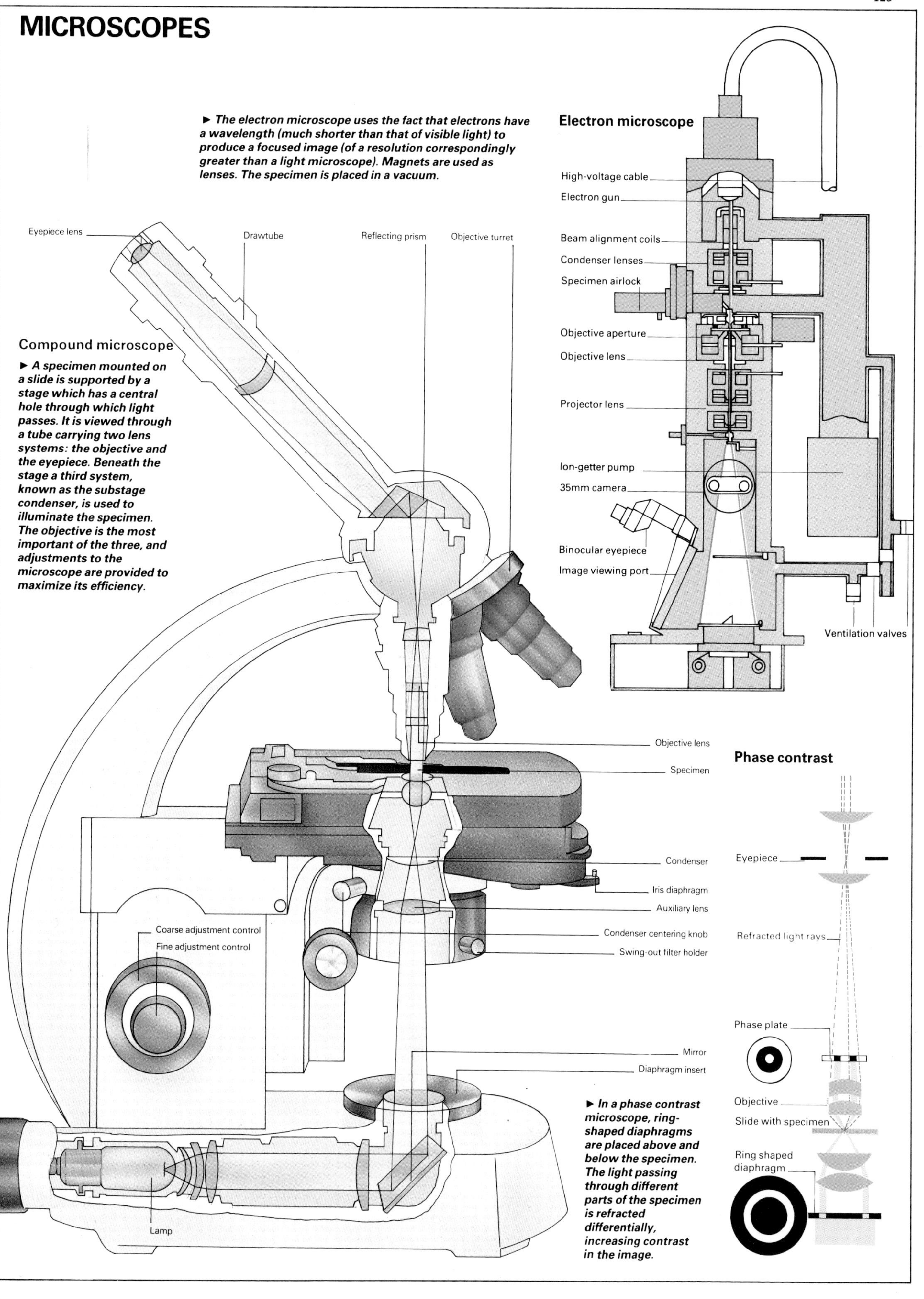

► *The electron microscope uses the fact that electrons have a wavelength (much shorter than that of visible light) to produce a focused image (of a resolution correspondingly greater than a light microscope). Magnets are used as lenses. The specimen is placed in a vacuum.*

► *A specimen mounted on a slide is supported by a stage which has a central hole through which light passes. It is viewed through a tube carrying two lens systems: the objective and the eyepiece. Beneath the stage a third system, known as the substage condenser, is used to illuminate the specimen. The objective is the most important of the three, and adjustments to the microscope are provided to maximize its efficiency.*

► *In a phase contrast microscope, ring-shaped diaphragms are placed above and below the specimen. The light passing through different parts of the specimen is refracted differentially, increasing contrast in the image.*

pipes; HYDROELECTRICITY, the conversion of water power to electric energy using hydraulic TURBINES; the design and construction of ditches, culverts and hydraulic jumps (a means of slowing down the flow of a stream by suddenly increasing its depth) for controlling and discharging flood water, and the treatment and disposal of industrial and human waste. Hydraulics applies the principles of HYDROSTATICS and HYDRODYNAMICS and is hence a branch of FLUID MECHANICS.

Hydrazine (N_2H_4)
Colorless liquid, a covalent HYDRIDE resembling AMMONIA, prepared by oxidizing UREA with hypochlorite in the presence of gelatin. It reacts with ALDEHYDES to form hydrazones ($RCH{=}NNH_2$); it is a weak BASE, forming hydrazonium salts. Hydrazine is a powerful reducing agent, and so is used as a rocket fuel (being oxidized by nitric acid), as well as a corrosion inhibitor in boilers. It is also used to cure rubber, and in the production of plastics, explosives and fungicides.

Hydrides
Binary compounds of HYDROGEN and another element. They fall into three classes.

COVALENT HYDRIDES are formed by the elements in Groups IB-VIIA of the PERIODIC TABLE, ie the nonmetals and some metals. They are mainly volatile, reactive compounds, though those of Groups IB and IIB, and aluminum, are nonvolatile polymers. (See the individual elements, and also BORANES; HYDROCARBONS; AMMONIA; HYDRAZINE; WATER; SULFIDES; HYDROGEN FLUORIDE; HYDROGEN CHLORIDE.)

IONIC HYDRIDES are formed by the ALKALI METALS and ALKALINE-EARTH METALS. They are crystalline solids containing the ion H^-, which is a very powerful BASE and reducing agent (see OXIDATION AND REDUCTION), and react violently with water to give hydrogen.

METALLIC HYDRIDES formed by the TRANSITION ELEMENTS (Groups IIIB-VIII), are mostly nonstoichiometric (see COMPOSITION, CHEMICAL) and electrically conducting. They resemble ALLOYS, and some have interstitial structures.

Hydrocarbons
Organic compounds composed of CARBON and HYDROGEN only. Like organic compounds in general, which are derived formally from hydrocarbons by adding FUNCTIONAL GROUPS, they are best divided into ALIPHATIC, ALICYCLIC and AROMATIC hydrocarbons; aliphatic hydrocarbons are further subdivided into ALKANES, ALKENES and ALKYNES. Some hydrocarbons, especially TERPENES, occur in plant oils, and solid, high-molecular-weight hydrocarbons occur as BITUMEN, but by far the largest sources of all sorts of hydrocarbons are PETROLEUM, NATURAL GAS and COAL GAS. They are used as FUELS, for LUBRICATION, and as starting materials for a wide variety of industrial syntheses.

Hydrocephalus
Enlargement of the normal cavities within the brain, called ventricles, with increased CEREBROSPINAL FLUID (CSF) within the SKULL. In children it causes a characteristic enlargement of the head. Brain tissue is attenuated and damaged by long-standing hydrocephalus. The condition may be caused by block to CSF drainage in the lower ventricles or brain stem aqueduct (for example by TUMOR or malformation), or by prevention of its reabsorption over the brain surface (for example, following MENINGITIS).

Hydrochloric acid
Solution of HYDROGEN CHLORIDE in water; a strong ACID of major industrial importance.

Hydrocyanic acid see CYANIDES.

Hydrodynamics
The branch of FLUID MECHANICS dealing with the FORCES, ENERGY and PRESSURE of FLUIDS in motion. A mathematical treatment of ideal frictionless and incompressible fluids flowing around given boundaries is coupled with an empirical approach in order to solve practical problems.

Hydroelectricity (hydroelectric power)
The generation of ELECTRICITY using water power, and the source of about a third of the world's electricity. Although the power station must usually be sited in the mountains and the electricity transmitted over long distances, the power is still cheap since water, the fuel, is free. Moreover, running costs are low. An exciting modern development is the use in coastal regions of the ebb and flow of the tide as a source of electric power. Hydroelectric power uses a flow of water to turn a TURBINE, which itself drives a GENERATOR.

The turbines are of two main types: impulse (eg the Pelton wheel) and reaction (eg the Francis and Kaplan wheels). The Pelton wheel has buckets about its edge, into which jets of water are aimed, so turning the wheel. The Francis wheel has spiral vanes: water enters from the side and is discharged along the axis. The Kaplan wheel is rather like a huge propeller immersed in the water.

Hydrogen (H)
The simplest and lightest element, a colorless, odorless gas. Hydrogen atoms make up about 90% of the UNIVERSE, and it is believed that all other elements have been produced by fusion of hydrogen (see STAR; FUSION, NUCLEAR). On Earth most hydrogen occurs combined with oxygen as WATER and mineral HYDRATES, or with carbon as HYDROCARBONS (see PETROLEUM). Hydrogen is produced in the laboratory by the action of a dilute ACID on zinc or other electropositive metals. Industrially it is made by the catalytic reaction of hydrocarbons with steam, or by the WATER GAS process, or as a byproduct of some ELECTROLYSIS reactions. Two-thirds of the hydrogen manufactured is used to make ammonia by the HABER PROCESS. It is also used in HYDROGENATION, PETROLEUM refining, and metal smelting. METHANOL and HYDROGEN CHLORIDE are produced from hydrogen. Being flammable, it has now been largely superseded by helium for filling BALLOONS and AIRSHIPS. Hydrogen is used in oxy-hydrogen WELDING; liquid hydrogen is used as fuel in rocket engines, in BUBBLE CHAMBERS, and as a refrigerant (see CRYOGENICS). Hydrogen is fairly reactive, giving HYDRIDES with most other elements on heating, and a moderate reducing agent. It belongs in no definite group of the PERIODIC TABLE, but has some resemblance to the HALOGENS in forming the ion H^-, and to the ALKALI METALS in forming the ion H^+ (see ACIDS); it is always monovalent. (See also HYDROGENATION; HYDROGEN BONDING.) A hydrogen atom consists of one ELECTRON orbiting a nucleus of one PROTON. A hydrogen molecule is two atoms combined (H_2). In parahydrogen both the protons have the same SPIN: in orthohydrogen the protons have opposite spin. They have slightly different properties. At room temperature, hydrogen is 75% orthohydrogen, 25% parahydrogen. DEUTERIUM (H^2) and TRITIUM (H^3) are ISOTOPES of hydrogen. (See also HYDROGEN BOMB.) AW 1.008, mp −259°C, bp −253°C. (♦ page 130, 147)

Hydrogenation
A reaction in which hydrogen is added to a compound. Hydrogenation converts unsaturated organic compounds (see BOND, CHEMICAL) into saturated ones. Catalysts (most commonly Raney nickel, palladium and platinum) are used. Hydrogenation is used notably to turn vegetable oils into margarine and in PETROLEUM refining.

Hydrogen bomb (thermonuclear bomb)
Very powerful bomb whose explosive energy is produced by nuclear FUSION of two DEUTERIUM atoms or of a deuterium and a TRITIUM atom. The extremely high temperatures required to start the fusion reaction are produced by using an ATOMIC BOMB as a fuze. Lithium 6 deuteride (Li^6D) is the explosive; neutrons produced by deuterium fusion react with the Li^6 to produce tritium. The end products are the isotopes of HELIUM He^3 and He^4. In warfare hydrogen bombs have the advantage of being far more powerful than atomic bombs, their power being measured in megatons of TNT, capable of destroying a large city. In defensive and peaceful uses they can be modified so that the radioactivity produced is minimal. Hydrogen bombs were first developed in the US (1949-52) by EDWARD TELLER and others, and have been tested also by the USSR, Great Britain, China and France.

Hydrogen bonding
The formation of a weak bond (see BOND, CHEMICAL) between a HYDROGEN atom (bound to a small electronegative atom, usually fluorine, oxygen, nitrogen or chlorine) and another such electronegative atom, in another or the same molecule. It is an electrostatic effect, but can be well described in molecular-orbital terms, especially when the three-atom system is symmetrical (eg HF_2^-), resembling the hydrogen-bridge bonding in BORANES. Hydrogen bonding leads to anomalous physical properties due to molecular association: high melting point and boiling point, low vapor pressure, high viscosity, etc. It is important in the hydrogen halides, WATER, ICE, ALCOHOLS, OXYACIDS, AMMONIA, AMINES and AMIDES, and hence in vital molecules such as AMINO ACIDS, PROTEINS and DNA. (♦ page 37)

Hydrogen chloride (HCl)
Colorless acrid gas fuming in air; a covalent HYDRIDE prepared by heating SALT with concentrated sulfuric acid or by direct combination of HYDROGEN and CHLORINE. It is unreactive when completely dry. mp −115°C, bp −85°C. Hydrochloric acid, a solution of hydrogen chloride in water, is a strong ACID, and reacts with active metals and bases to give chlorides (see HALIDES). It is used to make chlorine compounds and in the extraction and processing of metals. Dilute hydrochloric acid is produced in the stomach (see DIGESTIVE SYSTEM) but in excess causes gastric ULCERS. The concentrated acid is corrosive. (See also AQUA REGIA.)

Hydrogen fluoride (HF)
Colorless liquid, fuming in air; a covalent HYDRIDE, prepared by distilling FLUORITE with concentrated sulfuric acid. Its physical properties show typical anomalies due to HYDROGEN BONDING. It is a very strong ACID, and an ionizing solvent for many inorganic and organic compounds; it is used to make FLUORINE and its compounds, especially FREON and FLUOROCARBONS. mp −83°C, bp 20°C. Hydrofluoric acid, a solution of hydrogen fluoride in water, is (anomalously) a weak acid, but causes very severe burns and is toxic. It dissolves silica to give fluorosilicic acid (H_2SiF_6), and is therefore used to etch glass.

Hydrogen ion concentration
A measure of the acidity of a solution. It is frequently expressed as its negative logarithm, called pH. The hydrogen ion concentration in pure water is $1{\times}10^{-7}$M and its pH is thus 7. A pH less than 7 indicates an acid solution while a solution with a pH greater than 7 is basic (alkaline).

Hydrogen peroxide see PEROXIDES.

Hydrogen sulfide see SULFIDES.

Hydrography
Branch of hydrology dealing with bodies of water such as oceans, lakes and rivers on the Earth's

surface; and especially with the charting of their boundaries, currents, underwater contours and shipping hazards, as well as with the composition of their beds. (See also EARTH; HYDROLOGY; OCEANOGRAPHY.)

Hydroids
A group of CNIDARIANS, of the simple POLYP form in which the tentacles point upward. Some are solitary, others colonial. The best-known genus is *Hydra*.

Hydrologic cycle
The circulation of the waters of the Earth between land, oceans and atmosphere. Water evaporates from the oceans into the ATMOSPHERE, where it may form CLOUDS (see also EVAPORATION). Much of this water is precipitated as rain back into the ocean, but much also falls on land. Of this, some is returned to the atmosphere by the TRANSPIRATION of plants, some joins rivers and is returned to the sea, some joins the GROUNDWATER and eventually reaches a sea, lake or river, and some evaporates back into the atmosphere from the surface of the land or from rivers, streams, lakes, etc. Over 97% of the Earth's water is in the oceans; of the remaining fresh water, about 75% is in solid form (see GLACIER). (See also HYDROLOGY; HYDROSPHERE.) (◀ page 44)

Hydrology
The branch of geophysics concerned with the HYDROSPHERE (all the waters of the Earth), with particular reference to the HYDROLOGIC CYCLE. The science was born in the 17th century with the work of Pierre Perrault and EDME MARIOTTE.

Hydrolysis
A double decomposition effected by WATER, according to the general equation

$$XY + H_2O \rightarrow XOH + YH$$

If XY is a salt of a weak ACID or a weak BASE, the hydrolysis is reversible, and affects the pH of the solution (see BUFFER). Reactive organic compounds such as ACID CHLORIDES and ACID ANHYDRIDES are rapidly hydrolyzed by water alone, but others require acids, bases, or ENZYMES as catalysts (as in digestion).

Hydrometer
An instrument to measure the density of a liquid. In essence it consists of a closed glass tube calibrated along its stem and blown into a bulb, which is weighted, at the other. The hydrometer floats upright in the liquid to be tested: the denser the liquid, the more the instrument is buoyed up. The scale is read at the level of the liquid's surface. Though the instrument measures DENSITY directly, it is usual to calibrate the stem in terms of SPECIFIC GRAVITY, the ratio of the density of the liquid to the density of water at that temperature.

Hydrophobia see RABIES.

Hydrophone
An adaptation of the MICROPHONE for use underwater. As with the microphone, the device converts sound (pressure) waves into electrical impulses, which are passed by cable to an external amplifier. Its prime use is in SONAR.

Hydrophytes
Plants that live entirely submerged in water, whose upper leaves are floating, or which are entirely floating.

Hydroponics
The technique by which plants are grown without soil. All the minerals required for plant growth are provided by nutrient solutions in which the roots are immersed.

Hydroquinone (1,4-dihydroxybenzene)
The reduced form of quinone, used as a photographic developer and antioxidant. MW 110.11, mp 173-4°C, bp 285°C.

Hydrosphere
All the waters of the Earth, in whatever form: solid, liquid, gaseous. It thus includes the water of the ATMOSPHERE, water on the EARTH's surface (eg oceans, rivers and ice sheets) and GROUNDWATER. (See also HYDROGRAPHY; HYDROLOGY; LITHOSPHERE.) (▶ page 134)

Hydrostatics
The branch of FLUID MECHANICS dealing with FORCES and PRESSURES in stationary FLUIDS, and their effects on bodies immersed in them. The concept of fluid pressure (a normal force per unit area acting across any surface in the fluid or at its boundary) enables problems of flotation, buoyancy etc to be treated.

Hydrotherapy
System of treatment by use of water, usually by treading or bathing in special pools, often supplied from mineral springs historically credited with healing properties.

Hydroxides
Compounds containing the OH group, or the ion OH^-. Hydroxides of metals are generally BASES, and, if soluble, ionize to produce alkaline solutions (see ALKALIS) containing hydroxide ions. Nonmetals form ACID hydroxides, or OXYACIDS, which dissolve to produce hydrogen ions. Some metal hydroxides, such as zinc hydroxide, are amphoteric, that is, both basic and acidic. Hydroxides are formed by hydration of the oxide, or, if insoluble, by precipitation with an alkali. The OH^- ion acts as a LIGAND, forming hydroxo complexes. Organic compounds containing the OH group are ALCOHOLS, PHENOLS and CARBOXYLIC ACIDS.

Hygrometer
Device to measure HUMIDITY (the amount of water vapor the air holds). Usually, hygrometers measure relative humidity, the amount of moisture as a percentage of the SATURATION level at that temperature. There are several types: the hair hygrometer, though of limited accuracy, is common; the wet and dry bulb hygrometer (psychrometer) has two thermometers; the dewpoint hygrometer comprises a polished container cooled until the dew point is reached; the electric hygrometer measures changes in the electrical resistance of a hygroscopic (water-absorbing) strip.

Hymenoptera
An order of insects that includes the ants, wasps and bees. They are characterized by having a narrow constriction or "waist" between the thorax and abdomen. Most of the SOCIAL INSECTS belong to this order.

Hyperactivity
Inexact term used to describe ceaseless, purposeless, excessively energetic behavior in children.

Hyperbaric chamber
Chamber built to withstand and be kept at pressures above atmospheric. The high oxygen pressures achieved in them may destroy the anaerobic bacteria (*Clostridia*) responsible for gas gangrene; surgery may be done in the hyperbaric chamber. It is also used for aeroembolism in decompression.

Hyperidrosis
Excessive SWEATING.

Hyperon see SUBATOMIC PARTICLES.

Hyperopia (hypermetropia)
Far- or long-sightedness, a defect of VISION in which light entering the EYE from nearby objects comes to a focus behind the retina. As a result objects close to appear blurred.

Hypertension see BLOOD CIRCULATION.

Hyperthyroidism see THYROID GLAND.

Hypha
A hollow filament with a rigid cell wall found in FUNGI. Most hyphae are too small to be seen with the naked eye, but some are up to a millimeter in diameter. The network of hyphae is known as a MYCELIUM.

Hypnosis
An artificially induced mental state characterized by an individual's loss of critical powers and consequent openness to suggestion. It may be induced by an external agency or by the individual (autohypnosis). Hypnotism has been widely used in medicine (usually to induce ANALGESIA) and especially in PSYCHIATRY and PSYCHOTHERAPY. Here the particular value of hypnosis is that, while in trance, the individual may be encouraged to recall deeply repressed memories (see MEMORY; REPRESSION) that may be the heart of, for example, a COMPLEX.

Hypo (sodium thiosulfate) see SODIUM.

Hypochlorites
Salts containing the ClO^- ion derived from hypochlorous acid, HOCl, an unstable, weak acid. They are made by absorbing CHLORINE into an ALKALI, and are used for BLEACHING (see also BLEACHING POWDER) and as disinfectants.

Hypochondria (hypochondriasis)
A mental disorder involving undue ANXIETY about real or supposed ailments, usually in the belief that these are incurable.

Hypoglycemia
An abnormally low level of glucose in the blood.

Hypogonadism
Underdevelopment of the sexual organs. It is a symptom of several rare hormonal or developmental diseases.

Hypothalamus
Central part of the base of the brain, closely related to the PITUITARY GLAND. It contains vital centers for controlling the autonomic NERVOUS SYSTEM, body temperature and water and food intake. It also produces HORMONES for regulating pituitary secretion and two systemic hormones. (▶ page 126)

Hypothermia
Abnormally low body temperature. In old people it occurs because of their reduced ability to maintain their body temperature. It also occurs when people suffer from exposure, eg when forced to spend a winter night outdoors. It dulls the senses so that the person may not feel cold and hence does not realize the danger he is in.

Hypothesis see SCIENTIFIC METHOD.

Hypoxemia see HYPOXIA.

Hypoxia (anoxia)
Lack of oxygen in blood and body tissues. ASPHYXIA, LUNG disease, paralysis of respiratory muscles and some forms of COMA prevent enough oxygen reaching the blood. Disease of heart or circulation may also lead to tissue hypoxia. Irreversible brain damage follows prolonged hypoxia.

Hysterectomy
Surgical removal of the UTERUS, with or without the OVARIES and FALLOPIAN TUBES. It may be performed via either the abdomen or the vagina and is most often used for FIBROIDS, CANCER of the cervix or the body of the uterus, or for diseases causing heavy MENSTRUATION. If the ovaries are preserved HORMONE secretion remains intact, though periods cease and infertility is inevitable.

Hysteresis
A memory phenomenon found in ferromagnetic materials (see MAGNETISM), the magnetization depending on the MAGNETIC FIELD applied and on how it was applied. In electric motors, generators and transformers the refusal of the magnetization to follow the field directly often results in substantial ENERGY loss (hysteresis loss).

Hysteria
Psychiatric disorder characterized by exaggerated responses, emotional lability with excess tears and laughter, overactivity and often overbreathing, occasionally leading to TETANY. It is often a manifestation of attention-seeking behavior.

NERVOUS SYSTEM

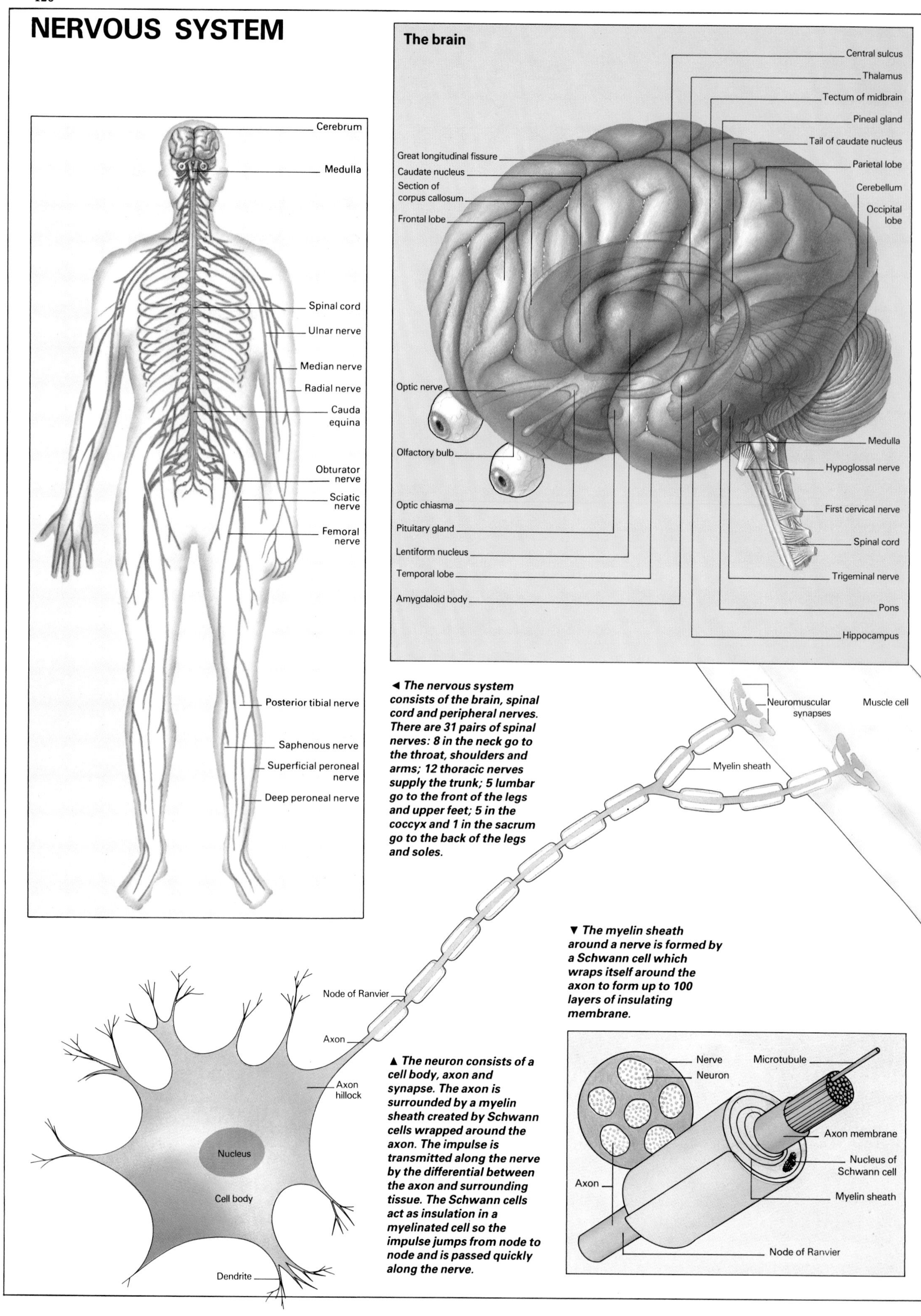

◀ *The nervous system consists of the brain, spinal cord and peripheral nerves. There are 31 pairs of spinal nerves: 8 in the neck go to the throat, shoulders and arms; 12 thoracic nerves supply the trunk; 5 lumbar go to the front of the legs and upper feet; 5 in the coccyx and 1 in the sacrum go to the back of the legs and soles.*

▼ *The myelin sheath around a nerve is formed by a Schwann cell which wraps itself around the axon to form up to 100 layers of insulating membrane.*

▲ *The neuron consists of a cell body, axon and synapse. The axon is surrounded by a myelin sheath created by Schwann cells wrapped around the axon. The impulse is transmitted along the nerve by the differential between the axon and surrounding tissue. The Schwann cells act as insulation in a myelinated cell so the impulse jumps from node to node and is passed quickly along the nerve.*

The autonomic nervous system

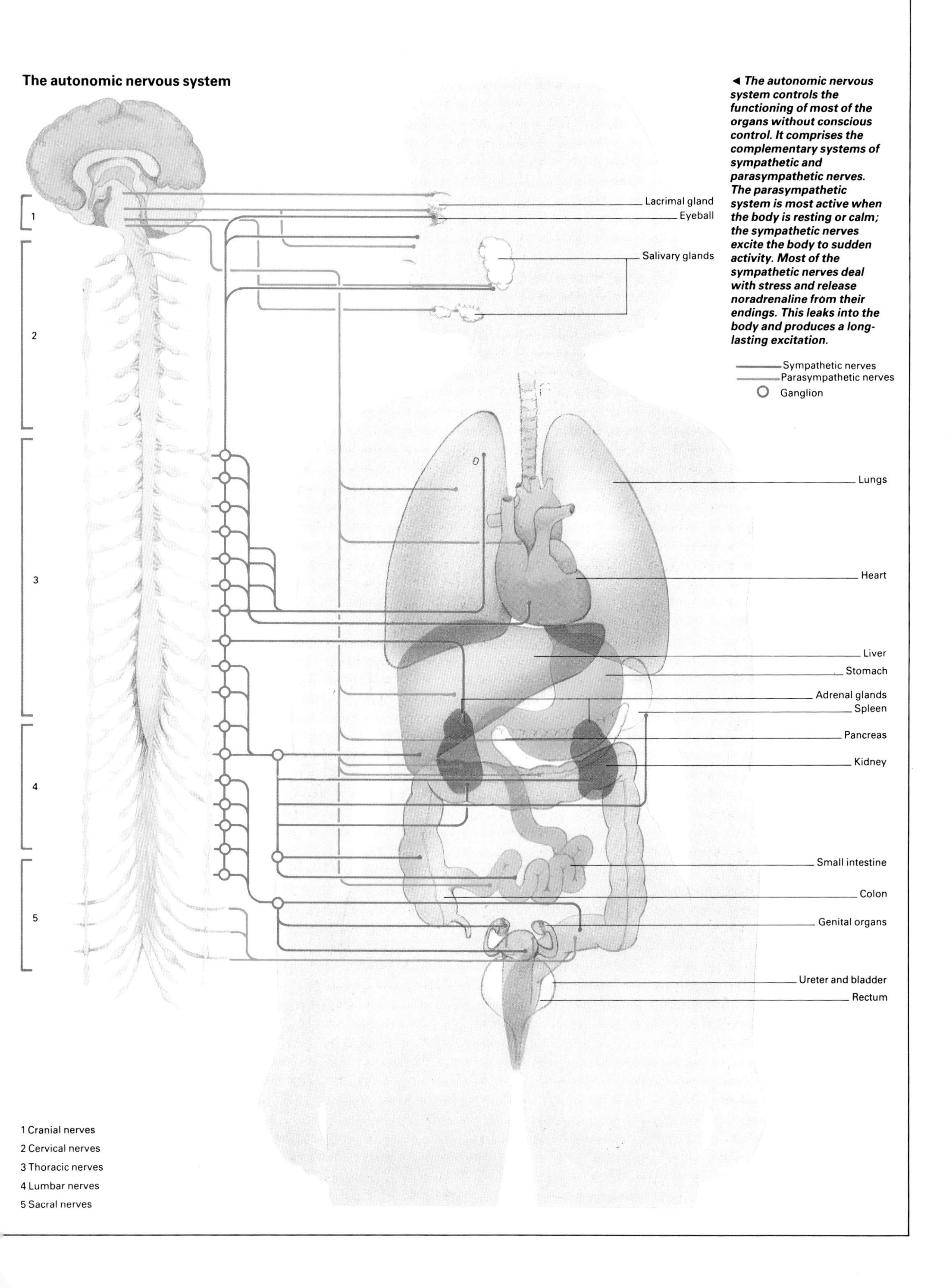

◀ ***The autonomic nervous system controls the functioning of most of the organs without conscious control. It comprises the complementary systems of sympathetic and parasympathetic nerves. The parasympathetic system is most active when the body is resting or calm; the sympathetic nerves excite the body to sudden activity. Most of the sympathetic nerves deal with stress and release noradrenaline from their endings. This leaks into the body and produces a long-lasting excitation.***

Iatrochemistry
An aspect of ALCHEMY that sought chemical treatments for disease, particularly as promoted in the 16th century by PARACELSUS and his followers. The analytic methods used in iatrochemistry were highly significant in the development of modern CHEMISTRY, and the search for new remedies led to the discovery of many new chemical substances.

Iatrogenic
Caused by medical treatment. As most drugs have some adverse effect, the term is used for conditions that persist after the original illness is cured.

Ice
Frozen WATER: a colorless, crystalline solid in which the strong, directional HYDROGEN BONDING produces a structure with much space between the molecules. Thus ice is less dense than water, and floats on it. The expansion of water on freezing may crack pipes and automobile radiators. Since dissolved substances lower the freezing point, seawater freezes at about −2°C. Ice has a very low coefficient of FRICTION. (See also FROST; GLACIER; HAIL; ICE AGES; ICEBERG; ICEBREAKER; SNOW.) mp 0°C, bp 100°C, sg 0.92 (0°C).

Ice ages
Periods when glacial ice covers large areas of the Earth's surface that are not normally covered by ice. Ice ages are characterized by fluctuations of climatic conditions: a cycle of several glacial periods contains interglacial periods, perhaps of a few tens of thousands of years, when the climate may be as temperate as between ice ages. It is not known whether the Earth is currently between ice ages or merely passing through an interglacial period.

There seem to have been several ice ages in the PRECAMBRIAN, and certainly a major one immediately before the start of the CAMBRIAN. There were a number in the PALEOZOIC, including a major ice age with a complicated cycle running through the CARBONIFEROUS (MISSISSIPPIAN and PENNSYLVANIAN) and early PERMIAN. The ice age that we know most about, however, is that of the QUATERNARY period, continuing through most of the PLEISTOCENE and whose last glacial period ended about 10,000 years ago, denoting the start of the HOLOCENE. (See GEOLOGY.) At their greatest, the Pleistocene glaciers covered about a third of the Earth's surface, or some 45 million km^2, and may have been up to 3km thick in places. They covered most of Canada, N Europe and N Russia, N parts of what is now the US, and, in the S Hemisphere, Antarctica, parts of South America, and some other areas.

Theories about the cause of ice ages include that the Sun's energy output varies, that the Earth moves with respect to its axis, that CONTINENTAL DRIFT may alter global climatic conditions, and that volcanic dust in the ATMOSPHERE could reduce the amount of solar heat received. (See also EARTH; GLACIER; VOLCANO.) (◆ page 106)

Iceberg
A large, floating mass of ice. In the S Hemisphere, the Antarctic ice sheet overflows its land support to form shelves of ice on the sea; huge pieces, as much as 200km across, break off to form icebergs. In the N Hemisphere, icebergs are generally not over 150m across. Most are "calved" from some 20 GLACIERS on Greenland's W coast. Small icebergs (growlers) may calve from larger ones. Some 75% of the height and over 85% of the mass of an iceberg lies below water. Northern icebergs usually float for some months to the Grand Banks, off Newfoundland, there melting in a few days. They endanger shipping, the most famous tragedy being the sinking of the *Titanic* (1912). The International Ice Patrol now keeps a constant watch on the area.

Icecap see GLACIER.

Iceland spar see CALCITE.

Ichthyology
The study of fish.

Ichthyosaurs
Extinct REPTILES that lived from the Triassic to the mid-Cretaceous periods, 248-80 million years ago. They were aquatic and appear almost fish-like, with streamlined bodies, limbs modified as fins and a fish-like tail. They gave birth to live young, an ability that is not found in most living reptiles.

Id
The part of the unconscious mind that consists of a system of primitive urges, and according to Freud persists unrecognized into adult life. (See also EGO; SUPEREGO.)

Identity (personal identity)
In psychology, the individual's sense of being a distinct, continuous entity, roughly corresponding to self-awareness (see CONSCIOUSNESS).

Igneous rocks
One of the three main types of ROCKS, those whose origin is related to heat. They crystallize from the MAGMA either at the Earth's surface (extrusion) or beneath (intrusion). There are two main classes. Volcanic rocks are extruded (see VOLCANISM), typical examples being LAVA and PYROCLASTIC ROCKS. Plutonic rocks are intruded into the rocks of the Earth's crust at depth, a typical example being GRANITE: those forming near to the surface are sometimes called hypabyssal rocks. Types of intrusions include BATHOLITHS, DIKES, SILLS and LACCOLITHS. As plutonic rocks cool more slowly than volcanic rocks they have a coarser texture, more time being allowed for crystal formation. The other two principal types of rocks are METAMORPHIC and SEDIMENTARY ROCKS. (◆ page 208)

Idiopathic
When used of diseases, those that are of spontaneous or constitutional origin, and have no apparent cause.

Ignis fatuus see WILL-O'-THE-WISP.

Ileitis
INFLAMMATION of the ileum, part of the small intestine (see ENTERITIS; GASTROINTESTINAL TRACT).

Ileostomy
A form of STOMA.

Ileum see GASTROINTESTINAL TRACT.

Illusion
An erroneous perception of reality, often the result of misinterpretation by the brain of information received by the SENSES.

Ilmenite
Hard, black OXIDE mineral, iron (II) titanium (IV) oxide ($FeTiO_3$), the chief ore of TITANIUM. It occurs widely in IGNEOUS ROCKS, notably in the USSR, Norway, Quebec and Wyoming. It crystallizes in the rhombohedral system.

Image, Optical
A representation of an object formed in an optical instrument. Although a virtual image has no physical existence – light only seems to come from its apparent position – light actually comes to a focus in a real image and these can be made visible by using a suitable screen.

Image processing
The processing of images using computer techniques. Data from the image are digitized, and mathematical operations are applied to the data, usually by a COMPUTER, to "manipulate" the imagery. It covers a range of processes including enhancing images, emphasizing particular features, or storing the image digitally for later retrieval or transmission. Image processing is used routinely to process photographs from data received via remote sensors on board spacecraft and SATELLITES.

Imago
A term referring to the adult insect emerging from a pupa in complete METAMORPHOSIS.

Imides
Organic compounds with the general formula R.CO.NH.CO.R′, obtained by reacting an AMIDE with an ACID CHLORIDE, or AMMONIA with an ACID ANHYDRIDE.

Immersion foot (trench foot)
Disease of the feet after prolonged immersion in water, due to a combination of vasoconstriction and waterlogging. It usually starts with red, cold and numb feet, which on warming develop through EDEMA and blistering to ulceration and sometimes skin GANGRENE.

Immune system see IMMUNITY.

Immunity
The ability to resist infection by pathogens (viruses, bacteria and other unicellular parasites) and by multicellular endoparasites. The term immunity is often used to refer to specific immunity against a particular organism, but the immune system also includes many non-specific defenses against invasion. In mammals these include the acid in the stomach, an antibacterial enzyme known as lysozyme in the tears and urine, and the physical barrier presented by the skin and mucous membranes. Backing up these defenses are cells known as polymorphonuclear neutrophils, PMNs (also known as polys, neutrophils or granulocytes), which ingest foreign organisms by PHAGOCYTOSIS and kill them. A second set of phagocytic cells, the acrophages, can also ingest foreign particles indiscriminately, but they play a part in specific immune responses as well.

Specific immune responses function by recognizing particular chemical structures – known as antigens – on the surface of invading cells. An antigen can be a protein, lipid, carbohydrate or any other molecule. These antigens interact with protein molecules produced by the host, the immunoglobulins, which bind the antigen in much the same way that an ENZYME binds its substrate. Specific immune responses involve several different types of cell. One type, the B-lymphocyte, or B-cell, is capable of producing free immunoglobulins, known as antibodies, when immunoglobulin receptors on its cell surface combine with antigen. The antibodies it releases have the same structure, and therefore the same specificity, as the immunoglobulin receptors on the cell surface. Each B-cell produces its own unique type of immunoglobulin due to random and irreversible rearrangements of the DNA in its chromosomes. Thus there are hundreds of thousands of different immunoglobulins available to fight off invasion by a new invading cell – one of them is bound to "fit" chemically with the antigens on that cell's surface. When it does, it stimulates the B-cell carrying it to divide rapidly, and its daughter cells release antibodies that can bind the antigen. (Once bound, the cell bearing the antigen is more easily ingested by phagocytic cells – PMNs and macrophages. Binding between the antigen and antibody also activates the complement system, described below.)

In order to release antibodies, however, the B-cell must interact with two other cells. One is a type of T-lymphocyte, or T-cell, known as a T-helper cell or Th. This also carries immunoglobulins that are highly variable, as in the B-cells. If the Th immunoglobulin "recognizes" antigens from the invading cell it binds with the antigens concerned and releases chemical messengers that stimulate the B-cell to action. The second is a macrophage or similar cell, which acts as an antigen-presenting cell, or APC. This cell ingests the foreign cell, partially digests it, and then "presents" fragments of antigen from the invader on its own cell surface, where the Th cell rcognizes it. Without an APC, the Th cell cannot become activated, and without an active Th cell the B-cell cannot begin to multiply and release antibodies. These elaborate multicell mechanisms are probably needed to maintain control over the immune system, which can be life-threatening if it begins attacking the host's own cells.

A second specific immune mechanism involves another type of T-cell, known as a cytotoxic T-cell, or Tc. This carries immunoglobulins on its surface, and when these bind to antigens on the surface of an invading cell, the Tc cell is activated to kill the invading cell. As with B-cells, a Th cell and an APC are needed to activate the Tc cell. The main function of Tc cells is to kill body cells that have been invaded by a virus – these begin to show the viral proteins on their cell surface soon after infection, and before the new viruses are released, so the Tc cells can obliterate the infected cell before the virus has had a chance to proliferate.

A third type of T-cell, known as Tdth, responds to antigens that its immunoglobulins recognize by releasing various chemical messengers. These attract other cells, particularly macrophages, to the site to fight off the infection. The observed effect is that the tissues become inflamed as cells pile into the area.

Following infection, B-cells, Tc cells and Tdth cells all leave dormant memory cells, with the same immunoglobulin specificity, which can spring into action again if the same invading organism is encountered. This is why the host becomes "immune" to many diseases after being infected once. Immunity to viral diseases, such as the common cold, is rarely effective against a subsequent infection because the virus can mutate so quickly, and a new, mutant form, with different coat proteins, is not recognized by the memory cells.

Another important factor in immunity is the complement system, a set of proteins that, when activated, form powerful enzymes. These attack and destroy the membranes of the invading cell. The complement system is activated by the formation of antigen-antibody complexes, or directly by the presence of foreign substances.

Several different control pathways operate on the immune system, one involving a fourth type of T-cell, the suppressor T, or Ts cell. When these control mechanisms fail the individual may suffer ALLERGIES to harmless substances such as pollen or food molecules, or AUTOIMMUNE DISEASES.

Immunoglobulins
Proteins that combine with foreign molecules (antigens) in the body (see ANTIBODIES AND ANTIGENS). Those that are released into the blood serum or other body fluids are known as antibodies. Others are carried on the surface of cells, and enable those cells to recognize antigens. Several types of immunoglobulin are recognized. Although they share basic structural features they differ in size, site, behavior and response to different antigens. Absence of all or some immunoglobulins causes disorders of IMMUNITY, increasing susceptibiity to infection, while the excessive formation of one type is the basis for myeloma, a disease characterized by bone pain, pathological fractures and liability to infection. Immunoglobulins are available for replacement therapy; a type from highly immune subjects is sometimes used to protect against certain diseases, eg serum hepatitis, TETANUS.

Immunosuppressive drugs
Drugs that suppress the immune response. They are used to prevent the body from rejecting transplanted tissues or organs, but also make the person more susceptible to infections and cancers.

Impedance
The ratio of the AC voltage applied to an electric circuit to the current it produces. It is a generalization of the concept of electric resistance (see ELECTRICITY) to include cases where the current oscillates ahead of or behind the voltage (ie out of PHASE with it) based on the mathematics of complex numbers. The term is also applied to the ratio between the driving force and response of other oscillatory or wave systems.

Impetigo
A skin infection caused by streptococcal bacteria. Clusters of pustules the size of pinheads form, and gradually merge with each other. They then burst, forming a yellow scab that is firmly attached at the center but raised at the margins.

Implantation
In mammals, the earliest stage of EMBRYO development in which the embryo becomes attached to the wall of the uterus. After fertilization by sperm, the egg divides into a small ball of cells, whose outer layer is specialized to invade the uterine wall, which is itself prepared for implantation. The interface between embryo and uterus develops into the PLACENTA. In some mammals implantation can be delayed by several weeks or months so that the overall GESTATION period is much longer. Delayed implantation is seen in many seals and sea-lions, for example, allowing them a gestation period of a year so that they can mate while on the breeding ground, after giving birth to the previous year's pup.

Implosion
A bursting inward, as opposed to explosion, a bursting outward. Implosion due to gravitational collapse is an important end-stage in the lives of STARS (see also BLACK HOLES).

Impotence
In men, inability to maintain sufficient erection and hardness of the penis to permit intromission into the vagina and to maintain it until ejaculation. It has a variety of causes, including alcohol, medicines, atherosclerosis, and psychological factors.

Imprinting
A form of learning that occurs principally in young animals. It has been studied only in birds and mammals, but may occur in other animals that show extended parental care. One type of imprinting enables a young animal to recognize its parent or parents by either their appearance, call or smell. Once imprinting has occurred, the young one will follow its parent, and be able to distinguish it from other adults. There is a fairly short period, usually a matter of weeks or days, in which imprinting can take place, known as the sensitive period. If the young are separated from the mother during this period and exposed to another object, they will imprint on this. Experiments have shown that young birds will imprint on a variety of objects, including human beings, moving cardboard boxes and flashing lights. The sensitive period for sexual imprinting is generally longer than for parental imprinting and involves recognition of the general species pattern, rather than an individual. It teaches the young to recognize potential mates in later life, and is much more common in birds than in mammals. If the eggs of one bird species are placed under a foster parent of a related species, the young birds will often court an individual of the foster species rather than one of their own when they reach maturity. After giving birth, adult females (including humans) undergo a period of imprinting in which they learn to recognize the newborn offspring.

Impulse
The integral of a FORCE over an interval of time. By NEWTON's second law, the impulse (a VECTOR quantity) equals the change of MOMENTUM produced by the force. It is a useful concept when a large and variable force acts for a short time, as in an impact.

Inbreeding
Breeding between close relatives. For some species inbreeding is the norm, as in self-fertilizing flowering plants, for example. For others, outbreeding is the norm, and if inbreeding occurs in such species the phenomenon of inbreeding depression may be seen. This usually occurs because the species has accumulated various deleterious mutations that are normally not seen. The fact that they are recessive, and that progeny produced by outbreeding are usually HETEROZYGOUS for most genes, means that these deleterious alleles are masked by the presence of normal, dominant genes (see DOMINANCE). When inbreeding occurs, however, individuals that are HOMOZYGOUS for the mutant allele may be produced, because the parents both carry the same mutation. In the homozygous condition, the recessive deleterious mutation is expressed, and inbreeding depression results.

Inch (in)
A unit of length in the US Customary and British Imperial systems. Since 1959 it has been defined exactly equal to 25.4mm.

Inclined plane see MACHINE.

Incontinence
The inability to control the release of urine, feces, or both. It is normal only in young children.

Incubation
The technique of growing bacteria (and sometimes viruses) in the laboratory. Also the period of time that elapses between infection with a microorganism and manifestation of the disease.

Incubator
A device used for incubating microorganisms. The term is also used for an enclosed cot in which a baby, particularly if premature, is placed to create an ideal protective and controllable environment. Temperature is regulated thermostatically and the possibility of infection is minimized; the air in the incubator may be enriched with controlled amounts of oxygen. Nursing is carried out through portholes. The use of incubators has been significant in reducing mortality in premature infants.

Indicator
Substance that indicates when the concentration of a chemical species has passed a certain threshold value, by a change of color, turbidity or fluorescence. They are generally used to find the end-point of a TITRATION. The indicator is a substance existing in two visibly different forms in an EQUILIBRIUM that is the same kind as that of the reaction being followed. Thus, to follow an acid-base titration, a conjugate ACID-base system is used as indicator which changes color over a narrow range of pH corresponding to the end-point. (For this to take place, the equilibrium

PERIODIC TABLE OF ELEMENTS

Ordering the elements

If the elements are placed in order of increasing atomic number, correlations are found between their chemical and physical properties. Elements with similar physical properties are found at definite intervals of atomic number and it is possible to draw a table to show these similarities.

In a periodic table, the elements are separated into horizontal periods and vertical groups. Elements within each group are very like one another. Across the periods, the properties of the elements change steadily. The table has allowed chemists to predict the existence of new elements.

▼ Elements arranged by atomic number repeat properties such as their ionization energies – the energy needed to pluck the outer electrons from an atom. Some elements are loath to give up electrons, especially if they have a full shell. Thus in the graph the pinnacles are capped by the gases in group 18. The elements of group 1 offer least resistance to losing their sole outer electrons, and so are at the low points.

▼▼ Electrons occupy shells around the nucleus. The inner shell can hold two electrons, the next eight, and so on, up to 32 for the outer shells. Within these shells are subshells of 2, 6, 10 or 14 electrons with the symbols s, p, d and f. The table is built up by a step-by-step addition of electrons to atoms, from the nucleus outwards. A full shell results in an atom of great chemical inertness, such as helium or neon.

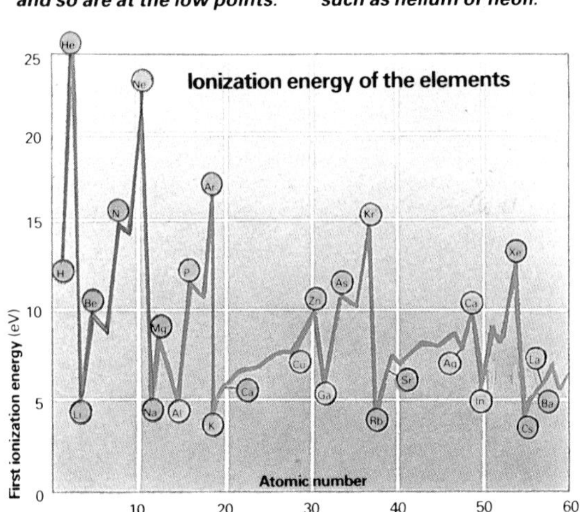

The "buildup" of electrons

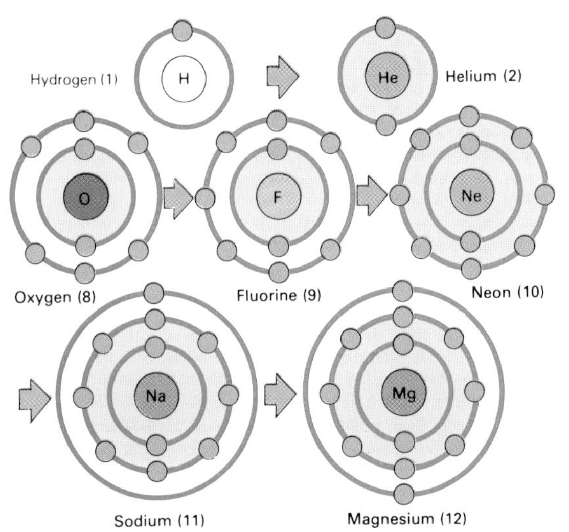

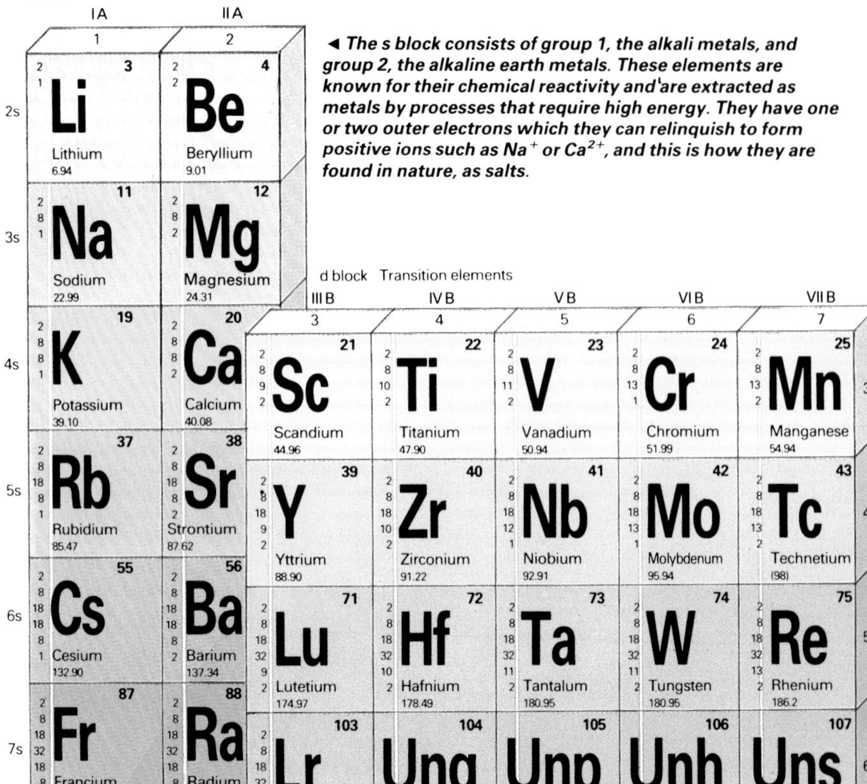

◄ The s block consists of group 1, the alkali metals, and group 2, the alkaline earth metals. These elements are known for their chemical reactivity and are extracted as metals by processes that require high energy. They have one or two outer electrons which they can relinquish to form positive ions such as Na^+ or Ca^{2+}, and this is how they are found in nature, as salts.

f block Inner transition elements

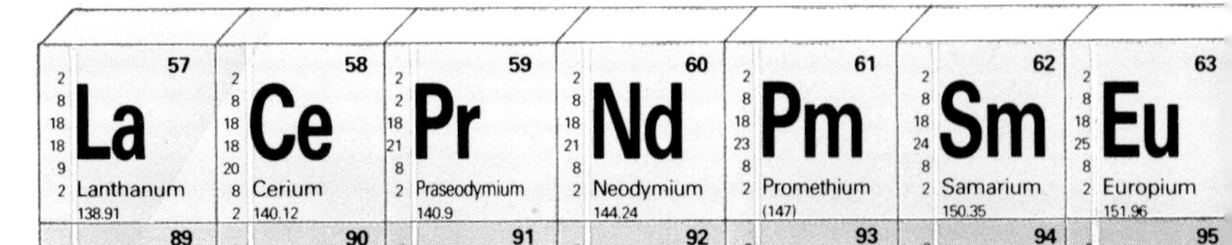

▼ Recognizing a metal is fairly easy. It is hard, dense, shiny and gives out a characteristic "ping" when struck. Chemists, however, define a metal by its ability to conduct electricity. Conductivity stems from the free moving electrons that are a feature of the chemical bonding in metals. A few elements are semi-conductors, and these straddle the boundary between the metals and the nonmetals.

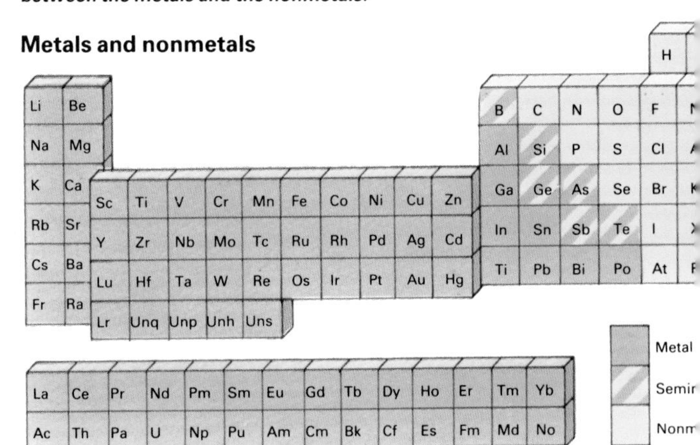

▼ *The d block elements are all metals, in which the d shell, underlying an outer s shell, is being filled with electrons. In their compounds these elements exhibit several oxidation states, depending on the number of their electrons transferred to other atoms. Thus manganese ranges from two [Mn(II)] as in Mn^{2+} salts, up to seven [Mn(VII)] as in potassium permanganate, $KMnO_4$.*

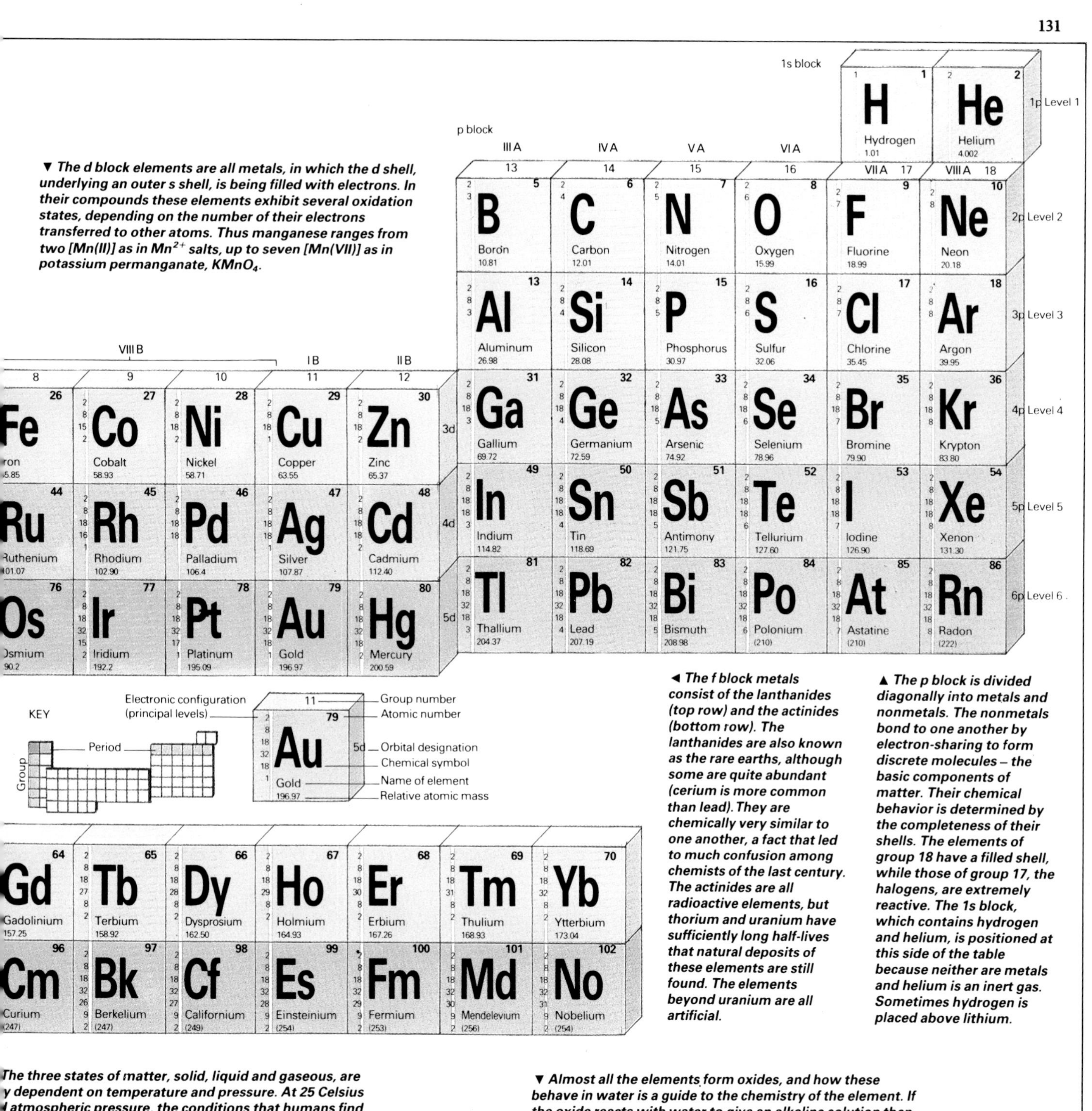

◀ *The f block metals consist of the lanthanides (top row) and the actinides (bottom row). The lanthanides are also known as the rare earths, although some are quite abundant (cerium is more common than lead). They are chemically very similar to one another, a fact that led to much confusion among chemists of the last century. The actinides are all radioactive elements, but thorium and uranium have sufficiently long half-lives that natural deposits of these elements are still found. The elements beyond uranium are all artificial.*

▲ *The p block is divided diagonally into metals and nonmetals. The nonmetals bond to one another by electron-sharing to form discrete molecules – the basic components of matter. Their chemical behavior is determined by the completeness of their shells. The elements of group 18 have a filled shell, while those of group 17, the halogens, are extremely reactive. The 1s block, which contains hydrogen and helium, is positioned at this side of the table because neither are metals and helium is an inert gas. Sometimes hydrogen is placed above lithium.*

The three states of matter, solid, liquid and gaseous, are y dependent on temperature and pressure. At 25 Celsius atmospheric pressure, the conditions that humans find st agreeable, only two elements are depicted as liquids – mine and mercury. At 30 Celsius, cesium and gallium uld also have been shown as liquids. These melt at 28·5 29·8 Celsius respectively.

▼ *Almost all the elements form oxides, and how these behave in water is a guide to the chemistry of the element. If the oxide reacts with water to give an alkaline solution then it is described as a basic oxide; if it reacts to form an acid solution it is described as an acidic oxide. Metals give basic oxides, nonmetals give acidic ones. Amphoteric oxides, such as aluminum oxide, show both kinds of behavior.*

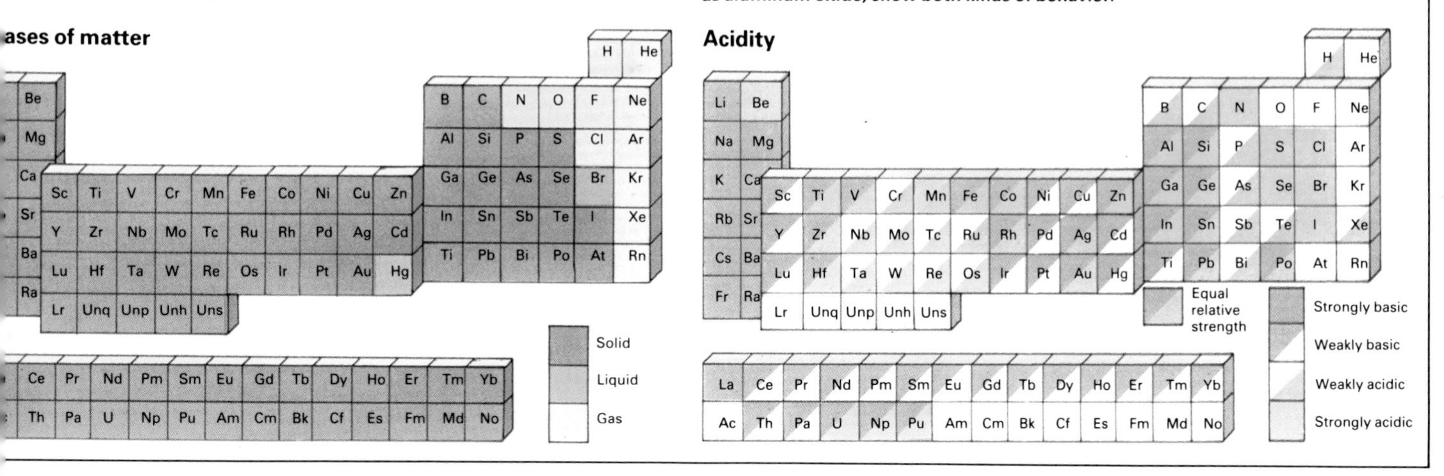

constant K of the indicator must approximately equal the hydrogen-ion concentration at the end-point.) A universal indicator is a mixture of indicators that changes color continuously over a wide pH range, used as a quick guide to acidity. To follow an OXIDATION-reduction reaction, an indicator is used that exists reversibly in an oxidized or reduced state, its oxidation potential being about the same as that of the reaction. For a precipitation or complexing reaction, an indicator is used that itself forms a colored precipitate or complex with excess added reagent. A good indicator must be visible at such low concentrations that it does not interfere in the reaction.

Indigestion see DYSPEPSIA.

Indium (In)

Rare metal, very soft and silvery white, in Group IIIA of the PERIODIC TABLE, resembling ALUMINUM. It is prepared from flue dust residues of zinc processing. Indium forms trivalent compounds and some monovalent ones. It is used in solders, low melting-point alloys, germanium TRANSISTORS, glass seals, bearing alloys, and (combined with Group VA elements) in SEMICONDUCTORS. AW 114.8, mp 156.6°C, bp 2050°C, sg 7.31 (20°C). (♦ page 130)

Inductance (Self-)

The ratio of the voltage induced in an electric circuit (see INDUCTION, ELECTROMAGNETIC) to the rate of change of the current in it. It depends on the circuit geometry, being large for coils and small for extended circuits, and is greatly increased by the presence of ferromagnetic materials (see MAGNETISM). Voltages induced by currents in a different circuit are measured in terms of mutual inductance. Inductors have an IMPEDANCE to AC currents proportional to the current, and are widely used in electronics. The SI unit of inductance is the HENRY (H).

Induction, Electromagnetic

The phenomenon in which an ELECTRIC FIELD is generated in an electric circuit when the number of MAGNETIC FIELD lines passing through the circuit changes, independently discovered by M. FARADAY and J. HENRY. The voltage induced is proportional to the rate of the change of the field, and large voltages can be produced by switching off quite small magnetic fields suddenly. Frequently, the magnetic field is itself generated by an electric current in a coil, in which case the voltage induced is proportional to the rate of change of the current (see INDUCTANCE).

The principle finds numerous applications in electric GENERATORS and MOTORS, MICROPHONES, TRANSFORMERS, and engine ignition systems (see INDUCTION COIL, MAGNETO). In the less familiar technique of induction heating, widely used in metal working, an object is heated by currents created in it by the voltage induced by a high-frequency current in a nearby coil; as the coil field will pass through insulators without heating them, the principle can be applied to produce "cold hob" electric stoves.

Induction hardening

The use of electromagnetic INDUCTION to heat metals rapidly in order to harden them (see METALLURGY; STEEL). A very high-frequency current is passed through an induction coil surrounding the workpiece.

Industrial melanism

The appearance of darkly pigmented varieties as a response to the blackening of surfaces by air pollution. Overproduction of the pigment melanin in certain individuals, due to mutation, is not uncommon, but as such individuals do not usually survive to breed the mutation dies out. If, however, a melanic individual has better camouflage on soot-blackened surfaces than the normal form, then the mutation can spread through a population. The phenomenon was first observed in the peppered moth, *Biston betularia*, in the industrial north of England in the late 19th century. It has since been found in other insect species and a few spiders. Pollution controls have produced a decline in the proportion of melanics in certain areas.

Inert gases

Former name for the NOBLE GASES. (♦ page 130)

Inertia

Property of all MATTER, representing its resistance to any alteration of its state of MOTION. The MASS of a body is a quantitative measure of its inertia; a heavy body has more inertia than a lighter one and needs a greater FORCE to set it in motion. NEWTON's laws of motion (see MECHANICS) depend on the concept of inertia. In EINSTEIN's theory of RELATIVITY, the inertial properties of matter are interrelated with its total ENERGY content.

Inertial guidance

An automatic navigational apparatus carried in guided missiles, airplanes, ships and submarines, which depends on the forces of INERTIA for sensing changes in the magnitude and direction of the vehicle's motion. ACCELEROMETERS are mounted on gyrostabilized platforms to isolate them from the vehicle's angular motion; by measuring the forces needed to keep a suspended mass stationary with respect to the moving vehicle, they sense changes in its motion and gravitational fields. The orientations of the accelerometers are found from reference directions provided by GYROSCOPES. A computer calculates velocities or distances from the instrument signals and can compare these with stored data. The accuracy of the system is improved by a method of FEEDBACK called Schuler tuning.

Infantile paralysis see POLIOMYELITIS.

Infarct

An area of localized tissue death, caused by blockage to the artery that supplied it with blood. Some tissues receive blood from a network of arteries and thus have alternative supplies; these are protected unless all their arteries become infarcted.

Infectious diseases

DISEASES caused by any microorganism, but particularly VIRAL and BACTERIAL DISEASES and PARASITIC DISEASES, in which the causative agent may be transferred from one person to another (directly or indirectly). Knowledge of the stages at which a particular disease is liable to infect others and of its route helps physicians to limit the spread of diseases in EPIDEMICS.

Inferiority complex

Term used by ADLER and now mainly by psychoanalysts to describe the COMPLEX of fears and EMOTIONS arising out of feelings of inferiority or inadequacy. (See also SUPERIORITY COMPLEX.)

Infertility

Inability to produce offspring for any reason. May be due to an inability to produce eggs or sperm, defects in the ducts that carry them, difficulty in achieving fertilization or, in the case of mammals, of carrying the fetus to maturity.

Infestation

A disease caused by an organism on the outside of the body; harmful organisms inside the body cause infections.

Inflammation

The complex of reactions triggered in body TISSUES by injury and infection. It is typified by redness, heat, swelling and pain in the affected part. The first change is in the CAPILLARIES, which dilate (causing ERYTHEMA), and become more permeable to cells and PLASMA (leading to EDEMA). White BLOOD cells accumulate on the capillary walls and pass into affected tissues; foreign bodies, dead tissue and bacteria are taken up and destroyed by phagocytosis and ENZYME action. Active substances produced by white cells encourage increased blood flow and white cell migration into the tissues. LYMPH drainage is important in removing edema fluid and tissue debris. ANTIBODY AND ANTIGEN reactions, ALLERGY and other types of IMMUNITY are concerned with the initiation and the perpetuation of inflammation.

Inflationary Universe see COSMOLOGY.

Inflorescence

Term applied to a plant shoot bearing two or more flowers. There are several types of inflorescence, classified according to the arrangement of the individual flowers within the inflorescence. In the type of inflorescence known as a raceme the flowers are attached to the main flower axis by short stalks of equal length – as, for example, in the hyacinth; while in the spike the flowers are directly attached to the main axis, as in the gladiolus. Plants such as lilac and oats have an inflorescence similar to a raceme, but the flower stalks bear more than one flower. This formation is called a panicle. In the corymb, the pedicels are of unequal length so that the inflorescence has a flat-topped appearance, for example hawthorn. In some plants, particularly those of the daisy family, all the flowers are bunched on a flat disk, this arrangement being known as a capitulum or head. In the umbel the flower stalks arise from a central point, as in cow parsley.

Influenza

A viral disease causing respiratory symptoms, fever, malaise, muscle pains and headache and often occurring in rapidly spreading EPIDEMICS. GASTROINTESTINAL TRACT symptoms may also occur. Rarely, it may cause a severe viral PNEUMONIA. A characteristic of influenza viruses is their property of changing their antigenic nature frequently, so that IMMUNITY following a previous attack ceases to be effective. This also limits the usefulness of influenza VACCINATION.

Information retrieval

A branch of technology of ever-increasing importance as man attempts to cope with the "information explosion". To store and have reference to the vast amount of printed matter produced annually is impossible for most libraries. The problem can be solved by microphotography. Pages are photographed at a reduction (typically to about 1/20 and stored on 35mm or 16mm film (microfilm), on transparent cards measuring about 100×150mm (microfiches) or as positive prints on slightly smaller cards (microcards)). Videotape is also used. Reference may be manual or by machine, usually computer. The information must be classified so that the user may gain rapid access either to a particular book or paper or to all the relevant material on a particular subject.

In COMPUTERS, information retrieval involves a reverse of those operations used for data storage. The operator inserts a classification which the computer matches with the classification in its memory. (See also DATABASE; NETWORK.)

Information technology (IT)

The convergence of COMPUTER technology, TELECOMMUNICATIONS and MICROELECTRONICS for the processing and manipulation of information. The computers' processing and memory power are used to enhance the communications medium to the benefit of the user. Examples of IT are found in NETWORKS of computer terminals or WORD-PROCESSORS, electronic mail services, stored numbers and call transfers in telephone systems, and on-line DATABASES. The term IT is often misused and rarely means precisely the same thing to different firms, research groups or even nations. It is even used, on occasion, to encompass ARTIFICIAL INTELLIGENCE.

Infrared astronomy
The study of INFRARED RADIATION emitted by celestial objects. By placing an infrared telescope in space, the IR background can be reduced a million times. IR sources include cool dust clouds and warm dust around young stars in our GALAXY, in addition to nearby galaxies, active galaxies and quasars.

Infrared radiation
ELECTROMAGNETIC RADIATION of wavelength between 780nm and 1mm, strongly radiated by hot objects and also termed heat radiation. Detected using PHOTOELECTRIC CELLS, BOLOMETERS and photographically, it finds many uses – in the home for heating and cooking and in medicine in the treatment of muscle and skin conditions. Infrared absorption spectroscopy is an important analytical tool in organic chemistry. Military applications (including missile-detection and guidance systems and night-vision apparatus) and infrared PHOTOGRAPHY (often FALSE-COLOR PHOTOGRAPHY) exploit the infrared window, the spectral band between 7.5 and 11μm in which the ATMOSPHERE is transparent. This and the high infrared reflectivity of foliage give infrared photographs their striking, often dramatic clarity, even when exposed under misty conditions. (♦ page 25, 118)

Inhibition
In biochemistry, see ENZYMES.

Inhibition
The action of a mental process or function in restraining the expression of another mental process or function, eg fear of social condemnation inhibiting fulfilment of sexual desire.

Injection
The administration of a substance, usually a DRUG or vaccine, by SYRINGE and needle into the body. This may be into the skin (intradermal), blood vessels (intravenous or occasionally intraarterial), muscles (intramuscular) or cerebrospinal fluid (intrathecal). Injections may also be placed in the fat layer just under the skin (subcutaneously). Injection techniques bypass the gastrointestinal tract, which may be poor at absorbing the substance or may destroy the agent.

Inkblot test (Rorschach test) see RORSCHACH, H.

Inoculation
The INJECTION or introduction of microorganisms or their products into living TISSUES or culture mediums.

Inorganic chemistry
Major branch of CHEMISTRY comprising the study of all the elements and their compounds, except carbon compounds containing hydrogen (see ORGANIC CHEMISTRY). The elements are classified according to the PERIODIC TABLE. Classical inorganic chemistry is largely descriptive, synthetic and analytical: modern theoretical inorganic chemistry is hard to distinguish from PHYSICAL CHEMISTRY.

Insanity
Term descriptive of an individual's mental state employed in legal and popular usage.

Insecticide
Any substance toxic to insects and used to control them in situations where they cause economic damage, or endanger the health of man and his domestic animals. There are three main types: stomach insecticides, which are ingested by the insect with their food; contact insecticides, which penetrate the cuticle, and fumigant insecticides, which are inhaled. Stomach insecticides are often used to control chewing insects like caterpillars and sap-sucking insects like aphids. They may be applied to the plant prior to attack and remain active in or on the plant for a considerable time. They must be used with considerable caution on food plants or animal forage. Included in this group are the systemic insecticides, which are absorbed by the plant and transported to all its parts. Contact insecticides include the plant products NICOTINE, derris and pyrethrum, which are quickly broken down, and synthetic compounds such as chlorinated HYDROCARBONS, organophosphates (malathion, parathion) and carbamates. Highly persistent insecticides may be concentrated in food chains and exert harmful effects on other animals, including humans.

Insectivores
An order of MAMMALS that includes the shrews, hedgehogs, moles and tenrecs. All are small and have pointed snouts, often with prominent whiskers and sharp teeth. The word is also used to mean any insect-eating animal.

Insectivorous plants (carnivorous plants)
Specialized plants whose leaves are adapted to trap and digest insects, which supplement their food supply. They normally live in boggy habitats or as epiphytes. The insects may be caught in vase-like traps (eg pitcher plant), by leaves that spring shut (eg venus fly trap), by a trapdoor (eg bladderwort) or on sticky leaves (eg sundew). The captured insects are broken down by ENZYMES secreted from the plants and the products absorbed.

Insects
The major group of terrestrial invertebrates, and one of the most successful animal groups. Once classified as ARTHROPODS, they are now placed in a new phylum, the Uniramia, which also includes the MYRIAPODS (centipedes and millipedes). The insects can be divided into primitive flightless forms (the APTERYGOTES) and winged forms, together with flightless species that are descended from them (the Pterygotes). The latter can be further divided into the ENDOPTERYGOTES, which undergo a pupal stage in their life cycle, and the more primitive EXOPTERYGOTES, which hatch as miniature adults, or nymphs. The insects and myriapods are thought to have evolved from the same stock as the modern annelid worms, and the ONYCHOPHORANS may represent an intermediate stage. (See also SOCIAL INSECTS.) (♦ page 16)

Instinct
An important concept in early theories of animal and human behavior, which is now regarded as somewhat misleading. Instinct can be defined as a stereotyped pattern of actions that is seen in all equivalent members of a species (eg all juveniles; all females with young) and that requires no learning or experience before it can be expressed; it is assumed to be inherited genetically. Closer study of actions described as instinctive shows there to be a learning component that refines the action – frequently the behavior is less stereotyped initially than it is following learning, the opposite of what the traditional view of instinct would predict. Although some elements of behavior are clearly innate and inherited, it is now evident that these are extensively modified by learning.

Instruments, Scientific
Devices used for measurement and hence for scientific investigation and control. They extend the observing faculties of the human senses, providing accuracy and a greater range. They can also detect and measure phenomena such as X-rays which humans cannot sense. Early instruments, used mainly in the fields of astronomy, navigation and surveying, measured the basic quantities of mass, length, time and direction (see BALANCE; COMPASS). With the rise of modern science came several instruments including the MICROMETER, MICROSCOPE, TELESCOPE and THERMOMETER. A few simple instruments, such as the ruler or balance, work by direct comparison, but most are transducers, representing the quantity measured by another sensible quantity (usually the position of a pointer on a scale). All instruments require initial CALIBRATION against a known or calculable standard. In general, an instrument interacts with the measured phenomenon, and the resultant change in its state is amplified if necessary, displayed by means of a pointer, pen, light beam, oscilloscope etc., and recorded, usually on chart paper or by photography. Although precision instruments are designed for high accuracy, inevitably errors are introduced: amplification produces NOISE, the slowness of the instrument's response results in lag and damping, and the intrinsic nature of the response may be defective owing to hysteresis or drift; moreover, the observer may misread the scale because of parallax or interpolation errors. Most fundamental of all, the act of measuring a system may significantly alter the state of the system (see also UNCERTAINTY PRINCIPLE).

Insulation, Electric
The containment of electric currents or voltage by materials (insulators) that offer a high resistance to current flow, will withstand high voltages without breaking down, and will not deteriorate with age. Resistance to sunlight, rain, flame or abrasion may also be important. The electrical resistance of insulators usually falls with temperature (paper and asbestos being exceptions) and if chemical impurities are present. The mechanical properties desired vary with the application: cables require flexible coatings, such as polyvinyl chloride, while glass or porcelain are used for rigid mountings, such as the insulators used to support power cables. In general, good thermal insulators are also good electrical ones.

Insulation, Thermal
Reduction of the transfer of heat from a hot area to a cold. Thermal insulation is used for three distinct purposes: to keep something hot; to keep something cold; and to maintain something at a roughly steady temperature. HEAT is transferred in three ways: CONDUCTION, CONVECTION and RADIATION. The vacuum bottle thus uses three different techniques to reduce heat transfer: a vacuum between the walls to combat conduction and convection; silvered walls to minimize the transmission of radiant heat from one wall and maximize its reflection from the other; and supports for the inner bottle made of cork, a poor thermal conductor. (See also POLYSTYRENE; REFRACTORY.)

Insulin
HORMONE important in METABOLISM, produced by the islets of Langerhans in the PANCREAS, which act as an ENDOCRINE GLAND. Insulin is the only hormone that reduces the level of sugar in the BLOOD, and is secreted in response to a rise in blood sugar (eg after meals, or in conditions of stress); the sugar is converted into GLYCOGEN in the cells of MUSCLE and the LIVER under the influence of insulin. Absence or a relative failure in secretion of insulin occurs in DIABETES MELLITUS, in which blood sugar levels are high and in which sugar overflows into the urine. The isolation of insulin as a pancreatic extract by F.G. BANTING and C.H. BEST in 1921 was a milestone in medical and scientific history. It is a PROTEIN made up of fifty AMINO ACIDS as two peptide chains linked by sulfur bridges. (♦ page 28, 111)

Integrated circuit
A single structure in which a large number of individual electronic components are assembled. (See ELECTRONICS.)

Intelligence
The ability to solve novel problems by reasoned thought or the application of prior experience. Also used to mean any mental abilities of a higher order than INSTINCT, simple learning (eg learning that one food item is good to eat while another is not), or imitation of others. The intelligence of

PLANETS Internal Structure

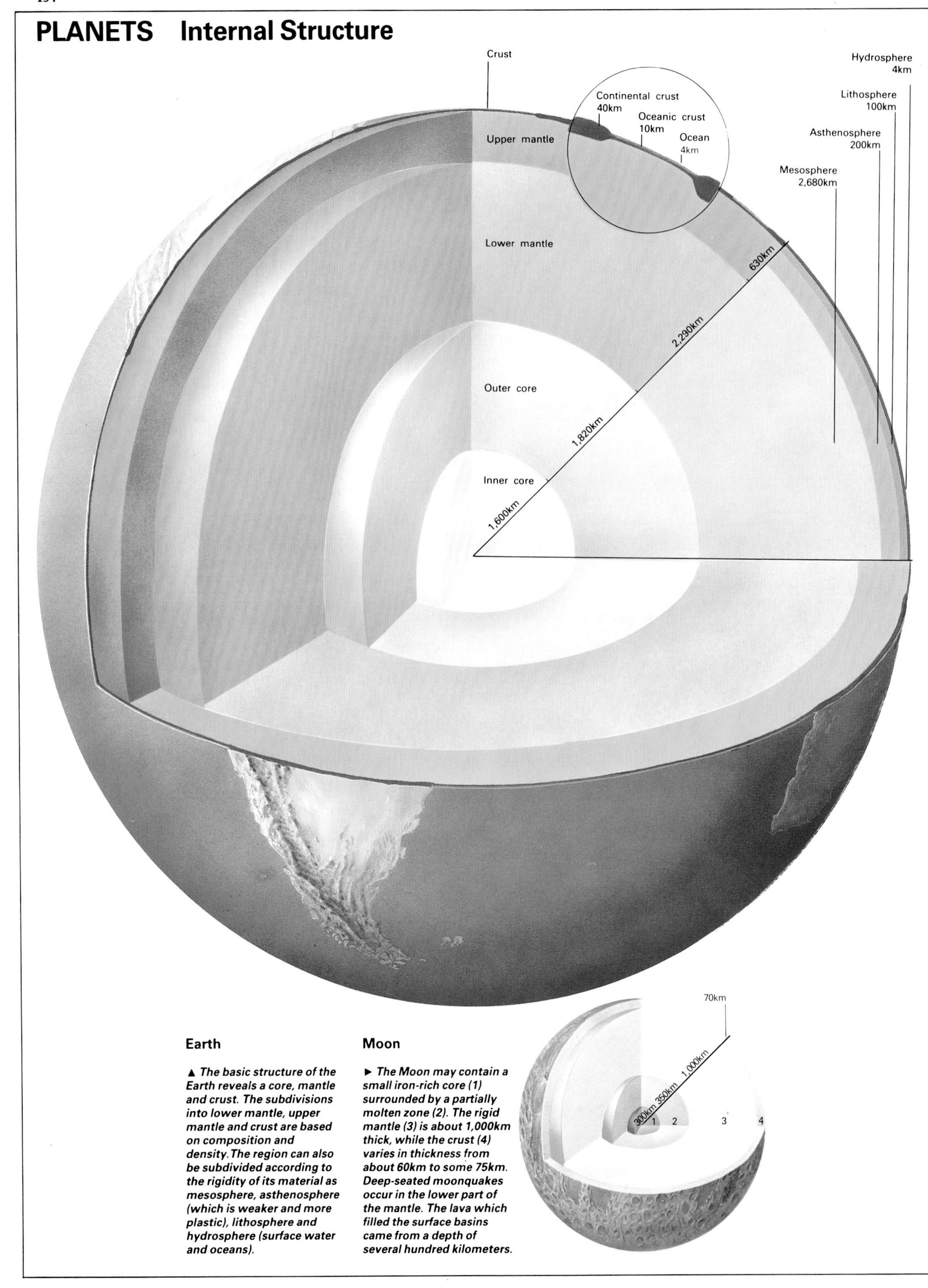

Earth

▲ *The basic structure of the Earth reveals a core, mantle and crust. The subdivisions into lower mantle, upper mantle and crust are based on composition and density. The region can also be subdivided according to the rigidity of its material as mesosphere, asthenosphere (which is weaker and more plastic), lithosphere and hydrosphere (surface water and oceans).*

Moon

► *The Moon may contain a small iron-rich core (1) surrounded by a partially molten zone (2). The rigid mantle (3) is about 1,000km thick, while the crust (4) varies in thickness from about 60km to some 75km. Deep-seated moonquakes occur in the lower part of the mantle. The lava which filled the surface basins came from a depth of several hundred kilometers.*

Of the nine planets, each has a distinct chemical composition and physical structure. The four low-density giants contain large amounts of hydrogen and helium, differing fractions of "ices" and small fractions of rocks and metals. (The outermost planet, Pluto, is probably a cosmic iceberg, made mainly of frozen water, ammonia and methane.) The four terrestrial planets (including Earth) consist almost entirely of rocky materials and metals – such as iron and nickel – in differing proportions. Each planet contains several distinct layers. The Moon and the seven planets on this page are all shown at the same size to illustrate the relative proportions of those layers common to all. (The Earth is shown in scale to the Moon.)

Mercury

◄ Despite its small size, Mercury has nearly the same density as the Earth, and this suggests that it must be about twice as rich in iron as the Earth. The iron-nickel core (1) probably extends to a radius of about 1,800km – 75 percent of the radius of the planet – and contains nearly 80 percent of its mass. Above this is a rocky mantle (2) and a lighter crust (3).

Venus

► The composition and internal structure of Venus is probably broadly similar to that of the Earth. Astronomers believe that the partially molten metallic core (1) is slightly smaller than its equivalent in the Earth, with a radius of some 2,940km. Above this lies the mantle (2) and a crust (3) some 60km thick – at least twice the thickness of the Earth's crust.

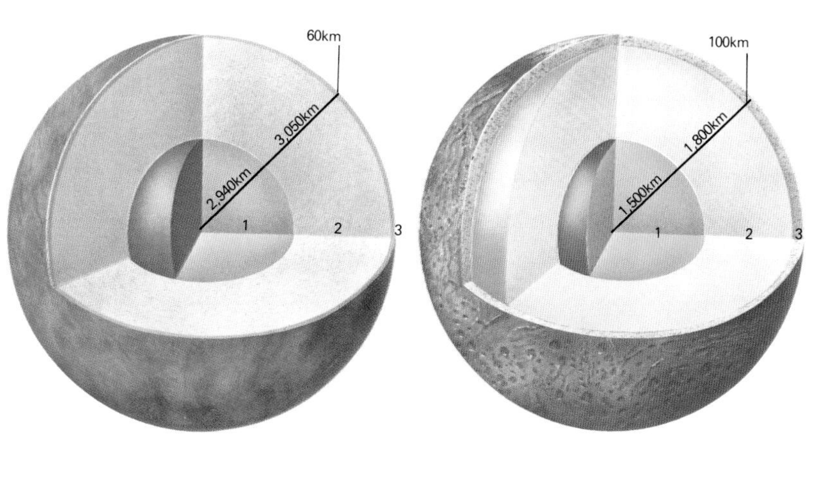

Mars

◄ The mean density of Mars is about 30 percent lower than that of Earth, and therefore the planet cannot have a large metallic core. Many alternatives have been suggested. The one illustrated here has a core (1) of iron and iron compounds some 1,500km in radius, a silicate mantle (2) about 1,800km thick, and a crust with a thickness of about 100km (3).

Jupiter

► Jupiter may contain a compact iron-silicate core (1) in a zone of liquid metallic hydrogen (2) which extends to a radius of about 45,000km. Above this there is a layer of liquid molecular hydrogen (3) 25,000km thick, then the 1,000km deep hydrogen-rich atmosphere (4). At the center temperature is about 30,000K and pressure is 100 million atmospheres.

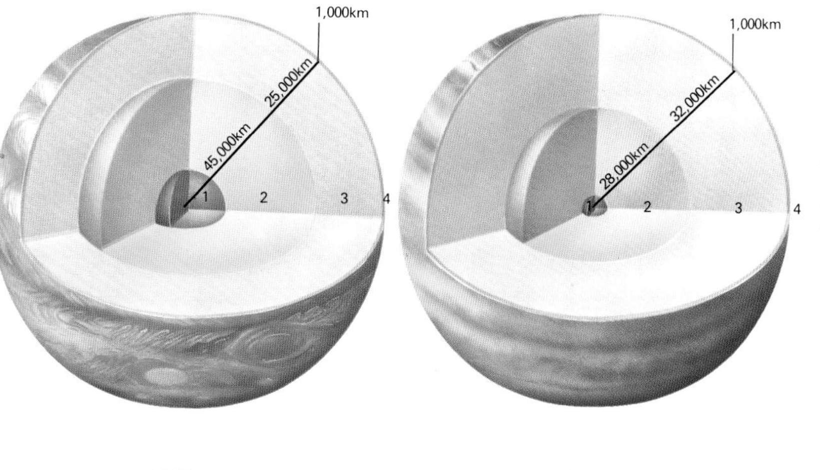

Saturn

◄ Like Jupiter, Saturn may contain a compact iron-silicate core (1) embedded in a liquid metallic hydrogen zone (2), up to a radius of some 28,000km. The liquid molecular hydrogen zone (3), probably 32,000km deep, is below the hydrogen-rich atmosphere (4). The temperature at Saturn's core may be 12,000K and pressure 8 million atmospheres.

Uranus

► Planetary geologists think that Uranus has an iron-silicate core (1) of about three Earth-masses which is somewhat larger than the Earth itself. This is probably surrounded by a mantle (2) of water, ammonia and methane ices extending to a radius of some 18,000km, and by a deep atmosphere (3) of hydrogen, helium and methane. The interior may be partly fluid.

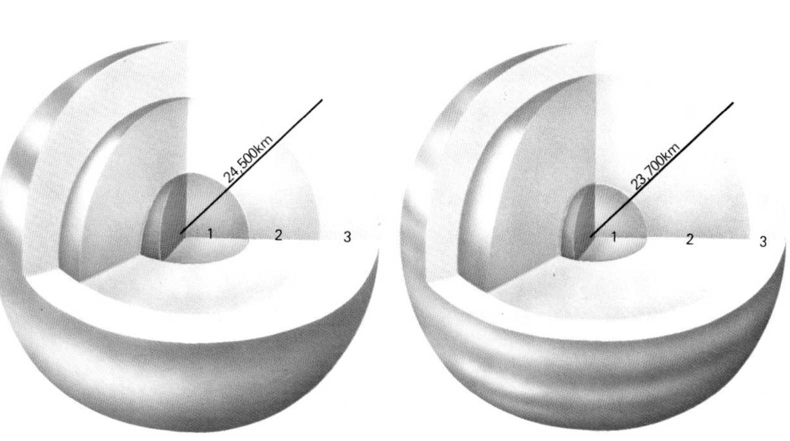

Neptune

◄ Neptune is denser than Uranus but probably has a similar general structure, with an iron-silicate core (1), an "icy" mantle (2) and a deep atmosphere (3) rich in hydrogen, helium and methane. Although the core is believed to be solid the planet emits sufficient heat to suggest that the mantle may be fluid, and that motions within it may generate a magnetic field.

animals is estimated in a variety of ways, including the use of mazes with a food reward at the end, and of levers that automatically deliver a food reward when the correct one (one of a particular color or in a particular location) is depressed. With animals such as chimpanzees, which can manipulate objects more skillfully, other tests can be devised. Some now employ specially adapted computers that the animal operates directly. The problem of measuring human intelligence is far more complex, especially when trying to compare people of different linguistic groups and cultures. No truly satisfactory, objective test has yet been devised.

Intelligence Quotient see IQ.

Intelligence test see PSYCHOLOGICAL TESTS.

Interference

The interaction of two or more similar or related WAVE MOTIONS establishing a new pattern in the AMPLITUDE of the waves. It occurs in all wave phenomena including SOUND, LIGHT and water waves. In most cases the resulting amplitude at a point is found by adding together the amplitudes of the individual interfering waves at that point. Interference patterns can result only if the interfering waves are of related wavelength and exhibit a definite PHASE relationship.

OPTICAL INTERFERENCE Light from ordinary sources is "incoherent" – there is no definite relationship between the phases of the waves associated with different PHOTONS. Until recently the only way to demonstrate optical interference was to use light from a single source that had been divided and led to the interference zone along paths of differing length, thus ensuring that the interfering beams were coherent at least with each other. In this way THOMAS YOUNG in 1801 first demonstrated optical interference, showing, because interference effects cannot be explained on either ray or particle models, that light was indeed to be regarded as a wave phenomenon. Young passed light from a single pinhole source through two parallel slits in an opaque screen and found that interference fringes – alternate bands of light and dark – were formed on another screen placed beyond the slits. The bright bands resulted from the constructive interference of the two beams, the wave amplitude of each reinforcing the other; the dark bands, destructive interference, the amplitude of one wave effectively canceling the effect of that of the other. Newton's rings, colored fringes seen in thin transparent films, are a similar interference effect. In recent years LASERS (which produce coherent light radiation having a uniform and controllable phase structure) have enabled physicists to produce optical interference effects more easily, an important application being HOLOGRAPHY. (See also INTERFEROMETER.)

Interferometer

Any instrument employing INTERFERENCE effects, used for measuring the wavelengths of LIGHT, RADIO, SOUND or other wave phenomena; for measuring the refractive index (see REFRACTION) of gases (Rayleigh interferometer); for measuring very small distances using radiation of known wavelength, or, in ACOUSTICS and RADIO ASTRONOMY, for determining the direction of an energy source. In most interferometers the beam of incoming radiation is divided in two, led along paths of different but accurately adjustable lengths and then recombined to give an interference pattern. Perhaps the best-known optical instrument is the Michelson interferometer, devised in 1881 for the MICHELSON-MORLEY EXPERIMENT. More accurate for wavelength measurements is the Fabry-Perot interferometer, in which the radiation is recombined after multiple partial reflections between parallel lightly silvered glass plates.

Interferon

A substance produced by cells in response to virus infections. It usually prevents other viruses from infecting simultaneously, so attempts have been made to produce it on a commercial scale to treat such infections.

Interhalogens

Group of binary compounds of the HALOGENS with each other, of general formula XX'_n where X' is the more electronegative halogen, and $n=1$, 3, 5 or 7. They are volatile, covalent, reactive substances. Bromine (III) fluoride (BrF_3) is a very powerful fluoridating agent, and a good solvent for fluorine compounds. Polyhalide salts, containing ions $XX'_{n+1}{}^-$, are formed by reacting halogens or interhalogens with HALIDES.

Internal-combustion engine

Type of engine now most commonly used, in which the fuel is burned inside the engine and the expansion of the combustion gases is used to provide the power. Because of their potential light weight, efficiency and convenience, internal-combustion engines largely superseded steam engines in the early 20th century. They are used industrially and for all kinds of transport, notably to power automobiles.

There are three classes of internal-combustion engine: reciprocating engines, which include the gasoline engine, the diesel engine and the free-piston engine; rotary engines, including the GAS TURBINE, the turbojet (see JET PROPULSION) and the WANKEL ENGINE; and rocket engines and non-turbine jet engines, working by reaction. Although originally coal gas and even powdered coal were used as fuel, now almost all fuels used are hydrocarbon products: diesel oil, gasoline, bottled gas and NATURAL GAS.

The first working (though not usable) internal-combustion engine was a piston engine made by HUYGENS (1680) that burned gunpowder. In 1794 Robert Street patented a practicable though inefficient engine into which the air had to be pumped by hand. In 1876 Nicholas August Otto (1832-1891) built the first four-stroke engine, using the principles stated earlier by Alphonse Beau de Rochas. The cycle is (1) intake of fuel/air mixture; (2) compression of mixture; (3) ignition and expansion of burned gases; (4) expulsion of gases as exhaust. Only the third stroke is powered, but the engine is highly efficient, and modern gasoline engines are basically the same. Generally four, six or eight cylinders are linked to provide balanced power. The engine is cooled by water circulating through pipes or by air from a fan. The fuel/air mixture is produced in the carburetor; greater power is given by supercharging, by which the proportion of air and the initial pressure of the mixture are increased. The two-stroke engine, giving greater power for a given size but less efficient in fuel use, does not usually have valves, but an inlet and an exhaust port in the cylinder, blocked and uncovered in turn by the piston. At the end of the powered stroke, the piston drives fresh fuel mixture from the crankcase into the cylinder, pushing out the exhaust gases.

The EFFICIENCY of an internal-combustion engine increases with the compression ratio; if this is too high, however, "knocking" occurs due to irregular burning and detonations. It is avoided by using fuel of high OCTANE number, and by using antiknock additives. The 1973 energy crisis sent engine designers in search of high efficiency. The results are the "fast-burn" and "lean-burn" engines, with high compression ratios but, more important, high air-fuel ratios. Most engines in the 1970s ran stoichiometric mixtures (14.7:1 air-fuel ratio). Fast burn is up to 16:1 and lean burn up to 22:1. These give better fuel consumption and lower emissions of toxic or acidic compounds. The engine of the foreseeable future is lean-burn. (See also AIR POLLUTION.)

International Date Line see DATE LINE, INTERNATIONAL.

International System of Units see SI UNITS.

Interstellar matter

Thinly dispersed matter, in the form of gases and dust, between the stars. The dust grains scatter and absorb starlight, causing distant stars to appear fainter than they otherwise would do. The amount of extinction due to the dust is greater at shorter wavelengths, so that blue light is affected more than red; consequently, distant stars appear redder in color than they really are (interstellar reddening). Dense dust clouds show up as dark patches (see NEBULA) against the background stars. Gas clouds containing high-temperature stars show up as luminous patches (emission nebulae). The interstellar gas also reveals its presence by the absorption lines (interstellar lines) that it superimposes on the SPECTRA of stars. Neutral hydrogen gas emits radio waves at a wavelength of 21cm (ie a frequency of 1420MHz). Radio astronomers can also detect emission lines from a variety of species of molecules, many of which are organic in nature, which exist mainly in dense, cool molecular clouds.

Intestine see GASTROINTESTINAL TRACT.

Intoxication

State in which a person is overtly affected by excess of a DRUG or poison. It is often used to describe the psychological effects of drugs, and particularly alcohol, in which behavior may become disinhibited, facile, morose or aggressive and in which judgment is impaired. Late stages of intoxication affecting the brain include stupor and COMA.

Intravenous feeding

Method of supplying directly into a vein the nutrients essential for health and maintenance of body mass when the GASTROINTESTINAL TRACT is unable to provide adequate nutrition in certain DISEASES of the gut and following SURGERY.

Introversion and extroversion

Terms coined by JUNG for two opposite character traits. Introverts are shy, introspective, "ingoing"; extroverts sociable, little concerned with their own inner thoughts and feelings, "outgoing". We all display both traits, one or other dominating at different times; Jung suggested that conflict between them was a cause of NEUROSIS.

Intussusception

A condition in which the intestine telescopes inside itself, causing blockage.

Invar

An alloy composed of 64% iron, 36% nickel and a trace of carbon. Having a very small coefficient of thermal expansion, it is used for pendulums, tuning-forks, measuring devices and other components whose dimensions must be independent of temperature.

Inverse square law

Relationship according to which the intensity of a spherical wave (see WAVE MOTION) varies inversely (see INVERSE VARIATION) with the SQUARE of its distance from the source. The law applies only where the source is small compared with the distance and the medium is unbounded, homogeneous, isotropic and nondissipative. The strengths of ELECTROMAGNETIC and GRAVITATIONAL FIELDS also vary according to inverse square law.

Inversion, Temperature

A condition of the lower part of the ATMOSPHERE in which temperature increases with increase in height above the surface. Normally, temperature decreases upward through most of the atmosphere, but certain atmospheric disturbances (eg a

FRONT) can create inversions. The condition also occurs on cold nights. Inversions sometimes aggravate AIR POLLUTION, as the cooler air trapped near the surface cannot rise and so carry away the pollutants.

Invertebrates
Animals without backbones, that is, all those not belonging to the VERTEBRATES. They are a miscellaneous collection of groups from simple sponges to highly specialized insects and spiders. Apart from the universal lack of an internal backbone, many of these groups have little in common. One phylum, the CHORDATA, is split between the invertebrates and the vertebrates.

Inverter see COMPUTER.

In-vitro fertilization
The FERTILIZATION of an ovum outside the body, used to overcome certain forms of infertility. The technique is employed in producing so-called "test-tube babies". An EGG is removed from a woman's OVARY, and mixed with human SPERM in a glass dish. The CELL starts to divide, and when the EMBRYO has reached the eight-cell stage (72 hours), the cells are implanted in a woman's UTERUS. Because there is a high failure rate doctors often implant several eggs, so multiple births may result. Scientists have also developed techniques to freeze the embryo after fertilization for later implanting. Most countries have strict guidelines for research on human embryos.

Iodine (I)
The least reactive of the HALOGENS, forming black lustrous crystals which readily sublime to pungent violet vapor. Most iodine is produced from calcium iodate ($Ca[IO_3]_2$), found in CHILE SALTPETER. In the US, much is recovered from oil-well brine, which contains sodium iodide (NaI). Chemically it closely resembles BROMINE, but has a greater tendency to covalency and positive oxidation states. It is large enough to form 6-coordinate oxy-anions. Most plants (especially seaweeds) contain traces of iodine; in the higher animals it is a constituent of the thryoxine hormone secreted by the THYROID GLAND. Iodine deficiency can cause GOITER. Iodine and its compounds are used as antiseptics, fungicides and in the production of dyes. The RADIOISOTOPE I^{131} is used as a tracer and to treat goiter. Silver iodide, being light-sensitive, is used in PHOTOGRAPHY. (See also HALIDES.) AW 126.9, mp 113.6°C, bp 184°C. sg 4.93 (20°C). (◆ page 130)

Ion
An ATOM or group of atoms that has become electrically charged by gain or loss of negatively charged ELECTRONS. In general, ions formed from metals are positive (CATIONS), those from nonmetals negative (ANIONS). Compound ions are usually anions derived from OXYACIDS, eg sulfate SO_4^{2-}. CRYSTALS of ionic compounds consist of negative and positive ions arranged alternately in the lattice and held together by electrical attraction (see BOND, CHEMICAL). Many covalent compounds undergo ionic DISSOCIATION in solution. Ions may be formed in gases by radiation or electrical discharge, and occur in the IONOSPHERE (see also ATMOSPHERE). At very high temperatures gases form PLASMA, consisting of ions and free electrons. In solution, many simple ions combine with LIGANDS to give complex ions, including HYDRATES. (See ELECTROLYSIS; ION EXCHANGE; IONIZATION CHAMBER; IONIZATION POTENTIAL; ION PROPULSION; ZWITTERION.) (◆ page 37, 118)

Ion counter see GEIGER COUNTER; SCINTILLATION COUNTER.

Ion exchange
Chemical reaction in which IONS in a solution are replaced by others of like charge. An insoluble solid is used that has an open, netlike molecular structure: a ZEOLITE, or a synthetic organic polymer called an ion-exchange resin, whose composition and properties can be tailored for the use required. The solid has attached anionic groups, which are neutralized by small mobile cations in the interstices. It is these cations that are exchanged for others when a solution is passed through. The principle of anion exchange is similar. Ion exchange is used for softening HARD WATER, for purifying sugar, and for concentrating ores of uranium and the NOBLE METALS. Ion-exchange CHROMATOGRAPHY is used to separate the RARE EARTHS, and in chemical ANALYSIS.

Ionization chamber
Instrument for measuring the amount of ionization created in it by radiation such as alpha particles, beta rays, gamma rays or X-rays. A gas-filled chamber contains two ELECTRODES with a variable potential difference between them. The ions produced move toward the oppositely charged electrode, forming an electric current which is a measure of the amount of incoming radiation.

Ionization potential
The ENERGY needed to remove an ELECTRON from the ground state of a given type of ATOM to infinity. It increases for removal of successive electrons, which are bound by the atom's positive charge. Ionization potentials can be determined by SPECTROSCOPY.

Ion microscope see FIELD-ION MICROSCOPE.

Ionosphere
The zone of the Earth's ATMOSPHERE extending outward from about 75km above the surface, in which most atoms and molecules exist as electrically charged IONS. The high degree of ionization is maintained through the continual ABSORPTION of high-energy solar radiation. Several distinct ionized layers, known as the D, E, F_1, F_2 and G layers, are distinguished. These are somewhat variable, the D layer disappearing and the F_1, F_2 layers merging at night. As the free ELECTRONS in these layers strongly reflect RADIO waves, the ionosphere is of great importance for long-distance radio communications. (◆ page 25)

Ion propulsion (ion drive)
Drive proposed for spacecraft on interstellar or longer interplanetary trips. The vaporized propellant (liquid CESIUM or MERCURY) is passed through an ionizer, which strips each atom of an ELECTRON. The positive IONS so formed are accelerated rearward by an ELECTRIC FIELD. The resultant thrust is low but in the near vacuum of space may be used to build up huge velocities by constant acceleration over a long period of time. The drive has been tested in orbit around the Earth.

IQ (Intelligence Quotient)
A measure of intelligence or scholastic ability that takes an average person as having a score of 100. IQ tests are relatively crude measures that take little account of cultural differences and certain handicaps.

Iridectomy
Cutting into the iris, used to drain the anterior chamber of the eye in glaucoma.

Iridescence
Production of colors of varied hue by INTERFERENCE of light reflected from front and back of thin films (as in soap bubbles) or from faults and boundaries within crystalline solids such as mica or opal. The colors of mother-of-pearl and some insects are due to iridescence.

Iridium (Ir)
Hard, white metal in the PLATINUM GROUP, the most resistant element to corrosion at room temperature. AW 192.2, mp 2447°C, bp 4530°C, sg 22.4 (20°C). (◆ page 130)

Iris
The colored part of the vertebrate EYE, which has the pupil at the center. (◆ page 213)

Iron (Fe)
Silvery gray, soft, ferromagnetic (see MAGNETISM) metal in Group VIII of the PERIODIC TABLE; a TRANSITION ELEMENT. Metallic iron is the main constituent of the EARTH'S core, but is rare in the crust; it is found in meteorites (see METEORS). Combined iron is found as HEMATITE, MAGNETITE, LIMONITE, SIDERITE, GOETHITE, TACONITE, CHROMITE and PYRITE. It is extracted by smelting oxide ores in a blast furnace to produce pig iron, which may be refined to produce cast iron or wrought iron, or converted to steel in the open-hearth process or the Bessemer process. Many other iron ALLOYS are used for particular applications. Pure iron is very little used; it is chemically reactive, and oxidizes to RUST in moist air. It has four allotropes (see ALLOTROPY). The stable oxidation states of iron are +2 (ferrous) and +3 (ferric), though +4 and +6 states are known. The ferrous ion (Fe^{2+}) is pale green in aqueous solution; it is a mild reducing agent, and does not readily form LIGAND complexes. Iron (II) sulfate ($FeSO_4.7H_20$), or green vitriol or copperas, is a green crystalline solid made by treating iron ore with sulfuric acid, used in tanning, in medicine to treat iron deficiency, and to make ink, fertilizers, pesticides and other iron compounds. mp 64°C. The ferric ion (Fe^{3+}) is yellow in aqueous solution; it resembles the ALUMINUM ion, being acidic and forming stable LIGAND complexes, especially with CYANIDES. Iron (III) oxide (Fe_2O_3), a red-brown powder used as a pigment and as jewelers' rouge (see ABRASIVES), occurs naturally as HEMATITE. mp 1565°C. (See also ALUM; SANDWICH COMPOUNDS.) In the human body, iron is a constituent of HEMOGLOBIN and the CYTOCHROMES. Iron deficiency causes ANEMIA. AW 55.8, mp 1540°C, bp 2760°C, sg 7.874 (20°C). (◆ page 130)

Irradiation
Exposure of a sample to RADIATION, usually for a definite purpose. Biological and pharmaceutical materials may have their properties altered by exposure to ULTRAVIOLET RADIATION; X-RAYS are widely used in medicine and industry. Materials may be irradiated directly with radiation of a given type and ENERGY by placing them in a particle ACCELERATOR or NUCLEAR REACTOR, but it is often more practical to use the radiation from manufactured radioactive ISOTOPES to change their physical and chemical properties as required. Neutrons and GAMMA RAYS are used to sterilize foodstuffs and control the reproduction of insect pests.

Irrigation
Artificial application of water to soil to promote plant growth. Irrigation is vital for agricultural land with inadequate rainfall. There are three main irrigation techniques: surface irrigation, in which the soil surface is moistened or flooded by water flowing through furrows or tubes; sprinkler irrigation, in which water is sprayed on the land from above; and subirrigation, in which underground pipes supply water to roots. The amount of water needed for a particular project is called the duty of water; this is expressed as the number of hectares irrigated by water supplied at a rate of $1m^3/s$.

Ischemia
Loss or deficiency of blood supply, producing poor function and eventually tissue death leading to gangrene.

Island
Comparatively small land area entirely surrounded by water, a result of the buildup of the cone of a submarine VOLCANO, EROSION by the sea or by GLACIERS of parts of coastal regions, DIASTROPHISM or other process. Island arcs are curving chains of islands. They are associated

with EARTHQUAKE activity, and have deep OCEAN trenches on the convex sides. (See also ATOLL; CORAL; PLATE TECTONICS.)

Islets of Langerhans see INSULIN; PANCREAS.

Isobar
Line drawn on a meteorological map joining points that are, at a given moment in time, experiencing the same air pressure (see ATMOSPHERE).

Isomerism
In nuclear physics, the existence of metastable states of an atomic nucleus (see ATOM), having the same atomic number and mass number as the ground state, but higher energy. Nuclear isomers are formed by bombardment or in a radioactive decay chain (see RADIOACTIVITY). They usually have a very short HALF-LIFE and decay by emitting GAMMA RAYS.

Isomers
Chemical compounds having identical chemical COMPOSITION and molecular FORMULA, but differing in the arrangement of atoms in their molecules, and having different properties. The two chief types are STEREOISOMERS, which have the same structural formula, and structural isomers, which have different structural formulas. The latter may be subdivided into positional isomers, which have the same FUNCTIONAL GROUPS occupying different positions on the carbon skeleton; and functional isomers, which have different functional groups. (See also CRACKING; TAUTOMERISM.)

Isometrics
Exercises in which MUSCLES are contracted against resistance, but without movement at JOINTS; the muscles remain at the same length but their tension is increased. It is used in some systems of physical training and in PHYSIOTHERAPY.

Isomorphism
The formation by different compounds or MINERALS of CRYSTALS having closely similar external forms and lattice structure. Isomorphous compounds have similar chemical composition – ions of similar size, charge, and electrical polarizability being substituted for each other – and form mixed crystals.

Isoprene (2-methyl-1,3-butadiene)
A conjugated diene (see ALKENES; DIELS-ALDER REACTION; RESONANCE). Isoprene is a colorless liquid made by destructive distillation of rubber or from PETROLEUM, and used to make synthetic RUBBER. It is the basic unit of plant products, including CAROTENOIDS, STEROLS and TERPENES.

Isostasy
The theoretical tendency of the Earth's crust to maintain equilibrium as it floats on the MANTLE; assumed to result from flows of the dense plastic ASTHENOSPHERE in the MANTLE in response to local changes in the pressure on it of the solid LITHOSPHERE above. Alternatively, the great weight of a mountain range may push lithosphere material down into the athenosphere, where it becomes asthenosphere material itself.

Isotherm
Line drawn on a meteorological map joining points that are, at a given moment, experiencing the same temperature. (See METEOROLOGY.)

Isotopes
ATOMS of a chemical ELEMENT that have the same number of PROTONS in the nucleus, but different numbers of NEUTRONS, ie having the same atomic number but different MASS NUMBER. Isotopes of an element have identical chemical and physical properties (except those determined by atomic mass). Most elements have several stable isotopes, being found in nature as mixtures. The natural proportions of the isotopes are expressed in the form of an abundance ratio. Because some isotopes have particular properties (eg 0.015% of HYDROGEN atoms have two neutrons and combine with oxygen to form HEAVY WATER, used in NUCLEAR REACTORS), mass-dependent methods of separating these out have been devised. These include MASS SPECTROSCOPY, DIFFUSION, DISTILLATION and ELECTROLYSIS. A few elements have natural radioactive isotopes (RADIOISOTOPES) and others of these can be made by exposing stable isotopes to RADIATION in a reactor. These are widely used therapeutically and industrially; their radiation may be employed directly, or the way in which it is scattered or absorbed by objects can be measured. They are useful as tracers of a process, as they may be detected in very small amounts and behave virtually identically to other atoms of the same element. They may also be used to "label" atoms in complex molecules, in attempts to work out chemical reaction mechanisms. (➧ page 147)

Isotropy
Property exhibited by a medium in which physical properties are independent of direction. Most liquids and materials composed of small randomly oriented crystals are isotropic in all properties, while crystalline materials are, in general, anisotropic.

Isthmus
Narrow strip of land joining two large landmasses, or a peninsula to the mainland. (See also STRAIT.)

IVF see IN-VITRO FERTILIZATION.

PLANT EVOLUTION

▼ ***Plants do not fossilize easily, so relatively little is known about how they evolved. Much of this evolutionary tree is based on comparisons of the structure, pigments, life cycles and reproductive organs of living plants.***

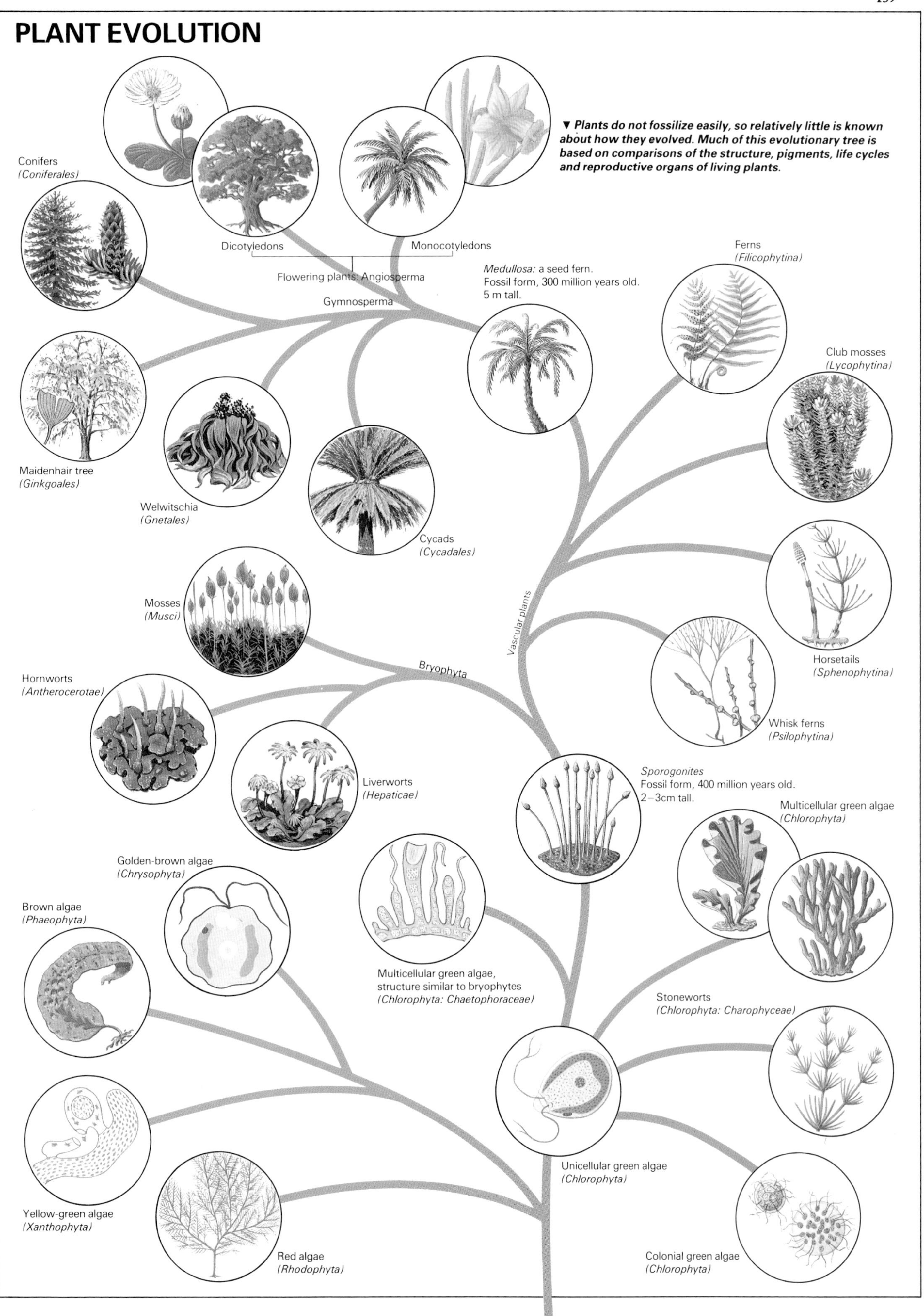

Jack o'lantern see WILL-O'-THE-WISP.

Jacob, François (b. 1920)
French biologist who shared with MONOD and LWOFF the 1965 Nobel Prize for Physiology or Medicine for his work with Monod on regulatory GENE action in BACTERIA.

Jade
Either of two tough, hard minerals with a compact interlocking grain structure, commonly green but also found as white, mauve, red-brown or yellow; used as a GEM stone to make carved jewelry and ornaments. Nephrite, the commoner form of jade, is an AMPHIBOLE, a combination of tremolite and actinolite, occurring in China, the USSR, New Zealand and the western US. Jadeite, rarer than nephrite and prized for its more intense color and translucence, is a sodium aluminum PYROXENE, found chiefly in upper Burma.

Janet, Pierre Marie Félix (1859-1947)
French psychologist and neurologist, best known for his studies of HYSTERIA and NEUROSIS, who played an important role in reconciling the theories of psychology and the practice of clinical treatment of mental disease.

Jansen, Zacharias (1580-1638)
Dutch spectacle-maker credited with inventing the compound MICROSCOPE (1590).

Jasper
Common variety of CHALCEDONY containing admixed HEMATITE or GOETHITE, normally red, brown or yellowish, often with banding and spotting. Good grades are used for semiprecious GEM stones.

Jaundice
Yellow color of the skin and sclera of the eye caused by excess bilirubin pigment in the BLOOD. HEMOGLOBIN is broken down to form bilirubin, which is excreted by the LIVER in the BILE. If blood is broken down more rapidly than normal (hemolysis), the liver may not be able to remove the abnormal amount of bilirubin fast enough. Jaundice also occurs with liver damage (HEPATITIS, late CIRRHOSIS) and when the bile ducts leading from the liver to the duodenum are obstructed by stones from the gall bladder or by cancer of the pancreas or bile ducts.

Jeans, Sir James Hopwood (1877-1946)
British physicist and mathematician best known for his contributions to astronomy and for his popular science books. He played a valuable role in proving the invalidity of the NEBULAR HYPOTHESIS, but his own theory of the formation of the SOLAR SYSTEM, that the planets were "drawn out" of the Sun by a star passing close by, has now in turn been largely discarded.

Jenner, Edward (1749-1823)
British pioneer of VACCINATION. He examined in detail the country maxim that dairymaids who had had COWPOX would not contract SMALLPOX. In 1796 he inoculated a small boy with cowpox and found that this rendered the boy immune from smallpox.

Jensen, Johannes Hans Danriel (1907-1973)
German physicist who shared with M.G. MAYER and E.H. WIGNER the 1963 Nobel Prize for Physics for his suggestion (independent of Mayer's) that the PROTONS and NEUTRONS of the atomic nucleus are arranged in concentric shells.

Jerne, Niels (b. 1911)
Danish immunologist who shared the 1984 Nobel Prize for Physiology or Medicine with MILSTEIN and KÖHLER for discovering the principle for production of MONOCLONAL ANTIBODIES.

Jet
Compact, hard variety of lignite COAL, deep black and polishable, mined at Whitby, England, and used as a GEM material.

Jet propulsion
The propulsion of a vehicle by expelling backward a jet of fluid whose MOMENTUM produces a reaction that imparts an equal forward momentum to the vehicle, according to NEWTON's third law of motion. The squid uses a form of jet propulsion. Jet-propelled boats, using water for the jet, have been built, and air jets have been used to power cars, but by far the chief use is to power airplanes and rockets, because to attain high speeds jet propulsion is essential. The first jet engine was designed and built in 1937 by Air Commodore Sir Frank Whittle (1907-1987), but the first jet-engine aircraft to fly was German (August 1939). Jet engines are INTERNAL-COMBUSTION ENGINES. The turbojet is the commonest form. Air enters the inlet diffuser and is compressed in the air compressor, a multistage device having sets of rapidly rotating fan blades. It then enters the combustion chamber, where the fuel (a kerosine/gasoline mixture) is injected and ignited, and the hot, expanding exhaust gases pass through a turbine that drives the compressor and engine accessories. The gases, sometimes heated further in an afterburner, are expelled through the jet nozzle to provide the thrust. The nozzle converges for subsonic flight, but for supersonic flight one that converges and then diverges is needed. The fanjet or turbofan engine uses some of the turbine power to drive a propeller fan in a cowling, for more efficient subsonic propulsion; the turboprop, similar in principle, gains its thrust chiefly from the propeller. The ramjet is the simplest air-breathing jet engine, having neither compressor nor turbine. When accelerated to supersonic speeds by an auxiliary rocket or turbojet engine, the inlet diffuser "rams" the air and compresses it; after combustion the exhaust gases are expelled directly. Ramjets are used chiefly in guided missiles.

Jet stream
A narrow band of very fast E-flowing winds, stronger in winter than in summer, found around the level of the tropopause (see ATMOSPHERE). Speeds average about 60km/h in summer, about 125km/h in winter, though over 300km/h has been recorded. (See also METEOROLOGY; WIND.)

Jewels see GEMS.

Jodrell Bank
Site of the Mullard Radio Astronomy Observatory in England, pioneered by Alfred Charles Bernard Lovell (b. 1913) and including one of the largest steerable RADIO TELESCOPES ("dish" 250ft (76.2m) across) (1957).

Joint
In vertebrates a specialized surface between BONES allowing movement of one on the other. Major joints, especially of limbs, are synovial joints which are lined by synovial membrane and CARTILAGE and surrounded by a fibrous capsule; they contain SYNOVIAL FLUID, which lubricates the joint surfaces. Parts of the capsule (eg in the ankle) or overlying TENDONS (eg in the knee) form LIGAMENTS important in joint stability, though at some joints (eg the SHOULDER) resting activity in MUSCLES ensures stability, while in others (eg the hip) it is due to the shape of the bony surfaces. Fibrous and cartilaginous joints between bones are relatively fixed except under special circumstances (eg the widening of the symphysis pubis in PREGNANCY). Joint disease causes ARTHRITIS, with pain, limitation of movement and sometimes increase in fluid. (➧ page 216)

Joliot-Curie, Irène (1897-1956)
French physicist, the daughter of Pierre and Marie CURIE. She and her husband, Jean Frédéric Joliot (1900-1958), shared the 1935 Nobel Prize for Chemistry for their discovery of artificial RADIOACTIVITY. Both later played a major part in the formation of the French atomic energy commission but, because of their communism, were removed from positions of responsibility there (Frédéric 1950, Irène 1951). Like her mother, Irène died from LEUKEMIA as a result of prolonged exposure to radioactive materials.

Josephson, Brian David (b. 1940)
British physicist awarded, with I. GIAEVER and L. ESAKI, the 1973 Nobel Prize for Physics for his discovery of the Josephson Effect, the passage of ELECTRICITY through an insulator between two superconductors (see SUPERCONDUCTIVITY). Pairs of ELECTRONS form in the superconductors and tunnel (see WAVE MECHANICS) through the insulating layer.

Joule (J)
The SI unit of ENERGY, defined as the WORK done when a FORCE of one NEWTON acts through the distance of one meter. Also equivalent to the energy dissipated by one WATT in one second, it equals 10^7 erg in CGS UNITS.

Joule, James Prescott (1818-1889)
British physicist who showed that HEAT energy and mechanical energy are equivalent and hinted at the law of conservation of ENERGY. From 1852 he and Thomson (later LORD KELVIN) performed a series of experiments in THERMODYNAMICS, especially on the Joule-Thomson effect (see CRYOGENICS). The JOULE (unit) is named for him. (See also JOULE'S LAW.)

Joule's Law
Law derived by J.P. JOULE that the heat evolved in a given time by passage of electricity through a conductor is proportional to the RESISTANCE of the conductor times the square of the electrical intensity. We now write it $H=I^2R$, where H is the rate of generation of heat in WATTS, I the current in AMPERES and R the resistance in OHMS.

Jugular veins
A pair of veins on each side of the neck which collect venous blood from the BRAIN (internal jugular vein) and the rest of the head (external jugular vein). Their proximity to the surface makes them liable to trauma with HEMORRHAGE and EMBOLISM. (➧ page 28)

Julian Calendar see CALENDAR.

Jung, Carl Gustav (1875-1961)
Swiss psychiatrist who founded analytical psychology. He studied PSYCHIATRY at Basel University, his postgraduate studies being of PARAPSYCHOLOGY. After working with BLEULER and JANET, he met FREUD (1907), whom he followed for some years. But he disagreed with, particularly, Freud's belief in the purely sexual nature of the LIBIDO, and in 1913 broke away completely. In *Psychological Types* (1921) he expounded his views on INTROVERSION AND EXTROVERSION. Later he investigated anthropology and the occult to form the idea of ARCHETYPES, the universal symbols present in the COLLECTIVE UNCONSCIOUS. (See also ANIMA AND ANIMUS; PERSONA; PSYCHOANALYSIS; PSYCHOLOGY.)

Jupiter
The largest and most massive planet in the Solar System (diameter about 143Mm, mass 317.8 times that of Earth), fifth from the Sun. Jupiter is larger than all the other planets combined and, with a mean solar distance of 5.20AU and a "year" of 11.86 Earth years, is the greatest contributor to the Solar System's angular MOMENTUM. Its atmosphere consists mainly of hydrogen, helium and hydrogen compounds such as methane and ammonia. Its disk is marked by prominent cloud-belts paralleling its equator, these being occasionally interrupted by turbulences, and particularly by the Great Red Spot, an elliptical area 40Mm long and 13Mm wide in the S Hemisphere that seems to be an anticyclonic high-pressure feature rotating counterclockwise in a period of six days. Another long-term feature, the South Tropical Disturbance, was first observed in 1901 and disappeared in 1939. The nature of these features is not yet known. When VOYAGER 1 detected two rings of particles, Jupiter became the third planet to be shown to have one or more rings. Jupiter's day is about 9.92h and this high rotational velocity causes a visible flattening of the poles: the equatorial diameter is some 6% greater than the polar diameter. Jupiter has 16 moons, the two largest of which, Callisto and Ganymede, are larger than MERCURY: Io has violently active volcanoes on its surface. Jupiter radiates twice as much heat as it receives from the Sun. The poles are as warm as the equator. (♦♦ page 135, 221, 240)

Jurassic
The middle period of the MESOZOIC era. lasting from about 213 to 144 million years ago. (See also GEOLOGY.) (♦ page 106)

Juvenile hormone
An insect HORMONE that maintains the presence of juvenile features in the LARVA by suppressing the development of the imaginal buds, precursors of the adult organs (see METAMORPHOSIS). In its absence, adult features appear on molting or following a PUPAL stage. It is also involved in normal egg production by female insects.

Kala-azar see LEISHMANIASIS.

Kalm, Peter (1716-1779)
Swedish botanist best known for his survey of North American natural history.

Kamerlingh-Onnes, Heike (1853-1926)
Dutch physicist awarded the 1913 Nobel Prize for Physics for his work on low-temperature physics (see CRYOGENICS). He discovered SUPERCONDUCTIVITY and was the first to liquefy HELIUM (1908).

Kaolin (china clay)
Soft, white CLAY composed chiefly of KAOLINITE, and mined in England, France, Saxony, Czechoslovakia, China and the southern US. It is used for filling and coating paper, filling rubber and paints, and for making pottery and porcelain.

Kaolinite
Prototypical member of the kaolinite group of CLAY minerals. It consists of hexagonal flakes of composition $Si_4Al_4O_{10}(OH)_8$. It is formed by alteration of other clays or FELDSPAR.

Kapitza, Peter Leonidovich (b. 1894)
Russian physicist best known for his work on low-temperature physics (see CRYOGENICS), for which he was corecipient of the 1978 Nobel Prize.

Kaposi's sarcoma
Cancer of the dermal layers of the skin. Formerly rare, it is relatively common in AIDS sufferers.

Karat see CARAT.

Karle, Jerome (b. 1918)
US scientist who jointly received with HERBERT HAUPTMANN the 1985 Nobel Prize for Chemistry for their development of direct methods for the determination of crystal structures.

Karrer, Paul (1889-1971)
Russian-born Swiss chemist awarded with W.N. HAWORTH the 1937 Nobel Prize for Chemistry for his work on the CAROTENOIDS and flavins, and on VITAMINS A and B_2.

Karst
A LIMESTONE topography typically including collapsed caverns, SINK HOLES where streams disappear underground, and areas of bare "limestone pavement".

Karyotype
The characteristic CHROMOSOMES of an individual organism, cell-line or species, arranged in a systematized form. The karyotype is obtained by microscopic examination of cells during meiosis, when the chromosomes become visible. The chromosomes are numbered by pairs and grouped by their appearance in descending order of size.

Kastler, Alfred (1902-1984)
French physicist awarded the 1966 Nobel Prize for Physics for his work on the structure of the ATOM, work which led eventually to the development of the LASER.

Katz, Sir Bernard (b. 1911)
German-born British biophysicist who shared with AXELROD and von EULER the 1970 Nobel Prize for Physiology or Medicine for their independent work on the chemistry of the transmission of nerve impulses.

Kekulé von Stradonitz, Friedrich August (1829-1896)
German chemist regarded as the father of modern ORGANIC CHEMISTRY. At the same time as Archibald Scott Couper (1831-1892) he recognized the quadrivalency of CARBON and its ability to form long chains. With his later inference of the structure of BENZENE (the "benzene ring"), structural organic chemistry was born.

Keloid
The abnormal but harmless ridge of connective tissue that sometimes forms on a scar, making it raised and lumpy. Its cause is unknown.

Kelvin (K)
The SI unit of thermodynamic TEMPERATURE, defined as 1/273.16 of the thermodynamic temperature of the triple point of WATER. It is used both as a unit of temperature difference (when the centigrade degree is defined equal to it) and for expressing ABSOLUTE temperatures (in kelvins above ABSOLUTE ZERO). Temperatures expressed in degrees Celsius equal temperatures expressed in kelvins less 273.15. The older terms "degree Kelvin" and "Kelvin scale" are obsolete.

Kelvin, William Thomson, 1st Baron (1824-1907)
British physicist who made important contributions to many branches of physics. In attempting to reconcile CARNOT's theory of heat engines and JOULE's mechanical theory of HEAT he formulated (independently of CLAUSIUS) the second law of THERMODYNAMICS and introduced the ABSOLUTE temperature scale, the unit of which is called KELVIN for him. His and FARADAY's work on ELECTROMAGNETISM gave rise to the theory of the electromagnetic field, and his papers, with those of Faraday, strongly influenced J. CLERK MAXWELL's work on the electromagnetic theory of LIGHT (though Kelvin himself rejected Maxwell's over-abstract theory). His work on wire-telegraphic signaling played an essential part in the successful laying of the first Atlantic cable.

Kendall, Edward Calvin (1886-1972)
US biochemist awarded with HENCH and REICHSTEIN the 1950 Nobel Prize for Physiology or Medicine for his work on the corticoids and isolation of cortisone (see STEROIDS), which was applied by Hench to the treatment of rheumatoid ARTHRITIS.

Kendrew, Sir John Cowdery (b. 1917)
British biochemist awarded with M. F. PERUTZ the 1962 Nobel Prize for Chemistry for his first determining the structure of a globular PROTEIN (myoglobin).

Kennelly, Arthur Edwin (1861-1939)
US electrical engineer who, independently of HEAVISIDE, proposed the existence of that layer of the IONOSPHERE (the E layer) now often called the Kennelly-Heaviside Layer.

Kenny, Sister Elizabeth (1886-1952)
Australian nurse best known for developing the treatment of infantile paralysis (see POLIOMYELITIS) by stimulating and reeducating the muscles affected.

Kepler, Johannes (1571-1630)
German astronomer who, using BRAHE's superbly accurate observations of the planets, advanced COPERNICUS' heliocentric model of the SOLAR SYSTEM in showing that the planets followed elliptical paths. His three laws (see KEPLER'S LAWS) were later the template about which NEWTON formulated his theory of GRAVITATION. Kepler also did important work in optics, discovering a fair approximation for the law of REFRACTION.

Kepler's Laws
Three laws formulated by JOHANNES KEPLER to describe the motions of the planets in the Solar System. (1) Each planet orbits the Sun in an

PLATE TECTONICS Volcanoes and Earthquakes

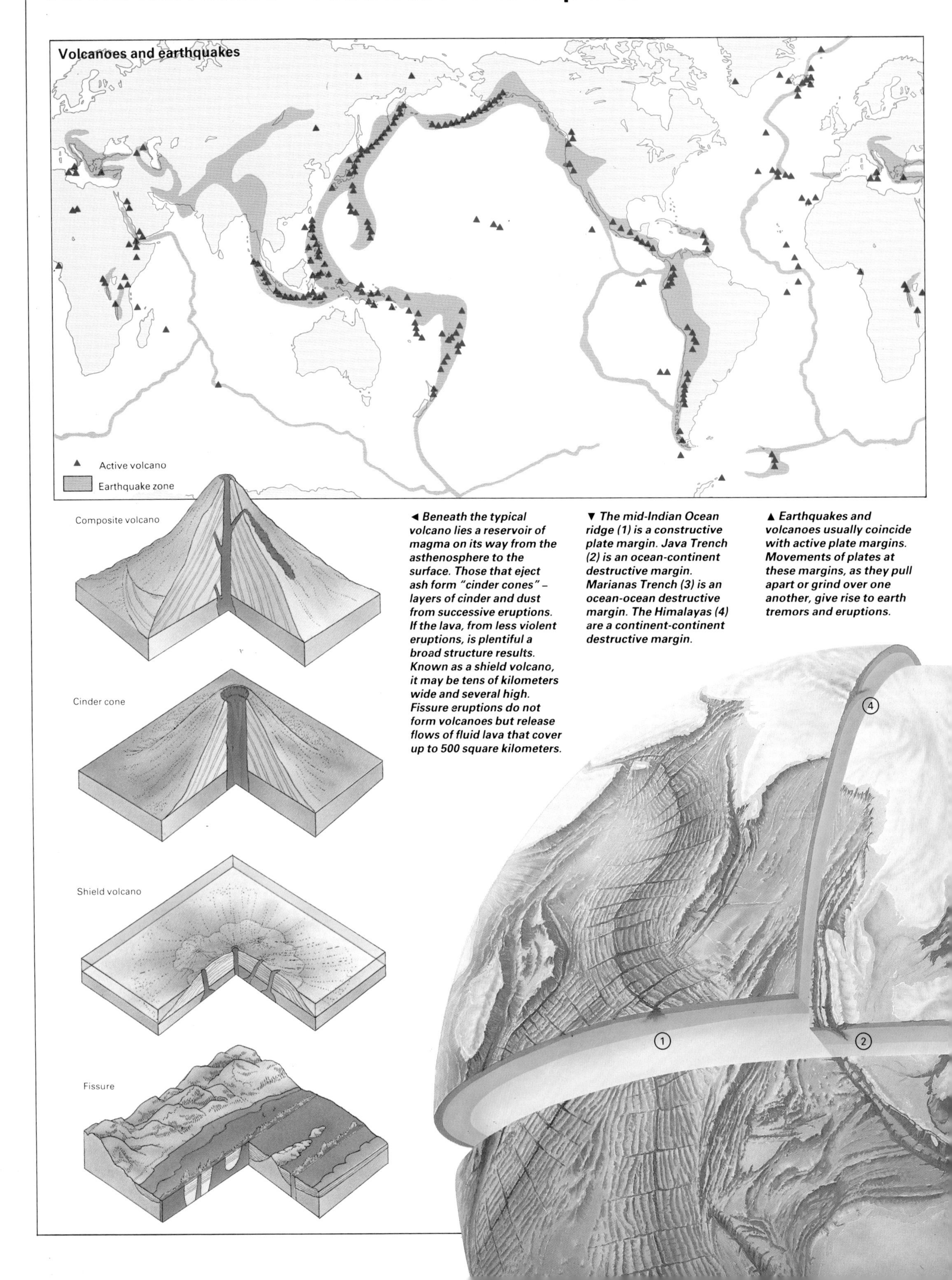

◀ *Beneath the typical volcano lies a reservoir of magma on its way from the asthenosphere to the surface. Those that eject ash form "cinder cones" – layers of cinder and dust from successive eruptions. If the lava, from less violent eruptions, is plentiful a broad structure results. Known as a shield volcano, it may be tens of kilometers wide and several high. Fissure eruptions do not form volcanoes but release flows of fluid lava that cover up to 500 square kilometers.*

▼ *The mid-Indian Ocean ridge (1) is a constructive plate margin. Java Trench (2) is an ocean-continent destructive margin. Marianas Trench (3) is an ocean-ocean destructive margin. The Himalayas (4) are a continent-continent destructive margin.*

▲ *Earthquakes and volcanoes usually coincide with active plate margins. Movements of plates at these margins, as they pull apart or grind over one another, give rise to earth tremors and eruptions.*

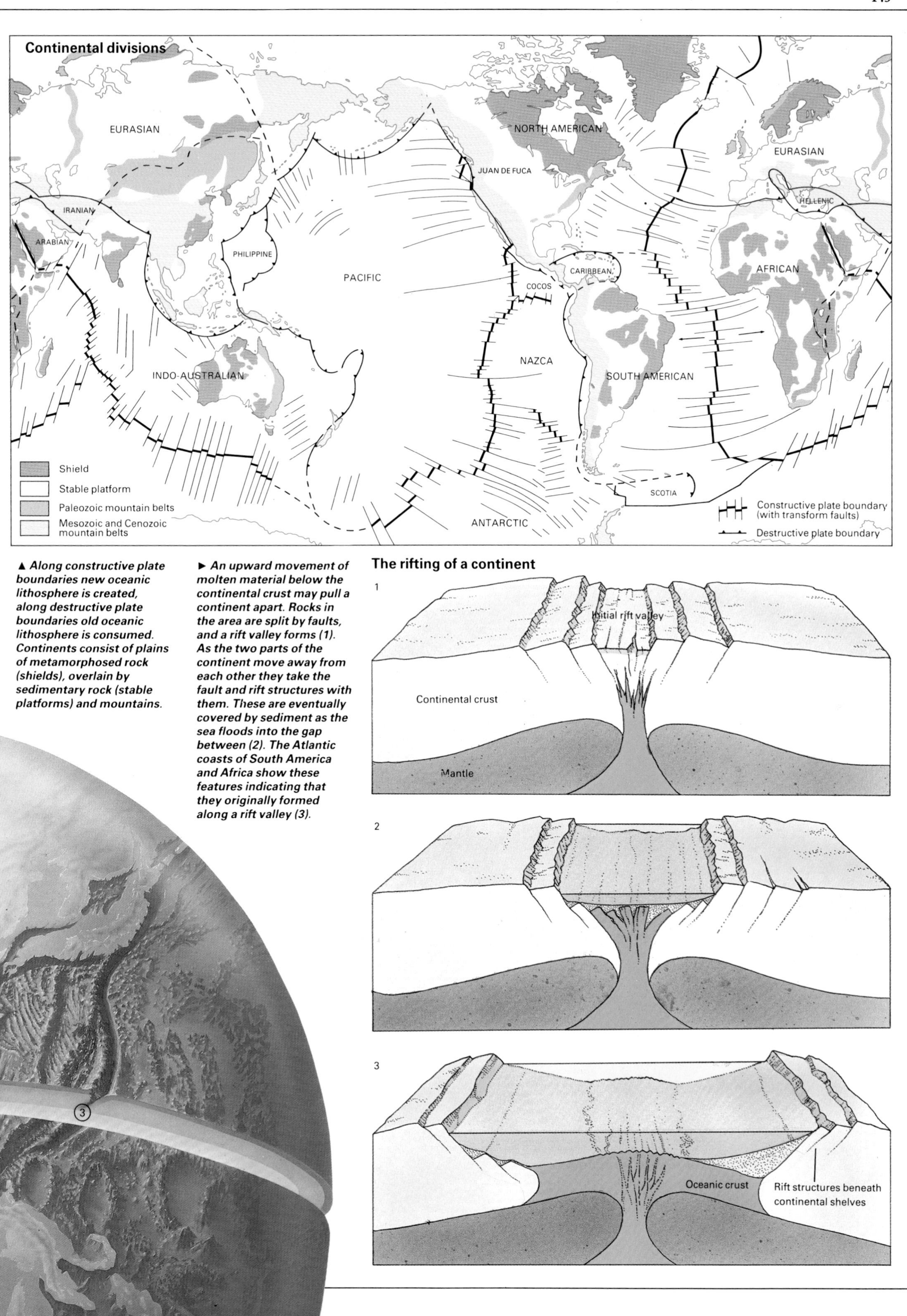

▲ *Along constructive plate boundaries new oceanic lithosphere is created, along destructive plate boundaries old oceanic lithosphere is consumed. Continents consist of plains of metamorphosed rock (shields), overlain by sedimentary rock (stable platforms) and mountains.*

► *An upward movement of molten material below the continental crust may pull a continent apart. Rocks in the area are split by faults, and a rift valley forms (1). As the two parts of the continent move away from each other they take the fault and rift structures with them. These are eventually covered by sediment as the sea floods into the gap between (2). The Atlantic coasts of South America and Africa show these features indicating that they originally formed along a rift valley (3).*

ELLIPSE of which the Sun is at one focus. (2) The line between a planet and the Sun sweeps out equal areas in equal times: hence the planet moves faster when closer to the Sun than it does when farther away. (3) The square of the time taken by a planet to orbit the Sun is proportional to the cube of its mean distance from the Sun.

Keratin
A fibrous insoluble PROTEIN found in the skin of vertebrates, and forming the major component of hair, feathers, nails, claws, horns and hooves.

Kernite
A mineral form of hydrated sodium tetraborate ($Na_2B_4O_7.4H_2O$), found associated with BORAX in Kern County, California. It forms colorless monoclinic crystals, and occurs as veins in clay shale beds. Kernite is a major source of BORON.

Ketones
Class of organic compounds of general formula RR′CO, containing a carbonyl group, but less reactive than ALDEHYDES, which in some ways they resemble. They are used as solvents and in industrial synthesis. The simplest and most important is ACETONE (see also CAMPHOR). Ketones are formed by dehydrogenation or oxidation of secondary ALCOHOLS, FRIEDEL-CRAFTS acylation of aromatic compounds, and by other methods. They may be reduced by hydrogen or metal HYDRIDES to secondary alcohols, and undergo addition and condensation reactions with NUCLEOPHILES (see also OXIMES). The presence of α-hydrogen yields greater reactivity because of keto-enol TAUTOMERISM.

Ketosis
Metabolic state in which breakdown of body FATS leads to the production of KETONE bodies (β-hydroxybutyrate and acetoacetate). These break down to ACETONE, which may be smelled in the breath.

Kettlehole
A depression in an area covered by glacial drift, formed where a mass of ice, submerged in the drift, has melted. Kettle lakes are water-filled kettleholes. (See also GLACIER.)

Khorana, Har Gobind (b. 1922)
Indian-born US biochemist who shared with HOLLEY and NIRENBERG the 1968 Nobel Prize for Physiology or Medicine for his major contributions toward deciphering the genetic code (see GENETICS).

Kidneys
Two organs concerned with the excretion of waste products in the urine and the balance of salt and water in the body. They lie behind the peritoneal cavity of the abdomen and excrete urine via the ureters, thin tubes passing into the pelvis to enter the BLADDER. The basic functional unit of the kidney is the nephron, consisting of a glomerulus and a system of tubules; these feed into collecting ducts, which drain into the renal pelvis and ureter. BLOOD is filtered in the glomerulus so that low-molecular-weight substances, minerals and water pass into the tubules; here most of the water, sugar and minerals are reabsorbed, leaving behind wastes such as urea in a small volume of salt and water. Tubules and collecting ducts are concerned with the regulation of salt and water reabsorption, which is partly controlled by two HORMONES (VASOPRESSIN and aldosterone). Some substances are actively secreted into the urine by the tubules, and the kidney is the route of excretion of many drugs. Hormones concerned with ERYTHROCYTE formation and regulation of aldosterone are formed in the kidneys, which also take part in protein METABOLISM. Diseases affecting the kidney may result in acute NEPHRITIS, including BRIGHT'S DISEASE, the nephrotic syndrome (EDEMA, heavy protein loss in the urine and low plasma albumin) or acute or chronic renal failure. In acute renal failure, nephrons rapidly cease to function, often after prolonged SHOCK, SEPTICEMIA, etc. They may, however, recover. In chronic renal failure, the number of effective nephrons is gradually and irreversibly reduced so that they are unable to excrete all body wastes. Nephron failure causes UREMIA. Disease of the kidneys frequently causes hypertension (see BLOOD CIRCULATION). (♦♦ page 111, 233)

Kieselguhr (diatomaceous earth)
A fine, porous, chalklike material (amorphous SILICA) formed by the accumulation on ocean floors of the shells of DIATOMS. It is used as an abrasive, a filter and an absorbent, especially in dynamite.

Kilo (k)
The SI prefix multiplying a unit one-thousandfold. Examples include kilohertz (kHz), kilometer (km), kilobyte (kb), kiloton and kilowatt (kW). (See SI UNITS.)

Kilogram (kg)
The base unit of MASS in SI units, defined as the mass of a platinum-iridium prototype kept under carefully controlled conditions at the International Bureau of Weights and Measures, near Paris, France.

Kilowatt-hour (kWh)
The commercial unit of electrical ENERGY, being the energy dissipated by a one-kilowatt device in one hour.

Kimberlite
Basic IGNEOUS ROCK, often altered and fragmented, which contains DIAMONDS formed in situ. It consists of OLIVINE with MICA, SERPENTINE, CALCITE and other minerals. Its chief occurrence is as pipes and DIKES at Kimberley, South Africa.

Kinematics
The branch of MECHANICS concerned with describing the motions of objects without consideration of the forces causing those motions. It thus deals with quantities such as distance, time, VELOCITY and ACCELERATION. With KINETICS it makes up DYNAMICS.

Kinesis
Random, undirected movement made in response to a stimulus. (See also TAXIS.)

Kinetic energy see ENERGY.

Kinetics, Chemical
Branch of PHYSICAL CHEMISTRY dealing with reaction rates and mechanisms. In a chemical system, several reactions may be possible according to THERMODYNAMICS, but in practice the fastest reaction predominates, not necessarily the most energetically favored. The reaction rate – the rate at which the concentration of one reactant decreases – is normally proportional to a certain POWER of the concentrations of the reactants, the sum of the exponents being called the reaction order. Thus for the reaction $A+B \rightarrow C+D$ it may be found that the rate is given by

$$\frac{d[A]}{dt}=k[A]^2[B]$$

(see EQUILIBRIUM, CHEMICAL): such a reaction is third order overall (second order in A, first order in B). The rate constant, k, depends exponentially (see EXPONENT) on the absolute TEMPERATURE (so that at room temperature most reactions double in rate for a 10K rise in temperature) and on the ACTIVATION ENERGY. CATALYSIS speeds up a reaction by providing an alternative mechanism with lower activation energy. Reaction rates are studied by measuring concentration as a function of time, regular or continuous chemical ANALYSIS being used. (See also FLASH PHOTOLYSIS.)

Kinetic theory
Widely used statistical theory based on the idea that matter is made up of randomly moving ATOMS or MOLECULES whose kinetic ENERGY increases with TEMPERATURE. It is closely related to statistical mechanics, and predicts macroscopic properties of solids, liquids and gases from motions of individual particles using MECHANICS and PROBABILITY theory. Gases are particularly suited to treatment by kinetic theory, and useful laws connecting their pressure, temperature, density, diffusion and other properties have been deduced with its aid.

Kingdom
The primary division of the living world in TAXONOMY. Originally just two kingdoms were recognized, Animalia and Plantae. The fungi and bacteria were placed with the plants on the basis of their rigid cell walls, while most non-photosynthetic unicellular organisms, other than bacteria and fungi, were designated as protozoa and placed with the animals. The artificiality of this two-kingdom system became apparent in the late 19th century, but a truly satisfactory alternative has still not been devised. Most biologists now use a five-kingdom system; Animalia, Plantae, Fungi, Protista or Protoctista, and Monera. The last of these includes all PROKARYOTES (bacteria and cyanobacteria) and is easily defined. The most problematic group is the Protista, which includes all protozoa, and is extended by some taxonomists to take in the slime molds and even the algae (on the basis of their structural simplicity), so that large kelps are grouped with tiny amebae in a new kingdom, the Protoctista. Unicellular organisms such as the diatoms and dinoflagellates are variously placed with the protists or with the algae in the plant kingdom.

Alternative schemes that create more kingdoms to accommodate the "difficult" minor groups have been proposed, some containing as many as 20 kingdoms.

King's Evil see SCROFULA.

Kinin
A chemical messenger released by ENZYME action, which causes contraction of smooth muscle and elevates blood pressure, thus playing an important role in INFLAMMATION and SHOCK. Kinin is a small PEPTIDE.

Kinorhynchs
Tiny worm-like animals, less than 1mm long, which are found in coastal sands. They are superficially segmented but do not appear to be related to the main group of segmented worms, the annelids.

Kinsey, Alfred Charles (1894-1956)
US zoologist best known for his statistical studies of human sexual behavior, published as *Sexual Behavior in the Human Male* (1948) and *Sexual Behavior in the Human Female* (1953).

Kirchhoff, Gustav Robert (1824-1887)
German physicist best known for his work on electrical conduction, showing that current passes through a conductor at the speed of light, and for deriving KIRCHHOFF'S LAWS. With BUNSEN he pioneered spectrum analysis (see SPECTROSCOPY), which he applied to the solar spectrum, identifying several elements and explaining the FRAUNHOFER LINES.

Kirchhoff's Laws
Two laws governing electric circuits involving Ohm's-law conductors and sources of electromotive force, stated by G.R. KIRCHHOFF. They assert that the sums of outgoing and incoming currents at any junction in the circuit must be equal, and that the sum of the current-resistance products around any closed path must equal the total electromotive force in it.

Kitasato, Shibasaburo (1856-1931)
Japanese bacteriologist who discovered, independently of YERSIN, the PLAGUE bacillus; and with BEHRING discovered that graded injections of

toxins could be used for immunization (see ANTITOXINS).

Klaproth, Martin Heinrich (1743-1817)
German chemist noted for his pioneering work in chemical ANALYSIS. He discovered the elements ZIRCONIUM (1789) and URANIUM (in fact, uranium oxide: 1789), and also rediscovered and named TITANIUM (1795).

Klebs, Edwin (1834-1913)
German pathologist and bacteriologist whose work permitted LÖFFLER to isolate the DIPHTHERIA bacillus (1884), often now called the Klebs-Löffler bacillus.

Kleptomania
An individual's obsessive urge to steal objects that he does not in general really wish to possess.

Klinefelter's syndrome
Chromosome abnormality associated with small testes.

Klitzing, Klaus von (b. 1943)
German scientist awarded the 1986 Nobel Prize for Physics for his discovery of the quantized Hall effect.

Klug, Aaron (b. 1926)
British physicist and molecular biologist who received the 1982 Nobel Prize for Chemistry in recognition of his work applying ELECTRON MICROSCOPY and for the light this threw on the structure and development of VIRUSES and the mechanisms of PROTEIN building in the CELL.

Knot (kn)
Unit of speed used at sea, defined as one nautical mile per hour. The international knot (1852m/h) used in the US differs slightly from the UK knot (6080ft/h). The name comes from the old practice of measuring speed at sea by counting the number of knots in a knotted line payed out over the stern in a given time.

Koch, Robert (1843-1910)
German medical scientist regarded as a father of BACTERIOLOGY, awarded the 1905 Nobel Prize for Physiology or Medicine for his work. He isolated the ANTHRAX bacillus and showed it to be the sole cause of the disease; devised important new methods of obtaining pure cultures; and discovered the bacilli responsible for TUBERCULOSIS (1882) and CHOLERA (1883).

Kocher, Emil Theodor (1841-1917)
Swiss surgeon awarded the 1909 Nobel Prize for Physiology or Medicine for his discovery of the relationship between the THYROID GLAND and CRETINISM.

Koffka, Kurt (1886-1941)
German-born US psychologist who, with KÖHLER and WERTHEIMER, was responsible for the birth of GESTALT PSYCHOLOGY.

Köhler, Georges (b. 1946)
German pharmacologist who received the 1984 Nobel Prize for Physiology or Medicine jointly with JERNE and MILSTEIN for making possible the production of MONOCLONAL ANTIBODIES.

Köhler, Wolfgang (1887-1967)
German-born US psychologist who, with KOFFKA and WERTHEIMER, was responsible for the birth of GESTALT PSYCHOLOGY.

Kornberg, Arthur (b. 1918)
US biochemist awarded with OCHOA the 1959 Nobel Prize for Physiology or Medicine for discovering an ENZYME (DNA polymerase) that could produce from a mixture of NUCLEOTIDES exact replicas of DNA molecules.

Kossel, Albrecht (1853-1927)
German biochemist awarded the 1910 Nobel Prize for Physiology or Medicine for his work on PROTEINS and NUCLEIC ACIDS, principally for showing that "nuclein" from cellular sources was not one substance but was composed of a protein and a nonprotein (nucleic acid) component.

Krafft-Ebing, Richard, Baron von (1840-1902)
German psychologist best known for his work on the psychology of sex. He also showed that there was a relation between syphilis (see SEXUALLY TRANSMITTED DISEASES) and general PARALYSIS.

Krebs, Sir Hans Adolf (1900-1981)
German-born British biochemist awarded (with LIPMANN) the 1953 Nobel Prize for Physiology or Medicine for his discovery of the CITRIC ACID CYCLE, or "Krebs cycle".

Krogh, Schack August Steenberg (1874-1949)
Danish physiologist awarded the 1920 Nobel Prize for Physiology or Medicine for his discovery that CAPILLARIES contract and expand so as to vary the amount of BLOOD-oxygen supplied to parts of the body in accordance with their requirements.

Krypton (Kr)
One of the NOBLE GASES, used to fill high-wattage electric light bulbs, flash lamps and electric-arc lamps. It combines with fluorine in an electric discharge to give krypton (II) fluoride (KrF_2), a highly reactive, colorless crystalline solid, which decomposes slowly at 20°C and is hydrolyzed by water. Other compounds have been claimed. AW 83.8, mp −157°C, bp −152°C. (♦ page 130)

Kuhn, Richard (1900-1967)
German chemist awarded the 1938 Nobel Prize for Chemistry for his work on the CAROTENOIDS and VITAMINS.

Kurchatov, Igor Vasilevich (1903-1960)
Russian nuclear physicist largely responsible for the development of Soviet nuclear armaments and for the first Soviet nuclear power station. The Soviets have named RUTHERFORDIUM kurchatovium for him.

Kurchatovium see RUTHERFORDIUM.

Kusch, Polykarp (b. 1911)
German physicist who shared with W.E. LAMB the 1955 Nobel Prize for Physics for showing that the ELECTRON had a magnetic moment (see MAGNETISM) greater than that theoretically calculated. This result inspired radical changes in nuclear theory.

Kwashiorkor
A form of malnutrition due to a diet deficient in protein and energy-producing foods. In affected children it causes EDEMA, skin and hair changes, loss of appetite, DIARRHEA, LIVER disturbance and apathy. It is common among some African tribes, and its name derives from its occurrence in children rejected from the breast at the birth of the next sibling.

Kyphosis
An arched curvature of the spine, as in the upper chest, the opposite of LORDOSIS. The term is usually used only when the arching is severely abnormal, producing a "hunchback" appearance.

Labia
The two pairs of skin folds, one fleshy and one thin, that surround the VAGINA. (➧ page 204)

Labradorite
Variety of plagioclase FELDSPAR, consisting of ALBITE and ANORTHITE, commonly occurring in BASALT and GABBRO. Gray to black in color, it often shows red, blue or green iridescence, and hence is used as a GEM stone and for building.

Labyrinthodonts
An extinct order of the AMPHIBIANS. They lived between the Carboniferous and Triassic periods, 360-213 million years ago. Some were large, up to 5m long, and all had teeth in which the dentine was folded into a pattern of wavy lines seen when the teeth are sectioned (hence "labyrinth-tooth").

Laccolith
A dome-like intrusion of igneous rock, usually arching the overlying strata and with an approximately flat floor. Phacoliths are similar but lens-shaped, with a concave side facing downward and a convex side facing upward.

Lacrimal
To do with tears; lacrimal glands constantly produce tears that lubricate the upper part of the inside of the nose, the tears becoming more copious during crying.

Lactation
The production of MILK by female mammals. Shortly before the birth of her young, hormonal changes in the mother result in increased development of the mammary glands and teats. Glandular cells in the body of the mammaries secrete milk, which is released to the young when the teats are stimulated.

Lactic acid ($CH_3.CHOH.COOH$)
A compound produced by the action of lactic acid bacteria on MILK (see FERMENTATION). It is responsible for the flavor of plain yoghurt. Lactic acid is also produced in muscle tissue that is being vigorously exercised and is unable to obtain sufficient oxygen from the blood to supply the energy it needs. Under these conditions additional energy can be obtained by GLYCOLYSIS, and the pyruvate produced is converted to lactic acid, or lactate. This lactate tends to build up in the muscle and is probably responsible for the sensation of tiredness that is experienced in vigorous exercise.

Lactose (milk sugar, $C_{12}H_{22}O_{11}$)
A disaccharide SUGAR forming about 4.5% of MILK. It yields GLUCOSE and galactose with the ENZYME lactase. (➧ page 40)

Lagomorphs
An order of MAMMALS, the hares, rabbits and pikas. They were at one time classified with the RODENTS because they too have constantly growing incisor teeth, but they are now placed in an order of their own. All are herbivores, with bacteria in the hindgut that aid the digestion of cellulose. The practice of coprophagy (eating primary fecal pellets) means that all food passes through the digestive system twice, so nutrients produced by hindgut bacteria can be absorbed in the stomach and small intestine.

Lagoon
Stretch of water separated from the sea by a bank or reef, a common phenomenon in CORAL reef areas (see also ATOLL). A lagoon may also be formed when the sea throws up a barrier beach at high tide. Haffs are lagoons created by a sandy spit at a river mouth.

Lake see RIVERS AND LAKES.

Lamarck, Jean Baptiste Pierre Antoine de Monet, Chevalier de (1744-1829)
French biologist who did pioneering work in taxonomy, especially that of invertebrates, which led him to formulate an early theory of evolution. Like CHARLES DARWIN, Lamarck believed that acquired characteristics could be inherited, but he differed from him in making it the cornerstone of the evolutionary process. He envisaged a desire for improvement in all living organisms that would give rise to new and better adapted forms through lifelong striving. The improvements thus acquired could be passed on to the offspring. Lamarck's best-known example was that of a giraffe stretching up toward higher and higher branches and thus gradually lengthening its neck. To account for the continuing existence of lowly forms, Lamarck espoused the doctrine of SPONTANEOUS GENERATION. Pasteur's refutation of spontaneous generation was a major blow to Lamarck; in the 20th century it has been established that acquired characteristics cannot be inherited (see CENTRAL DOGMA).

Lamb, Willis Eugene, Jr. (b. 1913)
US physicist who shared with KUSCH the 1955 Nobel Prize for Physics for his examinations of the hydrogen spectrum. He devised new techniques whereby he was able to show that the positions of certain lines differed from the theoretical predictions, thus necessitating a revision of atomic theory.

Lamellibranchs
Another name for the bivalve mollusks; see MOLLUSCA.

Laminates
Components where several laminae (thin sheets) of different substances are bonded together with RESINS. Laminated plastics comprise layers of cloth, paper, plastic, etc., impregnated with synthetic resin, bonded together by heat and pressure. Laminated glass is used in auto and airplane windows and as bulletproof glass; and laminated woods for many purposes (see also VENEER).

Laminectomy
Operation to expose the spinal cord by removing the back of one or more vertebrae.

Laparatomy
Operation of opening the abdomen, usually to make a diagnosis.

Lampreys
One of the two groups of jawless fish or AGNATHA. (➧ page 17)

Lampshells
A phylum of marine invertebrates once classified with the Molluska because of their resemblance to bivalves; this resemblance is now known to be a product of convergent evolution. Lampshells are enclosed in bivalved shells and are attached to the sea bed by a long, flexible stalk. There are about 260 living species, but at least 30,000 are known as fossils from rocks dating back to the Lower Cambrian period.

Lancelet
A fish-like invertebrate also called amphioxus.

Land, Edwin Herbert (b. 1909)
US physicist and inventor of Polaroid, a cheap and adaptable means of polarizing light (1932), and the Polaroid Land camera (1947). In 1937 he set up the Polaroid Corporation to manufacture scientific instruments and antiglare sunglasses incorporating Polaroid.

Landau, Lev Davidovich (1908-1968)
Soviet physicist who made important contributions in main fields of modern physics. His work on CRYOGENICS was rewarded by the 1962 Nobel Prize for Physics for his development of the theory of liquid HELIUM and his predictions of the behavior of liquid He^3.

Land reclamation
The transformation of useless land into productive land, usually for agricultural purposes. The major techniques are IRRIGATION, DRAINAGE, FERTILIZERS and DESALINIZATION.

Landsteiner, Karl (1868-1943)
Austrian-born US pathologist awarded the 1930 Nobel Prize for Physiology or Medicine for discovering the major BLOOD groups and developing the ABO system of blood typing.

Langley, Samuel Pierpont (1834-1906)
US astronomer, physicist, meteorologist and inventor of the BOLOMETER (1878) and of an early heavier-than-air flying machine. His most important work was investigating the Sun's role in bringing about meteorological phenomena.

Langmuir, Irving (1881-1957)
US physical chemist awarded the 1932 Nobel Prize for Chemistry for his work on thin films on solid and liquid surfaces (particularly oil on water), which gave rise to the new science of surface chemistry.

Language, Programming
A set of rules for giving instructions to a COMPUTER. In high-level languages, which use words similar to those in English and are therefore easier to learn, one statement may represent several machine instructions, and must be converted into machine code by a compiler. Low-level languages are more similar to machine code, and each statement represents one machine instruction.

Lanolin
Soft, yellow-white unctuous solid, a hydrated grease or wax from sheep's wool. It is a mixture of CHOLESTEROL and its ESTERS of FATTY ACIDS, and is used as a base for ointments and cosmetics.

Lanthanides
The 14 elements with atomic numbers (see ATOM) 58-71, immediately following LANTHANUM in the PERIODIC TABLE. They comprise cerium, praseodymium, neodymium, promethium, samarium, europium, gadolinium, terbium, dysprosium, holmium, erbium, thulium, ytterbium, and lutetium. (See also LANTHANUM SERIES.) (➧ page 130)

Lanthanum (La)
The second most abundant of the RARE EARTHS, and the prototypical member of the LANTHANUM SERIES of elements. AW 138.9, mp 920°C, bp 3453°C, sg 6.145 (25°C). (➧ page 130)

Lanthanum series
The 15 elements with atomic numbers (see ATOM) 57-71, comprising LANTHANUM and the LANTHANIDES (see PERIODIC TABLE). Their electronic structures are very similar, differing only in inner ORBITALS; hence their properties are very similar. This also produces a decrease (the lanthanide contraction) in ionic radii through the series, so the third-row TRANSITION ELEMENTS following the lanthanum series have ionic radii almost identical to those of their analogs in the second row, and hence have similar properties. The lanthanum series elements are all reactive metals resembling SCANDIUM, forming trivalent salts and LIGAND complexes. Cerium, praseodymium and terbium also form tetravalent compounds, and europium, ytterbium and samarium form divalent ones. All form divalent ionic hydrides and sulfides. (See also RARE EARTHS.)

Laparoscopy
The technique of introducing a fiber-optics tube

RADIOACTIVITY

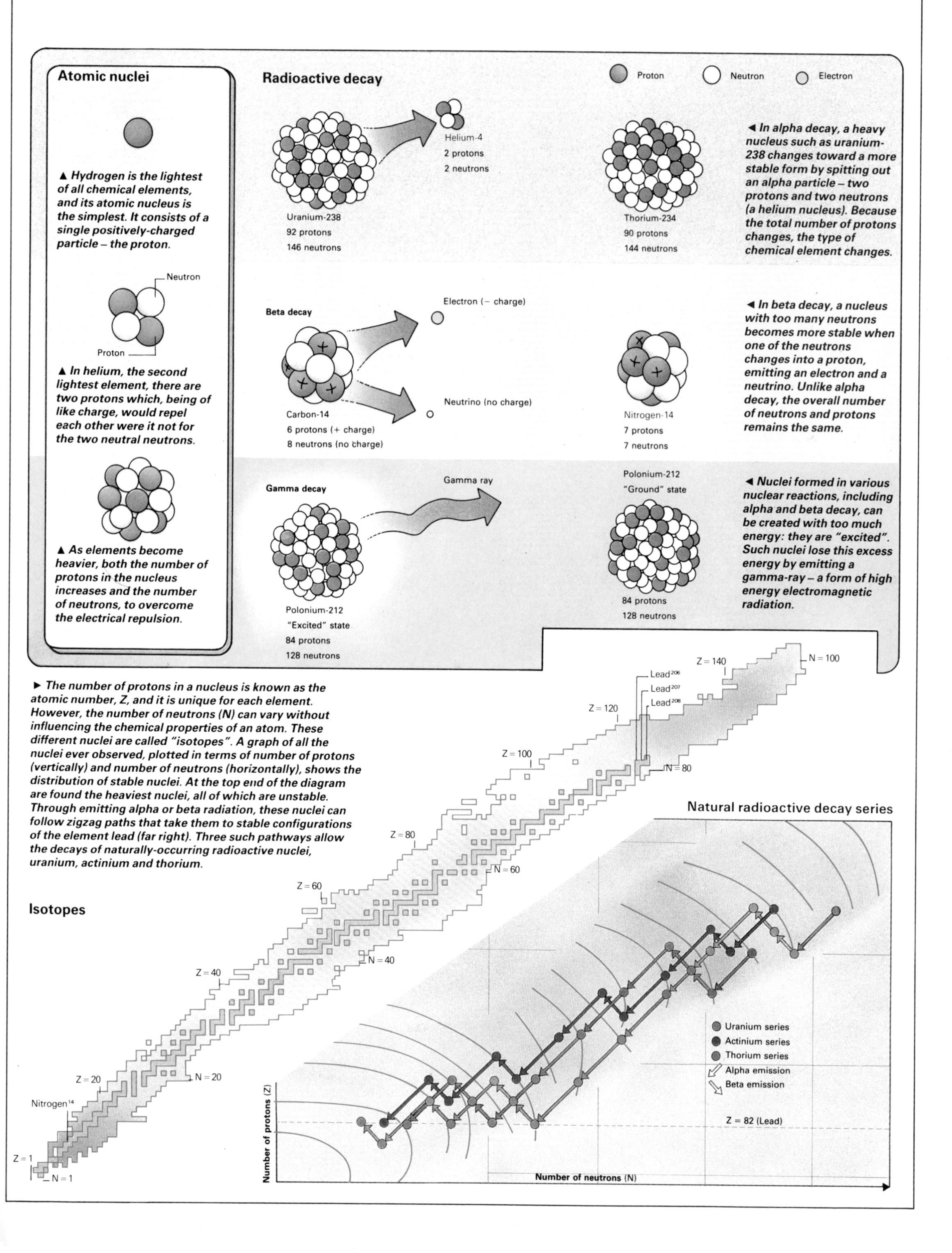

▲ *Hydrogen is the lightest of all chemical elements, and its atomic nucleus is the simplest. It consists of a single positively-charged particle – the proton.*

▲ *In helium, the second lightest element, there are two protons which, being of like charge, would repel each other were it not for the two neutral neutrons.*

▲ *As elements become heavier, both the number of protons in the nucleus increases and the number of neutrons, to overcome the electrical repulsion.*

◄ *In alpha decay, a heavy nucleus such as uranium-238 changes toward a more stable form by spitting out an alpha particle – two protons and two neutrons (a helium nucleus). Because the total number of protons changes, the type of chemical element changes.*

◄ *In beta decay, a nucleus with too many neutrons becomes more stable when one of the neutrons changes into a proton, emitting an electron and a neutrino. Unlike alpha decay, the overall number of neutrons and protons remains the same.*

◄ *Nuclei formed in various nuclear reactions, including alpha and beta decay, can be created with too much energy: they are "excited". Such nuclei lose this excess energy by emitting a gamma-ray – a form of high energy electromagnetic radiation.*

► *The number of protons in a nucleus is known as the atomic number, Z, and it is unique for each element. However, the number of neutrons (N) can vary without influencing the chemical properties of an atom. These different nuclei are called "isotopes". A graph of all the nuclei ever observed, plotted in terms of number of protons (vertically) and number of neutrons (horizontally), shows the distribution of stable nuclei. At the top end of the diagram are found the heaviest nuclei, all of which are unstable. Through emitting alpha or beta radiation, these nuclei can follow zigzag paths that take them to stable configurations of the element lead (far right). Three such pathways allow the decays of naturally-occurring radioactive nuclei, uranium, actinium and thorium.*

into the abdomen to enable the surgeon to view the inside of the body without making a large incision. The technique allows microsurgery, for example to remove ova from a woman's body for fertilization (see IN-VITRO FERTILIZATION) or to seal the Fallopian tubes for sterilization.

Lapis lazuli
Deep blue METAMORPHIC ROCK, found in crystalline LIMESTONE and consisting of LAZURITE mixed with other silicates, CALCITE and PYRITE. It chiefly occurs in Afghanistan and Chile, and has long been valued as a GEM stone and as the source of the pigment ultramarine.

Laplace, Pierre Simon, Marquis de (1749-1827)
French scientist known for his work on celestial mechanics, especially for his NEBULAR HYPOTHESIS; for his many fundamental contributions to mathematics, and for his PROBABILITY studies.

Larva
A pre-adult stage in the life history of many animals, differing structurally from the adult as well as being sexually immature. The possession of a larva usually enables a species to exploit a different food source from that used by the adult. Again, it may be important in dispersing individuals to new areas, or, in many parasites, as an infective phase. The larva undergoes a change of structure to adult form, known as METAMORPHOSIS.

Laryngitis
Inflammation of the larynx, usually due to either viral or bacterial infection or chronic voice abuse, and leading to hoarseness or loss of voice.

Larynx
Specialized part of the respiratory tract used in voice production. It lies above the trachea in the neck, forming the Adam's apple, and consists of several components linked by small muscles. Two folds, or vocal cords, lie above the trachea and may be pulled across the airway so as to regulate and intermittently occlude air flow. The movement of these cords produces audible sounds.

Laser
A device producing an intense beam of parallel LIGHT with a precise defined wavelength. The name is an acronym for "light amplification by stimulated emission of radiation", and the device is in fact a MASER operating as an oscillator at visible (as opposed to microwave) wavelengths.

The light produced by lasers is very different from that produced by conventional sources. In the latter all the source atoms radiate independently in all directions, whereas in lasers they radiate in step with each other and in the same direction, producing coherent light. Such beams spread very little as they travel, and provide very high capacity communication links. They can be focused into small intense spots, and have been used for cutting and welding – notably for refixing detached retinas in the human eye. Lasers also find application in distance measurement by INTERFERENCE methods, in SPECTROSCOPY and in HOLOGRAPHY.

The principles of laser operation are the same as those for the maser. The active material is enclosed between a pair of parallel mirrors, one of them half-silvered; light traveling along the axis is reflected to and fro and builds up rapidly by the stimulated emission process, passing out eventually through the half-silvered mirror, while light in other directions is rapidly lost from the laser.

In pulsed operation, one of the end mirrors is concealed by a shutter, allowing a much higher level of pumping than usual; opening the shutter causes a very intense pulse of light to be produced – up to 100MW for 30ns – while other pulsing techniques can achieve 10^{13}W in picosecond pulses.

Among the common laser types are ruby lasers (optically pumped, with the polished crystal ends serving as mirrors), liquid lasers (with RARE EARTH ions or organic dyes in solution), gas lasers (an electric discharge providing the high proportion of excited states), and the very small semiconductor lasers, which are based on electron-hole recombination. (♦ page 21)

Lassa fever
Severe, often fatal, viral fever endemic in parts of Africa.

Latent heat
The quantity of HEAT absorbed or released by a substance in an isothermal change of state, such as FUSION or vaporization. The temperature of a heated lump of ice will increase to 0°C and then remain at this temperature until all the ice has melted to water before again rising. The heat energy absorbed at 0°C overcomes the intermolecular forces in the ordered ice structure and increases the kinetic ENERGY of the water molecules.

Lateral line
A system of sensory cells embedded in pits or canals along the side of the body in fish and the larvae of amphibians. These organs probably detect the low-frequency vibrations in water which come from the movement of prey or shoaling companions.

Laterite
Residual red CLAY soil, usually soft and porous, consisting mainly of hydrated oxides of iron and aluminum. (Some is used as IRON ore.) It is formed from various iron-containing parent rocks by the process of laterization: secular weathering with powerful leaching out of silica, alkalis and alkaline earths, under oxidizing conditions. A tropical climate with heavy seasonal rainfall and good drainage is required.

Latex see RUBBER.

Latimeria see COELACANTH.

Latitude and longitude
The coordinate system used to locate points on the Earth's surface. Longitude "lines" are circles passing through the poles whose centers are at the center of the Earth; they divide the Earth rather like an orange into segments. Longitudes are measured 0°-180° E and W from the line of the GREENWICH OBSERVATORY. Assuming the Earth to be a sphere, we can think of the latitude of a point as the angle between a line from the center of the Earth to the point and a line from the center to the equator at the same longitude. Each pole, then, has a latitude of 90°, and so latitude is measured from 0° to 90° N and S of the EQUATOR, latitude "lines" being circles parallel to the equator that get progressively smaller towards the poles. (See CELESTIAL SPHERE.)

Lattice
Infinite three-dimensional periodic array of points in space, each point being surrounded in an identical way by its neighbors. An assembly of ATOMS placed in the same way at each lattice point makes up a CRYSTAL structure.

Laue, Max Theodor Felix von (1879-1960)
German physicist awarded the 1914 Nobel Prize for Physics for his prediction (and, with others, subsequent experimental confirmation) that X-rays can be diffracted by crystals (see X-RAY DIFFRACTION).

Laughing gas (nitrous oxide) see NITROGEN.

Laurasia
Ancient N Hemisphere supercontinent formed, with Gondwana to the S, after the splitting of Pangea (see CONTINENTAL DRIFT; GONDWANA; PANGEA). It appears to have comprised present Europe, North America and N Asia.

Lava
Both molten ROCK rising to the Earth's surface through VOLCANOES and other fissures, and the same after solidification. Originating in the MAGMA deep below the surface, most lavas (eg BASALT) are basic and flow freely for considerable distances. The acidic, SILICA-rich lavas such as RHYOLITE are much stiffer. (♦ page 134, 142)

Basic lavas solidify in a variety of forms, the commonest being *aa* (Hawaiian, rough) or block lava, forming irregular jagged blocks, and *pahoehoe* (Hawaiian, satiny) or ropy lava, solidifying in ropelike strands. Pillow lava, with rounded surfaces, has solidified under water, and slowly cooled basalt may form hexagonal columns.

Laval, Carl Gustaf Patrik de (1845-1913)
Swedish inventor best known for his pioneering work on high-speed steam TURBINES.

Laveran, Charles Louis Alphonse (1845-1922)
French physician awarded the 1907 Nobel Prize for Physiology or Medicine for his discovery of the MALARIA parasite, a protozoan of the genus *Plasmodium*. (See also PARASITE.)

Lavoisier, Antoine Laurent (1743-1794)
French scientist who was foremost in the establishment of modern CHEMISTRY. He applied gravimetric methods to the process of COMBUSTION, showing that when substances burned, they combined with a component in the air (1772). Learning from J. PRIESTLEY of his "dephlogisticated air" (1774), he recognized that it was with this that substances combined in burning. In 1779 he renamed the gas *oxygène*, because he believed it was a component in all acids. Then, having discovered the nature of the components in water, he commenced his attack on the PHLOGISTON theory, proposing a new chemical nomenclature (1787), and publishing his epoch-making *Elementary Treatise of Chemistry* (1789). In the years before his tragic death on the guillotine, he also investigated the chemistry of RESPIRATION, demonstrating its analogy with combustion.

Lawrence, Ernest Orlando (1901-1958)
US physicist awarded the 1939 Nobel Prize for Physics for his invention of the CYCLOTRON (1929; the first successful model was built in 1931).

Lawrencium (Lr)
A TRANSURANIUM ELEMENT; the final member of the ACTINIDE series, prepared by bombardment of lighter actinides. The most stable isotope, Lr^{256}, has a half-life of only 35s. (♦ page 130)

Lazear, Jesse William (1866-1900)
US physician, a member of W. REED's commission investigating C. FINLAY's theory that YELLOW FEVER is spread by the MOSQUITO. Lazear's death five days after a mosquito bite was a tragic demonstration of the truth of the theory.

Lazurite ($Na_4Al_3Si_3O_{12}S$)
Sulfur-bearing feldspathoid SILICATE mineral, forming deep blue granular masses: the chief constituent of LAPIS LAZULI.

Lazy eye
An old-fashioned and misleading name for STRABISMUS.

LCD (liquid crystal display)
A liquid crystal sealed between two pieces of glass with a conductive coating. Liquid crystals normally twist light passing through, but a voltage applied between the glass disrupts the order of the molecules, so that the liquid is darkened and forms visible characters. Used on watches, calculators etc.

LD50
A measure of the toxicity of chemical substances; it is the median lethal dose of a substance, which will kill 50% of the animals that receive that dose.

Leaching
The process whereby water, as it percolates through the soil, dissolves out various mineral salts. Rain water is slightly acidic, because of dissolved carbon dioxide from the atmosphere, and thus important in the leaching of soils.

Lead (Pb)
Soft, bluish gray metal in Group IVA of the PERIODIC TABLE, occurring as GALENA and also as CERUSSITE and anglesite (lead sulfate). The sulfide ore is converted to the oxide by roasting, then smelted with coke. Lead dissolves in dilute nitric acid, but is otherwise resistant to corrosion, because of a protective surface layer of the oxide, sulfate, etc. It is used in roofing, coverings for electric cables, RADIATION shields, ammunition, storage BATTERIES, and alloys, including solder, PEWTER, BABBITT METAL and type metal. Lead and its compounds are toxic (see LEAD POISONING). AW 207.2, mp 327.5°C, bp 1760°C, sg 11.35 (20°C).

Lead forms two series of salts; the lead (II) compounds are more stable than the lead (IV) compounds. Lead(II) oxide (PbO), or litharge, is a yellow crystalline solid made by oxidizing lead, used in lead acid storage batteries, glass and glazes. mp 888°C. Lead(IV) oxide (PbO_2), a brown crystalline solid, is a powerful oxidizing agent used in matches, fireworks and dyes; it decomposes at 290°C. Trilead tetroxide (Pb_3O_4), or red lead, an orange-red powder, is made by oxidizing litharge, and used in paints, inks, glazes and magnets. Lead tetramethyl ($Pb[C_2H_5]_4$), is a colorless liquid made by reacting a lead/sodium alloy with ethyl chloride. It is used as an antiknock additive to GASOLINE. (♦ page 130)

Lead poisoning
Condition caused by excessive levels of LEAD in tissues and blood. Lead may be taken in through the industrial use of the metal, through AIR POLLUTION due to lead-containing fuels or, in children, through eating old paint. Brain disturbance with COMA or CONVULSIONS, peripheral NEURITIS, ANEMIA and abdominal COLIC are important effects.

Leaf
The main outgrowths from the stems of plants and the principal site of PHOTOSYNTHESIS in most species. The form of leaves varies but the basic features are similar. Each leaf consists of a flat blade or lamina, attached to the main stem by a leaf stalk or petiole. Leaf-like stipules may be found at the base of the petiole. The green coloration is produced by CHLOROPHYLL, which is sited in the CHLOROPLASTS. Most leaves are covered by a waterproof covering or cuticle. Gaseous exchange takes place through small openings called STOMATA, through which water vapor also passes (see TRANSPIRATION). The blade of the leaf is strengthened by veins that contain the vascular tissue, which is responsible for conducting water around the plant and also the substances essential for metabolism. In some plants the leaves are adapted to catch insects (see INSECTIVOROUS PLANTS), while in others they are modified to reduce water loss (see SUCCULENTS, XEROPHYTE). Leaves produced immediately below the flowers are called bracts, and in some species, eg poinsettia, these are more highly colored than the flowers, taking over the role of petals. (♦ page 53)

Leap year see CALENDAR.

Learning
Almost everything that we do derives from learning. The concepts of learning and memory are closely related, though learning is usually considered to be the result of practice, and there is usually considered to be a particular stimulus that encourages such practice. The simplest learned response is the conditioned REFLEX. The most powerful learning stimulus is the satisfaction of instinctive drives (see INSTINCT).

Positive stimuli (rewards) are more effective encouragements to learning than are negative stimuli (punishments). All animals display the ability to learn, and even some of the most primitive have the ability to become bored with the tests of experimenters (where the reward is an inadequate stimulus). In humans, learning ability depends largely on INTELLIGENCE, though social and environmental factors also play a part. (See also CONDITIONING; HABIT; IMPRINTING.)

Leblanc, Nicolas (1742-1806)
French chemist who invented the Leblanc process for obtaining alkali (SODIUM carbonate) from common SALT. The salt was treated with SULFURIC ACID to give sodium sulfate, which was then heated with CHALK and CHARCOAL to give a "black ash" from which the sodium carbonate could be washed with water. The process was supplanted late in the 19th century by the Solvay process.

Le Châtelier, Henri Louis (1850-1936)
French chemist best known for formulating Le Châtelier's principle (1888), that if a change occurs in one of the conditions of a system initially in equilibrium, the system will adjust, tending to nullify the change and return to equilibrium.

LED (light emitting diode)
Tiny crystal diodes which emit light (usually red) when an electrical current is passed through them. Digits are made up from the diodes, and used as digital displays on calculators, watches, etc.

Lederberg, Joshua (b. 1925)
US geneticist awarded with G.W. BEADLE and E.L. TATUM the 1958 Nobel Prize for Physiology or Medicine for his work on bacterial genetics. With Tatum, he showed that the offspring of different mutants of *Escherichia coli* had genes recombined from those of the original generation, thus establishing the sexuality of *E. coli*. Later he showed that genetic information could be carried between *Salmonella* by certain bacterial viruses. (See also BACTERIA; GENETICS; VIRUS.)

Lee, Tsung Dao (b. 1926)
Chinese-born US physicist who shared with YANG the 1957 Nobel Prize for Physics for their investigations of violations of the principle of PARITY, which led to significant improvements in our understanding of SUBATOMIC PARTICLES.

Lee, Yuan T. (b. 1936)
Taiwanese-born US scientist, who received with D. HERSCHBACH and J. C. POLANYI the 1986 Nobel Prize for Chemistry for their work on the dynamics of chemical elementary processes.

Leeuwenhoek, Anton van (1632-1723)
Dutch microscopist who made important observations of CAPILLARIES, red BLOOD corpuscles and SPERM cells; best known for being the first to observe BACTERIA and PROTOZOA (1674-6), which he called "very little animalcules". (♦ page 123)

Left-handedness see HANDEDNESS.

Legionnaires disease
An infection of the lungs caused by the bacterium *Legionella pneumophila*. Symptoms include PNEUMONIA, dry cough, muscular pain and, sometimes, GASTROENTERITIS. The first reported outbreak was at an American Legion meeting in Philadelphia in 1976 – hence the name.

Legume
In botany, a type of FRUIT also known as a pod. It is a multiseeded dry fruit that releases its seeds by splitting along two margins – as in beans and peas. More generally, the name is used to refer to members of the Leguminosae, the pea and bean family, especially those grown as crops for their seed, which are often called grain legumes.

Leguminosae (leguminous plants)
Plants of the pea and bean family, the fruit of which are called LEGUMES (pods). In terms of number of species, this family is second in size only to the Compositae. Economically important species include alfalfa, chickpea, lentil, pea, peanut and soybean. The roots of leguminous plants produce nodules containing nitrogen-fixing bacteria (see NITROGEN FIXATION).

Lehn, Jean-Marie Pierre (b. 1939)
French scientist awarded with DONALD CRAM and CHARLES PEDERSEN the Nobel Prize for Chemistry in 1987 for their work on molecular recognition, or "host-guest" chemistry, opening up new fields of research also in biology, medicine and materials science.

Leiomyoma
Another name for a FIBROID.

Leishmaniasis (Kala-azar)
A chronic tropical disease, particularly of the young, caused by protozoa and carried by sandflies. It causes fever, systemic disturbance, ANEMIA, enlargement of the spleen and liver and susceptibility to infection. It also causes a chronic skin condition (oriental sore) with ulceration and crusting, which may affect mucous membranes of the mouth, nose and pharynx.

Leloir, Luis Federico (b. 1906)
Argentinian biochemist who won the 1970 Nobel Prize for Chemistry for his discovery of the existence and biological significance of the sugar NUCLEOTIDES.

Lemaître, Georges Edouard (1894-1966)
Belgian physicist who first proposed the Big Bang model of the Universe, explaining the RED SHIFTS of the galaxies as due to recession (see DOPPLER EFFECT), thereby inferring that the Universe is expanding. The theory holds that the origins of the Universe lie in the explosion of a primeval atom, the "cosmic egg". (See also COSMOLOGY.)

Lenard, Philipp Eduard Anton (1862-1947)
Hungarian-born German physicist awarded the 1905 Nobel Prize for Physics for his investigations of COSMIC RAYS, during which he showed that the ATOM is mainly empty space. He also made pioneering studies of the PHOTOELECTRIC EFFECT, showing that cathode rays are generated thereby.

Lens, Optical
A piece of transparent material having at least one curved surface and which is used to focus light radiation in cameras, glasses, MICROSCOPES, TELESCOPES and other optical instruments. The typical thin lens is formed from a glass disk, though crystalline minerals and molded plastics are also used and, as with spectacle lenses, shapes other than circular are quite common.

The principal axis of a lens is the imaginary line perpendicular to its surface at its center. Lenses that are thicker in the middle than at the edges focus a parallel beam of light traveling along the principal axis at the principal focus, a point on the axis on the far side of the lens from the light source. Such lenses are converging lenses. The distance between the principal focus and the center of the lens is known as the focal length of the lens; its focal power is the reciprocal of its focal length and is expressed in diopters (m^{-1}).

A lens thicker at its edges than in the middle spreads out a parallel beam of light passing through along its principal axis as if it were radiating from a virtual focus one focal length out from the lens center on the same side as the source. Such a lens is a diverging lens.

Lens surfaces may be either inward curving (concave), outward bulging (convex) or flat (plane), and it is the combination of the properties of the two surfaces that determines the focal power of the lens. In general, images of objects produced using single thin lenses suffer from various defects, including spherical and chromatic aberration (see ABERRATION, OPTICAL), coma (in which peripheral images of points are distorted into pear-shaped spots) and astigmatism. The effects of these are minimized by designing compound lenses in which simple lenses of different shapes and refractive indexes (see REFRACTION) are combined. ACHROMATIC LENSES reduce chromatic aberration; aplanatic lenses reduce this and coma, and anastigmatic lenses combat astigmatism. (See also LIGHT.) (◆◆ page 115, 123, 237)

Lenz's Law
Law of electromagnetic INDUCTION, stating that the ELECTROMOTIVE FORCE (emf) induced in a circuit is such as to oppose the flux change giving rise to it.

Leo
The Lion, a constellation on the ECLIPTIC and fifth sign of the ZODIAC. It contains the bright star REGULUS (apparent magnitude +1.35). Leo gives its name to the annual Leonid METEOR shower. (◆ page 240)

Lepidoptera
An order of insects, the butterflies and moths. They are characterized by their coiled proboscis, used in feeding, and by their wings, which are covered with minute, overlapping scales, giving them color and pattern.

Lepospondyls
An extinct order of AMPHIBIANS, known only from fossils. They lived during the Carboniferous and Permian periods, 360-248 million years ago, alongside the larger LABYRINTHODONTS. They were up to 60cm long. Some were snake-like, others displayed bizarre modifications of the skull. They have no living descendants.

Leprosy
Disease of low infectivity that damages nerve endings so that sufferers sustain cumulative injuries, especially of the fingers and toes.

Lepton see SUBATOMIC PARTICLES.

Lesbianism see HOMOSEXUALITY.

Leucippus (*c*490 BC)
Greek philosopher who, according to ARISTOTLE, originated the theory that matter is made up of indivisible, infinitely small atoms (see ATOMISM). His pupil Democritus further developed the theory.

Leukemia
Malignant proliferation of white blood cells in BLOOD or BONE MARROW. It may be divided into acute and chronic forms. In acute forms, primitive cells predominate and progression is rapid, with ANEMIA, bruising and infection. These tend to occur in childhood. Chronic forms present in adult life with mild systemic symptoms, susceptibility to infection and enlarged LYMPH nodes or SPLEEN and LIVER.

Leukocytes see BLOOD.

Leukoplakia
The presence of small white patches on a mucous membrane. Caused by infection on the tongue, or by hormone deficiency on the female genital organs.

Leukorrhea
Production of excess white vaginal discharge due to infection; from puberty a certain amount of discharge is normal.

Levee see FLOODS AND FLOOD CONTROL.

Level (spirit level)
A device used to determine whether a surface is level or not. Usually it is a glass tube, curved upward at the center, almost filled with alcohol or ether, leaving a bubble of vapor. This bubble floats to the highest point of the tube which is, when the level is on a perfectly horizontal surface, the center.

Lever
The simplest MACHINE, a rigid beam pivoted at a fulcrum so that an effort acting at one point of the beam may be used to shift a load acting at another point on the beam. There are three classes of lever: those with the fulcrum between the effort and the load; those with the load between the fulcrum and the effort; and those with the effort between the fulcrum and the load. The part of the beam between the load and the fulcrum is the load arm; that between the effort and the fulcrum, the effort arm. The effort multiplied by the length of the effort arm equals the load multiplied by the length of the load arm: a load of 50kg, 5m from the fulcrum, may be moved by any effort 10m from the fulcrum greater than 25kg (the longer the effort arm, the less effort required). Load divided by effort gives the mechanical advantage; in this case 2. A first-class lever (eg a crowbar) has a mechanical advantage greater, less than or equal to 1; a second-class (eg a wheelbarrow), always more than 1; a third-class (eg the human arm), always less than 1. (See also ARCHIMEDES; MECHANICS; MOMENT.)

Leverrier, Urbain Jean Joseph (1811-1877)
French astronomer whose calculations enabled Johann Gottfried Galle (1812-1910) to discover the planet NEPTUNE (1846). (See also ADAMS, J.C.)

Levi-Montalcini, Rita (b. 1909)
Italian-born scientist working in the US who received the 1986 Nobel Prize for Physiology or Medicine together with STANLEY COHEN for their discoveries of growth factors.

Lewin, Kurt (1890-1947)
Prussian-born US psychologist, an early member of the GESTALT PSYCHOLOGY school, best known for his development of the concept of group dynamics, especially field theory.

Lewis, Gilbert Newton (1875-1946)
US chemist who suggested that covalent bonding consisted of the sharing of valence-electron pairs. His theory of ACIDS and bases involved seeing acids (Lewis acids) as substances that are able to accept electron pairs from bases that are electron-pair donating species (Lewis bases). In 1933, Lewis became the first to prepare HEAVY WATER (D_2O).

Leyden jar
The simplest and earliest form of CAPACITOR, a device for storing electric charge. It comprises a glass jar coated inside and outside with unconnected metal foils, and a conducting rod which passes through the jar's insulated stopper to connect with the inner foil. The jar is usually charged from an electrostatic generator. The device is now little used outside the classroom.

Libby, Willard Frank (1908-1980)
US chemist awarded the 1960 Nobel Prize for Chemistry for discovering the technique of RADIOCARBON DATING (1947).

Libido
Popularly, the sexuality or general sex drive of the individual. In PSYCHOANALYSIS, following FREUD, the libido, with its source in the ID, is a type of mental energy (though it may, as in sexuality, generate physiological energy or activity) responsible for all human constructive action.

Libra
The Scales, an average sized constellation on the ECLIPTIC, the seventh sign of the ZODIAC.

Libration see MOON.

Lice
ARTHROPOD parasitic on mammals, living in the hair. They can carry typhus and other diseases. Their eggs are called nits, and are common in young children. Contrary to belief, they seem to prefer clean, straight hair.

Lichen
Name given to organisms that are an association between FUNGI and unicellular photosynthetic organisms, either CYANOBACTERIA or ALGAE. Their relationship may be a form of MUTUALISM in which the fungus prevents the alga or cyanobacterium from drying out, and benefits in return from food produced by photosynthesis. However, it has been shown that the algae of some lichens can survive without the fungus, while the fungus alone has difficulty in growing and reproducing, so the relationship may be one-sided, with the alga gaining little benefit.

Liebig, Baron Justus von (1803-1873)
German chemist who with WÖHLER proposed the radical theory of organic structure. This suggested that groups of atoms such as the benzoyl radical (C_6H_5CO-), now known as the benzoyl group (see BENZOIC ACID), remained unchanged in many chemical reactions. He also developed methods for organic quantitative analysis and was one of the first to propose the use of mineral fertilizers for feeding plants. (See also MIRROR.)

Life
A complex phenomenon that is still not fully understood, but is characterized by the ability to build up more orderly structures from less orderly ones, to maintain and repair those orderly structures, and to produce offspring of the same form and internal organization using only simple raw materials. Life on Earth is based on complex long-chain or cyclic molecules of carbon (known as organic compounds), and the ability to perform controlled transformations of these molecules, by means of ENZYMES, is a characteristic feature of living organisms. All living things have a definite boundary, marked by a MEMBRANE (often with a further boundary beyond this, such as a cell wall or a cuticle). This membrane is made up of fats and proteins and regulates movement of substances into and out of the organism. (The exceptions to these rules are the VIRUSES, which have no external membrane and in some cases may even lack their own enzymes, or enzyme-producing genes. Viruses are assumed to be degenerate organisms that have lost these features as a result of their parasitic way of life.) The ability of living things to repair and reproduce themselves is based on NUCLEIC ACIDS, which contain a record of the details of enzymes and other proteins needed to make a new organism. This information is stored in the order of bases along the nucleic acid molecule (usually DNA), which in turn determines the order of amino acids in the protein molecules, via the GENETIC CODE. Proteins (as enzymes) are responsible for the production of all the other complex constituents of a living body, such as fats and carbohydrates, while membrane proteins are able to regulate the entry of inorganic substances, such as water and salts, into the organism. By controlling the production of proteins, DNA thus controls the entire organism. Most living organisms are divided up into membrane-bound cells, or are small enough to consist of a single CELL. Notable exceptions are the fungi and some types of algae. One major division of living organisms is between PROKARYOTIC and EUKARYOTIC organisms. Taxonomically, living things are divided up into KINGDOMS.

Life cycle
The series of stages through which an individual organism passes in its progression through life. It may be simple, as in vertebrates – from the union of the gametes in fertilization to the death of the organism – or rather more complex, as in plants exhibiting ALTERNATION OF GENERATIONS or insects undergoing complete metamorphosis.

Life, Origin of
A subject whose study involves several different disciplines: chemistry, biochemistry, geology and paleontology. It is now widely accepted that life could have arisen in a marine environment on Earth, approximately 4000 million years ago. The atmosphere at that time contained very little free oxygen, so the conditions for chemical reactions were very different from those on Earth today, when any spontaneously arising organic compound would be rapidly oxidized and broken down. Experiments simulating conditions on the Earth at that time have produced compounds such as amino acids (the building blocks of protein) and purines and pyrimidines (the bases found in

RELATIVITY

The special theory

In 1905, Einstein proposed his special theory of relativity, based upon two simple statements. First, no physical experiment can distinguish between two identical laboratories, one of which is moving, the other stationary; so it is not possible to detect the absolute motion of the Earth. Second, all observers must obtain the same value for the speed of light through space, even though the light source may be approaching or receding.

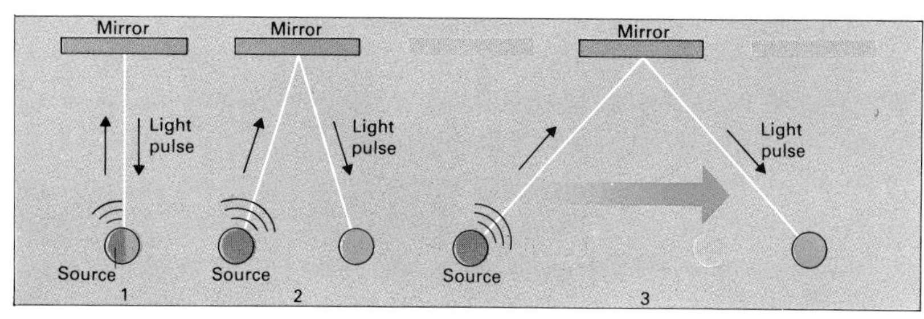

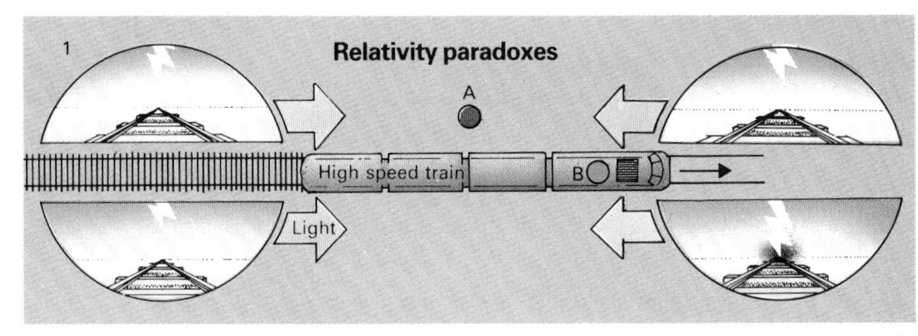

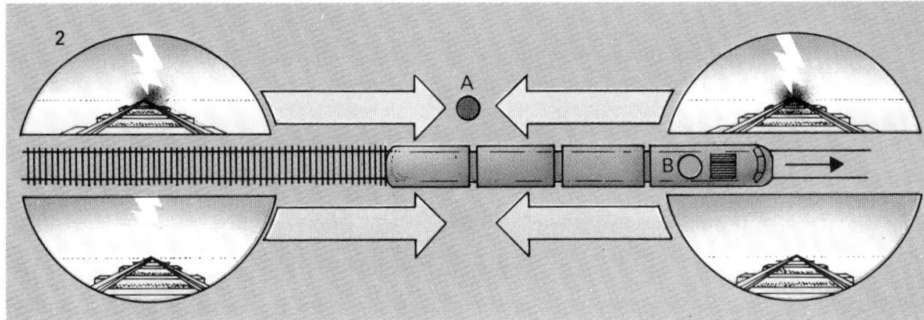

◄ *The special theory of relativity abolishes the idea of time as an absolute. In Einstein's own "thought experiment" two bolts of lightning strike opposite ends of a railroad track and are seen by an observer "A" on the embankment and a passenger "B" on a train. In situation (1), moving toward the light waves "B" sees the lightning strike the track only in front of him. In (2) "A" sees both bolts striking the track simultaneously. However, "B" has yet to see the second lightning bolt strike. The light has still to arrive from that source. Which view is then correct? Neither. Einstein said that measurements of time depend on the frame of reference – whether the observer is moving or not.*

▲ *The second postulate of the special theory is that the speed of light is independent of the velocity of the source. Suppose a moving clock sends out a time pulse while coincident with a stationary clock, and the light is reflected back from a mirror (1). From the viewpoint of the stationary clock the light will appear to have traveled farther to the moving clock, so that a stationary observer will say that the time interval for the moving clock has increased (2). The faster the clock is moving, the greater the dilation will seem (3). These effects will not be apparent to a moving observer (otherwise the principle of relativity would be violated and the observer would be able to detect the motion without reference outside).*

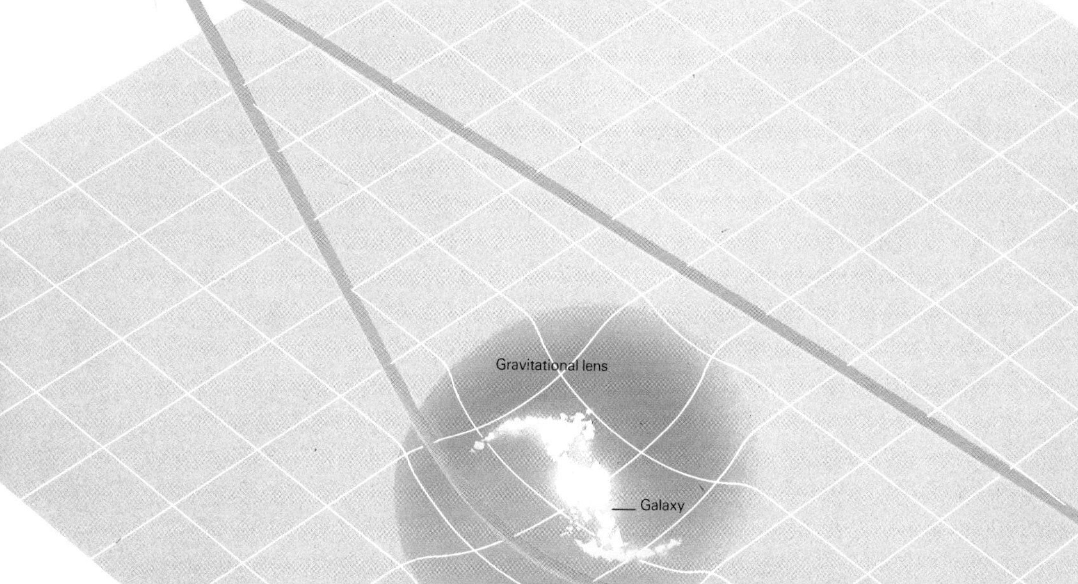

General theory of relativity

◄ *The general theory included gravity within relativity. Einstein proposed that space-time is curved in the presence of massive bodies. All particles, including light, travel on the shortest route, and where space-time is curved these routes are curved. Mass produces a curved space-time which makes particles move as if attracted. This has been summarized as, "Matter tells space how to curve and space tells matter how to move". Thus a large mass warps space so that light passing the object does not travel in a straight line but follows a bent path rather as when traveling through a lens. Such a "gravitational lens" is believed to have been seen in "double" quasars.*

nucleic acids), starting from inorganic gases and water only. It seems clear that all the basic constituents of life could have arisen in this way, but how they could have become organized into a living cell, and how the relationship between proteins and nucleic acids embodied in the GENETIC CODE could have become established is still unresolved.

Life may well have arisen more than once on Earth, but all present-day living forms are descended from a single common ancestor, as shown by the universality of the genetic code and the uniformity of basic biochemical reactions.

Ligament
Specialized fibrous thickening of a JOINT capsule, which helps to prevent the joint being forced beyond its normal range. Sudden twisting forces may cause ligamentous strain or tears (sprain). Ligaments are made up largely of COLLAGEN.

Ligand
An ION or molecule linked to a central metal ion by a coordinate bond (see BOND, CHEMICAL) to form a so-called complex compound. Almost any ion or molecule that can act as a BASE, having an atom able to donate an electron-pair, may act as a ligand – common examples include NH_3, H_20, Cl^-, OH^-, SO_4^{2-}, CO, NO^+, H^-, $C_5H_5^-$, CH_3COO^-. The complex formed may be cationic, uncharged or anionic. The coordination number of the central ion in the complex is the number of ligand-to-ion bonds; this equals the number of ligands unless they are polydentate – having more than one donating atom – when they may occupy more than one coordination site forming a CHELATE complex. Coordination numbers of 2 to 10 are known, but 6 (octahedral) and 4 (tetrahedral or square planar) are commonest. Many complexes with more than one kind of ligand have STEREOISOMERS. Complexes vary greatly in their lability, ie the rapidity with which the ligands are replaced by others: they are described as labile or inert. The bonding in complexes has been described by several theories: crystal field theory considers the effect that the electrostatic field due to the ligands has on the energies of the central ion d-ORBITALS; ligand field theory includes the mixing of ligand and ion orbitals.

Light
ELECTROMAGNETIC RADIATION to which the human EYE is sensitive. Light radiations occupy the small portion of the electromagnetic SPECTRUM lying between wavelengths 400nm and 770nm. The eye recognizes light of different wavelengths as being of different COLORS, the shorter wavelengths forming the blue end of the (visible) spectrum, the longer the red. The term light is also applied to radiations of wavelengths just outside the visible spectrum, those of energies greater than that of visible light being called ultraviolet light, those of lower energies, infrared. (See ULTRAVIOLET RADIATION; INFRARED RADIATION.) White light is a mixture of radiations from all parts of the visible spectrum, typified by the BLACKBODY RADIATION reaching the Earth from the Sun. Bodies which do not themselves emit light are seen by the light they reflect or transmit. In passing through a body or on reflection from its surface, particular wavelengths may be abstracted from white light, the body consequently displaying the colors that remain. Objects that reflect no visible light at all appear black.

For many years the nature of light aroused controversy among physicists. Although HUYGENS had demonstrated that REFLECTION and REFRACTION could be explained in terms of waves – a disturbance in the medium – NEWTON preferred to think of light as composed of material corpuscles (particles). YOUNG'S INTERFERENCE experiments reestablished the wave hypothesis and FRESNEL gave it a rigorous mathematical basis. At the beginning of the 20th century, the nature of light was again debated as PLANCK and EINSTEIN proposed explanations of blackbody radiation and the PHOTOELECTRIC EFFECT respectively, which assumed that light carried ENERGY in discrete quanta (see PHOTON). Today physicists explain optical phenomena in terms either of waves (reflection, refraction, DIFFRACTION, interference, polarization – see POLARIZED LIGHT – and SCATTERING) or quanta (blackbody radiation, photoelectric emission and the interaction of light with substantial MATTER). (See also WAVE MOTION; QUANTUM THEORY.)

Light from the Sun is the principal source of energy on Earth, being absorbed by plants in PHOTOSYNTHESIS. Many other chemical reactions involve light (see CHEMILUMINESCENCE; PHOTOCHEMISTRY; PHOTOGRAPHY) though few artificial light sources are chemical in nature. Most light sources employ radiation emitted from bodies which have become hot or have been otherwise energetically excited (see ENERGY LEVEL; LASER; LUMINESCENCE). Light can be converted into electricity using the PHOTOELECTRIC CELL. Light used for illumination is the subject of the science of PHOTOMETRY. (See also OPTICS.) (◆◆ page 21, 115, 118, 123, 151, 237)

Light meter
A device for measuring LIGHT levels, particularly in PHOTOGRAPHY where they are often coupled directly to the exposure controls of a camera. Most light meters employ either PHOTOVOLTAIC CELLS (eg selenium type) or PHOTOCONDUCTIVE DETECTORS (eg cadmium sulfide – "CdS" – type).

Lightning
A discharge of atmospheric electricity resulting in a flash of light in the sky. Most occur between two parts of a single cloud, some between cloud and ground, and a few between one cloud and another. Flashes range from a few km to about 150km in length, and typically have an energy of about 300kWh and an electromotive force of about 100MV.

Cloud-to-ground lightning usually appears forked. A relatively faint light moves towards the ground at about 125km/s in steps, often branching or forking. As this first pulse (leader stroke) nears the ground, electrical discharges (streamers) arise from terrestrial objects; where a streamer meets the leader stroke a brilliant, high-current flash (return stroke) travels up along the ionized (see ION) path created by the leader stroke at about 100Mm/s (nearly one-third the speed of light). Several exchanges along this same path may occur. If strong wind moves the ionized path, ribbon lightning results.

Sheet lightning occurs when a cloud either is illuminated from within or reflects a flash from outside, in the latter case often being called heat lightning (often seen on the horizon at the end of a hot day). Ball lightning, a small luminous ball near the ground, often vanishing with an explosion, and bead lightning, the appearance of luminous "beads" along the channel of a stroke, are rare.

Lightning results from a buildup of opposed electric charges in, usually, a cumulonimbus CLOUD, negative near the ground and positive on high (see ELECTRICITY). There are several theories that purport to explain this buildup. Understanding lightning might help us to probe the very roots of life, for lightning was probably significant in the formation of those organic chemicals that were to be the building blocks of life. (See also SAINT ELMO'S FIRE; THUNDER.)

Light year
In ASTRONOMY, a unit of distance equal to the distance traveled by light in a vacuum in one sidereal year, equal to 9461Tm (about 6 million million miles). The unit has largely been replaced by the PARSEC (1 ly=0.3069pc). (◆ page 240)

Lignin
A complex polymer made up of phenyl units (benzene rings with side-groups) joined together in a variety of ways. It gives strength and rigidity to the woody tissue of plants, and may account for 25% to 30% of the WOOD of some trees. Because it has a random structure it is difficult for enzymes to break down, and few organisms can digest it. Some fungi, however, and a few bacteria can do so. Wood-eating insects such as termites have symbiotic lignin-digesting bacteria in their gut.

Lignite (brown coal) see COAL.

Lilienthal, Otto (1848-1896)
German pioneer of aeronautics, credited with being the first to use curved, rather than flat, wings, as well as first to discover several other principles of AERODYNAMICS. He made over 2000 glider flights, dying from injuries received when one of his gliders crashed.

Lime (calcium oxide or hydroxide) see CALCIUM.

Limestone
SEDIMENTARY ROCK consisting mainly of calcium carbonate (see CALCIUM), in the forms of CALCITE and aragonite. Some limestones, such as CHALK, are soft but others are hard enough for use in building. Limestone may be formed inorganically (oolites) by evaporation of seawater or freshwater containing calcium carbonate, or organically from the shells of mollusks or skeletons of coral piled up on sea beds and compressed. In such limestone fossils usually abound.

Limnology
A branch of BIOLOGY that deals with the study of freshwater habitats and the plants and animals within them.

Limonene
A natural terpene hydrocarbon used as a flavoring and fragrance. It occurs in enantiomeric forms; d-limonene occurs in the oils of citrus fruits, while the l-form is found in spearmint and peppermint oils. MW 136.24, mp 74.4°C, bp 178°C.

Limonite
A dark brown, amorphous OXIDE mineral consisting of hydrated iron (III) oxide, formula $FeO[OH].nH_2O$. A major ION ore of widespread occurrence, often with GOETHITE, it is formed by alteration of other iron minerals.

Linear accelerator see ACCELERATORS, PARTICLE.

Linkage
The occurrence together on the same CHROMOSOME of two GENES. If the genes are close together they are said to be "closely linked", and are normally transmitted together from generation to generation. The more widely separated they are on a chromosome, the more likely it is that genes will be separated by CROSSING OVER.

Linnaeus, Carolus (later Carl von Linné) (1707-1778)
Swedish botanist and physician, the father of TAXONOMY, who brought system to the naming of living things. His classification of plants was based on their sexual organs (he was the first to use the symbols ♂ and ♀ in their modern sense), an artificiality dropped by later workers; but many of his principles and taxonomic names are still used today.

Linoleic acid (12-octadeca-cis,cis-dienoic acid)
A doubly unsaturated carboxylic acid containing 17 carbon atoms. It is found as a constituent of the triglycerides of plants and is an essential constituent of the human diet. AW 280.46, mp −5°C, bp 228°C.

Lipase
An enzyme that splits TRIGLYCERIDES into FATTY

ACIDS and GLYCEROL. As triglycerides are normally water-insoluble, lipases act relatively slowly on the surface of a fat globule. They occur widely in oil seeds and are involved in GERMINATION. They are also produced by the PANCREAS and secreted into the DUODENUM, where they digest the triglycerides in food. BILE SALTS produced by the liver emulsify these fats, making it easier for the lipases to attack them. (♦ page 40)

Lipids
A diverse group of organic compounds found in all living things, and characterized by their solubility in nonpolar organic solvents such as ETHER, CHLOROFORM and ETHANOL. Lipids include many heterogeneous substances, and unlike PROTEINS and CARBOHYDRATES have no characteristic type of building block, but they are mostly made up of carbon and hydrogen atoms only, which accounts for their general insolubility in water. Lipids are classified into TRIGLYCERIDES, PHOSPHOLIPIDS, WAXES, STEROIDS, TERPENES and other types, according to their structure.

Lipmann, Fritz Albert (1899-1986)
German-born US biochemist who shared with KREBS the 1953 Nobel Prize for Physiology or Medicine for his discovery in 1947 of coenzyme A (see ENZYMES).

Lipoma
Harmless fatty tumor of the skin.

Lippmann, Gabriel (1845-1921)
French physicist awarded the 1908 Nobel Prize for Physics for inventing (*c*1891) the first system of color PHOTOGRAPHY. His process required long exposures, and thus is now obsolete. He also invented the coelostat and predicted PIEZO-ELECTRICITY.

Lipscomb, William Nunn, Jr (b. 1919)
US physical chemist who won the 1976 Nobel Prize for Chemistry for his work on structure and bonding in BORANES. He proposed that these might contain three-center (hydrogen-bridge) bonds.

Liquid
One of matter's three states, the others being SOLID and GAS. Liquids take the shape of their container, but have a fixed volume at a particular temperature and are virtually incompressible (see COHESION). Nearly all substances adopt the liquid state under suitable conditions of temperature and pressure. (See also FLUID; KINETIC THEORY; VAPOR.) (♦ page 130)

Lissajous' figures (Bowditch curves)
Plane curves traced by a point moving in two SIMPLE HARMONIC MOTIONS that are at right angles to each other. They can most easily be formed by supplying different alternating voltages (see ELECTRICITY) to the *x*- and *y*-detection plates of an OSCILLOSCOPE. Only if the frequencies are commensurable will a true Lissajous figure (ie a closed curve) be formed.

Lister, Joseph, 1st Baron (1827-1912)
British surgeon who pioneered antiseptic SURGERY, perhaps the greatest single advance in modern medicine. PASTEUR had shown that microscopic organisms are responsible for PUTREFACTION, but his STERILIZATION techniques were unsuitable for surgical use. Lister experimented and, by 1865, succeeded by using carbolic acid (see PHENOL).

Liter (l)
A metric unit of volume, originally defined as that of 1kg of water at the temperature of its maximum density (=1.000028dm^3), but redefined in 1964 as exactly equal to one cubic decimeter (=1dm^3). The liter is not recommended for use alongside SI units.

Litharge (lead (II) oxide) see LEAD.

Lithification (diagenesis)
The process in which a loose and unconsolidated sediment, such as sand or mud, is turned into a sedimentary rock, such as sandstone or shale. It usually involves two processes. First the particles of sediment are compacted by the weight of the later sediments above them. This reduces the space between them and causes them to interlock. Groundwater percolating through the sediment then deposits crystals of minerals, such as calcite, on the surfaces of the particles, cementing them together into a solid whole.

Lithium (Li)
A white metallic element somewhat harder and less reactive than the other ALKALI METALS. Physically and chemically, lithium also resembles the ALKALINE EARTH METALS. It is the lightest element that is a solid at room temperature. It is made by ELECTROLYSIS of fused lithium chloride, and is useful in the treatment of MANIA. Lithium metal is used in heat transfer because of its high specific heat; the isotope Li^6 is important in thermonuclear processes. Lithium stearate is an additive to lubricating greases. AW 6.9, mp 180°C, bp 1347°C, sg 0.534 (20°C). (♦ page 130)

Lithosphere
The rocks of the Earth, as contrasted with the ATMOSPHERE and HYDROSPHERE. Today, use of the term is restricted to reference to the uppermost 100km depth of the substance of the Earth. In this definition it constitutes the solid mass of the plates (see PLATE TECTONICS) moving about on the more fluid ASTHENOSPHERE. (♦ page 134, 143)

Litmus
Mixture of colored compounds, extracted from LICHENS, used as an acid-base INDICATOR.

Little Dipper
Ursa Minor (the Little Bear), N Hemisphere circumpolar constellation containing POLARIS, the N polestar.

Littoral fauna
Term applied to animals that live on the seashore (the littoral or intertidal zone). They are uniquely adapted to being alternately covered and uncovered by the tide; on any one shore the organisms are "zoned" according to the amount of time they are covered by the sea.

Liver
The large organ lying on the right of the ABDOMEN beneath the DIAPHRAGM and concerned with many aspects of METABOLISM. It consists of a homogeneous mass of cells arranged round blood vessels and bile ducts. Nutrients absorbed in the GASTROINTESTINAL TRACT pass via the portal veins to the liver and many are taken up by it; they are converted into forms (eg GLYCOGEN) suitable for storage and release when required. PROTEINS, including ENZYMES, PLASMA proteins and CLOTTING factors, are synthesized from amino acids. The liver converts protein breakdown products into urea, and detoxifies or excretes other substances (including drugs) in the blood. Bilirubin, the HEMOGLOBIN breakdown product, is excreted in the BILE; this also contains bile salts, made in the liver from CHOLESTEROL and needed for the DIGESTIVE SYSTEM.

Diseases of the liver include CIRRHOSIS and HEPATITIS, while abnormal function is manifested as JAUNDICE, edema, ascites (excessive peritoneal fluid), and a variety of brain and nervous system disturbances including delirium and coma. Chronic liver disease leads to skin abnormalities, a bleeding tendency and alterations in routes of BLOOD CIRCULATION, which may in turn lead to HEMORRHAGE. Many drugs may damage the liver, causing disease similar to hepatitis, and both drugs and severe hepatitis can cause acute liver failure. (♦♦ page 28, 40, 233)

Liverworts
Small non-flowering plants growing in moist habitats and occasionally in water. They form the Class Hepaticae of the BRYOPHYTES, which also includes the MOSSES and HORNWORTS. The most familiar types have a flat, lobed thallus attached to the ground by root-like rhizoids. Other types have a thallus bearing leaf-like expansions. They exhibit ALTERNATION OF GENERATIONS. (♦ page 139)

Lizards see LACERTILIA.

Loam
Soil composd of about 30%-50% SAND particles, 30%-50% SILT particles, and less than 20% CLAY particles. (See SOIL.)

Lobe-finned fish see SARCOPTERYGII.

Lobotomy
Operation in which the FRONTAL LOBES are separated from the rest of the BRAIN, used in the past as treatment for refractory DEPRESSION. It leads to a characteristically disinhibited type of behavior and is now rarely used.

Lockjaw see TETANUS.

Lodestone see MAGNETITE.

Lodge, Sir Oliver Joseph (1851-1940)
British physicist best known for his work on the propagation of ELECTROMAGNETIC RADIATION, devising an eariy instrument (the coherer) for detecting it. He also did important work on PARAPSYCHOLOGY.

Loeb, Jacques (1859-1924)
German-born US biologist best known for his work on PARTHENOGENESIS, especially his induction of artificial parthenogenesis in sea urchins' and frogs' eggs, thereby highlighting the biochemical nature of FERTILIZATION.

Loess
Fine-grained, wind-deposited SILT found worldwide in deposits up up to 50m thick. Its main components are QUARTZ, FELDSPAR and CALCITE. Extremely porous, it forms highly fertile topsoil, often chernozem. It is able to stand intact in cliffs.

Loewi, Otto (1873-1961)
German-born US pharmacologist awarded (with SIR HENRY DALE) the 1936 Nobel Prize for Physiology or Medicine for his work showing the chemical nature of nerve impulse transmission. (See NERVOUS SYSTEM.)

Löffler, Friedrich August Johannes (1852-1915)
German bacteriologist who first isolated the DIPHTHERIA bacillus (1884), which had first been observed by KREBS the previous year. He also discovered the causative organism of GLANDERS; and, with others, showed that FOOT-AND-MOUTH DISEASE is caused by a VIRUS, developing a SERUM against it.

Lomonosov, Mikhail Vasilievich (1711-1765)
Russian scientist and man of letters, best known for his corpuscular theory of matter, in the course of developing which he made an early statement of the KINETIC THEORY.

Long, Crawford Williamson (1815-1878)
US physician who first discovered the surgical use of diethyl ETHER as an anesthetic (1842). His discovery followed an observation that students under the influence of ether at a party felt no pain when bruising or otherwise injuring themselves.

Longitude see LATITUDE AND LONGITUDE.

Lophophorates
A proposed phylum that would include all animals having a lophophore – a tentacular feeding organ found in the LAMPSHELLS, PHORONID WORMS, BRYOZOA and ENDOPROCTS. It is not certain at present if these organisms all share a common intermediate ancestor.

Lordosis
A hollowing of the spine, as in the small of the back; the opposite of KYPHOSIS. The term is often used to describe the posture of sexually receptive animals.

Lorentz, Hendrik Antoon (1853-1928)
Dutch physicist awarded with P. ZEEMAN the 1902

Nobel Prize for Physics for his prediction of the ZEEMAN effect. Basing his work on J. CLERK MAXWELL's equations, he explained the REFLECTION and REFRACTION of light; and proposed his electron theory, that LIGHT occurred through motion of electrons in a stationary electromagnetic ETHER. Thus the wavelength should change under the influence of a powerful magnetic field; and this was experimentally shown by Zeeman (1896).

The theory was, however, inconsistent with the results of the MICHELSON-MORLEY EXPERIMENT, and so Lorentz introduced the idea of "local time", that the rate of time's passage differed from place to place; and, incorporating this with the proposal of George Francis Fitzgerald (1851-1901) that the length of a moving body decreases in the direction of motion (the Fitzgerald contraction), he derived the Lorentz transformation, a mathematical statement which describes the changes in length, time and mass of a moving body. His work, with Fitzgerald's, laid the foundations for EINSTEIN's "special theory" of RELATIVITY.

Lorenz, Konrad (b. 1903)
Austrian zoologist and writer, the father of ETHOLOGY, awarded for his work the 1973 Nobel Prize for Physiology or Medicine with FRISCH and TINBERGEN. He is best known for his studies of bird behavior and of human and animal AGGRESSION. His best-known books are *King Solomon's Ring* (1952) and *On Aggression* (1966).

Lowell, Percival (1855-1916)
US astronomer and writer who predicted the existence of and initiated the search for PLUTO; but who is best known for his championing the theory (now discarded) that the "canals" of MARS were an irrigation system built by an intelligent race.

LSD (lysergic acid diethylamide)
A HALLUCINOGENIC DRUG based on ERGOT alkaloids. It may lead to psychotic reaction and bizarre behavior.

LSI (Large Scale Integration) see ELECTRONICS.

Lumbago
Popular term for low back pain or lumbar back ache. It may be of various origins, including chronic ligamentous strain, SLIPPED DISK (sometimes with SCIATICA), certain types of ARTHRITIS affecting the spine and congenital disease of the spine. Diagnosis and treatment may be difficult.

Lumen (lm)
SI unit of luminous flux (see PHOTOMETRY).

Luminance
In PHOTOMETRY, the brightness of an extended surface. In SI units luminance is measured in CANDELAS per square meter.

Luminescence
The nonthermal emission of ELECTROMAGNETIC RADIATION, particularly LIGHT, from a PHOSPHOR. Including both fluorescence and phosphorescence (distinguished according to how long emission persists after excitation has ceased, in fluorescence emission ceasing within 10ns but continuing much longer in phosphorescence), particular types of luminescence are named for the mode of excitation. Thus in photoluminescence, X-ray PHOTONS are absorbed by the phosphor and lower-energy radiations emitted; in CHEMILUMINESCENCE the energy source is a chemical reaction: cathodoluminescence is energized by cathode rays (ELECTRONS), and BIOLUMINESCENCE occurs in certain biochemical reactions. (See also ELECTROLUMINESCENCE.)

Lunar Society of Birmingham
The most illustrious of the British provincial scientific societies of the late 18th century. Its members included the industrialists Matthew Boulton and Josiah Wedgwood, the physician ERASMUS DARWIN, and the chemist and theologian JOSEPH PRIESTLEY.

Lungfish see DIPNOI.

Lungs
Any sac-like organ concerned with GAS EXCHANGE. In the terrestrial vertebrates, the two largely air-filled organs in the chest. Their surfaces are separated from the chest wall by two layers of pleura, with a little fluid betweeen them; this allows free movement of the lungs and enables the forces of expansion of the chest wall and DIAPHRAGM to fill them with air. Air is drawn into the TRACHEA via mouth or nose; the trachea divides into the BRONCHI, which divide repeatedly until the terminal air sacs or alveoli are reached. In the alveoli air is brought into close contact with unoxygenated BLOOD in lung CAPILLARIES. Lung diseases in humans include ASTHMA, BRONCHITIS, PNEUMONIA, PLEURISY, PNEUMOTHORAX, PNEUMOCONIOSIS, EMBOLISM, CANCER and TUBERCULOSIS; lungs may also be involved in several systemic diseases, eg sarcoidosis, SYSTEMIC LUPUS ERYTHEMATOSUS.

Lungs are also found in some fish (see LUNGFISH), and in the terrestrial snails, where there is a large air-filled cavity beneath the shell, which opens to the outside just above the mollusk's head. Some spiders breathe by means of lung-books – lung-like cavities with many parallel membranes within them, which increase the surface area of the lung in the same way as the alveoli do in the lungs of vertebrates. (♦ page 28)

Lupus vulgaris
TUBERCULOSIS of the SKIN.

Luria, Salvador Edward (b. 1912)
Italian-born US biologist who shared with DELBRÜCK and HERSHEY the 1969 Nobel Prize for Physiology or Medicine for research on BACTERIOPHAGES.

Luteinizing hormone (LH)
A pituitary gland GONADOTROPHIN which in female mammals causes release of eggs from the ovary and promotes the secretion of PROGESTERONE. In males it stimulates testicular ANDROGEN formation. Variations in LH secretion are involved in the onset of PUBERTY and in the menstrual cycle. (♦ page 111)

Luteotrophic hormone see PROLACTIN.

Lutetium (Lu)
The final member of the LANTHANUM SERIES of elements. AW 175.0, mp 1700°C, bp 3400°C, sg 9.840 (25°C). (♦ page 130)

Lux (lx)
SI unit of illuminance. (See PHOTOMETRY.)

Lwoff, André Michael (b. 1902)
French microbiologist awarded with F. JACOB and J. MONOD the 1965 Nobel Prize for Physiology or Medicine for his work on lysogeny, a process of genetic interaction between BACTERIOPHAGES and BACTERIA.

Lycopsida see CLUBMOSSES.

Lye
Any strong caustic ALKALI, especially POTASSIUM or SODIUM hydroxide.

Lyell, Sir Charles (1797-1875)
Scottish geologist and writer whose most important work was the promotion of geological UNIFORMITARIANISM (originally developed by JAMES HUTTON as an alternative to the CATASTROPHISM of CUVIER and others). The prime expression of these views came in his *Principles of Geology* (1830-33). His other works included the *Elements of Geology* (1838), and *Geological Evidence of the Antiquity of Man* (1863). Here he expressed guarded support for DARWIN's theory of evolution.

Lymph
A colorless fluid, similar in composition to blood but lacking erythrocytes (red blood cells). It originates from the capillaries, escaping through the capillary walls to bathe the cells of the body, then being drained off into the LYMPHATIC SYSTEM. (♦ page 40)

Lymphatic system
A circulatory system that carries LYMPH from the extracellular fluid via lymph vessels and nodes (sometimes wrongly called lymph glands) back to the blood system. Important node sites are in the neck, groin, chest and abdomen. Fine ducts carry lymph to the nodes, which are filled with lymphocytes (see IMMUNITY) and reticulum cells. These act as a filter, particularly for infected debris or CANCER cells. From the nodes lymph may drain to other nodes or directly into the major thoracic duct, which returns it to the blood. Specialized lymph ducts or lacteals carry fats absorbed in the GASTROINTESTINAL TRACT to the thoracic duct. In addition, there are several areas of lymphoid tissue at the portals of the body as a primary defense against infection (TONSILS, ADENOIDS, Peyer's patches in the gut). Lymph node enlargement may be due to inflammation following disease in the area drained, to development of an abscess in the node, or to secondary spread of cancer. (See also LYMPHOMA; LEUKEMIA.)

Lymphoma
Malignant proliferation of LYMPH tissue, usually in the lymph nodes, SPLEEN or GASTROINTESTINAL TRACT. The prototype is HODGKIN'S DISEASE, but a number of other forms occur with varying HISTOLOGY and behavior.

Lynen, Feodor (1911-1979)
German biochemist who shared with K.E. BLOCH the 1964 Nobel Prize for Physiology or Medicine for their independent work on the metabolism of CHOLESTEROL and the fatty acids.

Lyra
The Lyre, a medium-sized N Hemisphere constellation containing VEGA and the Ring Nebula (M57), a fine planetary NEBULA some 700pc from the Earth.

Lysenko, Trofim Denisovich (1898-1977)
Soviet agronomist whose antipathy for GENETICS and position of power under the Stalin regime led to the stifling of any progress in Soviet biological studies for 25 years or more. Refusing on ideological grounds to believe in GENES, he adopted a peculiar form of Lamarckism (see LAMARCK; MICHURIN), and forced other scientists to support his views. He was removed from power in 1964.

Lysis
Disintegration, eg of blood cells or bacteria.

Lysosome
A small vesicle found in the cytoplasm of many cells, consisting of degradative ENZYMES within a lipoprotein membrane. The enzymes are inactive until released into the cytoplasm, or a food vacuole formed by PHAGOCYTOSIS. (♦ page 13)

McClintock, Barbara (b. 1902)
US biologist who received the 1983 Nobel Prize for Physiology or Medicine for her theory of GENE regulation.

Mach, Ernst (1838-1916)
Austrian physicist and philosopher whose name is commemorated in MACH NUMBERS. His greatest influence was in philosophy, where he rejected from science all concepts which could not be validated by experience. This freed EINSTEIN from the absoluteness of Newtonian space-time (and thus helped him toward his theory of RELATIVITY) and helped inform the logical positivism of the Vienna Circle.

Machine
A device that performs useful work by transmitting, modifying or transforming motion, forces and ENERGY. There are three basic machines: the inclined plane, the lever, and the wheel and axle; from these, and adaptations of these, are built up all true machines, no matter how complex they may appear. There are two essential properties of all machines: mechanical advantage, which is the ratio of load to effort, and efficiency, the ratio of actual performance to theoretical performance. Mechanical advantage can be less than, equal to or greater than 1; while efficiency, owing to such losses as FRICTION, is always less than 100% (otherwise a perpetual motion machine would be possible). (See also EFFICIENCY; ENERGY; FORCE; LEVER; WORK.)

Simple machines derived from the three basic elements include: from the inclined plane, the wedge (effort at the top being translated to force at the sides) and the screw (an inclined plane in spiral form); from the lever, the wrench or spanner (the BALANCE also uses the principle of the lever), and from the wheel and axle, the pulley (which can also be viewed as a type of lever). (See also ENGINE; PUMP.)

Mach number
Ratio of the speed of an object or fluid to the local speed of SOUND, which is temperature dependent. Speeds are subsonic or supersonic depending on whether the mach number is less than or greater than one.

Macleod, John James Rickard (1876-1935)
British-born physiologist who shared with SIR F.G. BANTING the 1923 Nobel Prize for Physiology or Medicine for his role in the isolation of INSULIN.

Maclure, William (1763-1840)
British-born US geologist regarded as the father of American geology for his monumental *Observations on the Geology of the United States* (1809).

Mcmillan, Edwin Mattison (b. 1907)
US nuclear physicist who shared with SEABORG the 1951 Nobel Prize for Chemistry for his discovery of the first TRANSURANIUM ELEMENT, number 93, NEPTUNIUM (1940). Independently, he and the Soviet physicist Vladimir Veksler (1907-1966) developed the SYNCHROTRON, for which they shared the 1963 Atoms for Peace award.

Macrophages
Large white blood cells (leukocytes) that contribute to the non-specific IMMUNE SYSTEM by engulfing foreign particles, including infectious microorganisms, dust and dead body cells. They also participate in immune reactions directed against specific antigens, by interacting with T-cells and B-cells. Macrophages and their precursors, monocytes, are found in various body organs, such as the lungs, as well as in the blood and lymph where they originate. They engulf particles by PHAGOCYTOSIS.

Magellanic clouds
Two irregular GALAXIES that orbit the MILKY WAY, visible in S skies. The Large Magellanic Cloud (Nubecula Major), about 9kpc in diameter, has a well-marked axis suggesting that it may be an embryonic spiral galaxy. The Small Magellanic Cloud (Nubecula Minor) is about 5kpc across. Both are rich in CEPHEID VARIABLES. Their distances are about 50kpc (LMC) and 60kpc (SMC). The Large Magellanic Cloud contains a giant NEBULA, the Tarantula nebula, near to which a SUPERNOVA explosion occurred in 1987.

Magendie, François (1783-1855)
French physiologist, regarded as the father of experimental PHARMACOLOGY. He introduced the medical use of several DRUGS (eg MORPHINE); and first proved that in a spinal nerve (see NERVOUS SYSTEM; SPINAL CORD) the ventral (anterior) root has a motor function and the dorsal (posterior) root a sensory function.

Maglev (magnetic levitation)
The suspension of a vehicle above a track by magnetic forces. Eric Laithwaite's work on the LINEAR MOTION in Britain made maglev a real possibility for high-speed trains; low friction, and therefore high energy efficiency, were the main attractions. Laithwaite took his original idea a step further in the early 1970s by turning the winding sideways and calling it the transverse flux linear motor. He demonstrated that it could not only suspend the vehicle, but also propel it and guide it along a track. The difficulty is maintaining stability. The conventional linear motor is in service, however, in the US and the UK for mass-transit systems operation over short distances at low speeds.

Magma
Molten material formed in the upper mantle or crust of the EARTH, composed of a mixture of various complex SILICATES in which are dissolved various gaseous materials, including WATER. On cooling magma forms IGNEOUS ROCKS, though any gaseous constituents are usually lost during the solidification. Magma extruded to the surface forms LAVA. The term is loosely applied to other fluid substances (such as molten salt) in the Earth's crust. (See also HOT SPRINGS; VOLCANISM.) (♦♦ page 142, 208)

Magnesia (magnesium oxide) see MAGNESIUM.

Magnesite
Mineral form of magnesium carbonate ($MgCO_3$), found in the US, Austria and Manchuria. It forms hexagonal crystals, usually massive and white. Magnesite is mainly used for lining furnaces, as it decomposes to give magnesium oxide (see MAGNESIUM) on heating.

Magnesium (Mg)
A reactive, silvery white ALKALINE-EARTH METAL, the eighth most abundant element. Its chief ores are dolomite, brucite and magnesite. It is also found in many other minerals and in large quantities in the sea. Magnesium is light but strong, and forms useful alloys with aluminum and other metals. Magnesium is manufactured either by the ELECTROLYSIS of fused magnesium chloride or by the silicothermic process, in which mixed oxides, formed by the calcination of dolomite, are reduced at high temperature with ferrosilicon, forming magnesium crystals. When heated in air, divided magnesium burns readily with a dazzling white flame: hence its use in flashbulbs and flares. It is also used as a powerful reducing agent in the preparation of many other metals. AW 24.3, mp 650°C, bp 1100°C, sg 1.738 (20°C). (♦ page 130)

MAGNESIUM HYDROXIDE ($Mg(OH)_2$) occurs naturally as colorless hexagonal plates of brucite. It is formed as a gelatinous precipitate when ALKALI is added to magnesium salts. It is a BASE and loses water when heated to 350°C.

MAGNESIUM OXIDE (magnesia, MgO) is a white crystalline solid and also a BASE. Being highly refractory, it is used in furnace linings. When mixed with magnesium chloride solution, magnesia forms a durable cement used as a stucco finish on buildings.

MAGNESIUM SULFATE (epsom salts, $MgSO_4$) is a colorless crystalline solid. The aqueous solution is used as a purgative.

Magnet see MAGNETISM.

Magnetic field
What is said to exist where electric charges (see ELECTRICITY) experience a FORCE proportional to their VELOCITY but at right angles to it, or where magnetic dipoles (see MAGNETISM) experience a torque. The field is defined in the direction of zero torque, with a strength equal to the torque on a unit dipole at right angles to the field. Magnetic fields originate at magnetic dipoles or electric currents. (♦ page 118)

Magnetic lens
A device using a MAGNETIC FIELD to focus a beam of charged particles. The field is produced between two annular pole pieces around the beam driven by an electromagnet. Its imaging properties were first studied by Hans Busch in 1926.

Magnetic resonance detection
Electrons and all nuclei containing an odd number of nucleons possess SPIN, and therefore behave like tiny magnets. In the simplest case, these tend to align parallel or antiparallel to an applied magnetic field. This is true for electrons and for several interesting nuclei, in particular the nuclei of hydrogen (single protons). The application of an oscillating magnetic field, at right angles to the steady field, can induce the magnets to flip orientation, absorbing energy from the oscillating field. Once the oscillating field (typically the magnetic component of electromagnetic radiation) is removed, the magnets return to their original orientations, now emitting energy. The absorption and emission occur only at a characteristic frequency, which depends both on the size of the steady field and on the intrinsic size of the electron or nuclear magnet.

In electron spin resonance (ESR), frequencies of 28 GHz are typical in fields of around 1T. In nuclear magnetic resonance (NMR), hydrogen nuclei, for example, resonate at frequencies around 42.5 MHz in fields of 1T. In both cases, the exact value of the frequency for resonance depends on the molecular environment. Surrounding electrons and nuclei can influence the value of the steady magnetic field a particular electron or nucleus feels. NMR and ESR studies thus both reveal information about the molecular environment of electrons and nuclei.

In magnetic resonance imaging, used in medicine, hydrogen nuclei in the body are made to resonate in a magnetic field that varies across the body. The size of signal over a range of frequencies then reveals the distribution of hydrogen in the body, and can reveal, eg, tumors.

Magnetic storm
An occasional disturbance in the Earth's MAGNETIC FIELD, correlated with SUNSPOT activity. A high energy PLASMA ejected from a solar flare sets

up large currents in the MAGNETOSPHERE on reaching the Earth, causing a rapid rise of about 0.2% in the magnetic field at the surface. The plasma subsequently moves around the Earth, often accompanied by auroral displays, while the field drops to about 0.5% below its normal value, recovering over several days.

Magnetism

The phenomena associated with "magnetic dipoles", commonly encountered in the properties of the familiar horseshoe (permanent) magnet and applied in a multitude of magnetic devices.

Magnetism was first discovered through the properties of the lodestone, a shaped piece of MAGNETITE that had the property of aligning itself in a roughly north-south direction. Eventually a lodestone was used to magnetize a steel bar, thus making an artificial permanent magnet. The power of a magnet was discovered to be concentrated in two "poles", one of which always sought the north, and was called a north-seeking pole, or north pole, the other being a south-seeking pole, or south pole. It was early learned that, given two permanent magnets, the unlike poles were attracted to each other and the like poles repelled each other. Furthermore, dividing a magnet in two never resulted in the isolation of an individual pole, but only in the creation of two shorter two-poled magnets. Again, it was found that magnetic poles attracted or repelled each other according to an inverse square law. The explanation of these properties in terms of magnetic "lines of force" was an early achievement of the science of magnetostatics.

Today, physicists explain magnetism in terms of magnetic dipoles. Magnetic dipole moment is an intrinsic property of fundamental particles. ELECTRONS, for example, have a moment of 0.928×10^{-23} $A.m^2$ parallel or antiparallel to the direction of observation. The forces between magnetic dipoles are exactly analogous to those between electric dipoles (see ELECTRICITY). This leads scientists often to regard the dipoles as consisting of two magnetic charges of opposite type, the poles of traditional theory. But unlike electric charges, magnetic poles or MONOPOLES have never been found in isolation.

In ferromagnetic materials such as IRON and COBALT, spontaneous dipole alignment over relatively large regions known as magnetic domains occurs. Magnetization in such materials involves a change in the relative size of domains aligned in different directions, and can multiply the effect of the magnetizing field a thousand times. Other materials show much weaker, nonpermanent magnetic properties (see DIAMAGNETISM; PARAMAGNETISM).

Magnetism is intimately associated with electricity (see ELECTROMAGNETISM). Electric currents generate MAGNETIC FIELDS circulating around themselves – the EARTH's magnetic field is maintained by large currents in its liquid core – and small current loops behave like magnetic dipoles with a moment given by the product of the loop current and area. (◆ page 118)

Magnetism, Terrestrial see EARTH.

Magnetite

Hard, black OXIDE mineral of composition Fe_3O_4. As an IRON ore of widespread occurrence in igneous rocks, it is second only in importance to HEMATITE. Magnetite has the inverse SPINEL structure. It is strongly ferromagnetic (see MAGNETISM) and was used in the ancient world as a compass, under the name lodestone.

Magneto

A simple electrical AC GENERATOR based on a rotating permanent magnet which induces (see ELECTROMAGNETISM) a current in a coil. It is the basis of an ignition system used in INTERNAL-COMBUSTION ENGINES without batteries, the spark voltage being induced as usual in a large secondary coil when the current in a primary coil is interrupted, but the primary current itself being induced by the rotating magnet rather than being drawn from a battery.

Magnetohydrodynamics (MHD)

The DYNAMICS of conducting fluids such as liquid metals or PLASMAS, in ELECTRIC and MAGNETIC FIELDS. It is a macroscopic form of ELECTRODYNAMICS, deriving from fluid dynamics such concepts as magnetic pressure and magnetic viscosity. Its equations often defy exact solution. The most important applications are in magnetohydrodynamic GENERATORS and controlled nuclear FUSION processes. The extremely hot plasma produced by the fusion is contained by strong circulating magnetic fields; various designs are possible, the stability of each being the paramount consideration.

Magnetometer

A device measuring MAGNETIC FIELD strength. Various types exist, exploiting, for instance, the oscillation rate of a small, freely suspended bar magnet or the deflection of a magnet against its suspension or a reference field. Sensitive magnetometers used in space research include the proton precession magnetometer and the helium magnetometer.

Magnetosphere

The region of space around a body in which the magnetic field of that body is dominant.

Magnetostriction

The interaction between the physical dimensions of a ferromagnetic specimen and its magnetization. A long iron rod, for example, contracts slightly in a MAGNETIC FIELD. Magnetization by mechanical strain is exploited in high-frequency vibration detectors.

Magnifying glass

A simple MICROSCOPE, in which a converging LENS is used to form an enlarged image of an object. In normal use, the object is held within the focal length of the lens and an enlarged, upright virtual image is seen through the lens. A magnifying glass can also be used to form a real but inverted image of an object if the object is placed outside the focal length of the lens. A large converging lens can also be used as a burning glass, focusing light and heat from the Sun.

Magnitude, Stellar

A measure of a star's brightness. The foundations of the system were laid by HIPPARCHUS (*c*120 BC), who divided stars into six categories, from 1 to 6 in order of decreasing brightness. Later the system was extended to include fainter stars that could be seen only by telescope, and brighter stars, which were assigned negative magnitudes (eg Sirius, −1.5). Five magnitudes were defined as a 100-times increase in brightness. These apparent magnitudes depend greatly on the distances from us of the stars. Absolute magnitude is defined as the apparent magnitude a star would have were it at a distance of 10pc from us: Sirius then has magnitude +1.4. Absolute magnitudes clearly tell us far more than do apparent magnitudes. Stars are also assigned bolometric and photographic magnitude, and magnitudes measured at particular wavebands in the invisible and infrared regions of the spectrum. (◆ page 228-9)

Mainframe see COMPUTER.

Malabsorption

Failure to absorb essential parts of the diet, leading to deficiency diseases.

Malachite

A green mineral consisting of basic copper (II) carbonate ($Cu_2CO_3[OH]_2$). It is of widespread occurrence, usually with AZURITE, and is formed by weathering of other copper minerals. It is a minor ore of COPPER, and is used for ornamental stone and GEMS.

Malaria

Tropical PARASITIC DISEASE causing malaise and intermittent FEVER and sweating, either on alternate days or every third day; bouts often recur over many years. One form, cerebral malaria, develops rapidly with ENCEPHALITIS, COMA and SHOCK. Malaria is due to infection with *Plasmodium*, carried by mosquitoes of the genus *Anopheles*, from the blood of infected persons. The cyclic fever is due to the parasite's life cycle in the blood and liver; diagnosis is by examination of blood. Mosquito control, primarily by destroying their breeding places (swamps and pools), provides the best method of combating the disease. Resistance of the parasites to drugs and of the mosquitoes to insecticides is a major problem in malaria control.

Malic acid ($HOOC.CH_2.CHOH.COOH$)

Dibasic acid found in the juice of unripe apples and some other fruits. mp 133°C.

Malignancy see CANCER.

Malleability

The property of metals and alloys to be deformed by beating, rolling, etc without breaking. GOLD is the most malleable, and can be beaten into almost any shape. Malleability is not equivalent to DUCTILITY, the ability to be drawn out without breaking: LEAD is malleable but not ductile.

Malnutrition

Inadequate nutrition, especially in children. It may involve all the diet or just selected parts (eg total calories, or proteins or vitamins). Depending on the dietary deficiency different patterns of malnutrition are seen. (See also BERIBERI; KWASHIORKOR; MARASMUS; PELLAGRA; SCURVY.)

Malpighi, Marcello (1628-1694)

Italian physician and biologist, the father of microscopic ANATOMY, discoverer of the CAPILLARIES (1661), and a pioneer in several fields of medicine and biology.

Malthus, Thomas Robert (1766-1834)

English clergyman and economist best known for his *Essay on the Principle of Population* (1798; second, larger edition, 1803). In this he argued that the population of a region would always grow until checked by famine, pestilence or war. Even if agricultural production were improved, the only result would be an increase in population and the lot of the people would be no better. Although this pessimistic view held down the provision of poor relief in England for many decades, it also provided both C. DARWIN and A.R. WALLACE with a vital clue in the formulation of their theory of EVOLUTION by natural selection.

Maltose ("malt sugar", $C_{12}H_{22}O_{11}.H_2O$)

Disaccharide SUGAR, produced by the action of DIASTASE on STARCH and yielding GLUCOSE with the ENZYME maltase. (◆ page 40)

Mammals

A class of VERTEBRATES distinguished from all other animals by their ability to suckle their young with MILK produced by the mammary glands, and by their possession of hair. Like birds, they are HOMOIOTHERMS. The mammals are divided into three major groups: the MONOTREMATA (monotremes), METATHERIA (marsupials), and EUTHERIA (placental mammals). The first of these includes all the egg-laying mammals (the platypus and echidnas); they represent a link with the ancestral mammals, which evolved from egg-laying reptiles, and are sometimes referred to as prototherians. The second two groups both bear live young and are known collectively as therians, or modern mammals. These represent two distinct evolutionary lines within the mammals, which developed separately because of geographic isolation. (◆ page 17)

Mammal-like reptiles
Extinct VERTEBRATES that flourished during the Permian and Triassic periods, 286-213 million years ago, and that included the ancestors of the mammals. Some were large herbivores, but those that gave rise to the mammals were small carnivorous forms. Their fossils represent a continuous series, linking early reptiles with modern mammals.
Mammary glands see BREASTS; LACTATION.
Mandible
The upper jaw of vertebrate animals; the lower jaw is called the maxilla. The terms are also used for parts of the feeding apparatus of lower animals.
Manganese (Mn)
Hard, grayish metal in Group VIIB of the PERIODIC TABLE; a TRANSITION ELEMENT. It occurs naturally as PYROLUSITE and MANGANITE. Elementary manganese is obtained by the reduction of manganese (IV) oxide with aluminum in a furnace, or by ELECTROLYSIS. When smelted with iron ore, manganese ore gives the alloys spiegeleisen and ferromanganese, widely used in STEEL production. Manganese also forms useful ALLOYS with some nonferrous metals. Manganese is fairly reactive, resembling IRON chemically. Its main oxidation states are +2, +3, +4, +6 and +7. Manganese (IV) oxide (MnO_2), a black crystalline solid, is widely used as an oxidizing agent and as a depolarizer in electric dry cells. Permanganate (MnO_4^-) is used in nickel refining and tanning, and as a bleach, disinfectant and powerful oxidizing agent. Manganese (II) sulfate ($MnSO_4$) is a component of some fertilizers. AW 54.9, mp 1250°C, bp 2120°C, sg 7.20 (20°C). (◆ page 130)
Manganite
A black mineral consisting of hydrated manganese (III) oxide ($MnO[OH]$); an ore of MANGANESE, found in W Europe, Michigan and California. It occurs as bundles of prismatic crystals, and alters to PYROLUSITE.
Manhattan project
US project to develop an explosive device working by nuclear FISSION. It was established in August 1942, and research conducted at Chicago, California and Columbia universities, as well as at Los Alamos, New Mexico and other centers. By December 1942 a team headed by E. FERMI initiated the first self-sustaining nuclear chain reaction. On 16 July 1945 the first ATOMIC BOMB was detonated near Alamogordo, New Mexico, and similar bombs were the following month dropped on Hiroshima (6 August) and Nagasaki (9 August).
Mania
A PSYCHOSIS characterized by high elation, excitement and acceleration of physiological as well as mental processes. Homicidal mania is characterized by an uncontrollable urge to kill.
Manic-depressive psychosis
A PSYCHOSIS characterized by alternating periods of deep depression and MANIA. Periods of sanity may intervene.
Manometer
Device, usually consisting of a double-legged liquid column in a glass or metal tube, for determining the difference between two fluid pressures. In a simple U-tube manometer, the mercury (or other low-vapor-pressure liquid) rises in the lower pressure side and drops in the other, the difference in heights being measured on a suitably calibrated scale. Sensitivity is increased by inclining the tube or giving the legs different cross-sections.
Manson, Sir Patrick (1844-1922)
British medical scientist known for his pioneering researches in TROPICAL MEDICINE, especially for his naming the MOSQUITO as the transmitter of FILARIASIS and MALARIA. (See also ROSS, SIR RONALD.)
Mantle
The layer of the Earth lying between the CRUST and the CORE. (◆ page 134, 142-3)
Mantoux test
Injection of a small amount of tuberculosis bacillus, used to see whether immunity exists or if vaccination is necessary.
Map
A representation of the layout of features on the Earth's surface or part of it. Maps have many uses, including routefinding; marine or aerial navigation; administrative, political and legal definition, and scientific study. In cartography, or mapmaking, the techniques of SURVEYING and GEODESY are used to obtain the positional data to be represented. Since the EARTH is roughly spheroid – the GEOID being taken as the reference level – and as the surface of a sphere cannot be flattened without distortion, no plane map can perfectly represent its original, the distortion becoming worse the larger the area. But spherical maps or globes are impractical for large-scale work. Thus plane maps use various projections, geometrical algorithms for transforming spherical coordinates into plane ones. The choice of projection depends on the purpose of the map; one may aim for correct size or correct shape, but not both at once; a suitable compromise is generally reached. The scale of a map (assuming it to be constant) is the ratio of a distance on the map to the distance that it represents on the Earth's surface.
Marasmus
Extreme emaciation of malnourished infants, common in famine areas.
Marble
Metamorphic rocks formed by recrystallization of CALCITE or DOLOMITE. The finest Italian marble is pure white calcite and has been prized by sculptors throughout history (eg Carrara marble). Many types of marble contain impurities, causing discoloration.
Marconi, Guglielmo (1874-1937)
Italian-born inventor and physicist, awarded (with K.F. BRAUN) the 1909 Nobel Prize for Physics for his achievements. On learning of Hertzian (RADIO) waves in 1894, he set to work to devise a wireless TELEGRAPH. By the following year he could transmit and receive signals at distances of about 2km. He went to the UK to make further developments, and in 1899 succeeded in sending a signal across the English Channel. On 12 December 1901 in St John's, Newfoundland, he successfully received a signal sent from Poldhu, Cornwall, England, thus heralding the dawn of transatlantic radio communication.
Marijuana
A form of CANNABIS.
Mariner program
US unmanned space probes that have made close "fly-by" observations of VENUS, MARS and MERCURY. Mariner 9 orbited Mars (1971-72), sending back a detailed surface survey. Mariner 10 flew by Venus in 1974, and took the first pictures of Mercury's surface during three fly-bys (1974-75).
Mariotte, Edme (1620-1684)
French physicist who independently discovered BOYLE'S LAW (Mariotte's Law) and also discovered the blind spot of the EYE (1660).
Marl
Natural mixture of clay and calcium carbonate ($CaCO_3$: see CALCIUM; CLAY). If the former predominates, it is a mudstone; if the latter, it is termed calcareous. Greensand marls contain hardly any $CaCO_3$ (see also GLAUCONITE). Marls are used as soil conditioners and in making portland cement.
Marrow, Bone
The material in the center of bones, in which ERYTHROCYTES, white BLOOD cells and platelets are made. Mature cells only are released unless the marrow is diseased, as in LEUKEMIA, secondary CANCER or serious infections. Bone marrow aspiration or BIOPSY is often valuable in diagnosis.
Mars
The fourth planet from the Sun with a mean solar distance of 228Gm (about 1.52AU) and a "year" of 687 days. During the Martian day of about 24.62h the highest temperature at the equator is about 10°C, the lowest just before dawn being about −100°C. Mars has a mean diameter of 6775km, with a small degree of polar flattening, and at its closest to Earth (see CONJUNCTION) is some 56Gm distant. Its tenuous atmosphere is believed to consist mainly of carbon dioxide, nitrogen and NOBLE GASES, and the distinctive Martian polar caps are thought to be composed of frozen carbon dioxide and ice.

Telescopically, Mars appears as an ocher red disk marked by extensive dark areas: these latter have in the past been erroneously termed *maria* (seas). Several observers have reported sighting networks of straight lines on the Martian surface – the famous canals. However, photographs sent back by the MARINER probes revealed a network of subsidiary valleys. The Mariner scientists attributed these as dried river beds, but they are not the same as the "canals" seen by earlier observers. Mars is spotted with craters, rather as is the MOON. The VIKING spacecraft, which landed in 1976, detected no evidence of living organisms on Mars, but it has been suggested that such organisms might be able to survive near the edges of the polar ice caps. This remains to be tested. Mars has two moons, PHOBOS and DEIMOS. (◆◆ page 134, 221)
Marsh see SWAMP.
Marsh gas see METHANE.
Marsupials
A group of mammals; see METATHERIA.
Martin, Archer John Porter (b. 1910)
British biochemist awarded with R.L.M. SYNGE the 1952 Nobel Prize for Chemistry for their development of paper CHROMATOGRAPHY, a biochemical tool of great medical importance. He later helped to perfect gas chromatography.
Maser
A device used as a MICROWAVE oscillator or amplifier, the name being an acronym for "microwave (or molecular) amplification by stimulated emission of radiation". As OSCILLATORS they form the basis of extremely accurate atomic clocks; as amplifiers they can detect feebler signals than any other kind, and are used to measure signals from outer space.

Atoms and molecules can exist in various states with different energies; changes from one ENERGY LEVEL to another are accompanied by the emission or absorption of ELECTROMAGNETIC RADIATION of a particular frequency. Maser action is based on the fact that irradiation at the frequency concerned stimulates the process. If more atoms are in the higher energy (excited) state than in the lower state, incident waves cause more emission than absorption, resulting in amplification of the original wave.

The main difficulty is one of maintaining this arrangement of the states, as the EQUILIBRIUM configuration involves more atoms being in the lower than in the excited state. In the ammonia gas maser, molecules in the lower state are removed physically through their different response to an electric field, while in solid-state masers, often operated at low temperatures, a higher frequency "pumping" wave raises atoms into the excited state from some state not involved in the maser action.

Mass
A measure of the linear INERTIA of a body, ie of the extent to which it resists ACCELERATION when a FORCE is applied to it. Alternatively, mass can be thought of as a measure of the amount of MATTER in a body. The validity of this view seems to receive corroboration when one remembers that bodies of equal inertial mass have identical weights in a given gravitational field. But the exact equivalence of inertial mass and gravitational mass is only a theoretical assumption, albeit one strongly supported by experimental evidence. According to EINSTEIN's special theory of RELATIVITY, the mass of a body is increased if it gains ENERGY; according to the famous Einstein equation: $\Delta m = \Delta E/c^2$ where Δm is the change in mass due to the energy change ΔE, and c is the electromagnetic constant. It is an important property of nature that in an isolated system mass energy is conserved. The international standard of mass is the KILOGRAM. (♦ page 49, 151)

Massif
Plateau-like upland area, with abrupt margins and often complex geological structure. (See also MOUNTAIN.)

Mass number (A)
The total number of nucleons (PROTONS and NEUTRONS) in the nucleus of an ATOM, written as a number following its name after a hyphen (eg oxygen-16), or as a superscript following its chemical symbol (eg O^{16}). (♦ page 130)

Mass spectroscopy
Spectroscopic technique in which electric and magnetic fields are used to deflect moving charged particles according to their mass, employed for chemical analysis, separation, ISOTOPE determination or finding impurities. The apparatus for obtaining a mass spectrum (ie a number of "lines" of distinct charge-to-mass ratio obtained from the beam of charged particles) is known as a mass spectrometer or mass spectrograph, depending on whether the lines are detected electrically or on a photographic plate. In essence, it consists of an ion source, a vacuum chamber, a deflecting field and a collector. By altering the accelerating voltage and deflecting field, particles of a given mass can be focused to pass together through the collecting slit.

Mastectomy
Removal of part of or the entire BREAST, including the overlying skin and sometimes the nipple. LYMPH nodes from the armpit and some chest wall muscles may also be excised. It is used as a treatment for breast cancer.

Mastitis
Inflammation and soreness of the breasts. It is economically important in cattle; in humans it is painful but rarely serious.

Mastoid
Bone that projects down from the skull inside and behind the ear. Infection of the cavity in it can cause pain and deafness.

Mater
The dura mater is the membrane immediately under the skull, above the pia mater. The ARACHNOID MEMBRANE lies below both of these, over the brain. Together they form the MENINGES.

Materials, Strength of
A branch of MECHANICS concerned with the behavior of materials when subjected to loads. When force is applied to an object there is a tendency for it to deform: the internal forces resulting from the applied force are called stresses; the deformations are called strains.

STRESS The four main types of stress are: shearing (eg the forces set up in a rivet joining two plates that are pulling in opposite directions); bending; tension and compression (which tend to elongate or shorten the member); and torsion (twisting). Analysis of a structure, is concerned with the stresses that each component is called upon to resist, and its ability to do so without undue strain.

STRAIN Materials deform in different ways under load. Basic properties include ELASTICITY, where a material regains its original dimensions when load is removed; plasticity, where the deformation is permanent; brittleness, where deformation is negligible before fracture; and creep, deformation under a constant load over a period of time. (See also DUCTILITY; HARDNESS; MALLEABILITY.) Within limits, elastic materials deform in proportion to the stress:

ie $\frac{\text{stress}}{\text{strain}} = $ a constant (Hooke's Law).

The value of the constant depends on the material and on the type of stress. For tensile and compressive forces it is called Young's modulus, E (see YOUNG, THOMAS); for SHEARING forces, the shear modulus, S; and for forces affecting the VOLUME of the object the bulk modulus, B. There comes a point (the elastic limit), however, when further stress results in a permanent deformation. (See also TENSILE STRENGTH.)

Mathematics
The fundamental, interdisciplinary tool of all science. It can be divided into two main classes, pure and applied mathematics, though there are many cases of overlap between these. Pure mathematics has as its basis the abstract study of quantity, and thus includes the sciences of number – arithmetic and its broader realization, algebra – as well as the subjects described collectively as geometry and their extensions, the subjects described collectively as analysis (particularly calculus). In modern mathematics, many of these subjects are treated in terms of set theory. Applied mathematics deals with the applications of this abstract science. It thus has particularly close associations with physics and engineering.

Matter
Material substance, that which has extension in space and time. All material bodies have inherent INERTIA, measured quantitatively by their MASS, and exert gravitational attraction on other such bodies. Matter may also be considered as a specialized form of ENERGY. There are three physical states of matter: solid, liquid and gas. An ideal solid tends to return to its original shape after forces applied to it are removed. Solids are either crystalline or amorphous; most melt and become liquids when heated. Liquids and gases are both FLUIDS: liquids are only slightly compressible but gases are easily compressed. On the molecular scale, the state of matter is a balance between attractive intermolecular forces and the disordering thermal motion of the molecules. When the former predominate, MOLECULES vibrate about fixed positions in a solid crystal LATTICE. At higher temperatures, the random thermal motion of the molecules predominates, giving a featureless gas structure. The short-range intermolecular order of a liquid is an intermediate state between solid and gas. (♦ page 21)

Maupertuis, Pierre Louis Moreau de (1698-1759)
French mathematician and astronomer who showed the EARTH to be flattened at the poles (1738) and who formulated the principle of least action, which assumes that in nature phenomena such as the motion of bodies occur with maximum economy.

Maxwell, James Clerk (1831-1879)
British physicist whose contributions to science have been compared to those of Newton and Einstein. His most important work was in ELECTROMAGNETISM (he pointed to the electromagnetic nature of LIGHT) and THERMODYNAMICS. Most important of all was his derivation of the equations that bear his name, four equations that together describe in terms of the relevant VECTOR quantities the interrelation between electric and magnetic fields in a particular space. (See also ELECTROMAGNETIC RADIATION.)

Mayer, Julius Robert von (1814-1878)
German physician and physicist who contributed to the formulation of the law of conservation of ENERGY.

Mayer, Marie Goeppert (1906-1972)
German-born US physicist awarded, with J.H.D. JENSEN and E.P. WIGNER, the 1963 Nobel Prize for Physics for her proposal (independent of Jensen's) that the PROTONS and NEUTRONS of the atomic nucleus are arranged in concentric shells.

Meadow's syndrome
Form of MUNCHHAUSEN'S SYNDROME in which a parent fabricates symptoms in a child and submits them to medical procedures. A form of child abuse.

Measles
Common INFECTIOUS DISEASE, caused by a VIRUS. It involves a characteristic sequence of fever, headache and malaise followed by CONJUNCTIVITIS and RHINITIS, and then the development of a typical rash, with blotchy ERYTHEMA affecting the skin of the face, trunk and limbs. A cough may indicate infections in small BRONCHI, and this may progress to virus PNEUMONIA. Secondary bacterial infection may lead to middle ear infection or pneumonia. ENCEPHALITIS is seen in a small but significant number of cases and is a major justification for VACCINATION against this common childhood disease. In Africa and several other regions of the Third World measles is a major killer, particularly in children debilitated by malnutrition.

Mechanics
The branch of applied mathematics dealing with the actions of forces on bodies. There are three branches: kinematics, which deals with relationships between distance, time, velocity and acceleration; dynamics, dealing with the way forces produce motion; and statics, dealing with the forces acting on a motionless body.

KINEMATICS In kinematics distance and time, which are SCALAR quantities, and VELOCITY and ACCELERATION, which are VECTOR quantities, are studied. Velocity is the rate of change of position of a body in a particular direction with respect to time: it thus has both magnitude and direction. Its magnitude is the scalar quantity speed (S), related to distance (s) and time (t) by the equation $S = s/t$; similarly velocity ($\mathbf{v}$) is related to distance and time by $\mathbf{v} = ds/dt$. Acceleration is rate of change of velocity with respect to time: $\mathbf{a} = d\mathbf{v}/dt = d^2s/dt^2$. Thus, if velocity is measured in km/s, acceleration is measured in km/s^2.

Often we have to consider the combination of velocities in different directions. Consider a ship sailing due east at velocity $\mathbf{x}$, and carried north by a current with velocity $\mathbf{y}$. The resultant velocity $\mathbf{z}$ can be found by using a diagram where the arrows represent the velocities in both magni tude and direction or by trigonometry. Velocities can be resolved in a similar way (resolving means simply finding its components in different directions).

DYNAMICS is based on NEWTON's three laws of motion: that a body continues in its state of motion unless compelled by a force to act otherwise; that the rate of change of motion (acceleration) is proportional to the applied force and occurs in the direction of the force, and that every action is opposed by an equal and opposite reaction. The first gives an idea of INERTIA, which is proportional to the MASS and opposes the change of motion: combining the first with the second, we find that $\mathbf{F} \propto m\mathbf{a}$, where $\mathbf{F}$ is the force, m the mass of the body and $\mathbf{a}$ the acceleration produced by the FORCE. In practice we choose units such that $\mathbf{F}=m\mathbf{a}$. (See also MOMENTUM.)

Newton suggested that gravitational attraction existed between all bodies, and proposed a law to describe this: if two bodies, masses m_1 and m_2 are separated by distance d, the force of attraction, $\mathbf{F}$, between them is given by $\mathbf{F}=G(m_1m_2)/d^2$ where G is the universal gravitational constant (see GRAVITATION). Near the surface of the Earth we find for a body of mass m that $\mathbf{F}\propto m$ (G and the mass of the Earth are constant and the distance from the surface to the center is approximately so). We usually set $\mathbf{F}=m\mathbf{g}$ where $\mathbf{g}$ is another constant, the ACCELERATION due to gravity (usually denoted g although it is a vector).

STATICS Forces can be combined in much the same way as velocities. If forces $\mathbf{P}_1$ and $\mathbf{P}_2$, with resultant $\mathbf{R}$, act at a point, there must be a third force, $\mathbf{P}_3$, equal and opposite to $\mathbf{R}$, for equilibrium. We can combine forces similarly over more complex structures. For a bridge whose weight (w) acts as a downward force $\mathbf{W}$ at the center, the upward reactions $\mathbf{R}_1$ and $\mathbf{R}_2$ at the piers must be such that $\mathbf{R}_1+\mathbf{R}_2=-\mathbf{W}$ for it to be in EQUILIBRIUM. Similarly, the members of the bridge are examined to determine the stresses acting on each (see MATERIALS, STRENGTH OF).

Mechenikov, Ilya see METCHNIKOFF, ÉLIE.

Medawar, Sir Peter Brian (1915-1987)
British zoologist who shared with F.M. BURNET the 1960 Nobel Prize for Physiology or Medicine for their work on immunological tolerance. Inspired by Burnet's ideas, Medawar showed that if fetal mice were injected with cells from eventual donors, skin grafts made onto them later from those donors would "take", thus showing the possibility of acquired tolerance and hence, ultimately, organ TRANSPLANTS.

Medicine
The art and science of healing. Within the last 150 years or so medicine has become dominated by scientific principles; before this, healing was mainly a matter of tradition and magic.

The earliest evidence of medical practice is seen in Neolithic skulls in which holes have been bored, presumably to let evil spirits out, a practice called trepanning (see TREPHINE). Treatment in primitive cultures was either empirical or magical. Empirical treatment included bloodletting, dieting, primitive surgery and the administration of numerous potions, lotions and herbal remedies (some used in modern medicines). Faith healing and other traditional methods are still practiced in modern societies. ACUPUNCTURE and OSTEOPATHY, both ancient treatments, are also practiced today.

The growth of scientific medicine began with the Greek philosophy of nature. The great Greek physician HIPPOCRATES, with whose name is associated the Hippocratic oath, which codifies the physician's ideals of humanity and service, has justly been called the father of medicine. Galen of Pergamum, the encyclopedist of Classical medicine, clearly distinguished ANATOMY from PHYSIOLOGY. Medieval medicine was basically a corrupted Galenism. The 16th century saw the dawn of modern medicine. Men such as FABRICIUS, VESALIUS and WILLIAM HARVEY revived the critical, observational approach to medical research. Perhaps the most far-reaching advances since then have been in preventive medicine, anesthesia and drug therapy. Preventive medicine was attempted in medieval times when ships arriving in Europe during the Black Death were "quarantined" for 40 days. More recent major milestones have been EDWARD JENNER's work on VACCINATION, and the "germ theory of disease" proposed by LOUIS PASTEUR and developed by ROBERT KOCH. ANESTHESIA and ASEPSIS (see LISTER, JOSEPH) made possible great advances in SURGERY. CRAWFORD LONG and JAMES SIMPSON were both pioneers of their use. Drug therapy originated with herbal remedies, but perhaps the two most important discoveries in this field both came in the 20th century: that of INSULIN by FREDERICK BANTING and CHARLES BEST, and that of penicillin by ALEXANDER FLEMING. (See also ANTIBIOTICS; CHEMOTHERAPY; DRUGS.)

Medusa
One of the two body forms adopted by CNIDARIAN animals. It is typified by the jellyfish, and has an inverted saucer shape, with the tentacles hanging downward around the mouth. (See also POLYP.)

Meerschaum (sepiolite)
Fibrous CLAY mineral, light and porous, consisting of hydrated magnesium silicate. Occurring chiefly in Turkey, it is used to make tobacco pipes.

Meg-, mega- (M)
SI prefix multiplying a unit one millionfold. Examples include the megahertz (MHz), megameter (Mm) and megohm (MΩ). One megagram is equal to one tonne (t). (See SI UNITS.)

Meiosis
A special form of cell division seen in EUKARYOTES, and also known as reduction division because it halves the number of chromosomes. In DIPLOID organisms it results in the formation of a HAPLOID cell, which acts as a GAMETE. Meiosis is similar in all plant and animal cells; there are two divisions of the nucleus, in the course of which the CHROMOSOMES divide once so that one diploid cell gives four haploid daughter cells. The chromosomes form HOMOLOGOUS pairs as an initial step, and the daughter cells receive only one of each homologous pair. The distribution of the chromosomes to the daughter cells is a random process, so there is considerable genetic variety in the gametes produced. This is augmented by CROSSING OVER, in which genetic material is exchanged between homologous (and occasionally between non-homologous) chromosomes. The behavior of chromosomes during meiosis is controlled by the protein tubules of the spindle, as in MITOSIS. (♦ page 32)

Meitner, Lise (1878-1968)
Austrian physicist who worked with OTTO HAHN to discover PROTACTINIUM. Following their experiments bombarding URANIUM with NEUTRONS, Meitner collaborated with her nephew, Otto Robert Frisch (1904-1979), to discover nuclear FISSION and predict the chain reaction.

Melanin
A brownish black pigment that lies in various SKIN layers and is responsible for skin color. The distribution in the skin is altered by light and by certain hormones.

Melanism
An excessive development of the dark pigment MELANIN in an animal, as in the panther, a black variety of leopard. Animals without any pigmentation are termed albino (see ALBINISM). (See also INDUSTRIAL MELANISM.)

Melting point
The temperature at which FUSION occurs, so that a solid and the corresponding liquid are at equilibrium (see PHASE EQUILIBRIUM). For pure compounds it has a precise value equal to the FREEZING POINT.

Membranes
Layers that cover the surface of CELLS and enclose organelles in all living organisms. The cell membrane allows some substances into the cell, excludes others, and actively transports others into the cell even though the direction of movement may be against existing concentration gradients. Membranes are composed of two layers of LIPID, principally PHOSPHOLIPID, molecules with their water-soluble heads outermost on both sides, and their long fatty acid chains innermost. Various proteins are embedded in this lipid bilayer, and form channels that control the entry of substances into the cell. Molecules of cholesterol are also present in the bilayer, and help to stabilize the structure. (♦ page 13)

Memory
The capacity of living organisms to store and retrieve information about past events. Memory is observed in most animals, from very simple flatworms upward. It allows animals to find food more efficiently, to learn to avoid danger, and to find their own nest or burrow. Many animals display impressive feats of memory, as in some forms of NAVIGATION, for example, where complex routes and sets of landmarks are remembered. Memory can be studied in animals only by observing learned responses to stimuli, and is therefore difficult to separate from learning. Furthermore, laboratory experiments may not accurately reflect an animal's memory abilities in its natural habitat: both dogs and chimpanzees are good at finding caches of food in the wild, but perform poorly when similar tests are carried out in the laboratory. Nevertheless, certain discoveries have been made about memory from such experiments. One is that very young animals learn as quickly as adults, but do not retain these memories very long. This phenomenon, known as infant amnesia (loss of memory), may be adaptive, in that infant memories of what is dangerous or unpleasant are often irrelevant to the adult, and retaining these memories could be a handicap. Experiments with animals have also confirmed that there are at least two different types of memory: short-term and long-term. This is apparent from studies of human subjects, where a period of up to 30 minutes is required to convert short-term memories into long-term memories. Before that conversion the memory can be destroyed by electric shock treatment or by a blow to the head, producing amnesia of events leading up to car accidents and other traumatic events. The part of the brain responsible for processing new memories and converting them into long-term memories is known as the hippocampus, located on the lower inner margin of each cerebral hemisphere. If this portion of the brain is damaged or removed the person can still recall long-term memories stored before the hippocampus was lost, and still has short-term memory, but cannot convert new short-term memories to long-term memories. Studies of other brain-damaged individuals reveal intriguing patterns to memory formation. A patient with generalized viral brain damage may lose all memory of past events and be unable to create new long-term memories, while retaining the ability to read, write, play a musical instrument and perform other skilled tasks. Alternatively, a particular type of memory may be lost, such as that for a certain group of words. Not enough is yet understood about the brain for these observations to be explained.

The question of how simple memories are stored at a cellular level is being investigated in lower animals. It appears to involve changes in the SYNAPSES that connect the nerve cells; information passes across these synapses by means of chemicals known as neurotransmitters. The signal for the release of these neurotransmitters from a nerve cell is mediated by ion channels in the cell membrane, which open to allow ions to pass through. The memory of an adverse stimulus such as an electric shock will make the animal more responsive to later stimuli. This is now known to occur because certain ion channels open much more readily following the adverse stimulus and thus produce a greater release of neurotransmitters. These changes in the ion channels are produced by another nerve, known as a facilitator. It is likely that all memory formation involves changes to synapses to some extent, but whether this can account for all the complexities of human memory is unknown. Other possible mechanisms include the synthesis of "memory molecules", possibly RNA or protein, or the formation of new nerve cells and neural circuits. The question of how and why memories are lost in the normal person is also impossible to answer at present. Some investigators believe that memories are not lost – only the ability to retrieve them declines with time. (See also NERVOUS SYSTEM.)

Menarche
The age at first menstruation.

Mendel, Gregor Johann (1822-1884)
Austrian botanist and Augustinian monk who laid the foundations of the science of GENETICS. He found that self-pollinated dwarf pea plants bred true, but that under the same circumstances only about a third of tall pea plants did so, the remainder producing tall or dwarf pea plants in a ratio about 3:1. Next he crossbred tall and dwarf plants and found this without exception resulted in a tall plant, but one that did not breed true. Thus, in this plant, both tall and dwarf characteristics were present. He had found a mechanism justifying DARWIN's theory of EVOLUTION by NATURAL SELECTION; but contemporary lack of interest and his later, unsuccessful experiments with the hawkweeds discouraged him from carrying this further. It was not until 1900, when H. DE VRIES and others found his published results, that the importance of his work was realized. (See also GENETICS; POLLINATION.) (◆ page 32)

Mendelevium (Md)
A TRANSURANIUM ELEMENT in the ACTINIDE series, first made by bombarding einsteinium-253 with ALPHA PARTICLES. (◆ page 130)

Mendeleyev, Dmitri Ivanovich (1834-1907)
Russian chemist who formulated the Periodic Law, that the properties of elements vary periodically with increasing atomic weight, and so drew up the PERIODIC TABLE. (See also MEYER, J.L.)

Ménière's disease
Disorder of the cochlea and labyrinth of the EAR, causing brief acute episodes of VERTIGO, with nausea or vomiting, ringing in the ears and DEAFNESS. Ultimately permanent deafness ensues and vertigo lessens. It is a disorder of inner-ear fluid and each episode causes some destruction of receptor cells. Named for the French otologist Prosper Ménière (1801-1862).

Meningitis
INFLAMMATION of the membranes covering the brain (meninges), caused by BACTERIA or VIRUSES. Bacterial meningitis is of abrupt onset, with headache, vomiting, fever, neck stiffness and avoidance of light. Viral meningitis is a milder illness with similar signs in a less ill person. Tuberculous meningitis is an insidious chronic type, which responds slowly to antituberculous drugs. Some FUNGI, unusual bacteria and syphilis (see SEXUALLY TRANSMITTED DISEASES) may also cause varieties of meningitis.

Menopause see MENSTRUATION.

Menorrhagia
Excessive menstrual bleeding.

Menstruation
Specifically the monthly loss of blood (period), representing shedding of the uterine lining in women of reproductive age; in general, the whole monthly cycle of hormone, structural and functional changes in such women, punctuated by menstrual blood loss. After each period the uterus lining (endometrium) starts to proliferate and thicken under the influence of GONADOTROPHINS (follicle stimulating hormone) and ESTROGENS. In mid-cycle a burst of LUTEINIZING HORMONE secretion, initiated by the HYPOTHALAMUS, causes release of an egg from an ovarian follicle (ovulation). More PROGESTERONE is then secreted and the endometrium is prepared for IMPLANTATION of a fertilized egg. If the egg is not fertilized PREGNANCY does not ensue, and blood-vessel changes occur leading to the shedding of the endometrium and some blood; these are lost through the vagina for several days, sometimes with pain or colic. The cycle then restarts. During the menstrual cycle, changes in the BREASTS, body temperature, fluid balance and mood occur, the manifestations varying from person to person. Cyclic patterns are established at PUBERTY (menarche) and end in middle life (age 45-50) at the menopause, the "change of life". Disorders of menstruation include heavy, irregular or missed periods; bleeding between periods or after the menopause; and excessively painful periods. They are studied in GYNECOLOGY.

Mental illness (psychiatric disease)
Disorders characterized by abnormal function of the higher centers of the BRAIN, responsible for thought, perception, mood and behavior, in which organic disease has been eliminated as a possible cause. The borderline between disease and the range of normal variability is indistinct and may be determined by cultural factors. The first humane asylum for the mentally ill was founded in Paris by PINEL (1795). Originally only socially intolerable cases were admitted to such hospitals, but today voluntary admission is more common. The Viennese school of psychology, in particular SIGMUND FREUD and his pupils, emphasized the importance of past, especially childhood, experiences, sexual attitudes and other functional factors. Behavior therapy, PSYCHOANALYSIS and PSYCHOTHERAPY derive from this school. Others favored the influence of subtle organic factors (eg brain biochemistry); the use of this led to LOBOTOMY, SHOCK THERAPY and DRUGS.

Mental illness is generally classified into PSYCHOSIS, NEUROSIS and personality disorder. Schizophrenia is a psychosis causing disturbance of thought and perception in which mood is characteristically flat and behavior withdrawn. Features include: auditory hallucinations; delusions of person ("I'm the King of Spain"), of surroundings and other people (eg suspicion of conspiracy in PARANOIA); blocking, insertion and broadcasting of thought; and nonlogical sequence of ideas. Conversation lacks substance and may be in riddles and neologisms; speech or behavior may be imitative, stereotyped, repetitive or negative. In affective psychoses, disturbance of mood is the primary disorder. Subjects usually exhibit DEPRESSION with loss of drive and inconsolably low mood, either in response to situation (exogenous) or for no apparent reason (endogenous). Loss of appetite and characteristic sleep disturbance also commonly occur. In hypnomania or MANIA, excitability, restlessness, euphoria, ceaseless talk, flight of ideas and loss of social inhibitions occur. Financial, sexual and alcohol excesses may result. Neuroses include ANXIETY, a pathological exaggeration of a physiological response. This may coexist with depression. Obsessional and compulsive neuroses, manifested by extreme habits, rituals and fixations (which may be recognized as irrational); PHOBIAS, excessive and inappropriate fears of objects or situations; and HYSTERIA are other types. Psychopathy is a specific disorder of personality characterized by failure to learn from experience. Irresponsibility, inconsiderateness and lack of foresight result, and may lead to crime. Other personality disorders are exhibited by a variety of people, often with unstable backgrounds, who seem unable to cope with the realities of everyday adult life; attempted suicide is a common gesture. (See also ALCOHOLISM; ANOREXIA NERVOSA; DRUG ADDICTION.)

Mental retardation (mental handicap)
Low intellectual capacity arising not from MENTAL ILLNESS but from impairment of the normal development of the brain and NERVOUS SYSTEM. Causes include genetic defect (as in MONGOLISM); infection of the EMBRYO or FETUS; HYDROCEPHALUS or inherited metabolic defects (eg CRETINISM), and injury at birth including cerebral HEMORRHAGE and fetal ASPHYXIA. Disease in infancy such as ENCEPHALITIS may cause mental retardation in children with normal previous development. Retardation is initially recognized by slowness to develop normal patterns of social and learning behavior and confirmed through intelligence measurements.

Menthol (3-hydroxy-*p*-menthane)
A TERPENE alcohol, an ALICYCLIC COMPOUND. It is a white crystalline solid with a pungent odor and mint flavor, having a cooling, soothing effect on the throat and nasal passages, and used in medications, cosmetics, cigarettes and flavorings. Menthol is extracted from peppermint oil or synthesized.

Mercator, Gerardus (1512-1594)
Flemish cartographer and calligrapher, best known for Mercator's Projection (see MAP), which he first used in 1569 for a world map. The PROJECTION is from a point at the center of the Earth through the surface of the globe onto a cylinder that touches the Earth around the equator.

Mercury (quicksilver, Hg)
Silvery white liquid metal in Group IIB of the PERIODIC TABLE; an anomalous TRANSITION ELEMENT. It occurs as cinnabar, calomel and rarely as the metal, which has been known from ancient times. It is extracted by roasting cinnabar in air and condensing the mercury vapor. Mercury is fairly inert, tarnishing only slowly in moist air, and soluble in oxidizing acids only; it is readily attacked by the HALOGENS and sulfur. It forms Hg^{2+} and some Hg_2^{2+} compounds, and many important ORGANOMETALLIC COMPOUNDS. Mercury and its compounds are highly toxic. The metal is used to form AMALGAMS; for electrodes, and in barometers, thermometers, diffusion PUMPS, and mercury-vapor lamps. AW 200.6, mp −39°C, bp 357°C, sg 13.546 (20°C). (◆ page 130)

MERCURY(II) CYANATE (mercury fulminate, $Hg[ONC]_2$) is a white crystalline solid, sensitive to percussion, and used as a detonator.

MERCURY(II) CHLORIDE (corrosive sublimate, $HgCl_2$) is a colorless crystalline solid prepared by direct synthesis. Although highly toxic, it is used in dilute solution as an ANTISEPTIC, and also as a fungicide and a polymerization catalyst. mp 276°C, bp 302°C.

MERCURY(I) CHLORIDE (calomel, Hg_2Cl_2) is a white rhombic crystalline solid, found in nature. It is used in ointments and formerly found use as a laxative. A calomel/mercury cell with potassium chloride electrolyte (the Weston cell) is used to provide a standard ELECTROMOTIVE FORCE. mp 303°C, bp 384°C.

Mercury
The planet closest to the Sun, with a mean solar distance of 58Gm. Its highly eccentric ORBIT brings it within 46Gm of the Sun at perihelion and takes it 70Gm from the Sun at aphelion. Its diameter is about 4880km, its mass about 0.054 that of the Earth. It goes around the Sun in just under 88 days and rotates on its axis in about 59 days. The successful prediction by ALBERT EINSTEIN that Mercury's orbit would be found to

advance by 43″ per century is usually regarded as a confirmation of the general theory of RELATIVITY. Night surface temperature is thought to be about 100K, midday equatorial temperature over 700K. The planet's average density indicates a high proportion of heavy elements in its interior. Mercury has little or no atmosphere and no known moons. (♦♦ page 134, 221)

Mercury poisoning
A condition resulting in acute GASTROINTESTINAL TRACT and KIDNEY disease when mercury (II) salts are ingested. A chronic form, often from vapor inhalation, causes brain changes with tremor, ataxia, irritability and social withdrawal. The mental changes ensuing from the former use of mercury in making felt hats led to the phrase "mad as a hatter". Organic mercury from fish (eg tuna) living in contaminated water, or from cereals treated with antifungal agents, may cause ataxia, swallowing difficulty, abnormalities of vision and COMA. Nephrotic syndrome of the kidneys may also be seen.

Mercury program
First US manned space flights (1960-63), using the one-man Mercury capsule. In 1961 Alan Shepard and Virgil Grissom were launched on suborbital flights by the Redstone carrier missile. In 1962 John Glenn made the first US orbital flight, followed by M. Scott Carpenter, Walter Schirra and Leroy Cooper; orbital missions were launched by Atlas carriers. (See also SPACE EXPLORATION.)

Meridian
On the celestial sphere, the great circle passing through the celestial poles and the observer's ZENITH. It cuts his HORIZON N and S. (See also CELESTIAL SPHERE; TRANSIT.) The term is also used for a line of terrestrial longitude.

Meristem
In plants, a group of cells that is actively dividing, and thus provides new growth. Meristems are found at various points, notably at the tips of stems and roots (where they are known as apical meristems), and as a thin layer beneath the bark of trees, where the dividing cells produce expansion of the trunk. (♦ page 53)

Merrifield, Robert Bruce (b. 1921)
US biochemist who received the 1984 Nobel Prize for Chemistry for his work on synthesizing complex PEPTIDES, PROTEINS and oligonucleotides, which allows highly pure compounds to be easily separated from byproducts and starting materials. Vital to recent advances in BIOTECHNOLOGY and GENETIC ENGINEERING.

Mesa
Steep-sided, flat-topped area formed beneath a horizontal cap of hard rock where the surrounding softer rock has been worn away. Further EROSION of the sides produces a smaller hill, or butte.

Mescaline
HALLUCINOGENIC DRUG, derived from a Mexican cactus, whose use dates back to ancient times when "peyote buttons" were used in religious ceremonies among American Indians. The hallucinations experienced during its use were among the first to be described (by ALDOUS HUXLEY) and resemble those of LSD.

Mesentery
The membranous fold in which the GASTROINTESTINAL TRACT lies. It is suspended from the back wall of the ABDOMEN so that it lies relatively free and mobile in the peritoneal cavity. It consists of a double layer of peritoneum, and within it lie the blood vessels and lymph vessels that supply the gut.

Mesmer, Franz Anton (1734-1815)
German physician, controversy over whose unusual techniques and theories sparked in CHARCOT and others an interest in the possibilities of using "animal magnetism" (or mesmerism, ie HYPNOSIS) for psychotherapy.

Mesoderm
The middle layer of cells in animals with three body layers (all animals other than the sponges, cnidarians and ctenophores). The mesoderm layer develops between the ENDODERM and ECTODERM layers following gastrulation (see EMBRYOLOGY). It gives rise to most of the body's organ systems other than the nervous system (ectodermal) and the liver and inner layers of the respiratory organs and digestive system (endodermal).

Mesons see SUBATOMIC PARTICLES.

Mesosphere
The atmospheric zone immediately above the stratosphere, in which the temperature decreases with increasing altitude from about −10°C at 50km to a minimum of about −90°C at 85km. (See ATMOSPHERE.) (♦ page 134)

Mesozoa
A phylum of small parasitic invertebrates, of very unusual construction. They consist of two cell-layers which are not comparable to the endoderm and ectoderm of other animals. The mesozoans show no affinity with any other invertebrates.

Mesozoic
The middle era of the PHANEROZOIC, lasting from 248 until 65 million years ago. It has three periods: the TRIASSIC, JURASSIC and CRETACEOUS. (See GEOLOGY.) (♦ page 106)

Metabolic disorders
Disorders of the normal chemical functioning of the body. None can be cured but most can be chemically corrected.

Metabolism
The sum total of all chemical reactions that occur in a living organism. It can be subdivided into anabolism, which describes reactions that build up more complex substances from smaller ones, and catabolism, which describes reactions that break down complex substances into simpler ones. Anabolic reactions require ENERGY in the form of ATP, while catabolic reactions liberate energy. All metabolic reactions are catalyzed by ENZYMES in a highly integrated and finely controlled manner so that there is no overproduction or underutilization of any substance under normal circumstances.

Metal
An element with high specific gravity; high opacity and reflectivity to light (giving a characteristic luster when polished); that can be hammered into thin sheets and drawn into wires (ie is malleable and ductile); and is a good conductor of heat and electricity, its electrical conductivity decreasing with temperature. Roughly 75% of the chemical elements are metals, but not all of them possess all the typical metallic properties. Most are found as ores and in the pure state are crystalline solids (mercury, liquid at room temperature, is a notable exception), their atoms readily losing electrons to become positive IONS. ALLOYS are easily formed because of the nonspecific, nondirectional nature of the metallic bond. (♦ page 130)

Metalloid (semimetal)
An ELEMENT that has properties – physical and chemical – intermediate between those of METALS and those of NONMETALS. The metalloids – BORON, SILICON, GERMANIUM, ARSENIC, ANTIMONY, SELENIUM and TELLURIUM – form a diagonal band in the PERIODIC TABLE. They do not have high ELECTRONEGATIVITY or electropositivity, and form amphoteric OXIDES; they are SEMICONDUCTORS.

Metallurgy
The science and technology of METALS, concerned with their extraction from ores, the methods of refining, purifying and preparing them for use and the study of the structure and physical properties of metals and ALLOYS. A few unreactive metals such as silver and gold are found native (uncombined), but most metals occur naturally as MINERALS (ie in chemical combination with nonmetallic elements). Ores are mixtures of minerals from which metal extraction is commercially viable. Techniques for working ores and forming alloys have been developed over the last 5000 years, but only in the last two centuries have these methods been based on scientific theory. The production of metals from ores is known as process or extraction metallurgy; fabrication metallurgy concerns the conversion of raw metals into alloys, sheets, wires, etc, while physical metallurgy covers the structure and properties of metals and alloys, including their mechanical working, heat treatment and testing. Process metallurgy begins with ore dressing, using physical methods such as crushing, grinding and gravity separation to split up the different minerals in an ore. The next stage involves chemical action to separate the metallic component of the mineral from the unwanted nonmetallic part. The actual method used depends on the chemical nature of the mineral compound (eg if it is an oxide or sulfide, its solubility in acids, etc.) and its physical properties. Hydrometallurgy uses chemical reactions in aqueous solutions to extract metal from ore. ELECTROMETALLURGY uses electricity for firing a furnace or electrolytically reducing a metallic compound to a metal. Pyrometallurgy covers roasting, SMELTING and other high-temperature chemical reactions. It has the advantage of involving fast reactions and giving a molten or gaseous product that can easily be separated out. The extracted metal may need further refining or purifying: electrometallurgy and pyrometallurgy are again used at this stage. Molten metal may then be cast by pouring it into a mold, giving, for example, pig iron, or it may be formed into ingots, which are then hot or cold worked, as with wrought iron.

Mechanical working, in the form of rolling, pressing or forging, improves the final structure and properties of most metals; it tends to break down and redistribute the impurities formed when a large mass of molten metal solidifies. Simple heat treatment such as ANNEALING also tends to remove some of the inherent brittleness of cast metals. (See also BLAST FURNACE; SINTERING; STEEL.)

Metamorphic rocks
One of the three main types of rocks of the Earth's crust. They consist of rocks that have undergone change owing to heat, pressure or chemical action. SEDIMENTARY ROCKS undergo prograde metamorphism, by which they lose volatiles such as WATER and carbon dioxide, under conditions of heat and pressure beneath the Earth's surface. On exposure to the atmosphere this process may be reversed by weathering (see EROSION). LIMESTONE may be metamorphosed to give MARBLE, SHALE to give SLATE. IGNEOUS ROCKS and previous metamorphic rocks undergo retrograde metamorphism, absorbing volatiles, usually from nearby metamorphosing sediments. GRANITE, for example, may be metamorphosed to form a GNEISS. (♦ page 208)

Metamorphosis
In animals a marked and relatively rapid change in body form. This alteration in appearance is associated with a change in diet and way of life. Familiar examples are the changes that occur when a tadpole becomes a frog or a caterpillar enters a pupal stage (see PUPA) from which it emerges as a butterfly or moth. In the case of a tadpole new limbs and organs grow, and the larval ones are eventually resorbed, but the amphibian remains

functional throughout. In the case of the caterpillar most of its fatty tissues are broken down, and the adult organs formed from "imaginal buds" – clusters of cells that were present, but quiescent in the caterpillar; see JUVENILE HORMONE.

Metastases

Nodules of cancer that have split off from their original site and lodged elsewhere in the body. They are considered serious as they are often widely distributed around the body and hence are difficult to locate and treat.

Metatheria (marsupials)

A subclass of the MAMMALS that includes the opossums and kangaroos. There are 236 living species, limited to Australasia with a few in the Americas. Metatherians give birth to very immature young that are transferred to the mother's pouch, where they continue development while they are nourished on milk.

Metazoa

A rarely used term meaning multicellular animals. The word was originally introduced to distinguish between multicellular animals and unicellular animals, or PROTOZOA, at a time when the latter were included in the ANIMAL kingdom.

Metchnikoff, Élie (Ilya Mechnikov) (1845-1916)

Russian biologist who shared with PAUL EHRLICH the 1908 Nobel Prize for Physiology or Medicine for his discovery of phagocytes (in humans called leukocytes) and their role in defending the body from, for example, bacteria. (See BLOOD.)

Meteor

The visible passage of a meteoroid (a small particle of interplanetary matter) into the Earth's atmosphere. Because of friction it burns up, showing a glowing trail of ionized gas in the night sky. The velocity on entry lies in the range 11-72km/s.

Meteoroids are believed to consist of asteroidal and cometary debris. Although stray meteoroids reach our atmosphere throughout the year, for short periods at certain times of year they arrive in profuse numbers, sharing a common direction and velocity. In 1866 GIOVANNI SCHIAPARELLI showed that the annual Perseid meteor shower was caused by meteoroids orbiting the Sun in the same orbit as a comet observed some years before; moreover, as their period of orbit is unrelated to that of the Earth, the meteoroids must form a fairly uniform "ring" around the Sun for the shower to be annual. Other comet-shower relationships have been shown, implying that these streams of meteoroids are cometary debris.

Meteors may be seen by a nighttime observer on average five times per hour: these are known as sporadic meteors or shooting stars. Around twenty times a year, however, a meteor shower occurs and between 20 and 35,000 meteors per hour may be observed. These annual showers are generally named for the constellations from which they appear to emanate, eg Perseids (PERSEUS), Leonids (LEO). Large meteors are called FIREBALLS, and those that explode are known as bolides.

Meteorites are larger than meteors, and are of special interest in that, should they enter the atmosphere, they at least partially survive the passage to the ground. Many have been examined. They fall into two main categories: "stones", whose composition is not unlike that of the Earth's crust; and "irons", which contain about 80%-95% iron, 5%-20% nickel and traces of other elements. Intermediate types exist. Irons display a usually crystalline structure, which implies that they were initially liquid, cooling over long periods of time. Sometimes large meteorites shatter on impact, producing large craters like those in Arizona and Wolf Creek, Australia.

Meteorology

The study of the ATMOSPHERE and its phenomena, weather and climate. Based on atmospheric physics, it is primarily an observational science, whose main application is WEATHER FORECASTING AND CONTROL. The RAIN gauge and WIND vane were known in ancient times, and the other basic instruments – ANEMOMETER, BAROMETER, HYGROMETER and THERMOMETER – had all been invented by 1790. Thus accurate data could be collected; but simultaneous observations over a wide area were impracticable until the development of the telegraph. Since WWI observations of the upper atmosphere have been made, using airplanes, balloons, RADIOSONDE, and since WWII (when meteorology began to flourish) ROCKETS and artificial SATELLITES. RADAR has been much used. Meteorology may be classified by the type of phenomenon observed: CLOUDS, PRECIPITATION and HUMIDITY, WIND and air pressure, air temperature, and STORMS. More basic is the scale of the phenomena: the microscale deals with small, transient phenomena up to about 10km in size and lasting, say, an hour; the mesoscale, those up to 200km across and lasting a few hours; the synoptic scale is that of daily national and continental weather maps, while the macroscale treats of global, seasonal phenomena. The general circulation of the atmosphere is zonal by latitude (see JET STREAM; PREVAILING WESTERLIES; TRADE WINDS). Imposed on this are disturbances – chiefly CYCLONES and anticyclones – due to imbalance of pressure and temperature. An air mass is a large region of air, roughly homogeneous horizontally, that forms by stagnant contact with a land or sea surface and then moves elsewhere. When two air masses of different properties meet, a FRONT is formed. (See also ISOBAR; ISOTHERM.)

Meter (m)

The SI base unit of length, defined as the length equal to 1,650,763.73 times the wavelength of radiation corresponding to the transition between the ENERGY LEVELS $2p_{10}$ and $5d_5$ of the krypton-86 atom. It was originally intended that the meter should represent one ten-millionth of the distance from the N Pole to the equator on the meridian passing through Paris, but the surveyors got their sums wrong and for 162 years (to 1960) the meter was defined as an arbitrary distance marked on a metal bar (from 1889-1960, the "international prototype meter").

Methane (CH_4)

Colorless, odorless gas; the simplest ALKANE. It is produced by decomposing organic matter in sewage and in marshes (hence the name marsh gas), and is the "firedamp" of coal mines. It is the chief constituent of NATURAL GAS, occurs in COAL GAS and WATER GAS, and is produced in PETROLEUM refining. Methane is used as a FUEL and in the manufacture of carbon-black. MW 16.0, mp −182.5°C, bp −161.5°C. (◆ page 37)

Methanol (methyl alcohol or wood alcohol, CH_3OH)

Colorless liquid, the simplest ALCOHOL. Formerly made by destructive distillation of wood, it is now mostly made by catalytic reaction of carbon monoxide and hydrogen. It is used to make FORMALDEHYDE and other industrial chemicals, in antifreeze, rocket fuels and as a solvent. Being highly toxic, it is used to make ETHANOL undrinkable (when it is known as "denatured" or "methylated" spirit). MW 32.0, mp −97.7°C, bp 64.7°C.

Methedrine see AMPHETAMINES.

Methyl compounds

Organic compounds containing the methyl group, CH_3 (see ALKANES).

Metric system

A decimal system of weights and measures devised in Revolutionary France in 1791 and based on the METER. The original unit of MASS was the GRAM, the mass of a cubic centimeter of water at 4°C, the TEMPERATURE of its greatest DENSITY. Auxiliary units were to be formed by adding Greek prefixes to the names of the base units for their decimal subdivisions. The metric system forms the basis of the physical units systems known as CGS UNITS and MESA UNITS, the present International System of Units (SI UNITS) being a development of the latter.

Meyer, Adolf (1866-1950)

Swiss-born US psychiatrist best known for his concept of psychobiology, the use in psychiatry of both psychological and biological processes together.

Meyer, Julius Lothar (1830-1895)

German chemist who, independently from MENDELEYEV, drew up the PERIODIC TABLE, publishing his version in 1870. He showed the periodicity of atomic volume.

Meyerhof, Otto (1884-1951)

German physiologist who shared with A.V. HILL the 1922 Nobel Prize for Physiology or Medicine for their independent work on the biochemistry of MUSCLE action.

MHO(reciprocal OHM)

A unit of CONDUCTANCE.

Mica

Group of common SILICATE minerals composed of sheets of linked SiO_4 tetrahedra, with aluminum replacing silicon to some extent, and containing cations and hydroxyl groups between the layers. The three main types are: BIOTITE, MUSCOVITE and PHLOGOPITE; others include CHLORITE and GLAUCONITE. Micas occur widespread in many igneous, metamorphic and sedimentary rocks, and weather to CLAY minerals. They show perfect basal cleavage, producing thin, flexible flakes that are used as electrical insulators and as the dielectric in CAPACITORS; ground mica is used in paints, inks, wallpaper, rubber and waterproof coatings.

Michelson, Albert Abraham (1852-1931)

German-born US scientist awarded the 1907 Nobel Prize for Physics for his optical precision instruments and the spectroscopic and meteorological investigations carried out with their aid.

Michelson-Morley experiment

Important experiment whose results, by showing that the ETHER does not exist, substantially contributed to EINSTEIN's formulation of RELATIVITY theory. Its genesis was the development by ALBERT ABRAHAM MICHELSON (1852-1931) of an INTERFEROMETER (1881) whereby a beam of light could be split into two parts sent at right angles to each other and then brought together again. Because of the Earth's motion in space, the "drag" of the stationary ether should produce INTERFERENCE effects when the beams are brought together: his early experiments showed no such effects. With E.W. MORLEY he improved the sensitivity of his equipment, and by 1887 was able to show that there was no "drag", and therefore no ether. Michelson, awarded the Nobel Prize for Physics in 1907, was the first US Nobel prize-winner.

Michurin, Ivan Vladiurirovich (1855-1935)

Russian horticulturalist whose theories of heredity, including the inheritance of ACQUIRED CHARACTERISTICS (see also LAMARCK), officially displaced GENETICS in Soviet science from 1948 until LYSENKO's fall from power in 1964.

Microbe

A word for any microscopic organism, especially one that causes disease. The term covers bacteria, viruses, macrophages, chlamydia, fungi, molds and unicellular animals.

Microbiology

The study of MICROORGANISMS.

Microchemistry

A branch of CHEMISTRY in which very small

amounts (1μg to 1mg) are studied. Tracer methods, especially labeling with radioactive ISOTOPES, are used, as are instrumental methods of ANALYSIS.

Microchip see CHIP.

Microcline
Common mineral in the FELDSPAR group, occurring in igneous rocks as vitreous crystals of various colors. It has the same chemical composition as ORTHOCLASE, and differs from it only in the arrangement of the silicon and aluminum atoms.

Microcomputer
A small computer which uses one or more MICROPROCESSORS as its CENTRAL PROCESSING UNIT. (See COMPUTERS.)

Microelectronics
The technology of constructing electronic systems from small electronic parts, eg INTEGRATED CIRCUITS, CHIPS. (See also ELECTRONICS.)

Microgravity (free fall)
The preferred term for "zero gravity". Although weight seems to disappear in space, there is in fact always a small amount of gravity.

Micrometer
Instrument for accurately measuring dimensions or separations. Its basis is that when a screw is turned once, it advances or retreats a distance equal to its pitch. The hairlines in telescope and microscope eyepieces are adjusted by means of a precision micrometer screw to measure separations. The micrometer caliper has a G-shaped frame on whose "leg" is a scale; inside the leg runs a screw (of pitch usually 0.5mm) attached to a thimble, calibrated for fractions of a turn, which runs over the scale. An object is placed between the screw's spindle and an anvil at the far side of the G's opening, and the screw turned until the object is just held. For greater accuracy (of the order of 1μm) a VERNIER SCALE may also be used.

Micron (μ)
Unit of length in the CGS system (see CGS UNITS), equal to one millionth of a METER. In SI units it is replaced by the micrometer (μm).

Microorganisms
A term generally used to refer to the viruses, bacteria, cyanobacteria, protozoa, diatoms, dinoflagellates, unicellular algae, and yeasts and other microscopic fungi. The majority of microorganisms are unicellular, and most can be seen only with a microscope, but the term cannot be defined precisely; for example, the giant ameba is a unicellular protozoan yet it can be seen with the naked eye. The study of microorganisms is known as microbiology.

Microphone
Device for converting sound waves into electrical impulses. The carbon microphone used in telephone mouthpieces has a thin diaphragm behind which are packed tiny carbon granules. SOUND waves vibrate the diaphragm, exerting a variable pressure on the granules. This varies their RESISTANCE, so producing fluctuations in a DC current (see ELECTRICITY) passing through them. The crystal microphone incorporates a piezoelectric crystal in which pressure changes from the diaphragm produce an alternating voltage (see PIEZOELECTRICITY). In the electrostatic microphone the diaphragm acts as one plate of a CAPACITOR, vibration producing changes in capacitance. In the moving-coil microphone the diaphragm is attached to a coil located between the poles of a permanent magnet: movement induces a varying current in the coil (see INDUCTANCE). The ribbon microphone has, rather than a diaphragm, a metal ribbon held in a magnetic field; vibration of the ribbon induces an electric current in it.

Microprocessor
A SILICON chip which contains the arithmetic and logic function of a CENTRAL PROCESSING UNIT. It is an integral part of MICROCOMPUTERS (see also COMPUTER).

Microscope
An instrument for producing enlarged images of small objects. The simple microscope or MAGNIFYING GLASS, comprising a single converging LENS, was known in ancient times, but the first compound microscope is thought to have been invented by the Dutch spectacle-maker ZACHARIAS JANSSEN around 1590. However, because of the ABERRATION unavoidable in early lens systems, the simple microscope held its own for many years, ANTON VAN LEEUWENHOEK constructing many fine examples using tiny near-spherical lenses. Compound microscopes incorporating ACHROMATIC LENSES became available from the mid-1840s.

In the compound microscope a magnified, inverted image of an object resting on the "stage" is produced by the objective lens (system). This image is viewed through the eyepiece (or ocular) lens (system) which acts as a simple microscope, giving a greatly magnified virtual image. In most biological microscopy the object is viewed by transmitted light, illumination being controlled by mirror, diaphragm and "substage condenser" lenses. The near-transparent objects are often stained to make them visible. As this usually proves fatal to the specimen, phase-contrast microscopy, in which a "phase plate" is used to produce a DIFFRACTION effect, can alternatively be employed. Objects which are just too small to be seen directly can be made visible in dark-field illumination. In this an opaque disk prevents direct illumination and the object is viewed in the light diffracted from the remaining oblique illumination.

Although there is no limit to the theoretical magnifying power of the optical microscope, magnifications greater than about 2000× can offer no improvement in resolving power (see TELESCOPE) for light of visible wavelengths. The shorter wavelength of ultraviolet light allows better resolution and hence higher useful magnification. For yet finer resolution physicists turn to electron beams and electromagnetic focusing (see ELECTRON MICROSCOPE). The FIELD-ION MICROSCOPE offers the greatest magnifications. (♦ page 123)

Microtome
Device to prepare thin sections for the microscope. The specimen is embedded in a block of wax, then placed in the microtome. Turning a handle raises and lowers the block against a blade. At the top of its rise the block is advanced by a MICROMETER screw, which controls the thickness of the section cut on the descent.

Microtubules
Microscopic tubes of protein found within cells. They are made up of molecules of a protein, tubulin, which are arranged around a hollow core, rather like the kernels on a maize cob. Some microtubules are quite short, others up to a meter in length, as in the axons of motor nerves in mammals. Microtubules permeate the cytoplasm of EUKARYOTIC cells, forming part of the CYTOSKELETON and helping to transport vital molecules around the cell. Bundles of microtubules make up the FLAGELLA and CILIA, as well as the basal bodies that control these organs. The CENTRIOLES are also made up of microtubules, as is the spindle that controls the movement of chromosomes in MITOSIS and MEIOSIS. The bacterial flagellum, or undulipodium, is also made up of a tubular protein filament, but it is a single tubule rather than a bundle of tubules, and is composed of a slightly different type of protein. (♦ page 13, 126)

Microwaves
ELECTROMAGNETIC RADIATIONS with wavelengths between 1mm and 30cm; used in RADAR, telecommunications, SPECTROSCOPY and for cooking (microwave ovens). Their dimensions are such that it is easy to build antennas of great directional sensitivity and high-efficiency WAVEGUIDES for them. (♦ page 118)

Midnight Sun
Phenomenon observed N of the Arctic Circle and S of the Antarctic Circle. Each summer the Sun remains above the horizon for at least one 24-hour period (a corresponding period of darkness occurs in winter), owing to the tilt of the EARTH's equator to the ecliptic.

Mid-ocean ridge see OCEANS; PLATE TECTONICS.

Migraine see HEADACHE.

Migration
Long-distance mass movements made by animals of many different groups, both vertebrate and invertebrate, usually at regular intervals. Generally animals move from a breeding area to a feeding place, returning as the breeding season approaches the following year. This is the pattern of annual movements of migratory birds and fish. Migrations of this nature may be over great distances, up to 11,000km in some birds. In other cases migrations may follow cycles of food abundance: wildebeeste in E Africa follow in the wake of the rains, grazing on the new grass; caribou (reindeer) show similar movements. Certain carnivore species may follow these migrations. (See also NAVIGATION.)

Mildew
A general name for the superficial growth of many types of fungi often found on plants and material derived from plants. Also used more specifically for certain groups of fungi that infect plants. Powdery mildews are caused by fungi belonging to the Ascomycetes Order Erysiphales, the powdery effect being due to the masses of spores. These fungi commonly infest roses, apples, phlox, melons etc. Downy mildews are caused by Phycomycetes. They commonly infest many vegetable crops.

Mile (mi)
Name of many units of length in different parts of the world. The statute mile (st mi) is 1760 yards (exactly 1609.344m); the international (US) nautical mile is 1.15078st mi; the UK nautical mile, 1.15151st mi. The name derives from the Roman (Latin) *milia passuum*, a thousand paces.

Milk
A nutritious liquid secreted by the mammary glands of female mammals. The secretion of milk (LACTATION) is initiated immediately after birth by the hormone PROLACTIN. The milks produced by different mammals all have the same basic constituents but the proportion of fat, protein and other ingredients differs from species to species. In any species the milk produced is a complete food for the young until weaning.

A milk-like substance is also produced by certain birds, notably pigeons, flamingoes and some species of penguin. It is produced by the CROP and consists of nutrient-rich cells that are shed from the cell wall.

Milky Way
A hazy, milk-like band of stars encircling the night sky. Irregular dark patches are caused by intervening clouds of gas and dust. Its appearance is due to the view of the disk of our GALAXY, which we see from the Sun's location within it. The term is also used to name our Galaxy. It is a disk-shaped spiral galaxy containing some 100 billion stars, and has a radius of about 15kpc. Our SOLAR SYSTEM is in one of the spiral arms and is just over 9kpc from the galactic center, which lies in the direction of SAGITTARIUS. The Galaxy slowly rotates about a roughly spheroidal nuclear bulge (radius about 4.5kpc), though not at uniform

speed; the Sun circles the galactic center about every 230 million years. The Galaxy is surrounded by a spheroidal halo some 50kpc in diameter, composed of gases, dust, occasional stars and GLOBULAR CLUSTERS. This, in turn, seems to be immersed within an extended halo of dark matter, which cannot be seen directly but whose gravitational influence seems necessary to account for the high speeds at which gas and stars in the outer regions of the Galaxy are moving. (♦ page 240)

Millibar (mbar or mb)
CGS unit of PRESSURE. (See BAR.)

Millikan, Robert Andrews (1868-1953)
US physicist awarded the 1923 Nobel Prize for Physics for his determination of the charge on a single ELECTRON and his work on the PHOTOELECTRIC EFFECT. He also studied and named COSMIC RAYS.

Milstein, César (b. 1927)
Argentinian biochemist based in the UK who received the 1984 Nobel Prize for Physiology or Medicine jointly with NIELS JERNE and GEORGES KÖHLER for discovering the principle for the production of MONOCLONAL ANTIBODIES.

Mimicry
The close resemblance of one organism (the mimic) to another (the model). In most cases the model is unpalatable and conspicuous, and is avoided by certain predators. The mimic thus gains a degree of protection on the strength of the predator's avoidance of the model. Mimicry of this type, known as Batesian mimicry, is particularly well developed among insects. Where two or more equally unpalatable species resemble each other, for mutual protection, it is known as Mullerian mimicry. Some mimicry is also associated not with defense but with parasitism. Cuckoos, for example, lay eggs that mimic those of their hosts, and cuckoo bees resemble the bees in whose nests they lay their eggs.

Mind
Man's mental organ, with which he thinks, reasons, remembers and wills. The existence of the mind, as separate from the workings of the BRAIN, is not fully understood.

Minerals
Naturally occurring substances that can be obtained by mining, including COAL, PETROLEUM and NATURAL GAS; more specifically, in geology, substances of natural inorganic origin, of more or less definite chemical composition, CRYSTAL structure and properties, of which the ROCKS of the Earth's crust are composed. (See also GEMS; ORE.) Of the 3000 minerals known, fewer than 100 are common. They may be identified by their color (though this can often vary because of impurities), HARDNESS, luster, SPECIFIC GRAVITY, crystal forms and CLEAVAGE; or by chemical ANALYSIS and X-RAY DIFFRACTION. Minerals are generally classified by their ANIONS in order of increasing complexity: elements, SULFIDES, OXIDES, HALIDES, CARBONATES, NITRATES, SULFATES, PHOSPHATES and SILICATES. Others are classed with those they resemble chemically and structurally, eg arsenates with phosphates. A newer system classifies minerals by their topological structure (see TOPOLOGY). (♦ page 44)

Mining
The means for extracting economically important MINERALS and ORES from the Earth. Where the desired minerals lie near the surface, the most economic form of mine is the open pit. This usually consists of a series of terraces, which are worked back in parallel so that the mineral is always within convenient reach of the excavating machines. Strip mining refers to stripping off a layer of overburden to reach a usually thin mineral seam (often COAL). The excavating machines used in open-pit mining are frequently vast. Soft minerals such as KAOLIN can be recovered hydraulically, by directing heavy water jets at the pit face and pumping out the resulting slurry. Where a mineral is found in alluvial (river bed) deposits, bucket or suction dredgers may be used. Where minerals lie far below the surface, various deep mining techniques must be used. Sulfur is mined by pumping superheated water down boreholes into the mineral bed. This melts the sulfur, which is then pumped to the surface (see FRASCH PROCESS). Watersoluble minerals such as SALT are often mined in a similar way (solution mining). But most often, deep minerals and ores must be won from underground mines. Access to the mineral-bearing strata is obtained via a vertical shaft or sloping incline driven from the surface, or via a horizontal adit driven into the side of a mountain. The geometry of the actual mining area is determined by the type of mineral and the strength of the surrounding material. All underground mines require adequate ventilation and lighting, facilities for pumping out any groundwater or toxic gases seeping into the workings, and means (railroad or conveyor) for removing the ore and waste to the surface. As in open-pit mining, the rock is broken mechanically or with explosives. However, particular care must be exercised when using explosives underground. Several occupational diseases (eg PNEUMOCONIOSIS) are associated with mining and extraction metallurgy, particularly where high dust levels and toxic substances are involved.

Minot, George Richards (1885-1950)
US physician who shared with W. MURPHY and G. WHIPPLE the 1934 Nobel Prize for Physiology or Medicine for his work with Murphy showing daily consumption of large quantities of raw liver to be an effective treatment for pernicious ANEMIA.

Miocene
The penultimate epoch of the TERTIARY period, which lasted from 25 to 5 million years ago. (See GEOLOGY.) (♦ page 106)

Mirage
Optical illusion arising from the REFRACTION of light as it passes through air layers of different densities. In inferior mirages distant objects appear to be reflected in water at their bases: this is because light rays traveling initially toward the ground have been bent upward by layers of hot air close to the surface. In superior mirages objects seem to float in the air: this occurs where warmer air overlies cooler, bending rays downward.

Mirror
A smooth reflecting surface in which sharp optical images can be formed. Ancient mirrors were usually made of polished bronze but glass mirrors backed with tin amalgam became the rule in the 17th century. Silvered-glass mirrors were first manufactured in 1840, five years after LIEBIG discovered that a silver mirror was formed on a glass surface when an ammoniacal solution of silver nitrate was reduced by an aldehyde (now usually formaldehyde).

Undistorted but laterally reversed virtual IMAGES can be seen in plane (flat) mirrors. (Such images are "virtual" and not "real" because no light actually passes through the apparent position of the image.) Concave spherical mirrors form real inverted images of objects farther away than half the radius of curvature of the mirror and virtual images of closer objects. Concave mirrors (usually with an unglazed metallic surface) are used in astronomical TELESCOPES because of their freedom from many lens defects. Parabolic concave mirrors, which focus a parallel beam of light in a single point, also find use as reflectors for solar furnaces and searchlights. Convex spherical mirrors always form distorted virtual images but offer a wider field of view than plane mirrors. Half-silvered glass mirrors are used in many optical instruments and can also be used to give a one-way mirror effect. (See also LIGHT; REFLECTION.) (♦♦ page 115, 237)

Miscarriage
Strictly the term for any spontaneous abortion that occurs after the 28th week of pregnancy; popularly, any non-induced abortion.

Misch metal
An ALLOY composed of 50% CERIUM, 25% LANTHANUM, 15% NEODYMIUM, 10% other RARE EARTHS, and iron; used for "flints" for cigarette lighters and as a deoxidizer for vacuum tubes.

Missing mass
Matter that is not directly visible but is inferred to exist in order to provide sufficient gravitational attraction to account for several aspects of the large-scale motion of matter in COSMOLOGY. Stars and gas clouds in the outer regions of spiral galaxies (see GALAXY) move faster than would be the case if all their mass were contained in visible stars. Likewise, clusters of galaxies can hold themselves together only if they contain at least ten times as much mass as is visibly present in their member galaxies. If the Universe is closed (see COSMOLOGY), well over 90% of its total mass must consist of dark matter.

Mississippian
The antepenultimate period of the PALEOZOIC, lasting from about 360 to 320 million years ago. (See also CARBONIFEROUS; GEOLOGY.) (♦ page 106)

Mistral
Cold wind blowing S from the Central Plateau of France to the NW Mediterranean. It occurs mainly in winter, and speeds up to about 140km/h have been recorded. It is a hazard to air and surface transport, crops and buildings.

Mitchell, Peter Dennis (b. 1920)
British biochemist who received the 1978 Nobel Prize for Chemistry for his theory of how CELLS obtain energy from foodstuffs and sunlight.

Mitochondria
Elliptical, membrane-bound organelles found in the cells of EUKARYOTES. Internally they contain layers of membranes on which the enzymes involved in RESPIRATION are located. Mitochondria are particularly numerous in active cells, such as muscle cells, where the need for energy is high. They contain a small amount of DNA and their own protein-making machinery; this has led to the suggestion that they are derived from free-living cells (see ENDOSYMBIOSIS THEORY). (♦ page 13, 28)

Mitosis
The normal process by which a CELL divides into two. Initially the CHROMOSOMES become visible in the nucleus before longitudinally dividing into a pair of parallel chromatids. The chromosomes shorten and thicken and arrange themselves on a spindle across the equator of the cell. The cell then divides so that each daughter contains a full complement of chromosomes. (♦ page 32)

Mitral stenosis
Tightness of the heart's mitral valve, which makes it both hard for the heart to pump against the ventricle, and likely to leak blood back into it.

MKSA units (Giorgi system)
A metric system of units based on the METER (length), KILOGRAM (mass), SECOND (time) and AMPERE (electric current), forming the basis of the now internationally accepted SI UNITS. The system is "rationalized" in that, with the PERMEABILITY of free space set at $4\pi \times 10^{-7}$ henry/meter, equations contain factors reflecting the geometry of the situations they describe: 2π for cylindrical symmetries; 4π for spherical.

Modem
A device to convert binary signals into audio signals that can be sent to and received by a COMPUTER over the telephone network.

Modulation see RADIO.
Moho
Abbreviation for MOHOROVIČIĆ DISCONTINUITY.
Mohorovičić discontinuity
A layer of the Earth originally regarded as marking the boundary between CRUST and MANTLE (see EARTH), evidenced by a change in the velocity of seismic waves (see EARTHQUAKE). It is now regarded as of little physical significance. The US project Mohole, designed to drill through the "Moho", was abandoned in 1966. More important are the discontinuities between the SIAL and SIMA portions of the crust; the core and the mantle (Gutenberg or Oldham Discontinuity), with a radius of about 3500km; and the inner and outer cores, which have a radius of about 1200km to 1650km.
Mohs' scale see HARDNESS.
Moiré pattern
A family of curves formed by the intersections of one family of curves with another over which it has been superimposed. Moiré patterns may be seen by looking through the folds of a gauze or nylon curtain: motion of the curtain or observer will cause dramatic changes in the patterns observed. They are of particular note in color printing, where special techniques are employed to prevent their appearance in HALFTONES; and are used in industry to determine, for example, the degree of flatness of a surface. They are used also as mathematical models of physical phenomena, and occasionally in the solution of mathematical problems. Their disturbing optical properties are of interest in psychology.
Moissan, Ferdinand Frédéric Henri
(1852-1907)
French chemist awarded the 1906 Nobel Prize for Chemistry for isolating FLUORINE (1886) and for developing the electric arc furnace (see ELECTRIC FURNACE). He also claimed to have synthesized DIAMONDS (1893).
Mold
A general name for a number of filamentous FUNGI that produce powdery or fluffy growths on fabrics, foods and decaying plant or animal remains. Best known is the blue bread mold caused by *Penicillium*, from which the antibiotic PENICILLIN was first discovered.
Mole
Pigmented spot in the SKIN, consisting of a localized group of special cells containing MELANIN. Change in a mole, such as increase in size, change of color and bleeding should lead to suspicion of melanoma.
Mole (mol)
The SI base unit of amount of substance, defined as the amount of substance of a system which contains as many elementary entities (of a specified kind) as there are atoms in 0.012kg of carbon-12 (ie the AVOGADRO number). One mole of a compound is its MOLECULAR WEIGHT in grams. The molarity of a solution is its concentration in moles per liter; a solution of which the molarity is 1 is called molar.
Molecular biology
The study of the structure and function of the MOLECULES that make up living organisms. This includes the study of PROTEINS, ENZYMES, CARBOHYDRATES, LIPIDS and NUCLEIC ACIDS. (See also BIOCHEMISTRY.)
Molecular genetics
The study of the mechanisms by which genetic information is replicated and expressed at the molecular level, and the ways in which the expression is regulated. (See also DNA; GENETICS.)
Molecular weight
A term often (incorrectly) used to describe molecular mass, which is the sum of the ATOMIC MASSES of all the atoms in a MOLECULE. It is an integral multiple of the empirical FORMULA mass found by chemical analysis, and of the EQUIVALENT WEIGHT. Molecular masses may be found directly by MASS SPECTROSCOPY, or deduced from related physical properties including gas DENSITY; effusion; osmotic pressure (see OSMOSIS), and effects on solvents: lowering of VAPOR PRESSURE and freezing point, and raising of boiling point; for large molecules the ultracentrifuge is used. (See also MOLE.) (◀ page 130)
Molecule
Entity composed of ATOMS linked by chemical BONDS and acting as a unit; the smallest particle of a chemical compound which retains the COMPOSITION and chemical properties of the compound. The composition of a molecule is represented by its molecular FORMULA. Elements may exist as molecules, eg oxygen O_2, phosphorus P_4. Molecules range in size from single atoms to macromolecules – such as PROTEINS – with MOLECULAR WEIGHTS of 10,000 or more. The chief properties of molecules depend on their structure (bond lengths and angles) – determined by electron diffraction, X-RAY DIFFRACTION and SPECTROSCOPY – spectra, and DIPOLE MOMENTS. (See also ORBITAL; VAN DER WAALS, J.D.) (◀ page 130)
Mollusca
A phylum of soft-bodied invertebrate animals that typically have a chalky shell for protection. There are over 80,000 living species. These are classified into three main classes: Gastropoda, Bivalvia, and Cephalopoda. (◀ page 16)
GASTROPODS include slugs, snails, limpets, slipper limpets, ormers and abalones. They have bodies divided into a head, a visceral hump and a muscular foot. The head bears tentacles. During development, the visceral hump twists counterclockwise through 180°, a process known as torsion. This brings the anus and mantle cavity to the front of the body, so that the animal can withdraw its head into the mantle cavity when threatened. Slugs and sea slugs have lost their shells, but the majority still undergo torsion as a vestige of their evolutionary past. This can be seen in many garden slugs, where the pneumostome, or breathing pore, is situated at one side of the mantle, rather than centrally.
BIVALVES (lamellibranchs) include oysters, clams and mussels. They are enclosed in shells that form a pair of hinged valves, drawn together by an elastic ligament. Most of them are SESSILE and live by filtering food particles from the water, or by using their gills. (See also LAMPSHELLS.)
CEPHALOPODS include squids, octopuses and cuttlefish. They typically bear tentacles and, except in NAUTILUS, their shell is much reduced and enclosed within their bodies. The nervous system and brain are well developed and the eyes are extraordinarily similar to those found in vertebrate animals. Locomotion is by crawling on the sea bed or by jet propulsion using the sudden ejection of water through a siphon. There are also several minor groups of mollusks, including the CHITONS (Polyplacephora), the tusk shells (Scaphopoda), the segmented NEOPILINA (Monoplacophora) and the solenogasters (Aplacophora).
Molting
The shedding of the EXOSKELETON, skin, fur or feathers by an animal. It may be a seasonal occurrence, as a periodic renewal of fur or plumage, or it may be necessary for growth, as in insects or crustaceans, which have a rigid exoskeleton (see ECDYSIS).
Molybdenite (MoS_2)
Soft, gray SULFIDE mineral, the chief ore of MOLYBDENUM, mined mainly in Colorado. Purified molybdenite has properties very similar to those of GRAPHITE, and is used as a lubricant.
Molybdenum (Mo)
Silvery gray metal in Group VIB of the PERIODIC TABLE; a TRANSITION ELEMENT. It is obtained commercially by roasting MOLYBDENITE in air and reducing the oxide formed with carbon in an electric furnace or by the thermite process to give ferromolybdenum. Because of its high melting point, it is used to support the filament in electric lamps and for furnace heating elements. It also finds use in corrosion-resistant, high-temperature steels and alloys. Molybdenum is unreactive, but forms various covalent compounds. Some are used as industrial CATALYSTS. Molybdenum is a vital trace element in plants and a catalyst in bacterial NITROGEN FIXATION. AW 95.9, mp 2623°C, bp 4630°C, sg 10.2 (20°C). (◀ page 130)
Moment
The product of a quantity and its distance from some specific point connected with it. Statical moments such as the moment of a force (measuring its turning effect on a body by multiplying the force's magnitude by its perpendicular distance from the rotation axis) enter equations of static equilibrium. The moment of inertia, I, of a rotating body (the analog of MASS in the dynamics of translation) is the sum of the products of its mass elements, m_i, with the squares of their distances, r_i, from the rotation axis.

$$I=\sum_i^i m_i r_i^2$$

Momentum
The product of the MASS and linear VELOCITY of a body. Momentum is thus a VECTOR quantity. The linear momentum of a system of interacting particles is the sum of the momenta of its particles, and is constant if no external forces act. The rate of change of momentum with time in the direction of an applied force equals the force (Newton's second law of motion – see MECHANICS). In rotational motion, the analogous concept is angular momentum, the product of the moment of inertia and the angular velocity of a body relative to a given rotation axis. If no external forces act on a rotating system, the direction and magnitude of its angular momentum remain constant.
Monadnock
Isolated hill formed of erosion-resistant bedrock in an area otherwise well eroded (see EROSION; PENEPLAIN); named for Mt Monadnock, Cheshire County, New Hampshire.
Monazite
PHOSPHATE mineral of widespread occurrence, mined in India, Brazil, the US and South Africa as an ore of the RARE EARTHS and THORIUM. It usually forms small brown prismatic crystals in the monoclinic system.
Mond, Ludwig (1859-1909)
German-born British industrial chemist whose discovery (1889) of nickel carbonyl (see TRANSITION ELEMENTS) led him to devise the Mond process for refining NICKEL, and so to found the Mond Nickel Company.
Monel metal
Strong, corrosion-resistant ALLOY composed of 68% nickel, 29% copper, 3% iron manganese, silicon and carbon; used for turbine blades, propellers, etc. Originally made by smelting nickel/copper ore from Sudbury, Ontario.
Monera
In taxonomy, the KINGDOM that includes the BACTERIA and CYANOBACTERIA. All members of the Monera are unicellular organisms, though some live in simple colonies of cells. They are PROKARYOTES, with rigid or semi-rigid cell walls.
Mongolism see DOWN'S SYNDROME.
Moniz, Antonio Caetano de Abreu Freire Egas
(1874-1955)
Portuguese brain surgeon awarded with W.R. HESS the 1949 Nobel Prize for Physiology or Medicine

for his development of frontal LOBOTOMY as a treatment for mental disorders. The treatment is now little used (see PSYCHIATRY).

Monoamine oxidase
An ENZYME that inactivates ADRENALINE.

Monoclonal antibodies
Antibodies cloned (see CLONING) from a single antibody-producing cell, and therefore identical. "Pure" antibodies of this type have a number of potential applications in the diagnosis and treatment of disease. (See also ANTIBODIES AND ANTIGENS; IMMUNITY.)

Monocotyledons (monocots)
Flowering plants or ANGIOSPERMS that produce seeds with only one seed leaf (or COTYLEDON). They have parallel-veined leaves and the flowering parts are usually in threes or multiples of three, not fives as in DICOTYLEDONS. (◆ page 139)

Monod, Jacques Lucien (1910-1976)
French biochemist who, with F. JACOB and A. LWOFF, received the 1965 Nobel Prize for Physiology or Medicine for his work with Jacob on regulatory GENE action in BACTERIA.

Monoecious plants
Ones where the male and female organs are borne on the same plant, but in separate flowers; examples are oak, maize and walnut. (See also DIOECIOUS PLANTS.)

Mononucleosis (glandular fever)
Common virus infection of adolescence causing a variety of symptoms including severe sore throat, headache, fever, malaise and enlargement of lymph nodes and spleen. Diagnosis is usually made with a blood test. The infection may be transmitted from person to person.

Monosaccharides see CARBOHYDRATES; SUGARS.

Monosodium glutamate
White crystalline solid, the acid sodium salt of glutamic acid, an AMINO ACID. Obtained from GLUTEN or soybean protein, it is added to many foods to bring out their flavor.

Monotremates
A subclass of the MAMMALS that includes the echidnas or spiny anteaters and the platypuses. Monotremes differ from all other mammals in not bearing live young; like their reptile ancestors, they lay eggs, but they do suckle their young on milk. Living members are limited to Australasia.

Monsoon
Wind system where the prevailing WIND direction reverses in the course of the seasons, occurring where large temperature (hence pressure) differences arise between oceans and large landmasses. Best known is that of SE Asia. In summer moist winds, with associated HURRICANES, blow from the Indian Ocean into the low-pressure region of NW India, caused by intense heating of the land. In winter, cold dry winds sweep S from the high-pressure region of S Siberia.

Month
Name of several periods of time, mostly defined in terms of the motion of the MOON. The synodic month (lunar month or lunation) is the time between successive full moons; it is 29.531 DAYS. The sidereal month, the time taken by the Moon to complete one revolution about the Earth relative to the fixed stars, is 27.322 days. The anomalistic month, 27.555 days, is the time between successive passages of the Moon through perigee (see ORBIT). The solar month, 30.439 days, is one twelfth of the solar YEAR. Civil or calendar months vary in length through the year, lasting from 28 to 31 days (see CALENDAR). In popular usage the (lunar) month refers to 28 days.

Moon
A SATELLITE, in particular the Earth's largest natural satellite. The Moon is so large relative to the Earth (it has a diameter two-thirds that of MERCURY) that Earth and Moon are sometimes regarded as a double planet. The Moon has a diameter of 3476km and a mass 0.0123 that of the Earth; its ESCAPE VELOCITY is around 2.4km/s. The orbit of the Moon defines the several kinds of MONTH. The distance of the Moon from the Earth varies between 363Mm and 406Mm (perigee and apogee) with a mean of 384.4Mm. The Moon rotates on its axis every 27.322 days, hence keeping the same face constantly toward the Earth; however, in accordance with KEPLER'S second law, the Moon's orbital velocity is not constant and thus there is exhibited the phenomenon known as libration: to a particular observer on the Earth, marginally different parts of the Moon's disk are visible at different times. There is also a very small physical libration due to slight irregularities in its rotational velocity. The Moon is covered with craters, whose sizes range up to 250km diameter. These are sometimes seen in chains up to 1Mm in length. Other features include rilles, trenches a few kilometers wide and a few hundred kilometers long; the *maria* or great plains; the bright rays that emerge from the large craters, and the lunar mountains. There are also lunar hot spots, generally associated with those larger craters showing bright rays: these remain cooler than their surrounds during lunar daytime, warmer during the lunar night.

It is widely believed that the majority of craters and *mare* basins were formed by the impacts of meteorites and asteroidal-sized bodies, but some, too, may be the result of internal volcanic activity. Most of the giant impacts that formed the *maria* occurred between 4.5 and 3.8 billion years ago, and the basins so produced were subsequently filled by volcanic magma. Analysis of samples brought back to Earth by the Apollo missions has shown that lunar rocks contain a higher proportion of refractory elements (those with a high boiling point, such as titanium) and are depleted in volatiles (those with low boiling point), compared to terrestrial rocks. There is no consensus about how the Moon was formed; one theory suggests that Earth and Moon were formed simultaneously in close proximity. (◆◆ page 134, 221)

Moonstone
Translucent variety of alkali (or plagioclase) FELDSPAR, showing opalescence resulting from unmixing of sodium and potassium; a GEM stone.

Moore, Stanford (1913-1982)
US biochemist who shared with C.B. ANFINSEN and W.H. STEIN the 1972 Nobel Prize for Chemistry for his part in determining the structure of the ENZYME ribonuclease (see also NUCLEIC ACIDS).

Moraine
Accumulation of debris carried or dropped by a glacier. Ground moraine is DRIFT left in a sheet as the GLACIER retreats. Terminal moraines are ridges deposited when the ice is melting before the glacial retreat; a series of ridges may mark pauses in the retreat. Lateral moraines are formed of debris that falls onto the glacier: when two glaciers merge their lateral moraines may unite to form a medial moraine.

Mordants
Substances that fix DYES to fabrics by precipitating them in the fibers or by forming LIGAND complexes with them. Generally, metal salts and hydroxides (including ALUM), now chiefly CHROMIUM salts and dichromates, are used. They modify the hue and improve fastness.

Morgagni, Giovanni Battista (1682-1771)
Italian anatomist whose *Of the Seats and Causes of Diseases as Investigated by Anatomy* (1761) established him as the father of morbid anatomy.

Morgan, Thomas Hunt (1866-1945)
US biologist who, through his experiments with the fruit fly *Drosophila*, established the relation between GENES and CHROMOSOMES and thus the mechanism of HEREDITY. For his work he received the 1933 Nobel Prize for Physiology or Medicine.

Morley, Edward Williams (1838-1923)
US chemist who worked on the relative densities of OXYGEN and HYDROGEN, but is best known for his role in the MICHELSON-MORLEY EXPERIMENT.

Morphine
OPIUM derivative used as a NARCOTIC ANALGESIC and also commonly in DRUG ADDICTION. It depresses respiration and the cough reflex, induces sleep and may cause vomiting and constipation. It is valuable in heart failure as a premedication for anesthetics, and in terminal malignant disease (see also HEROIN).

Morton, William Thomas Green (1819-1868)
US dentist who pioneered the use of diethyl ETHER as an anesthetic (1844-46). In later years he engaged in bitter litigation over his refusal to recognize the work of former colleagues, especially C.W. LONG'S prior use of ether in this way.

Mosaic
A general name for several VIRUS diseases of plants, including tobacco, tomatoes, potatoes, soybeans and peas; these viral diseases produce a characteristic leaf-mottling and stunted growth. Transmission of the disease may be via aphids or by mechanical contact.

The term genetic mosaic is used of organisms made up of two or more genetically distinct cell types. Such organisms are produced in embryological experiments, for example, by combining the cells from two fertilized eggs. Another term used in embryology, mosaic development, refers to a pattern of differentiation in which the fate of each cell is determined at an early stage.

Moseley, Henry Gwyn Jeffreys (1887-1915)
British physicist who showed that an ELEMENT'S properties depend on what he called its atomic number (see ATOM), equivalent to its nuclear charge.

Moss animals
Another name for the sea mats or BRYOZOA.

Mössbauer, Rudolf Ludwig (b. 1929)
German-born US scientist awarded the 1961 Nobel Prize for Physics for his research into the resonance absorption of gamma radiation and his discovery of the related effect that bears his name.

Mössbauer effect
The recoilless emission of GAMMA RAYS from certain CRYSTALS, discovered by RUDOLF LUDWIG MÖSSBAUER in 1957. When gamma rays are emitted from most nuclei, the latter recoil to a variable extent, giving the emitted PHOTONS a broad ENERGY spectrum. Mössbauer found that certain crystals, eg Fe^{57}, recoiled as a whole, ie their effective recoil was negligible. Gamma rays of closely specified frequency are thus produced and can be used for nuclear clocks and for testing RELATIVITY theory predictions.

Motile
A term used to describe an animal or microorganism capable of movement.

Motion
Change in the aspect of one body relative to another by translation, rotation or revolution, or by combinations of these. (See also MECHANICS; PERPETUAL MOTION; RELATIVITY; VELOCITY.)

Motion sickness
Nausea and vomiting caused by rhythmic movements of the body, particularly the head, set up in automobile, train, ship or airplane travel. Not all people are equally susceptible.

Motor, Electric
A device converting electrical into mechanical energy. Traditional forms are based on the FORCE experienced by a current-carrying wire in a magnetic field (see ELECTROMAGNETISM). Motors

can be, and sometimes are, run in reverse as GENERATORS.

Simple direct-current (see ELECTRICITY) motors consist of a magnet or electromagnet (the stator) and a coil (the rotor) which turns when a current is passed through it because of the force between the current and the stator field. So that the force keeps the same sense as the rotor turns, the current to the rotor is supplied via a commutator – a slip ring broken into two semicircular parts, to each of which one end of the coil is connected, so that the current direction is reversed twice each revolution.

For use with alternating-current supplies, small DC motors are often still suitable, but induction motors are preferred for heavier duty. In the simplest of these, there is no electrical contact with the rotor, which consists of a cylindrical array of copper bars welded to end rings. The stator field, generated by more than one set of coils, is made to rotate at the supply frequency, inducing (see INDUCTION, ELECTROMAGNETIC) currents in the rotor when (under load) it rotates more slowly, these in turn producing a force accelerating the rotor. Greater control of the motor speed and torque can be obtained in "wound rotor" types in which the currents induced in coils wound on the rotor are controlled by external resistances connected via slip-ring contacts.

In applications such as electric clocks synchronous motors, which rotate exactly in step with the supply frequency, are used. In these the rotor is usually a permanent magnet dragged round by the rotating stator field, the induction-motor principle being used to start the motor.

The above designs can all be opened out to form linear motors producing a lateral rather than rotational drive. The induction type is the most suitable, a plate analogous to the rotor being driven with respect to a stator generating a laterally moving field. Such motors have a wide range of possible applications, from operating sliding doors to driving trains, being much more robust than rotational drive systems, and offering no resistance to manual operation in the event of power cuts. A form of DC linear motor can be used to pump conducting liquids such as molten metals, the force being generated between a current passed through the liquid and a static magnetic field around it.

Motor neurone disease
A rare, incurable, rapidly fatal disease of the motor cells in the brain and spinal cord.

Mott, Sir Nevill Francis (b. 1905)
UK physicist who shared the 1977 Nobel Prize for Physics with P.W. ANDERSON and J.H. VAN VLECK for contributions to the understanding of the electronic structure of magnetic and disordered systems.

Mottelson, Ben (b. 1926)
US-born Danish physicist who shared the 1975 Nobel Prize for Physics with A. BOHR and L.J. RAINWATER for their work on the physics and structure of the atomic nucleus.

Mould see MOLD.

Moulting see MOLTING.

Mountain
A landmass substantially elevated above its surroundings. The difference between a mountain and a hill is essentially one of size: the exact borderline is not clearly defined. Plateaus, or table mountains, unlike most other mountains, have a large summit area as compared with that of their base. Most mountains occur in groups, ranges or chains (see also MASSIF). The processes involved in mountain-building are termed orogenesis. OROGENIES can largely be explained in terms of the theory of PLATE TECTONICS. Thus the Andes have formed where the Nazca oceanic plate is being subducted beneath (forced under) the South American continental plate, and the Himalayas have arisen at the meeting of two continental plates.

Mountains are traditionally classified as volcanic, block or folded. Volcanic mountains occur where LAVA and other debris (eg PYROCLASTIC ROCKS) build up a dome around the vent of a VOLCANO. They are found in certain well-defined belts around the world, marking plate margins. Block mountains occur where land has been uplifted between FAULTS. Folded mountains occur through deformations of the EARTH's crust (see FOLD). EROSION eventually reduces all mountains to plains. But it may also play a part in the creation of mountains, as where most of an elevated stretch of land has been eroded away leaving a few resistant outcrops of rock (see MONADNOCK).

Mountain sickness see ALTITUDE SICKNESS.

Mucous membranes
Body surfaces that produce MUCUS.

Mucus
A viscid, aqueous solution of glycoproteins secreted by many cells. In mammals mucus is produced by the mucous membranes of the respiratory and gastrointestinal tracts, the salivary and other digestive system glands. It provides a nonliving protective layer which is constantly being renewed, allowing removal of any particulate matter absorbed onto it (in the BRONCHI) and lubrication of food and feces. Mucus contains some IMMUNOGLOBULINS and has a role in local IMMUNITY.

Mulch
Layer of usually organic material kept on the surface of soil in order to reduce surface evaporation of moisture, to protect the soil from wind erosion, or as a manure. Mulches may contain straw, peat or scattered topsoil.

Müller, Alex (b. 1927)
German scientist who received the 1987 Nobel Prize for Physics with GEORG BEDNORZ for their discoveries and research relating to high-temperature superconductivity.

Muller, Hermann Joseph (1890-1967)
US geneticist awarded the 1946 Nobel Prize for Physiology or Medicine for his work showing that X-RAYS greatly accelerate MUTATION processes.

Müller, Paul Hermann (1899-1965)
Swiss chemist awarded the 1948 Nobel Prize for Physiology or Medicine for his discovery of the effective insecticidal properties of DDT (1939), a major contribution to world health and food production.

Mulliken, Robert Sanderson (1896-1986)
US chemist and physicist awarded the 1966 Nobel Prize for Chemistry for his work on the nature of chemical bonding and hence on the electronic structure of molecules (see BOND, CHEMICAL).

Multiple births
The delivery of more than one child at the end of pregnancy. Twins, the commonest type of multiple birth, are of two distinct varieties. Identical (monozygotic) twins originate in a single fertilized egg (zygote) that divides, each half (containing identical genetic material) developing independently into EMBRYO and FETUS, though they may share a common PLACENTA. Nonidentical (ie dizygotic) twins originate in the release of two eggs at ovulation (see MENSTRUATION), each being fertilized, implanting and developing separately. There is no more relation between their GENES than between those of other siblings. Higher orders of multiple births (triplets, quadruplets, quintuplets, etc) usually arise from multiple ovulation and are rare unless ovarian follicle stimulants (eg GONADOTROPHINS) have been used in the treatment of infertility; here the dosage is critical. (See also IN-VITRO FERTILIZATION.) Multiple pregnancy may run in families.

Multiple sclerosis
A relatively common disease of the BRAIN and SPINAL CORD in which the MYELIN covering nerves is destroyed in plaques of INFLAMMATION. The precise cause is unknown, but in common with many other conditions both hereditary and environmental factors are believed to play a part. It may affect any age group, but particularly young adults. Symptoms and signs indicating disease in widely separate parts of the NERVOUS SYSTEM are typical. They occur episodically, often with intervening recovery or improvement. Blurring of vision, sometimes with eye pain; double vision; vertigo; abnormal sensations in the limbs; PARALYSIS; ATAXIA; and bladder disturbance are often seen, though individually these can occur in other brain diseases. The course of the disease is extremely variable, some subjects having but a few mild attacks, while others progress rapidly to permanent disability and dependency.

Mumps
Common VIRUS infection (a form of viral PAROTITIS) causing swelling of the parotid salivary gland, and occasionally inflammation of the pancreas, an ovary or a testis. Mild fever, headache and malaise may precede the gland swelling. Rarely, a viral MENINGITIS and even less often ENCEPHALITIS complicates mumps. A bilateral and severe testicular inflammation can occasionally cause STERILITY.

Munchhausen's syndrome
Deliberate faking of symptoms by people who want to have an operation. Little is known of their motives as they refuse psychiatric investigation, but they often undergo unpleasant procedures such as amputations. Named after Baron Munchhausen, a character in a novel by Rudolf Edgar Raspe, a habitual storyteller and liar of the 18th century. (See also MEADOW'S SYNDROME.)

Muon see SUBATOMIC PARTICLES.

Murphy, John Benjamin (1857-1916)
US surgeon who pioneered the use of immediate appendectomy (the surgical removal of the appendix) as a treatment for APPENDICITIS.

Murphy, William Parry (b. 1892)
US physician who shared the 1934 Nobel Prize for Physiology or Medicine with G. MINOT and G. WHIPPLE for his work with Minot showing that daily consumption of large quantities of raw liver is an effective treatment for pernicious ANEMIA.

Muscle
Contractile tissue that produces body movement. In humans and other vertebrates there are three types of muscle. Skeletal or striated muscle is the type normally associated with the movement of the body. Its action can either be initiated voluntarily, through the central NERVOUS SYSTEM, or it can respond to REFLEX mechanisms. Under the microscope this muscle is seen to be striped or striated. The stripes are produced by the overlapping bundles of actin and myosin molecules. These two proteins slide over each other, using energy supplied by ATP, and thus produce muscle contraction. If a muscle is starved of oxygen, a process termed GLYCOLYSIS provides the ATP. However, glycolysis involves lactic acid production, and the rapid buildup of lactic acid soon inhibits further muscle action.

Skeletal muscle functions by being attached via TENDONS to two parts of the SKELETON that move relative to each other. The larger attachment is known as the muscle's origin. Contraction of the muscle attempts to draw together the two parts of the skeleton. Muscles are arranged in antagonistic groups so that all movements involve the contraction of some muscles and the simultaneous relaxation of their antagonists. Smooth or involuntary muscle is under the control of the autonomic nervous system, and we are rarely aware of its action. Smooth muscle fibers are situated in hollow structures such as the gut, BRONCHI, uterus and blood vessels. Smooth muscle uses the property of "tone" (continual slight tension) to regulate the diameter of tubes such as blood vessels. Being responsive to HORMONES, notably EPINEPHRINE (adrenaline), it can thus decrease blood supply to nonessential organs during periods of stress. In the gut, smooth muscle propels the gut contents along by contracting along its length in waves (PERISTALSIS). Cardiac muscle, found only in the HEART, has the property of never resting throughout life. It combines features of both skeletal and smooth muscle, for it is striped yet involuntary. The fibers are branching and interlinked, not discrete, enabling it to act quickly and in unison when stimulated. (♦♦ page 40, 216)

Muscular dystrophy

A group of inherited diseases in which MUSCLE fibers are abnormal and undergo ATROPHY. Most develop in early life or adolescence. Duchenne dystrophy occurs in males, though the genes are carried by females. It starts in early life, and some swelling (pseudohypertrophy) of calf and other muscles may be seen. A similar disease can affect females. Other types, described by muscles mainly affected, include limb-girdle and facio-scapulo-humeral dystrophies. There are many diverse variants, largely due to structural or biochemical abnormalities in muscle fibers. Myotonic dystrophy occurs in older men, causing baldness, cataracts, testis atrophy and a characteristic myotonus, in which contraction is involuntarily sustained. Muscular dystrophies usually cause weakness and wasting of muscles, particularly of those close to and in the trunk; a waddling gait and exaggerated curvature of the lower spine are typical. The muscles of respiration may be affected, with resulting pneumonia and respiratory failure; heart muscle, too, can also be affected. These two factors in particular may lead to early death in severe cases.

Muskeg

A BOG or SWAMP almost completely filled with SPHAGNUM moss; found in the far north, particularly in TUNDRA.

Mustelids

Members of the Family Mustelidae, a group of long-bodied, short-legged mammals belonging to the Order CARNIVORA. The mustelids include the weasels, stoats, ermine, martens, polecats, ferrets, skunks, badgers, wolverine and otters.

Mutation

Any change to the hereditary material, DNA. The simplest type of mutation is a point mutation, which affects just one purine or pyrimidine base. In substitution one base is replaced by another during copying (replication) of the DNA strand. Such a change may produce a change in a protein molecule (see PROTEIN SYNTHESIS), but it will not necessarily do so. The area of the chromosome where the mutation has occurred may be a region of "junk DNA" that does not produce protein molecules; or the change in the base may give another codon that translates into the same amino acid – for example, CUC, CUU, CUA and CUG all code for leucine, so a mutation in the last base of these codons will have no effect. Even if the mutation does produce a change in the amino acid sequence of a protein, this may have no observable effect because not all amino acids play a part in protein action (see NEUTRAL MUTATIONS).

Other types of point mutation include deletion of a base and the insertion of a base. Larger scale mutations such as the deletion, duplication or inversion of a whole segment of a chromosome can also occur.

Mutations occur at random, and the nature of their effects cannot be directed by the body (see LAMARCK; CENTRAL DOGMA). The rate of mutation is increased by exposure to radiation and certain chemicals, accounting for the ability of these agents to produce cancers and birth defects. Most mutations, when passed on to the offspring, prove lethal, but a small minority improve or alter the performance of the organism in some way. These mutations form a reservoir of genetic VARIATION upon which NATURAL SELECTION can act.

Mute see DUMBNESS.

Mutualism

A kind of SYMBIOSIS in which both organisms benefit from the relationship. One example involves the RUMINANTS, which are able to digest the material they eat thanks to cellulose-digesting bacteria that live within their gut. The bacteria, for their part, live in a protected and constant environment.

Myasthenia gravis

A disease of the junctions between the peripheral NERVOUS SYSTEM and the MUSCLES, probably due to abnormal IMMUNITY, and characterized by the fatigability of muscles. It commonly affects eye muscles, leading to drooping lids and double vision, but it may involve limb muscles. Weakness of the muscles of respiration, swallowing and coughing may lead to respiratory failure and pneumonia. Speech is nasal, regurgitation into the nose may occur and the face is weak, lending a characteristic snarl to the mouth. It is associated with disorders of the THYMUS GLAND and THYROID GLAND.

Mycelium

The mass of intermeshed HYPHAE that permeate soil, rotting wood, or other substrate, and that constitute the "body" of a fungus. Fruiting bodies, such as toadstools and puffballs, grow from the mycelium, and are themselves composed of tightly packed hyphae.

Mycology

The scientific study of FUNGI.

Mycoplasmas

Minute organisms intermediate in size between BACTERIA and VIRUSES. They are believed to be degenerate bacteria, and lack the typical bacterial cell wall. They are the smallest living entities capable of an independent existence. Some live in soil or sewage, others live inside host organisms either as parasites or as commensals. Mycoplasmas can cause diseases in humans, animals and plants.

Mycorrhiza

An association between the roots of vascular plants and a fungus. Mycorrhizas are found in the majority of plant species, not just in forest trees, heath plants and orchids, as was once thought. Some plants grow very poorly in the absence of the their fungal partner. The relationship appears to be mutualistic, that is, of benefit to both plant and fungus. There are two main types of mycorrhiza, the ectotrophic and endotrophic. In the ectotrophic association the fungus forms a sheath around the roots, and in the endotrophic the fungus penetrates the cells within the cortex of the roots. (See SYMBIOSIS.)

Myelin

Specialized layer of membranes, up to a hundred membranes thick, formed by the Schwann cells of the peripheral NERVOUS SYSTEM. These cells wrap themselves around the axons of motor neurons (nerve cells carrying impulses to the muscles), producing an electrically insulating sheath, interrupted at intervals, which facilitates rapid nerve impulse conduction. Myelin sheaths are found only in vertebrates, and only on the larger-diameter axons. Myelin disorders include MULTIPLE SCLEROSIS. (♦ page 126)

Myeloma

Tumor of bone-marrow cells that tends to appear in several places at once. Some are harmless but others progressive, resembling leukemia.

Myocarditis see MYOCARDIUM.

Myocardium

The membrane lining the heart and forming its valves. It can become inflamed or suffer an INFARCT.

Myoglobin

A molecule related to HEMOGLOBIN that is found in MUSCLE cells and serves as a local store for oxygen. Its affinity for oxygen encourages the transfer of oxygen to it from the blood.

Myopia

Near- or shortsightedness, a defect of VISION in which light entering the EYE from distant objects is brought to a focus in front of the retina, making them appear blurred.

Myositis

Rare inflammation of muscle.

Myriapods

A group of invertebrates, the centipedes and millipedes. All members are terrestrial and have a single pair of antennae, a trunk composed of leg-bearing segments, and a long tubular gut. Zoologists classify the 11,000 species into four separate classes: Chilopoda (centipedes), Diplopoda (millipedes), Pauropoda and Symphyla. The Chilopoda have flattened bodies with one pair of legs per segment and poison claws; they are fast-moving carnivores. The Diplopoda have cylindrical bodies, usually with two legs per segment. The legs are much shorter than those of the centipedes, as they are adapted for burrowing. The last two groups are both made up of very small, rarely seen animals living in soil or under rotting logs. The myriapods are believed to be related to the insects, and are often classified with them in the Uniramia on the basis of their unbranched limbs. (See also ARTHROPODS.)

Myxedema

Failure of the THYROID gland in adult life; it causes skin coarsening, loss of hair and mental dullness.

Myxophyta see SLIME MOLDS.

Nadir
The point on the CELESTIAL SPHERE directly opposite an observer's ZENITH.

Nails
A keratinous covering (see KERATIN) protecting the tips of the digits in primates, including humans. They are derived from CLAWS.

Napalm
A SOAP consisting of the aluminum salt of a mixture of CARBOXYLIC ACIDS, with aluminum hydroxide in excess. When about 10% is added to GASOLINE it forms a GEL, also called napalm, used in flame throwers and incendiary bombs; it burns hotly and relatively slowly, and sticks to its target. Developed in WWII, it was used in the Vietnam War and caused great havoc.

Naphtha
Volatile mixture of liquid HYDROCARBONS boiling in the range 80°C to 180°C, used as a solvent and as an industrial chemical feedstock. It is obtained by distilling COAL TAR (yielding aromatic products) or shale oil, or from the refining and cracking of PETROLEUM.

Naphthalene ($C_{10}H_8$)
White crystalline solid, an AROMATIC HYDROCARBON consisting of two fused BENZENE rings, and more reactive than benzene. It is produced from COAL TAR, and is used to make phthalic anhydride (see ACID ANHYDRIDES), as an intermediate in DYE manufacture, and for mothballs. MW 128.2, mp 80.3°C, bp 218°C.

Narcissism
Exaggerated self-love, often at a sexual level, characteristic of early sexual development though sometimes continuing into adult life.

Narcolepsy
An inborn condition that makes a person drowsy, likely to fall asleep without warning, and have hallucinations on waking.

Narcotics
DRUGS that induce sleep; specifically, the OPIUM-derived ANALGESICS. These affect the higher brain centers, causing mild euphoria and sleep (narcosis). They may act as HALLUCINOGENIC DRUGS and are abused in DRUG ADDICTION.

Nathans, Daniel (b. 1928)
US scientist who shared the 1978 Nobel Prize for Physiology or Medicine with W. ARBER and H.O. SMITH for their work in MOLECULAR GENETICS.

National Oceanic and Atmospheric Administration (NOAA)
US government agency set up in 1970 to coordinate scientific research into atmosphere and oceans. Its specific aims are the monitoring and control of POLLUTION and the investigation of potential resources and weather control techniques. The NOAA is responsible for the work of several formerly independent agencies including the Coast and Geodetic Survey (founded in 1807) and the Weather Bureau (founded in 1870).

Natta, Giulio (1903-1979)
Italian chemist awarded (with ZIEGLER) the 1963 Nobel Prize for Chemistry for his synthesis of POLYMERS of propene (one of the ALKENES). These have industrially desirable properties such as high melting point and strength.

Natural gas
Mixture of gaseous HYDROCARBONS occurring in reservoirs of porous rock (commonly sand or sandstone) capped by impervious strata. It is often associated with PETROLEUM, with which it has a common origin in the decomposition of organic matter in sedimentary deposits. Natural gas consists largely of METHANE and ETHANE, with propane and butane (separated for bottled gas), some higher ALKANES (used for GASOLINE), nitrogen, oxygen, carbon dioxide, hydrogen sulfide, and sometimes valuable HELIUM. It is used as an industrial and domestic fuel, and also to make carbon-black and in chemical synthesis. Natural gas is transported by large pipelines or (as a liquid) in refrigerated tankers.

Natural resources, Conservation of see CONSERVATION; RECYCLING.

Natural selection
A process that produces ADAPTATION in living organisms, and is probably responsible for most evolutionary change. The idea of natural selection was first expounded by CHARLES DARWIN, who saw it as analogous to the artificial selection practiced by plant and animal breeders, who bred only from those individuals showing desirable characteristics. Natural selection occurs because all organisms produce more offspring than can possibly survive, and some must die or fail to breed. Although there will be an element of luck involved, on average it is the ones best suited to their environment that survive longest and leave most offspring. Thus their genes will become more numerous in subsequent generations, and the overall features of the population will change.

There are three main prerequisites for natural selection to occur: variability within the population, inheritance of the characteristics that improve survival rates, and production of more offspring than can survive. Selection may be directional, producing a steady change in one or more characteristics, or stabilizing, maintaining the present type. Directional selection generally occurs when the environment – either the physical environment or other organisms – is changing or when the species is spreading into new areas.

Natural selection has been simulated in the laboratory many times, and has also been demonstrated in the wild, as in the evolution of INDUSTRIAL MELANISM. Darwin believed it to be the major source of evolutionary change, though he proposed SEXUAL SELECTION as an additional force. More recently, some biologists have questioned the ability of natural selection to produce new taxonomic groups beyond the species level, and have suggested other mechanisms. None of these has any supporting evidence, however, and most biologists remain satisfied with natural selection as the main driving force of evolution.

Nautical mile see MILE.

Navel see UMBILICAL CORD.

Navigation
The art and science of directing a vessel from one place to another. Originally navigation applied only to marine vessels, but now air navigation and, increasingly, space navigation are also important. Although the techniques and applications of navigation have radically changed through time, the basic problems, and hence the principles, have remained much the same.

MARINE NAVIGATION Primitive sailors could not venture out of sight of land without the risk of getting lost. But soon they learned to use sunset and sunrise, the prevailing winds, the POLE STAR and so forth as aids to direction. Early on, the first fathometer, a weighted rope used to measure depth, was developed. Before the 10th century AD the magnetic COMPASS had appeared. But it was not until the 1730s, with the invention of the SEXTANT and the CHRONOMETER, that both LATITUDE AND LONGITUDE could be determined. (See also ASTROLABE; GREENWICH OBSERVATORY.)

Modern navigation uses electronic aids such as LORAN and the radiocompass; celestial navigation, the determination of position by sightings of celestial bodies; and dead reckoning where, by knowing one's position at a particular past time, the time that has elapsed since, the direction and speed, the present position can be estimated. (See also DIRECTION FINDER; MAP; SONAR.)

AIR NAVIGATION uses many of the principles of marine navigation. In addition, the pilot must work in a third dimension, using his altimeter to know his altitude, and in bad visibility must use aids like the Instrument Landing System (ILS). RADAR is also used.

SPACE NAVIGATION is a science in its infancy. Like air navigation, it works in three dimensions, but the problems are exacerbated by the motions both of one's source (the Earth) and one's destination, as well as by the distances involved. But, prior to developments in new areas, it seems that SPACE EXPLORATION has inaugurated a new era in navigation by the stars. (See also CELESTIAL SPHERE.)

Navigation in animals
The direction-finding ability of various animals, from bacteria to birds. Simple forms of navigation include following a set route about familiar territory, moving toward a landmark – which may be a chemical, visual or audible signal – and pursuing a particular compass direction. Setting the compass direction is often achieved by sensing the Earth's magnetic field or by means of the Moon or stars – migrating birds in the N Hemisphere fly away from the Pole Star (which they recognize by its stable position in the sky) to go south. A more complex, and less well understood, form of navigation allows certain birds to find their destination even though they are blown off course during migration. In experiments, such species can return to their nests after being removed to an unfamiliar location many thousands of miles away. This ability is assumed to be based on noting the position of the Sun in the sky and comparing it with the remembered position of the sun at that time of day at the nest site. This would require phenomenal powers of computation and a highly accurate BIOLOGICAL CLOCK, and considerable doubt remains about whether it is possible.

NDT
Non-destructive testing; the inspection of materials or machines without destroying or removing parts.

Neanderthal man see PREHISTORIC MAN.

Nearctic
A region in ZOOGEOGRAPHY.

Nearsightedness see MYOPIA.

Nebula
An interstellar cloud of gas or dust. The term was originally used to denote any fuzzy celestial object, including external GALAXIES: this practice has now been abandoned.

Emission nebulae are gas clouds that shine because they absorb ultraviolet radiation from very hot, highly luminous stars embedded within them. They are also known as HII regions because their hydrogen gas is ionized by the illuminating stars. Reflection nebulae are seen when light from a high-luminosity star is reflected, or scattered, from nearby dust. Dark nebulae, such as the Coal Sack or the Horsehead nebula, are due to relatively dense clouds of dust that obscure background stars; thus they show up as dark patches against the starry background. Star formation takes place within interstellar clouds of gas and dust. Heated cocoons of dust surrounding newly forming stars can be detected by infrared observations.

Planetary nebulae are shells of gas expelled from stars that are approaching their final stages of evolution. The central star is often visible and is usually a small, very hot object with a surface temperature of about 100,000K, probably representing the exposed core of a RED GIANT that, after expelling its outer envelope of gas, is evolving to become a WHITE DWARF. The Ring Nebula (M57) is an outstanding example.

Nebular hypothesis
Theory accounting for the origin of the Solar System put forward by LAPLACE. It suggested that a rotating NEBULA had formed gaseous rings that condensed into planets and moons, the nebula's nucleus forming the Sun.

Necrosis
The death of body cells, tissues or organs as a result of disease. (See DEATH.)

Nectar
A sweet viscous secretion containing from 55% to 80% SUGAR, produced by the flowers of higher plants. By attracting insects it facilitates POLLINATION. Nectars with over 15% sugar are used by honey bees to make honey. Several plants produce sugary exudates from their leaves, probably for excretion. These, too, are sometimes referred to as nectar, and it is probable that the nectar of flowers evolved from such exudates.

Néel, Louis Eugène Félix (b. 1904)
French physicist awarded (with ALFVÉN) the 1970 Nobel Prize for Physics for his work on the magnetic properties of solids. His researches permitted manufacture of products used in, eg, computers, and explained phenomena such as the recording by certain rocks of past geomagnetic fields (see EARTH; PALEOMAGNETISM).

Nekton
The fast-moving animals of the open sea, such as squid, cuttlefish, pelagic fish and whales. The term is used in contrast to PLANKTON, a pelagic assemblage of small animals with limited locomotor ability that float passively, usually at the ocean's surface.

Nematodes
A large phylum of unsegmented worms that are almost ubiquitous in the ecosystem. They are also known as eelworms, threadworms or roundworms, and many are parasites, or pests of crop-plants. A tough, multi-layered cuticle covers the body and makes them highly resistant to desiccation, injury or toxins. (♦ page 16)

Nemerteans
A phylum of soft-bodied, unsegmented, worm-like animals known as ribbonworms or proboscis worms. Nearly 600 species are known; most are marine, but some are freshwater, and a few live in the soil. Most species grow to a length of 20mm but one, the bootlace worm *Lineus longissima*, attains a length of 55m.

Neodymium (Nd)
One of the LANTHANUM SERIES of elements. AW 144.2, mp 1024°C, bp 3100°C, sg 6.80 (20°C). (♦ page 130)

Neon (Ne)
One of the NOBLE GASES. It is used in discharge tubes, low-wattage glow lamps, SPARK CHAMBERS, and in helium-neon LASERS. Liquid neon is used as a refrigerant in the range 25-40K (see CRYOGENICS), and is added to liquid-hydrogen BUBBLE CHAMBERS to provide a better particle target. AW 20.2, mp −248.6°C, bp −246°C. Neon, the earliest of the noble gases to be used in discharge tubes, glows orange when excited. (♦ page 130)

Neopilina
A member of the MOLLUSCA that was thought to have been extinct by the end of the Devonian period, 345 million years ago, but was discovered alive in 1952. Unlike other mollusks, it is segmented, and it has been suggested that this indicates a segmented ancestor for the whole group. But there is doubt about whether segmentation in *Neopilina* is a primitive trait or one that has been acquired secondarily. Some details of its body plan suggest the latter. Study of *Neopilina* larvae could answer this question, but unfortunately the animal lives in very deep water and larval forms have never been seen.

Neoplasm
New cell growth (tumor) that is cancerous.

Neoteny see PEDOMORPHOSIS.

Nephrite see JADE.

Nephritis
Inflammation of the KIDNEYS. When due to bacteria ascending from the bladder it affects the center of the kidney first: this is pyelonephritis. Sometimes, for unknown reasons, the glomeruli are infected; this is glomerulonephritis. Both may be acute or chronic, and both are serious.

Nephron see KIDNEYS

Neptune
The fourth largest planet in the SOLAR SYSTEM and the eighth in position from the Sun, with a mean solar distance of 30.07AU. Neptune was first discovered in 1846 by J.G. Galle using computations by LEVERRIER based on the perturbations of URANUS' orbit. The calculation had been performed independently by JOHN COUCH ADAMS in England, but vacillations on the part of the then Astronomer Royal had precluded a rigorous search for the planet. Neptune has two moons, Triton and Nereid, the former having a circular, retrograde orbit (see RETROGRADE MOTION), the latter having the most eccentric orbit of any moon in the Solar System. Neptune's "year" is 164.8 times that of the Earth, its day being about 18h. Its diameter is about 50 Mm and its mass 17.2 times that of the Earth. Observations when Neptune passed in front of a star in 1984 suggest that Neptune too may have rings (see SATURN), but this is not confirmed. Like Jupiter and Saturn, Neptune emits more heat than it receives from the Sun. Its internal structure is believed to comprise a rocky metallic core surrounded by an ice mantle and a deep, hydrogen-rich atmosphere. The VOYAGER spaceprobe will no doubt reveal more of Neptune's structure and constitution on its 1989 fly-by. (♦♦ page 134, 221, 240)

Neptunism
Late 18th-century geological theory propagated by the school of A.G. WERNER in which it was claimed that the rocks originally forming the crust of the Earth had been precipitated out of aqueous solution.

Neptunium (Np)
The first TRANSURANIUM ELEMENT; one of the ACTINIDES. It is produced in breeder NUCLEAR REACTORS as a by-product of PLUTONIUM production by neutron irradiation of URANIUM (U^{238}). The most stable isotope is Np^{237} (half-life 2.2×10^6yr). Chemically neptunium resembles uranium. (For the Neptunium Series see RADIOACTIVITY.) mp 640°C, bp 3900°C, sg 20.45 (α). (♦ page 130)

Nernst, Walther Hermann (1864-1941)
German physical chemist awarded the 1920 Nobel Prize for Chemistry for his discovery of the Third Law of THERMODYNAMICS.

Nervous breakdown
Popular term used to describe various kinds of acute MENTAL ILLNESS, often associated with fatigue or emotional stress, which drastically impair a person's normal efficiency and disturb social behavior.

Nervous system
The system of tissues that conducts electrical impulses around the body to coordinate responses to external stimuli, initiate movement and process sensory information. Its responses are generally rapid, whereas those of the endocrine system with which it shares its coordinating and integrating function are generally slow (see GLANDS; HORMONES).

The nervous system can be divided into two parts. The central nervous system (CNS), consisting of BRAIN and SPINAL CORD, stores and processes information and sends messages to muscles and glands. The peripheral nervous system, consisting of 12 pairs of cranial nerves arising in and near the medulla oblongata of the brain and 31 pairs of spinal nerves arising at intervals from the spinal cord, carries messages to and from the central nervous system to the various parts of the body.

The nervous system can also be divided up in another way – into the part over which voluntary control can be exercised, and the part that is beyond such control. The latter is known as the autonomic nervous system, and it controls involuntary actions such as heartbeat and digestion. It consists of two complementary parts: the sympathetic system comes into play in emergencies, and prepares the body for "fight or flight", whereas the parasympathetic system controls the body's normal day-to-day functions. Most internal organs are innervated by both parts.

The nervous system's basic anatomical and functional unit is the highly specialized nerve cell or NEURON, the shape of which varies greatly in different regions. It possesses two kinds of processes: dendrites, which together with the cell body receive impulses from other neurons, and an axon, which conducts impulses to other neurons. Axons vary greatly in length (up to a few meters) and speed of conduction (up to about 90m/s).

Sensory or afferent neurons carry information to the central nervous system from sensory receptors (such as skin receptors and muscle stretch receptors), whereas efferent neurons carry information away from it. Efferent neurons passing to muscle are called motor neurons. Nerves are formed from many axons, both afferent and efferent, surrounded by their associated sheaths, which insulate them from each other. Axons surrounded by a fat and protein sheath, called a MYELIN sheath, conduct fastest. Just before entering the spinal cord each spinal nerve divides into a dorsal root containing afferent axons only and a ventral root containing efferent axons only.

Neurons communicate through specialized contact points or synapses which are either excitatory or inhibitory. The elaborate neural circuitry arising from synaptic contact in the central nervous system is responsible for much of behavior, from simple reflex action to complex thought and communication patterns.

The nerve impulse, or action potential, is an electrical signal. An electrical potential difference of about 70mV, called the resting potential, exists between the inside and outside of the neuron due to the ionic concentration imbalance between inside and outside, and a metabolic pump moving IONS across the cell membrane. If the resting potential is reduced below a certain threshold level, as may occur when impulses are received from other neurons, an impulse is initiated. Impulses are all the same strength ("all-or-none" law), and travel to the end of the axon to the synapse where a chemical transmitter substance (see NEUROTRANSMITTER) is released, which initiates a new electrical signal in the next neuron.

In invertebrates, nervous systems are generally less centralized than in vertebrates. The simplest animals, such as sea anemones, just have a "nerve net", with no centralized control. More active invertebrates have one or two nerve cords running the length of the body, and concentrations of nerve tissue known as ganglia (see GANGLION). Those in the head region tend to be largest, and they take overall control. These cerebral ganglia are the forerunners of brains, yet the other ganglia still have considerable autonomy, an the degree of centralization is much less than in the vertebrates. The most highly developed invertebrate nervous systems are found in the flying insects and the cephalopod mollusks. (♦ page 16-17, 126)

Network
In computer science, a system of computers connected by TELECOMMUNICATIONS. (See also COMPUTER.)

Neuralgia
Pain originating in a nerve and characterized by sudden sharp, often electric shock-like pain or exacerbations of pain. Neuralgia may be due to INFLAMMATION or trauma.

Neuritis (peripheral neuropathy)
Any disorder of the peripheral NERVOUS SYSTEM that interferes with sensation, the nerve control of MUSCLE, or both. Its causes include DRUGS and heavy metals (eg gold); infection or allergic reaction to it (as with LEPROSY or DIPHTHERIA); inflammatory disease (rheumatoid ARTHRITIS); infiltration, systemic and metabolic disease (eg DIABETES or PORPHYRIA); VITAMIN deficiency (BERIBERI); organ failure (eg of the LIVER or KIDNEY); genetic disorders, and the nonmetastatic effects of distant CANCER. Numbness, tingling, weakness and PARALYSIS result, at first affecting the extremities. Diagnosis involves electrical studies of the nerves and nerve BIOPSY.

Neurofibromatosis
A partly hereditary condition of variable form that produces pale "liver" spots on the skin and lumpy nodules on nerves. Occasionally the upper jaw is also affected.

Neurology
Branch of medicine concerned with diseases of the BRAIN, SPINAL CORD and peripheral NERVOUS SYSTEM.

Neuron (nerve cell)
The basic unit of the NERVOUS SYSTEM (including the BRAIN and SPINAL CORD). Each has a long AXON, specialized for transmitting electrical impulses and releasing chemical transmitters (NEUROTRANSMITTERS) at the SYNAPSE, which lies at the end of the axon. These synapses act on MUSCLE or effector cells or other neurons. The neuron itself is controlled by other neurons that have synapses acting on its cell body. The sum of the effects of these neurons – some stimulatory, some inhibitory – determines whether the neuron fires or not. (♦ page 126)

Neuropathy
Any disease of a nerve.

Neurosis
Any of several mild mental disorders such as anxiety or depression that are an overreaction to stress, rather than insanity. Actual neurosis is based in disorders of current sexual behavior; psychoneurosis is rooted in the past life; anxiety neurosis is characterized by exaggerated ANXIETY. (See OBSESSIONAL NEUROSIS.)

Neurosurgery see SURGERY.

Neurotransmitter
A substance that plays an important part in the transmission of nerve impulses within the NERVOUS SYSTEM. On release from the end of one nerve cell, it diffuses across the SYNAPSE and stimulates the adjoining nerve cell before being rapidly broken down by enzymes. The neurotransmitter binds to the proteins that make up the ion channels in the second cell (postsynaptic cell), and this affects the passage of ions into and out of the cell. (♦ page 111)

Neutralization
A chemical reaction between an ACID and a BASE to give a SALT.

Neutral mutations (silent mutations)
Changes in the genetic material (see MUTATIONS) that do not manifest themselves in the PHENOTYPE of the organism. This may be because they occur in regions of the DNA that are not transcribed, or because they change one triplet to another that codes for the same amino acid in the GENETIC CODE. Even if the mutation does produce a change in the amino acid sequence of a protein this may not have any observable effect, as large parts of each protein molecule have a structural role only, and many amino acids in these parts can be altered without affecting the protein's function. Neutral mutations tend to accumulate because they are not acted on by NATURAL SELECTION. The rate at which they accumulate is fairly constant for each particular protein, and differences between homologous proteins of two species can be used to determine the time since those two species diverged. Such measurements are now an important part of TAXONOMY.

Neutrino see SUBATOMIC PARTICLES.

Neutron (n)
Uncharged SUBATOMIC PARTICLE with rest mass 1.6748×10^{-27}kg (slightly greater than that of the PROTON) and SPIN ½. A free neutron is unstable, decaying to a proton, an ELECTRON and an antineutrino with HALF-LIFE 925s:

$$n \rightarrow p^{+} + e^{-} + \bar{\nu}$$

But neutrons bound within the nucleus of an ATOM are stable. All nuclei apart from hydrogen contain neutrons, which contribute to the nuclear cohesive forces and separate the mutually repulsive protons. Free neutrons are produced in many nuclear reactions, including nuclear FISSION, and hence nuclear reactors and particle ACCELERATORS are used as sources. The neutron was discovered in 1932 by SIR JAMES CHADWICK who bombarded beryllium with ALPHA PARTICLES emitted by a radioisotope. Neutrons are highly penetrating, and are moderated (slowed down) by colliding with the nuclei of light atoms. They induce certain heavy atoms to undergo fission. Shielding requires thick concrete walls. Neutrons may be detected by counting the ionizing particles or GAMMA RAYS produced when they react with nuclei. Neutrons have wave properties, and their DIFFRACTION is used to study crystal structures and magnetic properties. (See also CROSS-SECTION, NUCLEAR.) (♦♦ page 49, 147)

Névé
Compacted snow that lasts from year to year, representing one of the earliest stages in the development of a GLACIER. Further compaction, until there is no air left in the SNOW, results in the formation of FIRN.

Nevus see BIRTHMARK.

Newton, Sir Isaac (1642-1727)
The most prestigious natural philosopher and mathematician of modern times, the discoverer of the CALCULUS and author of the theory of universal GRAVITATION. Newton went to Trinity College, Cambridge, in 1661, retiring to Lincolnshire during the Plague of 1665-66, but returning to Cambridge as a Fellow in 1667 and accepting the Lucasian Chair of Mathematics in 1669. He was elected Fellow of the ROYAL SOCIETY in 1672, on the strength of his optical discoveries. In Cambridge, Newton spent much time in alchemical experiments. Toward the end of the century he tired of academic life, and in 1669 became Master of the Royal Mint. He resigned his chair and entered Parliament in 1701, and two years later began his presidency of the Royal Society, which he retained until his death. His achievements were legion: the method of fluxions and fluents (calculus); the theory of universal gravitation and his derivation of KEPLER'S LAWS; his formulation of the concept of FORCE as expressed in his three laws of motion (see MECHANICS); the corpuscular theory of LIGHT, and the binomial theorem, among many others. These were summed up in his two greatest works: *Philosophiae Naturalis Principia Mathematica* (1687) – the "Principia", which established the mathematical representation of nature as the paradigm of what counted as "science" – and *Opticks* (1704). Newton's often bitter controversies with his fellow scientists (notably HOOKE and Leibniz) are famous, but his influence is undoubted, even if, in the cases of optical theory and the Newtonian calculus notation, it retarded rather than accelerated the advance of British science. (♦ page 237)

Newton's rings see INTERFERENCE.

Niacin see VITAMINS.

Niche
In ecology, the total environment exploited by a particular species, including its food supply, nest site and other resources utilized. It is a maxim of ecology that no two species can occupy exactly the same niche – if they did, one would inevitably do better than the other, which would eventually become extinct.

Nicholas of Cusa (1401-1464)
German cardinal best known for his advanced cosmological views: he held that the Earth rotates on its axis, that space is infinite, and that the Sun is a star like other stars (see also ASTRONOMY). He also suggested the use of concave lenses for the shortsighted (see LENS).

Nickel (Ni)
Hard, gray-white, ferromagnetic (see MAGNETISM) metal in Group VIII of the PERIODIC TABLE; a TRANSITION ELEMENT. Roasting the ore gives crude nickel oxide, refined by electrolysis or by the Mond process (see MOND, LUDWIG). Nickel is widely used in ALLOYS, including MONEL METAL, INVAR and GERMAN SILVER. In many countries "silver" coins are made from cupronickel (an alloy of copper and nickel). Nickel-chromium alloys ("nichrome"), resistant to oxidation at high temperatures, are used as heating elements in electric fires, etc. Nickel is used for nickel plating and as a catalyst for HYDROGENATION. Chemically nickel resembles IRON and COBALT, being moderately reactive, and forming compounds in the +2 oxidation state; the +4 state is known in LIGAND complexes. AW 58.7, mp 1455°C, bp 2900°C, sg 8902 (25°C). (♦ page 130)

Nickel silver see GERMAN SILVER.

Nicolle, Charles Jules Henri (1866-1936)
French bacteriologist awarded the 1928 Nobel Prize for Physiology or Medicine for his discovery that body lice are main transmitters of TYPHUS.

Nicol prism
Optical device for producing a beam of plane-POLARIZED LIGHT. Two pieces of CALCITE crystal are cemented together with Canada balsam (which has a refractive index similar to that of glass). Incident light is split into ordinary and extraordinary linearly polarized rays in the prism. The ordinary ray hits the balsam layer obliquely and is totally internally reflected; the other ray emerges plane-polarized for a certain range of incidence angles.

Nicotine
Colorless oily liquid, an ALKALOID occurring in tobacco leaves and extracted from tobacco refuse. It is used as an insecticide and to make nicotinic acid (see VITAMINS). Nicotine is highly toxic.

Nictitating membrane
The third eyelid, a membrane found in many vertebrates that can be drawn horizontally across the eye to lubricate and clean the eyeball. It is

present in a few mammals, eg cats, and in humans is represented by the pink triangle of flesh in the corner of the eye.

Niépce, Joseph Nicéphore (1765-1833)
French inventor who in 1826 made the first successful permanent photograph. The image, recorded in asphalt on a pewter plate, required an 8hr exposure in a camera obscura (see CAMERA LUCIDA AND CAMERA OBSCURA). The method derived from heliography, Niépce's photoengraving process. In 1829 Niépce went into partnership with DAGUERRE.

Night blindness (nyctalopia)
Inability of the eyes to accommodate in or adapt to darkness. It may be a hereditary defect or an early symptom of VITAMIN A deficiency in adults.

Niobium (columbium, Nb)
A soft, silvery white metal in Group VB of the PERIODIC TABLE; a TRANSITION ELEMENT. It occurs as COLUMBITE and pyrochlore. Niobium is unreactive and corrosion resistant, and is used in STEELS, high-temperature ALLOYS and superconducting alloys (see SUPERCONDUCTIVITY). Being permeable to neutrons, it is also used in nuclear reactors. At high temperatures it reacts with nonmetals to give pentavalent covalent compounds. AW 92.9, mp 2477°C, bp 4900°C, sg 8.57 (20°C). (♦ page 130)

Nirenberg, Marshall Warren (b. 1927)
US biochemist who shared with HOLLEY and KHORANA the 1968 Nobel Prize for Physiology or Medicine for his major contributions toward the decipherment of the genetic code (see GENETICS).

Nitrates
Salts of NITRIC ACID, containing the nitrate ion (NO_3^-). Almost all nitrates are soluble in water, and only SODIUM and POTASSIUM nitrate occur significantly in nature, others being made by the action of nitric acid on the metal or its salts. Nitrates are used in explosives, fertilizers and fireworks. (See also NITROGEN CYCLE; NITROGEN FIXATION.) ESTERS of nitric acid ($RONO_2$) are also called nitrates.

Nitrates derived from fertilizers are now a serious source of pollution worldwide. As much as a third of the quantity applied can leach from the soil in rainwater and enter lakes, rivers, reser voirs and groundwater. In lakes, nitrates cause EUTROPHICATION, damaging plant and animal life. Levels in drinking water have been steadily rising and are a threat to health, particularly for babies. There is a possible link between nitrates and stomach cancer.

Nitration
Important process in ORGANIC CHEMISTRY, in which a nitro group ($-NO_2$) is introduced into a compound. AROMATIC COMPOUNDS are nitrated with a mixture of concentrated SULFURIC and NITRIC ACIDS which contains the electrophile (see NUCLEOPHILES) NO_2^+; the production of TNT and nitrobenzene are important cases. Nitration is also used for the formation of NITRATE esters including NITROCELLULOSE and NITROGLYCERIN.

Nitric acid (HNO_3)
Strong mineral ACID, a colorless, fuming liquid when pure. Nitric acid is made industrially by the oxidation of ammonia. It is a powerful oxidizing agent, reacting with most metals to give NITRATES and oxides of NITROGEN; with many organic compounds NITRATION occurs. Nitric acid is used to make nitrates, plastics, explosives and dyes, and as a rocket fuel. mp −42°C, bp 83°C.

Nitriles (organic CYANIDES)
Class of organic compounds of general formula RCN, named for the CARBOXYLIC ACID to which they can be hydrolyzed. The simplest is acetonitrile (or methyl cyanide), CH_3CN. Nitriles are prepared by dehydration of AMIDES or by reaction of sodium cyanide with ALKYL HALIDES or aryl sulfonates. They may be catalytically hydrogenated to AMINES, and react with GRIGNARD REAGENTS to yield KETONES. Acrylonitrile is used to make POLYMERS; some other nitriles are used as softening agents for rubber, etc.

Nitrocellulose
Properly called cellulose nitrate, a mixture of highly inflammable NITRATE esters of CELLULOSE made by NITRATION of cotton or wood pulp. The degree of nitration depends on the conditions used. Highly nitrated nitrocellulose – GUNCOTTON – was used as a high EXPLOSIVE and is still used as a propellant. Nitrocellulose with less than 12% nitrogen – collodion – is used in making propellants, lacquers and CELLULOID.

Nitrogen (N)
Nonmetal in Group VA of the PERIODIC TABLE; a colorless, odorless gas (N_2) comprising 78% of the ATMOSPHERE, prepared by fractional distillation of liquid air. Combined nitrogen occurs mainly as NITRATES. As a constituent of AMINO ACIDS, it is vital (see also NITROGEN CYCLE). Molecular nitrogen is inert because of the strong triple bond between the two atoms, but it will react with some elements, especially the ALKALINE-EARTH METALS, to give nitrides; with oxygen, and with hydrogen (see HABER PROCESS); it also forms N_2 LIGAND complexes with Group VIII transition metals. Activated nitrogen, formed in an electric discharge, consists of nitrogen atoms and is much more reactive. Nitrogen is used in NITROGEN FIXATION and to provide an inert atmosphere; liquid nitrogen is a CRYOGENIC refrigerant. It forms mainly trivalent and pentavalent compounds. AW 14.0, mp −210°C, bp −196°C.

NITRIC OXIDE (NO) is a colorless gas formed in the now obsolete electric-arc process; it is readily oxidized further to nitrogen dioxide. The NO molecule is unusual in having an odd number of electrons, and so gives the nitrosyl ions NO^+ and NO^-. mp −164°C, bp −152°C.

NITRITES are salts (or esters) of nitrous acid (HNO_2) and are mild reducing agents.

NITROUS OXIDE (N_2O), sometimes known as laughing gas, is a colorless gas with a sweet odor, prepared by heating ammonium nitrate, and used as a weak anesthetic, sometimes producing mild hysteria. mp −91°C, bp −88°C.

NITROGEN DIOXIDE (NO_2), a red-brown toxic gas in equilibrium with its colorless dimer (N_2O_4), is a constituent of automobile exhaust and smog. A powerful oxidizing agent, it is also an intermediate in the manufacture of NITRIC ACID. mp −11°C, bp 21°C. (See also AMMONIA; OSTWALD PROCESS; CYANAMIDE PROCESS; HYDRAZINE.) (♦ page 25, 130, 147)

Nitrogen cycle
The cycle of chemical changes in which NITROGEN is cycled between the air and the soil, and through living organisms. NITROGEN FIXATION by microorganisms turns atmospheric nitrogen into combined nitrogen as AMMONIA and NITRATES. These can be absorbed from the soil by plants, which use them to make protein. Animals ingest nitrogen compounds when they eat these plants. Urine and feces, and animal and plant remains return nitrogen to the soil as complex compounds. These are converted by fungi and bacteria to ammonium salts, which may then be oxidized to nitrites (see NITROGEN) and nitrates by other bacteria. These are either reused by plants or converted to nitrogen (by denitrifying bacteria) which returns to the air. (See also ECOLOGY.)

Humans affect the nitrogen cycle by fixing additional nitrogen in the production of FERTILIZERS. Some of this enters the nitrogen cycle as plant protein, while the rest runs off into lakes, streams and groundwater. The effect in lakes is to boost the available nitrogen for algae, causing EUTROPHICATION. The buildup of NITRATES in drinking water is another adverse effect. (♦ page 44)

Nitrogen fixation
The conversion of NITROGEN gas into nitrogen compounds. Nitrogen usually reacts only at high temperatures and pressures. Some is fixed naturally by the electrical discharges that occur during thunderstorms. Industrially, nitrogen is fixed in the HABER PROCESS, the CYANAMIDE PROCESS and the electric-arc process.

Some bacteria (*Rhizobium*) living in the root nodules of LEGUMINOSAE, and also some of those living free in the soil (eg *Azotobacter*), can fix nitrogen from the air as ammonia, nitrates and nitrites. Cyanobacteria also fix nitrogen. (See also NITROGEN CYCLE.) (♦ page 44)

Nitroglycerin ($C_3H_5(ONO_2)_3$)
Properly called glyceryl trinitrate, the NITRATE ester of GLYCEROL, made by its NITRATION. Since it causes VASODILATION, it is used to relieve ANGINA PECTORIS. Its major use, however, is as a very powerful high EXPLOSIVE, though its sensitivity to shock renders it unsafe unless used in the form of DYNAMITE or blasting gelatin. It is a colorless, oily liquid. MW 227.1, mp 13°C.

Nitrous oxide see NITROGEN.

Nits
The eggs of LICE.

Nobel, Alfred Bernhard (1833-1896)
Swedish-born inventor of dynamite and other explosives. About 1863 he set up a factory to manufacture liquid NITROGLYCERIN, but when in 1864 this blew up, killing his younger brother, Nobel set out to find safe handling methods for the substance, so discovering DYNAMITE, patented 1867 (UK) and 1868 (US). Later he invented gelignite (patented 1876) and ballistite (1888). A lifelong pacifist, he wished his explosives to be used solely for peaceful purposes, and was much embittered by their military use. He left most of his fortune for the establishment of the Nobel Foundation, and this fund has been used to award Nobel prizes since 1901.

Nobelium (No)
A TRANSURANIUM ELEMENT in the ACTINIDE series, prepared by bombardment of lighter actinides. The most stable isotope, No^{255}, has a HALFLIFE of only 3min. (♦ page 130)

Noble gases
The elements in Group O of the PERIODIC TABLE, comprising HELIUM, NEON, ARGON, KRYPTON, XENON and RADON. They are colorless, odorless gases, prepared by fractional distillation of liquid air (see ATMOSPHERE), except helium and radon. Owing to their stable filled-shell electron configurations, the noble gases are chemically unreactive: only krypton, xenon and radon form isolable compounds. They glow brightly when an electric discharge is passed through them, and so are used in advertising signs: neon tubes glow red, xenon blue, and krypton bluish white; argon tubes glow pale red at 10w pressures, blue at high pressures. (♦ page 130)

Noble metals
The unreactive, corrosion-resistant precious metals, comprising the PLATINUM GROUP, SILVER and GOLD, and sometimes RHENIUM. (♦ page 130)

Nocebo
The opposite of a placebo – a substance that is believed, due to fear or to coincidental occurrence of symptoms, to be harmful.

Noguchi, Hideyo (1876-1928)
Japanese bacteriologist best known for his work on syphilis (see SEXUALLY TRANSMITTED DISEASES), and YELLOW FEVER.

Noise
Unwanted SOUND; its precise definition is highly subjective. In air, sound is radiated spherically

from its source as a compressional wave, being part reflected, absorbed or transmitted on hitting an obstacle. Noise is usually a nonperiodic sound wave, as opposed to a periodic pure musical tone or a sine-wave combination. It is characterized by its intensity (measured in DECIBELS or nepers), frequency and spatial variation; a sound level meter and frequency analyzer measure these properties.

In electronics, noise is often used to refer to an unwanted "background" signal, particularly where it arises from the random motion of electrons due to thermal energy. This electronic noise is distributed over a wide range of frequencies, and can be reduced by making the equipment sensitive to a narrower range of frequencies close to that of the desired signal, or by cooling to reduce the thermal motion. (♦ page 225)

Nonmetal

A substance – in particular, an ELEMENT – showing none of the properties characteristic of METALS. (See also METALLOID.) The 17 or so nonmetallic elements fill the top right-hand corner of the PERIODIC TABLE. Their atoms are in general relatively small, with nearly filled electron shells, and have high IONIZATION POTENTIALS. They have high ELECTRONEGATIVITIES, and tend to form covalent BONDS with each other, and to form ANIONS. (♦ page 130)

Nonverbal communication

The gestures and signals that humans use to assist in communication, including facial expressions, hand gestures, body posture, eye contact, nods of the head etc.

Noradrenaline see ADRENALINE.

Normality see EQUIVALENT WEIGHT.

Norrish, Ronald George Wreyford (1897-1978)

British chemist awarded, with MANFRED EIGEN and GEORGE PORTER, the 1967 Nobel Prize for Chemistry for studies of extremely fast chemical reactions, and, in particular, for the development of FLASH PHOTOLYSIS.

North Atlantic drift

Eastward-moving continuation of the GULF STREAM, notable for its warming effect on the climates of W Europe.

Northern lights see AURORA.

North Pole

The point on the Earth's surface some 750km N of Greenland through which passes the Earth's axis of rotation. It does not coincide with the Earth's N Magnetic Pole, which is over 1000km away. The Pole lies roughly at the center of the Arctic Ocean, which there is permanently ice covered, and experiences days and nights each of six months. It was first reached by Robert E. Peary on 6 April 1909. (See also CELESTIAL SPHERE; MAGNETISM; SOUTH POLE.)

Northrop, John Howard (b. 1891)

US biochemist who shared with W.M. STANLEY and J.B. SUMNER the 1946 Nobel Prize for Chemistry for his work crystallizing the ENZYMES pepsin (*c*1930), trypsin (*c*1932) and chymotrypsin (*c*1935). He was also the first to isolate a BACTERIOPHAGE (1938).

North star see POLARIS.

Notochord

A rod of stiff cartilage that runs the length of the body in AMPHIOXUS, acting as a primitive type of skeleton. The notochord characterizes the Phylum CHORDATA, as all chordates possess one at some time during their life. Though replaced by cartilage or bone in the adult VERTEBRATE, and absent in the adults of other chordate groups (eg Tunicates), it is well developed in the embryos or larvae of all these groups, confirming evolutionary relationships within the class.

Nova

A star that over a short period (usually a few days) increases in brightness by 100 to 1,000,000 times. This is thought to be due to the star undergoing a partial explosion: that is to say, part of the star erupts, throwing out material at a speed greater than the ESCAPE VELOCITY of the star. A nova is believed to occur in a binary system (see DOUBLE STAR) in which one component is a WHITE DWARF (or, exceptionally, a NEUTRON STAR). Material dumped from the larger star onto the compact one eventually undergoes explosive nuclear "burning" to produce the nova outburst. The initial brightness fades rapidly, though it is usually some years before the star returns to its previous luminosity, having lost about 0.0001 of its mass. At that time a rapidly expanding planetary NEBULA may be seen to surround the star. Recurrent novae are stars that flare up at irregular periods of a few decades. Dwarf novae are subdwarf stars that suffer nova-like eruptions every few weeks or months. Novae have been observed in other galaxies besides the MILKY WAY. (See also SUPERNOVA.)

Nuclear energy

Energy released from an atomic nucleus during a nuclear reaction in which the atomic number (see ATOM), mass number or radioactivity of the nucleus changes. Nuclear energy arises from the special forces (about a million times stronger than chemical bonds) that hold the protons and neutrons together in the small volume of the atomic nucleus (see NUCLEAR PHYSICS). Lighter nuclei have roughly equal numbers of protons and neutrons, but heavier elements are only stable with a neutron:proton ratio of about 1.5:1. If one could overcome the electrostatic repulsion between protons and assemble them with neutrons to form a stable nucleus, its mass would be less than that of the constituent particles by the mass defect Δm, of the nucleus, and the binding energy, BE, given by $BE=\Delta m c^2$ (where c is the electromagnetic constant), would be released. Because c is large, a vast amount of energy would be released, even for a very small value of the mass defect. The binding energy (equivalent to the work needed to split up the nucleus into separate protons and neutrons) is always positive – nuclei are always more stable than their separate nucleons (protons or neutrons) – but is greatest for nuclei of medium mass, decreasing slightly for lighter and heavier elements. The low binding energy of very light elements means that energy can be released by combining, eg two DEUTERIUM nuclei to form a helium nucleus. This combination of two protons and two neutrons is particularly stable (see FUSION, NUCLEAR). For heavy elements the decrease in binding energy indicates that the more positively charged the nucleus becomes, the less stable it is, even though it contains more neutrons than protons. This sets a limit on the number of elements, and also explains why the nuclear-fission process, in which a heavy nucleus splits into two or more medium-mass nuclei with higher total binding energy, releases energy. The first nuclear reaction was performed experimentally in 1919 by RUTHERFORD, who exposed nitrogen to ALPHA PARTICLES (helium nuclei) from the radioactive element radium, producing oxygen and hydrogen:

$$N^{14}+He^{4}\rightarrow O^{17}+H^{1}$$

However, because nuclei are positively charged and repel each other, it was found difficult to bring them close enough together to react with each other. The discovery of the neutron in 1932 helped overcome this problem. Being uncharged and heavy (on the atomic scale), the neutron has high energy even when moving slowly, and is good for initiating nuclear reactions. By 1939 many nuclear reactions had been studied, but none seemed feasible as an energy source. Although energy might be released in a reaction, more energy was expended in producing particles able to initiate the reaction than could be recovered from it. Moreover, only a small fraction of the reagent particles would react as desired and any product particles would have little chance of reacting again. The situation was like trying to set fire to a damp forest with a box of matches. A breakthrough came in about 1939 when the violent reaction of the heavy element uranium on bombardment with slow neutrons (first observed experimentally by Fermi in 1934) was successfully interpreted. It was realized that this was an example of nuclear fission, the slow neutrons delivering enough energy to the small proportion of U^{235} nuclei in natural uranium to split them into two parts. This split does not always occur in the same way, and many radioactive fission products are formed, but each fission is accompanied by the release of much energy and two or three neutrons (these because the lighter nuclei of the fission products have a lower neutron:proton ratio than uranium). These neutrons were the key to the large-scale production of nuclear energy; they could make the uranium "burn" by setting up a chain reaction. Even allowing for the loss of some neutrons, sufficient are left to produce other fissions, each producing two or three more neutrons, and so on, leading to an explosive release of energy. The first controlled chain reaction took place in Chicago in 1942, using pure graphite as a moderator to slow down neutrons and natural uranium as fuel. Rods of neutron-absorbing material kept the reaction under control by limiting the number of neutrons available to cause fissions. The possibilities of nuclear energy as a weapon were exploited at once, and WWII ended shortly after the United States dropped two atomic bombs on Japan. Later, more powerful bombs, exploiting nuclear fusion, were developed. An increasing quantity of the world's energy is produced in NUCLEAR REACTORS from nuclear fission, though the Earth's natural supplies of fissionable material are limited. Moreover, because the fission products from these reactors are radioactive with long HALF-LIVES, atomic waste disposal is a major environmental problem. (♦ page 147)

Nuclear magnetic resonance (NMR) see SPECTROSCOPY.

Nuclear physics

The study of the physical properties and mathematical treatment of the atomic nucleus and SUBATOMIC PARTICLES. The subject was born when RUTHERFORD postulated the existence of the nucleus in 1911. The short-range exchange forces that hold together the nucleus, acting between positively charged protons and neutral neutrons, can be explained in terms of quarks and gluons, in the theory known as quantum chromodynamics (see SUBATOMIC PARTICLES). Despite the special techniques required to produce nuclear reactions, the subject has rapidly grown with the technical exploitation of NUCLEAR ENERGY. (♦ page 147)

Nuclear reactor

A device containing sufficient fissionable material, arranged so that a controlled chain reaction may be started up and maintained in it. Many types of reactor exist: all produce neutrons, gamma rays, radioactive fission products and heat, but normally use is made of only one of these. Neutrons may be used in nuclear research or for producing useful radioisotopes. Gamma rays are dangerous to human beings and must be shielded against, but have some uses (see IRRADIATION). The fragments produced by fission of a heavy nucleus have a large amount of energy and the heat they produce may be used for carrying out a variety of high-temperature processes or for heating a working fluid (such as steam) to operate a turbine and produce electricity. This is the function of most commercial reactors, although a number are used to power ships and submarines, since a small amount of nuclear fuel gives these a very long range. In an electricity-generating reactor, the fuel is normally uranium pellets surrounded by a moderator and the cooling fluid, heavy water or liquid sodium (which in turn heats the turbine fluid). There is much insulation and radiation shielding. The fuel is expensive, but produces several thousand times the heat of the same weight of coal. After some time it must be replaced (although only partly consumed) because of the buildup of neutron-absorbing fission products. This replacement, and the reprocessing of the radioactive products, needs costly remote handling equipment. New fast breeder reactors with no moderator avoid this problem, since as well as producing fission of U^{235} they convert nonfissionable U^{238} to plutonium, which also undergoes fission chain reactions – in effect they breed fuel! Research is continuing into more efficient reactors as power sources for the future.

Nucleic acids

Vital chemical constituents of living things; a class of complex threadlike molecules comprising two main types: the deoxyribonucleic acids (DNA) and the ribonucleic acids (RNA). DNA is found almost exclusively in the nucleus of the living CELL, where it forms the chief material of the CHROMOSOMES. (Small amounts of DNA are also found in the MITOCHONDRIA and CHLOROPLASTS.) It is the DNA molecule's ability to duplicate itself (replicate) that makes cell reproduction possible; and it is DNA, by directing PROTEIN SYNTHESIS, that controls HEREDITY in almost all organisms. (The notable exceptions are the RNA VIRUSES, which contain only RNA.) RNA performs several tasks connected with protein synthesis, and is found throughout the cell.

In both DNA and RNA, the backbone of the molecule is a chain of alternate phosphate and sugar (pentose) groups. To each sugar group is bonded one or other of four nitrogenous side groups known as bases. The bases are of two types, PURINES or PYRIMIDINES. Each unit consisting of a base, a sugar and a phosphate is called a NUCLEOTIDE. DNA differs chemically from RNA in that its sugar group has one less oxygen atom and one of its side groups, thymine, is replaced in RNA by uracil. DNA molecules are usually very much longer than RNA and may contain a million or so phosphate-sugar links.

It is the sequence in which the bases are arranged along the DNA molecule that constitutes stored genetic information and so makes the difference between one inherited characteristic and another. This information is in the form of coded instructions for the synthesis of particular protein molecules (see GENETIC CODE). The instructions are carried outside the cell nucleus by molecules of "messenger RNA", each incorporating a base sequence determined by DNA. The process by which mRNA is made on the DNA template is known as transcription. In the cytoplasm outside the nucleus are the RIBOSOMES, where protein assembly takes place, AMINO ACIDS, the "building blocks" of proteins, and molecules of another, smaller kind of RNA, "transfer RNA". Each of these transfer RNA molecules is able to capture an amino acid molecule of a particular type, with the aid of a special enzyme also specific for that amino acid. Once the transfer RNA has picked up its correct amino acid it can locate it in its proper place in the sequence dictated by messenger RNA. This process – the production of a polypeptide chain using mRNA as a template – is known as translation. (♦ page 13)

Nucleons see SUBATOMIC PARTICLES.

Nucleophiles

Chemical species that donate ELECTRONS, or a share in electrons, to another molecule in a substitution reaction. They are thus in general BASES, especially ANIONS, which are attracted to partially positively charged atoms in a molecule; examples are CYANIDE, HYDROXIDE, HALIDE and AMMONIA. Electrophiles are the exact opposite: species that receive electrons, or a share in electrons, from another molecule; in general ACIDS, especially CATIONS, attracted to partially negatively charged atoms in a molecule.

Nucleosides

The base-sugar component of NUCLEOTIDES.

Nucleotides

Organic chemicals of central importance in the life chemistry of all plants and animals. Some nucleotides form the basic constituents of NUCLEIC ACIDS – DNA and RNA. Another nucleotide, adenosine triphosphate (ATP), is the "energy currency" of all living organisms – it provides a means of storing and releasing the energy needed to drive biochemical processes. Other adenine nucleotides are part of three major coenzymes: coenzyme A, FAD and NAD. Finally, cyclic AMP (adenosine monophosphate or cAMP), is a vital messenger substance, with multiple roles in hormone action and many other regulatory processes.

The nucleotide molecule is a three-part structure, comprising a phosphate group linked to a 5-carbon sugar group (pentose) linked in turn to a nitrogenous side group (base). The five commonest bases are the PURINES, adenine and guanine, and the PYRIMIDINES, cytosine, thymine and uracil. Adenine, guanine, cytosine and thymine serve in the DNA molecule as the "letters" of the GENETIC CODE. In RNA, uracil replaces thymine.

Nucleus, Atomic see ATOM; SUBATOMIC PARTICLES.

Nucleus, Cell see CELL.

Nut

Botanically, a dry, indehiscent FRUIT with one seed, similar to an ACHENE but having a hard outer shell or pericarp. Nuts are produced by, for example, hazel and oak. The word is commonly used in a much broader sense to include seeds such as the peanut and brazil nut, the "stone" of DRUPES, such as the almond and walnut, and many other edible items that are not true nuts.

Nutation, Astronomical

Irregularities in the PRECESSION of the equinoxes owing to variations in the torque produced by the gravitational attractions of the Sun and Moon on the Earth.

Nutation, Mechanical

A "bobbing" superimposed on the PRECESSION of a rigid spinning body such as a GYROSCOPE. With increasing spin rate there is an increase in the frequency and a decrease in the magnitude of the nutation.

Nutrition

The means by which living organisms obtain the substances required for growth and the maintenance of life. Vital substances that cannot be synthesized within the CELL and must be present in the food are termed "essential nutrients". AUTOTROPHS, such as green plants, can derive energy from sunlight or chemical energy, and synthesize all their nutritional requirements from simple inorganic chemicals present in the soil, water and air. HETEROTROPHS, on the other hand, depend on previously synthesized organic materials obtainable only by eating living organisms.

Human nutrition involves five main groups of nutrients: PROTEINS, LIPIDS, CARBOHYDRATES, VITAMINS and minerals. Fats and carbohydrates are the body's sources of energy, and are required in relatively large amounts. They yield this energy by oxidation in the body cells, and nutritionists measure it in heat units called food CALORIES (properly called kilocalories, each equaling 1000 gram calories). Polysaccharides (long-chain carbohydrates, also known as STARCHES) normally form the most important energy source, contributing at least half the calories in a well-balanced diet. Fats, which provide about twice as many calories for a given weight as carbohydrates, consist largely of FATTY ACIDS, which divide into two main classes: saturated and unsaturated. Animal fats contain more saturated fatty acids than vegetable and fish oils. Proteins supply the remaining energy needs, but their real importance lies in the fact that the body proteins need certain essential AMINO ACIDS, found in protein foods, for growth and renewal. The correct proportions of essential amino acids are found in animal proteins; vegetable proteins tend to be different from our own, and therefore have insufficient quantities of some amino acids. Fortunately, the amino acid composition of cereals complements that of grain legumes (peas, beans, lentils, chickpeas, etc) so eaten together they give a balanced protein intake.

Minerals (inorganic elements) and VITAMINS (certain complex organic molecules) are also needed, but in much smaller quantities. Some minerals are components of body structures. Calcium and phosphorus, for example, are essential to bones and teeth. Iron in the blood is vital for the transport of oxygen to the tissues: an iron deficiency results in ANEMIA.

Another important component of the diet is roughage or fiber, supplied by plant foods such as whole cereal grains and leaf vegetables. Green leaf vegetables reduce the risk of cancer and may improve general health.

Nyctalopia see NIGHT BLINDNESS.

Nyctinasty

Regular, periodic movements of plant organs in response to the alternation of day and night, eg the opening and closing of flowers. It is a type of circadian rhythm (see BIOLOGICAL CLOCKS).

Nylon

Group of POLYMERS containing AMIDE groups recurring in the chain. The commonest nylon is made by condensation of adipic acid and hexamethylene diamine. Nylon is chemically inert, heat-resistant, tough and very strong, and is extruded and drawn to make SYNTHETIC FIBERS, or cast and molded into bearings, gears, zippers etc.

Nymph

A pre-adult stage in certain insects; strictly, used to contrast with LARVA. Nymphs typically resemble the adult in structure but are sexually immature and lack wings. Their METAMORPHOSIS (known as an incomplete metamorphosis) is gradual, and adult characteristics are developed progressively with each molt. The insects showing this type of life cycle are known as the exopterygotes (because their lungs develop on the outside, rather than within a pupa); they include the dragonflies, mayflies, cockroaches, mantids, termites, earwigs, stick insects, grasshoppers and crickets.

Nystagmus

Oscillation of the EYES, usually with a relatively slow drift in one direction and a correcting flick in the other. Looking out of a moving vehicle at passing objects induces nystagmus. It can be caused by failure of VISION fixation (a hereditary defect), weakness of the eye muscles, or diseases of the EAR labyrinths, brain stem or cerebellum.

Oasis
A fertile area in the midst of a desert. The water source is generally a SPRING though in the Sahara many oases have sprung up around WELLS. They may be hundreds of square kilometers in area, or merely a few trees clumped together. (See also DESERT; IRRIGATION.)
Obesity
The condition of a person having excessive weight for his height, build and age because of a buildup of fat. It is common in Western society. Obesity predisposes to or is associated with numerous diseases including diabetes, arthritis, atherosclerosis and high blood pressure; here premature death is usual.
Oblateness
The degree of flattening of an OBLATE SPHEROID. It is the ratio of the difference between the large and small axes to the large axis.
Oblate spheroid
A spheroid (see ELLIPSOID) that has two of its axes of symmetry of equal length greater than that of the third. The Earth, in common with the other planets of the SOLAR SYSTEM, is oblate, its polar diameter being some 45km greater than that of its equator. A prolate spheroid is one with two axes of symmetry equal in length and shorter than the third.
Observatory
Place from which a variety of astronomical observations are made. Ancient observatories such as Stonehenge were used to predict SOLSTICES and EQUINOXES. With TYCHO BRAHE and the advent shortly after his death of the TELESCOPE, the modern observatory was born. Apart from the telescope, modern instruments used by observatories include the spectroscope (see SPECTROSCOPY), the transit instrument and the meridian circle (used to measure the right ascension and declination of stars: see CELESTIAL SPHERE), the coelostat and the coronagraph (for observing the Sun), the photometer (for measuring stellar brightnesses) and, in RADIO ASTRONOMY, the RADIO TELESCOPE. The basic instruments are complemented by electronic instrumentation for the detection and analysis of radiation. Computers play an increasingly important part in the control of the instruments, in image processing and in data analysis.
Obsessional neurosis
A NEUROSIS characterized by obsessions (inability to rid the consciousness of certain ideas despite the individual's desire to do so and recognition of their abnormality) and COMPULSIONS.
Obsidian
Volcanic GLASS formed by rapid cooling of LAVA, usually with the composition of GRANITE. Commonly jet-black, but sometimes red, brown or variegated, it may be used as a GEM stone.
Obstetrics
The care of women during pregnancy, delivery and the postnatal period, a branch of medicine and surgery usually linked with GYNECOLOGY. Antenatal care and the avoidance or control of risk factors for both mother and baby have greatly contributed to the reduction of maternal and fetal deaths. The monitoring and control of labor and birth, with early recognition of complications; induction of labor and the prevention of post-delivery HEMORRHAGE with OXYTOCIN; safe forceps delivery and CESARIAN SECTION, and improved ANESTHETICS are important factors in obstetric safety.
Occlusion (occluded front) see FRONT.
Occultation
The covering up of one celestial body by another. The term is usually applied to the passage of planets in front of stars or of the Moon in front of stars or planets.
Occupational therapy
The ancillary speciality to MEDICINE concerned with practical measures to circumvent or overcome disability due to DISEASE. It includes the design or modification of everyday items such as cutlery, dressing aids, bath and lavatory aids, and wheelchairs. Assessment and education in domestic skills and industrial retraining are also important. Diversional activities are arranged for long-stay patients.
Oceanography
The study of all aspects of, and phenomena associated with, the oceans and seas. Most modern maps of the sea floor are compiled by use of echo sounders (see also SONAR), the vessel's position at sea being accurately determined by RADAR or otherwise. Water sampling, in order to determine, for example, salinity and oxygen content, is also important. Sea-floor sampling, to determine the composition of the sea floor, is carried out by use of dredges, grabs, etc (see DREDGING), and especially by use of hollow drills that bring up cores of rock. Ocean currents can be studied by use of buoys, drift bottles, etc, and often simply by accurate determinations of the different positions of a ship allowed to drift. Further information about the sea bottom can be obtained by direct observation (see BATHYSCAPHE; BATHYSPHERE), by study of the deflections of seismic waves (see EARTHQUAKE) or by SATELLITE.
Oceans
The oceans cover some 71% of the Earth's surface and comprise about 97% of the water of the planet (see HYDROSPHERE). They provide food, chemicals, minerals and transportation; and, by acting as a reservoir of solar heat energy, they ameliorate the effects of seasonal and diurnal temperature extremes for much of the world. With the atmosphere, they largely determine the world's climate (see also HYDROLOGIC CYCLE.)

Oceanographers generally regard the world's oceans as a single, large ocean. Geographically, however, it is useful to divide this into smaller units: the Atlantic, Pacific, Indian, Arctic and Antarctic or Southern Oceans (though the Arctic is often considered as part of the Atlantic, the Antarctic as parts of the Atlantic, Pacific and Indian). Of these, the Pacific is by far the largest and, on average, the deepest. However, the Atlantic has by far the longest coastline: its many bays and inlets are ideal for natural harbors.

OCEAN CURRENTS Large-scale permanent or semi-permanent movements of water at or beneath the surface of the oceans. Currents may be divided into those caused by winds and those caused by differences in DENSITY of seawater. In the former case, FRICTION between the prevailing wind and the water surface causes horizontal motion, and this motion is both modified by and in part transferred to deeper layers by further friction. Density variations may result from temperature differences, differing salinities, etc. The direction of tow of all currents is affected by the CORIOLIS EFFECT. Best known, perhaps, are the GULF STREAM and Humboldt current. (See also TIDES; WHIRLPOOL.)

OCEAN TRENCHES Long, narrow depressions of V-shaped cross-section running, typically, roughly parallel to continental coastal MOUNTAIN ranges or volcanic island arcs (see VOLCANISM). They are caused by one tectonic plate being overridden by another (see PLATE TECTONICS). Ocean ridges are submarine mountain belts: the first to be discovered was that running roughly N-S in the Atlantic. They are important sites of earthquakes and volcanic activity (see SEA-FLOOR SPREADING).

OCEAN WAVES Undulations of the ocean surface, generally the result of the action of wind on the water surface. At sea, there is no overall translational movement of the water particles: they move up and forward with the crest, down and backward with the trough, describing a vertical circle. Near the shore, FRICTION with the bottom causes increased wave height, and the wave breaks against the land. Waves thus cause much coastal EROSION. (See also TSUNAMI.)
Ochoa, Severo (b. 1905)
Spanish-born US biochemist who shared with KORNBERG the 1959 Nobel Prize for Physiology or Medicine for his first synthesis of a NUCLEIC ACID (or RNA).
Octane (C_8H_{18})
Liquid ALKANE with 18 ISOMERS, constituents of GASOLINE. Normal octane occurs in PETROLEUM; the branched isomers, which have high antiknock values, are made by ALKYLATION. The octane number of a gasoline is the percentage of isooctane (2,2,4-trimethylpentane) in the mixture of isooctane with n-heptane which, in standard tests, knocks to the same degree as the gasoline. Antiknock additives and modern refining techniques can raise the octane number above 100.
Oculist
One who practices OPHTHALMOLOGY.
Odonata
An order of insects, the dragonflies, damselflies and mayflies. They are probably the oldest of the flying insect groups, and are little changed from their ancestors of the Carboniferous period. All have aquatic nymphs.
Oedema see EDEMA.
Oedipus complex
COMPLEX typical of infantile sexuality, sometimes retained in the adult, comprising mainly unconscious desires to exclude the parent of one's own sex and possess the other. In boys, mother FIXATION and consequent father rivalry may lead to a CASTRATION COMPLEX.
Oersted (Oe)
The unit of MAGNETIC FIELD strength in CGS electromagnetic units (see ABAMPERE). It is defined as the field at the center of a single-turn circular coil of radius 1cm and carrying a current of $1/(2\pi)$ abamperes.
Oersted, Hans Christian (1777-1851)
Danish physicist whose discovery that a magnetized needle can be deflected by an electric current passing through a wire (1820) gave birth to the science of ELECTROMAGNETISM.
Oesophagus see ESOPHAGUS.
Oestrogen see ESTROGEN.
Ohm (Ω)
The SI unit of electric RESISTANCE. A conductor has a resistance of one ohm when a potential difference of one volt across it gives rise to a current of one ampere.
Ohm, Georg Simon (1787-1854)
Bavarian-born German physicist who formulated OHM'S LAW. He also contributed to ACOUSTICS, recognizing the ability of the human ear to resolve

mixed SOUND into its component pure (sinusoidal-wave) tones.

Ohmmeter

Instrument for providing a rapid, if approximate, value for the RESISTANCE of part of an electric circuit. It consists of an AMMETER (reverse calibrated in ohms – see OHM) in series with a fixed resistor, a battery and a variable resistor. This last is adjusted to zero the meter, with the terminals of the instrument shorted. The test resistance is then introduced between these terminals. The reduction in the current so caused is a measure of the resistance.

Ohm's Law

The statement due to G.S. OHM in 1827 that the electric POTENTIAL difference across a conductor is proportional to the current flowing through it, the constant of proportionality being known as the RESISTANCE of the conductor. It holds well for most materials and objects, including solutions, provided that the passage of the current does not heat the conductor, but ELECTRON TUBES and SEMICONDUCTOR devices show a much more complicated behavior.

Oil

Any substance that is liquid, insoluble in water, soluble in ETHER and greasy to the touch. There are three main groups: mineral oils (see PETROLEUM); fixed vegetable and animal oils (see FATS; LIPIDS), and volatile vegetable oils. Oils are classified as fixed or volatile according to the ease with which they vaporize when heated. Mineral oils include GASOLINE and many other fuel oils, heating oils and lubricants. Fixed vegetable oils are usually divided into three subgroups depending on the physical change that occurs when they absorb oxygen: oils such as linseed and tung, which form a hard film, are known as "drying oils"; "semidrying oils", such as cottonseed or soybean oil, thicken considerably but do not harden; "nondrying oils," such as castor and olive oil, thicken only slightly. Fixed animal oils include the "marine oils", such as cod-liver and whale oil. Fixed animal and vegetable fats, such as butterfat and palm oil, are often also classified as oils. Examples of volatile vegetable oils, which usually have a very distinct odor and flavor, include such oils as bitter almond, peppermint and turpentine. When dissolved in alcohol they are called "essences".

Oil refining see PETROLEUM.

Oils, Natural see FATS.

Oil sands

Loose sand or sandstone containing viscous oil. Depending on the proportion of oil to sand, they may occur as ASPHALT lakes, such as those in Trinidad, or BITUMINOUS SANDS such as the Athabaska tar sands (Alberta, Canada). Despite extraction problems, they are a potentially important OIL source.

Oil shale

A fine-grained, dark-colored sedimentary rock from which oil suitable for refining can be extracted. The rock contains an organic substance called kerogen, which may be distilled to yield OIL (see also DISTILLATION). It is nowadays an important oil source. (See also SEDIMENTARY ROCKS; SHALE.)

Oil well see PETROLEUM.

Olbers, Heinrich Wilhelm Matthäus (1758-1840)

German astronomer who discovered the ASTEROIDS Pallas (1802) and Vesta (1807), rediscovered Ceres (1802), and found five comets, one of which bears his name. He also formulated a method of calculating cometary orbits (1779); proposed that the pressure of light is responsible for COMETS' tails always pointing away from the Sun (1811); and stated Olbers' Paradox that, in an infinite, isotropic universe, the night sky should be uniformly illuminated, whereas in fact it is not. This he explained by suggesting the existence of clouds of INTERSTELLAR MATTER, which were later discovered; but the "paradox" was not fully resolved until HUBBLE showed that the UNIVERSE is expanding.

Olefins see ALKENES.

Oleic acid ($C_{17}H_{33}COOH$)

An unsaturated CARBOXYLIC ACID used in lubricating oils and varnishes. Triolein, its ester with GLYCEROL, is present in many natural oils and fats, helping to keep them liquid at room temperature. mp 16°C, bp 360°C.

Oleum see SULFURIC ACID.

Olfactory system see SMELL.

Oligocene

The third epoch of the TERTIARY period, of duration about 38-25 million years ago. (See also GEOLOGY.)

Oligochaetes

A class of ANNELIDS that includes the earthworms.

Olivine

Group of SILICATE minerals, orthosilicates of magnesium and iron $(Mg,Fe)_2SiO_4$. It forms olive green crystals in the orthorhombic system, and occurs commonly in IGNEOUS ROCKS, chiefly BASALT, GABBRO and PERIDOTITE. The transparent variety, peridot, is used as a GEM stone.

Omnivore

Any animal that feeds on a variety of foods, including both plants and animals. (See also CARNIVORE; HERBIVORE.)

Onchocerciasis see FILARIASIS.

Oncogenes

Genes carried by viruses that are involved in transforming normal cells to cancerous ones (see CANCER). Oncogenes are sections of DNA that the viruses originally picked up from the human GENOME – they are present in every human cell, healthy and tumorous. But when oncogenes are returned to the human cell by a virus they are present in far greater quantities than is normal, and therefore produce an abnormal proliferation of cells. Natural selection has happened to favor viruses that pick up this type of gene, because the proliferation of the cells they infect multiplies the reproductive potential of the virus. Oncogenes have been shown to code for enzymes that modify proteins by adding a phosphate group to them. These enzymes are associated with the cell membrane, and under normal circumstances are part of the complex mechanism controlling cell division. When they are present in excess, the control of cell division is disrupted and a tumor results.

Onsager, Lars (1903-1976)

Norwegian-born US chemist awarded the 1968 Nobel Prize for Chemistry for his fundamental work on irreversible chemical and thermodynamic processes.

Ontogeny

The process of development of an organism, from the fertilized egg to the mature individual. The term is not often used except in the phrase "ontogeny repeats phylogeny", which encapsulates the idea that organisms retrace the evolutionary pathway of their species during their development. Although there is a grain of truth in this, it should not be taken too literally. Animal embryos do display features that correspond to certain stages in their ancestry: for example, human embryos have gill slits when very small, as do most mammalian embryos. These features are the precursors of gills in fish, and would have been found in embryos of the rhipidistina fish, our ancestors of about 380 million years ago. However, the human embryo does not develop gills, only gill slits, and it displays few other fish-like characteristics, so it cannot be said to be retracing its evolutionary pathway.

Onychophorans

A group of terrestrial invertebrates that appear to be intermediate between the annelid worms and the Uniramia – the myriapods and insects. They are soft-bodied like the annelids, and worm-like in shape, but they have TRACHEAE for breathing, ANTENNAE for sensory purposes, and peg-like legs similar to those of a caterpillar. (◆ page 16)

Onyx

Variety of CHALCEDONY with variegated bands, straight rather than curved as in AGATE. Sardonyx has white and brown bands; carnelian onyx white and red. It is used as a GEM stone, especially for cameos and intaglios.

Oolith

A more or less spherical particle of rock that has developed by the accretion of material about an initial nucleus. The accretion may be concentric (so that in cross-section circular bands of material may be seen) or radial, and combinations are known. Larger ooliths are termed pisoliths. Concentration of ooliths can form, for example, oolitic limestones, often called oolites.

Oomycetes see FUNGI.

Oozes

Sediments on the deep-sea plains, consisting of PLANKTON remains, wind-borne volcanic dust, etc. Beneath about 5000m no organic matter is present (red clays); between 5000 and 3900m Radiolaria predominate in the siliceous oozes (see also PROTOZOA); and between 3900 and 2000m Foraminifera are most common in the calcareous oozes.

Opal

Cryptocrystalline variety of porous hydrated SILICA, deposited from aqueous solution in all kinds of rocks, and also formed by replacement of other minerals. Opals are variously colored; the best GEM varieties are translucent, with milky or pearly opalescence and iridescence due to light scattering and interference from internal cracks and cavities. Common opal is used as an abrasive, filler and insulator.

Open clusters see GALACTIC CLUSTERS.

Open-hearth process

Technique that was until recently responsible for most of the world's STEEL production. It derives its alternative name, the Siemens-Martin Process, from the work of Sir William Siemens and Pierre Emile Martin in the 1850s and 1860s. The FURNACE has two ducts, each leading to a chamber of brick checkerwork. In use, the hot exhaust fumes are passed out through one duct, heating the brick to high temperature, while air intake is through the other. Periodically the streams are reversed, so that the incoming air is preheated by the hot brick before passing the burners, thus greatly increasing the flame temperature. Some furnaces are liquid-fueled, but if the fuel is gaseous it may be fed in with the air. In the basic process, the charge is of IRON ore, scrap steel and LIMESTONE. The impurities in the ore combine with the limestone to form a basic SLAG. In the less important acid process REFRACTORIES of SILICA result in an acid slag.

Operation see SURGERY.

Ophthalmia

INFLAMMATION of the EYE. This may be CONJUNCTIVITIS, as in neonatal ophthalmia (often gonococcal, see SEXUALLY TRANSMITTED DISEASES), or UVEITIS, as in sympathetic ophthalmia in which an inflammatory reaction in both eyes follows injury to one.

Ophthalmology

The branch of MEDICINE and SURGERY concerned with diseases of VISION and the EYE. In infancy, congenital BLINDNESS and STRABISMUS, and in adults, glaucoma, UVEITIS, CATARACT, retinal detachment and vascular diseases are common, as are ocular manifestations of systemic diseases such as DIABETES. Disorders of eye movement, lids and TEAR production, color vision, infection and injury are also seen.

Ophthalmoscope
Instrument for examining the RETINA and structures of the inner EYE. A powerful light and lens system, combined with the cornea and lens of the eye, allows the retina and eye blood vessels to be seen at high magnification. It is a valuable aid to diagnosis in OPHTHALMOLOGY and internal medicine.

Opium
NARCOTIC extract from the immature fruits of the opium poppy, *Papaver somniferum*, which is native to Greece and Asia Minor. The milky juice is refined to a powder which has a sharp, bitter taste. Drugs, some drugs of abuse (see DRUG ADDICTION), obtained from opium include the narcotic ANALGESICS, HEROIN, MORPHINE and CODEINE. (Synthetic analogs of these include methadone and pethidine.) Older opium preparations, now rarely used, include laudanum and paregoric. The extraction of opium outside the pharmaceutical industry is banned throughout the West.

Oppenheimer, Julius Robert (1904-1967)
US physicist whose influence as an educator is still felt today and who headed the MANHATTAN PROJECT, which developed the ATOMIC BOMB. His main aim was the peaceful use of nuclear power (he fought against the construction of the HYDROGEN BOMB but was overruled by Truman in 1949); but, because of his left-wing friendships, he was unable to pursue his researches in this direction after being labeled a security risk (1954). He also worked out much of the theory of BLACK HOLES.

Opposition
In ASTRONOMY, the situation in which the Earth lies directly between another planet (or the Moon) and the Sun.

Optical activity
The property, possessed by certain substances, of rotating the plane of polarization of plane-POLARIZED LIGHT passing through them. This optical rotation is measured by POLARIMETRY. Optical activity is shown by asymmetric CRYSTALS that have two mirror-image forms – the rotation being to the left or right respectively – and by compounds with asymmetric molecules showing optical STEREOISOMERISM.

Optical character recognition (OCR)
A technique for reading typed letters automatically with an optical scanner. The scanner detects the character's shape by reflected light and converts the information into binary digits which a COMPUTER can process. In theory it is possible to recognize handwritten characters, but the enormous variation in handwriting makes it very unreliable.

Optical disk
A disk, similar in size and shape to an audio disk, in which the information is recorded and replayed by a LASER. The disk's surface is divided into tracks and sectors to aid location of specific information. In the original optical disk, invented by Philips of Holland, a helium-neon laser detected the presence or absence of pits (holes) in the coating of the disk as it spun at about 1800 revolutions per minute. But the gas laser has given way to semiconductor lasers (far more compact and using less power), and the disk technology is being used for digital audio (compact disks), video films and mass storage. For mass storage, the laser can operate at two powers, higher for burning the pits during the recording process, and lower for reading it. Erasable disks are also on the way; a tellurium coating is used that can be made to flow so that it seals over the pits, ready for re-recording. (See also COMPUTER.)

Optical fiber
Extruded glass fiber of high purity, used to transmit a light signal; used widely for telephone systems.

Optical illusion see ILLUSION.

Optician
One who practices OPTOMETRY.

Optics
The science of light and vision. Physical optics deal with the nature of LIGHT (see also COLOR; DIFFRACTION; INTERFERENCE; POLARIZED LIGHT; SPECTROSCOPY). Geometrical optics consider the behavior of light in optical instruments (see ABERRATION, OPTICAL; DISPERSION; LENS, OPTICAL; MICROSCOPE; MIRROR; PRISM; REFLECTION; REFRACTION; SPECTRUM; TELESCOPE). Physiological optics are concerned with vision (see EYE). (◆◆ page 115, 123, 213, 237)

Optometry
Measurement of the acuity (popularly "power") of VISION and the degree of lens correction required to restore "normal vision" in subjects with refractive errors (MYOPIA, HYPEROPIA, ASTIGMATISM). Its principal instrument is a chart of letters that subtend specific angles to the EYE at a given distance, temporary lenses being used to correct each eye.

Orbit
The path followed by one celestial body revolving under the influence of gravity (see GRAVITATION) about another. In the SOLAR SYSTEM the planets orbit the Sun, and the moons the planets, in elliptical paths, though Triton's orbit of NEPTUNE is as far as can be determined perfectly circular. The point in the planetary, asteroidal or cometary orbit closest to the Sun is called its perihelion; the farthest point is termed aphelion. In the case of a moon or artificial satellite orbiting a planet or other moon, the corresponding terms are perigee and apogee. (See also APSIDES, LINE OF; KEPLER'S LAWS.) Celestial objects of similar masses may orbit each other, particularly DOUBLE STARS.

Orbital
In chemistry, the mathematical wave function (see QUANTUM MECHANICS) that describes the motion of an ELECTRON around the nucleus of an ATOM or several nuclei in a molecule. The orbital represents the probability distribution of the electron in space; in effect, for each point, the likelihood of finding the electron there. Orbitals are defined and characterized by three quantum numbers, representing the energy level (and hence the size), the angular momentum (and hence the shape), and the orientation. An orbital can be occupied by one or two electrons (of opposite spin), according to FERMI-DIRAC STATISTICS. The precise energy of each orbital depends on the local electromagnetic field and is found by SPECTROSCOPY. In the formation of a covalent BOND, molecular orbitals are formed by linear combination of the outer atomic orbitals. (See also AROMATIC COMPOUNDS; RESONANCE.) (◆ page 21)

Order
In TAXONOMY, a group of related families. Examples include the Order Chiroptera, the bats, and the Order Lepidoptera, the butterflies and moths. Related orders are grouped together in classes. (◆ page 21)

Ordovician
The second period of the PALEOZOIC era, which lasted from about 505 to 438 million years ago and immediately followed the CAMBRIAN period. (See GEOLOGY.) (◆ page 106)

Ore
Aggregate of minerals and rocks from which it is commercially worthwhile to extract minerals (usually metals). An ore has three parts: the country rock in which the deposit is found; the gangue, the unwanted rocks and minerals of the deposit; and the desired mineral itself. Ore deposits may be VEINS; infillings of breccia (consolidated TALUS); sedimentary formations, as of the EVAPORITES; certain DIKES; or, especially with the SULFIDES, hydrothermal replacement deposits (where hot or superheated water has dissolved existing rocks and deposited in their place minerals held in solution). Mining techniques depend greatly on the form and position of the deposit.

Organ
In biology, a functionally adapted part of an organism, such as the heart, brain or liver.

Organelle
A discrete subcellular body found in the cells of EUKARYOTES. (◆ page 13)

Organic chemistry
Major branch of CHEMISTRY comprising the study of CARBON compounds that contain hydrogen (simple carbon compounds such as carbon dioxide being usually deemed inorganic). This apparently specialized field is in fact wide and varied, because of carbon's almost unique ability to form linked chains of atoms to any length and complexity; far more organic compounds are known than inorganic. Organic compounds form the basic stuff of living tissue (see also BIOCHEMISTRY), and until the mid-19th century, when organic syntheses were achieved, a "vital force" was thought necessary to make them. The 19th-century development of quantitative ANALYSIS by J. LIEBIG and J.B.A. DUMAS, and of structural theory by S. CANNIZZARO and F.A. KEKULÉ, laid the basis for modern organic chemistry. Organic compounds are classified as ALIPHATIC, ALICYCLIC, AROMATIC and HETEROCYCLIC COMPOUNDS, according to the structure of the skeleton of the molecule, and are further subdivided in terms of the FUNCTIONAL GROUPS present.

Organometallic compounds
Class of compounds containing bonds from carbon atoms to metal (or metalloid) atoms, and thus at the crossroads of INORGANIC and ORGANIC CHEMISTRY. In the last 30 years the subject has expanded enormously, with the development of many medical, industrial and synthetic uses. There are three main types of organometallic compounds: (1) the alkyl derivatives of Group IA and IIA metals of the PERIODIC TABLE (including GRIGNARD REAGENTS), which have ionic BONDS, and are powerful BASES and reducing agents (see OXIDATION AND REDUCTION); (2) derivatives of other Main Group metals, which are volatile, covalently bonded compounds; and (3) TRANSITION ELEMENT derivatives (including SANDWICH COMPOUNDS) with special *d*-ORBITAL bonding. Organometallic compounds are prepared by reacting a metal with an ALKYL HALIDE or a reactive HYDROCARBON, or by substitution.

Orion
The Hunter, a large constellation on the celestial equator that is visible during winter in N skies, containing RIGEL, BETELGEUSE and the Orion NEBULA.

Ornithischia
The bird-hipped dinosaurs, a group of DINOSAURS having bird-like pelvic girdles, with four prongs to each side. This resemblance to birds is entirely coincidental; the birds actually evolved from another group, the SAURISCHIA. All ornithischians were herbivorous. Four-legged types include the stegosaurs, with triangular bony plates along the back, and the armadillo-like ankylosaurs. The two-legged duck-billed dinosaurs were well equipped for swimming.

Ornithology
The scientific study of birds.

Orthoclase
Common mineral in igneous rocks, consisting of potassium aluminum silicate ($KAlSi_3O_8$); vitreous crystals of various colors. It is one of the three end-members (pure compounds) of the FELDSPAR group. (See also MICROCLINE.)

Orthodontics
The correction of malformed and displaced teeth.

Orthopedics
Speciality within SURGERY, dealing with bone and soft-tissue disease, damage and deformity. Its name derives from 17th-century treatments designed to produce "straight children". Until the advent of anesthetics, ASEPSIS and X-RAYS, its methods were restricted to AMPUTATION and manipulation for dislocation, etc. Treatment of congenital deformity, FRACTURES and TUMORS of bone, OSTEOMYELITIS, ARTHRITIS, and JOINT dislocation are common in modern orthopedics. Methods range from the use of splints, PHYSIOTHERAPY and manipulation, to surgical correction of deformity, fixing of fractures and refashioning or replacement of joints. Suture or transposition of TENDONS, MUSCLES or nerves are also performed.

Orthoptera
An order of insects that includes the grasshoppers, crickets, katydids and locusts. Most have hind legs greatly enlarged for jumping.

Oscillating universe see COSMOLOGY.

Oscillator
A device converting direct to alternating current (see ELECTRICITY), used, for example, in generating radio waves. Most types are based on an electronic amplifier, a small portion of the output being returned via a feedback circuit to the input, so as to make the oscillation self-sustaining. The feedback signal must have the same PHASE as the input: by varying the components of the FEEDBACK circuit, the frequency for which this occurs can be varied, so that the oscillator is easily "tuned". "Crystal" oscillators incorporate a piezoelectric crystal (see PIEZOELECTRICITY) in the tuning circuit for stability; in "heterodyne" oscillators, the output is the beat frequency between two higher frequencies.

Oscilloscope
A device using a CATHODE RAY TUBE to produce line graphs of rapidly varying electrical signals. Since nearly every physical effect can be converted into an electrical signal, the oscilloscope is very widely used. Typically, the signal controls the vertical deflection of the beam while the horizontal deflection increases steadily, producing a graph of the signal as a function of time. For periodic (repeating) signals, synchronization of the horizontal scan with the signal is achieved by allowing the attainment by the signal of some preset value to "trigger" a new scan after one is finished. Most models allow two signals to be displayed as functions of each other; dual-beam instruments can display two as a function of time. Oscilloscopes usually operate from DC to high frequencies, and will display signals as low as a few millivolts.

Osler, Sir William (1849-1919)
Canadian-born physician and educator best known for his work on platelets (see BLOOD) and for the informality of his educational techniques.

Osmium (Os)
Silvery gray hard metal in the PLATINUM GROUP. It is slowly oxidized in air. AW 190.2, mp 3030°C, bp 5000°C, sg 22.48 (20°C). (♦ page 130)
OSMIUM (VIII) OXIDE (OsO_4) A toxic oxidizing agent, used to stain tissues for microscope slides.

Osmosis
The diffusion of a solvent, usually water, through a SEMIPERMEABLE MEMBRANE that separates two solutions of different concentration. The movement of the solvent is from the more dilute to the more concentrated solution, owing to the thermodynamic tendency to equalize the concentrations. The liquid flow may be opposed by applying pressure to the more concentrated solution: the pressure required to reduce the flow to zero from a pure solvent to a given solution is known as the osmotic pressure of the solution. Osmosis was studied by THOMAS GRAHAM, who coined the term (1858); in 1886 VAN'T HOFF showed that, for dilute solutions (obeying Henry's Law), the osmotic pressure varies with temperature and concentration, as if the solute were a GAS occupying the volume of the solution. This enables MOLECULAR WEIGHT to be calculated from osmotic pressure measurements, and degrees of ionic DISSOCIATION to be estimated. Osmosis is important in DIALYSIS and in water transport in living tissue.

Ossicles see EAR.

Ossification see BONE.

Osteichthyes
A class of fish, also known as the BONY FISH. It is divided into two subclasses, the ACTINOPTERYGII and the SARCOPTERYGII. (♦ page 17)

Osteitis deformans
Another name for PAGET'S DISEASE.

Osteoblast
Cell that deposits calcium phosphate crystals in bone. Bone is constantly undergoing reshaping, and also acts as the calcium "bank" for the body so that blood calcium levels are kept constant.

Osteoclast
Cell that removes calcium phosphate crystals from bone.

Osteology
The study of the structure, function and diseases of BONE.

Osteomalacia
Loss of calcium from bone due to a shortage of calcium or vitamin D; an adult variant of RICKETS.

Osteomyelitis
BACTERIAL infection of BONE, usually caused by STAPHYLOCOCCUS, STREPTOCOCCUS or SALMONELLA. It is carried to the bone by the blood, or gains access through open fractures. It commonly affects children, causing fever and local pain. If untreated or partially treated it may become chronic, with bone destruction and a discharging SINUS.

Osteopathy
System of treatment based on the theory that disease arises from the mechanical and structural disorder of the body skeleton. Prevention and treatment are practiced by manipulation, often of the spine. While it may have a role in treatment of chronic musculo-skeletal pain, some orthodox orthopedic surgeons consider its methods hazardous, especially to the SPINAL CORD. Furthermore, serious disease may be overlooked.

Osteoporosis
Loss of bone common in old age, especially in thin women after the menopause, caused by the gradual disappearance of the organic structure of bone so that it cannot hold calcium well.

Ostwald, Friedrich Wilhelm (1853-1932)
Latvian-born German physical chemist regarded as a father of physical chemistry, and awarded the 1909 Nobel Prize for Chemistry for his work on CATALYSIS. He also developed the Ostwald process.

Otitis
Inflammation of the ear, which can lead to partial deafness if untreated.

Otosclerosis see DEAFNESS.

Otoscope
Instrument for examining the outer EAR and eardrum. It has a light, a lens and a conical earpiece.

Ovary
The female reproductive organ. In plants it contains the ovules (see FLOWER; ♦ page 53); in humans, the FOLLICLES in which the eggs, or ova, develop. (♦♦ page 111, 204)

Overweight see OBESITY.

Oviparous animals
Animals that reproduce by laying EGGS. (See also VIVIPAROUS and OVOVIVIPAROUS ANIMALS.)

Ovoviviparous animals
Animals that retain their eggs within their bodies until they have hatched, and thus give birth to live young. Unlike VIVIPAROUS ANIMALS, they have no direct link with their young, such as a placenta, through which nutrients can be supplied. Instead, the young draw nourishment from the yolk of their eggs. In a few species the young supplement this, once they have reached a certain size, by feeding on cells from the lining of the mother's reproductive tract, or on their siblings.

Ovulation see MENSTRUATION.

Ovule see FLOWER.

Ovum
The egg cell of sexually reproducing organisms. (♦♦ page 32, 204)

Oxalic acid $(COOH)_2$
White crystalline solid, a toxic, dibasic CARBOXYLIC ACID occurring in many plants. It is made by heating sodium formate or by fusing sawdust with sodium hydroxide. Oxalic acid is a mild reducing agent used as a standard for VOLUMETRIC ANALYSIS, in the leather and dye industries, to remove ink and rust stains, and to dissolve radiator scale. MW 90.0, mp 189.5°C.

Oxbow lake
C-shaped lake formed when a river cuts through, and deposits SILT across, the narrow neck of a meander. (See RIVERS AND LAKES.)

Oxidation and reduction (redox reactions)
Large class of chemical reactions, including many familiar processes such as COMBUSTION, CORROSION and RESPIRATION. Oxidation was originally defined simply as the combination of an element or compound with oxygen, or the removal of hydrogen from a compound; and reduction as combination with hydrogen or removal of oxygen. In the modern theory this has been generalized: oxidation is defined as loss of electrons, and reduction as gain of electrons. The two always go together: there is an oxidizing agent which is reduced, and a reducing agent which is oxidized. Thus, in the reaction

$$Fe^{3+} + I^- \rightarrow Fe^{2+} + \tfrac{1}{2}I_2$$

the iron (III) ion gains an electron and is reduced to iron (II), and the iodide ion loses an electron and is oxidized to iodine. The strength of a redox reagent, expressing its tendency to react, is measured by the electrode potential of the half-reaction (see ELECTROCHEMISTRY), and so redox reagents may be ranked in an extended ELECTROCHEMICAL SERIES. In a covalent compound or complex ion, each atom is assigned an oxidation number (O.N.), which is the charge it would have if all the BONDS were ionic – the electrons in a covalent bond between two atoms are assigned to the atom with the higher ELECTRONEGATIVITY. Thus the sulfur atom in sulfur dioxide has an O.N. of +4, and each of the two oxygen atoms has an O.N. of −2. When sulfur is oxidized by oxygen, $S + O_2 \rightarrow SO_2$, its O.N. increases from 0 (for elements, by definition) to +4, corresponding to a virtual loss of electrons. (See also INDICATOR.) (♦ page 130)

Oxides
Binary compounds of OXYGEN with the other elements (see also PEROXIDES). All the elements

form oxides except helium, neon, argon and krypton. Metal oxides are typically ionic crystalline solids (containing the O^{2-} ion), and are generally BASES, though the less electropositive metals form amphoteric oxides with acidic and basic properties (eg ALUMINUM oxide). Nonmetal oxides are covalent and typically volatile, though a few are macromolecular refractory solids (eg SILICON dioxide); most are acidic (see ACIDS), some are neutral (eg CARBON monoxide), and WATER is amphoteric. Oxides may be prepared by direct synthesis, or by heating hydroxides, nitrates or carbonates. Many oxide minerals are known; simple oxides are binary metal oxides, complex oxides contain several cations, and SPINELS are intermediate between the two. Some metal oxides are nonstoichiometric (see COMPOSITION, CHEMICAL). (♦ page 130)

Oximes
Derivatives of ALDEHYDES or KETONES (termed aldoximes and ketoximes respectively) formed by condensation with hydroxylamine (NH_2OH), and containing the group $C{=}N{-}OH$. Being easily isolated and with characteristic melting points, they are used for identification. Ketoximes undergo the BECKMANN rearrangement.

Oxyacids
Those ACIDS that contain an acidic hydroxyl (see HYDROXIDES) group, ie whose acidic hydrogen is bound to oxygen. They include CARBOXYLIC ACIDS and PHENOLS, but are typically (eg sulfuric acid) the hydration products of acidic (nonmetal) OXIDES. Acids, such as hydrochloric acid, whose acidic hydrogen is bound to an element other than oxygen, are termed hydracids.

Oxygen (O)
Gaseous nonmetal in Group VIA of the PERIODIC TABLE, comprising 21% by volume of the ATMOSPHERE and about 50% by weight of the Earth's crust. It was first prepared by SCHEELE and PRIESTLEY, and named *oxygène* by LAVOISIER. Gaseous oxygen is colorless, odorless and tasteless, liquid oxygen is pale blue. Oxygen has two allotropes (see ALLOTROPY): OZONE (O_3), which is metastable; and normal oxygen (O_2),which shows PARAMAGNETISM because its diatomic molecule has two electrons with unpaired spins. Oxygen is prepared in the laboratory by heating mercuric oxide or potassium chlorate (with manganese dioxide catalyst). It is produced industrially by fractional distillation of liquid air. Oxygen is very reactive, yielding OXIDES with almost all other elements and in some cases PEROXIDES. Almost all life depends on chemical reactions with oxygen to produce energy. Animals receive oxygen from the air, as do fish from the water (see RESPIRATION): it is circulated through the body in the blood. The amount of oxygen in the air remains constant because of PHOTOSYNTHESIS in plants and the decomposition by the Sun's ultraviolet rays of water vapor in the upper atmosphere.

Oxygen is used in vast quantities in metallurgy, smelting and refining, especially of iron and steel. Oxygen and ACETYLENE are used in oxyacetylene torches for cutting and welding metals. Liquid oxygen is used in rocket fuels. Oxygen has many medical applications (see ANESTHESIA; OXYGEN TENT), and is used in mixtures breathed by divers and high altitude fliers. It is also widely used in chemical synthesis. AW 16.0, mp −218.8°C, bp −183°C. (♦ page 25, 28, 37, 130)

Oxygen tent
Enclosed space, often made of plastic, in which a patient may be nursed in an atmosphere enriched with OXYGEN. It is mainly used for small children with acute respiratory DISEASES, or in adults when the use of a face mask is impractical.

Oxyhemoglobin see HEMOGLOBIN.

Oxytocin
HORMONE secreted by the posterior PITUITARY GLAND (see also HYPOTHALAMUS). It causes uterine contraction and is used in OBSTETRICS. It is also concerned with milk secretion by the BREAST. (♦ page 111)

Ozone (O_3)
Triatomic allotrope of OXYGEN (see ALLOTROPY); blue gas with a pungent odor. It is a very powerful oxidizing agent, and yields ozonides with ALKENES. It decomposes rapidly above 100°C. Ozone is made by subjecting oxygen to a high-voltage electric discharge. It is used for killing germs, bleaching, removing unpleasant odors from food and sterilizing water.

Ozone is one of the gases contributing to modern AIR POLLUTION problems. Some is released by industrial processes, but most is formed by the action of sunlight on car exhaust fumes, producing a reaction between nitrogen oxides and hydrocarbons. Even in very small amounts, ozone has been shown to provoke bronchial spasm in asthmatics and to retard the growth of plants.

In the upper atmosphere, however, ozone serves a useful purpose. Here there is a layer of ozone, formed when ULTRAVIOLET RADIATION acts on oxygen; this layer protects the Earth from the Sun's ultraviolet rays. The ozone layer is an important stabilizer of climate. In the 1970s chemists realized that long-lived chemicals and gases released into the atmosphere might damage the ozone layer, and cause an increase in the ultraviolet reaching the Earth's surface. Some trace gases from industrial processes and CHLOROFLUOROCARBONS (CFCs) used as propellants in spray cans and refrigerators are most likely to cause this type of damage. CFCs, which have been the major focus of concern, eventually break down to release chlorine, and it is this that destroys the ozone. The extent of the damage is out of all proportion to the amount of pollutant, because a chain reaction can be set up by a single CFC molecule that results in the destruction of several hundred thousand molecules of ozone. Breaches in the ozone layer over the Antarctic were first detected in 1984. Apart from the probable climatic effects of increased ultraviolet radiation, there is also the likelihood of more skin cancers in the human population. Moves to curb CFCs began in the 1980s with the ban on their use in aerosols by the United States. A major international agreement to limit CFC production was signed in 1987, under the United Nations Environment Program. Despite these measures, the level in the atmosphere is set to rise for several decades, as CFCs already produced (for example in refrigerators) are slowly released. Some environmentalists believe that a total ban is needed to prevent further damage, and express concern that the rising standard of living in developing countries – leading to a great many more refrigerators – is a huge potential threat to the ozone layer. One possible solution to the problem is to find alternative chemicals that either do not release chlorine or release it more readily, before they have time to reach the upper atmosphere. Some such alternatives are already available but they are more expensive to produce.

Pacemaker see PROSTHETICS.

Paget's disease
Inflammation of bone leading to calcium loss and resulting deformity, which is similar to RICKETS.

Pahoehoe (ropy lava) see LAVA.

Pain
The detection by the nervous system of harmful stimuli. The function of pain is to warn the individual of imminent danger: even the most minor tissue damage will cause pain, so that avoiding action can be taken at a very early stage. The level at which pain can only just be felt is the pain threshold. This threshold level varies slightly between individuals, and can be raised by, for example, HYPNOSIS, anesthetics, ANALGESICS and the drinking of alcohol. In some psychological illnesses, especially the NEUROSES, it is lowered. The receptors of pain are unencapsulated nerve endings (see NERVOUS SYSTEM), distributed variably about the body: the back of the knee has about 230 per cm^2, the tip of the nose about 40. Deep pain, from the internal organs, may be felt as surface pain or in a different part of the body. This phenomenon, referred pain, is probably due to the closeness of the nerve tracts entering the SPINAL CORD. (♦ page 213)

Palade, Georg Emil (b. 1912)
Romanian-born US cell biologist awarded with A. CLAUDE and C. DE DUVE the 1974 Nobel Prize for Physiology or Medicine for their researches into intercellular structures. Palade discovered the nature of the granules found in cell cytoplasm that are now called RIBOSOMES.

Palate
Structure dividing the mouth from the nose and bounded by the upper gums and teeth; it is made of bone and covered by a membrane. At the back it is a soft, mobile connective-tissue structure that can close off the nasopharynx during swallowing and speech. (♦ page 213)

Palearctic
A region in ZOOGEOGRAPHY.

Paleobotany see PALEONTOLOGY.

Paleocene
The first epoch of the TERTIARY period, which extended between about 65 and 55 million years ago. (See also GEOLOGY.)

Paleoclimatology
The determination of climatic conditions of the geological past by study of the FOSSIL paleoecology and sedimentary (sedimentology) evidence.

Paleogeography
The construction from geological, paleontological and other evidence of maps of parts or all of the Earth's surface at specific times in the Earth's past. Paleogeography has proved of considerable importance in CONTINENTAL DRIFT studies.

Paleomagnetism
The study of past changes in the EARTH's magnetic field by examination of rocks containing certain iron-bearing minerals (eg HEMATITE, MAGNETITE). Reversals of the field and movements of the magnetic poles can be charted and information on CONTINENTAL DRIFT may be obtained. (See also SEA-FLOOR SPREADING.)

Paleontology (paleobiology)
Study of the remains of living organisms of past eras. The two branches are paleobotany and paleozoology, dealing with plants and animals respectively. Such studies are essential to STRATIGRAPHY, and provide important evidence for EVOLUTION and CONTINENTAL DRIFT theories. (See also FOSSILS; PALEOCLIMATOLOGY; RADIOCARBON DATING.)

Paleozoic
The earliest era of the PHANEROZOIC, comprising two sub-eras: the Lower Paleozoic, 590-408 million years ago, containing the CAMBRIAN, ORDOVICIAN and SILURIAN periods; and the Upper Paleozoic, 408-248 million years ago, containing the DEVONIAN, CARBONIFEROUS (MISSISSIPPIAN and PENNSYLVANIAN) and PERMIAN. (♦ page 106)

Palladium (Pd)
White, soft, ductile metal in the PLATINUM GROUP. In addition to the general uses of these metals, palladium is used in dental and other ALLOYS. It absorbs 900 times its volume of hydrogen, forming a metallic HYDRIDE, and, being permeable to hydrogen at high temperatures, it is used in hydrogen purifiers. AW 106.4, mp 1554°C, bp 3000°C, sg 11.97 (0°C). (♦ page 130)

Pallas, Peter Simon (1741-1811)
German naturalist best known for his *Travels through Various Provinces of the Russian Empire* (3 vols, 1771-76), an account of his 6-year expedition collecting extant and fossil plant and animal specimens.

Palmer, Daniel David (1845-1913)
Canadian-born US founder of CHIROPRACTIC (1895).

Palmitic acid ($C_{15}H_{31}COOH$)
CARBOXYLIC ACID whose glyceryl ESTER is a major component of FATS. Palmitates are used in SOAPS.

Palomar Observatory see HALE OBSERVATORIES.

Palpitations
Increased awareness of the beating of the heart, a sign of emotional stress or panic attack.

Palynology
The study of fossil pollen and spores. These are relatively resistant to decay, and can be used to reconstruct the flora and climate prevailing when they were laid down.

Pancreas
Organ consisting partly of exocrine gland tissue, secreting into the DUODENUM, and partly of ENDOCRINE GLAND tissue (the islets of Langerhans), whose principal HORMONES include INSULIN and GLUCAGON. The pancreas lies on the back wall of the upper abdomen, much of it within the duodenal loop. Powerful digestive-system ENZYMES (pepsin, trypsin, lipase, amylase) are secreted into the gut; this secretion is in part controlled by intestinal hormones (SECRETIN) and in part by nerve reflexes. Insulin and glucagon have important roles in glucose and fat METABOLISM (see DIABETES MELLITUS); other pancreatic hormones affect GASTROINTESTINAL TRACT secretion and activity. Acute INFLAMMATION of the pancreas due to VIRUS disease, ALCOHOLISM or duct obstruction by gallstones, may lead to severe abdominal pain with shock and prostration caused by the release of digestive enzymes into the abdomen. Chronic pancreatitis leads to functional impairment and malabsorption. CANCER of the pancreas may cause JAUNDICE by obstructing the BILE duct. (♦♦ page 28, 40, 111, 233)

Pandemic see EPIDEMIC.

Pangea
Primeval supercontinent that, under PLATE TECTONIC action, split up to form Laurasia in the N and Gondwana in the S Hemisphere (see GONDWANA; LAURASIA; PLATE TECTONICS). In turn these too split up, forming our modern continents. (See also CONTINENTAL DRIFT.)

Pantograph
Instrument for enlarging or reducing a geometric figure or motion. It consists of four hinged bars forming a parallelogram, one vertex being fixed. It is used in technical drawing and mapmaking. A spring-loaded pantograph linkage is used on electric locomotives to collect current from an overhead wire.

Pantothenic acid see VITAMINS.

Papanicolaou, George Nicholas (1883-1962)
Greek-born US anatomist largely responsible for the development of cytologic PATHOLOGY (see also CYTOLOGY). He devised the PAP SMEAR TEST, used to reveal early signs of CANCER.

Pap smear test (Papanicolaou test)
CANCER screening test in which cells scraped from the cervix of the uterus are examined for abnormality under the microscope using the method of G.N. PAPANICOLAOU.

Paracelsus, Philippus Aureolus (1493-1541)
Swiss alchemist and physician who channeled the arts of ALCHEMY toward the preparation of medical remedies (see IATROCHEMISTRY). Born Theophrastus Bombast von Hohenheim, he adopted the name Paracelsus.

Paraffins see ALKANES.

Parallax
The difference in observed direction of an object due to a difference in position of the observer. Parallax in nearby objects may be observed by closing each eye in turn so that the more distant object appears to move relative to the closer. The brain normally assembles these two images to produce a stereoscopic effect (see BINOCULAR VISION). Should the length and direction of the line between the two points of observation be known, parallax may be used to calculate the distance of the object. In astronomy, the parallax of a star is defined as half the greatest parallactic displacement when viewed from Earth at different times of the year (see PARSEC).

Paralysis
Temporary or permanent loss of MUSCLE power or control. It may consist of inability to move a limb or part of a limb or individual muscles, paralysis of the muscles of breathing, swallowing and voice being especially serious. Paralysis may be due to disease of the brain (eg STROKE; TUMOR); SPINAL CORD (POLIOMYELITIS); nerve roots (SLIPPED DISK); peripheral NERVOUS SYSTEM (NEURITIS); neuromuscular junction (MYASTHENIA GRAVIS), or muscle (MUSCULAR DYSTROPHY). Disturbance of blood ELECTROLYTE levels can also lead to paralysis.

Paramagnetism
Weak magnetization of a material in the same direction as an applied MAGNETIC FIELD. Normally stronger than DIAMAGNETISM, the effect varies inversely with TEMPERATURE, and involves the partial alignment of intrinsic or orbital ELECTRON dipoles.

Paranoia
A PSYCHOSIS characterized by delusions of persecution (hence the popular term, persecution mania) and grandeur, often accompanied by HALLUCINATIONS. The delusions may form a self-consistent system that replaces reality. (See also MENTAL ILLNESS; SCHIZOPHRENIA.)

Paraplegia
PARALYSIS involving the lower part of the body, particularly the legs. Injury to the SPINAL CORD is often the cause.

Parapsychology (psychic research)
A field of study concerned with the scientific evaluation of two distinct types of phenomena: those collectively termed ESP, and those concerned with life after death, reincarnation, etc,

particularly including claims to communication with souls of the dead (spiritism or, incorrectly, spiritualism). Tests of the former have generally been inconclusive, of the latter almost exclusively negative. In both cases many "believers" hold that such phenomena, being beyond the bounds of science, cannot be subjected to laboratory evaluation. In spiritism, the prime site of the alleged communication is the seance, in which one individual (the medium) goes into a trance before communicating with the souls of the dead, often through a spirit guide (a spirit associated particularly with the medium). The astonishing disparity between different accounts of the spirit world has led to the whole field being treated with skepticism; fraud has often been suspected or proved.

Parasite
An organism that is for some part of its life history physiologically dependent on another, the host, from which it obtains nutrition. There is no clear demarcation between parasites and COMMENSALS on the one hand, and parasites and predators on the other, but parasites usually impose some cost on their host (unlike commensals), and are usually smaller than their host, with a higher reproductive rate (unlike predators). They also differ from most predators in not killing the host outright.

Nearly all the major groups of organisms have some parasitic members, though they are rare among the vertebrates. The most important parasites, besides the viruses which are a wholly parasitic group, occur in the bacteria, protozoa, platyhelminthes and nematodes. Study of the parasitic worms is termed helminthology. Blood-sucking arthropods, such as mosquitoes, tsetse flies and ticks, are also important because they transmit many diseases and serve as vectors for other parasites.

Parasites that live inside their host, such as tapeworms and liver flukes, are known as endoparasites. Those that live externally, such as fleas and lice, are known as ectoparasites. (See also BROOD PARASITES.)

Parasitic diseases
Infestation or infection by PARASITES, usually referring to nonbacterial and nonviral agents (ie to PROTOZOA and helminths). MALARIA, LEISHMANIASIS, trypanosomiasis (see TRYPANOSOMES), CHAGAS' DISEASE, FILARIASIS, SCHISTOSOMIASIS, toxoplasmosis, amebiasis and TAPEWORM are common examples. Manifestations may depend on the life cycle of the parasite; animal or insect vectors are usual.

Parasympathetic nervous system see NERVOUS SYSTEM.

Parathyroid glands
A set of four small ENDOCRINE GLANDS lying behind the THYROID that regulate calcium metabolism. Parathyroid hormone releases calcium from bone and alters the intestinal absorption and kidney excretion of calcium and phosphorus. Disease or loss of parathyroid glands may lead to TETANY, CATARACT or mental changes, and may be associated with disorders of IMMUNITY. Parathyroid overactivity or TUMORS cause raised blood calcium leading to bone disease, kidney disease and mental abnormalities. (◀ page 111)

Paré, Ambroise (*c*1510-1590)
French surgeon whose many achievements (eg adopting ligatures or liniments in place of CAUTERIZATION; introducing the use of artificial limbs and organs) have earned him regard as a father of modern SURGERY.

Paresis
Muscular weakness of a part, usually used in distinction to PARALYSIS, which implies complete loss of muscle power. It may be due to disease of the BRAIN, SPINAL CORD, peripheral NERVOUS SYSTEM or MUSCLES.

Parity
Physical property of a wave function (see WAVE MECHANICS) in QUANTUM MECHANICS specifying the function's behavior when its spatial coordinates are simultaneously reflected through the origin. If the parity is even, the wave function is unchanged by changing the sign of its coordinates; if it is odd, the wave function's sign changes. Parity has no significance in classical physics, but essentially arises from the symmetry of space. Strong nuclear and electromagnetic interactions conserve parity because they are governed by physical laws which do not distinguish between a right- or left-handed coordinate system (as in nature, where an object and its mirror image are equally realizable). Parity is not conserved in weak nuclear interactions.

Parkinson's disease
A common condition of the nervous system, most often seen in the elderly, causing a characteristic mask-like facial appearance, shuffling gait, slowness to move, muscular rigidity and tremor at rest; mental ability is preserved in most cases. It is a disorder of the basal ganglia of the BRAIN.

Parotid glands see SALIVA.

Parotitis
Inflammation of the salivary glands; mumps is an epidemic form of parotitis, but one infection confers lifelong immunity.

Parrot fever see PSITTACOSIS.

Parsec (pc)
In astronomy, the distance at which 1 ASTRONOMICAL UNIT would subtend an angle of 1 second. Originally defined as the distance of a star with a PARALLAX of 1″ viewed from Earth and Sun, the parsec was introduced to replace the LIGHT YEAR. 1 pc=3.258 ly=206,265AU.

Parthenogenesis
The production of young from unfertilized eggs. It is a modification of the normal process of sexual reproduction and allows females to reproduce without males. Parthenogenesis is seen in various plants (eg dandelions) and insects (eg aphids). Among higher animals it occurs in some fish and lizards. In some species it alternates with normal sexual reproduction, while in others it is the sole form of reproduction.

Particles, Elementary see SUBATOMIC PARTICLES.

Pascal, Blaise (1623-1662)
French mathematician, physicist and religious philosopher. His experiments (performed by his brother-in-law) observing the heights of the column of a BAROMETER at different altitudes on the mountain Puy-de-Dôme (1646) confirmed that the atmospheric air had weight. He also pioneered HYDRODYNAMICS and HYDROSTATICS, in doing so discovering PASCAL'S LAW.

Pascal (Pa)
The SI unit of PRESSURE, being that due to a force of one NEWTON acting per square meter.

Pascal's Law
In HYDROSTATICS, states that the pressure in an enclosed body of fluid arising from forces applied to its boundaries is transmitted equally in all directions with unchanged intensity. This pressure acts at right angles to the surface of the fluid container.

Paschen, Friedrich (1865-1947)
German physicist and a pioneer of SPECTROSCOPY, best known for his experimental work on infrared spectra and for explaining the "Paschen-Back Effect" – which concerns the splitting of spectral lines in an intense MAGNETIC FIELD.

Passerines
Members of the largest order of birds, the Passeriformes, also known as the perching birds or songbirds. The order includes no fewer than 60 different families, and accounts for about half the living species of birds. It encompasses the wrens, shrikes, swallows, larks, flycatchers, tits, treecreepers, finches and sparrows, starlings, orioles, bowerbirds, birds of paradise and crows. They are characterized by the foot, which has four unwebbed, forward-pointing toes; this structure is well adapted for gripping thin twigs or other perches. The structure of the syrinx (the "vocal cords") is also characteristic, and allows them to produce complex songs.

Pasteur, Louis (1822-1895)
French microbiologist and chemist. In his early pioneering studies in STEREOCHEMISTRY he discovered optical ISOMERISM. His attentions then centered around FERMENTATION, in which he demonstrated the role of microorganisms. He developed PASTEURIZATION as a way of stopping wine and beer from souring, and experimentally disproved the theory of SPONTANEOUS GENERATION. His "germ theory" of DISEASE proposed that diseases are spread by living germs (ie BACTERIA), and his consequent popularization of the STERILIZATION of medical equipment saved many lives. While studying ANTHRAX in cattle and sheep he developed a form of VACCINATION rather different from that of JENNER: he found that inoculation with dead anthrax germs gave future IMMUNITY from the disease. Treating RABIES similarly, he concluded that it was caused by a germ too small to be seen – ie a VIRUS. The Pasteur Institute was founded in 1888 to lead the fight against rabies.

Patch test
Test used in investigation of skin and systemic ALLERGY, especially contact DERMATITIS. Patches of known or likely sensitizing substances are placed on the skin for a short period. Local ERYTHEMA or HIVES indicate allergy.

Pathogen
A disease-causing microorganism. In ecological terms a pathogen is simply a very small parasite: there is no real difference, in terms of their way of life or effects, between pathogens and many endoparasites, but the term parasite is usually reserved for multicellular organisms.

Pathology
Study of the ANATOMY of DISEASE. Morbid anatomy, the dissection of bodies after death with a view to discovering the cause of disease and the nature of its manifestations, is complemented and extended by HISTOLOGY. In addition to AUTOPSY, biopsies (see BIOPSY) and surgical specimens are examined; these provide information that may guide treatment. It has been said that pathology is to MEDICINE what anatomy is to PHYSIOLOGY.

Patina
The attractive thin film of CORROSION products formed on metals by weathering, especially the green tarnish on copper or BRONZE.

Pauli, Wolfgang (1900-1958)
Austrian-born physicist awarded the 1945 Nobel Prize for Physics for his discovery of the Pauli EXCLUSION PRINCIPLE, that no two fermions (see FERMI-DIRAC STATISTICS) in a system may have the same four quantum numbers. In terms of the ATOM, this means that at most two electrons may occupy the same ORBITAL (the two having opposite SPIN).

Pauling, Linus Carl (b. 1901)
US chemist and pacifist awarded the 1954 Nobel Prize for Chemistry for his work on the chemical BOND (see also ELECTRONEGATIVITY) and the 1962 Nobel Peace Prize for his support of unilateral nuclear disarmament.

Pavlov, Ivan Petrovich (1849-1936)
Russian physiologist best known for his work on the conditioned REFLEX. Regularly, over long periods, he rang a bell just before feeding dogs, and found that eventually they salivated as soon as

they heard the bell, even when there was no food forthcoming. He also studied the physiology of the DIGESTIVE SYSTEM, and for this received the 1904 Nobel Prize for Physiology or Medicine.

Peat
Partly decayed plant material found in layers, usually in marshy areas, where acid conditions inhibit the bacteria that normally decompose plant remains. It is composed mainly of the peat moss *Sphagnum*, but also of sedges, trees, etc. Under the right geological conditions, peat forms COAL.

Peck (pk)
Unit of capacity of dry substances, corresponding to 8 quarts or 8.81 liter (US Customary System) or 2 gallons or 9.092 liter (British Imperial System).

Peck order
The term given to a DOMINANCE HIERARCHY in birds. The top bird can peck all others; the second can peck all but the top bird, and so on down to the bottom bird, who is pecked by all but can peck none.

Pectin
A polysaccharide found in plant tissue, capable of forming thick GELS with strong, acid SUGAR solutions and extensively used in the food industry to set jellies and jams. Commercial quantities are obtained from citrus and apple wastes after removal of the juice.

Pedersen, Charles (b. 1904)
Korean-born US scientist who received the 1987 Nobel Prize for Chemistry together with DONALD CRAM and JEAN-MARIE LEHN for their study of molecular recognition ("host-guest" chemistry), which has opened up new areas of research in medicine, materials science and biology.

Pediatrics
Branch of MEDICINE concerned with the care of children. This starts with newborn, especially premature, babies for whom intensive care is required to protect the baby from and adapt it to the environment outside the uterus. An important aspect is the recognition and treatment of congenital diseases in which structural or functional defects occur due to inherited disease (eg MONGOLISM) or disease acquired during development of embryo or fetus (eg SPINA BIFIDA). Otherwise, INFECTIOUS DISEASE, failure to grow or develop in a normal way, MENTAL RETARDATION, DIABETES, ASTHMA and EPILEPSY form the bulk of pediatric practice.

Pedicel see FLOWER.

Pedomorphosis
The precocious development of sexual organs and breeding ability in the juvenile or larval form of an organism. The best-known example is the axolotl, a salamander that normally breeds in its larval form, but can be induced to take on the adult form by treatment with hormones. Pedomorphosis is believed to have been important in the evolution of certain plants and animals, including humans; in this context, it is often known as neoteny. (See also CEPHALOCHORDATES).

Pegasus
The Winged Horse, a large N Hemisphere constellation noticeable for its Great Square, formed by four bright stars. Pegasus contains a GLOBULAR CLUSTER (M15).

Pegmatite
Very coarse-grained IGNEOUS or METAMORPHIC ROCK formed by slow crystallization from a melt containing volatiles. They usually have the composition of GRANITE, but often contain unusual MINERALS, including GEMS, and rare elements.

Pellagra
Deficiency disease (due to lack of nicotinic acid, a B vitamin) often found in maize- or millet-dependent populations. A DERMATITIS, initially resembling sunburn but followed by thickening, scaling and pigmentation, is characteristic; internal EPITHELIUM is affected (sore tongue, DIARRHEA). Confusion, DELIRIUM, hallucination and ultimately dementia may ensue.

Peltier effect
The heating or cooling effect at a junction between two dissimilar METALS when an electric current is driven through a circuit containing the junction. The effect is named for Jean Charles Athanase Peltier (1785-1845), the French scientist who discovered it in 1834. (See also THERMOCOUPLE.)

Pelvic inflammatory disease
General term for a variety of infections of the female reproductive system. Often sexually acquired, they are a common cause of infertility because fibrous tissue laid down during inflammation blocks the FALLOPIAN TUBES.

Pelvis
Lowest part of the trunk in animals, bounded by the pelvic bones and in continuity with the abdomen. The principal contents are the bladder and lower gastrointestinal tract (rectum) and reproductive organs, particularly in females – the uterus, ovaries, Fallopian tubes and vagina. The pelvic floor is a powerful muscular layer that supports the pelvic and abdominal contents and is important in urinary and fecal continence. The pelvic bones articulate with the legs at the hip joints. (♦ page 216)

Pendulum
A rigid body mounted on a fixed horizontal axis that is free to rotate under the influence of gravity. Many types of pendulum exist (eg Kater's and FOUCAULT's pendulum), the most common consisting of a large weight (the bob) supported at the end of a light string or bar. An idealized simple pendulum, with a string of negligible weight and length l, the weight of its bob concentrated at a point and a small swing amplitude, executes SIMPLE HARMONIC MOTION. The time, T, for a complete swing (to and fro) is given by $T=2\pi\sqrt{l/g}$, depending only on the string length and the local value of the gravitational ACCELERATION, g. Actual physical or compound pendulums approximate this behavior if they have a small angle of swing. They are used for measuring absolute values of g or its variation with geographical position, and as control elements in clocks.

Penicillin see ANTIBIOTICS.

Penis
Male reproductive organ for introducing sperm and semen into the female vagina and uterus; its urethra also carries urine from the bladder. The penis is made of connective tissue and specialized blood vessels that become engorged with blood in sexual arousal and cause the penis to become stiff and erect; this facilitates the intromission of semen in sexual intercourse. (♦ page 204)

Pennsylvanian
The penultimate period of the PALEOZOIC era, stretching between about 320 and 286 million years ago, the upper part of the CARBONIFEROUS period. (See GEOLOGY.) (♦ page 106)

Pentlandite
Bronze SULFIDE mineral with metallic luster, of composition $(Fe,Ni)_9S_8$; the chief ore of NICKEL. Usually found with PYRRHOTITE, its major occurrence is at Sudbury, Ontario.

Penumbra see SHADOW.

Penzias, Arno Allan (b. 1933)
US astrophysicist who shared the 1978 Nobel Prize for Physics with R. WILSON and P. KAPITZA. Penzias and Wilson detected the cosmic microwave background radiation, which supported the Big Bang theory of the origin of the Universe (see COSMOLOGY).

Peptide
A compound containing two or more AMINO ACIDS linked through the amino group ($-NH_2$) of one acid and the carboxyl group ($-COOH$) of the other. The linkage $-NH-CO-$ is termed a peptide or AMIDE bond. Peptides containing two amino acids are called dipeptides; with three, tripeptides, and so on; those with many units are polypeptides or proteins.

There is no clear dividing line between polypeptides or peptides and PROTEINS, but the term protein is usually reserved for fully functional long-chain molecules. Polypeptide or peptide is used for the individual chains in proteins containing two or more separate chains (dimers, oligomers), for partially synthesized chains, and for the breakdown products of proteins. Hormones such as insulin are sometimes referred to as peptide hormones, rather than proteins, since they consist of relatively short chains. (♦ page 111)

Perchloroethylene (tetrachloroethene)
A non-flammable hydrocarbon used widely in dry cleaning and industrial degreasing. MW 165.83, mp −19°C, bp 121°C.

Perennation
In plants, survival from one year to the next. In herbaceous perennials the part of the plant that survives the winter – the bulb, root, rhizome or corm – is known as the perennating organ.

Perennial
Any plant that continues to grow for more than two years. Trees and shrubs are examples of perennials that have woody stems that thicken with age. The herbaceous perennials such as the peony and daffodil have stems that die down each winter and regrow in the spring from underground perennating organs such as TUBERS and BULBS. (See also ANNUAL; BIENNIAL.)

Perianth see FLOWER.

Pericarditis
Inflammation of the pericardium, the membrane that contains the fluid surrounding the heart.

Pericardium
Two thin connective-tissue layers covering the HEART surface. It may become inflamed as a result of virus or bacterial infection and in UREMIA.

Peridot see OLIVINE.

Peridotite
Dark IGNEOUS ROCK consisting mainly of OLIVINE with some PYROXENE and HORNBLENDE but little FELDSPAR; it alters to SERPENTINE. Some varieties bear chromium ore, platinum or diamonds (see KIMBERLITE).

Perigee see ORBIT.

Perihelion see ORBIT.

Periodic Table
A table of the ELEMENTS in order of atomic number (see ATOM), arranged in rows and columns to illustrate periodic similarities and trends in physical and chemical properties. Such classification of the elements began in the early 19th century, when Johann Wolfgang Döbereiner (1780-1849) discovered certain "triads" of similar elements (eg calcium, strontium, barium) whose atomic weights were in arithmetic progression. By the 1860s many more elements were known, and their atomic weights determined, and it was noted by John Alexander Reina Newlands (1838-1898) that similar elements recur at intervals of eight – his "law of octaves" – in a sequence in order of atomic weight. In 1869 MENDELEYEV published the first fairly complete periodic table, based on his discovery that the properties of the elements vary periodically with atomic weight. There were gaps in the table corresponding to elements then unknown, whose properties Mendeleyev predicted with remarkable accuracy. Modern understanding of atomic structure has shown that the numbers and arrangement of the electrons in the atom are responsible for the periodicity of properties; hence the atomic number, rather than

the atomic weight, is the basis of ordering. Each row, or period, of the table corresponds to the filling of an electron "shell"; hence the numbers of elements in the periods is 2, 8, 8, 18, 18, 32, 32. (There are n^2 ORBITALS in the nth shell). The elements are arranged in vertical columns or groups containing those of similar atomic structure and properties, with regular gradation of properties down each group. The longer groups, with members in the first three short periods, are known as the Main Groups, usually numbered IA to VIIA, and 0 for the NOBLE GASES. The remaining groups, the TRANSITION ELEMENTS, are numbered IIIB to VIII (a triple group), IB and IIB. The characteristic VALENCE of each group is equal to its number N, or to $(8-N)$ for some nonmetals. Two series of 14 elements each, the LANTHANIDES and ACTINIDES, form a hyper-transition block in which the inner f ORBITALS are being filled; their members have similar properties, and they are usually counted in Group IIIB. European chemists number the groups differently: the members of the Main Groups are numbered I-IIA, III-VIIB and 0 for the noble gases; and the transition elements are number IIIA-VIII, IB and IIB. Hence, in Europe the lanthanides and actinides are counted in Group IIIA. (See also TRANSURANIUM ELEMENTS.) (♦ page 130)

Periodontics
Branch of DENTISTRY concerned with the structures that fix the teeth in the jaw.

Periscope
Optical instrument that permits an observer to view his surroundings along a displaced axis, and hence from a concealed, protected or submerged position. The simplest periscope, used in tanks, has two parallel reflecting surfaces (prisms or mirrors). An auxiliary telescopic gunsight may be added. Submarine periscopes have a series of lenses within the tube to widen the field of view, crosswires and a range-finder, and can rotate and retract.

Perissodactyls
An order of herbivorous MAMMALS that includes the rhinoceroses, tapirs and horses. All have one or three toes (the name means "odd-toed"), the weight of the body being supported by the middle toe of each foot. (See also ARTIODACTYLS.)

Peristalsis
The movements of the GASTROINTESTINAL TRACT, which cause forward propulsion and mixing of the contents. It is effected by the autonomic nervous system acting on visceral muscle layers. The term is also used for the type of movement shown by earthworms, in which some segments elongate while those behind them contract, propelling the animal forward. (♦ page 40)

Peritoneum
Two thin layers of connective tissue lining the outer surface of the abdominal organs and the inner walls of the ABDOMEN. A small amount of fluid lies between them in an extensive potential space, allowing free movement of the organs over each other.

Peritonitis
INFLAMMATION of PERITONEUM, usually caused by BACTERIAL INFECTION or chemical irritation of peritoneum when internal organs become diseased (as with appendicitis) or when gastrointestinal tract contents escape (as with a perforated peptic ULCER). Characteristic pain, sometimes with shock, fever, and temporary cessation of bowel activity (ileus), is common.

Perkin, Sir William Henry (1838-1907)
English chemist who, in 1856, while studying under VON HOFFMAN, discovered mauve, the first synthetic dye (see DYES AND DYEING.) He manufactured this and other dyes until 1874, and then devoted his remaining years to research. In 1808 he synthesized coumarin, the first synthetic PERFUME.

Permafrost
Permanently frozen ground, typical of the treeless plains of Siberia (see TUNDRA), though common throughout polar regions to depths of as much as 60m.

Permalloy
An ALLOY of iron and nickel, often with 5% molybdenum; it has a very high magnetic PERMEABILITY and is used in TRANSFORMERS.

Permeability (μ)
The ratio of the electromagnetic INDUCTION in a material to the MAGNETIC FIELD producing it. Materials showing DIAMAGNETISM and PARAMAGNETISM have permeabilities just below and above the free space value ($\mu_0=4\pi\times10^{-7}$H/m); ferromagnets have a permeability a thousand times greater.

Permian
The last period of the PALEOZOIC era, stretching between about 286 and 248 million years ago. (See also GEOLOGY.) (♦ page 106)

Permittivity (ε)
A constant of proportionality between an electric charge and the ELECTRIC FIELD emanating from it. The factor by which it exceeds the free space value ($\varepsilon_0=8.85\times10^{-12}$F/m) in a given material is known as the relative permittivity, or DIELECTRIC constant for the material.

Peroxides
Compounds of OXYGEN containing the peroxy group (–O–O–). ALKALI METAL and ALKALINE-EARTH METAL peroxides, containing the peroxide ion O_2^{2-}, are formed by heating the metals or their OXIDES in excess air. Covalent peroxides include peracetic acid and peroxymonosulfuric acid ("Caro's acid"). They are powerful oxidizing agents, used in bleaching. (See also SUPEROXIDES.) Hydrogen peroxide (H_2O_2) is a colorless liquid, usually produced as aqueous solutions by electrolytic or organic oxidation processes, a powerful oxidizing agent which readily decomposes into water and oxygen on heating or with various catalysts. It is used in bleaching, organic synthesis, medicine and in rocket fuels. mp −0.4°C, bp 150°C.

Perrin, Jean Baptiste (1870-1942)
French physical chemist awarded the 1926 Nobel Prize for Physics for his studies of BROWNIAN MOTION in which, by examining colloidal particles, he was able to arrive at a good value of the AVOGADRO number.

Perseus
Large N Hemisphere constellation containing the eclipsing binary (see DOUBLE STAR) ALGOL, two GALACTIC CLUSTERS, one of which is a double cluster, and the bright star Mirfak. It gives its name to the Perseid METEOR shower. (♦ page 240)

Persona
Term used by JUNG to describe the individual's projection of himself; ie the role that he plays to conform with others' expectations of his personality.

Perspiration see SWEAT.

Perturbation
In the elliptical orbit of one celestial body around another, an irregularity caused by the gravitational attraction of a third.

Pertussis see WHOOPING COUGH.

Perutz, Max Ferdinand (b. 1914)
Austrian-born British biochemist who shared with KENDREW the 1962 Nobel Prize for Chemistry for their research into the structure of HEMOGLOBIN and other globular PROTEINS.

Pesticide
Any substance used to kill "pests" – living organisms responsible for economic damage to crops (either growing or under storage), or to ornamental plants or domestic wild animals. Pesticides are subdivided into INSECTICIDES (which kill insects); miticides (which kill mites); HERBICIDES (which kill plants); FUNGICIDES (which kill fungi), and rodenticides (which kill rats and mice). The efficient control of pests is of enormous economic importance for humans, particularly as farming becomes more intensive. A major question with all pesticides is the possibility of adverse environmental side-effects (see ECOLOGY; POLLUTION).

Petit, Alexis Thérèse (1791-1820)
French physicist who worked with P.L. DULONG to discover Dulong and Petit's Law.

Petit mal see EPILEPSY.

PETN (Pentaerythritol tetranitrate, $C(CH_2ONO_2)_4$)
Colorless crystalline solid made by NITRATION of pentaerythritol. It is a high EXPLOSIVE used in detonators and grenades.

Petri dish
Shallow glass dish with a loose lid, used for growing CULTURES, named for Julius Petri (1852-1921), an assistant to R. KOCH.

Petrification see FOSSILS.

Petrochemicals
Chemicals made from PETROLEUM and NATURAL GAS, ie many organic chemicals, and the inorganic substances carbon black, sulfur, ammonia and hydrogen peroxide. Polymers, detergents, solvents and nitrogen fertilizers are major products.

Petrol see GASOLINE.

Petroleum
Naturally occurring mixture of HYDROCARBONS, usually liquid "crude oil", but sometimes taken to include NATURAL GAS. (See also ASPHALT; BITUMEN.) Petroleum is believed to be formed from organic debris, chiefly of plankton and simple plants, which has been rapidly buried in fine-grained sediment in marine conditions unfavorable to oxidation. After some biodegradation, increasing temperature and pressure cause CRACKING, and oil is produced. As the source rock is compacted, oil and water are forced out, and slowly migrate to porous reservoir rocks, chiefly sandstone or limestone. Finally, secondary migration occurs within the reservoir and the oil coagulates to form a pool, generally capped by impervious strata, and often associated with natural gas. Some oil seeps to the Earth's surface. The petroleum industry supplies about half the world's energy, as well as the raw materials for PETROCHEMICALS. Modern technology has made possible oil-well drilling to a depth of 5km, and deep-sea wells in 150m of water. Rotary drilling is used, with pressurized mud to carry the rock to the surface and to prevent escape of oil. When the well is completed the oil rises to the surface, usually under its own pressure, though pumping may be required. After removing salt and water, the petroleum is refined by fractional DISTILLATION producing the fractions GASOLINE, KEROSINE, diesel oil, fuel oil, lubricating oil, and ASPHALT. Undesirable compounds may be removed by solvent extraction, treatment with sulfuric acid, etc, and less valuable components converted into more valuable ones by CRACKING, reforming, ALKYLATION and polymerization. The chemical composition of crude petroleum is mainly ALKANES, saturated ALICYCLIC COMPOUNDS, and AROMATIC COMPOUNDS, with some sulfur compounds, oxygen compounds (carboxylic acids and phenols), nitrogen and salt. (See also OIL SHALE.)

Petrology
Branch of geology concerned with the history, composition, occurrence, properties and classification of rocks. (See GEOLOGY; ROCKS.)

Pewter
Class of ALLOYS consisting chiefly of TIN, now hardened with copper and antimony, and usually

containing lead. Roman pewter was high in lead and darkened with age. Pewter has been used for bowls, drinking vessels and candlesticks.

pH
Measure of the acidity (see ACID) of an aqueous solution, where the thermodynamic activity of the hydrogen ions in the solution approximates to their concentration in MOLES/liter for dilute solutions. Pure water has a pH of 7 (ie contains 10^{-7} mol/l H^+); acidic solutions have pH less than 7, basic ones greater than 7. (See also BUFFER.)

Phaecolith see LACCOLITH.

Phaeophyta (brown algae) see ALGAE.

Phage see BACTERIOPHAGE.

Phagocyte
A white blood cell (leukocyte) that contributes to the IMMUNE SYSTEM by engulfing microorganisms. The two main types are MACROPHAGES and neutrophils.

Phagocytosis
A process by which cells engulf particles. It is seen in cells that also show AMEBOID MOVEMENT, and involves the cell extending pseudopodia on both sides of the particle, which then coalesce so that the particle, within a membrane-bound vacuole, is surrounded by the cytoplasm of the cell. LYSOSOMES then empty their contents into the vacuole, and the particle is digested by enzymes. Amebae and similar organisms use phagocytosis to obtain their food. Cells in the IMMUNE SYSTEM of vertebrates, such as macrophages, use phagocytosis to engulf and incapacitate invading microorganisms, and to clear up cell debris, dust particles in the lungs, etc.

Phanerozoic
The eon of visible life, represented by rock strata in which FOSSILS appear – from about 590 million years ago to the present, containing the PALEOZOIC, MESOZOIC and CENOZOIC eras. (See also GEOLOGY; PRECAMBRIAN.) (◀ page 106)

Pharmacology
The study of DRUGS, their chemistry, mode of action, routes of absorption, excretion and METABOLISM, drug interactions, toxicity and side-effects. New drugs, based on older drugs, traditional remedies, chance observations, etc are tested for safety and efficacy, and manufactured by the pharmaceutical industry. The dispensing of drugs is PHARMACY. Drug prescription is the cornerstone of the medical treatment of disease.

Pharmacopeia
A text containing all available drugs and pharmacological substances used in medicine. It lists drugs; their properties and formulation; routes and doses of administration; mode of action, metabolism and excretion; known interaction with other drugs; contraindications and precautions in particular diseases; toxicity and adverse effects.

Pharmacy
The preparation or dispensing of DRUGS and pharmacological substances used in MEDICINE; also, the place where this is practiced.

Pharynx
The back of the throat where the mouth (oropharynx) and NOSE (nasopharynx) pass back into the ESOPHAGUS. It contains specialized MUSCLE for swallowing. The food and air channels are kept functionally separate so that swallowing does not interfere with breathing and speech. (◀ page 40)

Phase
The proportion of a cycle aready executed by an oscillating system, expressed as an angle (360° or 2π radians corresponding to a full cycle). Thus if an AC voltage is at its maximum value while the current is passing through zero, there is said to be a 90° (π/2) phase difference between them.

Phase equilibria
In THERMODYNAMICS, EQUILIBRIA between substances in different phases (solid, liquid or gas). Phase diagrams, basic to engineering, metallurgy and mineralogy, are empirical graphs showing what phases exist at different pressures and temperatures. The phase rule, deduced by J.W GIBBS, states that, for a closed system, $F=C-P+2$, where P is the number of phases, C the number of independent chemical components and F the number of degrees of freedom, ie the number of variables (pressure, temperature, composition) whose values must be specified to define the system.

Phenobarbital (phenobarbitone) see BARBITURATES.

Phenol (carbolic acid, C_6H_5OH)
The simplest of the PHENOLS, a white, hygroscopic crystalline solid, isolable from coal tar, but made by acid hydrolysis of cumene hydroperoxide, or by fusion of sodium benzene sulfonate (see SULFONIC ACIDS) with sodium hydroxide. Formerly used as an ANTISEPTIC, phenol is now of commercial importance in the manufacture of BAKELITE and many other resins, plastics, dyes, detergents, drugs, etc. MW 94.1, mp 40.9°C, bp 182°C.

Phenolphthalein (2,2-bis (p-hydroxyphenyl) phthalide)
A widely used indicator which is colorless in acid and neutral solutions, but red in basic solution. MW 318.33, mp 262-3°C.

Phenols
Class of AROMATIC COMPOUNDS in which a HYDROXIDE group is directly bonded to an aromatic ring system. They are very weak ACIDS and, like ALCOHOLS, form ETHERS and ESTERS. They are very liable to undergo electrophilic substitution (see NUCLEOPHILES), and hence condense with formaldehyde to form resins. The main phenols are PHENOL itself, CRESOL, RESORCINOL, pyrogallol (see GALLIC ACID) and PICRIC ACID.

Phenotype
The appearance of, and characteristics actually present in an organism, as contrasted with its GENOTYPE (its genetic makeup). Because of genetic DOMINANCE two organisms with the same phenotype (such as "tall") can have a different genotype (TT – homozygous for the tall allele, or Tt – heterozygous, but with the tall allele, T, dominant over the short allele, t). Organisms may also have an identical genotype but differing phenotype as a result of environmental influences such as lack of nutrients or extreme temperatures.

Phenylketonuria (PKU)
Inherited disease in which phenylalanine metabolism is disordered due to lack of an enzyme. It rapidly causes MENTAL RETARDATION, as well as irritability and vomiting, unless dietary foods low in phenylalanine are given from soon after birth and indefinitely. Screening of the newborn by urine tests (with confirmation by blood tests) facilitates prompt treatment.

Pheromone
A chemical messenger substance produced by one organism that affects the growth, development or behavior of another organism of the same species. Pheromones may be used to attract a mate, to induce or repress sexual development, to suppress growth, or to control the behavior of social animals.

Phlebitis
Inflammation of the wall of a vein, usually leading to thrombosis and obstruction. It is most commonly seen in patients with VARICOSE VEINS.

Phloem
A vascular tissue responsible for the transport of dissolved food substances thrugh the roots, stems and leaves of higher PLANTS. In flowering plants (ANGIOSPERMS) phloem mainly consists of elongated living sieve tubes, which have perforated end plates.

Phlogiston
The elementary principle postulated by G.H. STAHL to be lost from substances when they burn. The phlogiston concept provided 18th-century CHEMISTRY with its unifying principle, and the phlogiston theory of COMBUSTION found general acceptance until it was displaced by its inverse – LAVOISIER's oxygen theory.

Phlogopite
A range of magnesium-rich varieties of MICA, grading into BIOTITE.

Phobia
NEUROSIS characterized by exaggerated ANXIETY on confrontation with a specific object or situation; or the anxiety itself. Phobia is sometimes linked with OBSESSIONAL NEUROSIS, sometimes with HYSTERIA; in each case the object of phobia is usually merely symbolic. Common phobias include fear of public places (agoraphobia), of enclosed spaces (claustrophobia), of spiders (arachnophobia), of heights, flying, dentists and particular social situations.

Phobos
The inner moon of MARS, orbiting in 7.65h at a distance of 9270km. Roughly ellipsoidal in shape, it measures about 20×23×28km. (▶ page 221)

Phon
In acoustics, a unit of loudness. Loudness in phons is given by the number of DECIBELS above the reference level, 20 μPa, of a pure 1kHz-frequency SOUND which is judged by listeners to be of equal loudness with the original.

Phonon
In SOLID STATE PHYSICS, the particle (quantum) counterpart of the SOUND wave or LATTICE vibration, considered to have an important role in the CONDUCTION OF HEAT in electrical insulators.

Phoronid worms (horseshoe worms)
A phylum of marine, tube-dwelling, worm-like animals generally less than 200mm long. Each individual lives with its tube buried in sand or attached to a rock or shell, in shallow temperate or tropical seas. There are only about 11 species, most of which are colonial. The horseshoe worms possess a lophophore, or ciliated crown of tentacles, which is used for filtering food particles from the water. This feature suggests a relationship with the lampshells and the moss animals (see LOPHOPHORATES).

Phosgene (carbonyl chloride, $COCl_2$)
Colorless, reactive gas, hydrolyzed by water, made by catalytic combination of CARBON monoxide and CHLORINE, and used to make RESINS and DYES. Highly toxic, it was a poison gas in WWI.

Phosphates
Derivatives of phosphoric acid (see PHOSPHORUS): either phosphate ESTERS, or salts containing the various phosphate ions. Like SILICATES, these are numerous and complex, the simplest being orthophosphate, PO_4^{3-}. Of many phosphate minerals, the most important is APATITE. This is treated with sulfuric or phosphoric acid to give calcium dihydrogenphosphate ($Ca[H_2PO_4]_2$), known as superphosphate – the major phosphate FERTILIZER. The alkaline trisodium phosphate (TSP, Na_3PO_4) is used as a cleansing agent and water softener. Phosphates are used in making GLASS, SOAPS AND DETERGENTS.

Phosphoglycerides
Organic compounds that have an important role in the MEMBRANES of living organisms. Phosphoglycerides are made up of two FATTY ACID molecules bound to glycerol, with a polar group also bound to the glycerol via a phosphate group. In lecithin, one of the most widespread phosphoglycerides, the polar group is choline, a short-chain alcohol. Other phosphoglycerides have as

the polar group serine (an amino acid), inositol (a sugar), or another molecule of glycerol – all these compounds have one or more hydroxyl groups and are therefore alcohols. The hydrophilic nature of these groups, and of the phosphate and glycerol to which they are attached, accentuates the bipolarity of the fatty acid molecules, in which the hydrocarbon chains repel water while the carboxyl groups (now bound to the glycerol molecule) are attracted to water. This property accounts for the role of phosphoglycerides in biological membranes: the molecules are arranged in a bilayer, with the polar head-groups pointing outward on both sides, and the hydrocarbon chains of each layer pointing inward, forming a water-repellent inner layer to the membrane. Phosphoglycerides are the major type of PHOSPHOLIPID found in living organisms. (See also TRIGLYCERIDES.)

Phospholipids
Organic compounds made up of one or more fatty acid molecules bound to an alcohol – either glycerol or sphingosine – which is in turn bound to a phosphate group and a polar group. The polar group is a short-chain alcohol such as choline. The most abundant types of phospholipid are those in which glycerol is the main alcohol, the PHOSPHOGLYCERIDES.

Phosphor
A substance exhibiting LUMINESCENCE, ie emitting LIGHT (or other ELECTROMAGNETIC RADIATION) on nonthermal stimulation. Important phosphors include those used in television picture tubes (where stimulation is by electrons) and those coated on the inside wall of fluorescent lamp tubes to convert ULTRAVIOLET RADIATION into visible light.

Phosphorescence see LUMINESCENCE.

Phosphorus (P)
Reactive nonmetal in Group VA of the PERIODIC TABLE, occurring naturally as APATITE. This is heated with silica and coke, and elementary phosphorus is produced. Phosphorus has three main allotropes (see ALLOTROPY): white phosphorus, a yellow waxy solid composed of P_4 molecules, spontaneously flammable in air, soluble in carbon disulfide, and very toxic; red phosphorus, a dark red powder, formed by heating white phosphorus, less reactive, and insoluble in carbon disulfide; and black phosphorus, a flaky solid resembling GRAPHITE, consisting of corrugated layers of atoms. Phosphorus burns in air to give the trioxide and the pentoxide, and also reacts with the halogens, sulfur and some metals. It is used in making matches, ammunition, pesticides, steels, phosphor bronze, phosphoric acid and phosphate fertilizers. It is also of great biological importance. AW 31.0, mp (wh) 44°C, bp (wh) 280°C, sg (wh) 1.82, (red) 2.20, (bl) 2.69. Phosphorus forms phosphorous (trivalent) and phosphoric (pentavalent) compounds.
PHOSPHINE (PH_3) is a colorless, flammable gas, highly toxic, and with an odor of garlic. It is a weak BASE, resembling AMMONIA, and forms phosphonium salts (PH_4^+).
PHOSPHORIC ACID (H_3PO_4) is a colorless, crystalline solid, forming a syrupy aqueous solution. It is used to flavor food, in dyeing, to clean metals, and to make PHOSPHATES.
PHOSPHORUS PENTOXIDE (P_4O_{10}), is a white powder made by burning phosphorus in excess air. It is deliquescent (forming phosphoric acid), and is used as a dehydrating agent. (♦ page 130)

Photochemistry
Branch of PHYSICAL CHEMISTRY dealing with chemical reactions that produce LIGHT (see CHEMILUMINESCENCE; COMBUSTION), or that are initiated by light (visible or ultraviolet). Examples include PHOTOSYNTHESIS, PHOTOGRAPHY and bleaching by sunlight. One PHOTON of light of suitable wavelength may be absorbed by a molecule, raising it to an electronically excited state. Re-emission may occur by fluorescence or phosphorescence (see LUMINESCENCE), the energy may be transferred to another molecule, or a reaction may occur, commonly DISSOCIATION to form FREE RADICALS. The quantum yield, or efficiency, of the reaction is the number of molecules of reactant used (or product formed) per photon absorbed; this may be very large for chain reactions. (See also FLASH PHOTOLYSIS; LASER; RADIATION CHEMISTRY.)

Photoconductive detector
An electrical component whose CONDUCTIVITY increases as more light falls on it. Used in light detectors, light-sensitive switches, light meters, and in the Vidicon television camera tube, most employ photoconductive SEMICONDUCTORS such as lead telluride or cadmium sulfide.

Photoelectric cell
A device with electrical properties which vary according to the light falling on it. There are three types: PHOTOVOLTAIC CELLS; PHOTOCONDUCTIVE DETECTORS and phototubes (see PHOTOELECTRIC EFFECT).

Photoelectric effect
Properly known as the photoemissive effect: the emission of ELECTRONS from a surface when struck by ELECTROMAGNETIC RADIATION such as LIGHT. In 1905 EINSTEIN laid one of the twin foundations of QUANTUM THEORY by explaining photoemission in terms of the action of individual PHOTONS. The Einstein photoelectric law reflects the fact that no electrons are photoemitted unless the energy of the incident photons exceeds a certain threshold value, known as the surface work function for photoemission from the material. The effect is used in phototubes (ELECTRON TUBES having a photoemissive cathode), often employed as "electric eye" switches. Special types are used in image intensifiers and in the Image Orthicon television camera. (♦ page 21)

Photogrammetry
The use of photographs in mapmaking. Series of overlapping air photographs are generally used. If exposed in stereo pairs at a known altitude, such photographs can be used to make detailed, accurate relief maps.

Photography
The use of light-sensitive materials to produce permanent visible images (photographs). The most familiar photographic processes depend on the light-sensitivity of the silver halides. A photographic emulsion is a preparation of tiny crystals of these salts suspended in a thin layer of gelatin coated on a glass, film or paper support. On brief exposure to light in a camera or other apparatus, a latent image in activated silver salt is formed wherever light has fallen on the emulsion. This image is made visible in development, when the activated silver halide crystals (but not the unexposed ones) are reduced to metallic silver (black) using a weak organic reducing agent (the developer). The silver image is then made permanent by fixing, in the course of which it becomes possible to examine the image in the light for the first time. Fixing agents (fixers) work by dissolving out the silver halide crystals that were not activated on exposure. The image made in this way is densest in silver where the original subject was brightest and lightest where the original was darkest; it is thus a "negative" image. To produce a positive image, the negative (which is usually made on a film or glass plate support) is itself made the original in the above process, the result being a positive "print", usually on a paper carrier. An alternative method of producing a positive image is to bleach away the developed image on the original film or plate before fixing, and reexpose the unactivated halide in diffuse light. This forms a second latent image, which on development produces a positive image of the original subject (reversal processing).

The history of photography from the earliest work of NIÉPCE, DAGUERRE and FOX TALBOT to the present has seen successive refinements in materials, techniques and equipment. Photography became a popular hobby after EASTMAN first marketed roll film in 1889. The silver halides themselves are sensitive to light only from the blue end of the spectrum, so that in the earliest photographs other colors appear dark. The color sensitivity of emulsions was improved from the 1870s onward as small quantities of sensitizing dyes were incorporated. "Orthochromatic" plates became available after 1884 and "panchromatic" from 1906.

New sensitizing dyes also opened up the way to infrared and color photography. Modern "tripack" color films have three layers of emulsion, one each sensitive to blue, green and red light from the subject. Positive color transparencies are made using a reversal processing method in which the superposed, positive, silver images are replaced with yellow, magenta and cyan dyes respectively.

Motion-picture photography dates from 1890, when EDISON built a device to expose Eastman's roll film, and motion pictures rapidly became an important art form. Not all modern photographic methods employ the silver-halide process; eg xerography and the blueprint and ozalid processes work differently. FALSE-COLOR PHOTOGRAPHY and the diffusion process used in the polaroid land camera are both developments of the original silver-halide process.

Photometry
The science of the measurement of LIGHT, particularly as it affects illumination engineering. Because the brightness experienced when light strikes the human eye depends not only on the power conveyed by the radiation but also on the wavelength of the light (the visual sensation for a given power reaching a maximum at 555nm), a special arbitrary set of units is used in photometric calculations. In SI units, the photometric base quantity is luminous intensity, which measures the intensity of light radiated from a small source. The base unit of luminous intensity is the CANDELA (cd). The luminous flux (the photometric equivalent of the power radiating) from a point source is measured in lumens, where 1 lumen (lm) is the flux radiating from a 1 cd source through a solid angle of a steradian. The illuminance falling on a surface (formerly known as its illumination) is measured in luxes, where 1 lux (lx) is the level of illuminance occurring when a luminous flux of 1 lm falls on each m^2 of the surface. (See also LUMINANCE.)

Photon
The quantum of electromagnetic energy (see QUANTUM THEORY), often thought of as the particle associated with LIGHT or other ELECTROMAGNETIC RADIATION. Its ENERGY is given by $h\nu$ where h is the PLANCK CONSTANT and ν the frequency of the radiation. (♦ page 21)

Photoperiodism
The timing of seasonal activities in living organisms by means of daylength. Thus some plants flower only when the days reach a certain length; known as long-day plants, their flowering season is in spring or early summer. Short-day plants, on the other hand, flower in late summer or autumn, when the days are shorter. The critical daylength is different for each species of plant. The majority of birds also time their breeding season by photoperiodism, and in insects the onset of DIAPAUSE is

controlled by daylength. In many animals there is an additional mechanism for controlling the seasonal cycle, an endogenous circannual rhythm (see BIOLOGICAL CLOCKS).

Photosphere
A layer of gas on the Sun, 125-190km thick, visible to us as the Sun's apparent surface, emitting most of the Sun's light. Its temperature is estimated at 6000K.

Photosynthesis
The process by which green plants, cyanobacteria and other photosynthetic bacteria convert the energy of sunlight into chemical energy, which is used to fix carbon dioxide gas and transform it into CARBOHYDRATES. Overall, the process may be written as:

$$6CO_2+6H_20 \xrightarrow{light} C_6H_{12}O_6+6O_2$$

Although in detail photosynthesis is a complex sequence of reactions, two principal stages can be identified. In the "light reaction", CHLOROPHYLL (the key chemical in the whole process) is activated by absorbing a quantum of LIGHT, initiating a sequence of reactions in which the energy-rich compounds ATP (adenosine triphosphate) and NADPH (the reduced form of NADP nucleotide) are made, water being decomposed to give free oxygen in the process. In the second stage, the "dark reaction", the ATP and NADPH provide the energy and reducing power, respectively, for the assimilation of carbon dioxide gas, yielding glucose or other sugars. From them other carbohydrates, including STARCH, can be built up.

In plants, photosynthesis occurs in the CHLOROPLASTS. These are lacking in bacteria and cyanobacteria (both PROKARYOTES), and in them photosynthesis occurs on infoldings of the cell membrane. The photosynthetic process is controlled by ENZYMES. (◀ page 44)

Photovoltaic cell
A device for converting light radiation into electricity, used in light meters and for providing spacecraft power supplies. The photovoltage is usually developed in a layer of SEMICONDUCTOR (eg selenium) sandwiched between a transparent electrode and one providing support.

Phthalic acid (1,2-benzenedicarboxylic acid)
A dibasic acid used widely in chemical synthesis, mostly as its anhydride. MW 166.14, mp 210-11°C. Phthalic anhydride, made by oxidation of *o*-xylene is also used as a plasticizer and as a monomer for alkyde and polyester resins.

Phthalocyanine (tetrabenzoporphyrazine)
A blue-green pigment, MW 514.55. Copper phthalocyanine, in which a copper atom is complexed to the phthalocyanine molecule, is blue. Related green pigments are made by replacing some of the hydrogen atoms in copper phthalocyanine by chlorine atoms or sulphate groups.

Phylogeny
The ancestry of an organism. The term is most often used in connection with the idea that "ontogeny repeats phylogeny" (see ONTOGENY).

Phylum
In TAXONOMY, a major division of living organisms. Examples include the Phylum Chordata, Phylum Mollusca and Phylum Cyanobacteria. In botanical classification the term "division" is often used instead of phylum. Similar phyla are grouped together in KINGDOMS; a phylum is divided into classes.

Physical chemistry
Major branch of CHEMISTRY, in which the theories and methods of PHYSICS are applied to chemical systems. Physical chemistry underlies all the other branches of chemistry and includes theoretical chemistry. Its main divisions are the study of molecular structure; COLLOIDS; CRYSTALS; ELECTROCHEMISTRY; chemical EQUILIBRIUM; GAS laws; chemical KINETICS; MOLECULAR WEIGHT determination; PHOTOCHEMISTRY; SOLUTION; SPECTROSCOPY, and chemical THERMODYNAMICS.

Physical therapy see PHYSIOTHERAPY.

Physician see MEDICINE.

Physics
Originally, the knowledge of natural things (=natural science); now, the science dealing with the interaction of MATTER and ENERGY (but usually taken to exclude CHEMISTRY). Until the "scientific revolution" of the Renaissance, physics was a branch of PHILOSOPHY dealing with the natures of things. The physics of the heavens, for instance, was quite separate from (and often conflicted with) the descriptions of mathematical and positional ASTRONOMY. But from the time of GALILEO, and particularly through the efforts of HUYGENS and NEWTON, physics became identified with the mathematical description of nature; occult qualities were banished from physical science. Firm on its Newtonian foundation, classical physics gathered more and more phenomena under its wing until, by the late 19th century, comparatively few phenomena seemed to defy explanation. But the interpretation of these effects (notably BLACKBODY RADIATION and the PHOTOELECTRIC EFFECT) in terms of new concepts due to PLANCK and EINSTEIN involved the thorough reformulation of the fundamental principles of physical science (see QUANTUM THEORY; RELATIVITY). Physics today is divided into many specialisms. The principal of these are ACOUSTICS; ELECTRICITY and MAGNETISM; MECHANICS; NUCLEAR PHYSICS; OPTICS; QUANTUM THEORY; RELATIVITY; and THERMODYNAMICS.

Physiology
The study of function in living organisms. Based on knowledge of ANATOMY, physiology seeks to demonstrate the manner in which organs perform their tasks, and in which the body is organized and maintained in a state of HOMEOSTASIS. Normal responses to various stresses on the whole or on parts of an organism are studied. Important branches of physiology deal with RESPIRATION, BLOOD CIRCULATION, the NERVOUS SYSTEM, the DIGESTIVE SYSTEM, the KIDNEYS, the fluid and electrolyte balance, the ENDOCRINE GLANDS and METABOLISM.

Physiotherapy
System of physical treatment for disease or disability. Active and passive muscle movement, electrical stimulation, balancing exercises, heat, ultraviolet or shortwave radiation, and manual vibration of the chest wall with postural drainage, are some of the techniques used. Rehabilitation after FRACTURE, SURGERY, STROKE or other neurological disease, and the treatment of lung infection (PNEUMONIA, BRONCHITIS), are among the aims.

Phytoplankton see PLANKTON.

Piaget, Jean (1896-1980)
Swiss psychologist whose theories of the mental development of children, though now often criticized, have been of paramount importance. His many books include *The Psychology of Intelligence* (1947).

Piazzi, Giuseppe (1746-1826)
Italian astronomer who discovered Ceres, the first ASTEROID (1801). Through illness he lost it again, and it was rediscovered the following year by OLBERS.

Piccard
Name of the Swiss twin brothers Auguste (1884-1962), a physicist, and Jean Félix (1884-1963), a chemist. Both made famous high-altitude BALLOON ascents in order to study COSMIC RAYS with a minimum of atmospheric interference, Auguste in 1931 and 1932, and Jean in 1936. In 1948 Auguste successfully conducted an unpiloted trial dive of the BATHYSCAPHE, a deep-sea diving device built to his own design; the first piloted dive – in a new bathyscaphe – followed in 1953.

Pickering
Name of two US astronomers, Edward Charles Pickering (1846-1919) and his brother William Henry Pickering (1858-1938). Edward made important contributions to stellar PHOTOMETRY and was the inventor of the meridian photometer. William, in 1898, discovered Phoebe, the ninth moon of the planet SATURN.

Picric acid (2,4,6-trinitrophenol)
Yellow crystalline solid, made by NITRATION of PHENOL or its derivatives. A moderately strong ACID, it has been used as a DYE, as an ANTISEPTIC and ASTRINGENT for treating burns, and as a high EXPLOSIVE. MW 229.1, mp 122.5°C.

Piezoelectricity
A reversible relationship between mechanical stress and electrostatic POTENTIAL exhibited by certain CRYSTALS with no center of symmetry, discovered in 1880 during investigations of pyroelectric crystals (these are also asymmetric and get oppositely charged faces when heated). When pressure is applied to a piezoelectric crystal such as QUARTZ, positive and negative electric charges appear on opposite crystal faces. Replacing the pressure by tension changes the sign of the charges. If, instead, an electric potential is applied across the crystal, its length changes; this effect is linear. A piezoelectric crystal placed in an alternating electric circuit will alternately expand and contract. Resonance occurs in the circuit when its FREQUENCY matches the natural vibration frequency of the crystal, this effect being applied in frequency controllers. This useful way of coupling electrical and mechanical effects is used in MICROPHONES, phonograph pickups and ULTRASONIC generators.

Pigments, Natural
Chemical substances imparting colors to living organisms. In animals the most important examples include MELANIN (black), RHODOPSIN (purple) and the respiratory pigments, HEMOGLOBIN (red) and HEMOCYANIN (blue). In plants, CHLOROPHYLL (green) is important as the key chemical in PHOTOSYNTHESIS. Other plant pigments include the carotenes and xanthophylls, collectively known as carotenids (red-yellow), the anthocyanins (red-blue) and the anthoxanthins (yellow-orange).

Pile, Atomic see NUCLEAR REACTOR.

Piles see HEMORRHOIDS.

Pineal body (pineal gland)
A structure situated over the BRAIN stem in vertebrates, which appears to be a vestigial remnant of a light-sensitive organ. In birds it acts as a light-sensor (though within the skull) involved in response to daylength, or PHOTOPERIODISM. In many vertebrates it also secretes a hormone, melatonin, which controls changes in pigmentation. The pineal body develops from a paired brain outgrowth, the pineal apparatus. The posterior outgrowth becomes the pineal body, while in some lizards the anterior outgrowth develops into a functional eye that operates through a gap in the skull covered only by thin, transparent skin. It seems likely that the pineal apparatus represents the vestige of a pair of eyes located on the top of the head in the vertebrate ancestor. It is not known whether the pineal body has any function in man, but it may help to regulate our BIOLOGICAL CLOCK. (◀ page 111)

Pinel, Philippe (1745-1826)
French pioneer of the scientific study of MENTAL ILLNESS and the humane treatment of mental patients, whose remarkably modern ideas have earned him regard as a father of psychiatry.

Pinene
A natural terpene hydrocarbon. It is found in

turpentine and widely used as a solvent and fragrance. MW 136.24, mp −55°C, bp 164°C.

Pinkeye

Common name for CONJUNCTIVITIS.

Pinnipeds

Members of the Order Pinnipedia, the seals, sea-lions and walrus. Unlike other sea mammals, such as the CETACEANS, they emerge onto land to breed, and therefore retain hind limbs and fur.

Pint (pt)

Name of various units of dry or liquid measure. To measure liquids, this unit corresponds to 4 gills or 473.2 ml (US Customary System) or 568.26 ml (Imperial). As a dry measure the pint is equivalent to 0.5 quart or 0.551 liter (Customary).

Pioneer probes

US space probe series started in 1958. Pioneers 1 to 3 studied the VAN ALLEN RADIATION BELTS. Pioneers 5 to 9 were launched into solar orbit to study interplanetary space and the Sun itself. Pioneers 10 and 11 were Jupiter "fly-by" probes. Eleven years after its launch, Pioneer 10 was the first space vehicle to leave the Solar System, after crossing the orbit of Neptune in 1983. The Pioneer-Venus program launched in 1978 consisted of an orbiter and five atmosphere probes on board two spacecraft.

Pisces

The Fishes, a large, faint constellation on the ECLIPTIC, the 12th sign of the ZODIAC. The vernal EQUINOX now lies in Pisces.

Pistil see FLOWER.

Pitch

Black solid BITUMEN; the residue from distilling coal tar, wood tar or PETROLEUM, sometimes occurring naturally. It is used in roadmaking, for waterproofing and for caulking seams.

Pitch, Musical

The FREQUENCY of the vibrations constituting a SOUND. The frequency associated with a given pitch name (eg Middle C) has varied considerably over the years. The present international standard sets Concert A at 440Hz. (♦ page 225)

Pitchblende (uraninite)

Brown, black or greenish mineral, the most important source of URANIUM, RADIUM and POLONIUM. The composition varies between UO_2 and $UO_{2.6}$; thorium, radium, polonium, lead and helium are also present. Principal deposits are in Zaire, Bohemia, at Great Bear Lake, Canada, and in the Mountain States of the USA.

Pitot tube

Device invented by Henri Pitot (1695-1771) in 1732 and widely used in fluid dynamics for measuring FLUID velocities. One open end of a cylindrical tube points directly into the flowing stream, and the other end is connected to a pressure-measuring device. This compares the pitot-tube pressure with the static stream pressure, the difference being a measure of the fluid velocity.

Pituitary gland

Major ENDOCRINE GLAND situated just below the BRAIN, under the control of the adjacent HYPOTHALAMUS and in its turn controlling other endocrine glands. The posterior pituitary is a direct extension of certain cells in the hypothalamus and secretes VASOPRESSIN and OXYTOCIN into the BLOOD stream. The anterior pituitary develops separately and consists of several cell types which secrete different HORMONES, including growth hormone, follicle stimulating hormone, luteinizing hormone, prolactin, thyrotrophic hormone (which stimulates the thyroid gland) and adrenocorticotrophic hormone (ACTH). The anterior pituitary hormones are controlled by releasing hormones secreted by the hypothalamus into local blood vessels; the higher centers of the brain and environmental influences act by this route. FEEDBACK from the organs controlled occurs at both the hypothalamic and pituitary levels. Pituitary TUMORS or loss of blood supply may cause loss of function, while some tumors may be functional and produce syndromes such as ACROMEGALY (due to growth hormone imbalance). Pituitary tumors may also affect VISION by compressing the nearby optic nerves. (♦ page 111, 126)

Placebo

A tablet, syrup or other form of medication which is inactive and is prescribed in lieu of active preparations, eg in experimental studies of drug effectiveness. A placebo effect is an improvement caused by the psychological effects of treatment rather than by the treatment itself. The opposite effect, or substance producing it, is called a nocebo.

Placenta

In placental mammals, including humans, a specialized structure derived from the lining of the uterus and part of the EMBRYO. It separates and yet ensures a close and extensive contact between the maternal (uterine) and fetal (umbilical) blood circulations. This allows nutrients and oxygen to pass from the mother to the embryo, and waste products to pass in the reverse direction. GONADOTROPHINS are produced by the placenta that prepare the maternal body for delivery and the mammary glands for LACTATION. The placenta is delivered after the child at BIRTH (the afterbirth). (See also IMPLANTATION.) (♦♦ page 111, 204)

Placental mammals see EUTHERIA.

Placer mining

The extraction of minerals such as gold, platinum and diamonds from ORE that has accumulated through the processes of weathering or EROSION. The earliest and best-known form of placer mining is gold panning.

Plague

A highly infectious disease caused by a bacterium carried by rodent fleas. It causes greatly enlarged LYMPH nodes (buboes, hence bubonic plague), SEPTICEMIA with fever, prostration and coma; plague PNEUMONIA, which can be transmitted from person to person, is particularly severe. If untreated, death is common; epidemics occur in areas of overcrowding and poverty. It still occurs on a small rural scale in the Far East; massive epidemics such as the Black Death, which perhaps halved the population of Europe in the mid-14th century, are rare. Rat and flea control are the mainstay of prevention.

Planck, Max Karl Ernst Ludwig (1858-1947)

German physicist whose QUANTUM THEORY (with the theory of RELATIVITY) ushered physics into the modern era. Initially influenced by CLAUSIUS, he made fundamental researches in THERMODYNAMICS before turning to investigate BLACK-BODY RADIATION. To describe the electromagnetic radiation emitted from a BLACK BODY he evolved the Planck Radiation Formula, which implied that ENERGY, like MATTER, is not infinitely subdivisible – that it can exist only as quanta (see PLANCK CONSTANT). Planck himself was unconvinced of this, even after EINSTEIN had applied the theory to the PHOTOELECTRIC EFFECT and BOHR in his model of the ATOM; but for his achievement he received the 1918 Nobel Prize for Physics.

Planck constant

h ($=6.6256\times10^{-34}$Js), a quantity fundamental to quantum physics, named for MAX PLANCK, who in 1900 solved a long-standing problem in radiation physics with the hypothesis that the energy of a system vibrating with frequency v had to be a whole-number multiple of hv. The Planck constant also governs the accuracy with which different properties can be measured simultaneously (see UNCERTAINTY PRINCIPLE), and the wavelength of the wave associated with a particle (see QUANTUM MECHANICS).

Planet

In the SOLAR SYSTEM, one of the nine major celestial bodies orbiting the Sun; by extension, a similar body circling any other star. The possible presence of massive planets around a few nearby stars has been inferred from small periodic irregularities in their PROPER MOTIONS. For example, Barnard's Star (see BARNARD, E.E.) may have at least two planets comparable in mass to Jupiter. In 1983 the IRAS infrared satellite discovered infrared radiation from disks of cool dust, possibly planet-forming material, surrounding several stars, including VEGA, Fomalhaut and Beta Pictoris. In 1984 the Beta Pictoris disk was detected at optical wavelengths. These observations imply that planets are by no means unique to the Sun. (♦♦ page 134, 221)

Planetarium

An instrument designed to represent the relative positions and motions of celestial objects. Originally a mechanical model of the SOLAR SYSTEM the planetarium of today is an intricate optical device that projects disks and points of light representing Sun, Moon, planets and stars on to the interior of a fixed hemispherical dome. The various cyclic motions of these bodies as seen from a given latitude on Earth can be simulated. Of great assistance to students of ASTRONOMY and celestial NAVIGATION, planetariums also attract large public audiences. The first modern planetarium, built in 1923 by the firm of CARL ZEISS, is still in use at the Deutsches Museum, Munich, West Germany.

Planetary nebula see NEBULA.

Planetesimal hypothesis

A discarded theory proposed by T.C. Chamberlin and F.R. Moulton (1872-1952) to explain the formation of PLANETS. It states that a passing star drew matter out of the Sun, some of which condensed to form small solid particles (planetesimals), which in turn coalesced to form planets. The term "planetesimal" is now applied to the kilometer-sized, or larger, bodies that according to the accretion theory of planetary formation were formed by the hierarchical accretion of smaller particles, and subsequently themselves coalesced to form planets.

Plankton

Microscopic animals and plants that live mainly at the surface of the sea. They drift more or less passively under the influence of ocean currents, and are vitally important links in the marine food chain (see ECOLOGY). A major part of the plankton are the minute plants (phytoplankton), including unicellular ALGAE, DINOFLAGELLATES and DIATOMS. Phytoplankton may be so numerous as to color the water and cause it to have a "bloom". They are eaten by animals in the zooplankton. This includes the eggs, larvae and adults of a vast array of animal types, from protozoa to fish. Zooplankton is an important food for large animals such as whales, and for fish such as herring. Phytoplankton is confined to the upper layers of the sea where light can reach, but zooplankton has been found at great depths. (See also BENTHOS; NEKTON.)

Plant

A member of the KINGDOM Plantae. All plants carry out PHOTOSYNTHESIS, with the exception of a few parasitic species that have lost this ability. As photosynthesis relies on the green pigment CHLOROPHYLL, all non-parasitic plants are green, unless this color is masked by some other pigment. Plants are non-motile, unlike most animals, and this allows them to have a far more plastic growth-form – they generally have no definite size or shape to which all members of the species conform. The cells of plants are EUKARYOTIC, but with certain features not seen in other eukaryotes (animals and fungi): CHLOROPLASTS and other PLASTIDS, a large central vacuole, and CELLULOSE cell walls. Other characteristic plant features include a waterproof CUTICLE, openings known as STOMATA through which gas exchange can take place, and areas where growth occurs, known as MERISTEMS. In higher plants, the plant body can be divided into ROOTS, STEMS, LEAVES and BUDS. Among simpler forms the plant body consists of a THALLUS, sometimes with root-like RHIZOIDS. Growth and other processes are controlled by plant HORMONES, though these do not act in exactly the same way as animal hormones. Almost all plants show at least two forms of reproduction: the asexual production of SPORES, and sexual reproduction involving gametes; in all terrestrial plants and in most algae an ALTERNATION OF GENERATIONS is seen, in which the sexual and asexual forms alternate in the life cycle. Many plants also produce offspring by a variety of other asexual methods, generally known as VEGETATIVE REPRODUCTION.

Exactly which organisms are to be included in the plant kingdom is still a matter of debate among taxonomists. All flowering plants (ANGIOSPERMS), conifers and their allies (GYMNOSPERMS), ferns, clubmosses and horsetails (PTERIDOPHYTES), and mosses and liverworts (BRYOPHYTES) are included. (These are often referred to as the "terrestrial plants"; though some, such as the waterlilies and duckweeds, have returned to an aquatic way of life.) Most taxonomists also include the ALGAE – the seaweeds, freshwater weeds, and unicellular algae such as *Chlamydomonas*. Other taxonomists place the algae in the Kingdom PROTOCTISTA on the basis of their simpler level of organization. In reality there is probably a strong evolutionary link between unicellular green algae, multicellular green algae (such as *Spirogyra* and the sea-lettuce *Ulva*) and the simplest of terrestrial plants, the liverworts and mosses; this makes the separation of the algae from the terrestrial plants rather artificial.

Plants are extremely important in the ECOSYSTEM as they are AUTOTROPHS, upon which other organisms (HETEROTROPHS) all ultimately depend for food. Their role in the CARBON CYCLE is crucial, and the present destruction of much of the world's remaining forests could have catastrophic long-term consequences. (See also GREENHOUSE EFFECT.) (♦ page 13)

Plant diseases
Abnormalities of plant growth and development, caused either by infection with microorganisms or by deficiency of a nutrient in the soil. Most plant diseases are caused by microorganisms that infect the tissues, the most important being FUNGI, including MILDEW, RUSTS and SMUTS. Control methods for crops and ornamental plants are based on FUNGICIDES. VIRUSES are the next most damaging group of plant pathogens. Most of them are carried by aphids and other sap-sucking insects, and control is largely a matter of controlling these insect carriers. BACTERIA are less important, though they do cause some plant diseases; their main role, however, is in secondary infection, causing the tissues to rot. Deficiency diseases are caused by a lack of available minerals such as manganese in the soil.

Plant hormones see HORMONES.

Plaque
In dentistry, the aggregates of bacterial waste matter that accumulate on teeth. They act as frameworks in which bacteria can ferment carbohydrates, and if not removed by regular brushing become "fossilized" by calcium in the saliva. This makes them harder to remove, and therefore a better place for harboring bacteria.

Plasma
Almost completely ionized GAS, containing equal numbers of free ELECTRONS and positive IONS. Plasmas such as those forming stellar atmospheres (see STAR) or regions in an electron discharge tube are highly conducting but electrically neutral, and many phenomena occur in them that are not seen in ordinary gases. The TEMPERATURE of a plasma is theoretically high enough to support a controlled nuclear FUSION reaction. Because of this, plasmas are being widely studied, particularly in MAGNETO-HYDRODYNAMICS research. Plasmas are formed by heating low-pressure gases until the ATOMS have sufficient energy to ionize each other. Unless the plasma can be successfully contained by electric or magnetic fields, rapid cooling and recombination occurs; indeed the high temperatures needed for thermonuclear reactions cannot as yet be maintained in the laboratory for sufficiently long.

Plasma
The part of the BLOOD remaining when all CELLS have been removed, and which includes CLOTTING factors.

Plasmids
Small pieces of genetic material found in addition to the normal chromosome of bacterial cells. Each plasmid is a circular loop of DNA. There are two main plasmid types: larger ones, which are present as just one or two copies in the cell, and small ones that are present in high numbers in the cell. The larger ones are known as self-transmissible plasmids as they include genes for producing sex pili – protein tubes by which the plasmids can be transferred from one bacterial cell to another. The smaller plasmids cannot initiate their own transfer but they may be transferred along with a self-transmissible plasmid. Characteristics coded for by plasmids include antibiotic-resistance and resistance to heavy metal salts and ultraviolet light. Some plasmids code for enzymes that allow the bacterium to utilize novel sources of food, fix nitrogen or break down the hydrocarbons of petroleum. Because of the ease with which they travel from one cell to another, plasmids are important in GENETIC ENGINEERING.

Plasmodium
Genus of PROTOZOA responsible for MALARIA. Four main types are recognized: *P. falciparum*, *P. vivax*, *P. ovale*, and *P. malariae*, which cause variants of malaria and are endemic in different areas. *P. falciparum* causes cerebral malaria.

Plaster of Paris see CALCIUM.

Plastics
Materials that can be molded (at least in production) into desired shapes. A few natural plastics are known, eg BITUMEN, RESINS and RUBBER, but almost all are man-made, mainly from PETROCHEMICALS, and are available with a vast range of useful properties: hardness, elasticity, transparency, toughness, low density, insulating ability, inertness and corrosion resistance, etc. They are all high POLYMERS with carbon skeletons, each molecule being made up of thousands or even millions of atoms. Plastics fall into two classes: thermoplastic and thermosetting. Thermoplastics soften or melt reversibly on heating; they include celluloid and other cellulose plastics, lucite, nylon, polyethylene, styrene polymers, vinyl polymers, polyformaldehyde and polycarbonates. Thermosetting plastics, although moldable when produced as simple polymers, are converted by heat and pressure, and sometimes by an admixed hardener, to a cross-linked, infusible form. These include bakelite and other phenol resins, epoxy resins, polyesters, silicones, urea-formaldehyde and melamine-formaldehyde resins, and some polyurethanes. Most plastics are mixed with stabilizers, fillers, dyes or pigments and with plasticizers if needed. There are several fabrication processes: making films by calendering (squeezing between rollers), casting or extrusion, and making objects by compression molding, injection molding (melting and forcing into a cooled mold) and casting. (See also LAMINATES.)

Plastic surgery
The branch of SURGERY devoted to reconstruction or repair of deformity, surgical defect or the results of injury. Using bone, cartilage, tendon and skin from other parts of the body, or artificial substitutes, function and appearance may in many cases be restored. In skin grafting, the most common procedure, a piece of skin is cut, usually from the thigh, and stitched to the damaged area. Bone and cartilage (usually from the ribs or hips), or sometimes plastic, are used in cosmetic remodeling and facial reconstruction after injury. Congenital defects such as HARELIP and CLEFT PALATE can be treated in infancy.

Plastids
A collective term for organelles found in the cytoplasm of plant cells, containing CHLOROPHYLL (chloroplasts), other pigments (chromoplasts) or unpigmented (leucoplasts). They are thought to be derived from cyanobacteria; see ENDOSYMBIOSIS THEORY.

Platelet see BLOOD.

Plate tectonics
Fundamental theory of modern geology, arising from studies of CONTINENTAL DRIFT, EARTHQUAKE and VOLCANO distributions, and SEA-FLOOR SPREADING, which phenomena it largely explains. The Earth's crust is viewed as consisting of a number of semirigid plates in relative motion. Where plates meet, one edge is subducted beneath (forced under) the other: in midocean, this results in OCEAN trenches, deep seismic activity and arcs of volcanic ISLANDS; at continental margins, similar subduction of the oceanic plate results also in mountain-building. Where lighter continental blocks are forced together, neither edge is subducted and more complex mountain-building occurs. Belts of shallow earthquakes define the ocean ridges where new material is emerging from below. (♦ page 142-3)

Platinum (Pt)
Soft, silvery white metal in the PLATINUM GROUP. In addition to the general uses of these metals, platinum is used as a catalyst. AW 195.1, mp 1769°C, bp 3820°C, sg 21.45 (20°C). (♦ page 130)

Platinum group
The six NOBLE METALS in Group VIII of the PERIODIC TABLE, ie RUTHENIUM, RHODIUM, PALLADIUM, OSMIUM, IRIDIUM and PLATINUM (see also TRANSITION ELEMENTS). They are found together in PYROXENE deposits in South Africa and in the copper and NICKEL ores of Canada and the USSR. All are highly inert and corrosion-resistant, though palladium, osmium and platinum dissolve in AQUA REGIA; the others can be dissolved by fused oxidizing alkalis. Palladium dissolves slowly in oxidizing acids. Ruthenium and osmium show chief oxidation states +3, 4, 6 and 8; the other metals seldom exceed +4. All six metals form numerous HALIDES and complex halogen ions, and many other LIGAND complexes, including carbonyls resembling those of iron, cobalt and NICKEL. The platinum group metals are used, usually as ALLOYS with each other, for jewelry, the tips of pen nibs, electrical contacts, THERMOCOUPLES, crucibles, surgical instruments, standard weights and measures, and (finely divided) as catalysts (see CATALYSIS).

Platyhelminthes (flatworms)
A phylum of simple invertebrates with flattened bodies, containing parasitic flukes (Trematoda) and tapeworms (Cestoda), together with the free-living turbellaria, which are found in marine or

freshwater habitats, or glide across the leaves of trees in tropical rainforests. (♦ page 16)

Playa
Found in undrained areas in arid regions, a level tract formed of deposits from a temporary lake that has formed as a result of flooding or heavy rainfall, and then evaporated. (See also ALKALI FLATS, EVAPORITES.)

Pleiades
A GALACTIC CLUSTER in the constellation TAURUS. Seven of the stars, named for the seven daughters of Atlas, can be seen by the naked eye. The Pleiades are about 153pc from the Sun and are surrounded by faint reflection nebulosity (see NEBULA).

Pleistocene
The earlier epoch of the QUATERNARY period, from about 2 million to 10,000 years ago. (See also GEOLOGY; HOLOCENE.) (♦ page 106)

Plesiosaurs
Extinct reptiles that lived during the Jurassic and Cretaceous periods, 213-65 million years ago. They died out just before the end of the Cretaceous, along with their relatives the ICHTHYOSAURS. They were aquatic, with long necks, and limbs modified as paddles.

Pleurisy
INFLAMMATION of the pleura, the two thin connective tissue layers covering the outer lung surface and the inner chest wall. It causes a characteristic chest pain, which may be localized and is made worse by deep breathing and coughing. It is usually caused by pneumonia in the underlying lung tissue.

Pliocene
The final epoch of the TERTIARY period, lasting from about 5 to 2 million years ago. (See also GEOLOGY.) (♦ page 106)

Pluto
The ninth planet of the SOLAR SYSTEM, orbiting the Sun at a mean distance of 39.44AU in 247.7 years. Pluto was discovered in 1930 following observations of PERTURBATIONS in NEPTUNE'S orbit. Pluto has one moon, Charon, which is larger in relation to its parent planet than any other in the Solar System. Because of its distance from us, little is known of Pluto, though the changes of light in a series of eclipses between Pluto and Charon during 1985-90 should reveal more about its size and diameter. At present, its mass is believed to be 0.002 that of Earth and its diameter about 2500km. Its surface is highly reflective, and its low density suggests that it is composed mostly of water ice and frozen gases. Pluto's orbit is very eccentric: indeed, it currently lies closer to the Sun than Neptune, and will do so until 1999. Because of its smallness and unusual orbit, some astronomers suggest that Pluto should be reclassified as an asteroid. (♦ page 221, 240)

Plutonism (vulcanism)
The geological theory, often associated with the followers of J. HUTTON, that the rocks of the Earth were volcanic in origin. In the early 19th century plutonism rivalled NEPTUNISM for acceptance as the fundamental geological principle.

Plutonium (Pu)
The most important TRANSURANIUM ELEMENT, used as fuel for NUCLEAR REACTORS and for the ATOMIC BOMB. It is one of the ACTINIDES and chemically resembles URANIUM. Pu^{239} is produced in BREEDER REACTORS by neutron irradiation of uranium (U^{238}); like U^{235}, it undergoes nuclear FISSION, and was used for the Nagasaki bomb in WWII. AW 244, mp 640°C, bp 3200°C, sg 19.84 (α; 25°C). (♦ page 130)

PMR
Short for polymyalgia rheumatica, an acute but relatively short-lasting rheumatic disease accompanied by muscle pains and fever.

Pneumoconiosis
Restrictive disease of the LUNGS caused by deposition of dusts in the lung substance, inhaled during years of exposure, often in extractive industries such as coal mining. SILICOSIS, anthracosis and asbestosis are the principal kinds, though aluminum, iron, tin and cotton fiber also cause pneumoconiosis. Characteristic X-RAY changes are seen in the lungs.

Pneumocystis carinii
Organisms causing a form of pneumonia, rare except in AIDS patients and patients on immunosuppressant therapy.

Pneumonia
Inflammation of the tissue of the lungs. This leads to consolidation, where the ALVEOLI become blocked with pus and mucus and thus exclude air. It is usually caused by bacteria, and rarely by pure viral infection. A pneumonia may also develop if foreign bodies or noxious chemicals are inhaled. Most commonly the inflammation centers on the bronchi (air-carrying tubes) giving rise to bronchopneumonia. In lobar pneumonia the condition tends to be restricted to a single lobe. Cough with yellow or green sputum (sometimes containing blood), fever, malaise and breathlessness are common. The involvement of the pleural surfaces causes PLEURISY.

Pneumothorax
Collapse of the lung due to the presence of air in the pleural space between the LUNG and the CHEST wall. This may result from trauma, rupture of lung bullae in EMPHYSEMA or in ASTHMA, TUBERCULOSIS, PNEUMOCONIOSIS, CANCER, etc, or, in tall thin athletic males, it may occur without obvious cause.

Pogonophora
Also known as beadworms, a phylum of about 100 species of worm-like marine animals that lack both mouth and gut. They absorb nutrients directly from the water through their body surface; some also have symbiotic bacteria living within their tissues, which supply them with nutrients. Pogonophorans are tube dwellers, some growing to a length of 1.5m, but always very narrow, with a diameter of only a few millimeters.

Poikilotherm
An animal whose body temperature is largely dependent on external circumstances because it cannot generate heat metabolically. Although they are often called "cold-blooded" this is a misleading description, because when the air temperature is high the temperature of the animal's body is also high. Even when the air is cold, poikilotherms can warm themselves up by various means, such as basking in the sun or shivering their muscles. (See also HOMOIOTHERMS.)

Poisoning
The taking, via ingestion or other routes, of substances that are liable to produce illness or death. Poisoning may be accidental, homicidal, suicidal or as a suicidal gesture. DRUGS and medications are often involved, taken either by children in ignorance of their nature from accessible places, or by adults in suicide or attempted suicide. Easily available drugs such as aspirin, paracetamol and mild sedatives are often taken, though in serious suicide attempts barbiturates and antidepressants are more common. Chemicals such as disinfectants and weedkillers, cosmetics and paints are frequently swallowed as drinks by children, while poisonous berries may appear attractive. Poisoning by domestic gas or carbon monoxide has been used for suicide and homicide. Heavy metals (see LEAD POISONING, MERCURY POISONING, ARSENIC), INSECTICIDES and CYANIDES are common industrial poisons as well as being a risk in the community. Poisons may act by damaging body structures (eg weedkillers); preventing OXYGEN uptake by HEMOGLOBIN (carbon monoxide); acting on the NERVOUS SYSTEM (heavy metals); interfering with essential ENZYMES (cyanides, insecticides); with HEART action (antidepressants), or with the control of RESPIRATION (barbiturates).

Polanyi, John C. (b. 1929)
Canadian scientist who received with D. HERSCHBACH and Y.T. LEE the 1986 Nobel Prize for Chemistry for their work on the dynamics of chemical elementary pressure.

Polarimetry
Measurement of OPTICAL ACTIVITY by means of a polarimeter or polariscope, an instrument having two NICOL PRISMS, one fixed (the polarizer) and one rotatable (the analyzer), with the sample between them (see POLARIZED LIGHT). It is used in chemical analysis (notably for measuring sugar concentrations) and to study molecular configurations. The polariscope is also used to study strain (see MATERIALS, STRENGTH OF) in materials showing DOUBLE REFRACTION.

Polaris
Alpha Ursae Minoris, a CEPHEID VARIABLE star in the LITTLE DIPPER. Because of its close proximity to the N celestial pole (see CELESTIAL SPHERE), Polaris is also known as the Polestar or North Star, and has been used in navigation for centuries: owing to PRECESSION, Polaris is moving away from the N celestial pole. (♦ page 228)

Polarized light
LIGHT in which the orientation of the wave vibrations displays a definite pattern. In ordinary unpolarized light the wave vibrations (which occur at right angles to the direction in which the radiation is propagated) are distributed randomly about the axis of propagation. In plane-polarized light (produced in reflection from a DIELECTRIC such as glass or by transmission through a NICOL PRISM or polarizing filter), the vibrations all occur in a single plane. Polaroid filters work by subtracting the components of light orientated in a particular plane; two filters in sequence with their transmission planes crossed transmit no light. In elliptically polarized light (produced when plane-polarized light is reflected from a polished metallic surface) and circularly polarized light (produced on transmission through certain CRYSTALS exhibiting double refraction), the electric vector of the radiation at any point describes an ellipse or a circle. Much of the light around us – that of the blue sky, or reflected from lakes, walls and highways – is partially polarized. Polarizing sunglasses reduce glare by eliminating the light polarized by reflection from horizontal surfaces. Polariscopes employing two polarizing filters have proved to be valuable tools in organic chemistry (see POLARIMETRY). (♦ page 115)

Polarography
Chemical ANALYSIS, particularly for metals or organic groups, by measuring the saturation current and threshold voltage (see ELECTRICITY) for ELECTROLYSIS of a solution of the substance.

Polar wandering
The observation that the Earth's north and south magnetic poles have been in different positions at different times in the Earth's history. Magnetic particles in the rocks can indicate the positions of the poles at the time the rocks formed. By examining rocks of successive ages geophysicists can plot the movement of the poles over long periods of time. The "polar wandering" curves so produced are often consistent between continents only when the continents are moved to different relative positions on the globe. This is regarded as one of the proofs for CONTINENTAL DRIFT.

Polestar see POLARIS.

Poliomyelitis (infantile paralysis)
VIRAL DISEASE causing muscle PARALYSIS as a

result of direct damage to motor nerve cells in the SPINAL CORD. The virus usually enters by the mouth or gastrointestinal tract and causes a mild feverish illness, after which PARESIS or paralysis begins, often affecting mainly those muscles that have been most used in preceding days. Current polio vaccine is a live attenuated strain taken by mouth that colonizes the gut and induces immunity. Poliomyelitis vaccination has been one of the most successful developments in preventive medicine.

Pollen see POLLINATION.

Pollen count
An estimate of atmospheric pollen each day during spring and summer, which gives a guide to the intensity of the stimulus to ASTHMA and HAY FEVER in people with pollen ALLERGY. It is affected by wind, rain and humidity.

Pollination
In flowering plants or ANGIOSPERMS, the transfer of the male GAMETES from the anthers of a flower to the stigma of the same or another flower, where subsequent growth of the pollen leads to the fertilization of the female gametes (or eggs) contained in the ovules. This in turn leads to the production of SEEDS and FRUIT. Wind-pollinated plants, such as grasses, produce inconspicuous flowers with large feathery stamens and stigmas and usually large quantities of pollen. Insect-pollinated flowers have large, conspicuous and colorful flowers to attract pollinators, produce NECTAR on which the insects feed and have small stigmas. Some plants are pollinated by birds or small mammals, such as the Australian honey possum, and these have much in common with insect-pollinated flowers. Pollination by birds occurs mainly in the tropics and involves specialist groups such as the hummingbirds and sunbirds. Bat-pollination is also a feature of the tropics, and such flowers are usually dull in color, with a musty smell. The same is true of the few flowers that are pollinated by flies. It is believed that the earliest pollination in flowering plants was by insects, which probably ate part of the flowering shoot and inadvertently transferred some pollen in the process. From this a process of COEVOLUTION produced the mutualistic pollination relationships seen today. Some of these are highly specific, with a plant being totally dependent on a single species of pollinator and the pollinator entirely reliant on the species of plant.

Gymnosperms, such as the conifers, also require pollination to set seed. All are pollinated by the wind, and produce vast quantities of pollen in their male cones. In the cycads the pollen grains germinate to produce motile sperm (a primitive feature), which swim to the ovule in the female cone. In conifers the pollen is drawn in by a drop of sticky fluid that is exuded by the ovary and then shrinks back into it as it dries.

Pollution
The contamination of any natural environmental resource on which life or the quality of life depends, by any substance or form of energy, at a rate resulting in abnormal concentrations of what is then termed the pollutant. Some pollution is due to natural causes, such as volcanic eruption, but most is caused by human activities. Air (see AIR POLLUTION), water and soil are the natural resources chiefly affected. Some forms of pollution, such as urban sewage and garbage or inshore petroleum spillage, pose an immediate and obvious environmental threat; other forms, such as those involving toxic substances found in industrial wastes and agricultural PESTICIDES, present a more insidious hazard: they can enter biological food chains, affect the metabolism of organisms, and by killing some organisms but not others create an ecological imbalance (see ECOLOGY). Many pollutants have long-term effects that cannot easily be remedied; for example, chlorofluorocarbons used in aerosols and refrigerators are affected by sunlight and release chlorine, which causes the breakdown of OZONE in the upper atmosphere. This allows more ultraviolet light to penetrate, which can cause skin cancers. At ground level, ozone is itself a pollutant (see AIR POLLUTION). THERMAL POLLUTION is the excessive heating of lakes and rivers by industrial effluents, ionizing radiation from radioactive wastes (see FALL-OUT; RADIOACTIVITY) and non-ionizing radiation from high-energy radio waves, radar signals or electricity cables, all of which can cause ill-health in humans and other animals. The need to control environmental pollution in all its aspects is now widely recognized. (See also ACID RAIN; GREENHOUSE EFFECT; RECYCLING.)

Pollux see GEMINI.

Polonium (Po)
Soft, gray metal in Group VIA of the PERIODIC TABLE, occurring in PITCHBLENDE; usually produced by neutron bombardment of bismuth. All its isotopes are highly radioactive; the commonest, Po^{210}, emits ALPHA PARTICLES (half-life 138.4 days), and is used in NEUTRON sources. AW 209, mp 254°C. bp 960°C. sg 9.32 (α). (♦ page 130, 147)

Polyamide
A polymer in which the links between monomers are amide (–CONH–) groups, formed by the condensation reaction between a carboxylic acid group and an amine. Proteins are polyamides, as are some important synthetic polymers, such as the nylons.

Polychaetes
A class of ANNELIDS that includes the ragworms and lugworms.

Polycythemia
An excessive number of ERYTHROCYTES in the BLOOD, which leads to itching and a tendency to THROMBOSIS. It may be primary or secondary to prolonged HYPOXIA (with LUNG disease) or certain TUMORS.

Polyester
A polymer in which the links between monomers are ester (–COOC–) groups, formed by the condensation reaction between a carboxylic acid group and a hydroxyl group. Many synthetic fibers are polyesters. They can be made from mixtures of dicarboxylic acids and diols, eg terylene (dacron) from terephthalic acid and ethylene glycol.

Polyethylene
White, translucent resin, a POLYMER of ETHYLENE made catalytically at high pressure. Tough, elastic and inert, it is used to make plastic film, molded items, and SYNTHETIC FIBERS.

Polymer
Substance composed of very large MOLECULES (macromolecules) built up by repeated linking of small molecules (monomers). Many natural polymers exist, including PROTEINS, NUCLEIC ACIDS, polysaccharides (see CARBOHYDRATES), RESINS, RUBBER, and many minerals (eg quartz). The ability to make synthetic polymers to order lies at the heart of modern technology (see PLASTICS; SYNTHETIC FIBERS). Polymerization, which requires that each monomer has two or more FUNCTIONAL GROUPS capable of linkage, takes place by two processes: CONDENSATION with elimination of small molecules, or simple addition. CATALYSIS is usually required, or the use of an initiator to start a chain reaction of FREE RADICALS. If more than one kind of monomer is used, the result is a copolymer with the units arranged at random in the chain. Under special conditions it is possible to form stereoregular polymers, with the groups regularly oriented in space; these have useful properties. Linear polymers may form crystals in which the chains are folded sinuously, or they may form an amorphous tangle. Stretching may orient and extend the chains, giving increased tensile strength useful in synthetic fibers. Some crosslinking between the chains produces elasticity; a high degree of crosslinking yields a hard, infusible product (a thermosetting PLASTIC).

Polymorphism
In chemistry, the existence of certain chemical compounds in more than one crystalline form (see CRYSTAL). Usually the various forms are stable under different conditions. In some cases one is always stable, the others being metastable; thus CALCIUM carbonate has a stable hexagonal form, CALCITE, and a metastable orthorhombic form, aragonite. (See also ALLOTROPY.)

Polymorphism
In zoology, the existence of two or more types of individual within the same species of animal, other than male and female. An example is seen in some social insects such as ants and bees, in which many different types of worker are structurally adapted for different tasks within the colony. Polymorphism may also be useful in avoiding predation, as it varies the appearance of the prey, making it more difficult to find.

Polyp
Benign TUMOR of EPITHELIUM extending above the surface, usually on a stalk. Polyps may cause nasal obstruction and some (as in the GASTROINTESTINAL TRACT) may develop into a CANCER.

Polyp
A sessile, column-shaped form of certain CNIDARIANS, typified in the corals, sea anemones and hydroids. The crown of tentacles at the top surround the mouth, which also acts as an anus. (See also MEDUSA.)

Polypeptide
A PEPTIDE containing a large number of amino acid residues. (♦ page 40)

Polyploidy
Having more than two sets of chromosomes, as in triploidy (three sets), tetraploidy (four sets) and hexaploidy (six sets). Polyploids with an odd number of sets, such as triploids, are invariably sterile, because the chromosomes cannot pair up successfully during MEIOSIS. Those with an even number, such as tetraploids and hexaploids, are usually fertile. In plants, doubling of the chromosomes can make inter-species hybrids fertile by giving the chromosomes of each species a partner. Hybridization followed by polyploidy has played an important part in the evolution of some plants, notably bread wheat, which is a hexaploid produced by the successive hybridization of three different grass species. Spontaneous polyploidy can also generate new species that are immediately isolated reproductively from their diploid forebears. Such polyploids generally have larger flowers and fruits; they are produced artificially by plant breeders using colchicine, a compound extracted from the autumn crocus (*Colchicum*).

Polypropylene
A hydrocarbon polymer made from polypropylene (propene). Unlike polythene, polypropylene can exist in different isomeric forms. When propylene polymerizes, the third carbon atom in each monomer molecule forms a side chain along the growing polymer chain. The arrangement of these side chains affects the properties of the polymer, and it was only after the discovery of Ziegler-Natta (stereospecific) catalysts that it became possible to make polypropylene with useful properties.

Polysaccharides
Long-chain molecules made up of many sugar (monosaccharide) units (see CARBOHYDRATES). (♦ page 40)

Polystyrene
Polymer of STYRENE ($C_6H_5CH{=}CH_2$) used as a rigid molded plastic and (for upholstery and thermal insulation) as a foam.

Polyurethane
A polymer made by a condensation reaction between an organic isocyanate and a polyhydric alcohol. Polyurethanes are used as fibers, coatings, rubbers and foams. The foams can be produced during the condensation reaction by choosing reactants that release carbon dioxide.

Polywater (anomalous water)
A liquid formerly supposed to be a polymeric form of WATER. First reported in 1962, it is made by condensing water in very fine glass or silica capillary tubes, and has unusual properties (mp −40°C, bp *c*500°C, sg 1.4). It is now thought to contain substances dissolved from the glass.

Polyzoa see BRYOZOA.

Pome
A false fruit, the fleshy part of which is derived from the receptacle of the FLOWER, and not from the ovary. The apple is a pome: only the core represents the ovary, that is, the true FRUIT.

Population
In ecology and genetics, a group of individuals of the same species that are not separated from each other by any geographical barriers, and are therefore actively interbreeding.

Porphyria
Metabolic disease due to disordered HEMOGLOBIN synthesis. It runs in families and may cause episodic abdominal pain, skin changes, NEURITIS and mental changes. Certain drugs can precipitate acute attacks. Porphyria may have been the cause of the "madness" of George III of England.

Porphyrins
Water-soluble, nitrogen-containing compounds that occur widely in nature. They consist of four carbon-nitrogen rings, themselves joined in a ring structure, often with a metal atom at the center enclosed by the carbon and nitrogen atoms, but only loosely bound to them. Combinations of porphyrins with metal ions include the functional group of CHLOROPHYLL (containing magnesium) and heme (containing iron); combinations of heme with proteins give HEMOGLOBIN and the CYTOCHROMES. The status of the metal ion is such that it can participate in other reactions, as when hemoglobin binds to oxygen, or cytochromes accept electrons in the ELECTRON TRANSPORT CHAIN. In PHOTOSYNTHESIS it is the magnesium that, when activated by sunlight, gives off excited electrons; these then reduce NADP to NADPH. This in turn can reduce carbon dioxide (CO_2) to sugars (CH_2O).

Porphyry
An IGNEOUS ROCK having many large crystals (phenocrysts) set in a very fine-grained matrix, occurring in DIKES and SILLS. More generally, rocks are said to have porphyritic texture if they contain some phenocrysts in a finer grained matrix, eg porphyritic granite.

Portal vein
Vein carrying blood from the intestines to the liver; also known as the portal system. (♦ page 28)

Porter, Sir George (b. 1920)
British chemist awarded the 1967 Nobel Prize for Chemistry with MANFRED EIGEN and RONALD NORRISH for their studies of extremely fast chemical reactions, and, in particular, for their development of FLASH PHOTOLYSIS.

Porter, Rodney Robert (1917-1985)
British biochemist who shared with G.M. EDELMAN the 1972 Nobel Prize for Physiology or Medicine for his work on the molecular structure of ANTIBODIES.

Portwine stain
A form of BIRTHMARK.

Positron
The antiparticle corresponding to the ELECTRON. (See ANTIMATTER.)

Potash (potassium carbonate) see POTASSIUM.

Potassium (K)
A soft, silvery white, highly reactive ALKALI METAL. It is the seventh most abundant element, and is extensively found as SYLVITE, carnallite and other mixed salts: it is isolated by ELECTROLYSIS of fused potassium hydroxide. Potassium is chemically very like sodium, but even more reactive. It has one natural radioactive isotope, K^{40}, which has a half-life of 1.28 billion yr. K^{40} decays into Ar^{40}, an isotope of argon; the relative amounts of each are used to date ancient rocks. Potassium salts are essential to plant life (hence their use as fertilizers), and are important in animals for the transmission of impulses through the nervous system. AW 39.1, mp 63°C, bp 770°C, sg 0.862 (20°C).

POTASSIUM CARBONATE (potash, K_2CO_3) is a hygroscopic colorless crystalline solid, made from potassium hydroxide and carbon dioxide, an ALKALI used for making glass.

POTASSIUM CHLORIDE (KCl) is a colorless crystalline solid, found as SYLVITE. Used in fertilizers and as the raw material for other potassium compounds.

POTASSIUM NITRATE (saltpeter, KNO_3) is a colorless crystalline solid, soluble in water, which decomposes to give off oxygen when heated to 400°C. It is made from sodium nitrate and potassium chloride by fractional crystallization, and is used in gunpowder, matches, fireworks, some rocket fuels, and as a fertilizer. (♦ page 130)

Potential, Electric
The work done against ELECTRIC FIELDS in bringing a unit charge to a given point from some arbitrary reference point (usually earthed), measured in VOLTS (ie joules per coulomb). Charges will tend to flow from points at one potential to those at a lower potential, and potential difference, or voltage, thus plays the role of a driving force for electric current. In inductive circuits, the work done in bringing up the charge depends on the route taken, and potential ceases to be a useful concept.

Potentiometer
A device for accurate measurement of electric POTENTIAL by comparison with a standard cell potential: numerous DC and AC variants exist, mostly depending on OHM'S LAW. Typically, a potential drop is established in a long wire by a battery, and a sliding contact used to tap a variable proportion of this drop, the lengths needed to balance the standard and unknown potentials being noted in turn. The ratio of these lengths is the ratio of the potentials. The same arrangement is also used to vary an applied voltage, for example, in the thin carbon-film "potentiometers" used as volume controls in transistor radios.

Pound (lb)
The name of various units of weight. The pound avoirdupois is defined as exactly 0.45359237kg. The "metric pound" commonly used in continental Europe is 500g.

Powell, Cecil Frank (1903-1969)
British physicist and pacifist awarded the 1950 Nobel Prize for Physics for his development of a direct means of photographing the tracks of SUBATOMIC PARTICLES and subsequent discovery of the π-meson.

Power
The rate at which WORK is performed, or ENERGY dissipated. Power is thus measured in units of work (energy) per unit time, the SI unit being the watt (=the joule/second) and other units including the horsepower (=745.70W) and the cheval-vapeur (=735.5W). Frequently in engineering (and particularly in transportation) contexts, what matters is the power that a given machine can deliver or utilize – the rate at which it can handle energy – and not the absolute energies involved. A high-power machine is one which can convert or deliver energy quickly. While mechanical power may be derived as a product of a FORCE and a VELOCITY (linear or angular), the electrical power utilized in a circuit is a product of the potential drop and the current flowing in it (volts× amperes=watts). Where the electrical supply is alternating, the root-mean-square (rms) value of the voltage must be used.

Prairies
The rolling GRASSLANDS of North America. There are three types: tallgrass, midgrass (or mixed-grass) and shortgrass, which is found in the driest areas. Typical prairie animals are the coyote, badger, prairie dog and jackrabbit, and the largely vanished bison and wolf. (See also STEPPES.)

Praseodymium (Pr)
An easily tarnished metal, one of the LANTHANUM SERIES of elements. AW 140.9, mp 935°C, bp 3000°C, sg 6.48 (20°C). (♦ page 130)

Precambrian
The whole of geological time, from the formation of planet Earth to the start of the PHANEROZOIC (marked by the appearance of FOSSILS in rock strata), lasting from about 4550 to 590 million years ago. (See also GEOLOGY.) (♦ page 106)

Precession
The gyration of the rotational axis of a spinning body, such as a GYROSCOPE, describing a right circular CONE whose vertex lies at the center of the spinning body. Precession is caused by the action of a torque (spinning force) on the body. Precession of the equinoxes occurs because the Earth is not spherical but bulges at the EQUATOR, which is at an angle of 23.5° to the ECLIPTIC. Because of the gravitational attraction of the Sun, the Earth is subject to a torque that attempts to pull the equatorial bulge into the same plane as the ecliptic, therefore causing the planet's poles, and hence the intersections of the equator and ecliptic (the equinoxes), to precess in a period of about 26,000 years. The Moon (see NUTATION) and planets similarly affect the direction of the Earth's rotational axis.

Precipitation
In meteorology, all water particles that fall from clouds to the ground, including rain and drizzle, snow, sleet and hail. Precipitation is important in the HYDROLOGIC CYCLE. (♦ page 44)

Pregl, Fritz (1869-1930)
Austrian chemist awarded the 1923 Nobel Prize for Chemistry for his pioneering work on the microanalysis of organic compounds (see ANALYSIS, CHEMICAL; MICROCHEMISTRY).

Pregnancy
In humans the nine-month period from the fertilization and IMPLANTATION of an EGG, the development of EMBRYO and FETUS until the BIRTH of a child. Interruption of MENSTRUATION and change in the structure and shape of the BREASTS are early signs; morning sickness, which may be mild or incapacitating, is a common symptom. Later an increase in abdominal size is seen and other abdominal organs are pushed up by the enlarging uterus. Multiple pregnancy, hydatidiform mole, spontaneous ABORTION, antepartum HEMORRHAGE, toxemia and premature labor are common disorders of pregnancy. The time following birth is known as the puerperium. (♦♦ page 111, 204)

Prehistoric man
Humans and human ancestors living before recorded history, that is, before about 3000 BC. The earliest remains of human-like creatures, or hominids, are found in Africa and date back to between 3½ and 4 million years ago. These earliest

hominids, known as *Australopithecus afarensis*, were capable of walking upright, as footprints found at Laetoli in Tanzania show. They probably spent quite a lot of time in trees, and may have slept there at night. Their diet is assumed to have consisted mainly of plant material, as does that of the African apes (chimpanzee and gorilla). Anatomical and biochemical comparisons show us to be closely related to the African apes, and these earliest hominids, though they had upright posture and bipedal walking, retained many apelike features of the head, including small brains, a projecting "muzzle", and very little nose or chin.

Between about 2½ million and 1 million years ago at least four different types of hominids lived in Africa. Two were very heavily built, with massive jaw muscles and teeth; these are known as *Australopithecus robustus* and *A. boisei*, and they are believed to be a side-shoot of the evolutionary line leading to modern man. The other two were much more lightly built, and are known as *A. africanus* and *Homo habilis* respectively. Some authorities believe these two to be the same species, while others hold that *H. habilis* is distinctive, and has a larger brain. The question of whether *A. africanus* was a human ancestor is still debated, but *H. habilis* is widely believed to have been the forerunner of modern man. Simple pebble tools were made by this hominid, who included meat, probably scavenged from the kills of other animals, in his diet.

Tool-making and other skills improved gradually over the next million years, as *H. habilis* evolved into *H. erectus*, a hominid with a larger brain and distinctive bony ridges over the eyes. Organized hunting of large animals, the building of huts and the use of fire are all attributed to this hominid. Although migrations out of Africa had probably begun with *H. habilis*, the major migratory movements were those of *H. erectus*, whose remains and characteristic tools are found as far afield as China (Peking man), Indonesia (Java man), Israel, Spain, France and Britain. Further increases in brain size among *H. erectus* populations produced *Homo sapiens*, a group that is divided into two subspecies, *H. sapiens sapiens* or "fully modern man", and *H. sapiens neanderthalensis*, Neanderthal man. The latter was a stocky, robust form, found mainly in Europe and western Asia, and probably well adapted to the harsh climate of the last Ice Age. This form disappeared – or merged with modern man – about 40,000 years ago. Evidence of ritual burials and religious rites are associated with Neanderthal man, and these traits were developed further by the *H. sapiens sapiens* populations of the late Ice Age, who produced the magnificent cave paintings found in southern Europe and Africa, as well as many carvings and statuettes. With the ending of the Ice Age, food plants and animals began to be domesticated, and agriculture began to replace hunting and gathering as the main way of life from about 10,000 years ago.

The last major migrations of prehistoric man were from Asia into North America and thence to South America, and from Southeast Asia to the islands of the Pacific, Australia and New Zealand. All these migrations occurred some time after 40,000 years ago and involved *H. sapiens sapiens*.

Prelog, Vladimir (b. 1906)
Yugoslav-born Swiss chemist who shared with J. CORNFORTH the 1975 Nobel Prize for Chemistry for his work on ENZYMES.

Premenstrual syndrome
Irritability and impaired working performance experienced by some women in the 4-6 days preceding menstruation. It is caused by the hormonal changes that take place when the CORPUS LUTEUM atrophies.

Presbyopia
A defect of VISION coming on with advancing age in which the LENS of the EYE hardens, causing loss of the ability to accommodate (focus) nearby and often distant objects.

Pressure
The FORCE per unit area acting on a surface. The SI unit of pressure is the PASCAL (Pa=newton/(meter)^2) but several other pressure units, including the atmosphere (101.325kPa), the bar (100kPa) and the millimeter of mercury (mmHg =133.322Pa), are in common use. In the Universe, the pressure varies from roughly zero in interstellar space to an atmospheric pressure of roughly 100kPa at the surface of the Earth and much higher pressures within massive bodies and in stars. According to the KINETIC THEORY of matter, the pressure in a closed container of gas arises from the bombardment of the container walls by gas molecules: it is proportional to the temperature and inversely proportional to the volume of the gas. Pressure is a stress characterized by its uniformity in all directions and usually produces a decrease in volume. The value of the pressure affects most physical, chemical and biological processes. Consequently many different types of pressure gauges have been developed (see MANOMETER).

Prevailing westerlies
The predominant WINDS that blow between latitudes 30° and 60° both N and S of the equator. In the N Hemisphere they blow from the SW; in the S Hemisphere from the NW.

Priapulid worms
A small phylum of invertebrate worms. They are unsegmented, with a large eversible proboscis for feeding. The larvae of these worms move about freely in the surface layers of the sea-floor sediment, but the adults are burrowing, sedentary animals.

Prickly heat (heat rash)
An uncomfortable itching sensation due to excessive sweating, mainly seen in Europeans visiting the tropics.

Priestley, Joseph (1733-1804)
British theologian and chemist. Encouraged and supported by BENJAMIN FRANKLIN, he wrote *The History and Present State of Electricity* (1767). His most important discovery was OXYGEN (1774; named later by LAVOISIER), whose properties he investigated. However, he never abandoned the PHLOGISTON theory of COMBUSTION. He later discovered many other gases – AMMONIA, CARBON monoxide, hydrogen SULFIDE – and found that green plants require sunlight and give off oxygen. He coined the name RUBBER. His association in the 1780s with the LUNAR SOCIETY brought him into contact with scientists such as JAMES WATT and ERASMUS DARWIN. His theological writings and activity were important to him as a source of inspiration for his scientific work, and indeed he is regarded as a principal architect of the Unitarian Church. Hostile opinion over this and his support of the French Revolution led to his emigration to the US (1794).

Prigogine, Ilya (b. 1917)
Russian-born Belgian thermodynamicist who was awarded the 1977 Nobel Prize for Chemistry for contributions to understanding the THERMODYNAMICS of systems that are not in equilibrium states.

Primates
An order of MAMMALS to which tarsiers, bush babies, pottos, lemurs, monkeys, apes and humans all belong. In many ways, primates are not among the most highly evolved mammals: for example, the collar bone and the full complement of five digits on each hand are retained. In other ways they are specialized: the hands and feet are modified for grasping branches, the claws have become flat nails and the eyes provide stereoscopic and color vision. Most importantly, the brain is extremely large, especially those regions concerned with the abstract association of sensory signals. The tarsiers, bush babies, pottos and lemurs are the least advanced of the primates, and are often known collectively as the PROSIMIANS. In most parts of the world they have been replaced by the more advanced monkeys and apes. Some of the prosimians are wholly insectivorous, and a few primates are wholly herbivorous (eg the gorilla) but most primates are omnivores, including plant material, insects and small vertebrates in their diet.

Printed circuit see ELECTRONICS.

Prism
In geometry, a solid figure having two faces (the bases) which are parallel equal polygons and several others (the lateral faces) which are parallelograms. Prismatic pieces of transparent materials are much used in optical instruments. In spectroscopes (see SPECTROSCOPY) and devices for producing monochromatic LIGHT, prisms are used to produce DISPERSION effects, just as Newton first used a triangular prism to reveal that sunlight could be split up to give a SPECTRUM of colors. In binoculars and single-lens reflex cameras reflecting prisms (employing total internal reflection – see REFRACTION) are used in preference to ordinary mirrors. The NICOL PRISM is used to produce POLARIZED LIGHT.

Proboscis worms see NEMERTEANS.

Procyon
Alpha Canis Minoris, a visual DOUBLE STAR; the brightest star in CANIS MINOR (absolute magnitude +2.7) and 3.53pc distant. (♦ page 228, 240)

Progesterone
Female sex HORMONE produced by the corpus luteum (part of the ovary formed after ovulation) under the influence of LUTEINIZING HORMONE. It prepares the uterus lining for IMPLANTATION and other body organs for the changes of PREGNANCY. It is used in some oral contraceptives to suppress ovulation or implantation. (♦ page 111)

Program
The set of instructions to be followed by a COMPUTER.

Prokaryote
An organism with small, simple cells, lacking a nucleus and membranous organelles such as mitochondria, chloroplasts or endoplasmic reticulum. The prokaryotes include the bacteria and cyanobacteria, all other living organisms being EUKARYOTES. The genetic material of prokaryotes exists as a simple loop of DNA, located near the center of the cell. The cell contains RIBOSOMES, but these are of a different type from those found in eukaryotic cells. Within the prokaryotes there is considerable variation in cell structure. The cyanobacteria, for example, have complex infoldings of the cell membrane on which the photosynthetic enzymes are located. All prokaryotes have a rigid or semi-rigid cell wall, and all are unicellular, although some colonial forms occur. Some have a simple flagellum, or UNDULIPODIUM. On the basis of their cell structure, the prokaryotes are placed in a KINGDOM of their own, the Monera.

Prokhorov, Aleksandr Mikhailovich (b. 1916)
Soviet physicist awarded, with BASOV and TOWNES, the 1964 Nobel Physics Prize for work with Basov leading to development of the MASER.

Prolactin (luteotrophic hormone)
HORMONE secreted by the PITUITARY GLAND, concerned with LACTATION. (♦ page 111)

Prolateness see OBLATENESS.

Promethium (Pm)
Radioactive element (see RADIOACTIVITY) in the

LANTHANUM SERIES. It has no naturally occurring isotopes, and is formed in nuclear reactors. It is used in miniature nuclear-powered batteries. AW 145, mp 1168°C, bp 3300°C. (◆ page 130)

Prominence, Solar see SUN.

Proof spirit
Term describing the proportion of alcohol (ETHANOL) in distilled liquor. In the US the proof value is twice the percentage of ethanol by volume. Thus 90 proof represents 45% ethanol. In the UK proof spirit is somewhat stronger, containing 57.1% ethanol by volume.

Propane (C_3H_8)
Colorless gas, an ALKANE found in NATURAL GAS and light PETROLEUM. Mixed with butane, it is sold as bottled gas. It is also used to make ETHYLENE by CRACKING, and is oxidized to ACETALDEHYDE. MW 44.1, mp −187.7°C, bp −42°C.

Proper motion
The rate of motion at right angles to our line of sight of a star, measured in seconds of arc per year. (See RADIAL VELOCITY.)

Prosimians
A collective term for the more primitive groups of PRIMATES – that is, the tarsiers, galagos, bushbabies, pottos, lorises and lemurs. The more advanced primates, the monkeys and apes, have replaced the prosimians in most parts of the world.

Prospecting
The hunt for MINERALS economically worth exploiting. The simplest technique is direct observation of local surface features characteristically associated with specific mineral deposits. This is often done by prospectors on the ground, but aerial photography is increasingly employed (see PHOTOGRAMMETRY). Other techniques include examining the seismic waves caused by explosions, which supply information about the structures through which they have passed; testing local magnetic fields to detect magnetic metals or the metallic gangues associated with nonmagnetic minerals; and, especially for metallic sulfides, testing electrical CONDUCTIVITY.

Prostaglandins
Family of active compounds that occur naturally in the body, made from polyunsaturated fats. The greatest concentration occurs in SEMEN. Prostacyclin is one of the most interesting in the family, because it is an anti-coagulant and could be used as a drug to prevent BLOOD clotting during surgery. Other prostaglandins reduce blood pressure, induce contractions in the uterus and, possibly, act as mediators in the immune system (see IMMUNITY).

Prostate gland
Male reproductive GLAND that surrounds the urethra at the base of the bladder and secretes some of the semen. Benign enlargement of the prostate in old age is very common and may cause retention of urine. CANCER of the prostate is also common in the elderly, but it responds to HORMONE treatment. (◆ page 204)

Prosthetics
Mechanical or electrical devices inserted into or onto the body to replace or supplement the function of defective or diseased organs. Artificial limbs designed for persons with amputations were among the first prosthetics; but metal or plastic joint replacements or bone fixations for subjects with severe ARTHRITIS, FRACTURE or deformity are now also available. Replacement teeth for those lost by caries or trauma are included in prosthodontics. The valves of the heart may fail as a result of rheumatic or congenital heart disease or bacterial endocarditis, and may need replacement with mechanical valves (usually of ball-and-wire or flap types) sutured in place of the diseased valves under cardiorespiratory bypass. If the pacemaker of the heart fails, an electrical substitute can be implanted to stimulate the heart muscle at a set rate.

Prosthodontics see PROSTHETICS.

Protactinium (Pa)
Rare metal in the ACTINIDE series, found in URANIUM ores. More than 12 isotopes are known, all radioactive; Pa^{231} is the longest lived (half-life 34,000yr). Protactinium is usually pentavalent, resembling NIOBIUM and TANTALUM. AW 231, mp 1200°C, bp 4000°C, sg 15.37 (calc.). (◆ page 130)

Protective coloration see CAMOUFLAGE; MIMICRY; WARNING COLORATION.

Protein
Large molecules found in all living organisms, made up of unbranched chains of AMINO ACIDS. Proteins play a central part in life processes in the form of ENZYMES, which activate and regulate the entire metabolism of an organism, and are responsible for synthesizing all non-protein molecules. Proteins embedded in MEMBRANES also control the movement of substances into and out of the cell, while immunoglobulins are responsible for specific IMMUNITY against disease organisms. Structural proteins, such as those in cartilage or skin, give the body its shape and strength, while muscle proteins are responsible for movement.

Proteins are able to fulfil so many different tasks because they are immensely varied chemically. There are twenty common amino acids, each with a side-chain of different size and chemical nature. The types of amino acids present, and their order in the chain, gives the protein its unique characteristics. Once formed, the protein chain (or the POLYPEPTIDE chains that make up the protein) becomes folded in a particular way, thanks to reactions between amino acids at different points along the chain. This folding – known as the protein's conformation – creates specific areas where two or more amino acid side-chains come together and create an environment that is chemically reactive and has a special affinity for a particular molecule – this may be another protein, a fat, a carbohydrate, a mineral ion, a cofactor or a small metabolic intermediate. This specific binding is the basis for the action of enzymes, membrane proteins, immunoglobulins, muscle proteins and hemoglobin. In the case of enzymes the "active site" binds the substrate and then transforms it chemically. Many membrane proteins bind the molecule and then undergo a change in their conformation that projects the molecule through the membrane. Immunoglobulins, in the form of antibodies, simply bind the molecule (an antigen), allowing the antigen, or the cell that carries it, to be disposed of by other agents of the immune system. Muscle proteins bind to each other and then undergo a change of conformation that makes them move in relation to each other, producing a contraction. Hemoglobin binds to oxygen or to carbon dioxide, releasing the molecule when it encounters lower concentrations of the same gas.

In structural proteins the conformation is usually much more regular and rigid, often involving many more helical regions than are found in enzymes. However, some structural proteins, such as elastin, which surrounds the lungs and blood vessels, are very loosely structured and can readily expand and contract.

The order of amino acids in the protein chain is determined by the order of bases in the DNA chain (see PROTEIN SYNTHESIS). The importance of proteins is illustrated by the fact that DNA controls the whole organism throughout its lifespan, by coding for its proteins and regulating the rate at which they are produced. (◆◆ page 40, 233)

Protein synthesis
The process by which proteins are built up from their building blocks, AMINO ACIDS, in the bodies of living organisms. All the PROTEIN in any living organism is undergoing a continual process of breakdown and resynthesis. The white rat replaces half its protein in 17 days; in humans this requires 80 days. However, not all proteins are broken down and resynthesized at the same rate: half the human blood-serum proteins are replaced in 10 days; liver protein requires 20-25 days, while replacement of bone protein is very slow.

Protein synthesis is very rapid; within minutes of injecting an animal with a radioactively labeled amino acid, labeled protein can be isolated. Protein synthesis takes place within the cytoplasm of a cell, on the RIBOSOMES. The first step is the activation of an amino acid by reaction with ATP (see NUCLEOTIDES). This activated amino acid is then bound to a specific transfer RNA molecule. There is at least one type of transfer RNA for each amino acid. The soluble tRNA-amino acid complex is added to other amino acids to form a polypeptide chain, which eventually becomes a protein (see PEPTIDE). The order in which the amino acids are added is determined by messenger RNA, each CODON on the mRNA pairing up with an ANTICODON on the tRNA. Messenger RNA contains the code for a particular protein and is used only a few times before being destroyed. Messenger RNA from one organism can be used with transfer RNAs and ribosomes from another to synthesize a protein: this is how VIRUSES take over the synthetic systems of a cell. (See also GENETIC CODE.)

Proterozoic
The portion of the PRECAMBRIAN running from about 2390 to 590 million years ago, distinguished from the preceding ARCHEAN by having some trace of early life. (See also GEOLOGY.) (◆ page 106)

Protists
Members of the Kingdom Protista. All are single celled or form simple colonies of cells. Unlike the Kingdom Monera (bacteria and cyanobacteria) they are EUKARYOTIC. The PROTOZOA, such as the amebae, euglenoids and ciliates, form the core of the Protista, but beyond that, different taxonomists disagree on exactly which organisms should be included. Photosynthetic organisms such as the diatoms and dinoflagellates are classed by some with the protists but by others with the algae. Some modern classifications enlarge the group with various simple multicellular organisms, such as the algae, to form the Kingdom PROTOCTISTA; the word "protist" is occasionally used to refer to members of this new kingdom. (See also KINGDOM.)

Protoctista
A novel KINGDOM found in certain taxonomic schemes. It is based on the kingdom PROTISTA but also includes the slime molds, the algae (including large forms such as seaweeds), and certain other minor groups.

Proton
Elementary particle found in the nucleus of all ATOMS. It has a positive charge, equal in magnitude to that of the ELECTRON, and rest mass of 1.67252×10^{-27}kg (slightly less than the NEUTRON mass but 1836.1 times the electron mass). It has a lifetime of at least 10^{33} years, and may be completely stable, although some forms of UNIFIED FIELD THEORY predict that it decays after about 10^{32} years. (◆ page 49, 147)

Protoplasm
The contents of a living CELL, contained within the plasma membrane. It is usually differentiated into the NUCLEUS and the CYTOPLASM. The term is less used now that the contents of cells have been examined in greater detail.

Protozoa
A group of unicellular EUKARYOTIC organisms

that are not closely related to any group of multicellular organisms – that is, they are not yeasts (fungi), nor unicellular algae – and that have various animal-like features, such as motility, heterotrophic nutrition, and the absence of a rigid cell wall. The group is undoubtedly polyphyletic (not sharing an immediate common ancestor), and is difficult to define as it includes some organisms that show photosynthesis, though they can also grow in the dark by ingesting food, and may need certain amino acids in their diet even when photosynthesizing, so they are not true AUTOTROPHS. The principal photosynthetic protozoa are members of the genus *Euglena*. They have as close relatives some non-photosynthetic protozoans, and it is generally believed that these evolved from the photosynthetic forms by loss of their chloroplasts. Despite being photosynthetic, *Euglena* has many animal-like features that earn it its place in the Protozoa. (◆ page 13)

Proust, Joseph Louis (1754-1826)
French chemist who established the law of definite proportions, or Proust's Law (see COMPOSITION, CHEMICAL).

Proxima Centauri
The closest star to the Sun (1.33pc distant), a red dwarf star orbiting ALPHA CENTAURI. (◆ page 240)

Prussic acid (hydrocyanic acid) see CYANIDES.

Pseudohalogens
Class of monovalent inorganic radicals that chemically resemble the HALIDES and HALOGENS. They include CYANIDE (CN^-) and CYANOGEN ($[CN]_2$), cyanate (OCN^-), thiocyanate (SCN^-), and azide (N_3^-).

Pseudopodium
The "false limb" by which certain PROTOZOA are able to move; see AMEBOID MOVEMENT.

Psilocybin
HALLUCINOGENIC DRUG derived from a Mexican fungus (*Psilocybe mexicanus*); related to LSD.

Psittacosis (parrot fever)
LUNG disease with fever, cough and breathlessness caused by an organism intermediate between BACTERIA and VIRUSES. It is carried by parrots, pigeons, domestic fowl and related birds.

Psoriasis
Chronic, non-infectious inflammatory skin condition characterized by patches of red, thickened and scaling skin. It often affects the elbows, knees and scalp but may be found anywhere. Several forms are recognized, and the manifestations may vary in each individual with time. It may be associated with severe arthritis.

Psyche
In psychology, the MIND.

Psychedelic drugs see HALLUCINOGENIC DRUGS.

Psychiatry
The branch of medicine concerned with the study and treatment of MENTAL ILLNESS. It has two major branches: one is PSYCHOTHERAPY, the application of psychological techniques to the treatment of mental illnesses where a physiological origin is either unknown or does not exist (see also PSYCHOANALYSIS); the other, medical therapy, where attack is made either on the organic source of the disease or, at least, on its physical or behavioral symptoms. (Psychotherapy and medical therapy are often used in tandem.) As a rule of thumb, the former deals with NEUROSES, the latter with PSYCHOSES. (See also PSYCHOLOGY.)

Psychical research see PARAPSYCHOLOGY.

Psychoanalysis
A system of psychology having as its base the theories of SIGMUND FREUD; also, the psychotherapeutic technique based on that system. The distinct forms of psychoanalysis developed by JUNG and ADLER are more correctly termed respectively analytical psychology and individual psychology. Freud's initial interest was in the origins of the NEUROSES. On developing the technique of FREE ASSOCIATION to replace that of HYPNOSIS in his therapy, he observed that certain patients could in some cases associate freely only with difficulty. He decided that this was due to the memories of certain experiences being held back from the CONSCIOUS mind (see REPRESSION) and noted that the most sensitive areas were in connection with sexual experiences. He thus developed the concept of the UNCONSCIOUS (later to be called the ID), and suggested (for a while) that ANXIETY was the result of repression of the LIBIDO. He also defined "resistance" by the conscious to acceptance of ideas and impulses from the unconscious, and TRANSFERENCE, the idea that relationships with people or objects in the past affect the individual's relationships with people or objects in the present.

Psychological tests
Experiments devised to elicit information about the psychological characteristics of individuals. Such characteristics may relate to the intelligence, personality or aptitudes of the individual.

Psychology
Originally the branch of philosophy dealing with the mind, then the science of mind, and now, considered in its more general context, the science of behavior, whether human or animal. It is intimately related to ANTHROPOLOGY (the science of man) and somatology (the science of the body). (See also ANIMAL BEHAVIOR.) Psychology is also closely connected to MEDICINE and to SOCIOLOGY. There are a number of closely interrelated branches of human psychology. Experimental psychology embraces all psychological investigations controled by the psychologist. His experiments may center on the individual or group, in which latter case statistics will play a large part in the research. In particular, in clinical psychology information is gained through the treatment of those suffering from MENTAL ILLNESS. Social psychologists use statistical and other methods to investigate the effect of the group on the behavior of the individual. In applied psychology the discoveries and theories of psychology are put to practical use. Comparative psychology deals with the different behavioral organizations of animals (including humans). In this century the most important offspring of psychology are PSYCHIATRY and PSYCHOANALYSIS. Both are concerned with the treatment of mental illness, but from radically different viewpoints (see also PSYCHOTHERAPY). The former, in particular, is helped by the discoveries of physiological psychology, which attempts to understand the NEUROLOGY and PHYSIOLOGY of behavior. Two rather different branches of psychology sprang originally from FREUD's psychoanalysis. They are analytical psychology, founded by CARL GUSTAV JUNG, and individual psychology, founded by ALFRED ADLER. (See also ALFRED BINET; FRANCIS GALTON; WILHELM WUNDT; and BEHAVIORISM; GESTALT PSYCHOLOGY; INTELLIGENCE; PARAPSYCHOLOGY; PSYCHOLOGICAL TESTS; PSYCHOPHARMACOLOGY.)

Psychopathy
Any specific mental condition characterized by abnormal, antisocial behavior.

Psychopharmacology
The study of the effects of DRUGS on the mind, and particularly the development of drugs for treating MENTAL ILLNESS.

Psychosis
In contrast with NEUROSIS, any MENTAL ILLNESS, whether of neurological (see NEUROLOGY) or purely psychological origins, which renders the individual incapable of distinguishing reality from unreality or fantasy.

Psychosomatic illness
Any illness in which some mental activity, usually ANXIETY or the INHIBITION of the EMOTIONS (see also REPRESSION), causes physiological malfunction.

Psychotherapy
The application of the theories and discoveries of PSYCHOLOGY to the treatment of MENTAL ILLNESS. Psychotherapy does not usually involve physical techniques, such as the use of drugs or surgery (see PSYCHIATRY). The term is sometimes used misleadingly to distinguish other forms of therapy from PSYCHOANALYSIS.

Pteridophytes
A division of the PLANT kingdom that includes ferns, horsetails and clubmosses. All these plants contain vascular tissues (see ALTERNATION OF GENERATIONS) differentiated into ROOTS, STEMS and LEAVES. They do not produce seeds or flowers, but the seed-bearing plants (GYMNOSPERMS and ANGIOSPERMS) are thought to have evolved from a fern ancestor. Pteridophytes were once much more widespread and successful than they are today; tree-like clubmosses made up the swamp-forests of the Carboniferous era.

Pterodactyls
Another name for the PTEROSAURS.

Pterosaurs
Extinct REPTILES that lived during the Jurassic and Cretaceous periods, 213-65 million years ago. Pterosaurs were able to fly, and had greatly elongated fourth fingers that supported membranous wings. They were HOMOIOTHERMS ("warm-blooded") and their bodies were covered in fur. They were not of the dinosaur group, and form a separate line of development from the birds.

Ptolemy (Claudius Ptolemaeus, 2nd century AD)
Alexandrian astronomer, mathematician and geographer. Most important is his book on astronomy, now called *Almagest* ("the greatest"), a synthesis of Greek astronomical knowledge, especially that of HIPPARCHUS: his geocentric cosmology dominated Western scientific thought until the Copernican Revolution of the 16th century (see COPERNICUS, N.). His *Geography* supported Columbus' belief in the westward route to Asia. In his *Optics* he attempted to solve the astronomical problem of atmospheric refraction. (See also EPICYCLE.)

Ptosis
Drooping of an eyelid. It is a harmless condition that often runs in families, and is more noticeable when the subject is tired.

Puberty
The time during the growth of a person at which sexual development occurs, commonly associated with a growth spurt. Female puberty involves several stages – the acquisition of breast buds and of sexual hair, and the onset of MENSTRUATION – that may each begin at different times. Male puberty involves sexual hair development, voice change, and maturation of the TESTES and PENIS.

Puerperal fever
Disease occurring in puerperal women (ie those who have just given birth), usually a few days after the birth of the child. It is caused by infection of the uterus, often with STREPTOCOCCUS. It causes fever, abdominal pain and discharge of pus from the uterus. The introduction of ASEPSIS in OBSTETRICS by I.P. Semmelweiss (1818-1865) greatly reduced its incidence. Today, antibiotics are required if it develops.

Pulmonary
Anything to do with the lungs. The pulmonary artery takes blood from the heart to the lungs for oxygenation.

Pulsar
Short for pulsating radio star, a celestial radio source emitting brief, extremely regular pulses of ELECTROMAGNETIC RADIATION (normally at radio

frequency, apart from a few examples detectable at X-ray or optical frequencies). Each pulse lasts a few hundredths of a second and the period between pulses is of the order of one second or less. Though the pulse frequency varies from pulsar to pulsar, for each pulsar the period is highly regular, in some cases as regular as can be measured. The first pulsar was discovered in 1967 by ANTONY HEWISH and S.J. Bell. The fastest pulsar yet observed has a period of 0.0016s. It is likely that there are some 10,000 pulsars in the MILKY WAY, though less than 400 have as yet been discovered. It is believed that pulsars are the neutron STAR remnants of SUPERNOVAE, rapidly spinning and radiating through loss of rotational energy.

Pulse
The palpable impulse conducted in the ARTERIES representing the transmitted beat of the HEART. A normal pulse rate is between 68 and 80, but athletes may have slower pulses as they have larger, more efficient hearts. FEVER, heart disease, HYPOXIA and ANXIETY increase the rate.

Pulses
Edible seeds produced by plants of the LEGUMINOSAE, such as peas and beans, and a general name for plants yielding such seeds. They are also known as "grain legumes".

Pumice
Porous, frothy volcanic glass, usually silica-rich; formed by the sudden release of vapors as LAVA cools under low pressures. It is used as an ABRASIVE, an AGGREGATE and a railroad ballast.

Pump
Device for taking in and forcing out a fluid, thus giving it kinetic or potential ENERGY. The heart is a pump for circulating blood around the body. Pumps are commonly used domestically and industrially to transport fluids, to raise liquids, to compress gases or to evacuate sealed containers. Their chief use is to force fluids along pipelines. The earliest pumps were waterwheels, endless chains of buckets, and the ARCHIMEDES screw. Piston pumps, known in classical times, were developed in the 16th and 17th centuries, the suction types (working by atmospheric pressure) being usual, though they were unable to raise water more than about 10.4m. The steam engine was developed to power pumps for pumping out mines.

Piston pumps – the simplest of which is the syringe – are reciprocating volume-displacement pumps, as are diaphragm pumps, with a pulsating diaphragm instead of the piston. One-way inlet and outlet valves are fitted in the cylinder. Rotary volume-displacement pumps have rotating gear wheels or wheels with lobes or vanes. Kinetic pumps, or fans, work by imparting momentum to the fluid by means of rotating curved vanes in a housing: centrifugal pumps expel the fluid radially outward, and propeller pumps axially forward. Air compressors use the turbine principle (see also JET PROPULSION). Air pumps use compressed air to raise liquids from the bottom of wells, displacing one fluid by another. If the fluid must not come into direct contact with the pump, as in a nuclear reactor, electromagnetic pumps are used: an electric current and a magnetic field at right angles induce the conducting fluid to flow at right angles to both (see MOTOR, ELECTRIC); or the principle of the linear induction motor may be used. To achieve a very high vacuum, a diffusion pump is used, in which atoms of condensing mercury vapor entrain the remaining gas molecules.

Pupa
A stage in the development of those insects that have a LARVA completely different in structure from the adult, and in which "complete" METAMORPHOSIS occurs. The pupa is a resting stage in which the larval structure is reorganized to form the adult. Feeding and locomotion are meanwhile suspended. (See also JUVENILE HORMONE.)

Purcell, Edward Mills (b. 1912)
US physicist who shared with F. BLOCH the 1952 Nobel Prize for Physics for his independent work on nuclear magnetic moment, discovering nuclear magnetic resonance (NMR) in solids (see SPECTROSCOPY).

Purine ($C_5H_4N_4$)
A heterocyclic compound, consisting of a 6-membered ring fused with a 5-membered ring. The 6-membered ring has the PYRIMIDINE structure. Purine itself is the parent compound of a class of organic bases of major biochemical importance. The purines adenine and guanine are present in NUCLEIC ACIDS. Other important purine derivatives include CAFFEINE and theobromine. The end product of purine metabolism is URIC ACID, which is excreted in the urine.

Purkinje, Johannes Evangelista (1787-1869)
Bohemian-born Czech physiologist and pioneer of HISTOLOGY, best known for his observations of nerve cells (see NERVOUS SYSTEM) and discovery of the Purkinje effect, that at different overall light intensities the eye is more sensitive to different colors (see VISION).

Pus
The yellowish green exudate from wounds and ulcers in SEPSIS. It consists mostly of dead MACROPHAGES.

PVC (polyvinyl chloride)
A polymer made from vinyl chloride. It has a linear structure similar to polythene, except that one in every four hydrogen atoms is replaced by a chlorine atom. It is widely used in plastic pipes, flooring, records, clothing and packaging.

Pycnogonids (sea spiders)
Marine animals that are classified as CHELICERATES but are not closely related to the land spiders (see ARACHNIDS). They have very slender bodies with eight, ten or twelve legs. The tops of the legs are as thick as the body itself, and they accommodate some of the internal organs, including parts of the gut and the reproductive organs. Sea spiders feed on colonial hydroids, sponges and other sedentary prey. The male accepts eggs from the female and carries them around, stuck to his legs, until they hatch. (◆ page 116)

Pyorrhea
A flow of pus, usually used to refer to the pus related to poor oral hygiene and exuding from the margins of the gums and teeth; it causes loosening of the teeth and halitosis.

Pyrex
A borosilicate GLASS used in chemical and industrial apparatus and in ovenware. It is inert, a good insulator and heat-resistant (having a low coefficient of expansion and a high softening temperature).

Pyridine (C_5H_5N)
A colorless heterocyclic aromatic base with an unpleasant smell, found in bone oil and coal tar. It is used as a solvent and to denature grain alcohol (ETHANOL). Important derivatives are pyridoxine (VITAMIN B_6) and niacin.

Pyrimidine ($C_4H_4N_2$)
HETEROCYCLIC COMPOUND with a 6-membered aromatic ring containing nitrogen atoms in the 1 and 3 positions. It is the parent compound of a class of organic bases of major biochemical importance. The pyrimidines thymine, uracil and cytosine are present in NUCLEIC ACIDS. (See PURINE.)

Pyrite (iron pyrites, FeS_2, iron (II) disulfide)
A hard, yellow SULFIDE mineral known as fool's gold from its resemblance to gold. Of worldwide occurrence, it is a major ore of SULFUR. It crystallizes in the isometric system, usually as cubes. It alters to GOETHITE and LIMONITE.

Pyroclastic rocks
Rocks made up of particles thrown into the air by volcanic eruptions. (See also LAVA; VOLCANO.)

Pyroelectricity see PIEZOELECTRICITY.

Pyrogen
Any substance that causes FEVER.

Pyrolusite
Soft, gray-black OXIDE mineral composed of manganese (IV) oxide (MnO_2); of widespread occurrence, it is the chief ore of MANGANESE. It is a secondary mineral of aqueous origin crystallizing in the tetragonal system.

Pyrolysis
Chemical DECOMPOSITION of a substance by heat. (See also CALCINATION; COMBUSTION.)

Pyrometer
A temperature measuring device used for high temperatures. Platinum resistance thermometers and pyrometers operating on the principle of the THERMOCOUPLE have the disadvantage that they must be in contact with the hot body, but optical and radiation pyrometers can be used at a distance. Optical pyrometers estimate temperature from the light intensity in a narrow band of the visible spectrum by optical comparison of a glowing filament with an image of the hot body. Radiation pyrometers focus the body's heat radiation on a responsive thermal element such as a thermocouple.

Pyroxenes
Major group of SILICATE minerals occurring in IGNEOUS ROCKS. They have a chain structure, with prismatic CLEAVAGE close to 90°, and are related to the AMPHIBOLES. They are monoclinic or orthorhombic. Jadeite (see JADE) and SPODUMENE are commercially important.

Pyrrhotite
Bronze-brown iron SULFIDE mineral with the NICCOLITE structure, occurring in basic igneous rocks in Scandinavia and Ontario. It is non-stoichiometric (see COMPOSITION, CHEMICAL) with composition $Fe_{1-x}S$. Improved smelting techniques now make it possible to extract the iron economically.

Pythagoras (*c*570-*c*500 BC)
Greek philosopher who founded the Pythagorean school. Attributed to the school are: the proof of PYTHAGORAS THEOREM; the suggestion that the Earth travels around the Sun, the Sun in turn around a central fire; observation of the ratios between the lengths of vibrating strings that sound in mutual harmony, and ascription of such ratios to the distances of the planets, which sounded the "harmony of the spheres", and the proposition that all phenomena may be reduced to numerical relations.

Q fever
INFECTIOUS DISEASE caused by *Coxiella burnetti*, causing fever, headache and often dry cough and chest pain. It is transmitted by ticks from various farm animals and is common among farm workers and veterinarians.

Quadrant
A simple astronomical and navigational instrument used in early times to measure the altitudes of the Sun and stars. It consisted typically of a pair of sights, a calibrated quadrant (quarter) of a circle, and a plumb line. (See also SEXTANT.)

Quadruplets see MULTIPLE BIRTHS.

Quantum chromodynamics
A theory of the strong interactions among quarks. (See also SUBATOMIC PARTICLES; UNIFIED FIELD THEORY.)

Quantum electrodynamics
The quantum theory of electromagnetism. (See SUBATOMIC PARTICLES; UNIFIED FIELD THEORY.)

Quantum mechanics
Fundamental theory of small-scale physical phenomena (such as the motions of ELECTRONS and nuclei within ATOMS), developed during the 20th century when it became clear that the existing laws of classical mechanics and electromagnetic theory were not successfully applicable to such systems. Because quantum mechanics deals with physical events that we cannot directly perceive, it has many concepts unknown in everyday experience. DE BROGLIE struck out from the old QUANTUM THEORY when he suggested that particles have a wavelike nature, with a wavelength $\lambda = h/p$ (h being the Planck constant and p the particle momentum). This wavelike nature is significant only for very small particles such as electrons. These ideas were developed by SCHRÖDINGER and others into the branch of quantum mechanics known as WAVE MECHANICS. HEISENBERG worked along parallel lines with a theory incorporating only observable quantities such as ENERGY, using matrix algebra techniques. The UNCERTAINTY PRINCIPLE is fundamental to quantum mechanics, as is PAULI'S EXCLUSION PRINCIPLE. DIRAC incorporated relativistic ideas into quantum mechanics.

Quantum theory
Theory developed at the beginning of the 20th century to account for certain phenomena that could not be explained by classical PHYSICS. PLANCK described the previously unexplained distribution of radiation from a BLACK BODY by assuming that ELECTROMAGNETIC RADIATION exists in discrete bundles known as quanta, each with an ENERGY $E = h\nu$ (ν being the radiation frequency and h a universal constant – the PLANCK CONSTANT). EINSTEIN also used the idea of quanta to explain the PHOTOELECTRIC EFFECT, establishing that electromagnetic radiation has a dual nature, behaving sometimes as a WAVE MOTION and sometimes as a stream of particle-like quanta. Measurements of other physical quantities, such as the frequencies of lines in atomic spectra and the energy losses of electrons on colliding with atoms, showed that these quantities could not have a continuous range of values, discrete values only being possible. With RUTHERFORD's discovery in 1911 that ATOMS consist of a small positively charged nucleus surrounded by ELECTRONS, attempts were made to understand this atomic structure in the light of quantum ideas, since classically the electrons would radiate energy continuously and collapse into the nucleus. BOHR postulated that an atom only exists in certain stationary (ie nonradiating) states with definite energies and that quanta of radiation are emitted or absorbed in transitions between these states; he successfully calculated the stationary states of hydrogen. Some further progess was made along these lines by Bohr and others, but it became clear that the quantum theory was fundamentally weak in being unable to calculate intensities of spectral lines. The new QUANTUM MECHANICS was developed *c*1925 to take its place. (♦ page 21)

Quarantine
Period during which a person or animal must be kept under observation in isolation from the community if it is known or suspected that they have been in contact with an INFECTIOUS DISEASE. The duration of quarantine depends on the disease concerned and its maximum length of INCUBATION.

Quark see SUBATOMIC PARTICLES.

Quart (qt)
Name of various units of liquid and dry measure, equivalent to 2 pints or 0.9463 liter (US Customary System) or 1.137 liter (British Imperial System).

Quartz
Rhombohedral form of SILICA, usually forming hexagonal prisms, colorless when pure ("rock crystal"). A common mineral, it is the chief constituent of SAND, SANDSTONE, QUARTZITE and FLINT, and also occurs as the GEMS CHALCEDONY, AGATE, JASPER, and ONYX. Quartz is piezoelectric (see PIEZOELECTRICITY) and is used to make oscillators for clocks, radio and radar; and also to make windows for optical instruments. Crude quartz is used to make glass, glazes and abrasives, and as a flux.

Quartzite
Common rock consisting of QUARTZ, formed by metamorphism of SANDSTONE. It is tough, resisting weathering, and is used as road metal.

Quasar
Short for quasi-stellar object, a telescopically star-like celestial object whose SPECTRUM shows an abnormally large RED SHIFT. The largest quasar red shift so far measured is 4.01. Quasars are widely believed to be extremely distant objects, receding from us at high velocities, though some research suggests that their red shifts may not be entirely due to their sharing in the expansion of the Universe, in which case they may not be as distant as they seem.

Quasars show variability in light and radio emission (although the first quasars were discovered by RADIO ASTRONOMY, not all are radio sources), which indicates that they are comparatively small objects less than 0.3pc across. If quasars are as remote as their red shifts imply, they must be 100-10,000 times more luminous than a normal galaxy such as the Milky Way system. The most popular hypothesis is that a quasar is the compact hyperactive nucleus of a galaxy, so luminous that in most cases it completely swamps the light from the surrounding galaxy within which it is embedded, but not all astronomers accept this view. If this picture is correct, the central "energy machine" may be a supermassive BLACK HOLE accreting material from its surroundings.

Quaternary
The period of the CENOZOIC whose beginning is marked by the advent of humans. It has lasted about 2 million years, up to and including the present. (See also TERTIARY; GEOLOGY.) (♦ page 106)

Quicksand
Sand saturated with water to form a sand-water SUSPENSION possessing the characteristics of a liquid. Quicksands may form at rivermouths or on sandflats, and are dangerous as they appear identical to adjacent SAND.

Quicksilver see MERCURY.

Quimby, Phineas Parkhurst (1802-1866)
US pioneer of mental healing, an early user of SUGGESTION as a therapy. A strong influence on Mary Baker Eddy, he is held as a father of the New Thought movement.

Quinine
Substance derived from cinchona bark from South America, long used in treating a variety of ailments. It was preeminent in early treatment of MALARIA but is now rarely used.

Quinone (1,4-benzoquinone)
An aromatic compound used in the manufacture of dyes and pesticides. Many complex quinones are used as dyes, being applied as a soluble colorless hydroquinone and then oxidized to the insoluble colored quinone during the dyeing process. MW 108.1, mp 115.7°C.

Quinsy
Acute complication of TONSILLITIS in which ABSCESS formation causes spasm of the adjacent jaw muscles, fever and severe pain. It is rarely seen nowadays.

Quintal, Metric (q)
Unit of MASS equal to 100kg.

Quintuplets see MULTIPLE BIRTHS.

Rabbit fever see TULAREMIA.

Rabi, Isidor Isaac (1898-1988)
Austrian-born US physicist whose discovery of new ways of measuring the magnetic properties of ATOMS and MOLECULES both paved the way for the development of the MASER and the ATOMIC CLOCK and earned him the 1944 Nobel Prize for Physics.

Rabies (hydrophobia)
Potentially fatal VIRUS disease resulting from the bite of an infected animal, usually a dog. Headache, fever, and an overwhelming fear, especially of water, are early symptoms following an INCUBATION period of 3-6 weeks; PARALYSIS, spasm of muscles of swallowing, respiratory paralysis, DELIRIUM, CONVULSIONS and COMA due to an ENCEPHALITIS follow. Wound cleansing, anti-rabies vaccine and hyperimmune serum must be instituted early in confirmed cases to prevent the onset of these symptoms. Fluid replacement and respiratory support may help, but survival is rare if symptoms appear. Infected animals must be destroyed.

Race
Within a SPECIES, a subgroup whose members have sufficiently different physical characteristics from those exhibited by most members of another subgroup for it to be considered as a distinct entity. In particular the term is used with respect to the human species, *Homo sapiens*, the three most commonly distinguished races being caucasoid, mongoloid and negroid. It is impossible to make unambiguous distinctions between races, and the validity of the race concept in *Homo sapiens* is now in doubt, but the categories continue to be used. Such racial differences as do exist are mostly the product of adaptation to different environments, and largely superficial. No differences in intelligence exist between the races, and all can interbreed.

Raceme see INFLORESCENCE.

Racemic mixture
A mixture of equal quantities of a pair of ENANTIOMERS. A racemic mixture does not have optical activity as the activities of the enantiomers cancel each other.

Rad
A unit used for expressing absorbed dose of ionizing RADIATION (expressed as rd where it might be confused with rad(ian)). It represents the dosage absorbed when 1kg of matter absorbs 0.01 joules of energy. The related unit, the GRAY, is now used in preference.

Radar (radio detection and ranging)
System that detects long-range objects and determines their positions by measuring the time taken for radio waves to travel to the objects, be reflected and return. Radar is used for NAVIGATION, air control, fire control, storm detection, in RADAR ASTRONOMY and for catching speeding drivers. It developed out of experiments in the 1920s measuring the distance of the ionosphere by radio pulses. R.A. WATSON-WATT showed that the technique could be applied to detecting aircraft, and from 1935 Britain installed a series of radar stations which were a major factor in winning the Battle of Britain in WWII. From 1940 the UK and the US collaborated to develop radar. There are two main types of radar: continuous-wave radar, which transmits continuously, the frequency being varied sinusoidally, and detects the signals received by their instantaneously different frequency; and the more common pulsed radar. This latter has a highly directional antenna which scans the area systematically or tracks an object. A cavity magnetron or klystron emits pulses, typically 400 per second, 1μs across, and at a frequency of 3GHz. A duplexer switches the antenna automatically from transmitter to receiver and back as appropriate. The receiver converts the echo pulses to an intermediate frequency of about 30MHz, and they are then amplified, converted to a video signal, and displayed on a CATHODE-RAY TUBE. A synchronizer measures the time-lag between transmission and reception, and this is represented by the position of the pulse on the screen. Various display modes are used: commonest is the plan-position indicator (PPI), showing horizontal position in polar coordinates.

Radar astronomy
The study of celestial objects and the Earth's atmosphere by RADAR pulses. It includes tracking meteors and radar pulses reflected from the Moon and planets.

Radial symmetry
In animals, symmetry about a single polar axis rather than BILATERAL SYMMETRY, as seen in humans and other vertebrates. It is a feature of simple invertebrates such as sponges, and the sea anemones, corals and their relatives. Less than perfect radial symmetry is evident in the five-fold symmetry of echinoderms (starfish, sea-urchins, brittlestars etc).

Radial velocity
The component of the motion of a celestial body in the direction of the line of sight; that is, toward or away from the observer.

Radiation
The emission and propagation through space of ELECTROMAGNETIC RADIATION or SUBATOMIC PARTICLES. Exposure to X-RAYS and GAMMA RAYS is measured in RÖNTGEN units, absorbed dose of any high-energy radiation in GRAYS. (◀ page 25)

Radiation belts see VAN ALLEN RADIATION BELTS.

Radiation chemistry
Study of the chemical effects produced by interaction of RADIATION with matter. These include the effects of light (see PHOTOCHEMISTRY) but are in general more complex; ions are formed. It is important in infrared and X-ray photography and the study of mutations and of the origin of life.

Radiation sickness
Malaise, nausea, loss of appetite and VOMITING occurring several hours after exposure to ionizing RADIATION in large doses. These symptoms may be merely the first signs of lethal exposure. Large doses of radiation may cause BONE MARROW depression with ANEMIA, AGRANULOCYTOSIS and bleeding, or gastrointestinal disturbance with distension and bloody DIARRHEA. Skin ERYTHEMA and ulceration, LUNG fibrosis, NEPHRITIS and premature ATHEROSCLEROSIS may follow radiation, and malignancy may develop.

Radio
The communication of information between distant points using radio waves, ELECTROMAGNETIC RADIATION of wavelength between 1mm and 100km. Radio waves are also described in terms of their FREQUENCY – measured in HERTZ (Hz) and found by dividing the velocity of the waves (about 300Mm/s) by their wavelength. Radio communications systems link transmitting stations with receiving stations. In a transmitting station a piezoelectric OSCILLATOR is used to generate a steady radio-frequency (RF) "carrier" wave. This is amplified and "modulated" with a signal carrying the information to be communicated. The simplest method of modulation is to pulse (switch on and off) the carrier with a signal in, say, Morse code, but speech and music, entering the modulator as an audiofrequency (AF) signal from tape or a microphone, is made to interact with the carrier so that the shape of the audio wave determines either the amplitude of the carrier wave (amplitude modulation – AM) or its frequency within a small band on either side of the original carrier frequency (frequency modulation – FM). The modulated RF signal is then amplified to a high power and radiated from an ANTENNA. At the receiving station, another antenna picks up a minute fraction of the energy radiated from the transmitter together with some background NOISE. This RF signal is amplified and the original audio signal is recovered (demodulation or detection). Detection and amplification often involve many stages including FEEDBACK and intermediate frequency (IF) circuits. A radio receiver must of course be able to discriminate between all the different signals acting at any one time on its antenna. This is accomplished with a tuning circuit which allows only the desired frequency to pass to the detector (see also ELECTRONICS). In point-to-point radio communications most stations can both transmit and receive messages but in radio broadcasting a central transmitter broadcasts program sequences to a multitude of individual receivers. Programs are often produced centrally and distributed to a "network" of local broadcasting stations by wire or microwave link. Because there are potentially so many users of radio communications – aircraft, ships, police and amateur "hams" as well as broadcasting services – the use of the RF portion of the electromagnetic spectrum is strictly controlled to prevent unwanted INTERFERENCE between signals having adjacent carrier frequencies. The International Telecommunication Union (ITU) and national agencies such as the US Federal Communications Commission (FCC) divide the RF spectrum into bands that are allocated to the various users. Public broadcasting in the US uses MF frequencies between 535kHz and 1605kHz (AM) and VHF bands between 88MHz and 108MHz (FM). VHF reception, though limited to line-of-sight transmissions, offers much higher fidelity of transmission and much greater freedom from interference. International broadcasting and local transmissions in other countries frequently use other frequencies in the LF, MF and HF (short wave) bands.

THE DEVELOPMENT OF RADIO The existence of radio waves was first predicted by JAMES CLERK MAXWELL in the 1860s, but it was not until 1887 that HEINRICH HERTZ succeeded in producing them experimentally. "Wireless" telegraphy was first demonstrated by SIR OLIVER LODGE in 1894, and GUGLIELMO MARCONI made the first trans-Atlantic transmission in 1901. Voice transmission was first achieved in 1900, but transmitter and amplifier powers were restricted before the advent of LEE DE FOREST's triode ELECTRON TUBE in 1906. Only the development of the TRANSISTOR after 1948 has had as great an impact on radio technology. Commercial broadcasting began in the US in 1920.

Radioactivity
The spontaneous disintegration of certain unstable nuclei, accompanied by the emission of ALPHA PARTICLES (weakly penetrating HELIUM nuclei), BETA RAYS (more penetrating streams of ELECTRONS) or GAMMA RAYS (ELECTROMAGNETIC

RADIATION capable of penetrating up to 100mm of LEAD). In 1896, BECQUEREL noticed the spontaneous emission of ENERGY from URANIUM compounds (particularly PITCHBLENDE). The intensity of the effect depended on the amount of uranium present, suggesting that it involved individual atoms. The CURIES discovered further radioactive substances such as THORIUM and RADIUM, and about 40 natural radioactive substances are now known. Their rates of decay are unaffected by chemical changes, pressure, temperature or electromagnetic fields, and each nuclide (nucleus of a particular ISOTOPE) has a characteristic decay constant or HALF-LIFE. RUTHERFORD and SODDY suggested in 1902 that a radioactive nuclide can decay to a further radioactive nuclide, so that a series of transformations takes place that ends with the formation of a stable "daughter" nucleus. It is now known that for radioactive elements of high ATOMIC WEIGHT, three decay series (the thorium, actinium and uranium series) exist. As well as the natural radioactive elements, a large number of induced radioactive nuclides have been formed by nuclear reactions taking place in ACCELERATORS or NUCLEAR REACTORS (see also IRRADIATION; RADIOISOTOPE). Some of these are members of the three natural radioactive series. Various types of radioactivity are known, but beta emission is the most common, normally caused by the decay of a NEUTRON, giving a PROTON, an electron and an antineutrino (see SUBATOMIC PARTICLES). This results in a unit change of atomic number (see ATOM) and no change in MASS NUMBER. Heavier nuclides often decay to a daughter nucleus with atomic number two less and mass number four less, emitting an alpha particle. If an excited daughter nucleus is formed, gamma-ray emission may accompany both alpha and beta decay. The ionizing radiations emitted by radioactive materials are physiologically harmful, so special precautions must be taken in handling them. (◀ page 147)

Radio astronomy
The study of the ELECTROMAGNETIC RADIATION emitted or reflected by celestial objects in the approximate wavelength range 1mm-30m, usually by use of a RADIO TELESCOPE. The science was initiated accidentally in 1932 by Karl Jansky, who found an INTERFERENCE in a telephone system he was testing: the source proved to be the MILKY WAY. In 1937 an American, Grote Reber, built a 9.5m radio telescope in his backyard and scanned the sky at a wavelength around 2m. After WWII the science began in earnest. Investigation of the sky revealed that clouds of hydrogen gas in the Milky Way were radio sources, and mapping of these confirmed our Galaxy's spiral form (see GALAXY).

The sky is very different for the radio astronomer than for the optical astronomer. Bright stars are not radio objects (our Sun is one solely because it is so close), while many radio objects are optically undetectable. Radio objects include QUASARS, PULSARS, supernova remnants (eg the CRAB NEBULA) and other galaxies. The work of MARTIN RYLE in the 1960s and 1970s has enabled radio galaxies that are possibly at the farthest extremities of the Universe to be mapped. The Universe also has an inherent radio "background noise" (see COSMOLOGY).

Radiocarbon dating
A technique of dating organic material up to 70,000 years old. The radioactive carbon ISOTOPE C^{14} is produced naturally by the impact of COSMIC RAYS on NITROGEN atoms in the atmosphere, and, in the form of $C^{14}O_2$, enters the ecological CARBON CYCLE. Assuming that the amount of C^{14} produced in this way is constant, one may deduce that the proportion of C^{14} atoms present in living material is uniform throughout time. Knowing that C^{11} decays into N^{14} with a HALF-LIFE of 5730±40 years, examination of the amount of C^{14} remaining in organic material provides a reasonably accurate method of dating. Recent advances in DENDROCHRONOLOGY have shown that production of C^{14} in the atmosphere is not constant, and corrections have accordingly been made to the radiocarbon system.

Radiochemistry
The use of RADIOISOTOPES in chemistry, especially in studies involving chemical ANALYSIS, where radioisotopes provide a powerful and sensitive tool. Tracer techniques, in which a particular atom in a molecule is "labeled' by replacement with a radioisotope, are used to study reaction rates and mechanisms. (See also KINETICS, CHEMICAL; RADIATION CHEMISTRY.)

Radiocompass see COMPASS; DIRECTION FINDER.

Radiograph
A photograph exposed with X-RAYS or GAMMA RAYS. Special plates having thick emulsions are used to increase sensitivity.

Radioisotope
Radioactive ISOTOPE of an element. A few elements, such as RADIUM or URANIUM, have naturally occurring radioisotopes, but because of their usefulness in science and industry, a large number of radioisotopes are produced artificially. This is done by IRRADIATION of stable isotopes with particles such as NEUTRONS in an ACCELERATOR or NUCLEAR REACTOR. Radioisotopes with a wide range of HALF-LIVES and activities are available by these means. Because radioisotopes behave chemically and biologically in a very similar way to stable isotopes, and their radiation can easily be monitored even in very small amounts, they are used to "label" particular atoms or groups in studying chemical reaction mechanisms and to "trace" the course of particular components in various physiological processes. The radiation emitted by radioisotopes may also be utilized directly for treating diseased areas of the body (see RADIATION THERAPY), sterilizing foodstuffs or controlling insect pests.

Radiolaria
A group of PROTOZOA classified with the amebae in the Class Sarcodina. They are spherical, with an internal skeleton made up of spicules of silica radiating out from the center. Like amebae they have pseudopodia (see AMEBOID MOVEMENT), but these are of a special type known as axopodia, and are very long and slender. They have a central axis composed of protein MICROTUBULES, which make the axopod contractile, or of a spicule, making it rigid. Contractile axopods can be used for pulling prey, captured on their sticky surfaces, toward the cell, while rigid axopods can be used for "walking" along a surface, or for "rowing" the cell along. Radioloarians have a very complex cell structure, with a central membrane-bound compartment, containing one or more nuclei, that is not penetrated by the skeleton, and an outer compartment that contains many vacuoles as well as the skeletal spicules. These vacuoles allow the cell to regulate its buoyancy. Most radiolarians float in the plankton, but some are found on the sea floor.

Radiology
The use of RADIOACTIVITY, GAMMA RAYS and X-RAYS in MEDICINE, particularly in diagnosis but also in treatment. (See also RADIOTHERAPY.)

Radiometer
Instrument for measuring the intensity of radiant ENERGY. The term is usually applied to a simple vane type instrument (Crookes' radiometer) consisting of an evacuated glass bulb containing a pivot supporting vertical metal vanes blackened on one side. Incident radiation is more strongly absorbed by the blackened side of the vanes, and the forces exerted on them by residual gas molecules initiate rotation proportional to the radiation intensity.

Radiosonde
Meteorological instrument package attached to a small balloon capable of reaching the Earth's upper ATMOSPHERE. The instruments measure the TEMPERATURE, PRESSURE and HUMIDITY of the atmosphere at various altitudes, the data being relayed back to Earth via a radio transmitter. Radiosondes provide a cheap and reliable method of getting information for weather forecasting. (See also WEATHER FORECASTING AND CONTROL.)

Radio telescope
The basic instrument of RADIO ASTRONOMY. The receiving part of the equipment consists of a large parabola, the big dish, which operates on the same principle as the parabolic mirror of a reflecting TELESCOPE. The signals that it receives are then amplified and examined. In practice, it is possible to build radio telescopes effectively far larger than any dish could be by using several connected dishes; this is known as an array. (▶ page 237)

Radiotherapy
Use of ionizing RADIATION, as rays from an outside source or from radium or other radioactive metal implants, in treatment of malignant DISEASE, ie CANCERS, including LYMPHOMA and LEUKEMIA. The principle is that rapidly dividing TUMOR cells are more sensitive to the destructive effects of radiation on NUCLEIC ACIDS and are therefore damaged by doses that are relatively harmless to normal tissues. It is focussed on affected tissues and not on the whole body. Certain types of malignancy do indeed respond to radiation therapy, but RADIATION SICKNESS may also occur.

Radium (Ra)
Radioactive ALKALINE-EARTH metal similar to BARIUM, isolated from PITCHBLENDE by MARIE CURIE in 1898. It has white salts which turn black as the radium decays, and which emit a blue glow due to ionization of the air by radiation. It has four natural ISOTOPES, the commonest being Ra^{226} with HALF-LIFE, 1622 years. Radium is used in industrial and medical radiography. Aw 226.0, mp 700°C, bp 1150°C, sg 5. (◀ page 130, 147)

Radon (Ra)
A radioactive NOBLE GAS formed in the radioactive decay of RADIUM, ACTINIUM or THORIUM. Found in some radioactive minerals, Rn^{222} has a HALF-LIFE of 3.8 days, making it suitable for use in RADIOTHERAPY. Other natural and synthetic ISOTOPES have shorter half-lives. RADON (II) fluoride (RnF_2) is the only radon compound known. AW 222, mp −71°C, bp −62°C. (◀ page 130)

Rain
Water drops falling through the atmosphere; the chief form of PRECIPITATION. Raindrops range in size up to 4mm diameter; if they are smaller than 0.5mm the rain is called drizzle. The quantity of rainfall (independent of the drop size) is measured by a rain gauge, an open-top vessel that collects the rain, calibrated in millimeters or inches and so giving a reading independent of the area on which the rain falls. Light rain is less than 25mm/h, moderate rain 25 to 75mm/h, and heavy rain more than 75mm/h. Rain may result from the melting of SNOW or HAIL as it falls, but is commonly formed by direct condensation. When a parcel of warm air rises, it expands approximately adiabatically, cooling about 1K/100m. Thus its relative HUMIDITY rises until it reaches saturation, when the water vapor begins to condense as droplets, forming CLOUDS. These droplets may coalesce into raindrops, chiefly through turbulence and nucleation by ice particles or by cloud seeding (see also WEATHER FORECASTING AND CONTROL). Moist air can be lifted by CONVECTION, producing

convective rainfall; by forced ascent of air as it crosses a mountain producing orographic rainfall; and by the forces within CYCLONES, producing cyclonic rainfall. (See also GROUNDWATER; HYDROLOGIC CYCLE; METEOROLOGY; MONSOON.)

Rainbow
Arch of concentric, spectrally colored rings seen in the sky by an observer looking at rain, mist or spray with his back to the Sun. The colors are produced by sunlight being refracted and totally internally reflected (see REFRACTION) by spherical droplets of water. The primary rainbow, with red on the outside and violet inside, results from one total internal reflection. Sometimes a dimmer secondary rainbow with reversed colors is seen, arising from a second total internal reflection.

Rain forest see FOREST.

Rain shadow
Area of low rainfall on the lee side of a mountain barrier, which shelters it from prevailing rain-bearing winds. Rainfall on the corresponding windward side of the barrier is extremely high. (See also CLIMATE; RAIN.)

Rainwater, Leo James (1917-1987)
US physicist who shared the 1975 Nobel Prize for Physics with AAGE BOHR and BEN MOTTELSON for their work on the physics and structure of the atomic nucleus.

Ram (the constellation) see ARIES.

Raman, Sir Chandrasekhara Vankata (1888-1970)
Indian physicist awarded the 1930 Nobel Prize for Physics for his discovery of the Raman effect: when a medium is exposed to a beam of ELECTROMAGNETIC RADIATION, light scattered at right angles to the beam has a range of frequencies characteristic to the medium. This is the basis for Raman SPECTROSCOPY.

Ramapithecus see PREHISTORIC MAN.

Ramjet see JET PROPULSION.

Ramón y Cajal, Santiago (1852-1934)
Spanish neurohistologist who shared with C. GOLGI the 1906 Nobel Prize for Physiology or Medicine for his work showing the NEURON to be the fundamental "building block" of all nervous structures.

Ramsay, Sir William (1852-1919)
British chemist awarded the 1904 Nobel Prize for Chemistry for his discovery, prompted by a suggestion from RAYLEIGH (1892), of all the NOBLE GASES, including (with FREDERICK SODDY) HELIUM, though it had been earlier detected in the solar spectrum (1868).

Range finder
Instrument for remote measurement of distance. RADAR and SONAR provide nonoptical range finders. Optical range finders used – in cameras for correct focusing, and in military applications – are of two types. Coincidence range finders measure the angles formed from each end of a base line to the object viewed, and calculate the distance by trigonometry. The images from each optical path are made to coincide. Stereoscopic range finders use stereobinoculars which are adjusted until the stereo image formed by reticles in the eyepieces appears to be at the same distance as the object.

Rank, Otto (1884-1939)
Austrian-born US psychoanalyst who suggested that the psychological TRAUMA of birth is the basis of later anxiety NEUROSIS; and who applied PSYCHOANALYSIS to artistic creativity.

Rankine scale
Scale that expresses absolute TEMPERATURES in Fahrenheit degrees, devised by Scottish engineer William Rankine (1820-1872).

Raoult, François Marie (1830-1901)
French physical chemist best known for his work on the theory of SOLUTIONS. Raoult's Law, in its most general form, states that the VAPOR PRESSURE above an ideal solution is given by the sum of the PRODUCTS of the vapor pressure of each component and its mole fraction (the number of MOLES of the component divided by the total number of moles of all the components). (See also DISTILLATION.)

Rare earths
The elements of the LANTHANUM SERIES, in Group IIIB of the PERIODIC TABLE, occurring widespread in nature as MONAZITE and other ores. They are separated by chromatography and ion-exchange resins. Rare earths are used in alloys; their compounds (mixed or separately) are used as abrasives, for making glasses and ceramics, as "getters", as catalysts (see CATALYSIS) in the petroleum industry, and to make phosphors, lasers and microwave devices. (◆ page 130)

Rare gases
Former name for the NOBLE GASES. (◆ page 130)

Raster
A series of parallel sweeps, such as the sweeps of an electron beam across the image on a television camera or receiver.

Ratites
The major group of running or flightless birds, including the ostrich, emu, cassowary and kiwi. The group once also included giant forms, the moas and the elephant bird, which are now extinct because of overhunting. These flightless birds are known to have evolved from different ancestors.

Ray, John (1627-1705)
British biologist and natural theologian who, with Francis Willughby (1635-1672), made important contributions to TAXONOMY, especially in *A General History of Plants* (3 vols, 1686-1704).

Ray-finned fish
Members of the ACTINOPTERYGII, the largest group of living fish. They are so named because their fins are supported by radiating spines, and lack the fleshy central portion characteristic of the lobe-finned fish or SARCOPTERYGII. (◆ page 17)

Rayleigh, John William Strutt, Third Baron (1842-1919)
British physicist awarded the 1904 Nobel Prize for Physics for his measurements of the DENSITY of the atmosphere and its component gases, work that led to his isolation of ARGON (see also RAMSAY, SIR WILLIAM). He worked in many other fields of physics, and is commemorated in the terms Rayleigh scattering, which describes the way that ELECTROMAGNETIC RADIATION is scattered by spherical particles of radius less than 10% of the wavelength of the radiation (see SCATTERING), and Rayleigh waves (see EARTHQUAKES).

Raynaud's disease
A condition in which the fingers (or toes) suddenly become white and numb, often on exposure to mild cold, subsequently becoming in turn blue and then red and painful. It is caused by digital artery spasm. Raynaud's disease usually occurs in children and otherwise fit young women. Raynaud's syndrome is the same symptom as a manifestation of an underlying disease (eg SYSTEMIC LUPUS ERYTHEMATOSUS).

RDX (cyclonite or cyclotrimethylenetrinitramine)
A high EXPLOSIVE made by NITRATION of the condensation product of FORMALDEHYDE and AMMONIA. First introduced in WWII, it is used in blasting caps and plastic explosive.

Reactance
The ratio of an AC voltage applied to a single component of an electric circuit (particularly an inductor or CAPACITOR) to the current produced, the maximum values of each being taken irrespective of their relative PHASE. (See also IMPEDANCE.)

Reaction, Chemical
The process by which the CHEMICAL BONDS in molecules are broken and/or formed to convert reactant molecules to different product molecules.

Realgar
Soft, bright red SULFIDE mineral, arsenic disulfide (As_2S_2), an ore of ARSENIC associated with ORPIMENT and found in central Europe, Nevada and Utah. It forms monoclinic CRYSTALS.

Réaumur, René Antoine Ferchault de (1683-1757)
French scientist whose most important work was in ENTOMOLOGY, but who is best remembered for devising the now little used Réaumur temperature scale, in which 0°R=0°C and 80°R=100°C.

Recapitulation
In embryology, see ONTOGENY and PHYLOGENY.

Receptacle see FLOWER.

Recessiveness
In genetics, the failure of one ALLELE of a gene to be expressed when it is paired with another colloid of the same gene (the dominant allele); see DOMINANCE. (◆ page 32)

Recombinant DNA
Hybrid DNA molecules created by isolating segments of DNA and splicing them together with other DNA fragments to create molecules with specific properties. (See GENETIC ENGINEERING.)

Recombination
Processes that lead to the presence in offspring of GENE combinations not found in either parent. Such new combinations may be formed by CROSSING OVER and by the random distribution of the CHROMOSOMES to different GAMETES during MEIOSIS, followed by the union of gametes at FERTILIZATION. (See also VARIATION.) (◆ page 32)

Rectifier
A device such as an ELECTRON TUBE or SEMICONDUCTOR junction which converts alternating electric current (AC – see ELECTRICITY) to direct current (DC) by allowing more current to flow through it in one direction than another. A half wave rectifier transmits only one polarity of the alternating current, producing a pulsating direct current; two such devices are combined in full-wave rectification, giving a continuous pulse train which may be smoothed by a filter.

Recycling
The recovery and reuse of any waste material. Of obvious economic importance where reusable materials are available more cheaply than fresh supplies of the same materials, the recycling principle is finding ever wider application in the conservation of the world's natural resources and in solving the problems of environmental POLLUTION. The recycling of the wastes of a manufacturing process in the same process – eg the resmelting and recasting of metallic turnings and offcuts – is commonplace in industry. So also is the immediate use of wastes or byproducts of one industrial process in another – eg the manufacture of cattle food from the grain-mash residues found in breweries and distilleries. These are often termed forms of "internal recycling", as opposed to "external recycling": the recovery and reuse of "discarded" materials, such as waste paper, scrap metal and used glass bottles.

Red corpuscles (erythrocytes) see BLOOD.

Redi, Francesco (1627-1697 or 1698)
Italian biological scientist who demonstrated that maggots develop in decaying meat not through SPONTANEOUS GENERATION but from eggs laid on it by flies.

Red giant see STAR.

Red lead see LEAD.

Red shift
An increase in wavelength of the light from an object, usually caused by its rapid recession (see DOPPLER EFFECT). The red shift is determined by measuring the difference between the wavelength

of a given line in the SPECTRUM of a star or galaxy and the laboratory wavelength of that line. The red shift is equal to the ratio between that difference and the laboratory wavelength of the line; for speeds that are a small fraction of the speed of light, the red shift is also equal to the ratio between the speed of recession of the source and the speed of light. The spectra of distant GALAXIES show marked red shifts, and this is interpreted as implying that they are rapidly receding from us. (See also COSMOLOGY.)

Red tide
Ocean coloration caused by population explosion among certain DINOFLAGELLATES. They also release an alkaloid, poisonous to fish, into the sea.

Reduction see OXIDATION AND REDUCTION.

Reed, Walter (1851-1902)
US Army pathologist and bacteriologist who, in 1900, demonstrated the role of the mosquito *Aëdes aegypti* as a carrier of YELLOW FEVER, so enabling the disease to be controlled.

Refining
The purification of crude substances, especially metals (see METALLURGY), ores, petroleum and sucrose. Methods used, include DISTILLATION, ELECTROLYSIS and FLOTATION.

Reflection
The bouncing back of energy waves (eg LIGHT radiation, SOUND or WATER WAVES) from a surface. If the surface is smooth, "regular" reflection takes place, the incident and reflected wave paths lying in the same plane as, and at opposed equal angles to, the normal (a line perpendicular to the surface) at the point of reflection. Rough surfaces reflect waves irregularly, so an optically rough surface appears matt or dull while an optically smooth surface looks shiny. Reflected sound waves are known as ECHOES. (See also MIRROR; PRISM; REFRACTION.) (♦♦ page 115, 237)

Reflex
A muscle contraction or secretion resulting from nerve stimulation by a pathway from a stimulus to the effector organ without the interference of the higher thought centers.

Refraction
The change in direction of energy waves on passing from one medium to another in which they have a different velocity. In the case of LIGHT radiation, refraction is associated with a change in the optical density of the medium. On passing into a denser medium the wave path is bent toward the normal (the line perpendicular to the surface at the point of incidence), the whole wave path and the normal lying in the same plane. The ratio of the sine of the angle of incidence (that between the incident wave path and the normal) to that of the angle of refraction (that between the normal and the refracted wave path) is a constant for a given interface (Snell's Law). When measured for light passing from a vacuum into a denser medium, this ratio is known as the refractive index of the medium. Refractive index varies with wavelength (see DISPERSION). On passing into a less dense medium, light radiation is bent away from the normal, but if the angle of incidence is so great that its sine equals or exceeds the index for refraction from the denser to the less dense medium, there is no refraction and total (internal) REFLECTION (applied in the reflecting PRISM) results. Refraction is principally used in the design of LENSES. (See also DOUBLE REFRACTION.) (♦♦ page 115, 237)

Refractometer
An instrument measuring the refractive index (see REFRACTION) of liquids for a particular color of light; the index is often very sensitive to impurities, besides allowing easy determination of the proportions present in two component mixtures such as water/ethanol.

Refractories
Substances able to resist high temperatures without melting, decomposing or reacting, and hence used for thermal INSULATION and to line furnaces. Often made into firebricks, refractories are composed of various substances (mostly oxides), including the acidic fireclay, silica and zircon; the basic chromite, dolomite and magnesite; and the neutral carborundum, corundum and graphite. (See also CERAMICS.)

Refrigeration
Removal of HEAT from an enclosure in order to lower its TEMPERATURE. It is used for freezing water or food, for food preservation and for air conditioning, and is important for low-temperature chemical processes and CRYOGENICS studies and applications.

Regelation
The refreezing of ice that has melted under pressure alone, once the pressure is released. Ice skating depends on regelation, but as ice melts readily under pressure only while its temperature is near its FREEZING POINT, skating may not be feasible in very cold weather.

Regeneration
The regrowing of a lost or damaged part of an organism. In plants this is a common process and includes the production of, for example, dormant buds and adventitious roots. All animals possess some power to regenerate, but it is best seen in simple invertebrates such as starfish, which can regrow an arm if one is lost. In higher animals such as mammals regeneration is usually limited to the healing of wounds. Some animals are able to shed part of the body if attacked by a predator, and then regenerate that part; many lizards can lose their tails in this way, thus distracting the predator and making their escape.

Regnault, Henri Victor (1810-1878)
German-born French chemist best known for work on the physical properties of gases (eg showing that BOYLE'S LAW works only for ideal gases), and for inventing an air THERMOMETER and a HYGROMETER.

Regulus
Alpha Leonis, the brightest star in LEO. Its apparent magnitude is +1.35 and its distance from the Sun 26pc. It is a visual triple star. (♦ page 228)

Reichstein, Tadeus (born 1897)
Polish-born Swiss chemist awarded (with E.C. KENDALL and P.S. HENCH) the 1950 Nobel Prize for Physiology or Medicine for his work, independent of Kendall's, on the corticoids and isolation of what is now known as cortisone (see STEROIDS).

Reiter's syndrome
A feverish rheumatic disease with painful joints, urethral inflammation, and sore eyes. Of unknown cause but associated with a number of SEXUALLY TRANSMITTED DISEASES.

Relapsing fever
INFECTIOUS DISEASE caused by a bacterium of the genus *Borrelia*, carried by lice or ticks on rodents and causing episodic fever; it occurs in epidemics in areas of poverty and overpopulation. Rash, bleeding and respiratory symptoms are common, and the central NERVOUS SYSTEM may be affected.

Relativity
A frequently referred to but less often understood theory of the nature of space, time and matter. EINSTEIN's "special theory" of relativity (1905) is based on the premise that different observers moving at a constant speed with respect to each other find the laws of physics to be identical, and, in particular, find the speed of LIGHT waves to be the same (the "principle of relativity"). Among its consequences are that events occurring simultaneously according to one observer may happen at different times according to an observer moving past the first (although the order of two causally related events is never reversed); that a moving object is shortened in the direction of its motion; that time runs more slowly for a moving object; that the velocity of a projectile emitted from a moving body is less than the sum of the relative ejection velocity and the velocity of the body; that a body has a greater MASS when moving than when at rest, and that no body can travel as fast as, or faster than, the speed of light (2.998×10^8m/s – at this speed, a body would have zero length and infinite mass, while time would stand still on it).

These effects are too small to be noticed at normal velocities; they have nevertheless found ample experimental verification, and are commonplace considerations in many physical calculations. The relationship between the position and time of a given event according to different observers is known (for H.A. LORENTZ) as the Lorentz transformation. In this, time mixes on a similar footing with the three spatial dimensions, and it is in this sense that time has been called the "fourth dimension". The greater mass of a moving body implies a relationship between kinetic ENERGY and mass; Einstein made the bold additional hypothesis that all energy was equivalent to mass, according to the famous equation $E = mc^2$. The conversion of mass to energy is now the basis of nuclear reactors, and is indeed the source of the energy of the Sun itself.

Einstein's "general theory" (1915) is of importance chiefly to cosmologists. It asserts the equivalence of the effects of ACCELERATION and gravitational fields (see GRAVITATION), and that gravitational fields cause space to become "curved", so that light no longer travels in straight lines, while the wavelength of light falls as the light falls through a gravitational field. The direct verification of these last two predictions, among others, has helped deeply to entrench the theory of relativity in the language of physics. (♦ page 151)

Remission
Period of stability or recovery in a chronic or progressive incurable disease.

Remote sensing
The use of distant sensing devices (normally carried on aircraft or spacecraft) to detect features on the Earth.

Renal failure
Failure of KIDNEY function. Acute renal failure, which often follows injury, usually reverses spontaneously before great harm is done, but chronic renal failure is a serious condition, eventually fatal if untreated.

Renewable energy
The energy that comes from the Sun (SOLAR ENERGY), wind, water (TIDAL POWER and HYDROELECTRICITY), living plants (BIOMASS) and the Earth (GEOTHERMAL ENERGY). With the growing awareness of the finite supplies of fossil fuels (see COAL, GAS, PETROLEUM), there has been an interest in developing more efficient ways of using natural resources to produce electricity and to power vehicles. Renewable energy sources will last indefinitely and all, except geothermal power, are based on sunlight – which delivers annually more than 10,000 times as much energy as humanity uses.

Replication see NUCLEIC ACIDS.

Repression
The DEFENSE MECHANISM whereby an impulse or idea is restricted by the EGO or SUPEREGO to the UNCONSCIOUS (primary repression), or in which derivatives of it are similarly restricted (secondary repression). FREUD considered primary repression essential to the development of the ego. (See also INHIBITION.)

Reproduction
The process by which an organism produces offspring. Asexual reproduction involves a single

individual generating offspring that are in most cases genetically identical to the parent. Methods of asexual selection include SPORE production (plants, fungi, bacteria); BINARY FISSION (some simple invertebrates, protozoa, unicellular algae, bacteria); BUDDING (some simple invertebrates, yeasts); fragmentation (some simple invertebrates and filamentous algae); various means of VEGETATIVE REPRODUCTION (plants); and PARTHENOGENESIS (some animals and plants).

In sexual reproduction special sex cells called GAMETES are produced – in animals, sperm by males in the TESTES and ova (eggs) by females in the OVARY. In diploid organisms the gametes contain only half the number of chromosomes (haploid), being produced by MEIOSIS. the joining of two gametes (ie FERTILIZATION) produces a ZYGOTE with the normal number of chromosomes (DIPLOID). Fertilization may take place inside the female (internal fertilization) or outside (external fertilization). Internal fertilization demands that sperm be introduced into the female, often, but not always, by copulation.

While most higher organisms have separate male and female individuals, many plants and some invertebrates are HERMAPHRODITE, and some protists produce gametes that all look the same (isogamy), though they may actually belong to different mating strains, which can fuse only with one of the opposite type. In most fungi there are no gametes at all; instead, special hyphae known as gametangia fuse, and the nuclei of the two parent mycelia join up within the hyphae. The probable advantage of sexual reproduction is that the bringing together of genes derived from two individuals produces variation in each generation, enabling populations to change and thus adapt themselves to changing environmental conditions. At the molecular level, the most important aspect of reproduction is the ability of the chromosome to duplicate itself (see NUCLEIC ACIDS). (See also ALTERNATION OF GENERATIONS.) (♦♦ page 16-17, 204)

Reptiles
A class of the VERTEBRATES that includes the crocodiles, alligators, snakes, amphisbaenids, lizards, turtles and tortoises as well as a number of extinct groups (see DINOSAURS; ICHTHYOSAURS; MAMMAL-LIKE REPTILES; PLESIOSAURS; PTEROSAURS. Living reptiles have scaly skin and typically lay large, yolky eggs. Being POIKILOTHERMS ("cold-blooded"), they are most numerous in the tropics. FOSSIL forms are more numerous than living forms and occur mainly in rocks of the Permian to Cretaceous periods, 286-65 million years ago. (♦ page 17)

Research
The use of appropriate methods in attempting to discover new knowledge or to develop new applications of existing knowledge or to explore relationships between ideas or events. Scientific discoveries, technological achievements and scholarly publications are all the fruits of research. Every discipline develops research methods and tools appropriate to its subject matter; but whether undertaken by scholar, technologist or scientist, research always involves three basic steps: the formulation of a problem; the collection and analysis of relevant information; and a concerted attempt to discover a solution or otherwise resolve the problem in a manner dictated by the available evidence. Quite different kinds of initial problem may be formulated. In the field of science and technology, for example, fundamental (or properly scientific) research aims at enlarging our understanding of observable phenomena; the search is for general explanatory principles. Unlike applied (or technological) research, fundamental research is not explicitly directed toward the solution of a practical problem, although its results may, and usually do, suggest new technological possibilities. Knowledge of atomic structure is a goal of fundamental research; possible applications of this knowledge – nuclear power plants and weapons – demand technological research and development. In practice, however, the distinction is less clear-cut: accidental scientific discoveries are often made by research workers pursuing a technological goal. (See also SCIENTIFIC METHOD.)

Resin
A high-molecular-weight substance characterized by its gummy or tacky consistency at certain temperatures. Naturally occurring resins include bitumen, shellac (from insects) and rosin (from pine trees). Synthetic resins include the wide variety of plastic materials available today. (See also PLASTICS.)

Resistance
The ratio of the voltage applied to a conductor to the current flowing through it (see ELECTRICITY; OHM'S LAW), measured in OHMS. It is characteristic of the material of which the conductor is made (the resistance presented by a unit cube of a material being called its resistivity) and of the physical dimensions of the conductor, increasing as the conductor becomes longer and/or thinner. Resistance rises with TEMPERATURE in METALS, but falls in SEMICONDUCTORS and SOLUTIONS. Its accurate measurement is performed by the WHEATSTONE BRIDGE method. (♦ page 118)

Resonance
The large response of an oscillatory mechanical, acoustical or electrical system driven near its natural FREQUENCY. The ENERGY dissipation (due to FRICTION, etc) of all practical systems is termed damping: the amount of damping controls both the size of the resonant response and the sharpness of the resonance as a function of frequency.

Resonance
In chemistry, the theory of molecular structure in which the actual state of the bonding in a molecule is expressed as a "resonance hybrid" between two or more valence-bond structures (see BOND, CHEMICAL) and is intermediate between them, but of lower energy. There is no actual oscillation, and the model is equivalent to molecular ORBITAL theory. First proposed for benzene, resonance stabilizes AROMATIC COMPOUNDS and conjugated double-bond systems such as 1,3-butadiene.

Resorcinol (m-$C_6H_4(OH)_2$)
One of the dihydric PHENOLS, a colorless crystalline solid made by sulfonation (see SULFONIC ACIDS) of benzene followed by fusion with sodium hydroxide. It is used to make adhesives, formaldehyde RESINS, DYES, EXPLOSIVES, photographic developers, and as an ANTISEPTIC. (See also FLUORESCEIN.) MW 110.1, mp 110°C, bp 276.6°C.

Respiration
The breakdown of foods to yield energy. Anaerobic respiration involves a limited set of chemical reactions, known as GLYCOLYSIS, and does not require oxygen, whereas aerobic respiration involves the CITRIC ACID CYCLE and the ELECTRON TRANSPORT CHAIN as well as glycolysis, and is dependent upon oxygen. Aerobic respiration breaks the food molecules down into smaller fragments and yields more energy. It is found in all living organisms other than a few primitive bacteria (see ARCHEBACTERIA).

Respiration is sometimes also used to mean breathing (or other methods of obtaining oxygen, as in the gills of fish) but this is better described as GAS EXCHANGE. (♦ page 16-17, 28, 44)

Respiration, Artificial see ARTIFICIAL RESPIRATION.

Respiratory distress syndrome
Difficulty of breathing and blue appearance of some newborn babies, particularly those born to diabetic mothers or who are premature or born by CESARIAN SECTION.

Restriction enzymes
Enzymes, also called restriction endonuclease, that recognize specific short sequences of bases in DNA and break the DNA chain where these sequences occur. About 300 different restriction enzymes have been identified, each recognizing a different base sequence. These enzymes are a vital tool in GENETIC studies. If a given set of restriction enzymes is applied to DNA from genetically identical organisms, it will produce the same set of DNA fragments (which can be separated and identified by biochemical techniques). Genetic variation in a family group, population or species will result in different DNA fragments being produced. This technique is known as restriction site mapping.

Restriction mapping
A method used in molecular GENETICS for characterizing DNA. It relies on RESTRICTION ENZYMES, which cut the DNA strand at particular points. For DNA of a given type, a characteristic set of fragments is produced using particular restriction enzymes. The method is sufficiently sensitive to detect genetic differences between individuals in the same population, and can be used to identify close relatives; it is sometimes known as "DNA fingerprinting". This technique has been used in populations of wild animals to study breeding behavior or to quantify the genetic variability of a population, and in forensic science to identify samples of semen. It can also be used to confirm that individuals are members of the same family (as in immigration control, for example) or to establish the paternity of a child.

Reticulo-endothelial system
Generic name for those cells in the body that take up foreign material from the blood and other body fluids; they are also known as macrophages. Blood monocytes are functionally part of the system, as are macrophages in the LYMPH nodes, SPLEEN, BONE MARROW, LIVER (Kupffer cells) and LUNG alveoli. When foreign material (eg BACTERIA) is introduced into the bloodstream, macrophages rapidly take it up by PHAGOCYTOSIS and destroy it with intracellular enzymes. This constitutes a primary defence system and plays a part in IMMUNITY. Similarly, particulate matter in the lungs or liver is cleared by local macrophages.

Retina
The part of the EYE responsible for conversion of light into nerve impulses; it contains the rod and cone receptor cells, and nerve cells that connect with the optic nerve and thus with the brain. (♦ page 213)

Retinol (vitamin A) see VITAMINS.

Retinopathy
Any disease of the retina; the term is used especially of damage due to the long-term effects of diabetes.

Retrograde motion
The apparent backward (ie westward) motion of a PLANET due to the Earth's own motion (see ORBIT): also, the motion of any SOLAR SYSTEM body rotating or orbiting in the opposite (clockwise as viewed from the N celestial pole) direction to the majority.

Retrolental fibroplasia
A form of blindness occurring in premature babies exposed to high oxygen concentrations during the treatment of respiratory distress syndrome. Moderation in the use of oxygen and prevention of prematurity have reduced its incidence.

Retrovirus
A type of RNA VIRUS that has the ability to produce a DNA copy of its RNA once it has infected the

host cell. It does this by means of a special enzyme, reverse transcriptase. Retroviruses are very simple life-forms, which carry only enough genetic information to code for three proteins. Most produce tumors in birds or rodents, but one is responsible for AIDS in humans.

Reye's syndrome
An acute, often fatal disease of babies and young children marked by fatty degeneration of the liver and neurological symptoms, including fits. Its onset sometimes follows the administration of aspirin, which for this reason is now considered unsuitable for children under 12.

Reynolds, Osborne (1842-1912)
British physicist best known for his important contributions in FLUID MECHANICS, in particular his derivation (1883-84) of the REYNOLDS NUMBER.

Reynolds number
Important dimensionless parameter in FLUID MECHANICS, given by $R=\rho ul/\mu$, where ρ is the density, u the velocity, l the length and μ the viscosity of the fluid. The value of R allows for the effects of fluid viscosity on motion and determines whether a given fluid flow is steady or turbulent. It is useful for evaluating the behavior of scale models.

Rhenium (Re)
Very hard, silvery white TRANSITION ELEMENT in Group VIIB of the PERIODIC TABLE. It is very rare, and is obtained as a byproduct of MOLYBDENUM extraction. Analogous to MANGANESE, it forms compounds of all oxidation states between 0 and +7, those of the higher states being volatile and stable. Its uses are similar to those of the PLATINUM GROUP metals. AW 186.2, mp 3180°C, bp 5600°C, sg 20.5 (20°C). (♦ page 130)

Rheology
Branch of physics concerned with the structure and behavior of flowing and deformed materials, such as the way the shape and size of a body alters with time when subjected to mechanical forces. Properties such as FRICTION, stickiness and roughness are treated, and the study finds application in many fields from engineering to plant physiology.

Rheostat
A variable resistor used to control the current drawn by an electric motor, to dim lighting, etc. It may consist of a resistive wire, wound in a helix, with a sliding contact varying the effective length, or of a series of fixed resistors connected between a row of button contacts. For heavy loads, electrodes dipped in solutions can be used, the resistance being controlled by the immersion depth and separation of the electrodes.

Rhesus factor (Rh factor) see BLOOD.

Rheumatic fever
A rare disease, formerly more common, affecting mainly children and young adults, that arises as a delayed complication of infection of the upper respiratory tract with STREPTOCOCCUS. Skin rash, subcutaneous nodules and a migrating ARTHRITIS are common. Involvement of the HEART may lead to palpitations, chest pain, cardiac failure, MYOCARDITIS and INFLAMMATION of the PERICARDIUM; murmurs may be heard and the ELECTROCARDIOGRAPH may show conduction abnormality. Late effects include chronic valve disease of the heart leading to stenosis or incompetence, particularly of the mitral or aortic valves. Such valve disease appears in young to middle age and may require surgical correction.

Rheumatism
Imprecise term describing various disorders of the JOINTS, including RHEUMATIC FEVER and rheumatoid ARTHRITIS.

Rhine, Joseph Banks (born 1895)
US parapsychologist whose pioneering laboratory studies of ESP attempted to provide a scientific demonstration of the occurrence of telepathy (see PARAPSYCHOLOGY).

Rhinitis
INFLAMMATION of the mucous membranes of the nose, causing runny nasal discharge and seen in the common COLD, INFLUENZA and HAY FEVER. Irritation in the nose and sneezing are common.

Rhizoids
The name given to the simple root-like structures that are found on some mosses and liverworts, and on the thallus of the GAMETOPHYTE stages in ferns. They are much simpler in structure than true roots, but have the same functions – absorbing water and minerals from the soil.

Rhizome (rootstock)
The swollen horizontal underground stem of certain plants that acts as an organ of perennation and vegetative propagation. They last for several years, new shoots appearing each spring from the axils of scale leaves on the rhizomes.

Rhodium (Rh)
Moderately hard metal, the whitest of the PLATINUM GROUP. In addition to the general uses of these metals, rhodium is used for mirror surfaces. AW 102.9, mp 1963°C, bp 3700°C, sg 12.4 (20°C). (♦ page 130)

Rhodophyta (red algae) see ALGAE.

Rhodopsin (visual purple)
A photosensitive pigment, derived from vitamin A, found in the RETINAS of many vertebrates, including humans.

Rhyolite
Light-colored volcanic IGNEOUS ROCK, of the same composition as GRANITE, and very common and widespread. Often banded from LAVA flow as it solidified, it is fine-grained and usually porphyritic (see PORPHYRY). (See also OBSIDIAN; PUMICE.)

Rhythms, Biological see BIOLOGICAL CLOCKS.

Rib see SKELETON.

Ribbonworms see NEMERTEANS.

Riboflavin (vitamin B_2) see VITAMINS.

Ribonucleic acid see NUCLEIC ACIDS; RNA.

Ribose ($C_5H_{10}O_5$)
A pentose monosaccharide; part of the NUCLEOTIDE units from which RNA is made. (♦ page 28)

Ribosomes
Tiny granules, of diameter about 10nm, found in CELL cytoplasm. They are composed of PROTEIN and a special form of ribonucleic acid (see NUCLEIC ACIDS) known as ribosomal RNA or rRNA. The ribosome is the site of PROTEIN SYNTHESIS. In eukaryotic cells ribosomes are located mostly on the ENDOPLASMIC RETICULUM. (♦ page 13)

Richards, Dickinson Woodruff (1895-1973)
US physiologist awarded with COURNAND and FORSSMAN the 1956 Nobel Prize for Physiology or Medicine for his work with Cournand using Forssman's CATHETER technique to probe the heart, pulmonary artery and lungs.

Richards, Theodore William (1868-1928)
US chemist awarded the 1914 Nobel Prize for Chemistry for his determination of the atomic weights of some 60 elements. In particular, his accurate work showed the existence of isotopes, predicted earlier by FREDERICK SODDY.

Richardson, Sir Owen Willans (1879-1959)
British physicist awarded the 1928 Nobel Prize for Physics for his pioneering work on THERMIONIC EMISSION. The Richardson equation relates the rate of thermionic emission to the absolute TEMPERATURE of the heated metal.

Richet, Charles Road (1850-1935)
French physiologist awarded the 1913 Nobel Prize for Physiology or Medicine for his studies of ANAPHYLAXIS, which term he also coined.

Richter, Burton (born 1931)
US physicist who shared the 1976 Nobel Prize for Physics with S. TING for his independent discovery of the massive psi(3095) meson, the first particle identified as containing a charmed quark (see SUBATOMIC PARTICLES).

Richter scale
Scale devised by C. F. Richter (1900-1985), used to measure the magnitude of EARTHQUAKES in terms of the amplitude and frequency of the surface waves. The scale runs from 0 to 10, about one quake a year registering over 8.

Rickets
Bowing of the long bones, especially of the legs, in children whose diet is deficient in calcium or vitamin D, which is necessary for calcium absorption. Vitamin D can be obtained either from the diet or by the action of sunshine on the skin, so dark-skinned children need either more sunshine or more vitamin D than fair-skinned children.

Rickettsia
A group of bacteria that can survive only inside a cell (obligate intracellular parasites). This has led to their erroneous classification as intermediate between viruses and bacteria. They are responsible for a number of diseases (often borne by ticks or lice) including TYPHUS, SCRUB TYPHUS and ROCKY MOUNTAIN SPOTTED FEVER; related organisms cause Q FEVER and PSITTACOSIS.

Rift valley (graben)
A valley formed by the relative downthrow of land between two roughly parallel FAULTS. The best-known are the Great Rift Valley of E Africa and the Rheingraben. (♦♦ page 142-3, 208)

Rigel
Beta Orionis, a blue supergiant which is the brightest star in ORION. A quadruple star consisting of a visual binary and a spectroscopic binary (see DOUBLE STAR), it is 276pc distant.

Right ascension see CELESTIAL SPHERE.

Rigor mortis
Stiffness of the body MUSCLES occurring some hours after death. It is caused by the binding of the muscle proteins, actin and myosin. In the normal, living body their interaction occurs in two stages, binding and release, with ATP, the energy currency of the cell, being used up in the release stage. After death the absence of energy-generating processes, and thus of ATP, means that the two proteins remain bound together, making the muscles rigid. The body is set in the position held at the onset of the changes.

Rime see FROST.

Ringworm
A common FUNGUS disease of the skin of humans and animals that may also affect the hair or nails. Ring-shaped raised lesions occur.

Rittenhouse, David (1732-1796)
American astronomer and mathematician who invented the DIFFRACTION grating and built two famous orreries. He also discovered the atmosphere of VENUS (1768) independently of LOMONOSOV (1761), and built what was probably the first American TELESCOPE.

Ritter, Johann Wilhelm (1776-1810)
Silesian-born German physical chemist, a pioneer of ELECTROCHEMISTRY, who first positively identified ULTRAVIOLET RADIATION (1801).

Rivers and lakes
Bodies of inland water. Rivers flow in natural channels to the sea, lakes or, as tributaries, into other rivers. They are a fundamental component of the HYDROLOGIC CYCLE (see DRAINAGE). Lakes are land-locked stretches of water fed by rivers, though the term may be applied also to temporary widenings of a river's course or to almost enclosed bays and LAGOONS. In many parts of the world rivers and lakes may exist only during certain seasons, drying up partially or entirely during DROUGHT (see also PLAYA; WADI).

The main sources of rivers are SPRINGS, lakes and GLACIERS. Near the source a river flows swiftly, the rocks and other abrasive particles that it carries eroding a steep-sided V-shaped VALLEY (see EROSION). Variations in the hardness of the rocks over which it runs may result in WATERFALLS. In the middle part of its course the gradients become less steep, and lateral sideways erosion becomes more important than downcutting. The valley is broader, the flow less swift, and meandering more common. Toward the rivermouth the flow becomes more sluggish and meandering prominent: the river may form OXBOW LAKES. Sediment may be deposited at the mouth to form a DELTA (see also ESTUARY). (See also CANYON; HYDROELECTRICITY.)

Most lakes are the result of glacial erosion during the ICE AGES. Glaciers hollowed out deep basins, often depositing MORAINE to form natural dams. Most lakes have an outflowing stream; where there is great water loss through EVAPORATION there is no such stream and the lake water is extremely saline (see also EVAPORITES), as in the Dead Sea. Lakes are comparatively temporary features on the landscape, as they are constantly being infilled by silt. (See also DIVIDE; FIRTH; FJORD; GROUNDWATER.)

RNA (ribonucleic acid)
A long-chain molecule like DNA. The two are known collectively as the NUCLEIC ACIDS. The sugars in the RNA molecule are slightly different from those in DNA, and it has a different base, uracil, in place of thymine (see GENETIC CODE). RNA is concerned with the translation of the genetic message, stored in DNA, into the structure of proteins. There are three main types of RNA: messenger RNA (mRNA), transfer RNA (tRNA) and ribosomal RNA (rRNA). Messenger RNA carries the genetic message from the DNA and acts as a template during PROTEIN SYNTHESIS. Transfer RNAs bind to specific amino acids, and carry an anticodon of three bases that bind to the CODON on the messenger RNA molecule. Ribosomal RNA is part of the structure of the RIBOSOMES and has a poorly understood role in organizing protein assembly. In a few living organisms, the RNA VIRUSES, RNA itself acts as the genetic material.

RNA viruses
Viruses that have RNA instead of DNA as their genetic material; it may be double-stranded or single-stranded. In some single-stranded forms, the RNA in the virus acts directly as messenger RNA in the cell. In others it must first be replicated, to produce a complementary RNA strand that acts as the messenger RNA molecule. This replication is achieved using an enzyme, RNA polymerase, that is carried by the virus. Another group of RNA viruses, the RETROVIRUSES, must convert their RNA to DNA before they can replicate.

Robbins, Frederick Chapman (born 1916)
US virologist who shared the 1954 Nobel Prize for Physiology or Medicine with J.F. ENDERS and T.H. WELLER for their cultivation of the POLIOMYELITIS virus in nonnerve tissues.

Robinson, Sir Robert (1886-1975)
British organic chemist awarded the 1947 Nobel Prize for Chemistry for his pioneering studies of the molecular structures of the ALKALOIDS and other vegetable-derived substances.

Robot
An automatic machine that does work, simulating and replacing human activity; known as an android, if humanoid in form (which most are not). Robots have evolved out of simpler automatic devices, and many are now capable of decision-making, self-programming, and carrying out complex operations. Some have sensory devices. They are widely used in industry for mass production, including welding, grinding, paint spraying and die casting. They can work in environments that are dangerous or unpleasant to humans, and can perform simple manipulation tasks faster and more accurately than humans without tiring.

Rock crystal see QUARTZ.

Rocket
Form of jet-propulsion engine in which the substances (fuel and oxidizer) needed to produce the propellant gas jet are carried internally. Working by reaction, and being independent of atmospheric oxygen, rockets are used to power interplanetary space vehicles (see SPACE EXPLORATION.) In addition to their chief use to power missiles, rockets are also used for supersonic and assisted-takeoff airplane propulsion, and sounding rockets are used for scientific investigation of the upper atmosphere. The first rockets – of the firework type, cardboard tubes containing gunpowder – were made in 13th-century China, and the idea quickly spread to the West. Their military use was limited, guns being superior, until they were developed by William Congreve (1772-1828). Later Congreve rockets mounted the guide stick along the central axis; and William Hale eliminated it altogether, placing curved vanes in the exhaust stream, thus stabilizing the rocket's motion by causing it to rotate on its axis. The 20th century saw the introduction of new fuels and oxidants, eg a mixture of nitrocellulose and nitroglycerin for solid-fuel rockets, or ethanol and liquid oxygen for the more efficient liquid-fuel rockets. The first liquid-fuel rocket was made by R.H. GODDARD, who also invented the multistage rocket. In WWII Germany, and afterward in the US, WERNHER VON BRAUN made vast improvements in rocket design. Other propulsion methods, including the use of nuclear furnaces, electrically accelerated plasmas and ion propulsion, are being developed.

Rocks
The solid materials making up the Earth's crust. They may be consolidated (eg sandstone) or unconsolidated (eg sand). The study of rocks is petrology. Strictly, the term applies only to those materials that, unlike MINERALS, are not homogeneous and have no definite chemical composition. The primary constituents of rocks are OXYGEN and SILICON, combined with each other to form SILICA (see QUARTZ), and with each other and further elements (eg aluminum, iron, calcium, potassium, sodium and magnesium) to form SILICATES. Together, silica and silicates make up about 95% of the Earth's rocks. There are three main classes of rocks: igneous, sedimentary and metamorphic. IGNEOUS ROCKS form from the MAGMA, a molten, subsurface complex of silicates. They are the primary source of all the Earth's rocks. SEDIMENTARY ROCKS are consolidated accumulations of fragmented inorganic and organic material. They are of three types: classic, formed of weathered (see EROSION) particles of other rocks (eg SANDSTONE); organic deposits (eg COAL, some LIMESTONES); and chemical precipitates (eg the EVAPORITES). (See also FOSSILS; STRATIGRAPHY.) METAMORPHIC ROCKS have undergone change within the Earth under heat, pressure or chemical action. Sedimentary, igneous and even previously metamorphosed rocks may change in structure or composition in this way. (See also EARTH; GEOLOGY.)

Rock salt (halite) see SALT.

Rocky Mountain spotted fever
Tick-borne rickettsial disease (see RICKETTSIA) seen in much of the US, especially the Rocky Mountain region. It causes fever, headache and a characteristic rash starting on the palms and soles, later spreading elsewhere.

Rodents
An order of MAMMALS that includes the rats and mice, beavers, squirrels, porcupines, guinea pigs, capybaras and agoutis. It is the largest order of mammals, with almost 1500 living species. All rodents have a single pair of incisor teeth, which grow throughout life and wear to form sharp chisel edges used for gnawing. They are predominantly eaters of seeds and tough vegetable matter.

Rods see VISION.

Roebuck, John (1718-1794)
British inventor of the lead chamber process for making sulfuric acid (1746), and patron of JAMES WATT.

Roemer, Ole or **Olaus** see RØMER, OLE or OLAUS.

Roentgensee RÖNTGEN.

Rogers, Carl Ransom (b. 1902)
US psychotherapist who instituted the idea of the patient determining the extent and nature of his course of therapy, the therapist following the patient's lead.

Rohrer, Heinrich (b. 1933)
Swiss scientist awarded half of the 1986 Nobel Prize for Physics jointly with GERD BINNING for their design of the scanning tunneling microscope.

Rømer, Ole or **Olaus** (1644-1710)
Danish astronomer who first showed that light has a finite velocity. He noticed that JUPITER eclipsed its moons at times differing from those predicted, and correctly concluded that this was due to the finite nature of light's velocity, which he calculated as 227,000km/s (a modern value is about 299,800km/s).

Röntgen (R)
A unit of radiation exposure named for W.C. RÖNTGEN. Its value in SI units is such that exposure to 1 röntgen of X-RAYS or GAMMA RAYS causes 0.000258 coulombs of ionization per kg of air.

Röntgen (or **Roentgen**), **Wilhelm Conrad** (1845-1923)
German physicist, recipient in 1901 of the first Nobel Prize for Physics for his discovery of X-RAYS. This discovery was made in 1895 when by chance he noticed that a PHOSPHOR screen near a vacuum tube through which he was passing an electric current fluoresced brightly, even when shielded by opaque cardboard. (See also RÖNTGEN unit.)

Roots
The part of a PLANT that absorbs water and nutrients from the soil and anchors the plant to the ground. Water and nutrients enter a root through minute hairs that are outgrowths of the outermost cells. Roots need oxygen to function, and plants growing in swamps have special adaptations to supply it, like the aerial roots of the mangrove. There are two main types of root systems: the taproot system, where there is a strong main root from which smaller secondary and tertiary roots branch out; and the fibrous root system, in which a mass of equal-sized roots are produced. In plants such as the sugar beet the taproot may become swollen with stored food material. Adventitious roots anchor the stems of climbing plants, such as ivy. Epiphytic plants such as orchids have roots that absorb moisture from the air (see EPIPHYTE). The roots of parasitic plants such as dodder absorb food from other plants. (See also RHIZOIDS.)

Rootstock see RHIZOME.

Rorschach, Hermann (1884-1922)
Swiss psychoanalyst who devised the Rorschach Test (c1920), in which the subject looks at a series of ten symmetrical inkblots, and describes what he sees there. It is intended that, from his description, details of his personality can be deduced.

Rosin see RESIN.

REPRODUCTION

The female genitals

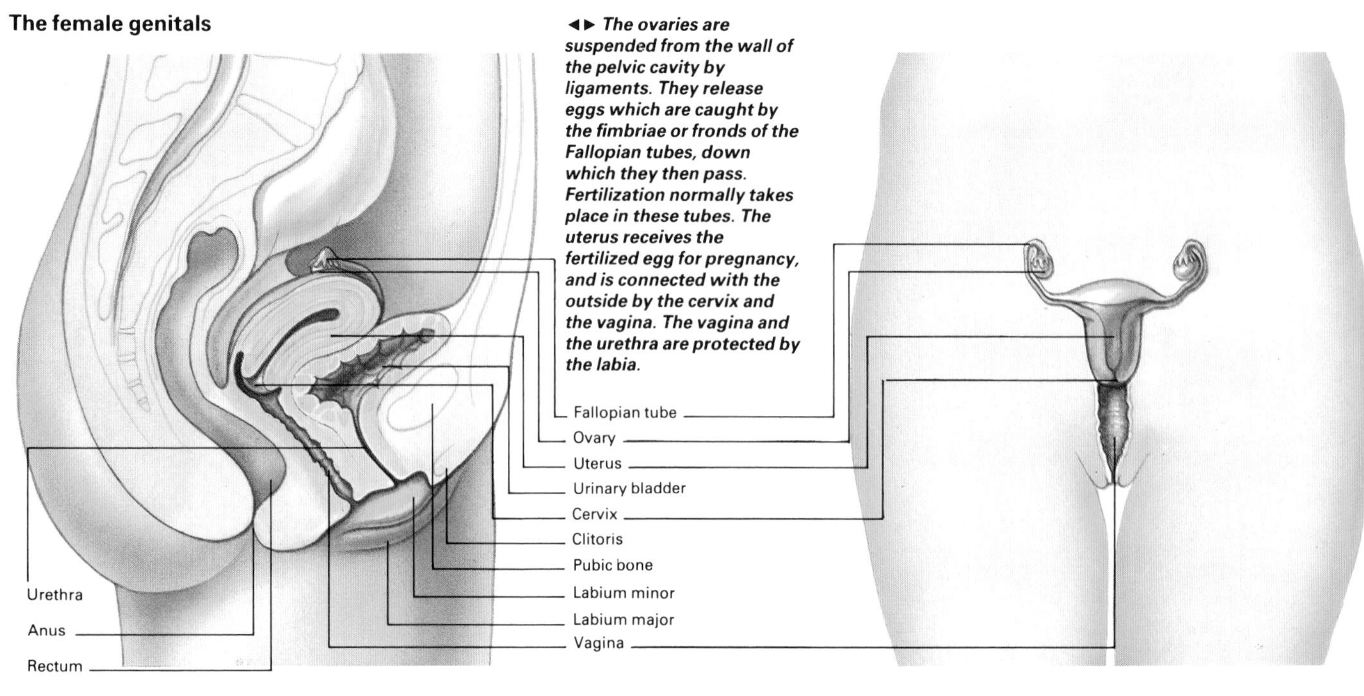

◄► *The ovaries are suspended from the wall of the pelvic cavity by ligaments. They release eggs which are caught by the fimbriae or fronds of the Fallopian tubes, down which they then pass. Fertilization normally takes place in these tubes. The uterus receives the fertilized egg for pregnancy, and is connected with the outside by the cervix and the vagina. The vagina and the urethra are protected by the labia.*

The human egg

The moment of fertilization

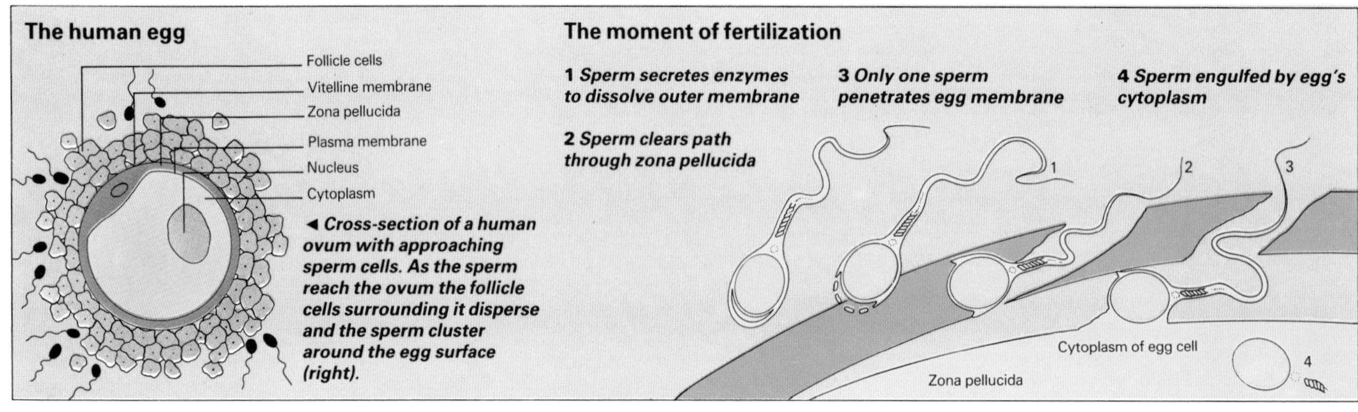

◄ *Cross-section of a human ovum with approaching sperm cells. As the sperm reach the ovum the follicle cells surrounding it disperse and the sperm cluster around the egg surface (right).*

The male genitals

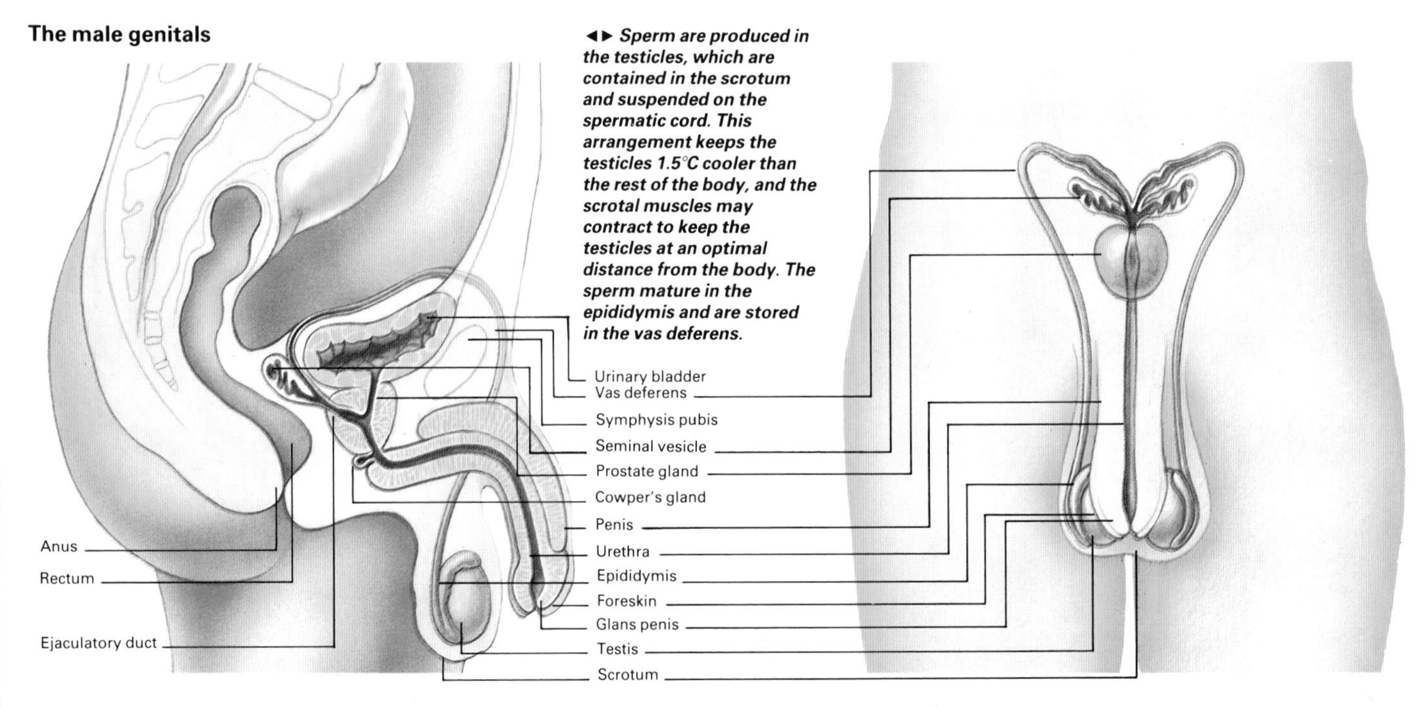

◄► *Sperm are produced in the testicles, which are contained in the scrotum and suspended on the spermatic cord. This arrangement keeps the testicles 1.5°C cooler than the rest of the body, and the scrotal muscles may contract to keep the testicles at an optimal distance from the body. The sperm mature in the epididymis and are stored in the vas deferens.*

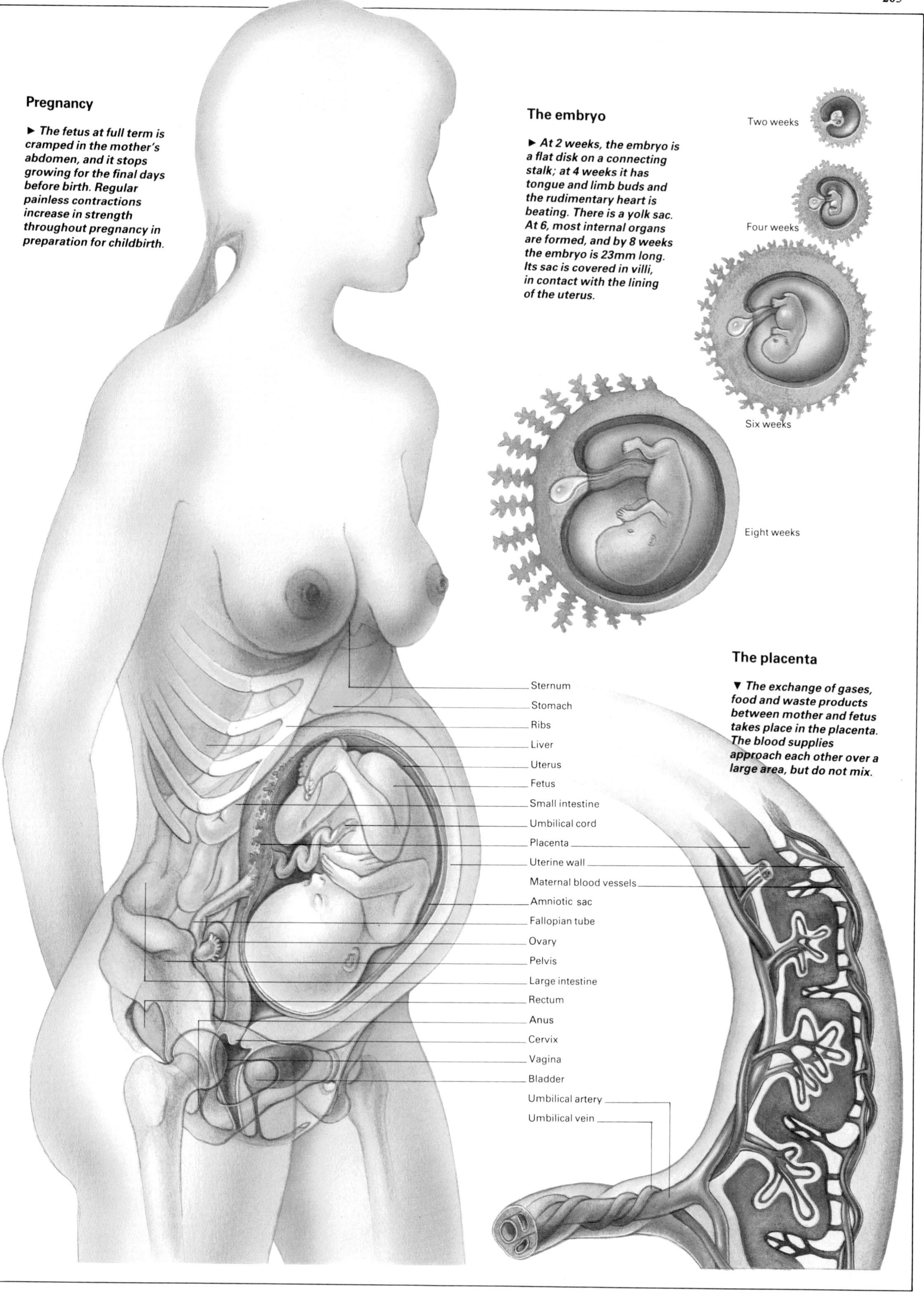

Pregnancy

► *The fetus at full term is cramped in the mother's abdomen, and it stops growing for the final days before birth. Regular painless contractions increase in strength throughout pregnancy in preparation for childbirth.*

The embryo

► *At 2 weeks, the embryo is a flat disk on a connecting stalk; at 4 weeks it has tongue and limb buds and the rudimentary heart is beating. There is a yolk sac. At 6, most internal organs are formed, and by 8 weeks the embryo is 23mm long. Its sac is covered in villi, in contact with the lining of the uterus.*

The placenta

▼ *The exchange of gases, food and waste products between mother and fetus takes place in the placenta. The blood supplies approach each other over a large area, but do not mix.*

Ross, Sir Ronald (1857-1932)
British physician awarded the 1902 Nobel Prize for Physiology or Medicine for his investigations, prompted by SIR PATRICK MANSON, of the *Anopheles* MOSQUITO in relation to the transmission of MALARIA, in the course of which he isolated malarial cysts in the mosquito's intestinal tract, later correctly identifying these with cysts he found in the bloodstreams of diseased birds (1897-98).

Rotifers
Microscopic invertebrate animals, the largest of which can just be seen with the naked eye. They are cup- or vase-shaped, with an organ known as a corona protruding from the upper opening. This is covered with cilia, which beat to create water currents that carry food to the mouth. At the base of the body is a "tail" (often referred to as the "foot" or "tailpiece"), which in many species ends in a gripping toe; this can grasp water plants and other surfaces when the rotifer is feeding. Rotifers are found in most freshwater habitats, from lakes to puddles, and even in the water that collects in the leaf axils of plants. They have extraordinary powers of survival, and can withstand being frozen for a century or more.

Rous, Francis Peyton (1879-1970)
US physician who shared (with C.B. HUGGINS) in the 1966 Nobel Prize for Physiology or Medicine for his discovery (*c*1910) of a VIRUS which causes TUMORS in chickens.

Roux, Pierre Paul Emile (1853-1933)
French bacteriologist noted for his work with PASTEUR toward a successful anthrax treatment, with METCHNIKOV on syphilis, and with YERSIN on diphtheria. Using Roux' and Yersin's results, VON BEHRING was able to develop the diphtheria antitoxin.

Rowland, Henry Augustus (1853-1901)
US physicist and engineer who developed the concave diffraction grating, in which the lines are ruled directly onto a concave spherical surface, thus eliminating the need for additional mirrors and lenses. (See also SPECTROSCOPY.)

Royal Institution
An English scientific society, founded in 1799 by Sir Benjamin Thompson (see COUNT RUMFORD) to encourage scientific study and the spread of scientific knowledge. It has been associated with many eminent scientists, including HUMPHRY DAVY and MICHAEL FARADAY.

Royal jelly
In social bees, a special food produced by the workers' bodies that is fed to a few larvae and stimulates them to develop into sexually reproducing adults rather than non-reproductive workers. (See SOCIAL INSECTS.)

Royal Society of London for the Improvement of Natural Knowledge
The premier English scientific society. Probably the most famous scientific society in the world, it also has a claim to be the oldest surviving. It had its origins in weekly meetings of scientists in London in the 1640s, and was granted a royal charter by Charles II in 1660. Past presidents include Samuel Pepys, SIR ISAAC NEWTON and LORD RUTHERFORD.

Rubber
An elastic substance; that is, one that quickly restores itself to its original size after it has been stretched or compressed. Natural rubber is obtained from many plants, and commercially from *Hevea brasiliensis*, a tree native to South America and cultivated also in SE Asia and W Africa.

Natural rubber is a chain POLYMER of ISOPRENE, known as caoutchouc when pure. It was a mere curiosity until the pioneer work of Thomas Hancock (1786-1865) and CHARLES MACINTOSH. Synthetic rubbers have been produced since WWI. They are long-chain polymers, elastomers; the main types are: copolymers of butadiene/styrene, butadiene/nitriles and ethylene/propylene; polymers of chloroprene (neoprene rubber), butadiene, isobutylene and SILICONES; and polyurethanes, polysulfide rubbers and chlorosulfonated polyethylenes.

Rubbia, Carlo (b. 1934)
Italian physicist who shared the 1984 Nobel Prize for Physics with VAN DER MEER for their discovery of the WEAK NUCLEAR FORCE – the charged W and the neutral Z particles (see SUBATOMIC PARTICLES).

Rubella (German measles)
Mild VIRUS infection, usually contracted in childhood and causing fever, skin rash, malaise and LYMPH node enlargement. Its importance lies in the fact that infection of a mother during the first three months of pregnancy leads to infection of the embryo via the placenta and is associated with a high incidence of congenital diseases including cataract, deafness and defects of the heart and esophagus. VACCINATION of intending mothers who have not had rubella is advisable.

Rubidium (Rb)
A soft, silvery white, highly reactive ALKALI METAL. It is fairly abundant, but is found only as a minor constituent of POTASSIUM and CESIUM minerals. It is more reactive than potassium, and reacts violently with water and ice. Metallic rubidium, prepared by ELECTROLYSIS of the chloride or reduction of the carbonate, is used in electron tubes, and its salts in making special glasses and ceramics. AW 85.5, mp 38.9°C, bp 705°C, sg 1532 (20°C). (◆ page 130)

Ruby
Deep red GEM stone, a variety of CORUNDUM colored by a minute proportion of chromium ions. It is found significantly only in upper Burma, Thailand and Sri Lanka, and is more precious by far than diamond. The name has been used for other red stones, chiefly varieties of garnet and spinel. Rubies have been synthesized by the Verneuil flame-fusion process (1902). They are used to make ruby LASERS.

Rumford, Sir Benjamin Thompson, Count (1753-1814)
American-born adventurer and scientist best known for his recognition of the relation between WORK and HEAT (inspired by observation of heat generated by FRICTION during the boring of cannon), which laid the foundations for JOULES's later work. He played a primary role in the founding of the ROYAL INSTITUTION (1799), to which he also introduced HUMPHRY DAVY.

Ruminants
Animals that ruminate or "chew the cud". They include the camels, deer, giraffes, antelopes, cattle, buffalo, bison, goats and sheep. All have a complex stomach with up to four chambers. The first, known as the rumen, contains bacteria that ferment the grass and other plant foods eaten by the animal, and digest the cellulose they contain. Mammals cannot digest cellulose for themselves, so this is a mutualistic relationship. The second chamber of the stomach, the omasum, which contains the stomach acid, is separated from the rumen by a very small orifice, as the acid would harm the bacteria. Only when the plant material is thoroughly digested can it pass on to the omasum. The breakdown of the food is helped by the animal regurgitating partially digested food from the rumen and chewing it: "chewing the cud".

Rupture
Common name for HERNIA.

Rush, Benjamin (1746-1813)
US physician, abolitionist and reformer. His greatest contribution to medical science was his conviction that insanity is a disease (see MENTAL ILLNESS): his *Medical Enquiries and Observations upon the Diseases of the Mind* (1812) was the first US book on PSYCHIATRY.

Ruska, Ernst (1906-1988)
German scientist who received half of the 1986 Nobel Prize for Physics for his fundamental work in electron optics, and for the design of the first electron microscope.

Russell, Henry Norris (1877-1957)
US astronomer who, independently of HERTZSPRUNG, showed the relation between a STAR's brightness and color – the resulting Hertzsprung-Russell diagram is important throughout astronomy and cosmology. (◆ page 228)

Rust see CORROSION.

Rusts
A group of FUNGI causing typical plant diseases. They form red or orange spots, which contain their spore-bearing organs, on the leaves of infected plants. Spores are carried by the wind to infect new plants. Some rusts are heteroecious: they alternate between two different host plants. The most important rust fungus is probably Puccinia graminis, which causes black stem rust of wheat. (See also SMUT.)

Ruthenium (Ru)
Hard metal in the PLATINUM GROUP. It is added to platinum or palladium to form hard ALLOYS. Ruthenium is a catalyst (see CATALYSIS) used in organic chemistry. AW 101.1, mp 2330°C, bp 4100°C, sg 12.30 (20°C). (◆ page 130)

Rutherford, Sir Ernest, 1st Baron Rutherford of Nelson (1871-1937)
New Zealand-born British physicist. His early work was with J.J. THOMSON on MAGNETISM and thus on RADIO waves. Following RÖNTGEN's discovery of X-RAYS (1895), they studied the CONDUCTIVITY of air bombarded by these rays and the rate at which the ions produced recombined. This led him to similar studies of the "rays" emitted by URANIUM (see RADIOACTIVITY). As a result of their work, he and FREDERICK SODDY were able in 1903 to put forward their theory of radioactivity. He was awarded the 1908 Nobel Prize for Chemistry. In 1911 he proposed his nuclear theory of the ATOM, on which BOHR based his celebrated theory two years later. In 1919 he announced the first artificial disintegration of an atom, NITROGEN being converted into OXYGEN and HYDROGEN by collision with an alpha particle. He was President of the ROYAL SOCIETY (1925-1930) and is commemorated by the naming of RUTHERFORDIUM.

Rutherfordium (Rf)
A TRANSURANIUM ELEMENT in Group IVB of the PERIODIC TABLE; atomic number 104. Soviet scientists claimed to have synthesized it, calling it kurchatovium, in 1964. American scientists claimed synthesis of rutherfordium in 1969. It is now known as unnilquadium.

Rutile
Red to black OXIDE mineral consisting of impure titanium (IV) oxide (TiO_2), a TITANIUM ore of widespread occurrence. It is used to color porcelain, and synthetic rutile is used for GEMS.

Ruzicka, Leopold (1887-1976)
Croatian-born Swiss chemist who shared with BUTENANDT the 1939 Nobel Prize for Chemistry for his work on the TERPENES and for his demonstration that the ring compounds muskone and civetone had respectively 16 and 17 carbon atoms in the ring (see ALICYCLIC COMPOUNDS): previously it had been thought that rings with more than 8 carbon atoms would be unstable.

Ryle, Sir Martin (1918-1984)
Radio astronomer, corecipient with A. HEWISH of the 1974 Nobel Prize for Physics. He was the first Professor of RADIO ASTRONOMY at Cambridge (1959), was knighted in 1966, and became British Astronomer Royal in 1972.

Sabatier, Paul (1854-1941)
French chemist who shared with GRIGNARD the 1912 Nobel Prize for Chemistry for his work on catalyst action in organic syntheses (see CATALYSIS), especially his discovery that finely divided nickel accelerates HYDROGENATION.
Sabin, Albert Bruce (b. 1906)
Russian-born US virologist best known for developing the oral POLIOMYELITIS vaccine, *Sabin* (1955). (See also SALK, J.E.).
Saccharides see CARBOHYDRATES; SUGARS.
Saccharin (*O*-benzosulfimide)
A SWEETENING AGENT, 550 times sweeter than sucrose, normally used as its soluble sodium salt. Not absorbed by the body, it is used by diabetics and in low-calorie DIETETIC FOODS. Some concern has been expressed over the continuing widespread use of saccharin because of fears that it may be carcinogenic.
Sachs, Julius von (1832-1897)
German botanist regarded as the father of experimental plant physiology. Among his many contributions are his discovery of what are now called CHLOROPLASTS; the elucidation of the details of the GERMINATION process; and his studies of plant TROPISMS.
Sacroiliac joint
The JOINT between the sacrum or lower part of the vertebral column and the iliac bones of the PELVIS. Little movement occurs about the joint but it may be affected by certain types of ARTHRITIS, such as ankylosing spondylitis.
Sadism
The derivation of erotic pleasure from inflicting pain on others; possibly a retention of infantile sexuality.
Sagittarius
The Archer, a constellation on the ECLIPTIC lying in the direction of the galactic center (see MILKY WAY). Sagittarius is the ninth sign of the ZODIAC.
Saint Elmo's fire
The glowing electrical discharge seen at the tips of tall, pointed objects – church spires, ships' masts, airplane wings, etc – in stormy weather. The negative electric charge on the storm clouds induces a positive charge on the tall structure. The impressive display is named (corruptly) for St Erasmus, patron of sailors.
Saint Vitus's Dance (Sydenham's chorea) see CHOREA.
Sakharov, Andrei Dimitrievich (b. 1921)
Soviet physicist who played a prominent part in the development of the first Soviet HYDROGEN BOMB. He subsequently advocated worldwide nuclear disarmament (being awarded the 1975 Nobel Peace Prize) and became a leading Soviet dissident.
Salam, Abdus (b. 1926)
Pakistani physicist who was awarded the 1979 Nobel Prize for Physics with S. GLASHOW and S. WEINBERG for their theory which united the WEAK NUCLEAR FORCE with the ELECTROMAGNETIC FORCE.
Sal ammoniac (ammonium chloride) see AMMONIA.
Salicylic acid (o-$C_6H_4(OH)COOH$)
White crystalline solid, made from PHENOL and carbon dioxide; used in medicine against calluses and warts, and to make aspirin and dyes. Its sodium salt is an analgesic and is used for rheumatism. MW 138.1, mp 159°C. Methyl salicylate, an ESTER, occurs in oil of WINTERGREEN, and is used as a liniment and a flavoring.
Saliva
The watery secretion of the salivary GLANDS which lubricates the mouth and food during chewing. It contains MUCUS, some immunoglobulins and digestive enzymes that break down starch. It is secreted in response to food in the mouth or by conditioned reflexes such as the smell or sight of food. Secretion is partly under the control of the parasympathetic autonomic NERVOUS SYSTEM. The various salivary glands – parotid, submandibular and sublingual – secrete slightly different types of saliva, varying in mucus and enzyme content. (♦ page 40)
Salk, Jonas Edward (b. 1914)
US virologist best known for developing the first POLIOMYELITIS vaccine, *Salk* (1952-54). (See also SABIN, A.B.)
Salmonella
A genus of bacteria that inhabit the intestines of animals and humans and cause disease, eg typhoid fever, food poisoning.
Salt
Common name for sodium chloride (NaCl), found in seawater and also as the common mineral, rock salt or halite. Pure salt forms white cubic CRYSTALS. Some salt is obtained by solar evaporation from salt pans, shallow depressions periodically flooded with seawater; but most is obtained from underground mines. The most familiar use of salt is to flavor food. (Magnesium carbonate is added to table salt to keep it dry.) It is also used in large quantities to preserve hides in leathermaking, in soap manufacture, as a food preservative and in keeping highways ice-free in winter. Rock salt is the main industrial source of chlorine and caustic soda. mp 801°C, bp 1465°C.
Salt, Chemical
An ionic compound (see BOND, CHEMICAL) formed by neutralization of an ACID and a BASE. The vast majority of MINERALS are salts, the best known being common SALT, sodium chloride. Salts are generally good electrolytes (see ELECTROLYSIS); those of weak acids or bases undergo partial HYDROLYSIS in water. Salts may be classified as normal (fully neutralized), acid (containing some acidic hydrogen, eg BICARBONATES), or basic (containing hydroxide ions). They may alternatively be classified as simple salts, double salts (two simple salts combined by regular substitution in the crystal lattice) including ALUMS, and complex salts (containing complex ions: see LIGAND). (♦ page 130)
Salt dome
A mass of EVAPORITE minerals that has pierced the strata above it and domed the strata near the surface. They often form natural traps for PETROLEUM, and occasionally penetrate to the surface to form salt glaciers.
Salt mat
Dried-up bed of an enclosed stretch of water that has evaporated, leaving the salts that it held in solution as a crust on the ground. Best known are the Lake Bonneville flats, near Salt Lake City, Utah. (See also EVAPORITES.)
Saltpeter (potassium nitrate) see POTASSIUM.
Sal volatile (ammonium carbonate, $(NH_4)_2CO_3$)
Colorless crystalline solid made by combining aqueous AMMONIA and CARBON dioxide. It is the main ingredient of smelling salts.
Samara
Botanically, a type of FRUIT; it is a winged ACHENE produced by, for example, maples and elms.
Samarium (Sm)
One of the LANTHANUM SERIES of elements. AW 150.4, mp 1050°C, bp 1600°C, sg 7.52 (α). (♦ page 130)
Samuelsson, Bengt Ingemar (b. 1934)
Swedish biochemist who shared the 1982 Nobel Prize for Physiology or Medicine with BERGSTROM and VANE for his work on the control of prostaglandins.
Sand
In geology, a collection of rock particles with diameters in the range 0.125-2.0mm. It can be graded according to particle size: fine (0.125-0.25mm); medium (0.25-0.5mm); coarse (0.5-1.0mm); and very coarse (1-2mm). Sands result from EROSION by GLACIERS, winds, or ocean or other moving water. Their chief constituents are usually QUARTZ and FELDSPAR. (See also BEACH; DESERT; DUNE; SANDSTONE.)
Sandstone
A SEDIMENTARY ROCK consisting of consolidated SAND, generally cemented by a matrix of CLAY minerals, CALCITE or HEMATITE. The sand grains are chiefly QUARTZ and FELDSPAR. The chief varieties are QUARTZITE, rich in silica; arkose, feldspar-rich; graywacke, coarse-grained and of varied composition; and subgraywacke, with more rounded grains and less feldspar than graywacke. Sandstone grades into CONGLOMERATE and SHALE. Sandstone beds may bear NATURAL GAS or PETROLEUM, and are commonly AQUIFERS. Sandstone is quarried for building, and crushed for use as aggregate.
Sandstorm (dust storm)
Windstorms in which clouds of SAND and DUST are driven across the land. Because of the high wind velocity, sandstorms are powerful factors in SOIL EROSION. (See also DUST BOWL; EROSION.)
Sandwich compounds
ORGANOMETALLIC COMPOUNDS in which the metal atom is sandwiched between two aromatic-ring LIGANDS – the whole of the ring electron system (see AROMATIC COMPOUNDS) interacting with the metal ORBITALS, giving great stability. FERROCENE and dibenzene-chromium are examples.
Sanger, Frederich (b. 1918)
British biochemist awarded the 1958 Nobel Prize for Chemistry for his work on the PROTEINS, particularly for first determining the complete structure of a protein, that of bovine INSULIN. In 1980 he became the fourth person to receive two Nobel Prizes when he was given the Chemistry award jointly with BERG and GILBERT for their work on the chemistry of NUCLEIC ACIDS. Sanger's contribution was to refine the techniques for determining the structure of more complex PROTEINS.
Saponification see SOAPS AND DETERGENTS.
Sapphire
All GEM varieties of CORUNDUM except those that, being red, are called RUBY; blue sapphires are best known, but most other colors are found. The best sapphires come from Kashmir, Burma, Thailand, Sri Lanka and Australia. Synthetic stones, made by flame-fusion, are used for jewel bearings, record styluses, etc.
Saprophytes
A term once used for the saprotrophic fungi and bacteria (see SAPROTROPHS), less commonly used nowadays because it implies, erroneously, that these organisms are plants (-phytes).
Saprotrophs
Organisms that obtain their food from dead matter – either the excreta or the corpses of other living organisms. Most fungi are saprotrophs (those that are not are parasites), and so are many

ROCK CYCLE Faulting and Folding

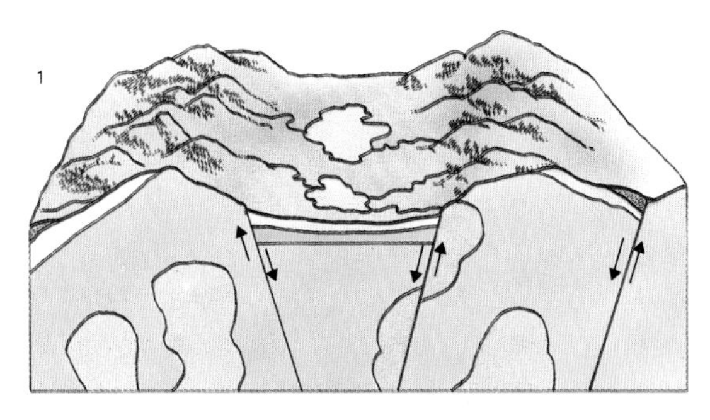

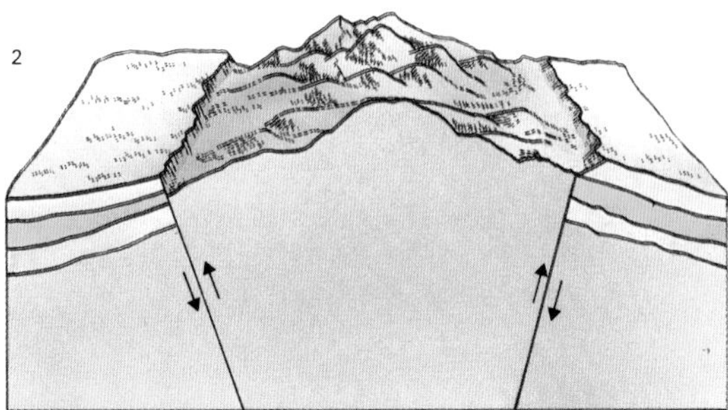

Rift valleys and block mountains

◄ Geological structures, such as folds and faults, are important in determining landscape. However, the surface of the land rarely follows the structures exactly, erosion shaping the final appearance. A block subsiding between two faults produces a structure called a graben (1). This may show as a rift valley on the surface, but erosion of the high areas and sedimentation in the low even out the topography so that the landscape feature is insignificant compared with the depth of the geological feature. The same is true of a horst block, uplifted between two faults, which may form block mountains (2).

Faulting

▼ There are many types of fault. When one block has slid down the fault face in relation to the other it produces a normal fault (1). When it appears to have moved up it is a reverse fault (2). A strike-slip fault (3) has moved the blocks sideways. An oblique fault (4) has both vertical and horizontal movements. Horsts (5) and grabens (6) result from blocks moving between faults.

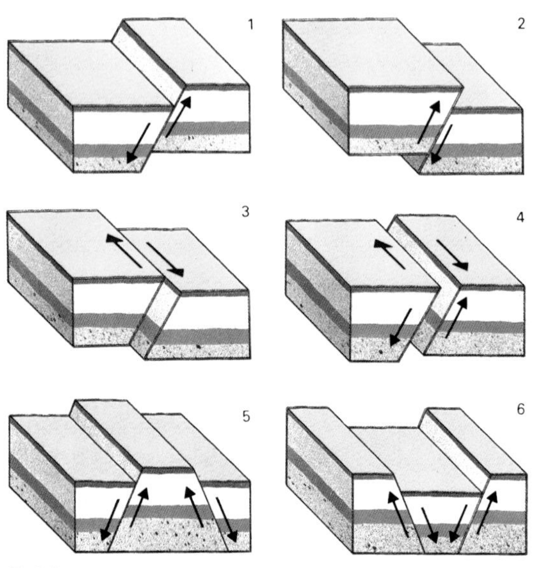

The rock cycle

► The rocks of the crust are constantly being destroyed and renewed. No sooner is a rock exposed to the elements at the surface than it is broken down by the weather and the debris washed away. This debris then collects and is eventually buried by more debris and turned into sedimentary rock. This may later be uplifted and exposed once more. Alternatively the rock may find its way to great depths where it melts and solidifies to become an igneous rock, or it is cooked and crushed to such an extent that its mineral content alters making a metamorphic rock. This change is known as the rock cycle.

Folding

▼ Folds occur when beds of compressed rock bend rather than break. The axial plane is the theoretical surface that passes through the "hinge line" in each successive bed. A vertical axial plane gives a symmetrical fold (1), a sloping one an inclined fold (2). An overturned fold (3) has both limbs sloping the same way. A fold pushed so far over that it has collapsed is a recumbent fold (4).

1 Axial plane · Syncline · Limb · Limb · Anticline

2 Axial plane

3 Axial plane

4 Axial plane

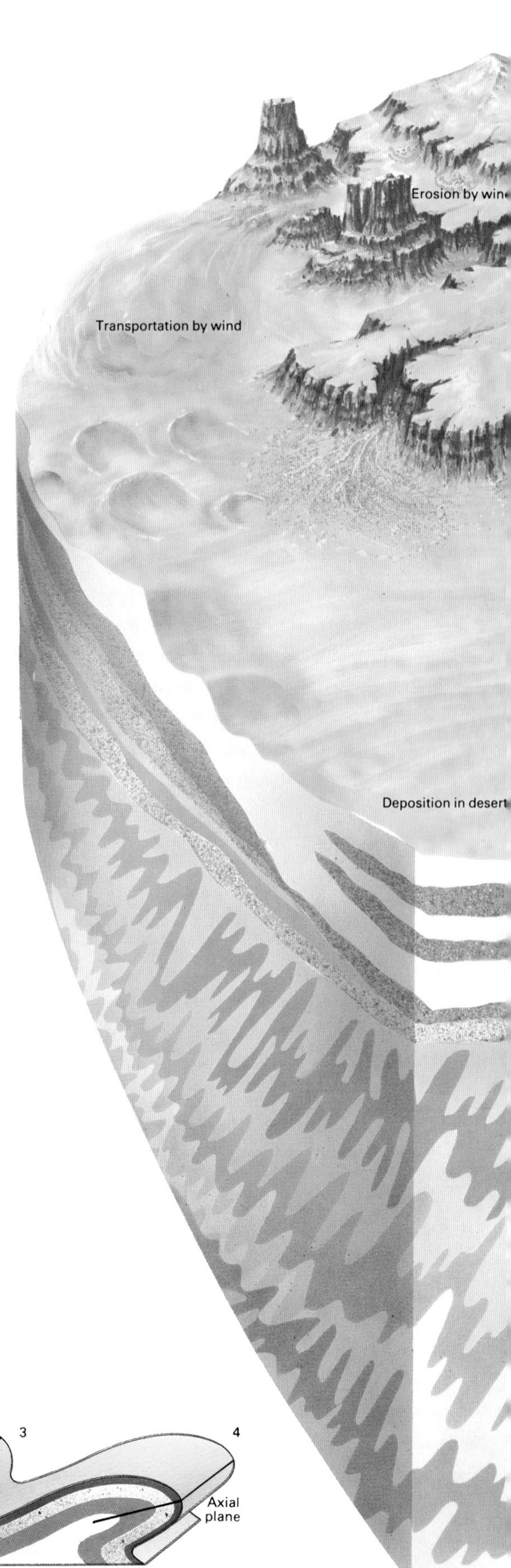

Erosion by frost
Erosion by rain
Transportation by rivers
Transportation by ice
Uplift
Emplacement of igneous rocks
Metamorphism
Melting
Deposition in river beds
Deposition from glaciers
Transportation by sea currents
Deposition by corals
Deposition in shallow seas
Transportation by turbidity currents
Deposition in deep seas
Igneous rock
Weathering, transport and deposition
Sediment
Cooling and solidification (crystalisation)
Heat and pressure (metamorphism)
Weathering, transport and deposition
Weathering, transport and deposition
Cementation and compaction (lithification)
Magma
Melting
Metamorphic rock
Heat and pressure (metamorphism)
Sedimentary rock

bacteria. Among animals, saprotrophs range from dung beetles to vultures. In ecology, saprotrophs are commonly referred to as "decomposers", and they play an important part in returning vital nutrients to soil and water. (page 44)

Sarcodina see PROTOZOA.

Sarcoma

A form of malignant TUMOR derived from ENDODERM. It commonly arises from BONE (osteosarcoma), fibrous tissue (fibrosarcoma) or CARTILAGE (chondrosarcoma).

Sarcopterygii (lobe-finned fish)

A group of fish whose only living representatives are the lungfish or DIPNOI and the coelacanth, last survivor of the CROSSOPTERYGII. The group is characterized by having fins with a fleshy central lobe, containing many small bones. In the living lungfish this feature has been much reduced, and the fin is reduced to a slender, whip-like organ; only in the coelacanth can the original type of fin be seen. This fleshy fin is believed to have evolved into the limb of the earliest land vertebrates, the ancestral amphibians.

Sard

A semiprecious stone, a brown variety of CHALCEDONY closely related to CARNELIAN.

Sardonyx see ONYX.

Sargasso Sea

Oval area of the N Atlantic, of special interest as the spawning ground of American eels, many of whose offspring drift across the Atlantic to form the European eel population. Bounded E by the Canaries Current, S by the N Equatorial Current, W and N by the GULF STREAM, it contains large masses of *Sargassum* weed.

Satellite

In astronomy, a celestial object that revolves with or around a larger celestial object. In our SOLAR SYSTEM this includes PLANETS, COMETS, ASTEROIDS and meteoroids (see METEOR), as well as the moons of the planets, though the term is usually restricted to this last sense. Of the 58 known moons, the largest is Ganymede (JUPITER III), the smallest Leda (Jupiter XIII). The MOON is one of the largest known satellites relative to its parent planet; indeed, the Earth-Moon system is sometimes considered a double planet. However, Pluto's satellite, Charon, is nearly half the size of its parent planet. (page 221)

Satellites, Artificial

Manmade objects placed in orbit as SATELLITES. First seriously proposed in the 1920s, they were impracticable until large enough ROCKETS were developed. The first artificial satellite, Sputnik I, was launched by the USSR in October 1957, and was soon followed by a host of others, mainly from the USSR and the US, but also from the UK, France, Canada, West Germany, Italy, Japan and China. They have many scientific, technological and military uses. Astronomical observations (notably X-RAY ASTRONOMY) can be made unobscured by the ATMOSPHERE. Studies can be made of the RADIATION and electromagnetic and gravitational fields in which the Earth is bathed, and of the upper atmosphere. Experiments have been made on the functioning of animals and plants in space (with zero gravity and increased radiation). Artificial satellites are also used for reconnaissance, surveying, meteorological observation, as navigation aids (position references and signal relays), and in communications for relaying television and radio signals.

Manned satellites, especially the historic Soyuz and Mercury series, have paved the way for space stations, which have provided opportunities for diverse research and for developing docking techniques; the USSR Salyut and US Skylab projects are notable. The basic requirements for satellite launching are determined by celestial mechanics. Launching at various velocities between that required for zero altitude and the escape velocity produces an elliptical orbit lying on a conic surface determined by the latitude and time of launch. To reach any other orbit requires considerable extra energy expenditure. Artificial satellites require: a power supply – solar cells, batteries, fuel cells or nuclear devices; scientific instruments; a communications system to return encoded data to Earth; and instruments and auxiliary rockets to monitor and correct the satellite's position. Most have computers for control and data processing, thus reducing remote control to the minimum. (page 237)

Saturated fats see FATTY ACIDS.

Saturation

Term applied in many different fields to a state in which further increase in a variable above a critical value produces no increase in a resultant effect. A saturated SOLUTION is one which will dissolve no more solute, EQUILIBRIUM having been reached; raising the temperature usually allows more to dissolve. Cooling a saturated solution may produce supersaturation, a metastable state, in which sudden crystallization depositing the excess solute occurs if a seed crystal is added. In organic chemistry, a saturated molecule has no double or triple bonds and does not undergo addition reactions.

Saturn

The second largest planet in the SOLAR SYSTEM and the sixth from the Sun. Until the discovery of URANUS (1781) Saturn was the outermost planet known. It orbits the Sun in 29.46 years at a mean distance of 9.54AU. Saturn does not rotate uniformly: its period of rotation at the equator is 10.23h, rather longer toward the poles. This rapid rotation causes a noticeable equatorial bulge: the equatorial diameter is 120.6Mm, the polar diameter 108.1Mm. Saturn has the lowest density of any planet in the Solar System, less than that of water, and may contain over 60% hydrogen by mass. Its total mass is about 95 times that of the Earth. Saturn has at least 21 moons. The largest, Titan, about the same size as MERCURY, is known to have an atmosphere. Of the other moons, four have diameters greater than 1000km, nine greater than 100km, and the rest are very small. The most striking feature of Saturn is its ring system, seven or more rings. The major rings are subdivided into thousands of ringlets, grooves and ripples. Until recently Saturn was thought to be the only ringed planet in the Solar System, but rings have since been identified around URANUS, JUPITER and possibly NEPTUNE. Rings may be a characteristic of the outer planets, representing partially consolidated material left over from their accretion. (page 134, 221, 240)

Saurischia

Lizard-hipped dinosaurs, a group of DINOSAURS having pelvic girdles resembling those of lizards, with three prongs to each side. They included the two-legged carnivorous theropods, eg *Tyrannosaurus* and *Allosaurus*, with enormous skulls and large teeth, and the four-legged herbivorous sauropods, eg *Brontosaurus* and *Diplodocus*, with small heads and long necks and tails. The other major group of dinosaurs was the ORNITHISCHIA.

Savanna

Tropical GRASSLANDS of South America and particularly of Africa, lying between equatorial FORESTS and dry DESERTS.

Scabies

Pubic lice, spread from person to person by sharing the same bed.

Scalds see BURNS AND SCALDS.

Scandium (Sc)

Silvery white metal in Group IIIB of the PERIODIC TABLE; a TRANSITION ELEMENT. It is widely distributed in low concentrations, and is extracted from thortveitite or as a byproduct of uranium extraction. Scandium forms trivalent ionic compounds and stable ligand complexes. The refractory scandium oxide (Sc_2O_3) is used in ceramics and as a catalyst; dilute scandium sulfate solution is used to improve germination of plant seeds. AW 45.0, mp 1500°C, bp 2800°C, sg 2.989 (25°C). (page 130)

Scar

Area of fibrous tissue that forms a bridge between areas of normal tissue as the end result of wound healing. The fibrous tissue lacks the normal properties of the healed tissue (eg, it does not tan). The size of a scar depends on the closeness of the wound edges during healing; excess stretching forces and infection widen scars. Scars sometimes develop harmless fibrous ridges called a keloid, for unknown reasons.

Scarlet fever (scarletina)

INFECTIOUS DISEASE caused by certain strains of *Streptococcus*. It was common in children, causing sore throat with TONSILLITIS, a characteristic skin rash and mild systemic symptoms.

Scarp (escarpment)

Steep slope or inland cliff, most often that of a CUESTA. The term is sometimes applied to similar slopes resulting from EROSION or faulting (see FAULT).

Scattering

The deflection of moving particles and energy waves (such as ELECTRONS, PHOTONS or SOUND waves) through collisions with other particles. RAYLEIGH scattering of sunlight gives rise both to the blue color of the ATMOSPHERE when the Sun is high in the sky and to the reds and yellows of the setting Sun, because blue light is scattered more strongly than red.

Schally, Andrew Victor (b. 1926)

Polish-born US physiologist who shared the 1977 Nobel Prize for Physiology or Medicine with R.S. YALOW and R. GUILLEMIN for his work identifying compounds stimulating peptide HORMONE release from the brain.

Schawlow, Arthur Z. (b. 1921)

US physicist who shared half of the 1981 Nobel Prize for Physics with BLOEMBERGEN for his work on laser SPECTROSCOPY.

Scheele, Karl (or **Carl**) **Wilhelm** (1742-1786)

Swedish chemist who discovered OXYGEN (*c*1773) perhaps a year before JOSEPH PRIESTLEY's similar discovery. He also discovered CHLORINE (1774).

Scheelite (calcium tungstate, $CaWO_4$)

A major ore of TUNGSTEN. It is of widespread occurrence, and forms tetragonal CRYSTALS of various colors.

Schiaparelli, Giovanni Virginio (1835-1910)

Italian astronomer who discovered the asteroid Hesperia (1861) and showed that meteor showers represent the remnants of comets. He is best known for terming the surface markings of Mars *canali* (channels). This was wrongly translated as "canals", implying Martian builders.

Schist

Common group of METAMORPHIC ROCKS that have acquired a high degree of schistosity – the tendency to split into layers along perfect CLEAVAGE planes. Their major constituents are flaky or platy minerals, especially MICA, TALC, AMPHIBOLES and CHLORITE.

Schistosomiasis (bilharzia)

A PARASITIC DISEASE caused by *Schistosoma* species of FLUKES. Infection is usually acquired by bathing in water where the parasite is carried by certain species of snail; the different species of parasite cause different manifestations. Infection of the bladder causes constriction, calcification and secondary infection, and can predispose to bladder CANCER. Another form leads to gastrointestinal tract disease with liver involvement.

Schizophrenia
A group of PSYCHOSES characterized by confusion of IDENTITY, HALLUCINATIONS, AUTISM, delusion and illogical thought. The three main types are CATATONIA; paranoid schizophrenia, which is similar to PARANOIA except that the intellect deteriorates; and hebephrenia, which is characterized by withdrawal from reality, bizarre or foolish behavior, delusions, hallucinations and self-neglect. (See also MENTAL ILLNESS.)

Schmidt, Bernhard Voldemar (1879-1935)
German optician best known for developing the Schmidt TELESCOPE, today one of the most-used tools of ASTROPHOTOGRAPHY. Its special advantage is that it avoids LENS coma. (♦ page 237)

Schrieffer, John Robert (b. 1931)
US physicist awarded with LEON COOPER and JOHN BARDEEN the 1972 Nobel Prize for Physics for their work on SUPERCONDUCTIVITY.

Schrödinger, Erwin (1887-1961)
Austrian-born Irish physicist and philosopher of science who shared with DIRAC the 1933 Nobel Prize for Physics for his elucidation of the Schrödinger wave equation, which is of fundamental importance in studies of QUANTUM MECHANICS (1926). It was later shown that his WAVE MECHANICS were equivalent to the matrix mechanics of HEISENBERG.

Schwann, Theodor (1810-1882)
German biologist who proposed the CELL as the basic unit of animal, as well as of plant, structure (1839), thus laying the foundations of HISTOLOGY. He also discovered the ENZYME pepsin (1836).

Schwarzschild radius
The radius of the SPHERE into which a given body of known MASS must be compressed in order to become a BLACK HOLE.

Schwinger, Julian Seymour (b. 1918)
US physicist who shared with FEYNMANN and TOMONAGA the 1965 Nobel Prize for Physics for his independent work in formulating the theory of quantum electrodynamics.

Sciatica
A characteristic pain in the distribution of the sciatic nerve in the LEG caused by compression or irritation of the nerve. The pain may resemble an electric shock and be associated with numbness and tingling in the skin area served by the nerve. One of the commonest causes is a SLIPPED DISK in the lower lumbar spine.

Scientific method
Science (from Latin *scientia*, knowledge) is too diverse an undertaking to be constrained to follow any single method. Yet from the time of FRANCIS BACON well into the 20th century the myth has persisted that true science follows a particular method – Bacon's celebrated "inductive method". This allegedly involved collecting a vast number of individual facts about a phenomenon, and then working out what general statements fitted those facts. After the 17th century nobody attempted to follow that program. In the 19th century, philosophers of science came to recognize the possible existence of the "hypothetico-deductive method". According to this model, the scientist studied the phenomena, dreamed up a hypothetical explanation, deduced some additional consequences of his explanation, and then devised experiments to see if these consequences were reflected in nature. If they were, he considered his theory (hypothesis) confirmed. But K. POPPER pointed to the logical fallacy in this last step – the theory had not been confirmed, but merely not falsified; it could, however, be worked with provisionally, so long as new tests did not discredit it. Philosophers of science now recognize that they cannot justly generalize about the psychology of scientific discovery; their role must be confined to the criticism of theories once they have been devised. Historians of science, meanwhile, have pointed to the importance in scientific discovery of "external factors" such as the contemporary intellectual context and the structures of the institutions of science. Once distinct terms – "theory", "model", "hypothesis", "explanation", "description" and "law" – are all now seen to represent different ways of looking at the same thing – the units in what constitutes scientific knowledge at any given time. Indeed there is still no general understanding of how scientists become dissatisfied with a once deeply entrenched theory and come to replace it with what, for the moment, seems a better version.

Scintillation counter
Instrument for detecting ionizing radiations. A brief localized light flash is produced in a PHOSPHOR when ionizing radiation such as X-rays or protons is incident on it. In early counters, the flashes were counted directly using a microscope, but now they are converted into electric impulses and counted electronically.

Scoliosis
A curvature of the spine to one side, with twisting. It occurs as a congenital defect or may be secondary to spinal diseases, including neurofibromatosis.

Scopolamine (hyoseine)
Anticholinergic drug related to ATROPINE and used widely in premedication for ANESTHESIA. It tends to be a central nervous system DEPRESSANT but otherwise resembles atropine in reducing secretions, gastrointestinal tract activity and vagus effects on the HEART (causing increased PULSE rate) and in dilating the pupils of the eyes. It may cause confusion in the elderly. Other common uses include treatment of MOTION SICKNESS, use as a mild SEDATIVE and in PARKINSON'S DISEASE.

Scoria
A vesicular form of LAVA, the vesicles having been formed by gases escaping from the lava while it was still hot.

Scorpius
The Scorpion, a medium-sized constellation on the ECLIPTIC; the eighth sign of the ZODIAC. Scorpius contains the bright star ANTARES.

Scree see TALUS.

Screw see ARCHIMEDES; MACHINE.

Scrub typhus (Tsutsugamushi disease)
A disease caused by RICKETTSIA carried by mites, and leading to ulceration at the site of INOCULATION, followed by FEVER, headache, lymph node enlargement and generalized rash. Cough and chest X-ray abnormalities are common. ENCEPHALITIS and MYOCARDITIS may occur with fatal outcome. It occurs mainly in the Far East and Australia and is generally seen in people who work on scrubland.

Scurvy
A nutritional disorder resulting from VITAMIN C deficiency, involving disease of the skin and mucous membranes, poor healing and ANEMIA; in infancy bone growth is also impaired. It may develop over a few months of low dietary vitamin C, beginning with malaise and weakness. Skin bleeding around hair follicles is characteristic, as are swollen, bleeding gums.

Sea see OCEANS.

Seaborg, Glenn Theodore (b. 1912)
US physicist who shared the 1951 Nobel Prize for Chemistry with E.W. MCMILLAN for his work in discovering several ACTINIDES (see TRANSURANIUM ELEMENTS): in 1944 AMERICIUM and CURIUM, and in 1949 BERKELIUM and CALIFORNIUM. Later discoveries were EINSTEINIUM (1952), FERMIUM (1953), MENDELEVIUM (1955) and NOBELIUM (1957).

Sea cucumbers
Amorphous, sausage-shaped animals that make up the Class Holothuroidea of the ECHINODERMS. The chalky exoskeleton, typical of the group, is reduced to tiny plates embedded in the skin, or is completely absent. Modified tube-feet around the mouth pick up sediment from the sea floor, from which the food is extracted. Sea cucumbers are most common in the ocean depths, where they crawl slowly over the bottom, picking up the organic detritus that rains down from the protective upper layers.

Sea-floor spreading
Key phenomenon supporting the theory of PLATE TECTONICS. Along ocean ridges (see OCEANS) material emerges from the EARTH's mantle to form new oceanic crust. This material, primarily BASALT, spreads out to either side of the ridges at a rate of the order of 10-50mm/yr. Newly laid down basalt is able to "fossilize" the prevailing geomagnetism (see PALEOMAGNETISM): the main evidence for sea-floor spreading comes from the symmetric pattern of alternately magnetized strips of basalt on either side of the ridges.

Sea gooseberry
A common type of comb jelly or CTENOPHORE.

Sea lilies
Common name given to some of the CRINOIDS.

Sea mats see BRYOZOA.

Sea pens
Type of coral; see CNIDARIANS.

Seasickness see MOTION SICKNESS.

Seasons
Divisions of the year, characterized by cyclical changes in the predominant weather pattern. In the temperate zones there are four seasons: spring, summer, autumn (fall) and winter. These result from the constant inclination of the Earth's polar axis (66½° from the ECLIPTIC) as the Earth orbits the Sun: during summer in the N Hemisphere the N Pole is tilted toward the Sun, in winter – when the solar radiation strikes the hemisphere more obliquely – away from the Sun. The summer and winter SOLSTICES (about 21 June and 22 December), popularly known as midsummer and midwinter, strictly speaking mark the beginnings of summer and winter respectively. Thus spring begins on the day of the vernal EQUINOX (about 21 March), autumn at the autumnal equinox (about 23 September). (♦ page 25)

Sea spiders see PYCNOGONIDS.

Sea squirts see TUNICATES.

Sebaceous glands
Small GLANDS in the SKIN that secrete sebum, a fatty substance that acts as a protective and water repellent layer on skin and allows the epidermis to retain its suppleness. Sebum secretion is fairly constant but varies from individual to individual. Obstructed sebaceous glands become blackheads, which are the basis for acne.

Second (s)
In SI units, the base unit of TIME, defined as the duration of 9,192,631,770 periods of the radiation corresponding to the transition between the two hyperfine levels of the ground state of the CESIUM-133 atom.

Secretin
A HORMONE of the gastrointestinal tract secreted by cells in the duodenum in response to the presence of food, and increasing the secretion of enzymes and bile.

Secretion see GLANDS.

Sedatives
DRUGS that reduce ANXIETY and induce relaxation without causing SLEEP; many are also hypnotics, drugs that in adequate doses may induce sleep. BARBITURATES were among the earlier drugs used in sedation, but they have fallen into disfavor because of the problems of addiction, side-effects and dangers of overdosage that they cause, and because of the availability of newer and safer

alternatives. Benzodiazepines (eg Valium, Librium) are now the most often used and have proved safe and effective.

Sedimentary rocks
One of the three main ROCK types of the Earth's crust. They consist of weathered (see EROSION) particles of igneous, metamorphic or even sedimentary rock transported, usually by water, and deposited in distinct strata. They may also be of organic origin, as in COAL, or of volcanic origin, as are PYROCLASTIC ROCKS. Most common are SHALE, SANDSTONE and LIMESTONE. Sedimentary rocks frequently contain FOSSILS, as well as most of the Earth's MINERAL resources. (♦ page 208)

Sedimentation
The processes whereby particles of solid material are transported and deposited elsewhere. The particles are the product of weathering (see EROSION); the transporting agent may be wind, water, GLACIERS or an AVALANCHE or landslide. Products of sedimentation thus include DRIFT, OOZES, TALUS and SEDIMENTARY ROCKS. (♦ page 208)

Seebeck, Thomas Johann (1770-1831)
German physicist who discovered but could not explain the Seebeck effect (1821). This was the first thermoelectric effect to be discovered (see THERMOCOUPLE).

Seed
A reproductive body produced by the ANGIOSPERMS and GYMNOSPERMS. In many species it also represents a resting stage that enables the plants to survive through unfavorable conditions. Seeds are generated by sexual REPRODUCTION, and develop from the fertilized ovule. Each seed is covered with a tough coat called a testa, and contains a young plant or embryo. In most seeds three main regions of embryo can be recognized: a radicle, which gives rise to the root; a plumule, which forms the shoot; and one or two seed leaves or COTYLEDONS. The seed contains enough stored food (in the cotyledons, or the ENDOSPERM) to support embryo growth during and after GERMINATION. Flowering plants (angiosperms) produce their seeds inside a FRUIT, which develops from the ovary wall, but the seeds of conifers (gymnosperms) lie naked on the scales of the CONE. Distribution of seeds is usually by wind, animals or water, the form of the seeds being adapted to a specific means of dispersal. (See also POLLINATION.) (♦ page 53)

Segmentation
The serial repetition of structures or organs along the long axis of the animal body. Segmentation is characteristic of many groups, notably the ANNELIDS.

Segrè, Emilio Gino (b. 1905)
Italian-born US nuclear physicist who shared with O. CHAMBERLAIN the 1959 Nobel Prize for Physics for their discovery (1955) of the antiproton (see ANTIMATTER). Earlier he had discovered TECHNETIUM, the first artificially produced ELEMENT (1937).

Seiche
A standing wave system (see WAVE MOTION) occurring in a lake or bay, set up by a disturbance such as wind, ocean swell, an earth tremor or a sudden change in air pressure. The oscillation period depends on the size and shape of the basin.

Seismograph
Instrument used to record seismic waves caused by EARTHQUAKES, nuclear explosions, etc. The record it produces is a seismogram. The simplest seismograph has a horizontal bar, pivoted at one end and with a recording pen at the other. The bar, supported by a spring, bears a heavy weight. As the ground moves, the bar remains roughly stationary owing to the inertia of the weight, while the rest of the equipment moves. The pen traces the vibrations on a moving belt of paper. Seismographs are also used in prospecting.

Seismology
The study of EARTHQUAKE phenomena.

Selachii
A subclass of the CHONDRICHTHYES that contains the sharks, rays and skates.

Selection see NATURAL SELECTION; SEXUAL SELECTION.

Selenite
Well-developed crystals of GYPSUM.

Selenium (Se)
Metalloid in Group VIA of the PERIODIC TABLE; it occurs as rare selenides with heavy metal sulfides, and is obtained as a byproduct of copper refining. Selenium has three allotropes (see ALLOTROPY), the most stable being the gray, metallic form. Its chemistry is analogous to that of SULFUR. It is used to make photoelectric cells, solar cells, rectifiers, in xerography and as a semiconductor; also to make ruby glass and to vulcanize rubber. AW 79.0, mp 220°C (gray), bp 685°C (gray), sg 4.79 (gray). (♦ page 130)

Semen
Fluid secreted by the TESTES containing SPERM. (See also REPRODUCTION.) (♦ page 204)

Semicircular canals see EAR.

Semiconductor
A material whose electrical CONDUCTIVITY is intermediate between that of an insulator and conductor at room temperature and increases with rising temperature and impurity concentration. Typical intrinsic semiconductors are single crystals of GERMANIUM or SILICON. At low temperatures their valence electron energy levels are filled and no electrons are free to conduct ELECTRICITY, but with increasing temperature, some electrons gain enough energy to jump into the empty conduction band, leaving a hole behind in the valence band. Thus, there are equal numbers of moving electrons and holes available for carrying electric current. Practical extrinsic semiconductors are made by adding a chosen concentration of a particular type of impurity atom to an intrinsic semiconductor (a process known as doping). If the impurity atom has more valence electrons than the semiconductor atom, it is known as a donor and provides spare conduction electrons, creating an n-type semiconductor. If the impurity atom has fewer valence electrons, it captures them from the other atoms and is known as an acceptor, leaving behind holes which act as moving positive charge carriers and enhance the conductivity of the p-type semiconductor that is formed. An n- and p-type semiconductor junction acts as a RECTIFIER; when it is forward biased, holes cross the junction to the negative end and electrons to the positive end, and current flows through it. If the voltage connections are reversed, the carriers will not cross the junction and no current flows. Semiconductor devices, such as the TRANSISTOR, based on the p-n junction, have revolutionized ELECTRONICS since they were first demonstrated in 1949, leading on to very small-scale INTEGRATED CIRCUITS with massive numbers of individual components. (♦ page 130)

Semimetal see METALLOID.

Semipermeable membrane
A MEMBRANE that allows certain substances, usually small molecules or ions, to pass through its pores, but keeps back others. Examples are cell membranes, parchment and cellophane. (See DIALYSIS; OSMOSIS.)

Semyonov, Nikolai Nikolaevich (1896-1986)
Russian physical chemist who shared with SIR CYRIL HINSHELWOOD the 1956 Nobel Prize for Chemistry for his work on chemical KINETICS, especially concerning chemical chain and branched-chain reactions.

Senility
The state of old age, usually referring to the general mental and physical deterioration, often (but not always) seen in the elderly. Failure of recent memory, dwelling on the past, episodic confusion and difficulty in absorbing new information are common.

Senses
Receptors that monitor the environment of an organism and provide information that helps the organism to survive (external senses); also, the internal senses that report on the internal state of the organism, such as its temperature, food requirement, water needs, etc. The organs of sense, the eye, ear, skin etc, all contain specialized cells that translate external stimuli into nervous impulses and nerve endings that communicate with centers in the BRAIN. Sense organs may be stimulated by pressure (in TOUCH, hearing and balance – see EAR), chemical stimulation (SMELL; TASTE), or electromagnetic radiation (VISION; heat sensors). Some animals also respond to gravity, magnetic forces, electrical fields, or the reflection of sound waves that they themselves have emitted (ECHOLOCATION).

Sensitivity
The ability of a measuring instrument to respond to small variations in the input signal.

Sensors
Devices that can detect specific properties and convert them quantitatively into signals (electrical, mechanical or thermal). For example, a thermometer uses mercury in a tube as a sensor; its position in the tube tells the viewer the temperature. A thermoelectric cell is another temperature sensor: it produces an electric current proportional to the temperature that can drive a meter.

Sensory deprivation
Condition of perceptual isolation – which may result in HALLUCINATIONS, thought or emotional disorders or spatiotemporal disorientation – experienced by people confined in highly unstimulating environments.

Sepals
In plants, the outer ring of green leaf-like organs that surrounds the petals of the FLOWER. The sepals are known collectively as the calyx. In some plants the calyx is petal-like and colored.

Sepsis
The condition in which bacteria are being attacked by MACROPHAGES, forming PUS.

Septicemia
Circulation of infective BACTERIA in the blood. Bacteria may transiently enter the blood normally but these are removed by the immune system. If this system fails and bacteria continue to circulate and multiply, a series of reactions takes place that leads to shock, with warm extremities, fever or hypothermia. Septic EMBOLISM may occur, causing widespread ABSCESSES.

Sere see SUCCESSION.

Serotonin
Aromatic AMINE, also known as 5HT or 5-hydroxytryptamine, found in serum and other tissues. It acts on BLOOD vessels and is also involved in PERISTALSIS and in the central NERVOUS SYSTEM as a transmitter at SYNAPSES. It may be involved in allergy ANAPHYLAXIS in some species; its action in increasing CAPILLARY permeability indicates its role in the reactions producing INFLAMMATION.

Serpentine
Common magnesium SILICATE mineral, formula $Mg_3Si_2O_5(OH)_4$, whose structure resembles that of KAOLINITE. There are three varieties: CHRYSOTILE, antigorite and lizardite. Gray, green or yellow with an attractive texture and easily polished, serpentine is used as an ornamental stone.

SENSES

Anatomy of the eye

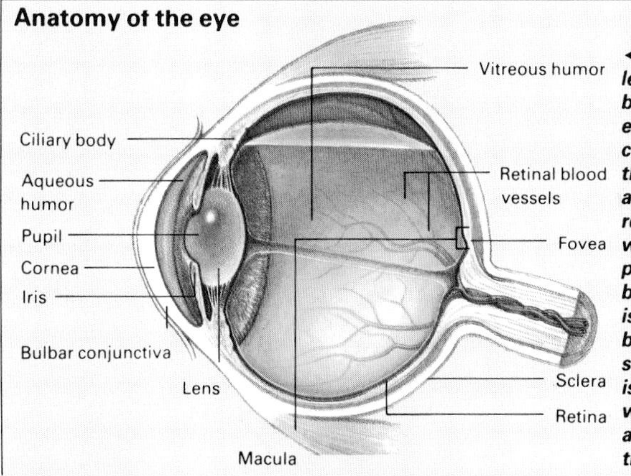

◄ *The eye consists of a lens (which can be focused by being squeezed or extended); an aperture to control the light entering the eye (the iris) and an area in which the image is resolved (the retina), from which light-sensitive cells pass information to the brain. Part of the focusing is done by the transparent, bulging cornea. The fovea, surrounded by the macula, is the part of the retina with most receptor cells, and so the part that forms the best image.*

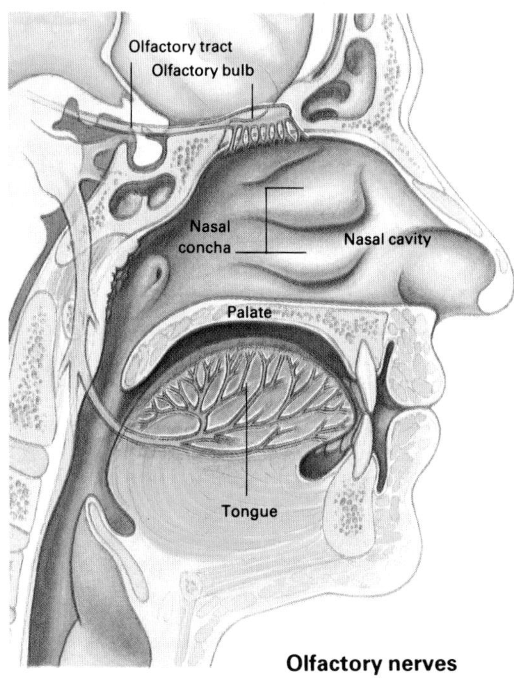

Olfactory nerves

◄ *The sense of smell depends on the olfactory organs; the receptor cells have cilia which project down into the nasal cavity. Taste is sensed by buds on the tongue and cheeks. Four kinds of buds respond to the four basic tastes.*

► *The outer ear collects sound waves and directs them to the middle ear. Here they are transmitted and amplified by three bones, the auditory ossicles, and passed to the inner ear where sensory receptors pick them up. The balance mechanism consists of three semi-circular canals in the inner ear, at right angles to one another. As the head moves, fluid in each canal disturbs receptors. The brain combines information from each plane to produce a sense of three-dimensional movement.*

Anatomy of the ear

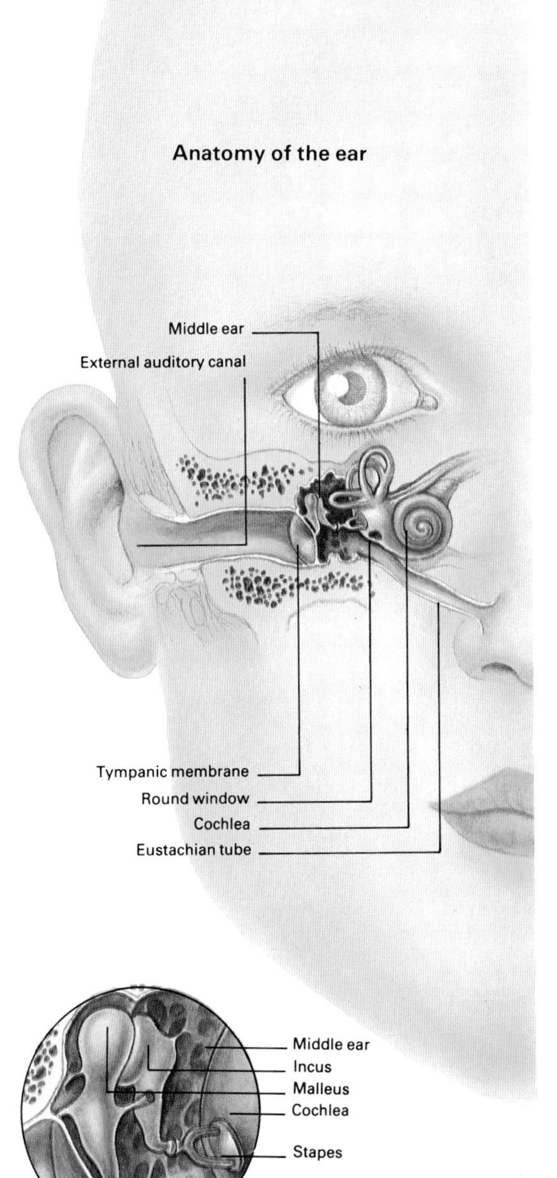

The skin's sense receptors

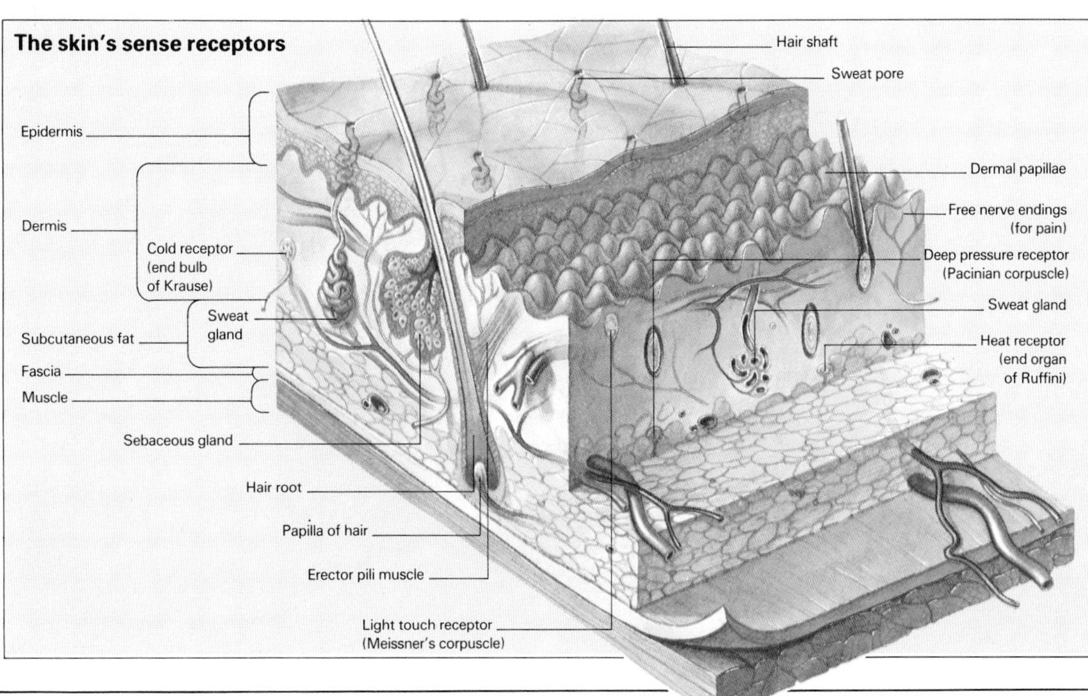

◄ *The skin contains a number of different sense receptors. Pain receptors have free ends which can be in the epidermis. Other sensory receptors have encapsulated ends. Meissner's corpuscles are small oval masses of connective tissue, each with one or more nerve fibers; they are most common in hairless regions, and are sensitive to motion and light touch. Pascinian corpuscles are large structures in the subcutaneous layers, especially of hands, feet, sexual organs, tendons and joints. Organs of Ruffini detect heavy touch and heat. Merkel's disks are dendrite formations in the epidermis. Bulbs of Krause, in the dermis, may be involved in the detection of cold.*

Serum
The clear, yellowish fluid that separates from BLOOD, LYMPH and other body fluids when they clot. It contains water, PROTEINS, fat, minerals, HORMONES and UREA. Serum therapy involves injecting serum-containing ANTIBODIES that can destroy particular pathogens.

Servetus, Michael or **Miguel Serveto**
(1511-1553)
Spanish biologist and theologian. In *Christianity Restored* (1553) he mentioned in passing his discovery of the pulmonary circulation (see BLOOD CIRCULATION). For heretical views expressed in this book he was denounced by the Calvinists to the Catholic Inquisition; escaping, he foolishly visited Geneva where he was seized by the Protestants, tried for heresy and burned alive.

Servomechanism
An automatic control device which controls the position, velocity or acceleration of a high-power output device by means of a command signal from a low-power reference device. The servomotor drive may be electrical, hydraulic or pneumatic.

Sessile animals
A term used to describe an animal or microorganism that does not move around, or does so only rarely. Also used in botany to mean a leaf, flower or fruit that lacks a stalk and therefore sits directly on a stem.

Sex chromosomes
The chromosomes that determine the sex of an individual (as well as carrying some genetic information not related to sex determination). Sex chromosomes are inherited in the same way as the other chromosome pairs; see MEIOSIS. In humans, individuals are either XX (female) or XY (male). The Y chromosome carries little genetic information, and it is largely the properties of the X chromosome that determine the "sex-linked" characteristics in males. Sex chromosomes are just one means of sex determination. (◆ page 32)

Sex hormones see ANDROGENS; ESTROGENS; GLANDS; HORMONES.

Sextant
Instrument for navigation, invented in 1730 and superseding the ASTROLABE. A fixed telescope is pointed at the horizon, and a radial arm is moved against an arc graduated in degrees until a mirror which it bears reflects an image of a known star or the Sun down the telescope to coincide with the image of the horizon. The angular elevation of the star, with the exact time (see CHRONOMETER), gives the latitude. The air sextant is a similar instrument, usually periscopic, designed for use in aircraft, and has an artificial horizon, generally a bubble level.

Sexual reproduction
A form of REPRODUCTION involving two organisms as parents. (◆ page 32, 204)

Sexual selection
An evolutionary process that CHARLES DARWIN suggested as an adjunct to NATURAL SELECTION. It depends on the fact that success in obtaining a mate is crucial for propagation of an individual's genes. Thus any factor that improves mating success will be selected for, regardless of whether it improves the individual's overall ADAPTATION to its environment or not. Indeed, it is possible that sexual selection could produce a characteristic that in some respects made the individual less well adapted. In the majority of animals sexual selection operates on the males rather than the females, because the females invest considerable energy resources in the young whereas the males often invest very little: thus the males compete for the female's favors, and in polygamous species, where sexual selection operates most strongly, the males try to inseminate as many females as possible. Sexual selection may operate through mate-choice by the females or through fighting between rival males. In the former case, ornaments such as the lyrebird's tail or the extravagant plumage of the birds of paradise result; while the latter produces weaponry such as the antlers of a deer stag or the enlarged claw of a fiddler crab. Often, both female choice and male rivalry are important in the selection of a particular feature.

Once sexual selection is established it is likely to be perpetuated, as males that lack the preferred feature will fail to breed, and their genes will quickly die out. Any female who fails to exert choice over her mate will risk producing offspring in whom the preferred feature is poorly developed, and they in turn may fail to breed, so that her genes will also die out. But how and why sexual selection gets started is less obvious. It may be that certain features shown by the male – such as glossy plumage in a bird – are indicative of good health, and therefore become a focus for sexual selection. Once they have become important, sexual selection automatically exaggerates them.

Sexually transmitted diseases
Those INFECTIOUS DISEASES transmitted mainly or exclusively by sexual contact, usually because the organism responsible is unable to survive outside the body and the close contact of genitalia provides the only means for transmitting viable organisms.

GONORRHEA An acute BACTERIAL DISEASE that is frequently asymptomatic in females, who therefore act as carriers, though they may suffer mild cervicitis or urethritis. In males it causes a painful urethritis with urethral discharge of PUS. ARTHRITIS, SEPTICEMIA and other systemic manifestations may also occur, and urethral stricture follow, though this is now rare in developed countries. Infection of an infant's eyes by mothers carrying the gonococcus causes neonatal OPHTHALMIA, previously a common cause of childhood blindness.

SYPHILIS Caused by *Treponema pallidum*, a SPIROCHETE, syphilis is a disease with three stages. A painless genital ULCER or chancre (a highly infective lesion) develops in the weeks after contact; this is usually associated with LYMPH node enlargement. Secondary syphilis, starting weeks or months after infection, involves systemic disease with fever, malaise and a characteristic rash, mucous membrane lesions and occasionally MENINGITIS, HEPATITIS or other organ disease. Tertiary syphilis takes several forms: eg gummas – chronic granulomas affecting SKIN, EPITHELIUM, BONE or internal organs – may develop. Largely a disease of blood vessels, tertiary syphilis causes disease of the aorta with aneurysm and aortic valve disease of the HEART, with incompetence. Syphilis of the NERVOUS SYSTEM may cause TABES DORSALIS, primary eye disease, chronic meningitis, multifocal vascular disease resembling STROKE or general PARESIS with mental disturbance, personality change, failure of judgment and muscular weakness. Congenital syphilis is disease transmitted to the fetus during pregnancy and leads to deformity and visceral disease. Syphilis is now relatively rare.

Other sexually transmitted diseases include non-gonoccal urethritis (NGU), a condition with symptoms very similar to those of gonorrhea. It is often caused by the bacterium *Chlamydia*; women may be asymptomatic carriers. Genital herpes is a viral disease caused by a type of *Herpes simplex* virus. It is characterized by the sporadic appearance of infective vesicles on the genitalia, usually at a time when bodily resistance is already low. AIDS and HEPATITIS B are also diseases that can be transmitted sexually.

Shadow
A nonilluminated region or area, shielded from receiving radiation from a light source by an extended object. A point source gives a sharply defined shadow, but an extended source gives rise to a region of full shadow (the umbra), surrounded by one of partial shadow (the penumbra).

Shale
Fine-grained SEDIMENTARY ROCK formed by cementation of SILT particles usually also containing fragments of other minerals. Shales are rich in FOSSILS and are laminated (they split readily into layers, or laminae). Their metamorphism (see METAMORPHIC ROCKS) produces SLATE. (See also OIL SHALE.)

Shapley, Harlow (1885-1972)
US astronomer who suggested that CEPHEID VARIABLES are not eclipsing binaries (see DOUBLE STAR) but pulsating stars. He was also the first to deduce the structure and approximate size of the Milky Way galaxy, and the position of the Sun within it.

Shearing
Type of deformation in which parallel planes in an object tend to slide over one another. Shear forces are always present in beams and whenever an object is subjected to bending stresses.

Shell
Any calcareous external covering secreted by an invertebrate, enclosing and protecting the body. The term is used particularly for the shells of mollusks, but also refers to those of foraminiferans, and may be used loosely to describe the EXOSKELETON of crustaceans.

Shelter belt (windbreak)
Natural or specially planted barrier of vegetation arranged to protect crops and agricultural land from erosion by the wind (see SOIL EROSION).

Sherrington, Sir Charles Scott (1857-1952)
British neurophysiologist who shared with E.D. ADRIAN the 1932 Nobel Prize for Physiology or Medicine for studies of the NERVOUS SYSTEM that form the basis of our modern understanding of its action.

Shingles
A VIRUS disorder characterized by development of pain, a vesicular rash and later scarring, often with persistent pain, over the skin of part of the face or trunk. The virus seems to settle in or near nerve cells following CHICKENPOX, which is caused by the same virus (*Herpes zoster*), and then becomes activated, perhaps years later and sometimes by disease. It then leads to the acute skin eruption described, which appears in the distribution of the nerve involved.

Shivering
Fine contractions of MUSCLES, causing slight repetitive movements that increase heat production by the body, thus raising body temperature in conditions of cold or when disease induces fever. Uncontrollable shivering with gross movements of the whole body is a rigor only seen in some fevers.

Shock
Specifically refers to the development of low blood pressure, inadequate to sustain BLOOD CIRCULATION, usually causing cold, clammy, gray skin and extremities, faintness and mental confusion and decreased urine production. It is caused by acute blood loss; burns with PLASMA loss; acute HEART failure; massive pulmonary EMBOLISM, and SEPTICEMIA. If shock is untreated, the person will die.

Shockley, William Bradford (b. 1910)
US physicist who shared with JOHN BARDEEN and WALTER BRATTAIN the 1955 Nobel Prize for Physics for their researches on semiconductors and their discovery of the TRANSISTOR effect.

Shock therapy (electroconvulsive therapy, ECT)
A form of treatment used in MENTAL ILLNESS, particularly DEPRESSION, in which carefully regulated electric shocks are given to the brains of

anesthetized patients. (Muscular relaxants are used to prevent injury through forceful muscle contractions.) The mode of action is unknown but rapid resolution of severe depression may be achieved. (See also PSYCHIATRY.)

Shooting star see METEOR.

Shortsightedness see MYOPIA.

Shoulder
Joint between the upper arm (humerus) and the upper trunk (scapula, clavicle and rib cage). It is an open ball-and-socket joint which is stable only by virtue of the numerous powerful muscles around it; this leads to increased maneuverability. (♦ page 216)

Shrub
Any small woody plant shorter than a TREE and with more than one main stem.

Sial
Collective term (silica-aluminum) for the rocks, lighter and more rigid than the SIMA and composed to a great extent of SILICA and ALUMINUM, that form the upper portion of the EARTH's crust. (See also ISOSTASY.)

Siamese twins
Twins (see MULTIPLE BIRTHS) that are physically joined at some part of their anatomy due to a defect in early separation. A variable depth of fusion is seen, most commonly at the head or trunk. SURGERY may be used to separate the twins if no vital organs are shared.

Sidereal time
Time referred to the rotation of the Earth with respect to the fixed stars. The sidereal DAY is about four minutes shorter than the solar day, as the Earth moves each day about 1/365 of its orbit about the Sun. Sidereal time is used in astronomy when determining the locations of celestial bodies.

Siderite
Brown or gray-green mineral consisting of iron(II) carbonate ($FeCO_3$), often with some magnesium, calcium and manganese. Of widespread occurrence in sedimentary or hydrothermal rocks, it is a major IRON ore. It has the CALCITE structure.

Siegbahn, Kai Manne Georg (1886-1978)
Swedish physicist who was awarded the 1924 Nobel Prize for Physics for his pioneer work in X-ray spectroscopy. He devised a way of measuring X-ray wavelengths with great accuracy, and developed an account of X-rays consistent with BOHR's theory of the atom. His son, Kai Siegbahn (b. 1918) shared the 1981 Nobel Prize for Physics for his contribution to high-resolution electron spectroscopy, which improved our understanding of chemical bonds between atoms.

Siemens (S)
The SI unit of electric CONDUCTANCE, equal in value to the reciprocal OHM (or MHO).

Siemensmar process see OPEN-HEARTH PROCESS.

Sievert (Sv)
The SI unit of dose equivalent, which modifies the absorbed dose to allow for the radiobiological effectiveness of the radiation concerned. Dose equivalent in sieverts is equal to the absorbed dose in grays multiplied by a radiation dependent quality factor. The sievert replaces the older unit, the REM, and 1 Sv=100 rem.

Signal
The variable impulse by which information is carried through a system.

Silica (silicon dioxide, SiO_2)
A very common mineral with three crystalline forms, the most common being QUARTZ. Silica is refractory and inert, though it dissolves in hydrogen fluoride and reacts with bases to form SILICATES. It is used to make glass, ceramics, concrete and CARBORUNDUM. (See also SILICON; KIESELGUHR.)

Silica gel
Amorphous form of SILICA made by acidifying a SILICATE and dehydrating the silicic acid formed. It is widely used as an adsorbent (see ADSORPTION) and drying agent (see DEHYDRATION). (See also GEL.)

Silicates
Salts of silicic acids. Discrete silicate anions include orthosilicates (SiO_4^{4-}), metasilicates (SiO_3^{2-}), and groups of SiO_4 units linked by Si–0–Si bonds; such condensation also produces infinite anions in chains, layers or three-dimensional arrays. Silicates (including aluminosilicates) are the most important class of minerals, forming 90% of the Earth's crust. (See also WATER GLASS.)

Silicon (Si)
Nonmetal in Group IVA of the PERIODIC TABLE; the second most abundant element (after oxygen), occurring as SILICA and SILICATES. It is made by reducing silica with coke at high temperatures. Silicon forms an amorphous brown powder, or gray semiconducting crystals, metallic in appearance. It oxidizes on heating, and reacts with the halogens, hydrogen fluoride, and alkalis. It is used in alloys, and to make TRANSISTORS and SEMICONDUCTORS. AW 28. 1, mp 1410°C, bp 3200°C, sg 2.42 (20°C). Silicon is tetravalent in almost all its compounds, which resemble those of CARBON, except that it does not readily form multiple bonds, and that chains of silicon atoms are relatively unstable. Silanes are series of volatile silicon hydrides, analogous to ALKANES, spontaneously flammable in air and hydrolyzed by water. (♦ page 130)

SILICON TETRACHLORIDE is a colorless fuming liquid, made by reacting chlorine with a mixture of silica and carbon, the starting material for preparing organosilicon compounds, including SILICONES. mp −70°C, bp 58°C. (For silicon carbide, see CARBORUNDUM; for silicon dioxide, see SILICA).

Silicon chip see CHIP.

Silicones
POLYMERS with alternate atoms of SILICON and oxygen, and organic groups attached to the silicon. They are resistant to water and oxidation, and are stable to heat. Liquid silicones are used for waterproofing, as polishes and anti-foam agents. Silicone greases are high- and low-temperature lubricants, and resins are used as electrical insulators. Silicone rubbers remain flexible at low temperatures.

Silicosis
A form of PNEUMOCONIOSIS, or fibrotic LUNG disease, in which long- standing inhalation of fine silica dusts in mining causes a progressive reduction in the functional capacity of the lungs. The normally thin-walled alveoli and small bronchioles become thickened with fibrous tissue and the lungs lose their elasticity. Characteristic X-RAY appearances and changes in lung function occur.

Silk
A fiber produced by certain insects and spiders to make cocoons and webs. It is a glandular secretion hardened into a filament on exposure to air. Commercial textile silk comes from the domestic silkworm.

Sill
Tabular body of IGNEOUS ROCK, under 1cm to over 100m thick and perhaps hundreds of kilometers wide, lying parallel to the beds of surrounding rocks. It has been suggested that sills are an extreme form of LACCOLITH or phacolith.

Silliman, Benjamin (1779-1864)
US chemist and geologist who founded *The American Journal of Science* (1819). The mineral sillimanite (a form of aluminum SILICATE, Al_2SiO_5) is named for him.

Silt
Soil composed of particles whose diameters range from 1/256 to 1/16mm. SOIL containing over 80% silt particles and less than 12% CLAY particles is often termed silt. In particular loess, accumulations of wind-blown dust, has particles of silt-size in the range 1/32 to 1/16mm.

Silurian
The third period of the PALEOZOIC era, which lasted between about 438 and 408 million years ago. (See also GEOLOGY.) (♦ page 106)

Silver (Ag)
Soft, white NOBLE METAL in Group IB of the PERIODIC TABLE, a TRANSITION ELEMENT. Silver has been known and valued from earliest times and used for jewelry, ornaments and coinage since the 4th millennium BC. It occurs as the metal, notably in Norway; in COPPER, LEAD and ZINC sulfide ores; and in ARGENTITE and other silver ores. It is concentrated by various processes including cupellation and extraction with CYANIDE (see also GOLD), and is refined by electrolysis. Silver has the highest thermal and electrical conductivity of all metals, and is used for printed circuits and electrical contacts. Other modern uses include dental ALLOYS and AMALGAM, high output storage batteries, and for monetary reserves. Although the most reactive of the noble metals, silver is not oxidized in air, nor dissolved by alkalis or nonoxidizing acids; it dissolves in nitric and concentrated sulfuric acid. Silver tarnishes by reaction with sulfur or hydrogen sulfide to form a dark silver sulfide layer. Silver salts are normally monovalent. Ag^+ is readily reduced by mild reducing agents, depositing a silver mirror from solution. AW 107.9, mp 962°C, bp 2170°C. sg 10.5 (20°C). (♦ page 130)

SILVER HALIDES (AgX) are crystalline salts used in photography. The chloride is white, the bromide pale yellow and the iodide yellow. On exposure to light, a crystal of silver halide becomes activated, and is preferentially reduced to silver by a mild reducing agent (the developer).

SILVER NITRATE ($AgNO_3$) is a transparent crystalline solid, used as an antiseptic and astringent, especially for removing warts.

Sima
Collective term (silica-magnesium) for the rocks, denser and more plastic than the SIAL and composed to a great extent of SILICA and MAGNESIUM, that form the lower portion of the Earth's crust. (See also ISOSTASY.)

Simoom (simoon)
Hot, dry wind or whirlwind occurring in the deserts of Arabia and N Africa. It usually carries much sand and greatly reduces visibility.

Simple harmonic motion (SHM)
A form of WAVE MOTION in which a moving particle traces a path symmetrical (see SYMMETRY) about a midpoint or equilibrium position, and through which it passes at regular intervals. The force responsible for the motion is always directed towards the midpoint, its magnitude proportional to the displacement of the particle. If the displacement of the particle is plotted as a FUNCTION of time the result is a sinusoidal CURVE

$$y=A \sin 2\pi ft$$

where A is the AMPLITUDE, f the FREQUENCY, y the displacement and t the time elapse from a particular zero-point. SHM derives its name from the fact that the vibrations produced by musical instruments (eg a string of a violin, the legs of a tuning fork), and hence the SOUND waves they propagate, approximate to it. In fact these, as all other vibrations and wave motions, may be treated as compounded of a number of SHMs.

Simpson, Sir James Young (1811-1870)
Scottish obstetrician who pioneered the use of

SKELETON AND MUSCLES

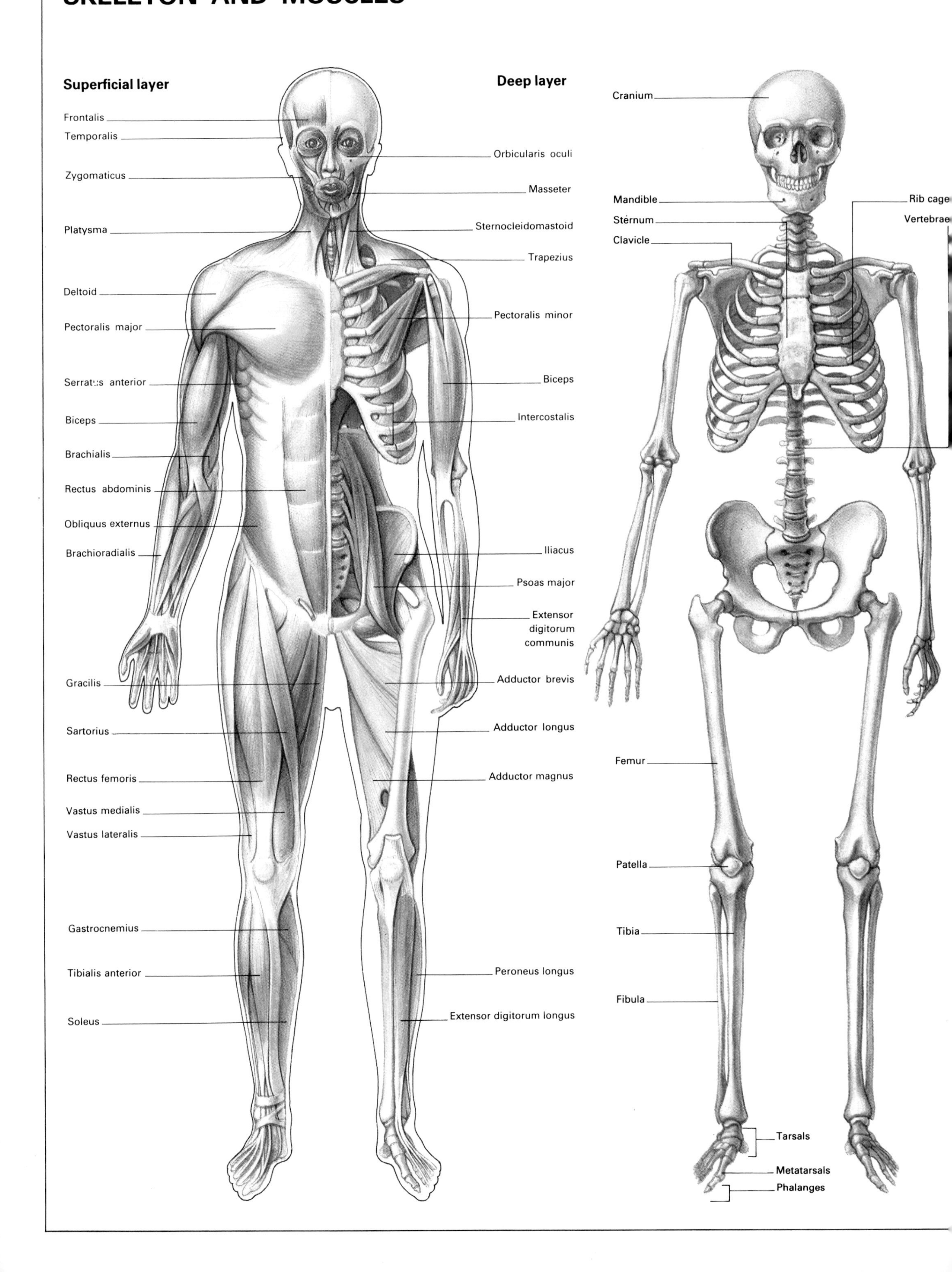

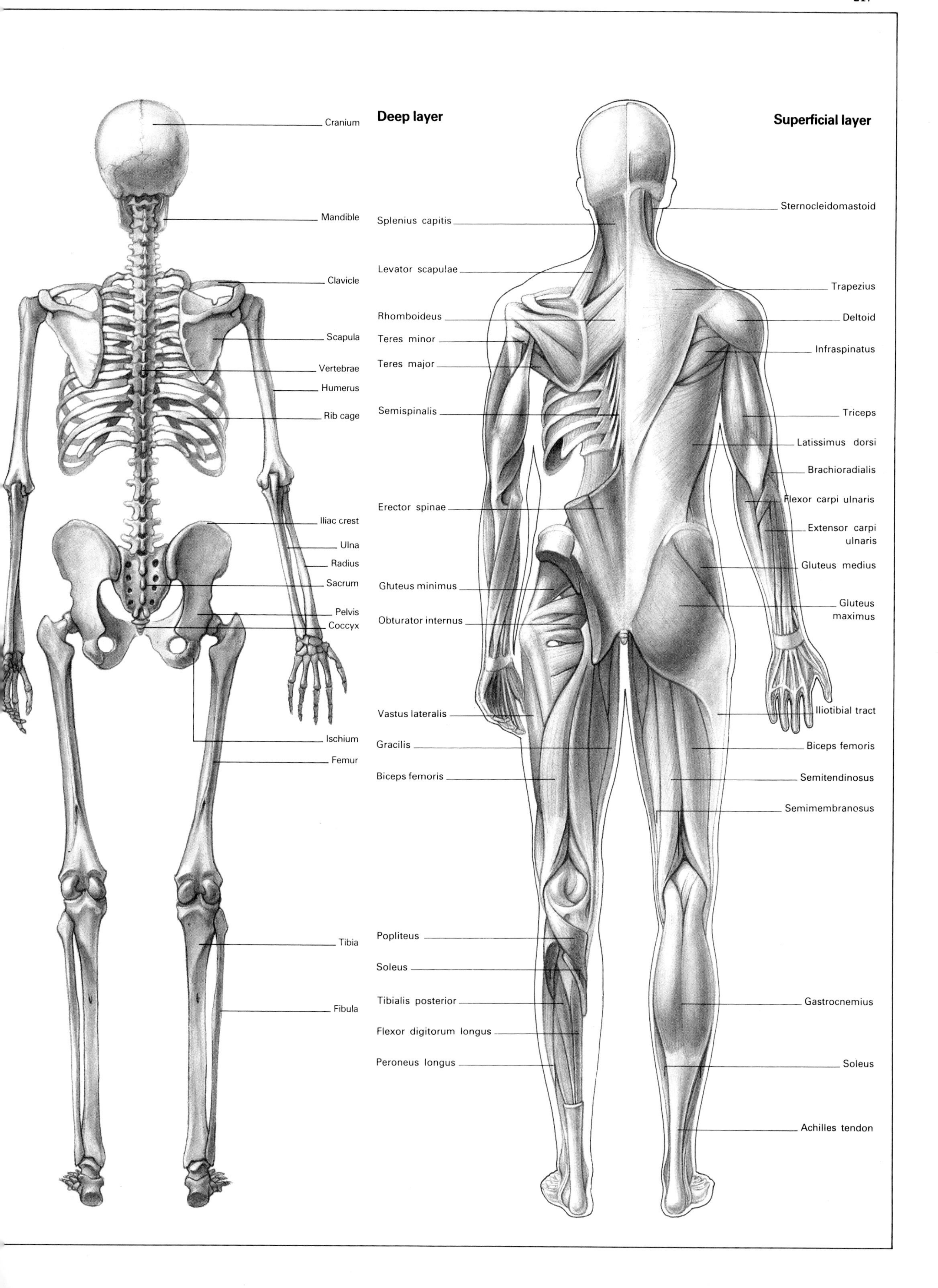
Cranium
Mandible
Clavicle
Scapula
Vertebrae
Humerus
Rib cage
Iliac crest
Ulna
Radius
Sacrum
Pelvis
Coccyx
Ischium
Femur
Tibia
Fibula
Deep layer
Splenius capitis
Levator scapulae
Rhomboideus
Teres minor
Teres major
Semispinalis
Erector spinae
Gluteus minimus
Obturator internus
Vastus lateralis
Gracilis
Biceps femoris
Popliteus
Soleus
Tibialis posterior
Flexor digitorum longus
Peroneus longus
Superficial layer
Sternocleidomastoid
Trapezius
Deltoid
Infraspinatus
Triceps
Latissimus dorsi
Brachioradialis
Flexor carpi ulnaris
Extensor carpi ulnaris
Gluteus medius
Gluteus maximus
Iliotibial tract
Biceps femoris
Semitendinosus
Semimembranosus
Gastrocnemius
Soleus
Achilles tendon

CHLOROFORM as an anesthetic, especially for mothers during childbirth (see ANESTHESIA; BIRTH).

Singularity
In astronomy, the point or ring within a BLACK HOLE at which the gravitational field is of infinite strength.

Sink hole (doline, swallow-hole)
Well or funnel-shaped hole, typical of KARST landscapes, formed when GROUNDWATER dissolves an underground cavity in LIMESTONE, followed by slump of the surface.

Sintering
The bonding together of compacted powder particles at temperatures below the melting point. The driving force is the decrease in surface energy that occurs as the particles merge and their total surface area lessens. The smaller the powder particles, the faster is the sintering. It is used to consolidate ORES, in powder METALLURGY, and in making CERAMICS and CERMETS.

Sinus
Large air space connected with the nose that may become infected and obstructed after upper respiratory infection and cause facial pain and fever (sinusitis). Also, a blind-ended channel that may discharge PUS or other material onto the skin or other surface. These may be embryological remnants or arise from a foreign body or deep chronic infection (eg OSTEOMYELITIS). The term is also used to describe a large venous channel, as in the liver and in the large vessels draining blood from the brain.

Siphonophores
Colonial relatives of the HYDROIDS and some of the most complex members of the phylum CNIDARIA. Unlike most colonial organisms they are not sessile, but float in the surface waters or actively swim by jet propulsion. Siphonophores are made up of many hundreds of individuals, some being POLYPS, others MEDUSAE. They all develop from a single founder individual by budding, so are genetically identical. Some are specialized for capturing prey, others for absorbing food or producing gametes for sexual reproduction. The enlarged, gas-filled float of some species is also formed from highly modified colony members. One of the best-known siphonophores is the Portuguese man o'war.

Sipunculids
A phylum of sausage-shaped marine worms that attain a length of 25-300mm. They are related to the ANNELIDS, but differ from them by having an unsegmented body and lacking the bristles called chaetae. There are about 320 species.

Sirenians
An order of MAMMALS, the sea cows, that is now largely extinct except for the dugong and manatees. They are found in the Atlantic and Indo-Pacific Oceans and are distantly related to the elephants.

Sirius
Alpha Canis Majoris (the Dog Star), the brightest STAR in the night sky, with an apparent magnitude of −1.46. It is 2.7pc distant, 20 times more luminous than the Sun, and has absolute magnitude +1.4. A DOUBLE STAR, its major component is twice the size of the Sun; its minor component (the Pup), the first white dwarf star to be discovered, has a diameter only 50% greater than that of the Earth, but is extremely dense, with a mass just less than that of the Sun. (▸ page 228, 240)

Sirocco
In S Europe, warm, humid WIND from the S or SE, originating over the Sahara and gaining humidity from the Mediterranean.

Sitter, Willem de (1872-1934)
Dutch astronomer who helped to get EINSTEIN'S theory of RELATIVITY widely known and who proposed a modification to it allowing for a gradual expansion of the Universe.

SI units
The internationally adopted abbreviation for the Système International d'Units (International System of Units), a modification of the system known as rationalized MKSA UNITS adopted by the 11th General Conference of Weights and Measures (CGPM) in 1960 and subsequently amended. SI units are the legal standard in many countries and find almost universal use among scientists.

Skeleton
A rigid or semi-rigid structure found in many animals, which gives support and protection to the body and provides a point of attachment for the muscles, against which they can pull during movement. An ENDOSKELETON occurs inside the body and may be made of BONE or CARTILAGE in the vertebrates, while in the invertebrates it can be of calcium carbonate (chalk) or silica. An EXOSKELETON occurs outside the body and usually consists of a mixture of the polysaccharide CHITIN and proteins, sometimes reinforced with calcium carbonate, as in most crustaceans. (◂ page 216)

Skin
The TISSUE that forms a sensitive, elastic, protective and waterproof covering for the body, together with its specializations (eg NAILS, HAIR, HORNS). In the adult human, it weighs 2.75kg, covers an area of 1.7m^2 and varies in thickness from 1mm (in the eyelids) to 3mm (in the palms and soles). It consists of two layers: the outer epidermis, and the inner dermis, or true skin. The outermost part of the epidermis, the stratum corneum, contains a tough protein called KERATIN. Consequently it provides protection against mechanical trauma, a barrier against microorganisms, and waterproofing. The epidermis also contains cells that produce MELANIN,whichis responsible for skin pigmentation and also provides protection against the Sun's ultraviolet rays. The unique pattern of skin folding on the soles and palms provides a gripping surface, and is the basis of identification by fingerprints. The dermis is usually thicker than the epidermis and contains blood vessels, nerves and sensory receptors, SEBACEOUS GLANDS (blockage of which leads to blackhead formation), hair follicles, fat cells and fibers. Human skin also contains sweat glands. Temperature regulation of the body is aided by the evaporative cooling of sweat (see PERSPIRATION), regulation of the skin blood flow, and the erection of hairs that trap an insulating layer of air next to the skin. The rich nerve supply of the dermis is responsible for the reception of touch, pressure, pain and temperature stimuli. Leading into the hair follicles are sebaceous glands that produce the antibacterial sebum, a fluid that keeps the hairs oiled and the skin moist. The action of sunlight on the human skin initiates the formation of VITAMIN D, which helps to prevent RICKETS. (◂ page 213)

Skinner, Burrhus Frederic (b. 1904)
US psychologist and author whose staunch advocacy of BEHAVIORISM has done much to gain it acceptance in 20th-century PSYCHOLOGY. His best-known books are *Science and Human Behavior* (1953) and *Beyond Freedom and Dignity* (1971).

Skull
In vertebrates the bony structure of the head, situated at the top of the vertebral column. It forms a thick bony protection for the BRAIN with small apertures for blood vessels, the SPINAL CORD, eyes, ears, etc., and the thinner framework of facial structure. (◂ page 216)

Slate
A dark gray, low-grade metamorphic rock. Because of the comparatively low temperatures and pressures under which it was formed, slate still retains the texture and cleavage properties of the SHALE from which it is derived, though the direction of cleavage is usually different from the direction of the original bedding. For this reason, it is widely used as a roofing material.

Sleep
A state of relative unconsciousness and inactivity. The need for sleep recurs periodically in all animals. If deprived of sleep humans initially experience HALLUCINATIONS and acute ANXIETY and become highly suggestible, and eventually COMA and sometimes DEATH result. During sleep, the body is relaxed and most bodily activity is reduced. Cortical, or higher, brain activity (as measured by the ELECTROENCEPHALOGRAPH), blood pressure, body TEMPERATURE, rate of HEART BEAT and breathing are decreased. However, certain activities, such as gastric and alimentary activity, are increased. Sleep tends to occur in daily cycles, which exhibit up to 5 or 6 periods of orthodox sleep – characterized by its deepness – alternating with periods of paradoxical, or rapid-eye-movement (REM), sleep – characterized by its restlessness and jerky movements of the eyes. Paradoxical sleep occurs only when we are dreaming and occupies about 20% of total sleeping time. Sleepwalking (SOMNAMBULISM) occurs only during orthodox sleep when we are not dreaming. Sleeptalking occurs mostly in orthodox sleep. Many theories have been proposed to explain sleep but none is completely satisfactory. Separate sleeping and waking centers in the HYPOTHALAMUS cooperate with other parts of the BRAIN in controlling sleep. Sleep as a whole, and particularly paradoxical sleep when dreaming occurs, is essential to health and life. Sleep learning experiments have so far proved ineffective. A rested brain and concentration are probably the most effective basis for LEARNING.

Sleeping pills
Drugs that induce SLEEP are properly termed hypnotics (see SEDATIVES). They all tend to suppress REM sleep. (See also ANESTHESIA; NARCOTICS.)

Sleeping sickness
INFECTIOUS DISEASE caused by TRYPANOSOMES occurring in Africa and carried by tsetse flies. It initially causes fever, headache, often a sense of oppression and a rash; later the characteristic somnolence follows and the disease enters a chronic, often fatal stage.

Sleepwalking see SOMNAMBULISM.

Sleet
PRECIPITATION consisting of small ICE pellets (diameter 5mm or less) formed by the freezing of raindrops or of partially melted snowflakes. A mixture of rain and snow is often termed sleet. (See also HAIL.)

Slime molds
Simple organisms that live in rotting logs, soil or leaf litter, ingesting bacteria, yeasts or other organic material. Some species live as parasites of plants, algae or fungi. The best-known forms are the cellular slime molds, which have a unicellular and a multicellular phase to the life cycle. In the unicellular phase the cells are indistinguishable from protozoan AMEBAE, and live by ingesting bacteria (see PHAGOCYTOSIS). When food runs short, these amebae aggregate to form a slug-like mass of cells, known as a pseudoplasmodium, which moves around until it finds a suitable place for the release of spores. It then develops into a stalked fruiting body, which releases spores, these later germinating to produce new amebae. Other slime molds exist as acellular networks of protoplasm, rather like the MYCELIUM of a fungus. Known as plasmodial slime molds, they too produce fruiting bodies that release spores. It is among the plasmodial slime molds that parasitic forms occur, including those that cause clubroot

in brassica crops. The slime molds were once regarded as fungi, but they have many features that suggest an independent origin, and modern taxonomists usually classify them separately from the fungi, for example in the Kingdom PROTOCTISTA.

Slipped disk
A common condition in which the intervertebral disks of the spinal column degenerate with extension of the central soft portion through the outer fibrous ring. The protruding material may cause back pain, or may press upon the SPINAL CORD or on nerves as they leave the spinal cord (causing SCIATICA). (♦ page 216)

Sloane, Sir Hans, 1st Baronet (1660-1753)
British physician and natural historian who served as President of the ROYAL SOCIETY (1727-41) after NEWTON. His collection of books and specimens, left to the nation (for a fee of £20,000 to be paid to his family) formed the basis of the British Museum (founded 1759).

Slow virus
Type of virus believed, but not proved, to exist; thought to be responsible for some diseases (possibly including MULTIPLE SCLEROSIS) that develop slowly and seem to have a long dormant period between the probable time of infection and the onset of symptoms.

Smallpox
A severe INFECTIOUS DISEASE caused by a VIRUS, and often fatal. It caused FEVER, headache and general malaise, followed by a rash. Major and minor forms of smallpox existed, with high fatality rate in major cases, often with extensive skin HEMORRHAGE. Immunization against smallpox was the earliest form practiced, initially through self-inoculation with the minor form. Later JENNER introduced VACCINATION with the related cowpox virus (later vaccinia were used). In 1980 the World Health Organization declared that the disease had been eliminated. QUARANTINE and vaccination were crucial to this. Today, smallpox viruses are found only in laboratories for research purposes.

Smell
A SENSE for detecting and recognizing airborne chemical substances. Detection of environmental odors is of vital importance in recognizing individuals of the same species or group, assessing food, detecting other animals or objects of danger, and in sexual behavior and attraction. Smell reception in insects is localized in the antennae and detection is by specialist (PHEROMONE) receptors and generalist (other odor) receptors. In terrestrial vertebrates the nostrils contain a specialized receptor surface, the olfactory epithelium. Receptor cells detect the tiny concentrations of odors in the air stream and stimulate nerve impulses that pass to olfactory centers in the BRAIN for coding and perception. Aquatic animals such as crustaceans and fish have a comparable sense detecting waterborne chemicals. In fish the receptors are often distributed all over the body, but may be concentrated in chin barbels or around the mouth. (♦ page 213)

Smelting
In METALLURGY, process of extracting a metal from its ore by heating the ore in a BLAST FURNACE or reverberatory furnace (one in which a shallow hearth is heated by radiation from a low roof heated by flames from the burning fuel). A reducing agent (see OXIDATION AND REDUCTION), usually coke, is used, and a flux is added to remove impurities. Sulfide ores are generally roasted to convert them to oxides before they are smelted.

Smith, Hamilton O. (b. 1931)
US molecular biologist who shared the 1978 Nobel Prize for Physiology or Medicine with ARBER and NATHANS for their research into MOLECULAR GENETICS.

Smith, William "Strata" (1769-1839)
The "father of stratigraphy". He established that similar sedimentary rock strata in different places may be dated by identifying the fossils each level contains, and made the first geological map of England and Wales (1815).

Smithson, James (1765-1829)
Earlier known as James Lewis and as Louis Macie, British chemist and mineralogist who left £100,000 (then about $500,000) for the foundation of the SMITHSONIAN INSTITUTION.

Smithsonian Institution
US institution of scientific and artistic culture, located in Washington, D.C., and sponsored by the US Government. Founded with money left by JAMES SMITHSON, it was established by Congress in 1846. It is governed by a board of regents comprising the US Vice-president and Chief Justice, three Senators, three representatives and six private citizens appointed by Congress. Although it undertakes considerable scientific research, it is best known as the largest US collection of museums.

Smithsonite
White, yellow or green mineral, formerly called CALAMINE. Smithsonite consists of zinc carbonate ($ZnCO_3$). It is of widespread occurrence, an ore of ZINC formed by alteration of other zinc minerals; it has the CALCITE structure. It was named for JAMES SMITHSON.

Smoking
The habit of inhaling or taking into the mouth the smoke of dried tobacco or other leaves from a pipe or wrapped cylinder; it has been practiced for many years in various communities, often using leaves of plant with hallucinogenic or other euphoriant properties. Since the rise in cigarette consumption, epidemiology has demonstrated an unequivocal association with lung CANCER, chronic BRONCHITIS and EMPHYSEMA and with ATHEROSCLEROSIS and heart disease. Smoking appears to play a part in other forms of cancer and in other diseases such as peptic ULCER. It is not yet clear what part of smoke is responsible for disease. It is now known that nonsmokers may be affected by environmental smoke (passive smoking). A minor degree of physical and a large degree of psychological addiction occur.

Smut
A group of FUNGI that cause plant diseases, so named for the masses of sooty spores formed on the surface of the host plant. Smuts require only one host plant to complete their life cycle, unlike RUSTS. *Ustilago maydis* causes an important disease of maize; *U. tritici* is the loose smut of wheat. (See also PLANT DISEASES.)

Sneeze
Explosive expiration through the nose and mouth stimulated by irritation or INFLAMMATION in the nasal EPITHELIUM. It is a REFLEX attempt to remove the source of irritation.

Snell, George D. (b. 1903)
US geneticist who was corecipient of the 1980 Nobel Prize for Physiology or Medicine for his research pinpointing the way in which the mouse's immune system recognizes itself and rejects TRANSPLANTS. (See also ANTIBODIES AND ANTIGENS; IMMUNITY.)

Snell's Law see REFRACTION.

Snow
PRECIPITATION consisting of flakes or clumps of ICE crystals. The crystals are plane hexagonal, showing an infinite variety of beautiful branched forms; needles, columns and irregular forms are also found. Snow forms by direct vapor-to-ice condensation from humid air below 0°C. On reaching the ground, snow crystals lose their structure and become granular. Fresh snow is very light (sg about 0.1), and is a good insulator, protecting underlying plants from severe cold. In time, pressure, sublimation and melting and refreezing lead to compaction into NÉVÉ. The slow melting of mountain snow is important in natural irrigation.

Snow blindness
Temporary loss of VISION with severe pain, tears and EDEMA due to excessive ultraviolet light reflected from snow.

Soaps and detergents
Substances that, when dissolved in water, are cleansing agents. Soap has been known since 600 BC; it was used as a medicine until its use for washing was discovered in the 2nd century AD. Until about 1500 it was made by boiling animal fat with wood ashes (which contain the alkali, potassium carbonate). Then caustic soda (see SODIUM), a more effective ALKALI, was used; vegetable FATS and oils were also introduced. Saponification, the chemical reaction in soap-making, is an alkaline HYDROLYSIS of the fat (an ESTER) to yield GLYCEROL and the sodium salt of a long-chain CARBOXYLIC ACID. The potassium salt is used for soft soap. Synthetic detergents, introduced in WWI, generally consist of the sodium salts of various long-chain SULFONIC ACIDS, derived from oils and PETROLEUM products. The principle of soaps and detergents is the same: the hydrophobic long-chain hydrocarbon part of the molecule attaches itself to the grease and dirt particles, and the hydrophilic acid group makes the particles soluble in water, so that by agitation they are loosed from the fabric and dispersed. Detergents do not (unlike soaps) form scum in HARD WATER. Their persistence in rivers, however, causes pollution problems, and biodegradable detergents have been developed. Household detergents may contain several additives: bleaches, brighteners, and ENZYMES to digest protein stains (egg, blood, etc).

Soapstone (steatite)
METAMORPHIC ROCK consisting of compacted TALC with SERPENTINE and carbonates, formed by alteration of PERIDOTITE. Soft and soapy to the touch, when fired it becomes hard and is used for insulators.

Social insects
Insects that live in complex, ordered societies. The term is generally used to mean those that have cooperative care of the young, overlap of at least two generations within the colony, and some non-reproductive individuals who rear or defend the young of others. Technically, such species should be called the eusocial or "truly social" insects. Included within this group are all termites and ants, and a large proportion of the bees and wasps. Ants, bees and wasps all belong to the Order Hymenoptera, and the eusocial ones are often referred to collectively as the "social Hymenoptera". In the most advanced of the social insects there is a single reproductive female, known as a "queen", who is of great size compared to other colony members, lays eggs at a prodigious rate and is often immobile. The non-reproductives are divided into two or more castes, some acting as workers and mainly concerned with feeding the young, others acting as soldiers and defending the nest. Caste membership may be determined at the larval stage, by the feeding regime or be based on age. Some soldier castes are so greatly modified for defence, with massive biting or piercing mouthparts, that they are unable to eat. In the Hymenoptera the worker and soldier castes are all-female; in the termites they contain individuals of both sexes. At certain times of the year individuals capable of reproduction are reared, and these fly away to found new colonies. Such reproductives

are all winged, in contrast to the wingless workers and soldiers of the ant and termite colonies. Once they have landed on the ground, however, the wings snap off, so the mature ant or termite queen is wingless like her offspring.

Social sciences
Group of studies concerned with humans in relation to their cultural, social and physical environment.

Sociobiology
A relatively new scientific discipline that attempts to elucidate the biological principles governing social behavior and social organization in animals. It brings together the ideas of evolutionary theory, genetics and population biology, and applies them to the observations of ethology in an effort to understand the genetic advantages of particular behavioral strategies. Sociobiology also draws upon "game theory" (which was originally developed for the analysis of military tactics) to understand how different behavioral strategies interact. Originally the proponents of sociobiology, such as Edward O. Wilson of Harvard University, regarded human behavior as within the scope of the subject, but as human behavior is so diverse there are now doubts about the validity of such studies.

Sociology
Systematic study that seeks to describe and explain collective human behavior – as manifested in cultures, societies, communities and subgroups – by exploring the institutional relationships that hold between individuals and so sustain this behavior.

Soda (sodium carbonate) see SODIUM.

Soddy, Frederick (1877-1956)
British chemist awarded the 1921 Nobel Prize for Chemistry for his work with RUTHERFORD on radioactive decay, and particularly for his formulation (1913) of the theory of ISOTOPES. He also worked with SIR WILLIAM RAMSAY to discover HELIUM.

Sodium (Na)
A soft, reactive, silvery white ALKALI METAL. It is the sixth most common element, occurring naturally in common salt and many other important minerals such as cryolite and Chile saltpeter. It is very electropositive, and is produced by ELECTROLYSIS of fused sodium chloride (Downs process). Sodium rapidly oxidizes in air and reacts vigorously with water to give off hydrogen, so it is usually stored under kerosine. Most sodium compounds are highly ionic and soluble in water, their properties being mainly those of the anion. Sodium forms some organic compounds, such as alkyls. It is used in making sodium cyanide, sodium hydride and the antiknock additive tetraethyl lead. Its high heat capacity and conductivity make molten sodium a useful coolant in some nuclear reactors. AW 23.0, mp 98°C, bp 900°C, sg 0.971 (20°C). (♦ page 130)

SODIUM BICARBONATE ($NaHCO_3$) is a white crystalline solid, made from sodium carbonate and carbon dioxide. It gives off carbon dioxide when heated to 270°C or when reacted with acids, and is used in baking powder, fire extinguishers and as an antacid.

SODIUM BORATES are sodium salts of BORIC ACID, differing in their degree of condensation and hydration; BORAX is the most important. They are white crystalline solids, becoming glassy when heated, and used in the manufacture of detergents, water softeners, fluxes, glass and ceramic glazes.

SODIUM CARBONATE (washing soda, Na_2CO_3) is a white crystalline solid obtained primarily from natural deposits. It is used in making glass, other sodium compounds, soap and paper. The alkaline solution is used in disinfectants and also in water softeners. mp 851°C.

SODIUM HYDROXIDE (caustic soda, NaOH) is a white deliquescent solid, usually obtained as pellets. It is a strong alkali, and absorbs carbon dioxide from the air. It is made by electrolysis of sodium chloride solution or by adding calcium hydroxide to sodium carbonate solution. Caustic soda is used in the production of cellulose, plastics, soap, dyestuffs, paper and in oil refining. mp 318°C, bp 1390°C.

SODIUM NITRATE (soda niter, $NaNO_3$) is a colorless crystalline solid, occurring naturally in CHILE SALTPETER. Its properties are similar to those of potassium nitrate (see POTASSIUM), but as it is hygroscopic it is unsuitable for gunpowder. mp 307°C.

SODIUM THIOSULFATE (hypo, $Na_2S_2O_3$) is a colorless crystalline solid. It is a mild reducing agent, used to estimate iodine and as a photographic fixer, dissolving the silver halides that have remained unaffected by light. (For sodium chloride, see SALT.)

Software
Term used in the COMPUTER industry to refer to all the non-hardware elements of a computer system, principally the programs.

Softwood
Timber produced from CONIFERS, accounting for about 80% of world production. The wood contains a different type of conducting element from that of ANGIOSPERM (broadleaved) trees such as oak and maple, whose timbers are known as hardwoods; while they are on the whole harder than softwoods, some are very soft, eg balsa. Similarly, softwoods include woods that are physically both soft and hard.

Soil
The uppermost surface layer of the Earth, in which plants grow and on which, directly or indirectly, all life on Earth depends. Soil consists, in the upper layers, of organic material mixed with inorganic matter as a result of weathering (see EROSION; HUMUS). Soil depth, where soil exists, may reach to many meters. Between the soil and the bedrock is a layer called the subsoil. Mature soil may be described in terms of four soil horizons: A, the uppermost layer, containing organic matter, though most of the soluble chemicals have been leached (washed out); B, strongly leached and with little or no organic matter (A and B together are often called the topsoil); C, the subsoil, a layer of weathered and shattered rock; and D, the bedrock. Three main types of soil are commonly distinguished: pedalfers, associated with temperate, humid climates, have a leached A-horizon but contain IRON and ALUMINUM salts with clay in the B-horizon; pedocals, associated with low-rainfall regions, contain soluble substances such as CALCIUM carbonate (soluble in rainwater, which contains carbon dioxide) and other salts; and laterites, tropical red or yellow soils, heavily leached and rich in iron and aluminum. Soils may also be classified in terms of texture (see CLAY; SILT; SAND); LOAMS, with roughly equal proportions of sand, silt and clay, together with humus, are among the richest agricultural soils. (See also PERMAFROST.)

Soil erosion
The wearing away of soil, a primary cause of concern in agriculture. There are two types: geological erosion denotes those naturally occurring EROSION processes that constantly affect the Earth's surface features; it is usually a fairly slow process and naturally compensated for. Accelerated erosion describes erosion hastened by the intervention of humans. Sheet erosion usually occurs on plowed fields. A fine sheet of rich topsoil (see SOIL) is removed by the action of rainwater. Repetition over the years may render the soil unfit for cultivation. In rill erosion, heavy rains may run off the land in streamlets: swiftly moving water cuts shallow trenches that may be plowed over and forgotten until, after years, the soil is found to be poor. In gully erosion, trenches are cut by repeated or heavy flow of water. Wind erosion is important in exposed, arid areas. (See CONSERVATION; LAND RECLAMATION.) (♦ page 208)

Sol
A COLLOID, usually in liquid form, in which the dispersed phase is initially a solid, the continuous phase initially a liquid. (See also GEL.)

Solanaceae
A family of flowering plants (ANGIOSPERMS) that includes the potato, tomato, peppers, tobacco, aubergine (egg plant) and nightshades. It is noted for the production of ALKALOIDS.

Solar cell
Device for converting the ENERGY of the Sun's radiation into electrical energy. The commonest form is a large array of SEMICONDUCTOR *p-n* junction devices in series and in parallel. By the PHOTOELECTRIC EFFECT each junction produces a small voltage when illuminated. Solar cells are chiefly used to power artificial SATELLITES. Their low efficiency (about 12%) makes them uncompetitive on Earth except for mobile or isolated devices. (♦ page 237)

Solar energy
The ENERGY given off by the SUN as ELECTROMAGNETIC RADIATION. In one year the Sun emits about 1.2×10^{34}J of energy, of which half of one-billionth (6×10^{24}J) reaches the Earth. Of this, most is reflected away, only 35% being absorbed. The power reaching the ground is at most 1.2kW/m^2, and on average 0.8kW/m^2. Solar energy is naturally converted into WIND power and into the energy of the HYDROLOGIC CYCLE, increasingly exploited as hydroelectric power. Plants convert solar energy to chemical energy by PHOTOSYNTHESIS, normally at only 0.1% efficiency; the cultivation of ALGAE in ponds can be up to 0.6% efficient, and is being developed to provide food and fuel. Solar heat energy may be used directly in several ways. Solar evaporation is used to convert brine to SALT and distilled water. Flat-plate collectors – matt black absorbing plates with attached tubes through which a fluid flows to collect the heat – are beginning to be used for domestic water heating, space heating, and to run air-conditioning systems. Focusing collectors, using a parabolic mirror, are used in solar furnaces, which can give high power absorption at high temperatures. They are used for cooking, for high-temperature research, to power heat engines for generating electricity, and to produce electricity more directly by the SEEBECK effect. Solar energy may be directly converted to electrical energy by SOLAR CELLS.

Solar plexus
The GANGLION of nerve cells and fibers situated at the back of the ABDOMEN that subserve autonomic NERVOUS SYSTEM function for much of the GASTROINTESTINAL TRACT. A blow on the abdomen over the plexus causes pain and "winding".

Solar system
The Sun and all the celestial objects that move in ORBIT around it, including the nine known planets (MERCURY; VENUS; EARTH; MARS; JUPITER; SATURN; URANUS; NEPTUNE; PLUTO), their 58 known moons, the ASTEROIDS, COMETS, meteoroids (see METEOR) and a large quantity of gas and dust. The planets all move in their orbits in the same direction and, with the exceptions of Venus and Uranus, also rotate on their axes in this direction: this is known as direct motion. Most of the Moons of the planets have direct orbits, with the exception of four of Jupiter's minor moons, the outermost moon of Saturn and the inner moon of Neptune, whose orbits are retrograde (see RETRO-

SOLAR SYSTEM

	Mean distance (AU)	Mean distance (millions of km)	Eccentricity	Inclination to ecliptic	Sidereal period (days)	Equatorial diameter (km)	Polar diameter (km)	Equatorial rotation	Mass (kg)	Density (water = 1)
Sun	—	—	—	—	—	1,392,530	1,392,530	24.6d	1.9891×10^{30}	1.41
Moon	—	—	0.0549	5°09′	27.322	3,476	3,476	27.32d	7.3483×10^{22}	3.34
Mercury	0.3871	59.91	0.2056	7°00′	87.969	4,878	4,878	58.65d	3.3022×10^{23}	5.43
Venus	0.7233	108.21	0.0068	3°23′	224.701	12,104	12,104	243d	4.8689×10^{24}	5.24
Earth	1.0000	149.60	0.0167	—	365.256	12,756	12,714	23.93hr	5.9742×10^{24}	5.52
Mars	1.5237	227.94	0.0934	1°50′	686.980	6,794	6,759	24.62hr	6.4191×10^{23}	3.94
Jupiter	5.2028	778.34	0.0485	1°18′	4332.59	142,800	134,200	9.8hr	1.899×10^{27}	1.32
Saturn	9.5388	1,427.01	0.0556	2°29′	10,759.20	120,000	108,000	10.2hr	5.684×10^{26}	0.70
Uranus	19.1818	2,869.6	0.0473	0°46′	30,684.8	51,800	49,000	16.3hr	8.6978×10^{25}	1.27
Neptune	30.0580	4,496.7	0.0086	1°46′	60,190.5	49,500	47,400	18.2hr	1.028×10^{26}	1.77
Pluto	39.44	5,900.0	0.250	17°12′	90,465.0	2,500	2,500	6.3d	1.6×10^{22}	1–2

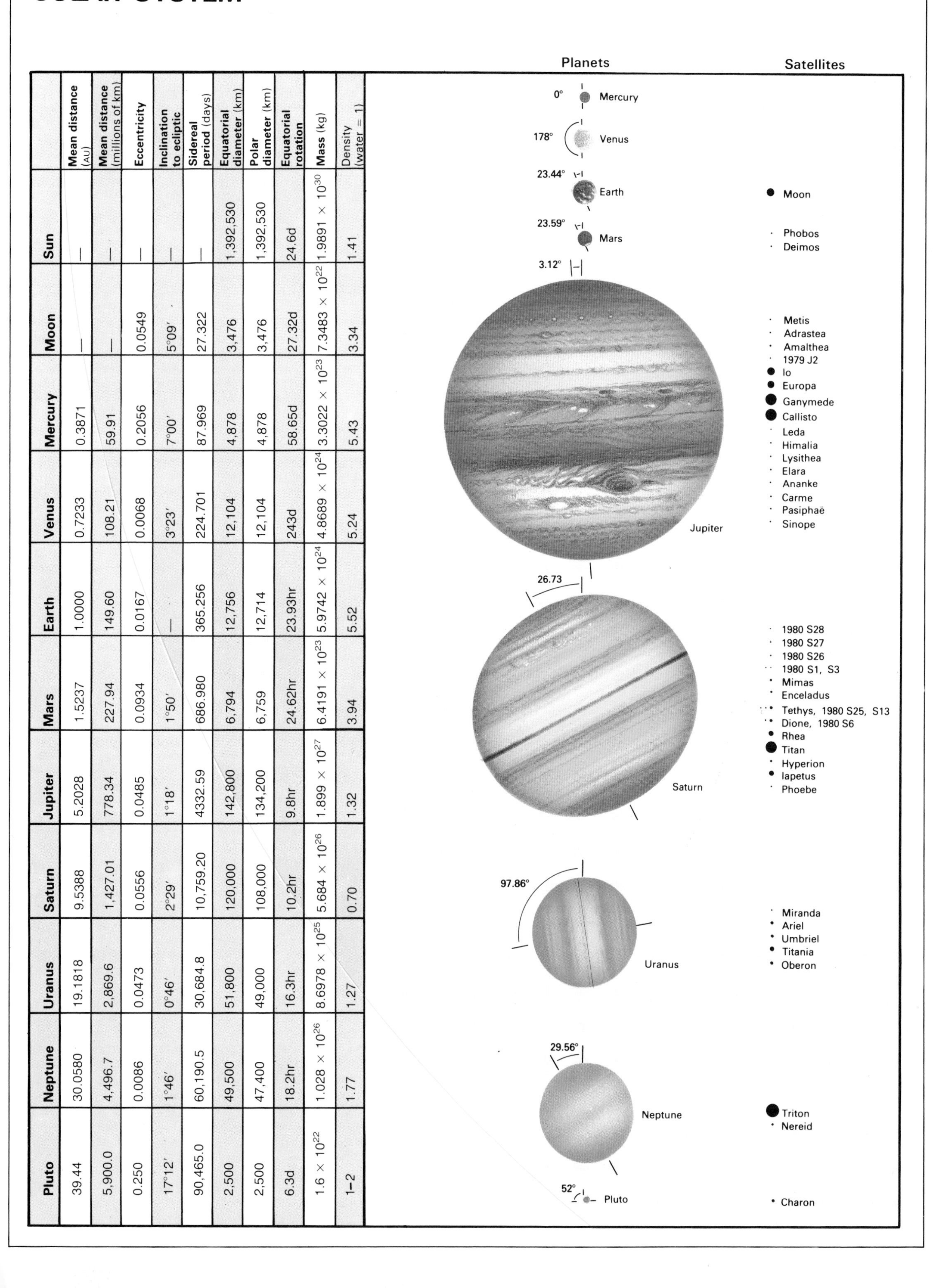

GRADE MOTION). Most of the planets move in elliptical, near circular orbits, and roughly in the same plane. The origin of the Solar System is not known, though various theories have been proposed (see NEBULAR HYPOTHESIS; PLANETESIMAL HYPOTHESIS). The most popular current theory suggests that the Sun and planets formed from the same cloud of gas and dust, which, as it contracted, formed a central condensation (the Sun) surrounded by a disk of matter that broke up to form the planets. It would not appear to be unique among the stars (see BETA PICTORIS; PLANET; VEGA). (◆◆ page 134, 221, 240)

Solar wind
The electrically charged material, mostly electrons and protons, thrown out by the Sun at an average speed of 400km/s. The "quiet" component is a continuous stream to which is added an "active" component produced by bursts of activity on the Sun's surface. The solar wind affects the magnetic fields of the planets, and causes the tails of COMETS.

Solenoid
Device used in CIRCUIT BREAKERS for producing a short lateral movement of a sliding iron core. This is attracted by the MAGNETIC FIELD produced when an electric current flows in one of the coils surrounding either end of the slider. Also, an elongated coil used to produce a region of uniform magnetic field.

Solid
One of the three physical states of matter, characterized by the property of cohesion: solids retain their shape unless deformed by external forces. True solids have a definite melting point and are crystalline, their molecules being held together in a regular pattern by stronger intermolecular forces than exist in liquids or gases. Amorphous solids (eg GLASS) are not crystalline, melt over a wide temperature range and are effectively supercooled liquids. (◆ page 130)

Solid state physics
Branch of physics concerned with the nature and properties of solid materials, many of which arise from the association and regular arrangement of atoms or molecules in crystalline solids. The term is applied particularly to studies of SEMICONDUCTORS and solid-state electronic devices.

Solstices
The two times each year when the Sun is on the points of the ECLIPTIC farthest from the equator (see CELESTIAL SPHERE). At the summer solstice in late June the Sun is directly overhead at noon on the Tropic of Cancer; at winter solstice, in late December, it is overhead at noon on the Tropic of Capricorn. (◆ page 25)

Solution
A homogeneous molecular mixture of two or more substances, commonly of a solid, liquid or gas in a liquid, though solid/solid and gas/gas solutions also exist. The major liquid component is usually termed the SOLVENT, the other component, which is dissolved in it, the solute. The solubility of a solute in a given solvent at a particular temperature is usually stated as the mass which will dissolve in 100g of the solvent to give a saturated solution (see SATURATION). Solubility generally increases with temperature. For slightly soluble ionic compounds, the solubility product – the product of the individual ionic solubilities – is a constant at a given temperature. Most substances are solvated when dissolved: that is, their molecules become surrounded by solvent molecules acting as LIGANDS. Ionic crystals dissolve to give individual solvated ions, and some good solvents of high dielectric constant (such as water) cause certain covalent compounds to ionize, wholly or partly (see also ACID). Analogous to an ideal gas, the hypothetical ideal solution is one which is formed from its components without change in total volume or internal energy: it obeys RAOULT's Law and its corollaries, so that the addition of solute produces a lowering of the freezing point, elevation of the boiling point and increase in osmotic pressure (see OSMOSIS), all proportional to the number of MOLES added. (See also DISSOCIATION; ELECTROLYSIS; EQUIVALENT WEIGHT.)

Solvay process
A process for the manufacture of sodium carbonate (see SODIUM). Salt, ammonia, carbon dioxide and water react to give precipitated sodium bicarbonate, which on heating gives sodium carbonate, and carbon dioxide for RECYCLING.

Solvent
A liquid capable of dissolving a substance to form a SOLUTION. Generally "like dissolves like"; thus a nonpolar covalent solid such as naphthalene dissolves well in a hydrocarbon solvent. The best solvents are those with polar molecules and high dielectric constant; water is the most effective.

Somnambulism (sleepwalking)
State in which the body is able to walk and perform other automatic tasks while consciousness is diminished. Often seen in anxious children, it is said to be unwise to awaken them as intense fear may be felt.

Sonar
Sound navigation and ranging, a technique used to detect and determine the position of underwater objects (eg submarines; shoals of fish) and to find the depth of water under a ship's keel. Sonar works on the principle of echolocation: high-frequency SOUND pulses are beamed from the ship, and returning ECHOES are analyzed to give the direction and range of the reflecting objects.

Sound
Mechanical disturbance, such as a change of pressure, particle displacement or stress, propagated in an elastic medium (eg air or water), that can be detected by an instrument or by an observer who hears the auditory sensation it produces. Sound is a measurable physical phenomenon and an important stimulus. It forms a major means of communication in the form of spoken language, and both natural and manmade sounds (of traffic or machinery) contribute largely to our environment. The EAR is very sensitive and will tolerate a large range of sound energies, but enigmas remain as to exactly how it produces the sensation of hearing. The Greeks appreciated that sound was connected with air motion and that the PITCH of a musical sound produced by a vibrating source depended on the vibration FREQUENCY. Attempts to measure the velocity of sound in air date from the 17th century. Sound is carried as a longitudinal compressional wave in an elastic medium: part of the medium next to a sound source is compressed, but its elasticity makes it expand again, compressing the region next to it and so on. The velocity of such waves depends on the medium and the temperature, but is always much less than that of light. Sound waves are characterized by their wavelength and frequency. Humans cannot hear sounds of frequencies below 16Hz and above 20kHz, such sounds being known as infrasonic and ULTRASONIC respectively. The sound produced by a TUNING FORK has a definite frequency, but most sounds are a combination of frequencies. The amount of motion in a sound wave determines its loudness or softness and the intensity falls off with the square of distance from the source. Sound waves may be reflected from surfaces (as in an ECHO), refracted or diffracted, the last property enabling us to hear around corners. The intensity of a sound is commonly expressed in DECIBELS above an arbitrary reference level; its loudness is measured in PHONS. (◆ page 225)

Southern Cross
Crux, a small, bright constellation about 30° distant from the S celestial pole. The four bright stars forming the cross are Acrux (alpha), Mimosa (beta), Gacrux (gamma) and Delta Crucis.

Southern Lights see AURORA.

South Pole
The point in Antarctica through which the Earth's axis of rotation passes, first reached by Roald Amundsen (14 December 1911). It does not coincide with the EARTH's S Magnetic Pole. (See also CELESTIAL SPHERE; MAGNETISM; NORTH POLE.)

Space exploration
At 10.56pm EDT on 20 July 1969 Neil Armstrong became the first man to set foot on the MOON. This was the climax of an intensive US space program sparked off by the successful launch of the Russian artificial SATELLITE Sputnik 1 in 1957, and accelerated by Yuri Gagarin's flight in Vostok 1, the first manned spacecraft, in 1961. Later that year Alan Shepard piloted the first American manned spacecraft, and President Kennedy set the goal of landing a man on the Moon and returning him safely within the decade. On 20 February 1962 John Glenn orbited the Earth three times in the first MERCURY craft to be boosted by an Atlas rocket, but it was Valery Bykovsky who set the one-man endurance record with a five-day mission in June 1963. His Vostok 5 craft was accompanied in orbit for three days by Vostok 6, which carried Valentina Tereschkova, the first woman cosmonaut. Alexei Leonov completed the first spacewalk in March 1965, but then it was the turn of the GEMINI MISSIONS to break all records. Both countries lost men, on the ground and in space: among them V.I. Grissom, E.H. White and R.B. Chaffee – in a fire on board Apollo during ground tests – and the crew of Soyuz 11, killed during reentry in 1971, though earlier Soyuz had docked successfully with the first space station, Salyut 1, and set new records. Unmanned probes such as Orbiter, Ranger and Surveyor were meanwhile searching out Apollo landing sites, while Russian Luna and Lunokhod craft were also studying the Moon. In 1968 Apollo 7 carried out an 11-day Earth-orbit flight, and at Christmas Apollo 8 made 10 lunar orbits. The lunar landing craft was tested on the Apollo 9 and Apollo 10 missions, leaving the way clear for the triumphant landing of Apollo 11. Apollo 12 was equally successful, landing only 550m from the lunar probe Surveyor 3, but the Apollo 13 mission was a near disaster: an explosion damaged the craft on its way to the Moon, and re-entry was achieved only with great difficulty. Apollo 14 had no such problems in visiting Fra Mauro, and on the Apollo 15 mission a lunar Roving Vehicle allowed collection of a very wide range of samples. Apollo 16 brought back over 90kg of moon rock; in December 1972 Apollo 17 made the last lunar landing. In 1973 the Skylab missions returned attention to the study of world resources from space, and movements toward cooperation in this field were demonstrated by the joint Apollo-Soyuz mission in 1975. Exploration of the planets has been carried out by unmanned probes: the MARINER series to Mars, Venus and Mercury; the PIONEER and VOYAGER missions to outer space, and a number of Soviet contributions, such as VENERA, the Zond bypass probe to Venus and Mars soft-landing craft. Results from the two American VIKING probes that soft-landed on Mars in 1976 did not reveal the existence of life there. By 1986 all the known planets, apart from Neptune and Pluto and two comets, Giacobini-Zinner and HALLEY'S, had been inspected at close range by spacecraft. Voyager 2 is scheduled to encounter Neptune in 1989.

The launch of the first test flight space shuttle in

1981 saw the introduction of reusable spacecraft. In 1982, two satellites were launched into orbit from the first operational shuttle, and the following year a shuttle flight carried the first American woman, Sally Rider, into space. In 1984 shuttle astronauts set a new record for working in space when they captured and repaired the faulty Solar Maximum Satellite.

Spacelab, a manned, reusable space laboratory built by the European Space Agency, took its first flight in the cargo bay of a shuttle in 1983. It proved that a scientist can carry out experiments in space, including processing materials in MICROGRAVITY and high-vacuum conditions.

The shuttle program brought a change in emphasis from exploration to operation in space. The intention is for crews, consisting of astronauts, and mission and payload specialists, routinely to launch and repair satellites and to carry out research. The program received a tragic setback in January 1986 when the shuttle Challenger exploded after liftoff, killing all seven crew.

Both Russia and America are now working toward permanently manned space stations. The Soviets launched their program in 1971, with the Salyut series of space stations. They are equipped for Earth, solar and astronomical observations and material processing experiments and, although not permanently manned, are visited regularly by crews flown up in Soyuz spacecraft. Some crews have stayed on Salyut for over 6 months at a time. The most recent version of the space station, Mir, has been enlarged by add-on modules to provide a basis for what may become a permanently manned facility.

The US space station program for the 1990s will use shuttles to ferry modules into orbit, where they can be assembled. Permanently manned in low orbit, the space station will consist of living quarters, solar power sources, computer, work space and docking bay. Space stations could be used for processing materials that are difficult to manufacture on Earth, launching and servicing satellites, space telescopes, building spacecraft, and launching missions to outer space.

The next century could see permanently occupied outposts on the Moon to mine minerals for manufacturing satellites, space stations and planetary probes, and a manned mission to Mars. Another possibility is electricity generated in GEOSTATIONARY ORBIT by large arrays of solar cells, and transmitted back to Earth.

Space factories
Envisaged to process materials in space. These would take advantage of MICROGRAVITY and atmospheric pressure to manufacture materials and products, on a large scale, that are difficult or impossible to produce on Earth; eg pharmaceuticals, electronics, optics and advanced alloys. Some preliminary investigations have taken place on Salyut space stations and space shuttles (see also SPACE EXPLORATION.)

Space medicine
The specialized branch of medicine concerned with the special physical and psychological problems arising from space flight. In particular, the effects of prolonged weightlessness and isolation are studied, simulated space flight forming the basis for much of this work.

Space shuttlesee SPACE EXPLORATION.

Space station see SPACE EXPLORATION.

Space-time
A way of describing the geometry of the physical Universe arising from EINSTEIN's special theory of RELATIVITY. Space and time are considered as a single 4-dimensional continuum rather than as a 3-dimensional space with a separate infinite 1-dimensional time. Time thus becomes the "fourth dimension". Events in space-time are analogous to points in space, and invariant space-time intervals to distances in space. (♦ page 151)

Spallanzani, Lazzaro (1729-1799)
Italian biologist who attacked the contemporary belief in the SPONTANEOUS GENERATION of life by demonstrating that organisms did not appear in vegetable infusions if the infusions were boiled and kept from contact with the air.

Spark
Momentary electric discharge in a gas.

Spastic paralysis
Form of PARALYSIS due to disease of the brain (eg STROKE) or SPINAL CORD (eg MULTIPLE SCLEROSIS), in which the involved muscles are in a state of constantly increased tone (or resting contraction). Spasticity is a segmental motor phenomenon where muscle contraction occurs without voluntary control.

Speciation
The process by which a new species is formed. It depends principally upon barriers to interbreeding developing between the new species and the parent species, so that reproductive isolation occurs. This may come about in a great many different ways, including the separation of the two populations by a geographical barrier; the development of a new breeding season, so the two are not fertile at the same time; a change to the courtship ritual or typical call in the new species; an alteration in the genital organs of the new species so the two cannot mate; a change to the chromosomes so that, although they mate, their two gametes cannot combine successfully, or they produce infertile hybrids. An example of a change to the chromosomes that instantly produces a new species is POLYPLOIDY, a fairly common occurrence in plants.

Species
A group of living organisms that can all interbreed, either actually or potentially, to produce fertile offspring. Unless the PHENOTYPE is strongly affected by the environment (as in some plants, for example), members of a species all look the same, or belong to a few recognizable groups – eg males and females, different castes in social insects, different morphs in POLYMORPHIC species, different subspecies in species with wide geographic distribution. The formation of a new species occurs by a process of SPECIATION. Related species are grouped together in a GENUS, and in the scientific name the first word denotes the genus and the second the species. Although different species are usually reproductively isolated from each other, two closely related species can sometimes breed to produce HYBRIDS.

Specific gravity (relative density, sg)
Ratio of the density of a substance to that of a reference material at a given temperature, usually water at 4°C. If the sg of an inert substance is less than unity (1), it will float in water at 4°C. The sg of liquids is measured with a HYDROMETER.

Specific heat
The HEAT required to raise the temperature of 1kg of a substance through one KELVIN; expressed in J/K.kg, and measured by CALORIMETRY. The concept was introduced by JOSEPH BLACK; PIERRE DULONG and ALEXIS PETIT showed that the specific heat of elements is approximately inversely proportional to their ATOMIC WEIGHTS, which could thus be roughly determined.

Spectroheliograph
A device used to obtain spectroheliograms, composite photographs of the Sun in light of a single wavelength. An optical image of the Sun is scanned by a slit admitting light to a PRISM which forms a SPECTRUM. A second slit is used to select a particular wavelength and expose a photographic plate passing behind it at the same rate as the first slit is scanning the image.

Spectrometer see SPECTROSCOPY.

Spectrophotometry
The measurement of the intensity of different colors (wavelengths) present in a beam of light, normally using a PRISM or DIFFRACTION grating, and a PHOTOELECTRIC CELL. It is used for COLOR comparison and in chemical ANALYSIS.

Spectroscopy
The production, measurement and analysis of SPECTRA; an essential tool of astronomers, chemists and physicists. All spectra arise from transitions between discrete energy states of matter, as a result of which photons of corresponding energy (and hence characteristic frequency or wavelength) are absorbed or emitted. From the energy levels thus determined, atomic and molecular structure may be studied. Moreover, by using the observed spectra as "fingerprints", spectroscopy may be a sensitive method of chemical analysis. Most of the different kinds of spectroscopy, corresponding to the various regions of ELECTROMAGNETIC RADIATION, relate to particular kinds of energy-level transitions. Gamma-ray spectra arise from nuclear energy-level transitions; X-ray spectra from inner-electron transitions in atoms; ultraviolet and visible spectra from outer (bonding) electron transitions in molecules (or atoms); infrared spectra from molecular vibrations; and microwave spectra from molecular rotations. There are several more specialized kinds of spectroscopy. Raman spectroscopy, based on the effect discovered by C.V. RAMAN, scans the scattered light from an intense monochromatic beam. Some of the scattered light is at lower (and higher) frequencies than the incident light, corresponding to vibration/rotation transitions. The technique thus supplements infrared spectroscopy. Mössbauer spectroscopy, based on the MÖSSBAUER EFFECT, gives information on the electronic or chemical environments of nuclei; as does nuclear magnetic resonance spectroscopy (NMR), based on transitions between nuclear SPIN states in a strong magnetic field. Electron spin resonance spectroscopy (ESR) is similarly based on electron spin transitions when there is an unpaired electron in an orbital, and so is used to study free radicals. The instrument used is a spectroscope, called a spectrograph if the spectrum is recorded photographically all at once, or a spectrometer if it is scanned by wavelength and calibrated from the instrument.

Spectrum
The array of colors produced on passing LIGHT through a PRISM; by extension, the range of a phenomenon displayed in terms of one of its properties. ELECTROMAGNETIC RADIATION arranged according to wavelength thus forms the electromagnetic spectrum, of which that of visible light is only a minute part. Similarly the mass spectrum of a particular collection of ions displays their relative numbers as a function of their masses. (See SPECTROSCOPY; MASS SPECTROSCOPY.) (♦ page 21)

Speech and speech disorders
Speech may be subdivided into conception, or formulation, and production, or phonation and articulation, of speech (see VOICE). Speech development in children starts with associating sounds with persons and objects, comprehension usually predating vocalization by some months. A phase of babbling speech, where the child toys with sounds resembling speech, is probably essential for development. READING is closely related to speech development, involving the association of auditory and visual symbols. Speech involves coordination of many aspects of brain function (HEARING, VISION, etc), but three areas particularly concerned with aspects of speech are located in the dominant hemisphere of right-handed

people and in either hemisphere of left-handed people (see HANDEDNESS). Disease of these parts of the brain leads to characteristic forms of dysphasia or APHASIA, ALEXIA, etc. Developmental DYSLEXIA is a childhood defect of visual pattern recognition. Stammering or stuttering, with repetition and hesitation over certain syllables, is a common disorder, in some cases representing frustrated left-handedness. Dysarthria is disordered voice production and is due to disease of the neuromuscular control of voice. In speech therapy, attempts are made to overcome or circumvent speech difficulties, this being particularly important in children. (See also DEAFNESS.)

Speed see VELOCITY; also AMPHETAMINES.

Spemann, Hans (1869-1941)
German embryologist awarded the 1935 Nobel Prize for Physiology or Medicine for his researches into the development of the EMBRYO, showing that specific CELLS adopted specific functions not through any predetermination of form but because of the action of local chemical "organizers" (in fact, HORMONES).

Spencer, Herbert (1820-1903)
English philosopher, social theorist and early evolutionist. In his multivolume *System of Synthetic Philosophy* (1862-96), he expounded a world view based on a close study of physical, biological and social phenomena, arguing that species evolve by a process of differentiation from the simple to the complex. His political individualism deeply influenced US social thinking. (See also SURVIVAL OF THE FITTEST.)

Sperm
The male GAMETE or sex cell (see REPRODUCTION). Sperm are usually motile, having a single flagellum. (◆ page 32, 204)

Spermatophytes
A division of the plant kingdom that includes the GYMNOSPERMS and ANGIOSPERMS.

Sperry, Roger (b. 1913)
US psychobiologist who received half of the 1981 Nobel Prize for Physiology or Medicine for his research on the organization of nerve CELLS in the BRAIN.

Sphalerite (blende)
The low-temperature (β) form of zinc sulfide (ZnS); the chief ore of ZINC, occurring worldwide with GALENA. It forms lustrous crystals of the isometric system, white when pure, but usually brown to black with iron impurity. (See also WURTZITE.)

Sprenopsida see HORSETAILS.

Sphincter
Muscle or group of muscles that can tighten temporarily to prevent movement of the contents of hollow viscera. Sphincters may be under autonomic or voluntary control.

Sphygmomanometer
Instrument for measuring blood pressure by determining the pressure in a cuff attached around a limb needed to prevent blood flow. A stethoscope is used to hear the changing sounds over the ARTERY as the cuff deflates.

Spin
Intrinsic angular MOMENTUM of a nucleus or SUBATOMIC PARTICLE, as if about an axis within itself. Spin, s, has definite values given by $nh/2\pi$, where $n=\frac{1}{2}$, 1, $\frac{3}{2}$, 2, ..., and h is the PLANCK CONSTANT. Particles or nuclei with n equal to an integer (1, 2,...) are bosons, and obey BOSE-EINSTEIN STATISTICS; those with half-integer values of n ($\frac{1}{2}$, $\frac{3}{2}$,...) are called fermions and obey FERMI-DIRAC STATISTICS.

Spina bifida
Congenital defect of the SPINAL CORD and spinal canal leading to a variable degree of leg PARALYSIS and loss of urine and feces SPHINCTER control; it may be associated with other malformation – particularly HYDROCEPHALUS. It is an embryological disorder due to failure of fusion of the neural tube.

Spinal column see VERTEBRAE.

Spinal cord
The part of the central NERVOUS SYSTEM outside the skull. It joins the BRAIN at the base of the skull, forming the medulla oblongata, and extends downward in a bony canal enclosed in the VERTEBRAE. Between the bone and cord are three sheaths of connective TISSUE called the meninges. A section of the cord shows a central core of gray matter (containing the cell bodies of nerve fibers running either to the muscles or within the cord itself), completely surrounded by white matter (composed solely of nerve fibers). There is a central canal containing CEREBROSPINAL FLUID, which opens into the cavities of the brain.

Spinal tap (lumbar puncture)
Procedure to remove CEREBROSPINAL FLUID (CSF) from the lumbar spinal canal using a fine needle. It is used in diagnosis of MENINGITIS, ENCEPHALITIS, MULTIPLE SCLEROSIS and TUMORS. In NEUROLOGY it is used in treatment, by reducing CSF pressure or allowing insertion of DRUGS. (◆ page 126)

Spine (backbone)
In VERTEBRATES, the column of bony (or cartilaginous) disks known as VERTEBRAE that runs the length of the body and contains the SPINAL CORD. (◆ page 216)

Spinels
Group of OXIDE minerals of general formula $M_2^{III}M^{II}O_4$, formed in high-temperature IGNEOUS or METAMORPHIC ROCKS. The chief members are spinel itself, aluminum magnesium oxide, with some GEM varieties; CHROMITE; and MAGNETITE. Free substitution occurs. Synthetic spinels are used as refractories and in solid-state components.

Spirochete
Spiral-shaped BACTERIA, species of which are responsible for RELAPSING FEVER, YAWS and syphilis (see SEXUALLY TRANSMITTED DISEASES).

Spleen
Spongy vascular lymphoid organ (see LYMPH) between the STOMACH and DIAPHRAGM on the left side of the ABDOMEN. A center for the RETICULO-ENDOTHELIAL SYSTEM, it also eliminates worn-out red BLOOD cells, recycling their iron. Most of its functions are duplicated by other organs of the body.

Spodumene
A lithium-bearing PYROXENE, $LiAlSi_2O_6$; a major LITHIUM ore found in PEGMATITES, forming clear, often large monoclinic crystals, and when colored used as GEM stones.

Spondylosis
Degeneration in the structure of a vertebra. Although it is often blamed for aching and stiffness, most people over 30 have X-ray evidence of it without symptoms.

Sponges
Simple, sessile animals that feed by filtering particles of food from the water. Most are marine, but some live in fresh water. Water currents are created by flagella, and each cell within the sponge absorbs its food independently of the others. Sponges are in effect little more than assemblages of cells, some with specialized roles, such as the production of gametes, but making relatively few demands upon each other and with no coordinating or circulatory systems. An internal skeleton supports most sponges, and this may be made up of calcium carbonate (chalk), silica, or a protein, spongin. (◆ page 16)

Spontaneous combustion
COMBUSTION occurring without external ignition, caused by slow OXIDATION or FERMENTATION which (if heat cannot readily escape) raises the temperature to burning point. It may occur when hay or small coal is stored.

Spontaneous generation (abiogenesis)
A theory, dating from the writings of ARISTOTLE, that living creatures can arise from non-living matter. The idea remained current even after it had become clear that higher orders of life could not be created in this way, and it was an essential part of LAMARCK's theory of evolution: if evolution was powered by a drive for self-improvement in all organisms, then the continued existence of the "lower orders" could only be explained by spontaneous generation. It was only with the work of REDI, who showed that maggots did not appear in decaying meat to which flies had been denied access, and PASTEUR, who proved that the equivalent was true of microorganisms (ie BACTERIA), that the theory was finally discarded. Life cannot arise spontaneously on Earth today, but it could well have done so in Precambrian times, when conditions on Earth were very different; see LIFE, ORIGIN OF.

Spore
A small, often unicellular entity produced by plants, bacteria and fungi, from which a new organism or organisms can grow. The word is used to describe a great variety of disparate items. Most spores have a reproductive purpose, and are therefore generated in huge numbers. However, some bacteria produce spores with thick cell walls to resist adverse conditions; these are formed from a single cell, and later develop into a single cell, so they are not reproductive. Plant spores are generated by asexual means (see ALTERNATION OF GENERATIONS), as are bacterial spores, but some fungal spores originate from a process of sexual reproduction.

Sporophyte
Spore-producing phase of the life-cycle of a plant representing the diploid generation. (See ALTERNATION OF GENERATIONS.)

Sporozoa see PROTOZOA.

Sprain see LIGAMENT.

Spring
A naturally occurring flow of water from the ground. This may be, for example, an outflow from an underground stream; but most often a spring occurs where an AQUIFER saturated with GROUNDWATER intersects with the Earth's surface. Such an aquifer, if confined above and below by aquicludes, may travel for hundreds of kilometers underground before emerging to the surface, there, perhaps, in desert areas giving rise to OASES. Spring water is generally fairly clean, as it has been filtered through permeable rocks; but all spring water contains some dissolved MINERALS. (See also GEYSER; HOT SPRINGS; WELL.)

Squall
A sudden increase in WIND speed of 8m/s or more, raising the wind speed to at least 11m/s and lasting for one minute or longer. Commonly associated with thunderstorms and heavy rain, squalls may do great damage. A squall line is a line of thunderstorms often hundreds of kilometers long with squalls along its advancing edge.

Squint see STRABISMUS.

Stahl, Georg Ernst (1660-1734)
Bavarian-born German physician and chemist who developed the PHLOGISTON theory to explain combustion.

Stainless steel
Corrosion-resistant STEEL containing more than 10% chromium, little carbon, and often nickel and other metals. There are four main types: ferritic, martensitic, austenitic and precipitation-hardening. Stainless steel is used for cutlery and many industrial components.

Stalactites and stalagmites
Rocky structures found growing downward from

SOUND

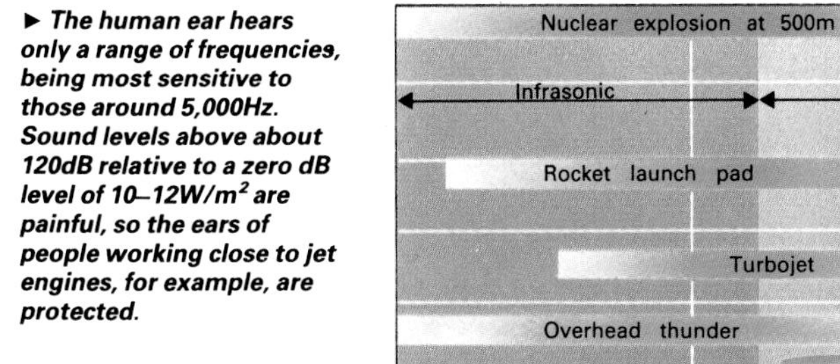

► *The human ear hears only a range of frequencies, being most sensitive to those around 5,000Hz. Sound levels above about 120dB relative to a zero dB level of 10–12W/m² are painful, so the ears of people working close to jet engines, for example, are protected.*

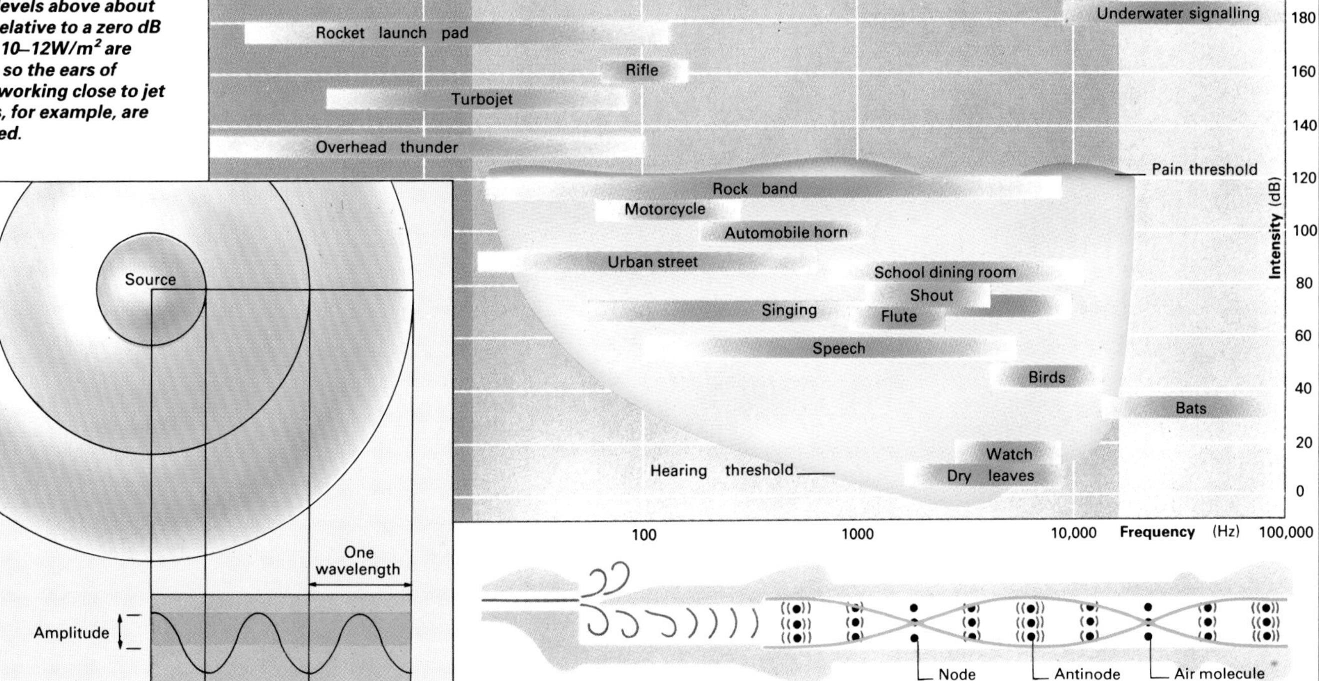

▲ *Sound waves spread out like ripples on a pond, but the ripples are variations in pressure that spread in three dimensions. "Crests" correspond to regions of increased pressure; "troughs" occur where the pressure is lower. Wavelength is the distance between crests; frequency the number of crests that pass a point each second.*

▼ *A familiar wave phenomenon of sound is the change in pitch of the noise from a passing aircraft. This is an example of the Doppler effect. As the source of the sound moves closer to the listener, each successive compression is emitted closer to the previous one. The wave arriving at the listener is thus squeezed together, so that its frequency appears higher as the source approaches. As soon as the source has passed, successive compressions are emitted at increasing intervals. The pitch of the sound drops.*

The Doppler effect

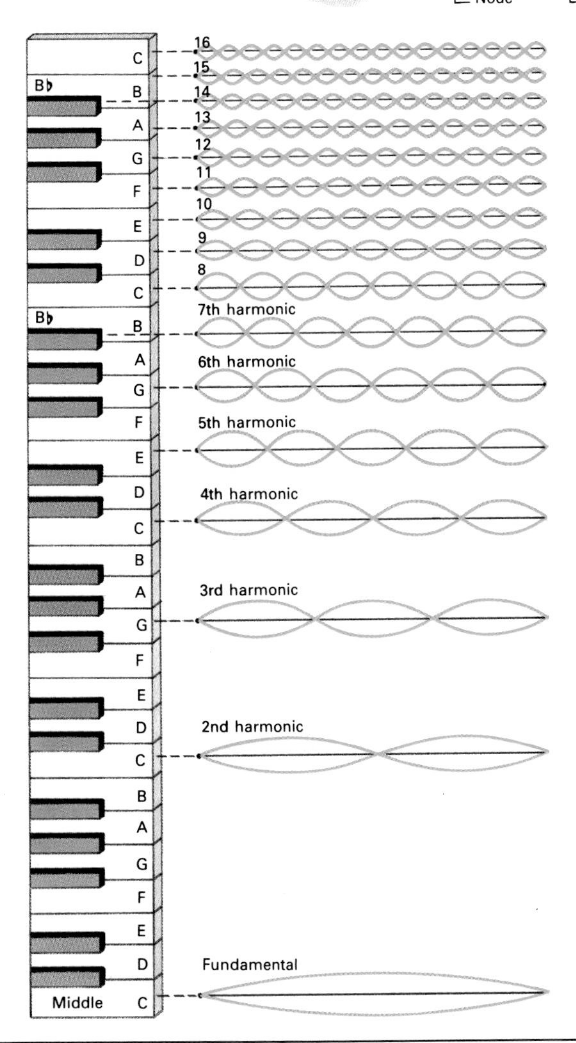

▲ ◄ *When a musician blows into a wind instrument such as a recorder (above), the air vibrates, setting up a "standing wave" in the pipe. This wave does not move along the tube, but consists of a stationary pattern of air moving by varying amounts. Positions where there is no movement are called nodes, while movement is greatest at the antinodes, for example at the ends of the pipe. In the simplest standing wave, one wavelength fits within the tube; this corresponds to the fundamental frequency of this note. Notes of higher fundamental frequency are made by shortening the tube – removing fingers covering holes along the tube. But each note contains overtones. These are weaker waves of higher frequency which also have antinodes at the open ends. Similar standing waves are set up when strings are plucked or struck, as in a piano (left). Here in the fundamental mode the ends of the string are held fixed, while the center vibrates. The profile of the vibrating string maps out half a wave pattern. The keyboard shows how notes of higher frequency correspond to the overtones, or harmonics, of the fundamental middle C.*

the roof (stalactites) and upward from the floor (stalagmites) of CAVES formed in LIMESTONE. Rainwater percolates through the rocks above the cave and, as it contains atmospheric carbon dioxide, can dissolve calcium carbonate en route. On reaching the cave, the water drips from the roof to the floor; as a drop hangs carbon dioxide is lost to the atmosphere, reducing the solubility and leaving a little calcium carbonate as CALCITE on the roof. Repetition forms a stalactite, and the impact of water falling on the floor deposits more calcite and forms a stalagmite. On occasion, the rising stalagmite and descending stalactite fuse to form a pillar.

Stamen see FLOWER.

Stammer see SPEECH AND SPEECH DISORDERS.

Standard time
The practice of defining the hour of the day throughout a specified geographical area time zone as being so many hours behind or ahead of an international standard, GREENWICH MEAN TIME (GMT). The US and Canada are covered by five such zones, the USSR by 11.

Stanford-Binet test
An adaptation of the Binet-Simon test for INTELLIGENCE, introduced by TERMAN (1916, revised 1937), and used primarily to determine the IQs of children. (See also BINET.)

Stanley, Wendell Meredith (1904-1971)
US biochemist who shared with J. NORTHROP and J. SUMNER the 1946 Nobel Prize for Chemistry for his first crystallization of a virus.

Staphylococcus
BACTERIUM responsible for numerous SKIN, soft tissue and BONE infections, less often causing SEPTICEMIA, PNEUMONIA, bacterial endocarditis and enterocolitis. BOILS, CARBUNCLES, IMPETIGO and OSTEOMYELITIS are all commonly due to staphylococci.

Star
A large incandescent ball of gases held together by its own gravity. The SUN is a fairly normal star in its composition, parameters and color. The lifespan of a star depends upon its mass and luminosity: a very luminous star may have a life of only one million years, the Sun a life of ten billion years, the faintest main sequence stars a life of ten thousand billion years. Stars are divided into two categories, Populations I and II. The stars in Population I are slower moving, generally to be found in the spiral arms of GALAXIES, and believed to be younger. Population II stars are faster moving and mainly to be found in the spheroidal halo of stars around a galaxy and in the GLOBULAR CLUSTERS. Compared to Population I stars they have a smaller proportion of metals in their composition. Many stars are DOUBLE STARS.

It is believed that stars originate as condensations out of INTERSTELLAR MATTER. In certain circumstances a protostar will form, slowly contracting under its own gravity, part of the energy from this contraction being radiated, the remainder heating up the core: this stage may last several million years. At last the core becomes hot enough for thermonuclear reactions (see FUSION, NUCLEAR) to be sustained, and stops contracting. Eventually the star as a whole ceases to contract, and radiates entirely by the thermonuclear conversion of hydrogen into helium: it is then said to be on the main sequence. When all the hydrogen in the core has been converted into helium, the now purely helium core begins to contract while the outer layers continue to "burn" hydrogen: this contraction heats up the core and forces the outer layers outward, so that the star as a whole expands for some 100-200 million years until it becomes a red giant star. Although the outer layers are comparatively cool, the core has become far hotter than before, and thermonuclear conversions of helium into carbon begin. The star contracts once more when its nuclear "fuels" become exhausted, and ends its life as a white dwarf star. It is thought that more massive stars become neutron stars, whose matter is so dense that its PROTONS and ELECTRONS are packed together to form NEUTRONS; were the Sun to become a neutron star, it would have a radius of less than 20km. Finally, when the star can no longer radiate through thermonuclear or gravitational means, it ceases to shine. Some high-mass stars may at this stage undergo ultimate gravitational collapse to form BLACK HOLES. (See also CARBON CYCLE; CEPHEID VARIABLES; CONSTELLATION; COSMOLOGY; GALACTIC CLUSTER; MAGELLANIC CLOUDS; MILKY WAY; NEBULA; NOVA; PULSAR; QUASAR; SOLAR SYSTEM; STAR CLUSTER; SUPERNOVA; UNIVERSE; VARIABLE STAR.) (➧ page 228-9, 240)

Starch
A CARBOHYDRATE consisting of chains of GLUCOSE arranged in one of two forms to give the polysaccharides amylose and amylopectin. Amylose consists of an unbranched chain of 200-500 glucose units, whereas amylopectin consists of chains of 20 glucose units joined by cross links to give a highly branched structure. Most natural starches are mixtures of amylose and amylopectin; eg potato and cereal starches are 20-30% amylose and 70-80% amylopectin. Starch is found in plants, occurring in grains scattered throughout the CYTOPLASM. The grains from any particular plant have a characteristic microscopic appearance, and an expert can tell the source of a starch by its appearance under the microscope. Starches in the form of rice, potatoes, wheat and other cereals supply about 70% of the world's food. (➧ page 40)

Star cluster
A cluster of stars sharing a common origin. There are two distinct types of star cluster: see GALACTIC CLUSTER; GLOBULAR CLUSTER.

Stark, Johannes (1874-1957)
Bavarian-born German physicist awarded the 1919 Nobel Prize for Physics for discovering the Stark effect (1913), the splitting of degenerate spectral lines through the application of a powerful ELECTRIC FIELD. The explanation of this was an early triumph of QUANTUM THEORY. (See also ZEEMAN, P.)

Starling, Ernest Henry (1866-1927)
British physiologist who worked with SIR WILLIAM MADDOCK BAYLISS on the study of hormones. He also did research into the heart, demonstrating the Starling equilibrium in 1896 – the balance between hydrostatic pressure, causing fluids to flow out of the capillary membrane, and osmotic pressure, causing the fluids to be absorbed from the tissues into the capillary.

Static
An accumulation of electric charge (see ELECTRICITY) responsible, eg, for the attractive and repulsive properties produced in many plastics and fabrics by rubbing. It leaks away gradually through warm damp air, but otherwise may cause small sparks (and consequent RADIO interference) or violent discharges such as LIGHTNING.

Statics
Branch of MECHANICS dealing with systems in EQUILIBRIUM, ie in which all FORCES are balanced and there is no motion.

Staudinger, Hermann (1881-1965)
German chemist awarded the 1953 Nobel Prize for Chemistry for showing that the long chains of MOLECULES known as POLYMERS are really giant molecules held together by normal chemical BONDS.

Steady State theory see COSMOLOGY.

Steam
The vapor formed from WATER at or above the boiling point (100°C). It is colorless, but appears white when it contains droplets of condensed water. Steam is used to make WATER GAS; it is a valuable industrial heat carrier because the latent heat of water is very high (40.65kJ/mol) and steam has a high specific heat: it is thus used in central heating and in pressure cookers. Because steam occupies about 1700 times the volume of the water producing it, when water is heated in a boiler pressure is built up, which is used to drive turbines and steam engines. Steam occurs naturally in springs, geysers and volcanoes.

Stearic acid ($CH_3(CH_2)_{16}COOH$)
White solid CARBOXYLIC ACID obtained from its ESTER, glyceryl tristearate (tristearin), which is found in many natural fats and oils. Sodium stearate is a principal constituent of many soaps.

Steatite see SOAPSTONE.

Steel
An alloy of iron and up to 1.7% carbon, with small amounts of manganese, phosphorus, sulfur and silicon. These are termed carbon steels; those with other metals are termed alloy steels; low-alloy steels if they have less than 5% of the alloying metal, high-alloy steels if more than 5%. Carbon steels are far stronger than iron, and their properties can be tailored to their uses by adjusting composition and treatment. Alloy steels – including STAINLESS STEEL – are used for their special properties. The main processes are the BESSEMER PROCESS, the Linz-Donawitz process and the similar electric-arc process used for making highest-quality steel, and the OPEN-HEARTH PROCESS.

Stefan-Boltzmann Law see BLACKBODY RADIATION.

Stegosaurs
A group of DINOSAURS, some of which grew to a length of about 5.5m, characterized by large bony plates on their backs. Like other ORNITHISHIANS, they were herbivorous and quadripedal.

Stein, William Howard (1911-1980)
US biochemist who shared with C.B. ANFINSEN and S. MOORE the 1972 Nobel Prize for Chemistry for his part in determining the structure of the ENZYME ribonuclease (see also NUCLEIC ACIDS).

Steinmetz, Charles Proteus (1865-1923)
Formerly Karl August Rudolf Steinmetz. German born US electrical engineer who first worked out the theory of HYSTERESIS (c1892), but who is best remembered for working out the theory of alternating current (1893 onward), so making it possible for AC to be used rather than DC in most applications.

Stem
The part of the PLANT that supports the LEAVES, FLOWERS and FRUITS. It may be short or creeping, as in low-growing plants, or tall as in TREES, or even underground (see RHIZOME). Stems also conduct nutrients, other substances and water between the various organs of the plant. Green stems carry out PHOTOSYNTHESIS.

Steno, Nicolaus or **Stensen, Niels** (1638-1686)
Danish geologist, anatomist and bishop. In 1669 he published the results of his geological studies: he recognized that many rocks are sedimentary (see SEDIMENTARY ROCKS); that fossils are the remains of once-living creatures and that they can be used for dating purposes, and established many of the tenets of modern crystallography.

Steppes
Extensive level GRASSLANDS of Europe and Asia (equivalent to the North American prairies and South American pampas). They extend from SW Siberia to the lower reaches of the River Danube.

Stereochemistry
The study of the arrangement in space of atoms in molecules, and of the properties that depend on such arrangements. The two chief branches are the study of STEREOISOMERS and stereospecific

reactions (which involve only one isomer); and CONFORMATIONAL ANALYSIS, including the study of steric effects on reaction rates and mechanisms.

Stereoisomers

Isomers having the same molecular structure, but differing in the spatial arrangement of their atoms. There are two main types.

OPTICAL ISOMERS are asymmetric molecules – usually having an asymmetric carbon atom with four different groups bonded to it – which hence have two mirror-image forms (enantiomers) and show OPTICAL ACTIVITY. Absolute spatial configurations have now been found for many isomers, and may be represented by projection formulae. Resolution, ie separation of the two enantiomers, is achieved by combining them with a single optical isomer, thus producing a pair of diastereoisomers which, not being mirror-images, have different properties and are separable. Inversion of configuration often occurs in substitution reactions.

GEOMETRICAL ISOMERS contain groups which are differently oriented with respect to a double bond or ring where rotation is impossible; they have different properties.

Stereoscope

Optical instrument that simulates binocular vision by presenting slightly different pictures to the two eyes so that an apparently three-dimensional image is produced. The simplest stereoscope, invented in the 1830s, used a system of mirrors and prisms (later, converging lenses) to view the pictures. In the color separation method the left image is printed or projected in red and seen through a red filter, and likewise for the right image in blue. A similar method uses images projected by polarized light and viewed through polarizing filters, the polarization axes being at right angles. The pictures are produced by a stereoscopic camera with two lenses a small distance apart. The stereoscope is useful in making relief maps by aerial photographic survey.

Sterility

The condition in which an organism is unable to produce offspring. Its many possible causes include failure to produce GAMETES, the production of abnormal gametes, the inability of males to introduce sperm into females, or inability of the embryo to develop normally. In humans, sterility develops naturally in old age.

Sterilization

Surgical procedure in which the FALLOPIAN TUBES are cut and tied to prevent eggs reaching the uterus, thus providing permanent CONTRACEPTION. (See also VASECTOMY.)

Stern, Otto (1888-1969)

German-born US physicist awarded the 1943 Nobel Prize for Physics for his development of the molecular-beam method of studying the magnetic properties of ATOMS, and especially for measuring the magnetic moment of the PROTON.

Steroid

An organic compound in which the four-ring carbon skeleton of the STEROLS makes up the central part of the molecule. Various side-chains may be attached to this sterol "nucleus". Steroids are widely distributed in living organisms and many act as hormones, including the insect hormone ecdysone (which controls molting), the various mammalian steroid hormones such as ESTROGENS and ANDROGENS (sex hormones), and the adrenocortical hormones, or corticosteroids, such as cortisol and aldosterone (which regulate glucose metabolism, and water and salt balance). Other steroids include the glycosides, which control the heart, the bile salts (breakdown products of cholesterol and other steroids), which aid the digestion of fats, and ALKALOIDS such as caffeine and nicotine, produced by plants.

In medicine, steroids are used for a variety of purposes. Sex hormones may be administered to correct deficiencies or as a means of contraception, while corticosteroids are often used to reduce inflammation or suppress the action of the immune system. The latter have adverse effects if taken for long periods. Many synthetic steroids – which have a slightly different action from that of the naturally occurring hormones – are available for medical use.

Sterols

Naturally occurring secondary ALCOHOLS with a fused ring structure of three six-membered carbon rings and one five-membered carbon ring, all of which are hydrogenated and contain in total one or more double bonds. Sterols are generally colorless, crystalline nonsaponifiable compounds. Important sterols include CHOLESTEROL, the major sterol found in most animals, and sitosterol, found in plants.

Stethoscope

Instrument devised by René T.H. Laennec (1781-1826) for listening to sounds within the body, especially those from the HEART, LUNGS, ABDOMEN and blood vessels.

Stevinus, Simon or **Stevin, Simon** (1548-1620)

Dutch mathematician and engineer who made many contributions to hydrostatics; disproved, before GALILEO, ARISTOTLE's theory that heavy bodies fall more swiftly than light ones; introduced the decimal system into popular use; and first used the triangle of forces in MECHANICS.

Stibnite

Soft, gray mineral, antimony(III) sulfide (Sb_2S_3), the chief ore of ANTIMONY, used in the manufacture of fireworks and matches; found in China, Czechoslovakia, Idaho, Nevada, and California.

Stigma see FLOWER.

Stimulant

DRUG that stimulates an organ. NERVOUS SYSTEM stimulants range from HALLUCINOGENIC DRUGS to drugs liable to induce CONVULSIONS. CARDIAC stimulants include digitalis and ADRENALINE and are used in cardiac failure and resuscitation respectively. Bowel stimulants have a laxative effect. Uterus stimulants (OXYTOCIN and ergometrine) are used in OBSTETRICS to induce labor and prevent postpartum HEMORRHAGE.

Stochastic process

Any process governed by the laws of PROBABILITY: for example, the BROWNIAN MOTION of submicroscopic particles or the decay of an atomic nucleus in RADIOACTIVITY. Most stochastics involve time: in the case of the particles, the state of the system at a time t is a random variable, $x(t)$.

Stoma

A surgically made connection between the intestine and the skin. When sections of the small and large intestine are removed surgically, usually because of cancer, the remaining intestine is connected with the outside of the body via a stoma. The feces produced through the stoma are collected in a disposable bag.

Stomach

The large distensible hopper of the DIGESTIVE SYSTEM. It receives food from the ESOPHAGUS and mixes it with hydrochloric acid and the stomach ENZYMES; fats are partially emulsified. After some time, the pyloric SPHINCTER at the base of the stomach relaxes and food enters the DUODENUM and the rest of the GASTROINTESTINAL TRACT. Diseases of the stomach include gastric ULCER, CANCER and pyloric stenosis, causing pain, anorexia or VOMITING. (◆◆ page 40, 111, 233)

Stomata

Tiny pores in the surface of leaves that are automatically opened and closed by a pair of guard cells. They allow exchange of gases and water vapor between the PLANT and the atmosphere.

Stoneworts (Charophyta)

Members of the green ALGAE, some species of which have a calcareous (chalky) outer covering to their cells. They also show an unusual growth-form, with whorls of cells arising at nodes on the main stem. Between these nodes the stem consists of a long coenocytic cell (one having many nuclei). Stoneworts are found in a variety of freshwater habitats. (◆ page 139)

Storm

A transient but often violent atmospheric disturbance with high WINDS, accompanied by SQUALLS, PRECIPITATION, and often thunder and lightning (see THUNDERSTORMS). Varying in type with latitude and season, storms are associated with CYCLONES. (See HURRICANE; TORNADO.)

Strabismus (cross-eye or squint)

A disorder of the EYES in which the alignment of the two ocular axes is not parallel, impairing binocular VISION; the eyes may diverge or converge. It is often congenital. Acquired squints are usually due to nerve or muscle disease and cause double vision or so-called lazy eye, when the brain learns to ignore the image from one eye because it conflicts with that from the other.

Strain see MATERIALS, STRENGTH OF.

Strain gauge

A device for measuring strain at the surface of a material by determining variations in the current through a piezoelectric device attached to it.

Strait

A narrow strip of sea joining two large areas of sea, possibly the result of marine EROSION of an ISTHMUS. (See also FIRTH.)

Strangeness number

Nonzero integral quantum number originally assigned to certain "strange" SUBATOMIC PARTICLES (K mesons and hyperons) because of their unusually long lifetimes. (Other particles have zero strangeness number.) It is now known to be a fundamental property of one type of QUARK – the strange quarks – so that strange particles all contain one or more strange quarks. Strangeness is conserved in strong nuclear interactions.

Strassmann, Fritz (1902-1980)

German physicist who worked on uranium fission with OTTO HAHN after LISE MEITNER had fled Germany (1938). All three shared the 1966 Fermi Award.

Stratigraphy

The branch of GEOLOGY concerned with the chronological sequence and the correlation of rock strata in different districts. (See also PALEONTOLOGY; ROCKS; SEDIMENTARY ROCKS.)

Stratosphere

The atmospheric zone immediately above the tropopause, including the OZONE layer. (See ATMOSPHERE.) (◆ page 25)

Strawberry mark

A type of BIRTHMARK.

Streptococcus

A group of disease-causing bacteria of the coccus shape (see COCCI).

Streptomycin see ANTIBIOTICS.

Stress see MATERIALS, STRENGTH OF.

Stroboscope

Instrument that produces regular brief flashes of intense light, used to study periodic motion, to test machinery and in high-speed photography. When the flash frequency exactly equals that of the rotation or vibration, the object is illuminated in the same position during each cycle, and appears stationary. A gas discharge lamp is used, with flash duration about 1μ and frequency from 2 to 3000Hz.

Stroke (cerebrovascular accident)

The sudden loss of some aspect of BRAIN function

STARS Classification and Evolution

In 1914 a Danish astronomer, Ejnar Hertzsprung (1873-1967), and an American, Henry Russell (1877-1957), published the first of a series of diagrams that revolutionized the study of stellar evolution. The two astronomers had discovered that there was a correlation between the brightness (absolute luminosity) and color (spectral class) – and hence surface temperature – of stars. Plotting these two values on a graph gives the diagram known as the Hertzsprung-Russell (or H-R) diagram. Drawing such a diagram for a large number of stars shows that most lie within a broad band of decreasing brightness and surface temperature. This band is known as the "main sequence". In their lifetimes most stars move across the main sequence, or on to it and back again the same way; once on they tend to stay there for a long time.

► The Hertzsprung-Russell diagram describes the relationship between stellar luminosity and temperature. The vertical scale shows luminosity (or an equivalent quantity such as absolute magnitude); the horizontal scale gives temperature (or an equivalent quantity such as spectral class). A star's luminosity is the amount of energy radiated from its entire surface in a second. A star's color is a guide to its temperature. Stars like the Sun emit at the middle of the visible spectrum and so appear yellowish. Cooler stars appear red, and hotter stars appear white or blue.

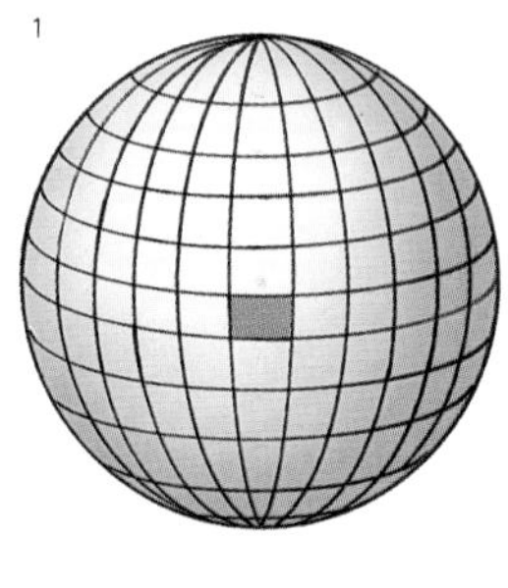

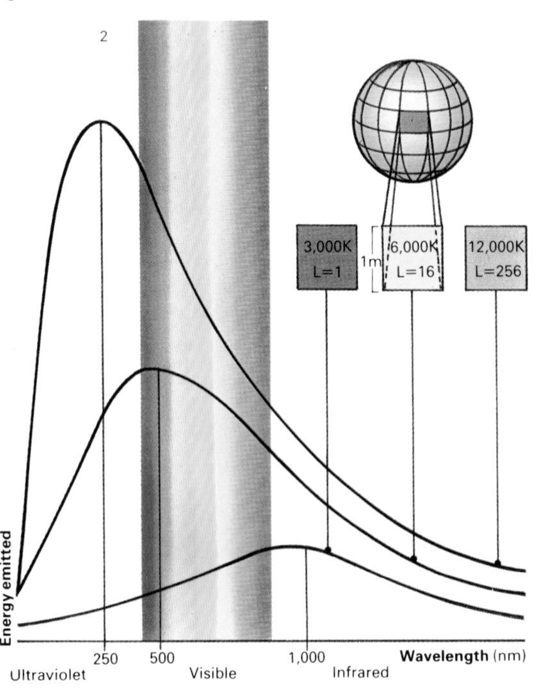

▲► If two stars have the same radius, the hotter one will be the more luminous of the two. If two stars of different sizes have the same temperature, the larger of the two will be the more luminous (L). The Stefan-Boltzmann law (1): the amount of energy emitted from equal areas depends on the power of their temperatures. The Wein law (2): the higher the temperature, the shorter the wavelength of the peak emission of radiation.

Stellar properties

Type of star	Diameter (Sun = 1)	Luminosity (Sun = 1)	Surface temperature (K)
Red Supergiant (MOI)	500	30,000	3,000
Red giant (K5III)	25	200	3,000
Main sequence stars:			
05	18	500,000	40,000
B0	7	20,000	28,000
A0	2.5	80	9,000
F0	1.35	6.3	7,400
G0	1.05	1.5	6,000
G2 (Sun)	1	1	5,800
K0	0.85	0.4	4,900
M0	0.63	0.06	3,500
M5	0.32	0.008	2,800
White dwarf	0.01	0.001	10,000
Neutron star	10^{-5}	–	10^{6}

◄► Not all stars lie on the main sequence. Some of low temperature but high luminosity lie above it and to the right. Known as red giants, they are cool red stars of great diameter. White dwarfs lie below it and to the left. Despite their temperatures – 10,000K or more – they have less than one thousandth of the Sun's luminosity. A typical white dwarf is comparable in size to the Earth and a million times denser than the Sun. Neutron stars are only 10-20km in diameter and are so dense that a teaspoonful of their material could weigh up to one billion tonnes. By contrast, a red supergiant's mean density is about one ten-thousandth of the density of air at sea-level.

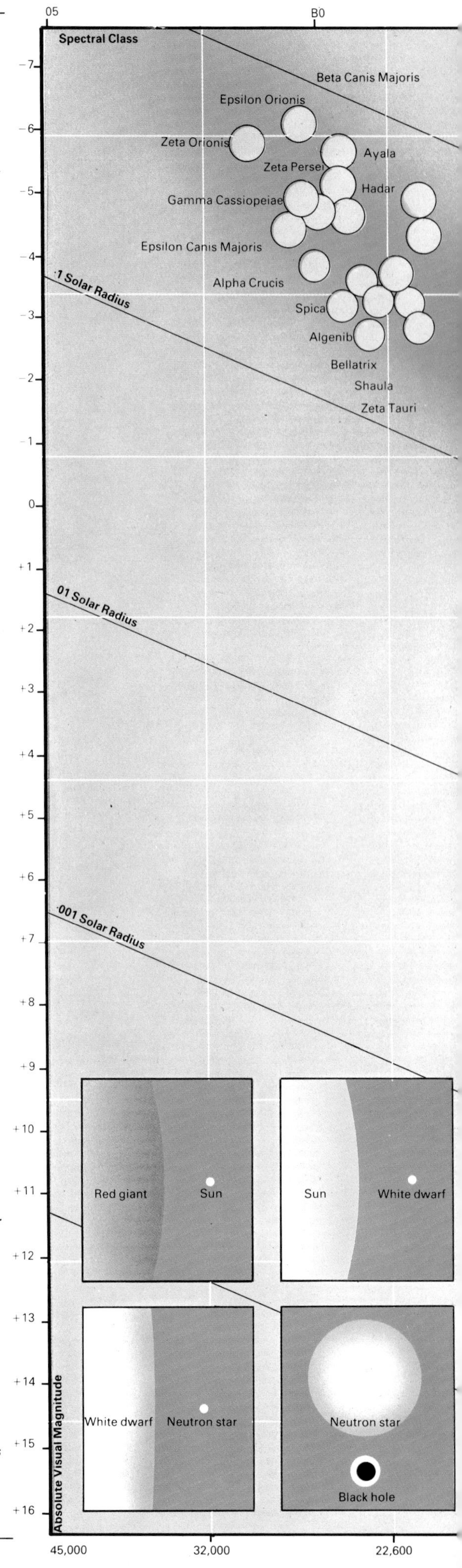

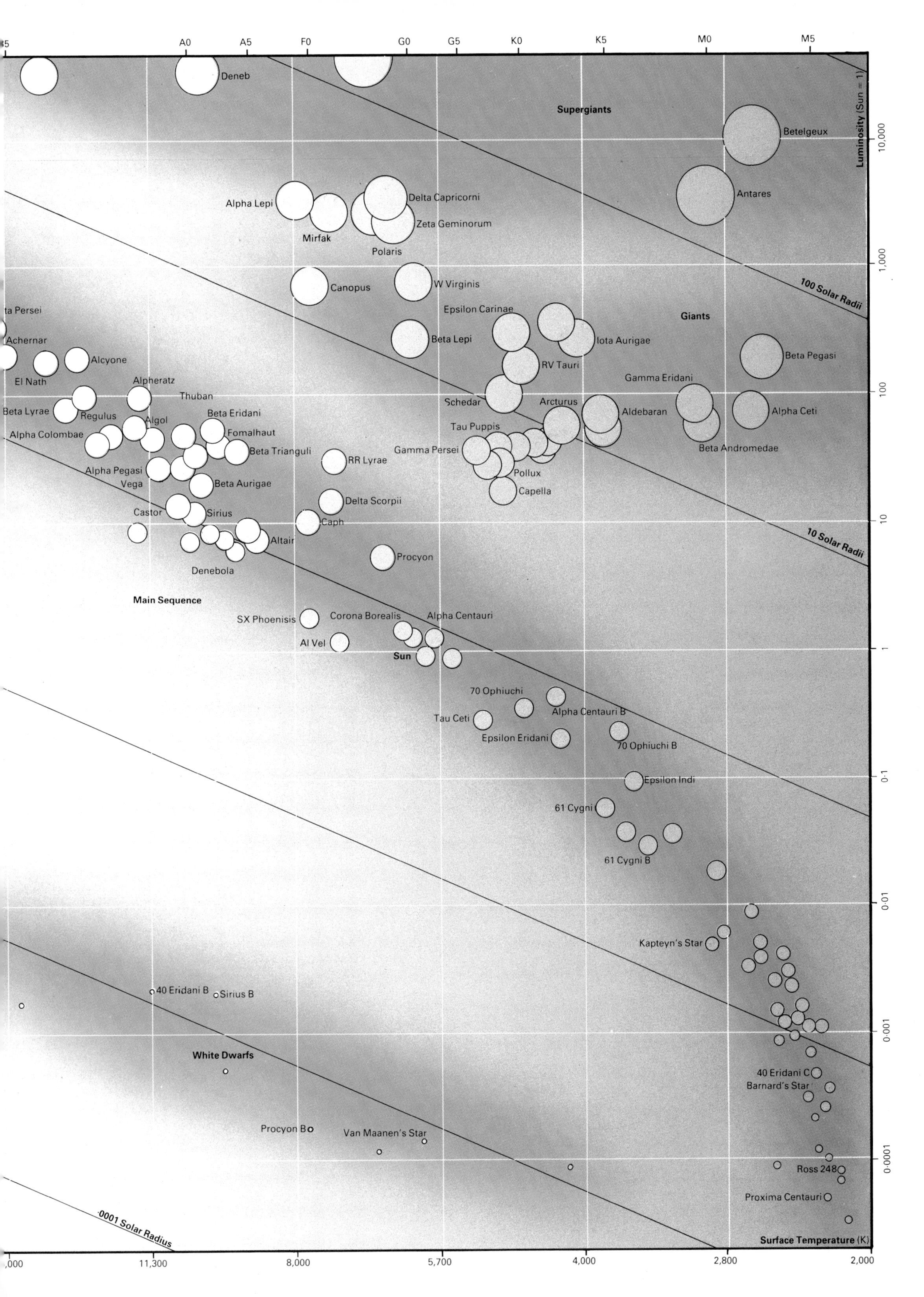

A0
A5
F0
G0
G5
K0
K5
M0
M5
Deneb
Supergiants
Betelgeux
Luminosity (Sun = 1)
10,000
Alpha Lepi
Delta Capricorni
Antares
Zeta Geminorum
Mirfak
Polaris
1,000
100 Solar Radii
Canopus
W Virginis
Epsilon Carinae
Giants
ta Persei
Beta Lepi
Iota Aurigae
Achernar
Alcyone
Beta Pegasi
RV Tauri
El Nath
Gamma Eridani
Alpheratz
Thuban
100
Schedar
Arcturus
Beta Lyrae
Regulus
Aldebaran
Alpha Ceti
Algol
Beta Eridani
Tau Puppis
Alpha Colombae
Fomalhaut
Gamma Persei
Beta Trianguli
Beta Andromedae
RR Lyrae
Alpha Pegasi
Pollux
Vega
Beta Aurigae
Capella
Delta Scorpii
Castor
Sirius
10
Caph
10 Solar Radii
Altair
Procyon
Denebola
Main Sequence
SX Phoenisis
Corona Borealis
Alpha Centauri
Al Vel
1
Sun
70 Ophiuchi
Alpha Centauri B
Tau Ceti
Epsilon Eridani
70 Ophiuchi B
Epsilon Indi
0·1
61 Cygni
61 Cygni B
0·01
Kapteyn's Star
40 Eridani B
Sirius B
0·001
White Dwarfs
40 Eridani C
Barnard's Star
Procyon B
Van Maanen's Star
0·0001
Ross 248
Proxima Centauri
·0001 Solar Radius
Surface Temperature (K)
,000
11,300
8,000
5,700
4,000
2,800
2,000

due to lack of BLOOD supply to a given area; control of limbs on one side of the body, APHASIA or dysphasia, loss of part of the visual field or disorders of higher function are common. Stroke may result from EMBOLISM, ATHEROSCLEROSIS and THROMBOSIS, or HEMORRHAGE. Areas with permanent loss of blood supply do not recover but other areas may to some extent take over their function.

Stromatolite
A hummock of limy mud with a concentric structure, formed in tropical intertidal areas when fine suspended particles are trapped by mats of blue-green algae. A mat of filamentous algae will entangle a layer of mud when it is covered at high tide. The algae will then regrow on the outside and trap yet another layer the next time the tide covers it. This eventually builds a dome or pedestal shape a meter or so in diameter. Today they are found on the west coast of Australia; fossil stromatolites are among the earliest known indications of life on Earth, some being 2900 million years old.

Strong nuclear force
The force that binds QUARKS together to form the larger SUBATOMIC PARTICLES called HADRONS. It is also responsible for binding protons and neutrons within the atomic NUCLEUS. (See also UNIFIED FIELD THEORY.) (♦ page 49)

Strontium (Sr)
Reactive, silvery white ALKALINE-EARTH METAL, occurring as strontianite ($SrCO_3$) and celestite ($SrSO_4$). Strontium is made by ELECTROLYSIS of the chloride or reduction of the oxide with aluminum. It resembles calcium physically and chemically. The radioactive isotope Sr^{90} is produced in nuclear FALLOUT, and is used in nuclear electric power generators. Strontium compounds are used in fireworks (imparting a crimson color), and to refine sugar. AW 87.6, mp 770°C, bp 1390°C. sg 2.54. (♦ page 130)

Strutt, John William see RAYLEIGH, BARON.

Struve, Otto (1897-1963)
Russian-born US astronomer known for work on stellar evolution (see STAR) and primarily for his contributions to astronomical SPECTROSCOPY, especially his discovery thereby of INTERSTELLAR MATTER (1938).

Stutter see SPEECH AND SPEECH DISORDERS.

Sty see BOIL.

Style see FLOWER.

Styrene (vinylbenzene)
Colorless liquid, an AROMATIC COMPOUND found in coal tar and essential oils, and made by dehydration of ethylbenzene. It is polymerized to make PLASTICS and RUBBERS, and especially polystyrene, a molding plastic which (expanded to a solid foam) is used for heat insulation. MW 104.2, mp −30.6°C, bp 145°C.

Subarachnoid hemorrhage
Bleeding under the ARACHNOID membrane, often following head injury, and leading to a harmful increase in intracranial pressure.

Subatomic particles
Small packets of matter-energy which are constituent of ATOMS or are produced in nuclear reactions or in interactions between other subatomic particles. The first such particle to be discovered was the (negative) ELECTRON (e^-), the constituent of CATHODE RAYS. Next, the nucleons were discovered; first the (positive) PROTON (p^+); then, in 1932, the (neutral) NEUTRON (n^o). The same year saw the discovery of the first antiparticle, the positron (or antielectron, $\bar{e}^+$ – see ANTIMATTER), and from that time the number of known subatomic particles, found in COSMIC RAYS or detected using particle ACCELERATORS, grew rapidly, until by the early 1970s about 100 were known, most of them highly unstable with very short HALF-LIVES, ranging from 10^{-8}s down to 10^{-23}s. A first division of particles classifies them according to whether they obey BOSE-EINSTEIN STATISTICS (bosons), or FERMI-DIRAC STATISTICS (fermions). Among the bosons, an important group are the gauge bosons, which transmit the four fundamental forces of physics. The familiar massless PHOTON carries the ELECTROMAGNETIC FORCE between electrically charged particles; the heavy intermediate vector bosons (the charged W and the neutral Z) are responsible for the WEAK NUCLEAR FORCE; massless gluons transmit the STRONG NUCLEAR FORCE; the as yet hypothetical graviton is the particle associated with GRAVITATION. The fermions can be divided into two groups: the hadrons, which are built from smaller particles called quarks, and the leptons, which are not built from quarks. Leptons, quarks and gauge bosons are, as far as scientists can tell, truly elementary particles, with no structure of their own. The leptons are the electron, the muon, the tau and the three types of neutrino. They do not feel the strong nuclear force because they do not contain quarks. The neutrinos have little or no rest mass, and are produced in various decay processes. The charged muon and tau seem to behave simply as heavier versions of the electron. The hadrons, including the mesons, nucleons and hyperons, interact via the strong force, which holds the atomic nucleus together in spite of the mutual repulsion of its constituent protons. Boson hadrons are known as mesons; they are built from a quark combined with an antiquark. The mesons include the pions (pi-mesons) and the heavier kaons (K-mesons). Fermion hadrons are known as baryons; they are built from combinations of three quarks. They include the nucleons (protons and neutrons) and the heavier hyperons. To explain the great variety of observed hadrons, the quarks must occur in at least six varieties, known as up, down, strange, charm, bottom and top. Only up and down quarks are necessary to build the familiar nucleons and the pions, but the others are required for the many short-lived particles found in COSMIC RAYS and at accelerators. The quarks are held together by passing gluons between themselves, in processes that are described by the theory known as quantum chromodynamics. (♦ page 49)

Subconscious
The area between CONSCIOUSNESS and UNCONSCIOUSNESS; in PSYCHOANALYSIS, a rarely used synonym for the UNCONSCIOUS.

Sublimation
In psychoanalysis, process whereby energies derived from instinctive drives, particularly the sexual and aggressive, are channeled into noninstinctive behavior, through inhibition or otherwise.

Sublimation
Transformation of a substance from the solid to the vapor state without its becoming liquid. All solids will sublime below their triple point (at which solid, liquid and vapor are all in EQUILIBRIUM), but in only a few cases – including DRY ICE, IODINE, NAPHTHALENE and SULFUR – is this at a high enough temperature and rate to be useful for purification. Freeze-drying (see DEHYDRATION) is by sublimation.

Subliminal perception see SUGGESTION.

Submarine canyon
Suboceanic canyon cutting across the CONTINENTAL SHELF, sometimes the continental slope. They are thought to be of comparatively recent origin.

Succession
The progressive changes in a plant population during the development of vegetation on a bare rock or soil surface, such as occurs in an abandoned field or following a volcanic eruption or a landslide. The land is colonized by opportunistic annuals and perennials ("weeds"), then is progressively invaded by woody species until a CLIMAX VEGETATION is achieved. This is commonly woodland. The term sere is often used to describe the sequence of plant communities; for example, lithosere describes the succession that starts on bare rock; hydrosere, one that starts with water.

Succulents
Plants that have swollen leaves or stems and are thus adapted to living in arid regions. Cacti are the most familiar, but representatives occur in other families, notably the Crassulaceae (stonecrops and houseleeks) and Aizoaceae (living stones, mesembryanthemum). Many succulents have colorful, though often short-lived, flowers. Succulents are just one type of XEROPHYTE.

Sucrose (cane sugar, $C_{12}H_{22}O_{11}$)
Disaccharide CARBOHYDRATE, commercially obtained from sugar beet, sugar cane and sweet sorghum. As table sugar, sucrose is the most important of the SUGARS. It comprises a GLUCOSE unit joined to a FRUCTOSE unit. Sucrose, glucose and fructose all exhibit OPTICAL ACTIVITY, and when sucrose is hydrolyzed the rotation changes from right to left. This is called inversion, and an equimolar mixture of glucose and fructose is called invert sugar. (♦ page 40)

Suffocation see ASPHYXIA.

Sugars
Sweet, soluble CARBOHYDRATES (of general formula $C_x(H_2O)_y$, comprising the monosaccharides and the disaccharides. Monosaccharides cannot be further degraded by HYDROLYSIS and contain a single chain of CARBON atoms. They normally have the suffix -ose and a prefix indicating the length of the carbon chain; thus trioses, tetroses, pentoses, hexoses and heptoses contain 3, 4, 5, 6 and 7 carbon atoms respectively. The most abundant natural monosaccharides are the hexoses, $C_6H_{12}O_6$ (including GLUCOSE), and the pentoses, $C_5H_{10}O_5$ (including xylose). Many different isomers of these sugars are possible, and often have names reflecting their source, or a property, eg FRUCTOSE is found in fruit, arabinose in gum arabic and the pentose, xylose, in wood. Disaccharides contain two monosaccharide units joined by an oxygen bridge. Their chemical and physical properties are similar to those of monosaccharides. The most important disaccharides are SUCROSE (cane sugar), LACTOSE and MALTOSE. (Table sugar consists of sucrose.) The most characteristic property of sugars is their sweetness. If we accord sucrose an arbitrary sweetness of 100, then glucose scores 74, fructose 173, lactose 16, maltose 33, xylose 40 (compare SACCHARIN 55,000). (♦ page 40)

Suggestion
Process whereby an individual loses his critical faculties and thus accepts IDEAS and beliefs that may be contrary to his own. People under HYPNOSIS are particularly suggestible (see also BRAINWASHING), as are those in a state of exhaustion.

Sulfates
Salts of SULFURIC ACID, containing the sulfate ion (SO_4^{2-}); formed by reaction of the acid with metals, their oxides or carbonates, or by oxidation of SULFIDES or sulfites (see SULFUR). Most sulfates are soluble in water, the main exceptions being calcium, strontium, barium and lead sulfates. They decompose at high temperatures to give sulfur trioxide and dioxide. Sulfates form LIGAND complexes and double salts (see ALUM). Many sulfate minerals occur in nature, often as evaporites or from oxidation of SULFIDES (See ANHYDRITE; BARITE; EPSOM SALTS; GYPSUM). Bisulfates contain the ion HSO_4^-; they are acid, and are converted to pyrosulfates ($S_2O_7^{2-}$) on heating. ESTERS of sulfuric acid, $(RO)_2SO_2$, are also called sulfates.

Sulfides
Binary compounds of SULFUR. (For organic sulfides see THIOETHERS.) Nonmetal sulfides, formed by direct synthesis, include CARBON disulfide and several sulfides of nitrogen and phosphorus. Metal sulfides (S^{2-}) are mostly insoluble (except those of the ALKALI METALS and ALKALINE-EARTH METALS), and are prepared by precipitation with hydrogen sulfide. Soluble sulfides are readily hydrolyzed to the soluble bisulfides (HS^-), and are used as reducing agents and in making dyes and pesticides. Many sulfide minerals are important ores: see ARGENTITE; ARSENOPYRITE; BORNITE; CHALCOCITE; CHALCOPYRITE; CINNABAR; REALGAR; SPHALERITE; STIBNITE; WURTZITE.
HYDROGEN SULFIDE (H_2S) is a colorless, highly toxic gas with a foul odor of rotten eggs, occurring in volcanoes; a covalent HYDRIDE. It is obtained industrially as a byproduct of petroleum refining, and prepared in the laboratory by reacting a sulfide with an acid. Hydrogen sulfide burns to give sulfur dioxide and water; in aqueous solution it is a very weak acid, forming sulfides with most metal salts. It is a good reducing agent. With sulfur, hydrogen sulfide gives hydrogen polysulfides (H_2S_n, n=2-9), whose salts are also known. mp −85.5°C, bp −60.3°C.
nSulfonic acids
Organic compounds with the general formula RSO_2OH; strong, water-soluble ACIDS often used as their sodium salts, sulfonates. AROMATIC sulfonic acids are made by sulfonation with fuming SULFURIC ACID. They are used to make detergents (see SOAPS AND DETERGENTS), DYES, SULFA DRUGS and ION-EXCHANGE resins. They are useful in synthesis, since the sulfonate group is readily replaced.
Sulfur (S)
Nonmetal in Group VIA of the PERIODIC TABLE. It is recovered from deposits and from natural gas and petroleum. Combined sulfur occurs as SULFATES and SULFIDES. There are two main allotropes of sulfur (see ALLOTROPY): the yellow, brittle rhombic form is stable up to 95.6°C, above which monoclinic sulfur (almost colorless) is stable. Both forms are soluble in carbon disulfide; they consist of eight-membered rings (S_8). Plastic sulfur is an amorphous form made by suddenly cooling molten sulfur. Sulfur is reactive, combining with most other elements. It is used in gunpowder, matches, as a fungicide and insecticide, and to vulcanize rubber. AW 32.1, mp 115°C (rh), 119°C (mono), bp 445°C, sg 2.07 (rh, 20°C). (◆ page 37, 130)
SULFUR DIOXIDE (SO_2) is a colorless, acrid gas, formed by combustion of sulfur. It is an oxidizing and reducing agent and is important as an intermediate in the manufacture of sulfur trioxide and SULFURIC ACID. It is also used in petroleum refining and as a disinfectant, food preservative and bleach. It reacts with water to give sulfurous acid (H_2SO_3), which is corrosive. Thus sulfur dioxide in flue gases is a harmful cause of POLLUTION. mp −73°C, bp −10°C.
SULFITES are salts containing the ion SO_3^{2-}, formed from sulfur dioxide and BASES; readily oxidized to SULFATES. Bisulfites are acid sulfites, containing the ion HSO_3^-.
SULFUR TRIOXIDE (SO_3) is a volatile liquid or solid formed by oxidation of sulfur dioxide (see CONTACT PROCESS). It reacts violently with water to give SULFURIC ACID. mp 17°C (α), bp 43°C (α).
THIOSULFATES are salts containing the ion $S_2O_3^{2-}$, usually prepared by dissolving sulfur in an aqueous sulfite solution. They are mild reducing agents, and form LIGAND complexes; in acid solution they decompose to give sulfur and sulfur dioxide. (For sodium thiosulfate, see SODIUM.)
Sulfuric acid (H_2SO_4)
An oily, colorless liquid, made in large quantities by the CONTACT PROCESS. It is an oxidizing agent, reacting with metals, sulfur and carbon on heating, and is a powerful dehydrating agent (see DEHYDRATION). In aqueous solution it is a strong ACID, and reacts with BASES and most metals to give SULFATES. Fuming sulfuric acid or oleum is 100% sulfuric acid containing dissolved sulfur trioxide; it is used to make SULFONIC ACIDS. Sulfuric acid is used to make fertilizers, pigments, explosives, dyes and detergents, to refine petroleum and coal tar, and in lead-acid storage batteries. mp 10°C, bp 33.7°C (98%). (◆ page 118)
Sullivan, Harry Stack (1892-1949)
US psychiatrist who made important contributions to SCHIZOPHRENIA studies and originated the idea that PSYCHIATRY depends on study of interpersonal relations (including that between therapist and patient).
Sulphide see SULFIDE.
Sulphur see SULFUR.
Sulphuric acid see SULFURIC ACID.
Sumner, James Batcheller (1887-1955)
US biochemist who shared with J.H. NORTHROP and W.M. STANLEY the 1946 Nobel Prize for Chemistry for his first crystallization of an ENZYME (urease, 1926), showing it was a PROTEIN.
Sun
The star about which the Earth and the other planets of the SOLAR SYSTEM revolve. The Sun is an incandescent ball of gases, by mass 69.5% hydrogen, 28% helium, 2.5% carbon, nitrogen, oxygen, sulfur, silicon, iron and magnesium altogether, and traces of other elements. It has a diameter of about 1393Mm, and rotates more rapidly at the equator (24.65 days) than at the poles (about 34 days). Although the Sun is entirely gaseous, its distance creates the optical illusion that it has a surface: this visible edge is called the PHOTOSPHERE. It is at a temperature of about 6000K, cool compared to the center of the Sun (15,000,000K) or the corona (1,000,000K); the photospheres of other stars may be at temperatures of less than 2000K or more than 500,000K. Above the photosphere lies the chromosphere, an irregular layer of gases between 1.5Mm and 15Mm in depth. It is in the chromosphere that FLARES and prominences occur: these last are great plumes of gas that surge out into the corona and occasionally off into space. The corona is the sparse outer atmosphere of the Sun. During solar ECLIPSES it may be seen to extend several thousand megameters and as bright as the full Moon, though in effect it extends beyond the orbit of NEPTUNE. The Earth lies within the corona, which at this distance from the Sun is termed the SOLAR WIND. The Sun is a very normal STAR, common in its characteristics though rather smaller than average. It lies in one of the spiral arms of the MILKY WAY. (◆◆ page 221, 228-9, 240)
Sunburn
Burning effect on the SKIN following prolonged exposure to ULTRAVIOLET RADIATION from the Sun, common in travelers from temperate zones to hot climates. First-degree BURNS may occur, but usually only a delayed ERYTHEMA is seen, with extreme skin sensitivity. Systemic disturbance occurs in severe cases. Fair-skinned persons are most susceptible.
Sunspots
Apparently dark spots visible on the face of the SUN, regions about 2000K cooler than the adjacent photosphere that appear dark merely by contrast with their surroundings. Single spots are known, but mostly they form in groups or pairs. They are never seen at the Sun's poles. They are associated with concentrated localized magnetic fields, but their precise cause is not fully understood. Their prevalence reaches a maximum about every 11 years.
Sunstroke (heatstroke)
Rise in body temperature and failure of sweating in hot climates, often following exertion. DELIRIUM, COMA and CONVULSIONS may suddenly develop.
Superconductivity
A condition occurring in many metals, alloys, etc, usually at low temperatures, involving zero electrical RESISTANCE and perfect DIAMAGNETISM. In such a material an electric current will persist indefinitely without any driving voltage, and applied MAGNETIC FIELDS are exactly canceled out by the magnetization they produce. In type I superconductors both these properties disappear abruptly when the temperature or applied magnetic field exceed critical values (typically 5K and 10^4A/m), but in type II superconductors the diamagnetism decay is spread over a range of field values. Large ELECTROMAGNETS sometimes use superconducting coils, which will carry large currents without overheating; the exclusion of fields by superconducting materials can be exploited to screen or direct magnetic fields. Superconductivity was discovered by H. KAMERLINGH-ONNES in 1911, and is due to an indirect interaction of pairs of ELECTRONS via local elastic deformations of the metal CRYSTAL.
Supercooling and superheating
A liquid cooled below its FREEZING POINT without the solid phase separating out is in a metastable supercooled state. Addition of a small amount of solid or shaking may cause the liquid to freeze. A liquid heated above its BOILING POINT or a saturated vapor heated after all traces of liquid have evaporated is superheated. This state is also metastable.
Superego
According to FREUD, the third and last part of the psychic apparatus to develop, a part of the EGO containing self-criticism, inhibitions, etc. Unlike the conscience, its strictures may date from an earlier stage of the individual's development, even clashing with his current values.
Superfluidity
The property whereby "superfluids" such as liquid HELIUM below 2.186K exhibit apparently frictionless flow. The effect requires QUANTUM MECHANICS for its explanation.
Supergiant star see STAR.
Superiority complex
Overevaluation by an individual of his abilities, usually a DEFENSE MECHANISM countering an INFERIORITY COMPLEX.
Supernova
A catastrophic stellar explosion in which a substantial fraction of the star's mass is blown out into space, or in some cases in which the star is completely destroyed. During the event, the star's luminosity may increase by a factor of millions or even billions, so that it may shine briefly as brightly as several billion suns. There are two principal types of supernova, Type I and Type II. Type I supernovae are typically about ten times more luminous than Type II; Type II supernovae show hydrogen lines in their spectra, while Type I supernovae do not. A Type I event is widely believed to occur in a binary system within which one star has dumped sufficient material onto the surface of a WHITE DWARF companion to cause it to be squeezed and heated gravitationally to such an extent that explosive thermonuclear reactions blow the white dwarf to pieces. A Type II event is believed to involve a massive Population I star that has exhausted the nuclear fuel in its core. The core is thought to collapse to form a NEUTRON STAR; the rest of the star (the envelope) is blasted into space to form an expanding supernova remnant

nant such as the CRAB NEBULA. Supernovae are rare events; on average only about three supernovae per century occur in a typical galaxy. The most recent supernova in the Milky Way Galaxy occurred in 1604. The most recent naked-eye supernova occurred in the Large MAGELLANIC CLOUD in 1987.

Superphosphate see PHOSPHATES.

Superposition, Principle of
Law of STRATIGRAPHY, first stated by WILLIAM SMITH, that when SEDIMENTARY ROCK strata are undisturbed the younger ROCKS lie above the older.

Supersaturation see SATURATION.

Supersonics
The study of fluid flow at velocities greater than that of SOUND, usually with reference to the supersonic flight of AIRPLANES and MISSILES when the relative velocity of the solid object and the air is greater than the local velocity of sound propagation. (♦ page 225)

Suprarenal glands see ADRENAL GLANDS.

Surface tension
FORCE existing in any boundary surface of a liquid such that the surface tends to assume the minimum possible area. Surface tension arises from the cohesive forces between liquid molecules and makes a liquid surface behave as if it had an elastic membrane stretched over it. Thus, the weight of a needle floated on water makes a depression in the surface. Surface tension governs the wetting properties of liquids, CAPILLARITY and detergent action.

Surfactant
A type of molecule which reduces the interfacial tension of an aqueous solution, usually by having both a polar and a non-polar region in its structure. They are used as detergents, emulsifiers and wetting agents.

Surgery
The branch of MEDICINE chiefly concerned with manual operations to remove or repair diseased, damaged or deformed body tissues. Otolaryngology deals with the EAR, LARYNX (voicebox) and upper respiratory tract: tonsillectomy is one of its most common operations. Colon and rectal surgery deals with the large intestine. Urological surgery deals with the urinary system (KIDNEYS, ureters, BLADDER, urethra) and male reproductive system. Neurosurgery deals with the NERVOUS SYSTEM (BRAIN, SPINAL CORD, nerves). Thoracic surgery deals with structures within the chest cavity.

Surrogate mother
A woman who carries the baby of another woman, after a fertilized egg is implanted into her. The process causes legal and ethical problems as there is often dispute about who should keep the child after it is born. (See IN-VITRO FERTILIZATION.)

Surveying
The accurate measurement of distances and features on the Earth's surface. The science began to attain modern accuracy in the 17th century with the introduction of Gunter's chain (1620) – 66ft (20.3m) long and the standard for measuring distance until superseded by the steel or INVAR tape – and of the VERNIER SCALE, the telescopic sight and the spirit level. For making MAPS and charts, the latitude and longitude of certain primary points are determined from astronomical observations. Geodetic surveying, for large areas, takes the Earth's curvature into account (see GEODESY). Much modern surveying is done by PHOTOGRAMMETRY, using the STEREOSCOPE to determine contours. Surveying is also important to chart land, to fix boundaries and to plan transportation routes, dams, etc.

Survival of the fittest
Term first used by HERBERT SPENCER in his *Principles of Biology* (1864) and adopted by CHARLES DARWIN to describe his theory of EVOLUTION by NATURAL SELECTION. (♦ page 16)

Suspension
System of macroscopic particles dispersed in a fluid in which settlement is hindered by intermolecular collisions and by the fluid's VISCOSITY. (See also COLLOID.)

Sutherland, Earl Wilbur, Jr. (1915-1974)
US physiologist awarded the 1971 Nobel Prize for Physiology or Medicine for demonstrating the role of cyclic adenosine 3′,5′-monophosphate in the way that HORMONES affect bodily organs.

Svedberg, Theodor (1884-1971)
Swedish chemist awarded the 1926 Nobel Prize for Chemistry for inventing the ultracentrifuge, important in studies of COLLOIDS and large MOLECULES. (See also CENTRIFUGE.)

Swammerdam, Jan (1637-1680)
Dutch microscopist whose precision enabled him to make many discoveries, including red BLOOD cells (before 1658).

Swamp
A poorly drained, low-lying area of land permanently saturated with water. Swamps usually develop where the surface is flat enough for rainwater runoff to be very slow, or where a lake basin has become filled in; vegetation helps to retain the swampiness. Marshes have standing surface water, and are usually only temporary. (See also BOG; MUSKEG.)

Sweat
Fluid secreted by the glands of the human skin as a method of cooling the body by latent heat of evaporation. Sweat also contains salts; excessive sweating may deplete the body of these. Sweating is the main method of cooling of hairless animals such as humans and pigs.

Sweetening agents
Substances used to sweeten food and drink (see TASTE). The commonest are the SUGARS, especially SUCROSE (table sugar) and GLUCOSE, which are themselves FOODS. Artificial sweeteners, with no food value and up to several thousand times sweeter than sugar, are used by diabetics and in DIETETIC FOODS, toothpaste, etc. They include ASPARTAME, CYCLAMATE and SACCHARIN. Some have been banned because of possible harmful effects.

Swimbladder
Organ found in modern TELEOST fish that has evolved from the lungs of earlier forms. Control of the gas pressure inside the bladder enables the fish to remain buoyant in water at any depth. In other, less advanced groups of ray-finned fish (ACTINOPTERYGII) the swimbladder can also act as a lung, as in the bowfin, a holostean, and *Polypterus*, a chondrostean. In the DIPNOI, or lungfish, the lungs have retained their original function.

Sydenham, Thomas (1624-1689)
"The English Hippocrates", who pioneered the use of QUININE for treating MALARIA and of laudanum as an anesthetic, wrote an important treatise on gout, and first described Sydenham's CHOREA (St Vitus' Dance).

Sylvite
HALIDE mineral, POTASSIUM chloride (KCl); white, cubic crystals. Occurs in West Germany and New Mexico in EVAPORITE deposits.

Symbiosis
A close relationship between two organisms of different species. It may be a relationship where both benefit (MUTUALISM), where one benefits without harming the other (COMMENSALISM), or where one lives inside the other, but without drawing nutrients from it or damaging it (inquilinism). Relationships where one draws nutrients directly from the other (PARASITISM) are sometimes also included.

Sympathetic nerves see NERVOUS SYSTEM.

Synapse
The point of connection between two NEURONS (nerve cells) or between a nerve and a muscle. An electrical nerve impulse, arriving at the end of the AXON in the pre-synaptic cell, releases a chemical transmitter (a NEUROTRANSMITTER). This crosses a small gap, the synaptic cleft, and affects ion channels on the dendrites and cell body of the post-synaptic cell. Its effect will be either to inhibit stimulation of the post-synaptic cell, or to excite it. The sum of all the effects from the various synapses impinging on the cell body determines whether or not it fires a nerve impulse. (♦ page 126)

Synchrocyclotron
Type of CYCLOTRON where the accelerating field is frequency-modulated to give charged particles, such as PROTONS, increased energies as their mass increases due to the effects of RELATIVITY. Pulses of particles start at the center of the accelerator under the influence of a magnetic field and move in circles of increasing radius as they are accelerated.

Synchrotron
Particle accelerator for accelerating pulses of PROTONS, deuterons or ELECTRONS to high energies. The particles move in a circular path of almost constant radius, guided by a ring of electromagnets while the particles are accelerated by an alternating electric field.

Syncline see FOLD.

Syncope
Another word for FAINTING.

Synergism
The working together of two or more agencies (eg synergistic muscles) to greater effect than both would have working independently.

Synge, Richard Laurence Millington (b. 1914)
British scientist who was awarded the 1952 Nobel Prize for chemistry with A.J.P. MARTIN for developing paper CHROMATOGRAPHY, a tool of considerable importance in modern biochemical research.

Synovial fluid
The small amount of fluid that lubricates JOINTS and the synovial sheaths of TENDONS.

Synthesis, Chemical
The process by which chemical compounds are made, often by multistage reactions, from readily available starting materials.

Synthesis gas
A mixture of carbon monoxide and hydrogen, used in the manufacture of ammonia, methanol and other organic molecules. It can be made by the reaction between coke (carbon) and steam (water) or by steam cracking of natural gas.

Syphilis see SEXUALLY TRANSMITTED DISEASES.

Systemic lupus erythematosus (SLE)
A rare disorder mainly affecting young women and causing a characteristic SKIN rash with butterfly distribution over the face. It also causes LUNG, KIDNEY and BLOOD disease, largely due to abnormal IMMUNITY directed against substances in cell nuclei.

Systole see BLOOD PRESSURE.

Szent-Györgyi von Nagyrapolt, Albert (1893-1986)
Hungarian-born US biochemist awarded the 1937 Nobel Prize for Physiology or Medicine for work on biological COMBUSTION processes, especially in relation to VITAMIN C.

Szilard, Leo (1898-1964)
Hungarian-born US physicist largely responsible, with FERMI, for the development of the US atom bomb (see MANHATTAN PROJECT). In 1945 he was a leader of the movement against using it. Later he made contributions in the field of molecular biology.

STORAGE AND EXCRETION

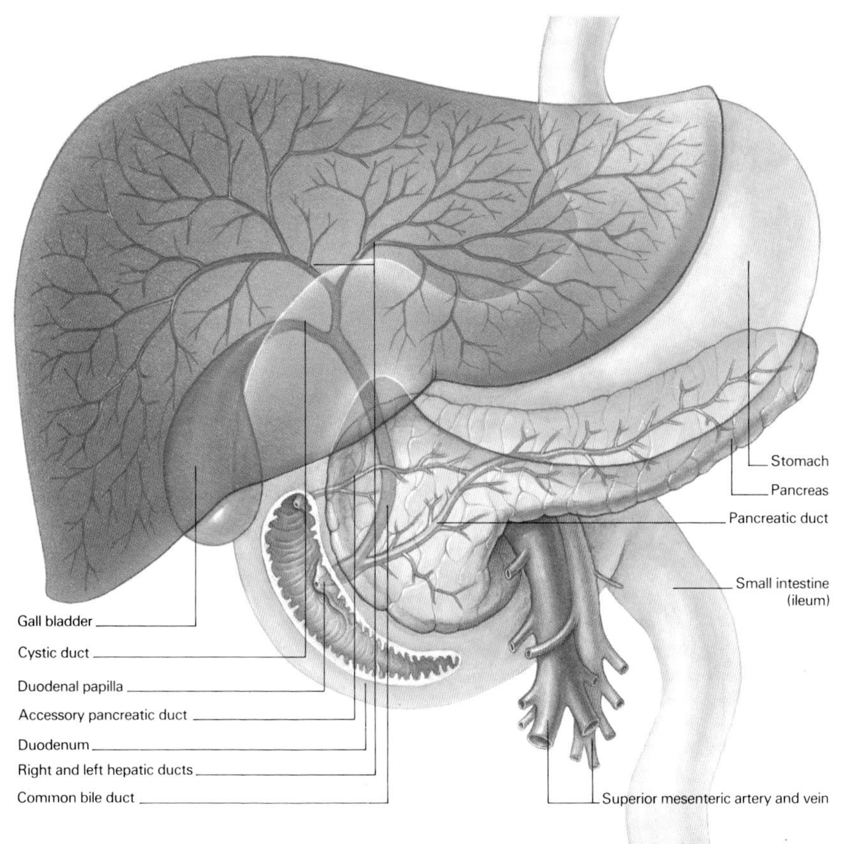

Liver

◄ ***Among the liver's many metabolic functions are: breaking down old blood cells; producing bile salts; storing fat-soluble vitamins; making blood-clotting substances; breaking down poisons; converting fats and proteins into carbohydrate; turning ammonia into urea; and milk sugar into glucose.***

The liver's blood supply

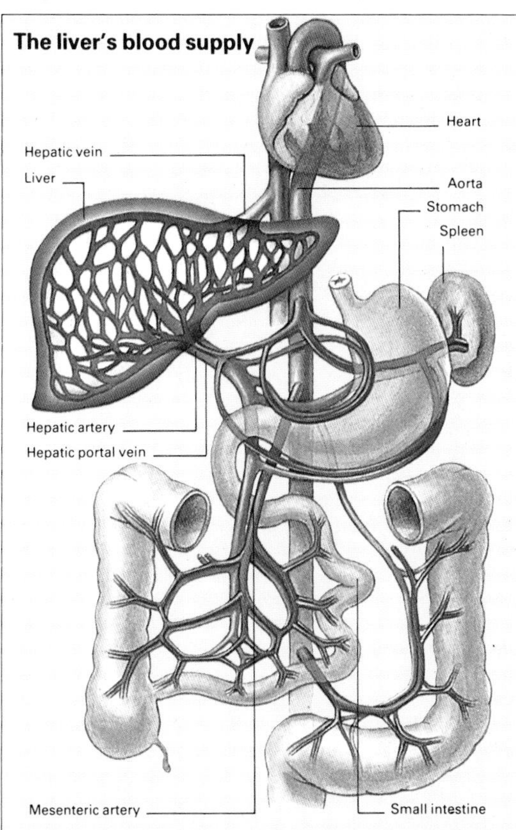

▲ ***The blood from the stomach and intestines, carrying the nutrients absorbed from the food, passes to the liver via the hepatic portal vein. The liver is also provided with oxygenated blood via the hepatic artery. Within the liver, these two blood supplies pass via the system of capillaries into the liver lobules. Blood finally leaves the liver in the hepatic veins.***

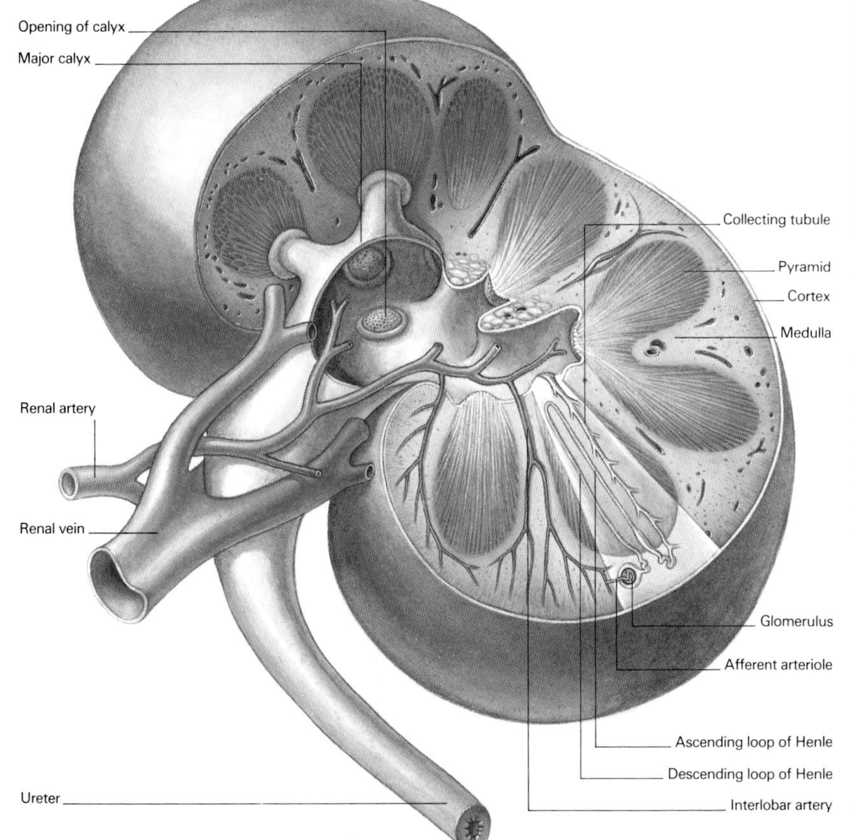

Kidney

◄ ***The kidney consists of an outer cortex, and a striped medulla. Blood is taken to the cortex where it enters a nephron. In the glomerulus, most of the blood's water and plasma, containing many minerals, are filtered out, but much of the water and valuable substances are reabsorbed in the long tubules of the nephron.***

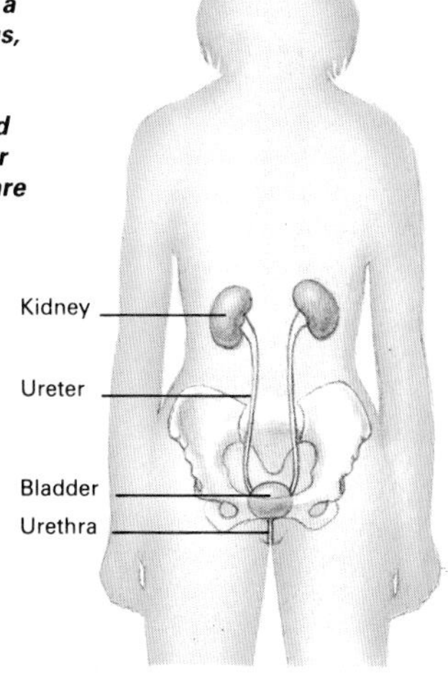

► ***The urine passes from the kidney to the bladder via the ureter.***

Tabes dorsalis
Form of tertiary syphilis (see SEXUALLY TRANSMITTED DISEASES) in which certain tracts in the SPINAL CORD – particularly those concerned with position sense – degenerate, leading to a characteristic high-stepping gait, sensory abnormalities and sometimes disorganization of joints, abdominal pain and abnormal pupil reactions.

Tachometer
Instrument for measuring the angular velocity of a rotating shaft. The simplest is a timed revolution counter. Other mechanical tachometers include the centrifugal tachometer, similar to the flyball governor; the vibrating-reed tachometer, a group of reeds of different lengths which is held against the shaft housing so that the reed whose natural vibration frequency equals the rotation frequency of the shaft vibrates by resonance; and the velocity-head tachometer, in which a pump or fan on the shaft produces a measured air pressure. Electrical tachometers are usually electric generators or electric impulse counters. The eddy-current tachometer is used as a speedometer.

Tachycardia see HEART.

Taconite
An unleached, low-grade IRON ore. It consists of fine-grained FLINT containing HEMATITE, MAGNETITE and several silicates. Taconite must be concentrated by leaching or magnetic processes before smelting.

Tadpoles
The larvae of frogs and toads. An aquatic larva is characteristic of all the AMPHIBIANS, but in salamanders and newts it is similar to the adult. In frogs and toads, the tadpole is globular with a long muscular tail. A full METAMORPHOSIS must be undergone to reach adult form.

Taiga
A vegetation zone lying between the northern coniferous forests (BOREAL FOREST) and the TUNDRA. It consists of clumps of trees separated by open areas, often boggy or rocky in character, where lichens and other low-growing plants predominate. The term is often used for the virgin coniferous forests of the northern (boreal) zone.

Talbot, William Henry Fox (1800-1877)
British astronomer, mathematician and pioneer of PHOTOGRAPHY who first produced photographic images on silver chloride paper (1838), and published the first photographically illustrated book (1844).

Talc
Basic magnesium SILICATE mineral, with formula $Mg_3Si_4O_{10}(OH)_2$, occurring in METAMORPHIC ROCKS, chiefly in the US, USSR, France and Japan. It has a layer structure resembling that of MICA, and is extremely soft (see HARDNESS). Talc is used in ceramics, roof insulation, cosmetics, as an insecticide carrier and as a filler in paints, paper and rubber. Compacted talc forms SOAPSTONE.

Talipes
Another name for clubfoot, a deformity sometimes present at birth.

Talus (scree)
Accumulation of rocky debris at the base of a cliff or steep mountain slope, the result of mechanical weathering (see EROSION) of the rocks above. A breccia is a SEDIMENTARY ROCK formation of consolidated talus.

Tamm, Igor Yevgenevich (1895-1971)
Soviet physicist awarded, with P.A. CHERENKOV and I.M. FRANK, the 1958 Nobel Prize for Physics for work with Frank interpreting the Cherenkov effect (1937).

Tannins (tannic acid)
A group of complex, bitter-tasting organic substances occurring in many plants, especially oak galls, tea, and the bark of certain trees. Tannins are used in curing hides to make leather.

Tantalum (Ta)
Hard, silvery gray metal in Group VB of the PERIODIC TABLE; a TRANSITION ELEMENT. It is found in COLUMBITE with NIOBIUM, which it closely resembles. Tantalum oxide is separated by solvent extraction and reduced to the metal. Highly inert, it is used in laboratory ware, capacitors, surgical instruments and as a "getter". AW 180.9, mp 3000°C, bp 5500°C, sg 16.6 (20°C). (◆ page 130)

Tapeworms
Intestinal parasites, so named because they are long and flat. They are classifed in the Class Cestoda of the flatworm Phylum PLATYHELMINTHES. A scolex, or head, only 1.5-2mm in diameter is attached to the gut, and behind this the body, which may be well over a meter long, consists of a ribbon of identical flat segments, or proglottids, each containing reproductive organs. Mature proglottids containing eggs are budded off at the end of the body and pass out with the feces. From there the larval stages can infect intermediate hosts. Several different tapeworms infect humans, and they have different intermediate hosts, including pigs and fish. Eating undercooked pork or fish can lead to infestation.

Tar
Dark, odorous liquid obtained by destructive distillation of coal or wood, especially from conifers. (See also CREOSOTE; PITCH.)

Tarnishing see CORROSION.

Tartaric acid (dihydroxysuccinic acid, HOOC.CHOH.CHOH.COOH)
A CARBOXYLIC ACID with three STEREOISOMERS, used in foods and soft drinks, as a metal cleaner, and in dyeing and photography. It is obtained by treatment of maleic anhydride with hydrogen peroxide and from the lees of wine fermentation, in which it occurs as potassium hydrogen tartrate, known as argol, or cream of tartar when pure, used in baking powder and as a sequestrant. From argol are made Rochelle salt, potassium sodium tartrate, used in making processed cheese, mirrors and cathartics; and tartar emetic, antimony potassium tartrate, used as an emetic, insecticide, and a mordant in dyeing.

Taste
A chemical sense, akin to that of SMELL but concerned with assessing food once it is in the mouth. Receptors are distributed over the surface of the TONGUE, and in humans are able to distinguish salt, sweet, sour, bitter and possibly water as primary tastes. Much of what is termed taste is actually perception of odors reaching the nose. Receptors for sweet are concentrated at the tip of the tongue, for salt and sour along the sides, with bitter mainly at the back. (◆ page 213)

Tatum, Edward Lawrie (1909-1975)
US biochemist awarded the 1958 Nobel Prize for Physiology or Medicine with G.W. BEADLE and J. LEDERBERG for work with Beadle showing that individual GENES control production of particular ENZYMES (1937-40).

Taube, Henry (b. 1915)
US chemist who was awarded the 1983 Nobel Prize for Chemistry for his work on the mechanisms by which transition metal complexes transfer ELECTRONS, which forms the basis of modern INORGANIC CHEMISTRY and gave insights into the biochemical reactions that maintain life.

Taurus
The Bull, a large constellation on the ECLIPTIC; the second sign of the ZODIAC. It contains the CRAB NEBULA, the GALACTIC CLUSTERS the Hyades and PLEIADES, and the bright star Aldebaran.

Tautomerism
The existence of two interconvertible ISOMERS of a compound, usually in labile EQUILIBRIUM, though in some cases isolable. It may be demonstrated by spectroscopy or by the exhibition of properties characteristic of both tautomers. Most tautomerism is by hydrogen transfer, as with carbonyl compounds, in which the keto form is in equilibrium with the enol form. SUGARS display tautomerism between cyclic and straight-chain forms.

Taxis
Directed movement, away from or toward a directional stimulus. (See also KINESIS.)

Taxonomy
The classification of living organisms into hierarchical groups, representing the relationships between them. The fundamental group in all taxonomy is the SPECIES, and related species are grouped together in a genus. Thus horses and zebras are classed together in the genus *Equus* owing to their obvious similiarities. Together with the asses and onagers, they are placed in the Family Equidae. This family is grouped with two others, the rhinoceroses and the tapirs, on the basis that all have an odd number of toes on each hoof (one or three); these three families make up the Order Perissodactyla, the odd-toed ungulates. Together with all the other orders of hair-covered, milk-producing animals, this order constitutes the Class Mammalia. Mammals can be grouped with other backboned animals in the Subphylum Vertebrata, which is part of the Phylum Chordata, containing all animals with a notochord at some time in their life cycle. Finally, the chordates are placed with all other multicellular, heterotrophic organisms that lack a rigid cell wall into the Kingdom Animalia.

Methods of modern taxonomy include cladistic analysis, in which similarities between the organisms to be classified are recorded and a cladogram, or standard branching diagram, is chosen from among various alternatives to produce the best "fit" with the data. Aother method is numerical taxonomy, in which dissimilarities are recorded as well as similarities. As many factors as possible are noted, and all the data is entered into a computer that calculates the probable relatedness of the different species involved.

Technetium (Tc)
Radioactive metal (see RADIOACTIVITY) in Group VIIB of the PERIODIC TABLE; a TRANSITION ELEMENT. It does not occur naturally, but was discovered in 1937 by EMILIO SEGRE and Carlo Perrier in bombarded molybdenum – the first element to be made artificially. It is now recovered from the fission products of nuclear reactors. Technetium is chemically very like RHENIUM. AW 98, mp 2200°C, bp 4600°C, sg 11.5 (20°C). (◆ page 130)

Teeth
The specialized hard structures used for biting and chewing food. Each tooth consists of a crown, or part above the gum line, and a root, or insertion into the bone of the jaw. The outer surfaces of the crowns are covered by a thin layer of enamel, the hardest animal tissue. This overlies the dentine, a substance similar to bone; in the center of each

tooth is the pulp, which contains blood vessels and nerves. The incisors at the front of the mouth are developed for biting off food with a scissor action, while the canines or "eye teeth" are particularly developed in some species for stabbing or slicing into prey. The molars and premolars, at the back of the jaw, are adapted for chewing and macerating food, which partly involves side-to-side movement of one jaw over the other. In some mammals, notably the rodents, the incisor teeth grow throughout life, as an adaptation to a very abrasive diet. In elephants the worn-out teeth are replaced by others that move forward from the back of the jaw. Similar adaptations are found in other groups, such as the sharks, which have replacement teeth that develop below and behind the existing ones and gradually move up to take their place.

Tektites
Glassy objects, usually of less than 100mm diameter, found only in certain parts of the world. Most are rich in SILICA: they resemble OBSIDIAN, though have less water. Despite suggestions that they are of extraterrestrial, particularly lunar, origin, it seems most likely that they have resulted from meteoritic impacts in SEDIMENTARY ROCK in the remote past.

Telecommications
The transmission of information – voices, data or pictures – using ELECTROMAGNETIC RADIATION. The radio, telephone and television are forms of telecommunications. One of the key technologies of the second half of the 20th century, a cornerstone of INFORMATION TECHNOLOGY.

Telemetry
The transmission of data from distant automatic monitoring stations to a recording station for analysis. It is of immense importance in SPACE EXPLORATION. (See also RADIO.)

Teleosts
Fish belonging to the Infraclass Teleosti, one of the three groups in the ACTINOPTERYGII, or ray-finned fish. They are by far the largest group of fish, with at least 20,000 living species. Most of the familiar fish, apart from sharks and rays (CHONDRICHTHYES) are teleosts. Their most distinctive feature is that the upper jaw is not fused to the skull, giving the mouth much more mobility – for example, the jaws can be pushed forward to suck in small food items. (♦ page 17)

Telepathy see ESP.

Telescope, Optical
Instrument used to detect or examine distant objects. It consists of a series of lenses and mirrors capable of producing a magnified IMAGE and of collecting more light than the unaided eye. The refracting telescope essentially consists of a tube with a LENS system at each end. Light from a distant object first strikes the objective lens, which produces an inverted image at its focal point. In the terrestrial telescope the second lens system, the eyepiece, produces a magnified, erect image of the focal image, but in instruments for astronomical use, where the image is usually recorded photographically, the image is not reinverted, thus reducing light losses. The reflecting telescope uses a concave MIRROR to gather and focus the incoming light, the focal image being viewed using many different combinations of lenses and mirrors in the various types of instrument, each seeking to reduce different optical aberrations. The size of a telescope is measured in terms of the diameter of its objective. Up to about 30cm diameter the resolving power (the ability to distinguish finely separated points) increases with size, but for larger objectives the only gain is in light gathering. A 500cm telescope can thus detect much fainter sources but resolve no better than a 30cm instrument. Because mirrors can be supported more easily than large lenses, the largest astronomical telescopes are all reflectors. (See also ASTRONOMY; OBSERVATORY; SCHMIDT.) (♦ page 237)

Teller, Edward (b. 1908)
Hungarian-born US nuclear physicist who worked with FERMI on nuclear FISSION at the start of the MANHATTAN PROJECT, but who is best known for his fundamental work on, and advocacy of, the HYDROGEN BOMB. In recent years he has been deeply involved in promoting the Strategic Defense Initiative (SDI) of the United States.

Tellurium (Te)
Silvery white metalloid in Group VIA of the PERIODIC TABLE, occurring as heavy-metal tellurides, and extracted as a byproduct of copper refining. It resembles SELENIUM in its chemistry, but is rather more metallic; tellurium (IV) compounds resemble PLATINUM (IV). It is added to metals to make them easier to machine, and to lead for greater corrosion resistance; bismuth telluride is used in thermoelectric devices. AW 127.6, mp 450°C, bp 1000°C, sg 6.24 (20°C). (♦ page 130)

Temin, Howard Martin (b. 1934)
US virologist who shared with R. DULBECCO and D. BALTIMORE the 1975 Nobel Prize for Physiology or Medicine for their work on cancer-forming VIRUSES.

Temperature
The degree of hotness or coldness of a body, as measured quantitatively by THERMOMETERS. The various practical scales used are arbitrary: the FAHRENHEIT scale was originally based on the values 0°F for an equal ice-salt mixture, 32°F for the freezing point of water and 96°F for normal human body temperature. (See also CELSIUS, ANDERS.) Thermometer readings are arbitrary also because they depend on the particular physical properties of the thermometric fluid, etc. There are now certain primary calibration points corresponding to the triple points, boiling points or freezing points of particular substances, whose values are fixed by convention. The thermodynamic, or ABSOLUTE, temperature scale is not arbitrary; starting at ABSOLUTE ZERO and graduated in KELVINS, it is defined with respect to an ideal reversible heat engine working on a CARNOT cycle between two temperatures T_1 and T_2. If Q_1 is the heat received at the higher temperature T_1, and Q_2 the heat lost at the lower temperature T_2, then T_1/T_2 is defined equal to Q_1/Q_2. Such absolute temperature is independent of the properties of particular substances, and is a basic THERMODYNAMIC function, arising out of the zeroth law. It is an intensive property, unlike HEAT, which is an extensive property – that is, the temperature of a body is independent of its mass or nature; it is thus only indirectly related to the heat content (internal energy) of the body. Heat flows always from a higher temperature to a lower. On the molecular scale, temperature may be defined in terms of the statistical distribution of the kinetic energy of the molecules.

Temperature Humidity Index (THI)
Formerly known as the discomfort index. An empirical measure of the discomfort experienced in various warm weather conditions, and used to predict how much power will be needed to run air-conditioning systems. It is given by

$$\mathrm{THI}=0.4(T_1+T_2)+15$$

where T_1 is the dry-bulb temperature and T_2 the wet-bulb temperature in degrees Fahrenheit (see HYGROMETER). When the index is 70 most people feel comfortable; at 80 or more, no one does.

Tendon
Tough, fibrous structure, made principally of COLLAGEN, formed at the ends of most MUSCLES, which transmits the force of contraction to the point of action (usually a bone). They allow bulky power muscles to be situated away from small bones concerned with fine movements, as in the hands. (♦ page 216)

Tensile strength
The resistance of a material to tensile stresses (those that tend to lengthen it). The tensile strength of a substance is the tensile force per unit area of cross-section that must be applied to break it.

Terbium (Tb)
One of the LANTHANUM SERIES of elements. AW 158.9, mp 1360°C, bp 2500°C, sg 8229 (25°C). (♦ page 130)

Terman, Lewis Madison (1877-1956)
US psychologist best known for developing the STANFORD-BINET TEST.

Terpenes
HYDROCARBONS that are oligomers or POLYMERS of ISOPRENE, and their derivatives. Most are odorous liquids, reactive and unstable, and are found in ESSENTIAL OILS, especially TURPENTINE. They contain double bonds and usually one or more rings. They include CAMPHOR, CAROTENOIDS, MENTHOL and VITAMIN A. Latex is a polyterpene.

Terramycin (oxytetracycline) see ANTIBIOTICS.

Territory
An area that an animal (or pair of animals, or group of animals) defends against other members of the same species. It may be used solely for feeding, for courtship, or for nesting, but in many species the territory serves all three purposes. A few species of birds have separate feeding and nesting territories, however. Territories are defended by calls, scent-marking, visual signals such as damaged vegetation, and sometimes by actual fighting. Territory size usually reflects the size of the animal, but it will depend also on population density – the more animals there are, the more pressure there will be on boundaries and the smaller territories will become. In many species an animal that does not secure a territory cannot breed, and where this occurs characteristics that help an animal to defend a territory are a product of SEXUAL SELECTION.

Tertiary
The period of the CENOZOIC era before the advent of humans, lasting from about 65 to 2 million years ago. Sometimes the Tertiary is regarded as synonymous with the Cenozoic, the QUATERNARY being merely a subperiod. (♦ page 106)

Tesla, Nikola (1856-1943)
Croatian-born US electrical engineer whose discovery of the rotating magnetic field permitted his construction of the first AC induction motor (*c*1888). Since it is easier to transmit AC than DC over long distances, this invention was of great importance.

Test-tube baby
Baby conceived by IN-VITRO FERTILIZATION.

Testes
Pair of male GONADS, which in humans lie in the scrotum suspended from the perineum below the PENIS. This position allows a lower temperature than in the abdomen, thus favoring SPERM production, the principal function of the testes. ANDROGEN hormones (mainly TESTOSTERONE) are secreted by the testes under the control of the HYPOTHALAMUS and PITUITARY GLAND. The testes develop at the back of the abdomen and descend in the fetus and infant. If the testes fail to descend in childhood, surgical correction may be required. (♦ page 111, 204)

Testosterone
ANDROGEN STEROID produced by the interstitial cells of the TESTES, and to a lesser extent by the ADRENAL cortex in both sexes, under the control of LUTEINIZING HORMONE. It is responsible for most male sexual characteristics – voice change,

hair distribution and the development of sex organs. (◆ page 111)

Tetanus (lockjaw)
BACTERIAL DISEASE in which a TOXIN produced by anaerobic tetanus bacilli growing in contaminated wounds causes MUSCLE spasm due to nerve toxicity. Minor cuts may be infected with the bacteria, whose spores are common in soil. The first symptom may often be painful contraction of jaw and neck muscles; trunk muscles, including those of respiration, and muscles close to the site of injury are also frequently involved. Untreated, many cases are fatal. Regular VACCINATION and adequate wound cleansing are important in prevention.

Tetany
Involuntary MUSCLE contractions, with excessive muscular irritability due mostly to lack of ionic CALCIUM in the BLOOD and tissues. True hypocalcemia (low blood calcium level) may be due to PARATHYROID GLAND insufficiency or pancreatitis, while ALKALOSIS may transiently reduce ionization of calcium compounds.

Tethys
Primeval sea that lay between the supercontinents GONDWANA and LAURASIA, separating what is now Africa from what is now S Eurasia. As the continents evolved toward their present form the Tethys narrowed, leaving only the present Mediterranean. The sediments of the Tethys GEOSYNCLINE are to be found in folded MOUNTAIN ranges such as the Himalayas. (See also PLATE TECTONICS.)

Tetracyclines see ANTIBIOTICS.

Tetraethyl lead see LEAD.

Tetrapods
Vertebrate animals having four legs, and their descendants: that is, the amphibians, reptiles, birds and mammals. In birds the front pair of legs have become modified into wings, while in snakes the limbs have been lost entirely, as has also happened in the legless lizards, amphisbaenids and caecilians. In marine animals the front limbs have developed into flippers and the hind ones are either lost or greatly modified.

Thalamus
Two nuclei of the upper BRAIN stem involved in the transmission of impulses to and from the cerebral cortex, especially in sensory pathways. (◆ page 126)

Thalassemia
An inherited form of ANEMIA found mainly in people of Mediterranean and Eastern origin.

Thalidomide
Mild sedative introduced in the late 1950s and withdrawn a few years later when it was found to be responsible for congenital deformities in children due to an effect on the EMBRYO in early pregnancy; in particular it caused defective limb bud formation. Thalidomide is still used in the treatment of LEPROSY.

Thallium (Tl)
Soft, bluish gray metal in Group IIIA of the PERIODIC TABLE, resembling LEAD. Like INDIUM, it is obtained as a byproduct of zinc processing. It forms monovalent (thallous) compounds which may be oxidized to the less stable trivalent (thallic) compounds. Thallous sulfide is used in photocells, and mixed crystals of the bromide and iodide in infrared detectors. Thallium compounds are dangerously toxic. AW 204.4, mp 304°C, bp 1460°C, sg 11.85 (20°C). (◆ page 130)

Theiler, Max (1899-1972)
South African-born US microbiologist awarded the 1951 Nobel Prize for Physiology or Medicine for his discovery that an attenuated strain of YELLOW FEVER could be prepared by infecting mice, so making possible the first yellow fever vaccine.

Theodolite
Surveying instrument comprising a sighting TELESCOPE whose orientation with respect to two graduated angular scales, one horizontal, the other vertical, can be determined. It represents a development of the transit, which traditionally included only the horizontal scale.

Theophrastus (*c*370-*c*285 BC)
Greek philosopher of the Peripatetic School, generally considered the father of modern BOTANY. His *Enquiry into Plants* deals with description and classification, his *Plant Etiology* with physiology and structure.

Theorell, Axel Hugo Teodor (1903-1982)
Swedish biochemist awarded the 1955 Nobel Prize for Physiology or Medicine for his studies of ENZYME action, specifically the roles of enzymes in biological OXIDATION AND REDUCTION processes.

Thermal analysis (thermoanalysis)
Group of methods for detecting and studying physical and chemical changes in substances heated at a standard rate through a temperature range; sometimes used for chemical ANALYSIS. In thermogravimetric analysis (TGA), the sample is weighed in a thermobalance – a sensitive BALANCE with the sample pan inside a furnace – and its weight is plotted against temperature. Weight loss is due to giving off gases or vapors; weight gain to reaction with the atmosphere. In differential thermal analysis (DTA), the sample is heated simultaneously with an inert reference substance (usually aluminum oxide), and the temperature difference between them is plotted against temperature. This deviates from zero in one direction when an exothermic reaction occurs, and in the other direction when an endothermic reaction occurs (see THERMOCHEMISTRY).

Thermionic emission
Spontaneous emission of ELECTRONS from metal or oxide-coated metal surfaces at temperatures between 1000K and 3000K. It supplies the electrons in electron- and cathode-ray tubes.

Thermistor
A SEMICONDUCTOR device the electrical resistance of which falls rapidly as its temperature rises. It is used as a sensor in electronic circuits measuring or regulating temperature and also in time-delay circuits.

Thermite
Mixture of powdered ALUMINUM and iron oxide (Fe_3O_4) in equivalent amounts, used in welding and incendiary bombs. On ignition with a barium peroxide or magnesium fuze, a violently exothermic OXIDATION reaction occurs, producing molten iron at 2500°C and alumina slag. It thus supplies both the heat and the metal for welding, and can be used to join large parts in a preheated refractory mold.

Thermochemistry
Branch of PHYSICAL CHEMISTRY that deals with HEAT changes accompanying chemical reactions. Practical thermochemistry is mainly by CALORIMETRY, which yields standard heats of reaction or enthalpy values ($\Delta H°$) (see THERMODYNAMICS). If this is negative, the reaction is termed exothermic (heat-producing); if positive, endothermic (heat-absorbing). Hess' Law, or the law of constant heat summation, a corollary of the first law of THERMODYNAMICS, states that the overall heat change in a chemical reaction is the same whether it takes place in one or several steps. Thus, by algebraic addition of chemical equations and their ΔH values, inaccessible heats of reaction may be calculated, including the heat of formation of a compound, which is the heat change when one mole of the compound is formed from its constituent elements in their standard states.

Thermocouple
An electric circuit involving two junctions between different METALS or SEMICONDUCTORS; if these are at different temperatures, a small ELECTROMOTIVE FORCE (emf) is generated in the circuit (Seebeck effect). Measurement of this emf provides a sensitive, if approximate THERMOMETER, typically for the range 70K-1000K, one junction being held at a fixed temperature and the other providing a compact and robust probe. Semiconductor thermocouples in particular can be run in reverse as small refrigerators. A number of thermocouples connected in series with one set of junctions blackened form a thermopile, measuring incident radiation through its heating effect on the blackened surface. Thermoelectricity embraces the Seebeck and other effects relating to heat transfer, thermal gradients, ELECTRIC FIELDS and currents.

Thermodynamics
Division of PHYSICS concerned with the interconversion of HEAT, WORK and other forms of ENERGY, and with the states of physical systems. It is basic to ENGINEERING, parts of GEOLOGY, METALLURGY and PHYSICAL CHEMISTRY. Building on earlier studies of the thermodynamic functions TEMPERATURE and heat, SADI CARNOT pioneered the science by his investigations of the cyclic heat ENGINE (1824), and in 1850 CLAUSIUS stated the first two laws. Thermodynamics was further developed by J.W. GIBBS, H.L.F. VON HELMHOLTZ, LORD KELVIN and J.C. MAXWELL.

In thermodynamics, a system is any defined collection of matter: a closed system is one that cannot exchange matter with its surroundings; an isolated system can exchange neither matter nor energy. The state of a system is specified by determining all its properties such as pressure, volume, etc. A system in stable EQUILIBRIUM is said to be in an equilibrium state, and has an equation of state (eg the general GAS law) relating its properties. (See also PHASE EQUILIBRIA.) A process is a change from one state A to another B, the path being specified by all the intermediate states. A state function is a property or FUNCTION of properties which depends only on the state and not on the path by which the state was reached; a differential dX of a function X (not necessarily a state function) is termed a perfect differential if it can be integrated between two states to give a value $X_{AB} = \int_A^B dX$ which is independent of the path from A to B. If this holds for all A and B, X must be a state function.

There are four basic laws of thermodynamics, all having many different formulations that can be shown to be equivalent. The zeroth law states that, if two systems are each in thermal equilibrium with a third system, then they are in thermal equilibrium with each other. This underlies the concept of temperature. The first law states that for any process the difference of the heat Q supplied to the system and the work W done by the system equals the change in the internal energy U: $\Delta U = Q - W$. U is a state function, though neither Q nor W separately is. Corollaries of the first law include the law of conservation of ENERGY, Hess' Law (see THERMOCHEMISTRY), and the impossibility of PERPETUAL MOTION machines of the first kind. The second law (in Clausius' formulation) states that heat cannot be transferred from a colder to a hotter body without some other effect, ie without work being done. Corollaries include the impossibility of converting heat entirely into work without some other effect, and the impossibility of PERPETUAL MOTION machines of the second kind. It can be shown that there is a state function ENTROPY, S, defined by

$\Delta S = \int dQ/T$, where T is the absolute temperature.

The entropy change ΔS in an isolated system is zero for a reversible process and positive for all

TELESCOPES

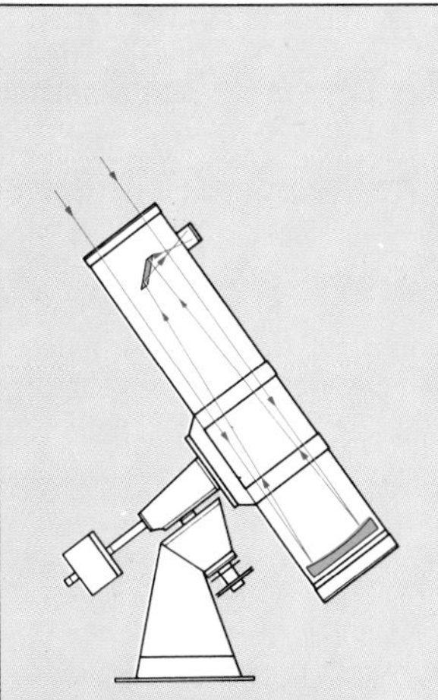

▲ *In the Newtonian design light reflected from the concave primary mirror is then reflected from a flat secondary to the side of the tube, where the eye piece is located. This avoids chromatic aberration.*

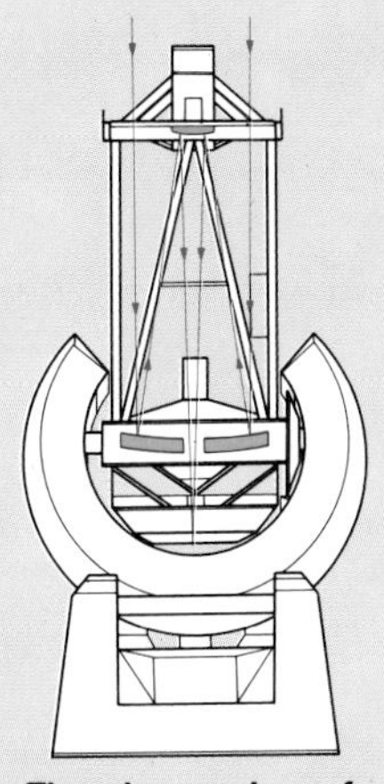

▲ *The primary mirror of a Cassegrain reflector is of parabolic cross-section. The converging cone of light is reflected from a convex secondary mirror, through a hole in the primary, to the eyepiece or instrument platform.*

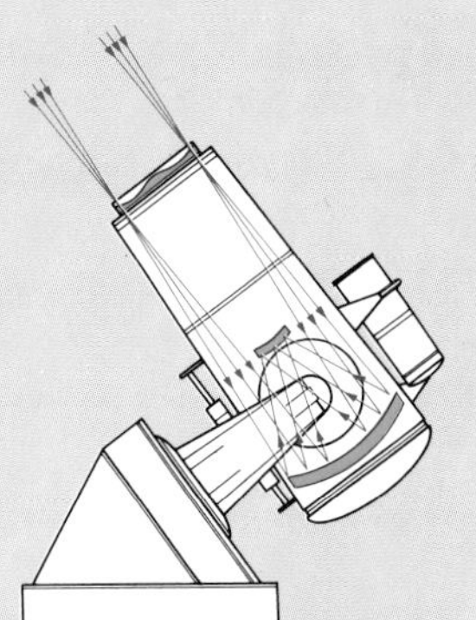

▲ *A Schmidt telescope has a specially shaped lens at the front of the tube to ensure that light rays reflected from the concave primary mirror are focused in the same plane, where a photographic plate is placed.*

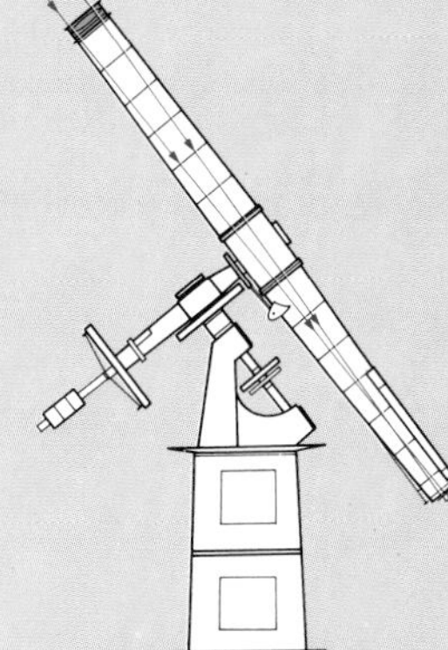

▲ *A refracting telescope has a large objective (a lens facing the object being watched) of long focal length, which brings light to focus where the image may be magnified by an eyepiece. This arrangement may be unwieldy.*

▲ *A radio telescope collects long-wave radiation which the human eye does not detect. In one type a dish antenna reflects radio waves to a receiver placed at the focus. The dish is pointed towards the area of sky to be studied.*

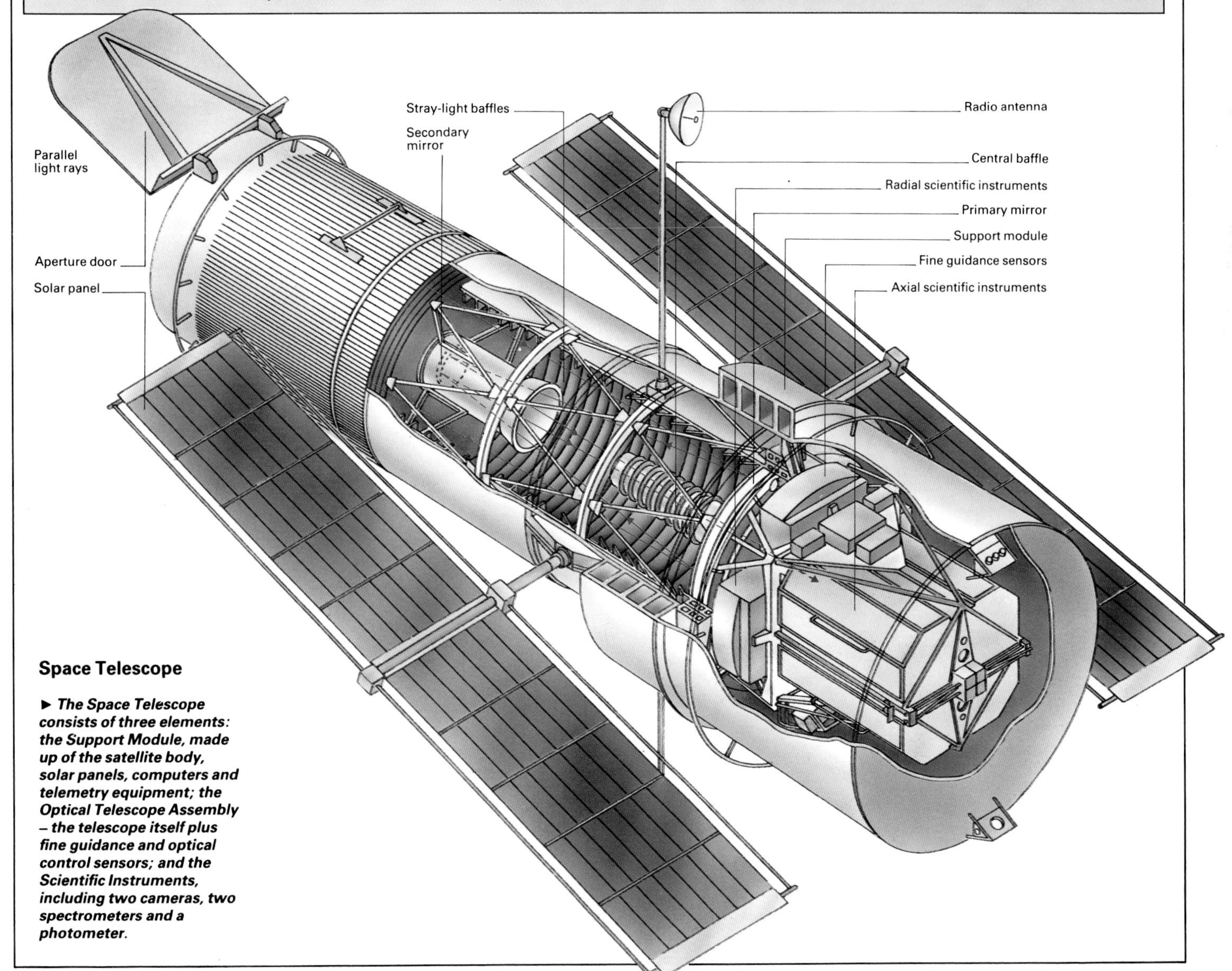

Space Telescope

► *The Space Telescope consists of three elements: the Support Module, made up of the satellite body, solar panels, computers and telemetry equipment; the Optical Telescope Assembly – the telescope itself plus fine guidance and optical control sensors; and the Scientific Instruments, including two cameras, two spectrometers and a photometer.*

irreversible processes. Thus entropy tends to a maximum (see HEAT DEATH). It also follows that a heat ENGINE is most efficient when it works on a reversible CARNOT cycle between two temperatures T_1 (the heat source) and T_2 (the heat sink), the EFFICIENCY being $(T_1-T_2)/T_2$. The third law states that the entropy of any finite system in an equilibrium state tends to a finite value (defined to be zero) as the temperature of the system tends to absolute zero. The equivalent NERNST heat theorem states that the entropy change for any reversible isothermal process tends to zero as the temperature tends to zero. Hence absolute entropies can be calculated from specific heat data. Other thermodynamic functions, useful for calculating equilibrium conditions under various constraints, are: enthalpy (heat content) $H=U+pV$; the Helmholtz free energy $A=U-TS$; and the Gibbs free energy $G=H-TS$. The free energy represents the capacity of the system to perform useful work. Quantum statistical thermodynamics, based on QUANTUM MECHANICS, has arisen in the 20th century. It treats a system as an assembly of particles in quantum states. The entropy is given by $S=k \ln P$ where k is the BOLTZMANN constant and P the statistical probability of the state of the system. Thus entropy is a measure of the disorder of the system.

Thermoelectricity see THERMOCOUPLE.

Thermograph
Any type of THERMOMETER that is self-registering, recording variations of temperature with time on a graph. A bimetallic strip is often used as the temperature-sensitive element, its deflection being recorded on a rotating drum via a system of levers. Thermographs are widely used in meteorology and atmospheric investigations.

Thermoluminescence
Emission of light from a steadily heated material that has previously been excited by exposure to radiation. It is a type of LUMINESCENCE and arises from electron displacements within the material's crystal lattice. A long time delay may elapse between excitation and subsequent light emission; this is utilized in thermoluminescent dating in archeology.

Thermometer
Instrument for measuring the relative degree of hotness of a substance (its TEMPERATURE) on some reproducible scale. Its operation depends upon a regular relationship between temperature and the change in size of a substance (as in the mercury-in-glass thermometer) or in some other physical property (as in the platinum resistance thermometer). The type of instrument used in a given application depends on the temperature range and accuracy required.

Thermonuclear reactions
The reactions used in nuclear FUSION devices such as the HYDROGEN BOMB.

Thermopile see THERMOCOUPLE.

Thiamine (aneurin)
Alternative name for VITAMIN B_1.

Thioethers
Organic compounds that are structurally similar to ETHERS but in which sulfur substitutes for the oxygen bridging the two alkyl groups.

Thiosulfates see SULFUR.

Thirst and hunger
Desires for water and food respectively, which have a role in regulating their intake. Thirst is the end result of a mixture of physical and psychological effects including dry mouth, altered blood mineral content, and the sight and sound of water; hunger, those of stomach contractions, low blood sugar levels, habit, and the smell and sight of food. Repleteness with either inhibits the sensation. Food and water intake are regulated by the HYPOTHALAMUS, and are closely related to the control of HORMONE secretion and other vegetative functions, being part of the system preserving the HOMEOSTASIS (constancy) of the body's internal environment. Drugs, smoking, systemic disease and local brain damage are among the many factors influencing thirst and hunger. Excessive thirst may be a symptom of DIABETES or kidney failure (UREMIA), but organic excessive hunger is rare.

Thompson, Benjamin see RUMFORD, COUNT.

Thomson, Sir George Paget (1892-1975)
British physicist awarded with C. DAVISSON the 1937 Nobel Prize for Physics for showing that ELECTRONS can be diffracted, thus demonstrating their wave nature.

Thomson, Sir Joseph John (1856-1940)
British physicist generally regarded as the discoverer of the electron. It had already been shown that cathode rays could be deflected by a magnetic field; in 1897 Thomson showed that they could also be deflected by an electric field, and could thus be regarded as a stream of negatively charged particles. He showed their mass to be much smaller than that of the hydrogen atom – this was the first discovery of a subatomic particle. His model of the atom, though imperfect, provided a good basis for RUTHERFORD's more satisfactory later attempt. Thomson was awarded the 1906 Nobel Prize for Physics.

Thomson, William see KELVIN, BARON.

Thorax
The middle part of the body in insects, lying between the head and the ABDOMEN and bearing the wings and legs. It houses many of the viscera. Crustaceans and chelicerates show a similar division of the body, but in spiders the head and thorax are fused together to form a "cephalothorax".

Thorium (Th)
Silvery white radioactive metal, one of the ACTINIDES. Its chief ore is MONAZITE. Thorium is tetravalent, resembling ZIRCONIUM and HAFNIUM. The metal is used in magnesium ALLOYS and to produce uranium-233 for atomic fuel. The refractory thorium (IV) oxide was used to make incandescent gas mantles; it is added in small amounts to the tungsten filaments in electric lamps. AW 232.0, mp 1700°C, bp 4500°C, sg 11.66 (17°C). (♦ page 130, 147)

Thrombosis
The formation of a clot (thrombus) in a blood vessel. It is commonly associated with VARICOSE VEINS, where thrombosis occurs in the tortuous superficial veins in the leg. It is more serious, even fatal, if it occurs in one of the major arteries supplying the heart or brain.

Thrush (monilia or candidiasis)
Mucous membrane infection with the yeast *Candida*, seen as multiple white spots, most often affecting the mouth or vagina. Patients with DIABETES or KIDNEY failure and those on STEROIDS and/or ANTIBIOTICS are particularly at risk.

Thulium (Tm)
Least abundant of the RARE EARTHS; one of the LANTHANUM SERIES of elements. AW 168.9, mp 1550°C, bp 1900°C, sg 9.321 (25°C). (♦ page 130)

Thunder
The acoustic shock wave caused by the sudden expansion of air heated by a LIGHTNING discharge. Thunder may be a sudden clap, or a rumble lasting several seconds if the lightning path is long and thus varies in distance from the hearer. It is audible up to about 15km away; the distance in kilometers can be roughly estimated as one-third the time in seconds between the lightning and thunder.

Thunderstorm
A STORM accompanied by THUNDER and LIGHTNING, heavy PRECIPITATION and SQUALLS. Usually short-lived, thunderstorms are generally associated with cumulonimbus CLOUDS.

Thurstone, Louis Leon (1887-1955)
US psychologist whose application of the techniques of statistics to the results of PSYCHOLOGICAL TESTS permitted their more accurate interpretation and demonstrated that a plurality of factors contributed to an individual's score.

Thymus
A ductless two-lobed gland lying just behind the breast bone and mainly composed of lymphoid cells (see LYMPH). It plays a part in setting up the body's IMMUNITY system. Autoimmunity is thought to result from its pathological activity. After PUBERTY it declines in size. (♦ page 111)

Thyroid gland
A ductless two-lobed gland lying in front of the trachea in the neck. The principal HORMONES secreted by the thyroid are thyroxine and triiodothyroxine; these play a crucial part in regulating the rate at which cells oxidize fuels to release energy, and strongly influence growth. The release of thyroid hormones is controlled by thyroid stimulating hormone (TSH), released by the PITUITARY GLAND when blood thyroid-hormone levels are low. Deficiency of thyroid hormones (hypothyroidism) in adults leads to myxedema, with mental dullness and cool, dry and puffy skin. Oversecretion of thyroid hormones (known as hyperthyroidism or thyrotoxicosis) produces nervousness, weight loss and increased heart rate. GOITER, an enlargement of the gland, may result when the diet is deficient in iodine. (See also CRETINISM.) (♦ page 111)

Tibia (shin bone)
The principal bone of the hind leg (human lower leg), paralleled by the FIBULA. (♦ page 216)

Tic
A stereotyped movement, habit spasm or vocalization that occurs irregularly, but often more under stress, and that is outside voluntary control. Its cause is unknown. Tic doloureux is a condition in which part of the face is abnormally sensitive, any touch provoking intense pain.

Tidal power
Form of HYDROELECTRICITY produced by harnessing the ebb and flow of the TIDES. Barriers containing reversible TURBINES are built across an estuary or gulf where the tidal range is great. The Rance power plant in the Gulf of St Malo, Brittany, the first to be built (1961-67), produces 240MW power, mostly at ebb tide.

Tidal wave
Obsolete term for TSUNAMI.

Tides
The periodic rise and fall of land and water on the Earth. Tidal motions are primarily exhibited by water: the motion of the land is barely detectable. As the Earth-Moon system rotates about its center of gravity, which is within the Earth, the Earth bulges in the direction of the Moon and in the exactly opposite direction, because of the Moon's gravitational attraction and the centrifugal forces resulting from the system's revolution. Toward the Moon, the lunar attraction is added to a comparatively small centrifugal force; in the opposite direction it is subtracted from a much larger centrifugal force. As the Moon orbits the Earth in the same direction as the Earth rotates, the bulge "travels" round the Earth each lunar day (24.83h); hence most points on the Earth have a high tide every 12.42h. The Sun produces a similar though smaller tidal effect.

Exceptionally high tides occur at full and new moon (spring tides), particularly if the Moon is at perigee (see ORBIT); exceptionally low high tides (neap tides) at first and third quarter. The friction of the tides causes the day to lengthen 0.001s per century.

Till (boulder clay)
The unsorted material left behind on the land after the retreat of a GLACIER. (See also DRIFT; DRUMLIN; MORAINE.)
Tillite see CONGLOMERATE.
Time
A concept dealing with the order and duration of events. If two events occur nonsimultaneously at a point, they occur in a definite order with a time lapse between them. Two intervals of time are equal if a body in equilibrium moves over equal distances in each of them; such a body constitutes a clock. The Sun provided man's earliest clock, the natural time interval being that between successive passages of the Sun over the local meridian – the solar DAY. For many centuries the rotation of the Earth provided a standard for time measurements, but in 1967 the SI unit of time, the SECOND, was redefined in terms of the frequency associated with a cesium energy-level transition. In everyday life, we can still think of time in the way Newton did, ascribing a single universal time-order to events. We can neglect the very short time needed for light signals to reach us, and believe that all events have a unique chronological order. But when velocities close to that of light are involved, relativistic principles become important; simultaneity is no longer universal and the time scale in a moving framework is "dilated" with respect to one at rest – moving clocks appear to run slow (see RELATIVITY). (♦ page 151)
Tin (Sn)
Silvery white metal in Group IVA of the PERIODIC TABLE, occurring as CASSITERITE in SE Asia, Bolivia, Zaire and Nigeria. The ore is reduced by smelting with coal. Tin exhibits ALLOTROPY: white (β) tin, the normal form, changes below 13.2°C to gray (α) tin, a powdery metalloid form resembling GERMANIUM, and known as "tin pest". Tin is unreactive, but dissolves in concentrated acids and alkalis, and is attacked by HALOGENS. It is used as a protective coating for steel, in alloys including solder, and in type metal. AW 118.7, mp 232°C, bp 2720°C, sg (β) 7.31. Tin forms organotin compounds, used as biocides, and also inorganic compounds: tin (II) and tin (IV) salts.
TIN (IV) OXIDE (SnO_2) White powder prepared by calcining CASSITERITE or burning finely divided tin; used in glazes and as an abrasive. subl 1800°C. (♦ page 130)
TIN (II) CHLORIDE ($SnCl_2$) White crystalline solid, prepared by dissolving tin in hydrochloric acid, used as a reducing agent, in tinplating, and as a mordant for dyes. mp 246°C, bp 652°C.
Tinbergen, Nikolaas (b. 1907)
Dutch ethologist awarded with K. LORENZ and K. VON FRISCH the 1973 Nobel Prize for Physiology or Medicine for their individual, major contributions to the science of ANIMAL BEHAVIOR.
Ting, Samuel C. C. (b. 1936)
US physicist who shared the 1976 Nobel Prize for Physics with B. RICHTER for his independent discovery of the J particle (also known as the psi (3095) particle).
Tinnitus
A ringing or other sounds in the EAR, found in certain diseases of the ear.
Tiselius, Arne Wilhelm Kaurin (1902-1971)
Swedish chemist awarded the 1948 Nobel Prize for Chemistry for his development of new techniques and equipment in order to apply ELECTROPHORESIS to the study of PROTEINS, notably those of the BLOOD.
Tissues
Similar CELLS grouped together in certain areas of the body in multicellular organisms. These cells are usually specialized for a single function, as with nervous tissue or muscle tissue. Groups of tissues, each with its own functions, make up ORGANS. (See also CONNECTIVE TISSUE; HISTOLOGY.) (♦ page 13)
Tissue typing
Testing tissues of a person or animal to see what immune genes they carry; people of similar tissue types can tolerate transplants from each other, with the aid of IMMUNOSUPPRESSIVE DRUGS.
Titanium (Ti)
Silvery gray metal in Group IVB of the PERIODIC TABLE; a TRANSITION ELEMENT. Titanium occurs in RUTILE and in ILMENITE, from which it is extracted by conversion to titanium (IV) chloride and reduction by magnesium. The metal and its alloys are strong, light, and corrosion- and temperature-resistant, and, although expensive, are used for construction in the aerospace industry. Titanium is moderately reactive, forming tetravalent compounds, including titanates (TiO_3^{2-}), and less stable di- and trivalent compounds. (♦ page 130)
TITANIUM (IV) OXIDE (TiO_2) is used as a white pigment in paints, ceramics, etc.
TITANIUM (IV) CHLORIDE ($TiCl_4$) is used as a catalyst. AW 47.9, mp 1670°C, bp 3300°C, sg 4.54.
Titration
Common technique of VOLUMETRIC ANALYSIS in which a standard solution of one reagent is added little by little from a BURETTE to a second reagent whose amount is to be determined. The end point, at which an exactly equivalent amount of reagent has been added, may be determined by using an INDICATOR, or by measurements of color, resistance, current flow or potential, whose variation with added reagent changes abruptly when the end point is reached.
TNT (trinitrotoluene)
Pale yellow crystalline solid made by NITRATION of TOLUENE. It is the most extensively used high EXPLOSIVE, being relatively insensitive to shock, especially when melted by steam heating and cast. MW 227.1, mp 80.8°C.
Tobacco
Dried and cured leaves of the tobacco plant *Nicotiana tabacum*, used for smoking, chewing and as snuff. Tobacco contains an alkaloid, nicotine, that makes it addictive. It also produces tarry substances when burned, which are a direct cause of lung cancer. Smoking increases the risk of heart disease and other illnesses, and chewing tobacco can result in cancer of the mouth and throat.
Tocopherol (Vitamin E) see VITAMINS.
Todd, Alexander Robertus, Baron Todd of Trumpington (b. 1907)
British organic chemist awarded the 1957 Nobel Prize for Chemistry for his work on the structure and synthesis of NUCLEOTIDES.
Toit, Alexander Logie du (1878-1948)
South African geologist, a disciple of WEGENER, whose work did much to validate Wegener's theories about CONTINENTAL DRIFT and was to a great extent responsible for their eventual acceptance.
Tokamak
A ring-shaped device in which a plasma is contained for experiments in FUSION reaction.
Toluene (methylbenzene, $C_6H_5CH_3$)
Colorless liquid HYDROCARBON, an AROMATIC COMPOUND produced mainly by catalytic reforming of PETROLEUM hydrocarbons. It is used as a solvent, in GASOLINE, and for making TNT, BENZALDEHYDE, BENZOIC ACID, etc. MW 92.2, mp −95°C, bp 110.6°C.
Tombaugh, Clyde Wilhelm (1906-1987)
US astronomer who discovered the planet PLUTO (1930).
Tomography
Computer-assisted tomography (CAT) is a technique for reassembling X-RAY information so as to construct a three-dimensional image of solid opaque bodies. The main use of the CAT scanner is in medicine, but it can also be used to observe the internal structures of nonliving objects.
Tomonaga, Shinichiro (1906-1979)
Japanese physicist who shared with FEYNMAN and SCHWINGER the 1965 Nobel Prize for Physics for their independent work on quantum electrodynamics.
Ton
Name of various actual or nominal units of weight. The short ton commonly used in the US is 2000lb (907kg), the long ton 2240lb (101.6kg). The metric ton or tonne (t) is 1000kg (2204.62lb). The ton used for measuring ships' cargoes is 40cu ft (1.13 cu m) in the US, 42cu ft (1.19 cu m) in the UK.
Tonegawa, Susumu (b. 1939)
Japanese biologist based in the US who was awarded the Nobel Prize for Physiology or Medicine for his pioneering work in immunology, establishing that random grouping of genes on a chromosome thereby achieve the necessary multiformity of antibodies against disease.
Tongue
In vertebrates, a muscular organ in the floor of the mouth that is concerned with the formation of food boluses and self-cleansing of the mouth, TASTE sensation and voice production. Its mobility allows it to move substances around the mouth, and in humans to modulate sound production in speech. In certain animals the tongue is extremely protrusile and is used to draw food into the mouth from a distance. (♦ page 40, 213)
Tonne (t)
1000kg, the metric TON.
Tonsillitis
INFLAMMATION of the TONSILS due to VIRUS or BACTERIAL infection. It may follow sore throat or other pharyngeal disease or it may be a primary tonsil disease. Sore throat and red swollen tonsils, which may exude PUS or cause swallowing difficulty, are common; LYMPH nodes at the angle of the jaw are usually tender and swollen. QUINSY is a rare complication.
Tonsils
Areas of LYMPH tissue aggregated at the sides of the PHARYNX. They provide a basic site of body defense against infection via the mouth or nose and are thus particularly susceptible to primary infection (TONSILLITIS). As with the ADENOIDS, they are particularly important in children first encountering infectious microorganisms in the environment.
Tooth see TEETH.
Topaz
Aluminum SILICATE mineral with composition $Al_2SiO_4(F,OH)_2$, which forms prismatic crystals (orthorhombic) that are variable and unstable in color, and valued as GEM stones. The best topazes come from Brazil, Siberia and the US.
Tornado
The most violent kind of STORM; an intense WHIRLWIND of small diameter, extending downward from a convective cloud in a severe THUNDERSTORM, and generally funnel-shaped. Air rises rapidly in the outer region of the funnel but descends in its core, which is at very low pressure. The funnel is visible because of the formation of cloud droplets by expansional cooling in this low pressure region. Very high winds spiral in toward the core. These, and explosions due to the low pressure, account for the almost total devastation and loss of life in the path of a tornado – which itself may move at up to 200m/s. Though generally rare, tornadoes occur worldwide, especially in the US and Australia in spring and early summer. (See also WATERSPOUT.)

UNIVERSE Scale and Structure

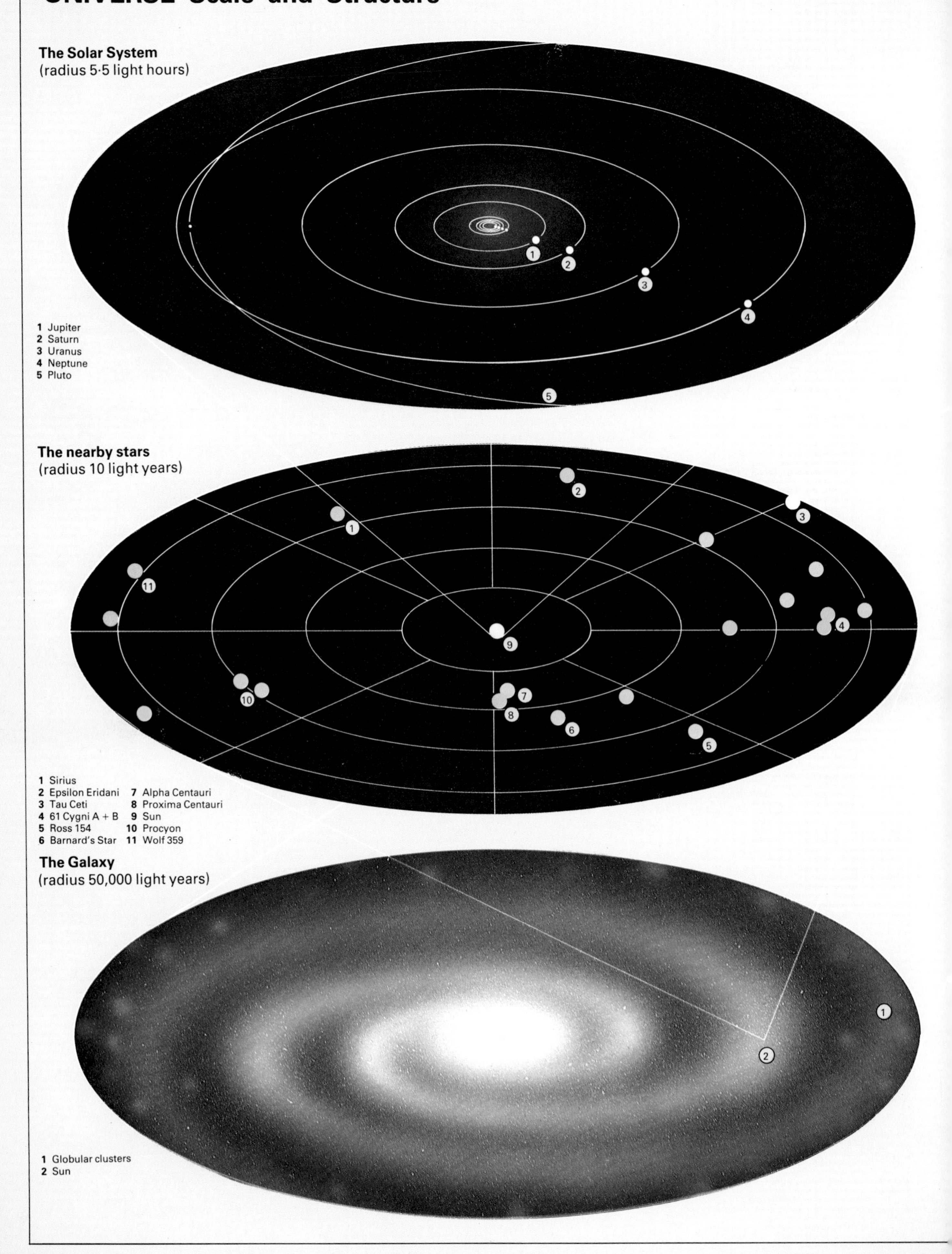

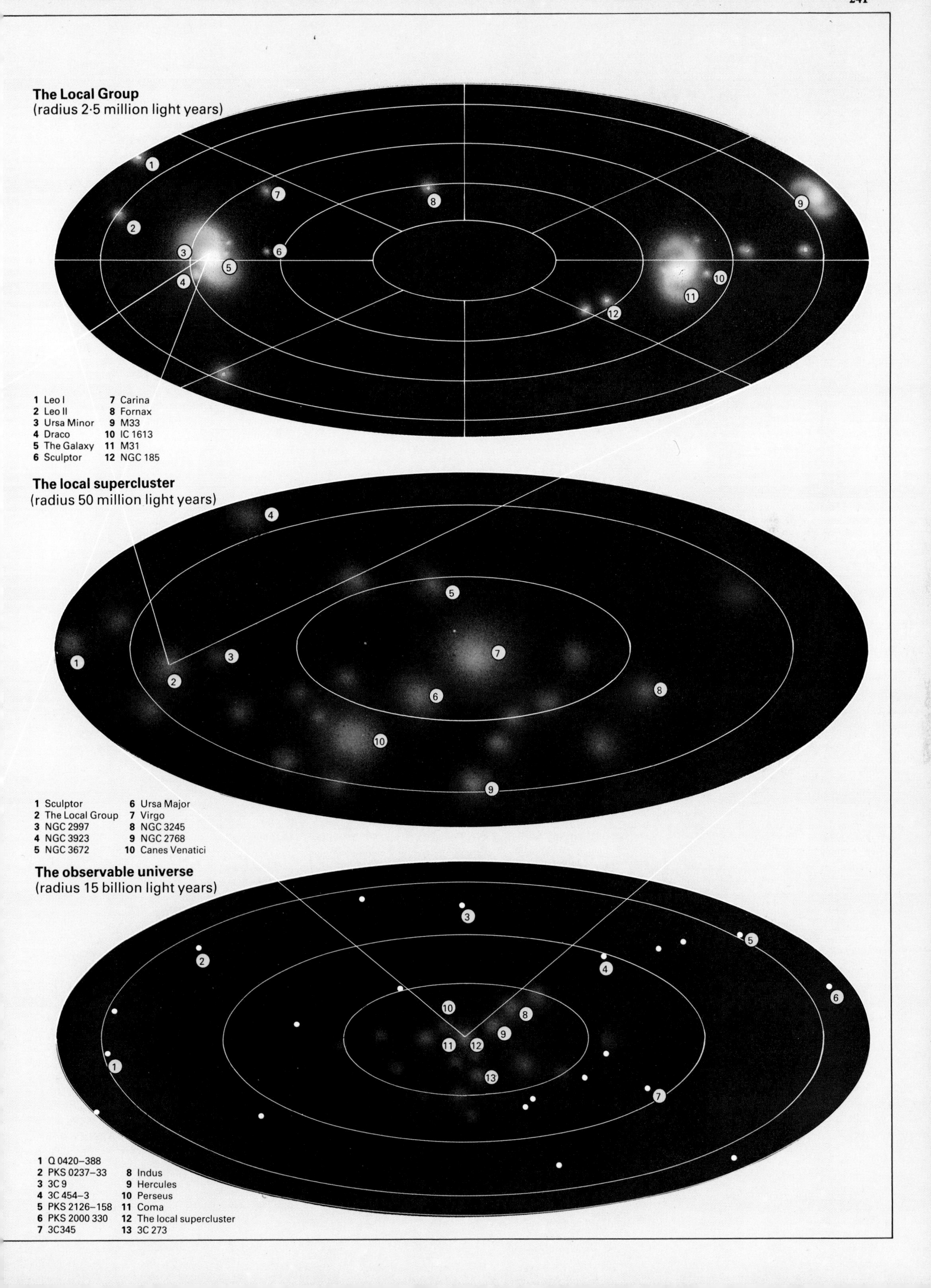
The Local Group
(radius 2·5 million light years)
1 Leo I
2 Leo II
3 Ursa Minor
4 Draco
5 The Galaxy
6 Sculptor
7 Carina
8 Fornax
9 M33
10 IC 1613
11 M31
12 NGC 185
The local supercluster
(radius 50 million light years)
1 Sculptor
2 The Local Group
3 NGC 2997
4 NGC 3923
5 NGC 3672
6 Ursa Major
7 Virgo
8 NGC 3245
9 NGC 2768
10 Canes Venatici
The observable universe
(radius 15 billion light years)
1 Q 0420–388
2 PKS 0237–33
3 3C 9
4 3C 454–3
5 PKS 2126–158
6 PKS 2000 330
7 3C345
8 Indus
9 Hercules
10 Perseus
11 Coma
12 The local supercluster
13 3C 273

Torque
A measure of the effectiveness of a FORCE or MOMENT in setting a body in rotation. In mechanics, a torque is a twisting moment or couple which tends to twist a fixed object such as a shaft about a rotation axis. If the shaft starts to rotate, the POWER it transmits is given by the product of the rotational speed and the torque.

Torricelli, Evangelista (1608-1647)
Italian physicist and mathematician, a one-time assistant of GALILEO, who improved the telescope and microscope and invented the (mercury) BAROMETER (1643).

Torsion
Strain produced by a twisting motion about an axis (a TORQUE), such as a couple applied perpendicular to a cylinder axis. The resistance of a bar of given material to torsion is a measure of its rigidity and elasticity.

Touch
The sensory system concerned with surface sensation, found in all external body surfaces including the SKIN and some mucous membranes. Functional categories of touch sensation include light touch (including movement of hairs), heat, cold, pressure and pain sensation. These are to some degree physiologically distinct. Receptors for all the senses are particularly concentrated and developed over the face and hands. When the various types of skin receptor are stimulated, they activate nerve impulses in cutaneous nerves; these impulses pass via the SPINAL CORD and brain stem to the BRAIN, where coding and perception occur. With painful stimuli, REFLEX withdrawal movements may be induced. (◆ page 213)

Tourmaline
Borosilicate mineral of variable composition (in general $XY_3Al_6(BO_3)_3Si_6O_{18}(OH)_4$ where X=Na,Ca and Y=Al,Fe^{III},Li,Mg), found in PEGMATITES as trigonal/hexagonal crystals used as GEM stones. Tourmaline crystals exhibit DOUBLE REFRACTION and PIEZOELECTRICITY, and hence are used in polarizers and in pressure-sensing devices.

Townes, Charles Hard (b. 1915)
US physicist awarded the 1964 Nobel Prize for Physics with N. BASOV and A. PROKHOROV for independently working out the theory of the MASER and, later, the LASER. He built the first maser in 1951.

Toxemia of pregnancy
Old name for pre-ECLAMPSIA.

Toxic shock syndrome
Condition of menstruating women, characterized by high fever, nausea, vomiting and watery diarrhea, possibly leading to severe illness. It is thought to be caused by a toxin produced by *Staphylococcus aureus* bacteria proliferating in tampons.

Toxin
A poisonous substance produced by a living organism. Many species produce chemical substances that are poisonous to some other organism; the toxin may be released continuously into the immediate environment, released only when danger is imminent, or contained within the body, so that it affects only an animal that makes a direct attack. Examples include poisonous plants whose toxins deter herbivores; fungi, which secrete substances that destroy bacteria (as ANTIBIOTICS these are of great value to humans); and poisonous spiders and snakes, which deliver their toxin via fangs. Toxins play an important part in defense and in the killing of prey. The symptoms of many INFECTIOUS DISEASES in humans, eg CHOLERA, DIPHTHERIA and TETANUS, are due to the release of toxins by the bacteria concerned. Many animals that produce toxins for defense also have WARNING COLORATION. (See also ANTITOXINS.)

Toxoid see ANTITOXINS.

Trace elements
Minerals required in minute quantities in an adequate human or animal diet (see NUTRITION), or for the optimum growth and yield of plants (see FERTILIZERS).

Tracer, Radioactive see RADIOCHEMISTRY, RADIOISOTOPE.

Trachea
In air-breathing vertebrates, the primary tube of the breathing system, leading from the throat to the two bronchi, which lead in turn to the lungs. In insects, air-filled tubes that ramify through the body, taking oxygen to all the cells. (See also GAS EXCHANGE.) (◆ page 28)

Tracheophyta see VASCULAR PLANTS.

Trachoma
INFECTIOUS DISEASE due to an organism (*Chlamydia*) intermediate in size between BACTERIA and VIRUSES. It is the commonest cause of blindness in the world. It causes acute or chronic CONJUNCTIVITIS and corneal INFLAMMATION with secondary blood-vessel extension over the cornea, resulting in loss of translucency. Eyelid deformity with secondary corneal damage is also common. It is transmitted by direct contact.

Trade winds
Persistent warm, moist WINDS that blow westward from the high-pressure zones at about 30°N and S latitude toward the DOLDRUMS (intertropical convergence zone) at the equator. They are thus northeasterlies in the N Hemisphere and southeasterlies in the S Hemisphere. They are stronger and displaced toward the equator in winter.

Tranquilizers
Agents that induce a state of quietude in anxious or disturbed patients. Minor tranquilizers are SEDATIVES valuable in the anxious. In certain types of psychotic disorder (see MENTAL ILLNESS), especially schizophrenia and (hypo-)mania, major tranquilizers are required to suppress abnormal mental activity as well as to sedate.

Transducers
Devices that convert power levels or signals carried in one energy mode to equivalent signals in another mode; eg electric motors; microphones; loudspeakers; turbines.

Transduction
The transfer of genetic material (DNA) from one bacterium to another by a bacteriophage (a VIRUS). The phages concerned are "temperate" ones, which do not kill the host bacterium immediately, but live within the cell for a while, often with their own DNA integrated into that of the host. When they eventually detach themselves they occasionally take a small portion of the host DNA with them as a result of a faulty excision process, and this can then become integrated into the DNA of the next cell that is infected. Transduction is used in genetic studies of bacteria to reveal linkage between genes.

Transfer RNA see NUCLEIC ACIDS; RNA.

Transformer
A device for altering the voltage of an AC supply (see ELECTRICITY), used chiefly for converting the high voltage at which power is transmitted over distribution systems to the normal domestic supply voltage, and for obtaining from the latter voltages suitable for electronic equipment. It is based on INDUCTION: the "primary" voltage applied to a coil wound on a closed loop of a ferromagnetic core creates a strong oscillating MAGNETIC FIELD which in turn induces in a "secondary" coil wound on the same core an AC voltage proportional to the number of turns in the secondary coil. The core is laminated to prevent the flow of "eddy" currents which would otherwise also be induced by the magnetic field and would waste some ENERGY as HEAT.

Transfusion, Blood
Means of BLOOD replacement in ANEMIA, SHOCK or HEMORRHAGE by intravenous infusion of blood from donors. It is the simplest and most important form of transplant and of enormous value, though it carries certain risks. Blood group compatibility based on ANTIBODY AND ANTIGEN reactions is of critical importance, as incompatible transfusion may lead to life-threatening shock and KIDNEY failure. Infection (eg HEPATITIS and AIDS) may be transmitted by blood, and FEVER or ALLERGY are common.

Transistor
Electronic device made of semiconducting materials used in a circuit as an AMPLIFIER, RECTIFIER, detector or switch. Its functions are similar to those of an electron tube, but it has the advantage of being smaller, more durable and consuming less power. The early and somewhat unsuccessful point-contact transistor has been superseded by the junction transistor, invented in 1948 by BARDEEN, BRATTAIN and SHOCKLEY. The junction transistor is a layered device consisting of two p-n junctions (see SEMICONDUCTOR) joined back to back to give either a p-n-p or n-p-n transistor. The three layers are formed by controlled addition of impurities to a semiconductor crystal, usually silicon or germanium. The thin central region (p-type in an n-p-n transistor and n-type in a p-n-p one) is known as the base, and the two outer regions (n-type semiconductor in an n-p-n transistor) are the emitter and collector, depending on the way an external voltage is connected. To act as an amplifier in a circuit, an n-p-n transistor needs a negative voltage to the collector and base. If the base is sufficiently thin, it attracts electrons from the emitter, which then pass through it to the positively charged collector. By altering the bias applied to the base (which need be only a few volts), large changes in the current from the collector can be obtained and the device amplifies. A collector current up to a hundred times the base current can be obtained. This type of transistor is analogous to a triode, the emitter and collector being equivalent to the cathode and anode respectively and the base to the control grid. The functioning of a p-n-p transistor is similar to the n-p-n type described, but the collector current is mainly holes rather than electrons. Transistors revolutionized the construction of electronic circuits, but are being replaced by INTEGRATED CIRCUITS in which they and other components are produced in a single semiconductor wafer.

Transit
The passage of a star across an observer's meridian (the great circle on the CELESTIAL SPHERE passing through his ZENITH and the north point of his horizon). The term is also applied to the passage of the inferior planets, Mercury and Venus, across the disk of the Sun.

Transition elements
The elements occupying the short groups in the PERIODIC TABLE – ie Groups IIIB to VIII, IB and IIB – in which the *d*-ORBITALS are being filled. The transition elements are all metals, including most of the technologically important ones. In general they are dense, hard and of high melting point. Their electronic structures, with many loosely bound unpaired *d*-electrons, account for their properties: they exhibit many different VALENCE states, form stable LIGAND complexes, mostly colored and paramagnetic, and are generally good catalysts. They form many stable ORGANOMETALLIC COMPOUNDS as well as carbonyls (compounds in which carbon monoxide, CO, acts as a ligand) with specially stable "push-pull" bonding. The second and third row transition elements are less reactive than the first row, and stable in higher valence states. (◆ page 130)

Transpiration
The loss of water by evaporation from the aerial parts of plants. Considerable quantities of water are lost in this way, far more than is needed to provide a flow of water for the upward movement of solutes. Transpiration is a necessary corollary of PHOTOSYNTHESIS, because in order to obtain sufficient sunlight considerable areas of leaf surface are needed, from which high loss of water by evaporation is inevitable. Plants have many means for reducing this, the waterproof CUTICLE that exchanges gas only through STOMATA playing an important part – stomata can be closed when evaporation rates are high. XEROPHYTES in particular are adapted for minimizing transpiration. (◆ page 44)
Transplants
Organs or tissues that are removed from one person and surgically implanted in another to replace lost or diseased organs. Autotransplantation is the moving of an organ from one place to another within a person where the original site has been affected by local disease (eg skin grafting). Blood TRANSFUSION was the first practical form of transplant. The nature of blood allows free transfusion between those with compatible blood groups. The next most important, and now most successful, of organ transplants is that of the KIDNEY. Here a single kidney is transplanted (from a live donor or from a person who has recently suffered sudden death) into a person who suffers from chronic renal failure. The kidney is placed beneath the skin of the abdominal wall and plumbed into the major ARTERIES and VEINS in the PELVIS and into the BLADDER. High doses of STEROIDS and IMMUNITY suppressants are used to minimize the body's tendency to reject the foreign tissue of the graft. Tissue typing methods are used in addition to blood grouping to minimize rejection. HEART transplantation has been much publicized, and is now becoming an accepted treatment for severe heart conditions. LIVER and LUNG transplants have also been attempted. Corneal grafting is a more widespread technique in which the cornea of the EYE of a recently dead person replaces that of a person with irreversible corneal damage leading to BLINDNESS. The lack of blood vessels in the cornea reduces the problem of rejection. Grafts from nonhuman animals are occasionally used (eg pig skin as temporary cover in extensive burns). Both animal and human heart valves are used in cardiac surgery.
Transsexual
A person who has undergone surgery to change the outward signs of gender. Most have their GONADS removed and are treated with hormone injections to make them resemble their new sex.
Transuranium elements
The elements with atomic numbers greater than that of URANIUM (92 – see PERIODIC TABLE; ATOM). None occurs naturally except in trace quantities: they are prepared by bombardment (usually with NEUTRONS or ALPHA PARTICLES) of suitably chosen lighter ISOTOPES. All are radioactive (see RADIOACTIVITY), and those of higher atomic number tend to be less stable. Those so far discovered are the ACTINIDES from neptunium through lawrencium, RUTHERFORDIUM and HAHNIUM. Only neptunium and plutonium have been synthesized in large quantity; most of the others have been produced in weighable amounts, but some with very short HALF-LIVES can be studied only by special tracer methods. (◆ page 130)
Trauma
Any sudden wound to the body or mind, often closely followed by SHOCK. In PSYCHOANALYSIS, a trauma is an immediate cause of ANXIETY that may develop into NEUROSIS. An infantile trauma is one that occurred in childhood but that affects the adult.
Travertine
Compact, banded LIMESTONE, usually light in color, evaporated or deposited from hot springs; sometimes applied also to STALACTITES AND STALAGMITES. Taking a high polish, it is used for interior decoration. Tufa, or calcareous sinter, is a porous equivalent.
Tree
A woody perennial PLANT with a well-defined main stem, or trunk. The trunk of a typical tree consists of thick-walled, water-conducting cells or xylem, which carry water and minerals upward, and phloem, which carries a solution of sugars downward, from the leaves to the roots. These conducting tissues are renewed every year as the trunk increases in girth, giving rise to the familiar ANNUAL RINGS. The older wood in the center of the tree (the heartwood) is much denser and harder than the younger, outer sapwood. The outer skin or BARK insulates and protects the trunk. Most living trees belong to the two most advanced groups of plants, the GYMNOSPERMS and the ANGIOSPERMS (the flowering plants). The former include the CONIFERS or cone-bearing trees such as the pine, spruce and cedar; they are nearly all evergreens and form the predominant vegetation in the cooler regions of the world (see BOREAL FOREST). The angiosperms have broader leaves and a different type of wood (see HARDWOOD; SOFTWOOD); in tropical climates they are mostly EVERGREEN, but in temperate regions the majority are DECIDUOUS. (See also FORESTS; WOOD.)

A tree-like growth form has evolved many separate times in the angiosperm group, with trees being found in dozens of different plant families. Most are DICOTYLEDONS, but the palm trees are MONOCOTYLEDONS; their structure is very different from that of other trees, the trunk being built up from old leaf-bases. A similar growth form is seen in the cycads and tree-ferns of the tropics and subtropics. These are now rare, but tree-like ferns were once widespread, as were giant tree-like clubmosses.
Trematoda
Parasitic flukes included in the PLATYHELMINTHES. They are similar to free-living flatworms (TURBELLARIA) but have a thick cuticle, and suckers enabling them to cling to their hosts. Trematode worms are responsible for diseases such as SCHISTOSOMIASIS.
Trephrine
Surgical instrument used for trepanning – making a circular hole in the skull or an internal organ in order to drain fluid, PUS, BLOOD or air.
Triassic
The first period of the MESOZOIC era, which lasted from about 248 to 213 million years ago. (See also GEOLOGY.) (◆ page 106)
Triceps
The major ARM muscle concerned with straightening the elbow. It has three heads or bellies at its upper end with separate attachments. (◆ page 216)
Trichinosis
Infestation with the larva of a worm (*Trichinella*), contracted from eating uncooked pork, etc., causing a feverish illness. Edema around the eyes, muscle pains and diarrhea occur early; later the lungs, heart and brain may be involved. It is avoided by the adequate cooking of pork.
Triglycerides
Organic compounds containing three long-chain FATTY ACIDS bound to a molecule of GLYCEROL. They are produced by animals as energy storage compounds. Triglycerides are insoluble in water, and occur naturally as either liquids or solids; those liquid at 20°C are normally termed oils and are generally found in plants and fish. Oils often contain triglycerides of the unsaturated fatty acid OLEIC ACID, which can be converted to the more saturated, and therefore solid, STEARIC ACID by HYDROGENATION in the presence of finely divided nickel. This process is basic to the manufacture of margarine. Triglycerides and other lipids are the most concentrated sources of energy in the human diet, giving over twice the energy of STARCHES. Diets containing high levels of animal fat (rich in saturated fatty acids and CHOLESTEROL) have been implicated as causative factors in heart disease, and replacement of animal fat by plant oils (eg peanut oil, sunflower oil) is recommended.
Trilobites
A group of extinct invertebrates. They were segmented, with a pair of legs on each segment, and looked something like woodlice (sowbugs), but with a broad shield, known as a cephalon, acros the head region. Trilobites were very common from about 590 to 350 million years ago, and did not finally become extinct until about 260 million years ago. They are not closely related to any living organisms, but may have shared a common ancestor with the CRUSTACEANS. (See also ARTHROPODS.)
Trinitrotoluene see TNT.
Triode
ELECTRON TUBE with positive ANODE, electron-emitting CATHODE and negatively biased control grid, used as an AMPLIFIER or OSCILLATOR.
Triplets see MULTIPLE BIRTHS.
Tritium (T or $_1H^3$)
The heaviest isotope of HYDROGEN, whose nucleus has one proton and two neutrons, produced by neutron irradiation of lithium (Li^6). Tritium is weakly radioactive and emits beta rays. It is used for tracer studies, in luminous paints, and together with deuterium in HYDROGEN BOMBS. (See also FUSION, NUCLEAR.) AW 3.0, mp −252.5°C, bp −248.1°C.
Tropical medicine
Branch of MEDICINE concerned with the particular diseases encountered in and sometimes imported from the tropics. These largely comprise INFECTIOUS DISEASES due to VIRUSES (eg YELLOW FEVER, SMALLPOX, lassa fever), BACTERIA (eg CHOLERA), protozoa (eg MALARIA, TRYPANOSOME diseases) and worms (eg FILARIASIS) that are generally restricted to tropical zones. The diseases of MALNUTRITION also fall within the province of tropical disease, as do SUNSTROKE and SNAKE BITES.
Tropics
The lines of latitude lying about 23½°N (Tropic of Cancer) and S (Tropic of Capricorn) representing the farthest northerly and southerly latitudes where the Sun is, at one time of the year, directly overhead at noon. This occurs at the summer SOLSTICE in each hemisphere. The term is also used of the area between the tropics. (◆ page 25)
Tropism
Movement of plants in response to external directional stimuli. If a plant is laid on its side, the stem will soon start to bend upward again. This movement (geotropism) is a response to the force of gravity. The stem is said to be negatively geotropic. Roots are generally positively geotropic and grow downward. Phototropisms are bending movements in response to the direction of illumination. Stems are generally positively phototropic (bend toward the light). Most roots are negatively phototropic, though some appear unaffected by light. Some roots exhibit positive hydrotropism: they bend toward moisture. This response is more powerful than the response to gravity; roots can be deflected from their downward course if the plants are watered only on one side. Tropisms are controlled by differences in concentration of growth HORMONES. (See also AUXINS.)
Troposphere
The innermost zone of the Earth's ATMOSPHERE, from the surface up to the tropopause. (◆ page 25)

Troy weight
A system of weights used for precious metals and stones, named for the French town of Troyes, famed for its medieval trade fairs. Today the price of gold is still quoted in dollars per troy ounce. The troy ounce, equivalent to 1.0971 ounces avoirdupois, is equal to the apothecaries' ounce.

Trypanosomes
Protozoa responsible for trypanosomiasis of the African (SLEEPING SICKNESS) and South American (CHAGAS' DISEASE) varieties, carried by the tsetse fly and certain bugs respectively.

Trypsin
ENZYME catalyzing the breakdown of PROTEINS in the DIGESTIVE SYSTEM. (♦ page 40)

Tsiolkovsky, Konstantin Eduardovich (1857-1935)
Russian physicist who pioneered ROCKET science, but who is perhaps most important for his role in educating the Soviet government and people into acceptance of the future potential of SPACE EXPLORATION. He also built one of the first WIND TUNNELS (*c*1892). A large crater on the far side of the MOON is named for him; and the timing of Sputnik I's launch commemorated the centenary of his birth.

Tsunami
Fast-moving ocean wave, formerly called a tidal wave, caused by submarine EARTHQUAKES, volcanic eruptions, etc, found mainly in the Pacific, and often taking a high toll of lives in affected coastal areas. In midocean, the wave height is usually under 1m, the distance between succeeding crests of the order of 200km, and the velocity about 750km/h. Near the coast, FRICTION with the sea bottom slows the wave, so that the distance between crests decreases, the wave height increasing to about 25m or more.

Tubers
Swollen underground stems and roots that are organs of perennation and VEGETATIVE REPRODUCTION and contain stored food material. The potato is a stem tuber. It swells at the tip of a slender underground stem (or stolon) and gives rise to a new plant the following year. Dahlia tubers are swollen roots.

Tuberculin
PROTEIN derivative of the mycobacteria responsible for TUBERCULOSIS. This may be used in tests of cell-mediated IMMUNITY to tuberculosis, providing evidence of previous disease (often subclinical) or immunization (BCG). The substance was originally isolated by KOCH.

Tuberculosis (TB)
A group of INFECTIOUS DISEASES caused by the BACILLUS *Mycobacterium tuberculosis*, which kills some 3 million people every year throughout the world. TB may invade any organ but most commonly affects the respiratory system, where it has been called consumption or phthisis. The disease is spread in three ways: inoculation via cuts, etc; inhalation of infected sputum; and ingestion of infected food. In pulmonary TB there are two stages of infection. In primary infection there are usually no significant symptoms; dormant small hard masses called tubercles are formed by the body's defenses. In postprimary infection the dormant BACTERIA are reactivated as a result of the weakening of the body's defenses, and clinical symptoms become evident. Symptoms include fatigue, weight loss, persistent cough with green or yellow sputum and possibly blood.

The TUBERCULIN skin test can show whether a person has some IMMUNITY to the disease, though the detection of the disease in its early stages, when it is readily curable, is difficult. Control of the disease is accomplished by preventive measures such as BCG vaccination, isolation of infectious people and food sterilization.

Tufa see TRAVERTINE.

Tularemia (rabbit fever)
INFECTIOUS DISEASE due to BACTERIA, causing fever, ulceration, LYMPH node enlargement and sometimes PNEUMONIA. It is carried by wild animals, particularly rabbits, and insects.

Tumor
Strictly, any swelling on or in the body, but more usually used to refer to an abnormal overgrowth of tissue (neoplasm). These may be harmless proliferations such as fibroids of the uterus, or they may be forms of CANCER, LYMPHOMA or SARCOMA, which are generally malignant. The rate of growth, the tendency to spread locally and to distant sites via the BLOOD vessels and LYMPH system, and systemic effects determine the degree of malignancy of a given tumor. Tumors may present as a lump, by local compression effects (especially with BRAIN tumors), by bleeding (GASTROINTESTINAL TRACT tumors) or by systemic effects including ANEMIA, weight loss, false HORMONE actions, NEURITIS, etc.

Tundra
The vegetation zone nearest the polar regions, where the subsoil is permanently frozen (permafrost), preventing surface water from draining away and creating a landscape of pools and mires. The vegetation consists principally of lichens, mosses, grasses and low-growing shrubs such as dwarf willow. During the brief summer a great many insects breed in the pools, and migrant birds nest and rear their young. In BIOGEOGRAPHY, the term tundra is often extended to include the vegetation of mountainous regions above the tree-line. In these areas permafrost is rarely found, but the climate and vegetation are similar to those of the polar tundra.

Tungsten (wolfram, W)
Hard, silvery gray metal in Group VIB of the PERIODIC TABLE; a TRANSITION ELEMENT. Its chief ores are SCHEELITE and WOLFRAMITE. The metal is produced by reduction of heated tungsten dioxide with hydrogen. Its main uses are in tungsten steel ALLOYS for high-temperature applications, and for the filaments of incandescent lamps. It is relatively inert, and resembles MOLYBDENUM. Cemented tungsten carbide (WC) is used in cutting tools. AW 183.9, mp 3422°C, bp 5700°C, sg 19.3 (20°C). (♦ page 130)

Tunicates
A subphylum of the CHORDATES containing the sea squirts. They are sessile marine organisms, and have sac-shaped bodies with two openings, a mouth and an atrial port. Water is sucked in through the mouth, filtered for food particles, and passed out through the atrial port. The tunicate larva is a small, free-swimming, tadpole-like animal, with a NOTOCHORD, similar to AMPHIOXUS. It is believed that amphioxus evolved from a tunicate member by NEOTENY, and that the earliest vertebrate may have arisen by a similar route. (♦ page 17)

Tunnel diode
SEMICONDUCTOR device with a high impurity concentration and negative RESISTANCE over part of its operating range, used in amplifying, oscillating and switching circuits. Its operation depends on quantum-mechanical tunneling of charges through a narrow p-n junction at zero voltage which ceases at increased forward voltages.

Turbine
Machine for directly converting the kinetic and/or thermal ENERGY of a flowing FLUID into useful rotational energy. The working fluid may be air, hot gas, steam or water. This either pushes against a set of blades mounted on the drive shaft (impulse turbines) or turns the shaft by reaction when the fluid is expelled from nozzles (or nozzle-shaped vanes) around its circumference (reaction turbines). Water turbines were the first to be developed. They now include the vast inward-flow reaction turbines used in the generation of HYDROELECTRICITY, and the smaller scale tangential-tow "Pelton wheel" impulse types used when exploiting a very great "head" of water. In the 1880s Charles Algernon Parsons (1854-1931), a British engineer, designed the first successful steam turbines, having realized that the efficient use of high pressure steam demanded that its energy be extracted in a multitude of small stages. Steam turbines thus consist of a series of vanes mounted on a rotating drum with stator vanes redirecting the steam in between the moving ones. They are commonly used as marine engines and in thermal and nuclear power plants. Gas turbines are not as yet widely used except in airplanes (see JET PROPULSION) and for peak-load electricity generation.

Turbulence
Type of irregular flow of FLUIDS in which the motion at any point varies rapidly in magnitude and direction. The value of REYNOLDS NUMBER determines whether fluid flow is laminar (smooth and well-defined), or turbulent. Most natural fluid motion is turbulent.

Turquoise
Hydrated copper aluminum PHOSPHATE mineral, $CuAl_6(PO_4)_4(OH)_8.4H_2O$; it is used as a semi-precious GEM stone, blue in color. Deposited from water, it occurs in veinlets and as masses. The finest turquoise comes from Iran.

Twins see MULTIPLE BIRTHS; SIAMESE TWINS.

Tyndall, John (1820-1893)
British physicist who, through his studies of the scattering of light by colloidal particles or large molecules in SUSPENSION (the Tyndall effect), showed that the daytime sky is blue because of the Rayleigh SCATTERING of impingent sunlight by dust and other colloidal particles in the air (see COLLOID).

Typhoid fever
INFECTIOUS DISEASE due to a SALMONELLA species causing fever, a characteristic rash, LYMPH node and SPLEEN enlargement, also GASTROINTESTINAL TRACT disturbance with bleeding and ulceration, and usually marked malaise or prostration. It is contracted from other cases or from disease carriers, the latter often harboring asymptomatic infection in the GALL BLADDER or urine, with contaminated food and water as major vectors.

Typhoon see HURRICANE.

Typhus
INFECTIOUS DISEASE caused by RICKETTSIA and carried by lice, leading to a feverish illness with a rash. Severe headache typically precedes the rash, which may be erythematous or may progress to skin HEMORRHAGE; mild respiratory symptoms of cough and breathlessness are common. Death ensues in a high proportion of untreated adults, usually with profound SHOCK and KIDNEY failure. Recurrences may occur in untreated patients who recover from their first attack, often after many years (Brill-Zinsser disease). A similar disease due to a different but related organism is carried by fleas (Murine typhus).

Ulcer
Pathological defect in SKIN or other EPITHELIUM, caused by INFLAMMATION secondary to infection, loss of blood supply, failure of venous return or CANCER. Various skin lesions can cause ulcers, including infection, arterial disease, varicose veins and skin cancer. Aphthous ulcers in the mouth are painful epithelial ulcers of unknown origin. Peptic ulcers include gastric and duodenal ulcers, though the two have different causes; they may cause characteristic pain, acute HEMORRHAGE, or lead to perforation and PERITONITIS. Severe scarring or EDEMA around the pylorus may cause stenosis with vomiting and stomach distension.
Ultracentrifuge see CENTRIFUGE.
Ultra high frequency waves (UHF) see ELECTROMAGNETIC RADIATION; RADIO.
Ultramicroscope
A MICROSCOPE for studying liquid suspensions of particles too small for direct microscopy (10nm-1μm), using light scattered by the particles at right angles. It allows their number and position to be determined, their motion to be followed, and their size to be estimated, though no structural detail can be discerned.
Ultrasonics
Science of SOUND waves with frequencies above those that humans can hear (720 kHz). With modern piezoelectric techniques, ultrasonic waves having frequencies above 24 kHz can readily be generated with high efficiency and intensity in solids and liquids, and exhibit the normal wave properties of REFLECTION, REFRACTION and DIFFRACTION. They can thus be used as investigative tools or for concentrating large amounts of mechanical energy. Low-power waves are used in thickness gauging and HOLOGRAPHY, high-power waves in surgery, for industrial homogenization, cleaning and machining. (♦ page 225)
Ultraviolet astronomy
The study of ULTRAVIOLET radiations emitted by celestial objects. As UV radiation above 310nm can penetrate to the Earth's surface, the sky in the longer UV can be studied with optical telescopes. UV astronomy is therefore taken to be the study of wavelengths between 10 and 310nm, with UV telescopes on balloons, rockets and satellites. A star that is hotter than 10,000K shines most brightly at UV wavelengths, and it is these young, hot stars that UV telescopes see.
Ultraviolet radiation
ELECTROMAGNETIC RADIATION of wavelength between 0.1nm and 380nm, produced using gas discharge tubes. Although it constitutes 5% of the energy radiated by the Sun, most falling on the Earth is filtered out by atmospheric OXYGEN and OZONE, thus protecting life on the surface from destruction by the solar ultraviolet light. This also means that air must be excluded from optical apparatus designed for ultraviolet light, similar strong absorption by glass necessitating that lenses and prisms be made of QUARTZ or FLUORITE. Detection is photographic or by using fluorescent screens. The principal use is in fluorescent tubes (see LIGHTING) but important medical applications include GERMICIDAL LAMPS, the treatment of RICKETS and some skin diseases and VITAMIN D enrichment of milk and eggs. (♦ page 25, 118)
Umbellifers
Members of the plant family Umbelliferae, which are characterized by their lacy, umbrella-like flowerheads. They include such plants as cow parsley and hemlock, as well as various crops – carrot, parsley, fennel, coriander, parsnip and angelica.
Umbilical cord
Long tubelike structure linking the developing EMBRYO or FETUS to the PLACENTA through most of PREGNANCY. It consists of BLOOD vessels taking blood to and from the placenta, and a gelatinous matrix. In humans the cord is clamped at birth to prevent blood loss and is used to assist delivery of the placenta. It undergoes ATROPHY and becomes the navel.
Umbra see SHADOW; ECLIPSE.
Uncertainty principle (indeterminacy principle)
A restriction, first enunciated by W.K. HEISENBERG in 1927, on the accuracy with which the position and MOMENTUM of an object can be established simultaneously: the product of the accuracies attainable in each cannot be less than the PLANCK CONSTANT. Likewise, time and energy cannot be known simultaneously with arbitrary accuracy. Relevant only near the atomic level, the principle arises from the wave nature of matter: a particle consists of a superposition of waves with slightly different speeds producing a localized disturbance of which neither the position nor the speed is precisely defined.
Unconformity
A surface between two contiguous rock strata representing a break in the normal succession; usually owing to EROSION having removed layers of rock before the deposition of the younger stratum. They may be parallel (strata parallel), angular (strata not parallel), heterolithic (sediment over intruded IGNEOUS ROCKS) or non-depositional (a genuine break in the deposition pattern). (See also SEDIMENTARY ROCKS.)
Unconscious
The part of the mind in which events that the individual is unaware of take place; ie the part of the mind that is not the conscious. Unconscious processes can, however, alter the behavior of the individual (see also DREAMS; INSTINCT). FREUD renamed the unconscious the ID. (See also COLLECTIVE UNCONSCIOUS.)
Unconsciousness
Lack of awareness, the commonest example of which is sleep; or the lack of self-awareness displayed by most, if not all, animals (see CONSCIOUSNESS; COLLECTIVE UNCONSCIOUS).
Undulant fever see BRUCELLOSIS.
Undulipodium
The flagellum of bacterial cells. This alternative name has been advocated because of the great structural differences between it and the flagellum of EUKARYOTES.
Ungulates
General name for all hoofed mammals, including the odd-toed and even-toed groups, PERISSODACTYLS and ARTIODACTYLS respectively.
Unified field theory
Theory which tries to incorporate the four fundamental forces of nature, and explain them in terms of one underlying force. The electroweak theory, developed in the 1970s, successfully combines the ELECTROMAGNETIC FORCE with the WEAK NUCLEAR FORCE. The so-called grand unified theories attempt to combine the theory of the STRONG NUCLEAR FORCE (quantum chromodynamics) with electroweak theory; one outcome of this unification may be that the PROTON is not stable. The final stage or total unification will be to include GRAVITY, in particular as it is described by the general theory of RELATIVITY.
Uniformitarianism
The principle, due to J. HUTTON and C. LYELL, that the same agencies are at work in nature today, operating at the same intensities, as they have always done throughout geological time. It was originally opposed to CATASTROPHISM.
Uniramia
The insects and myriapods; see ARTHROPODS. Their limbs are unbranched, whereas in the CRUSTACEANS the limbs are branched (biramous). (♦ page 16)
Universe
The system of all that exists or happens (see also HEAT DEATH). In COSMOLOGY, the term is applied to our Universe, ie all that we can observe and the presumably homogeneous and isotropic extension thereof. It is possible that other universes may exist or that, if we inhabit an oscillating universe, there may have been prior universes and there may be subsequent universes. (♦ page 240)
Unnilquadium (Unq)
One of four newly postulated elements, none of which has yet been isolated. The others are unnilpentium (Unp); unnilhexium (Unh); and unnilseptium (Uns). (♦ page 130)
Unsaturated fats see FATTY ACIDS.
Uracil see NUCLEIC ACIDS; NUCLEOTIDES.
Uraninite see PITCHBLENDE.
Uranium (U)
Soft, silvery white radioactive metal in the ACTINIDE series; the heaviest natural element. It occurs widely as PITCHBLENDE (uraninite), CARNOTITE and other ores, which are concentrated and converted to uranium (IV) fluoride, from which uranium is isolated by electrolysis or reduction with calcium or magnesium. The metal is reactive and electropositive, reacting with hot water and dissolving in acids. Its chief oxidation states are +4 and +6, and the uranyl (UO_2^{2+}) compounds are common. Uranium has three naturally occurring ISOTOPES: U^{238} (HALF-LIFE 4.5×10^9yr); U^{235} (half-life 7.1×10^8yr) and U^{234} (half-life 2.5×10^5 yr). More than 99% of natural uranium is U^{238}. The isotopes may be separated by fractional DIFFUSION of the volatile uranium (VI) fluoride. Neutron capture by U^{235} leads to nuclear FISSION, and a chain reaction can occur which is the basis of NUCLEAR REACTORS and of the ATOMIC BOMB. U^{238} also absorbs neutrons and is converted to an isotope of PLUTONIUM (Pu^{239}) which (like U^{235}) can be used as a nuclear fuel. Uranium is the starting material for synthesis of the TRANSURANIUM ELEMENTS. AW 238.0, mp 1135°C, bp 4000°C, sg 19.05(α). (♦ page 130, 147)
Uranus
The third largest planet in the SOLAR SYSTEM and the seventh from the Sun. Physically very similar to NEPTUNE, but rather larger (equatorial diameter 52Mm) it orbits the Sun every 84.02 years at a mean distance of 19.2AU, rotating in about 17h. The plane of its equator is tilted 98° to the plane of its orbit, such that the rotation of the planet and the revolution of its five moons, which orbit closely parallel to the EQUATOR, are retrograde (see RETROGRADE MOTION). In March 1977 an occultation of a star by Uranus led to the discovery of a series of rings similar to SATURN's. VOYAGER 2's fly-by in January 1986 greatly enhanced our knowledge of Uranus revealing faint atmospheric cloud belts, a magnetic field tilted by 55° to the rotation axes, and 15 moons. (♦ page 134, 221, 240)
Urea ($CO(NH_2)_2$)
The end-product of protein breakdown in many

mammals, excreted in the URINE. It was the first organic compound to be synthesized – by Wöhler in 1828 from ammonium cyanate – in a historic demonstration that disproved the idea that organic compounds could not be synthesized from inorganic precursors. Urea is now prepared by heating ammonia and carbon dioxide under pressure. When purified it is a white crystalline solid. MW 60.1, mp 135°C. (See also EXCRETION.) (♦ page 233)

Uremia
The syndrome of symptoms and biochemical disorders seen in KIDNEY failure, associated with a rise in blood UREA and other nitrogenous waste products of PROTEIN metabolism. Nausea, vomiting, malaise, itching, pigmentation, ANEMIA and acute disorders of fluid and mineral balance are common presentations, but the manifestations depend on the type of disease, rate of waste buildup, etc.

Ureters
The pair of tubes that carry urine from the KIDNEYS to the BLADDER. (♦ page 233)

Urethra
The tube that carries urine from the BLADDER for EXCRETION. (♦ page 204)

Urey, Harold Clayton (1893-1981)
US chemist awarded the 1935 Nobel Prize for Chemistry for his discovery of DEUTERIUM, an isotope of HYDROGEN having one proton and one neutron in its nucleus, and who played a major part in the MANHATTAN PROJECT. He is also important as a cosmologist: his researches into geological dating using oxygen isotopes enabled him to produce a model of the atmosphere of the primordial planet Earth; and hence to formulate a theory of the planets' having originated as a gaseous disk about the Sun (see SOLAR SYSTEM).

Uric acid (2,6,8-trihydroxypurine)
The end-product of protein breakdown in birds, invertebrates and snakes, and of PURINE breakdown in many insects, reptiles, birds, primates (including humans) and the Dalmatian dog. Sufferers from GOUT have a high blood level of uric acid.

Urine
A waste product comprising a dilute solution of excess salts and unwanted nitrogenous material, such as UREA or URIC ACID, excreted by many animals. The wastes are filtered from the blood in the KIDNEYS or equivalent structures, and the urine stored in the BLADDER until it can be eliminated. The passage of urine serves not only to eliminate wastes, but also provides a mechanism for maintaining the water and salt concentrations of the blood. While all mammals excrete their nitrogenous wastes in urine as urea, some animals excrete them as ammonia gas or in solid crystals of uric acid. In birds the urine is not watery but a thick white paste that is discharged with the feces.

Urochordates see TUNICATES.

Ursa major see GREAT BEAR.

Ursa minor see LITTLE DIPPER.

Urticaria see HIVES.

Uterus (womb)
Female reproductive organ which is specialized for IMPLANTATION of the EGG and development of the EMBRYO and FETUS during PREGNANCY. The regular turnover of its lining under the influence of ESTROGEN and PROGESTERONE is responsible for MENSTRUATION. Disorders of the uterus include malformation, abnormal position, and disorders of menstruation, of which benign tumors or fibroids are a common cause. CANCER of the uterus or its cervix is relatively common and may be detected by the use of regular smear tests. Removal of the uterus for cancer, fibroids, etc is HYSTERECTOMY. (♦ page 204)

Uveitis
Inflammation of the iria and choroid membrane of the eye, which together form the uveal tract.

Uvula
Soft central portion of the soft PALATE that hangs at the back of the PHARYNX, forming part of the occluding mechanism that can functionally separate the nasopharynx from the oropharynx.

Vaccination
Method of inducing IMMUNITY to INFECTIOUS DISEASE due to BACTERIA or VIRUSES. Based on the knowledge that second attacks of diseases such as SMALLPOX were uncommon, early methods of protection consisted in inducing immunity by deliberate inoculation of material from a mild case. Starting from the observation that farm workers who had accidentally acquired cowpox by milking infected cows were resistant to smallpox, EDWARD JENNER in the 1790s inoculated cowpox material into nonimmune persons, who then showed resistance to smallpox. LOUIS PASTEUR extended this work to experimental chicken CHOLERA, human ANTHRAX and RABIES. The term vaccination became general for all methods of inducing immunity by inoculation of products of the infectious organism. ANTITOXINS were soon developed in which specific immunity to disease TOXINS was induced. Vaccination leads to the formation of antibodies and the ability to produce large quantities rapidly at a later date (see ANTIBODIES AND ANTIGENS); this gives protection equivalent to that induced by an attack of the disease. It is occasionally followed by a reaction resembling a mild form of the disease, but rarely by its serious manifestations.

Vacuole
A small, usually fluid-filled, cavity in a cell. (♦ page 13)

Vacuum
Any region of space devoid of ATOMS and MOLECULES. Such a region will neither conduct HEAT nor transmit SOUND waves. Because all materials that surround a space have a definite VAPOR PRESSURE, a perfect vacuum is an impossibility, and the term is usually used to denote merely a space containing air or other gas at very low PRESSURE. Pressures less than 0.1μPa occur naturally about 800km above the Earth's surface, though pressures as low as 0.01nPa can be attained in the laboratory. The low pressures required for many physics experiments are obtained using various designs of vacuum PUMP.

Vacuum tube
An evacuated ELECTRON TUBE.

Vagina
The canal connecting the UTERUS to the outside of the female body, which receives the PENIS during sexual intercourse and acts as a canal during birth. Vaginitis is a condition in which it becomes inflamed, often due to hormone loss after the menopause. Vaginismus is painful contraction of the vagina during intercourse. (♦ page 204)

Valence (valency)
The combining power of an ELEMENT, expressed as the number of chemical BONDS that one atom of the element forms in a given compound. In general, the characteristic valence of an element in Group N of the PERIODIC TABLE is N or $(8-N)$. (♦ page 37)

Valley
Long, narrow depression in the Earth's surface, usually formed by GLACIER or river EROSION. Young valleys are narrow, steep-sided and V-shaped; mature valleys are broader, with gentler slopes. Some (RIFT VALLEYS) are the result of collapse between FAULTS. Hanging valleys, of glacial origin, are side valleys whose floor is considerably higher than that of the main valley.

Valve
Mechanical device which, by opening and closing, enables the flow of fluid in a pipe or other vessel to be controlled. Common valve types are generally named after the shape or mode of operation of the movable element, eg cone, or needle, valve; gate valve; globe valve; poppet valve; and rotary plug cock. Self-acting valves include: safety valves, usually spring-loaded and designed to open at a predetermined pressure; nonreturn valves, which permit flow in one direction only; and float-operated valves, set to shut off a feeder pipe before a container overflows.

Valve, Electronic see ELECTRON TUBE.

Vanadium (V)
Silvery white, soft metal in Group VB of the PERIODIC TABLE; a TRANSITION ELEMENT. It is widespread, the most important ores being CARNOTITE and roscoelite; it is isolated by reduction of vanadium (V) oxide with calcium. Most is used in ALLOYS to make hard and wear-resistant STEELS. Vanadium is fairly unreactive. It forms compounds in oxidation states +2, +3, +4 and +5. Vanadium (V) oxide is used in ceramics and as a catalyst in the CONTACT PROCESS. AW 50.9, mp 1920°C, bp 3400°C, sg 5.96 (20°C). (♦ page 130)

Van Allen, James Alfred (b. 1914)
US physicist responsible for the discovery of the VAN ALLEN RADIATION BELTS (1958).

Van Allen radiation belts
The belts of high-energy charged particles, mainly PROTONS and ELECTRONS, surrounding the Earth, named for VAN ALLEN, who discovered them in 1958. They extend from a few hundred to about 50,000km above the Earth's surface, and radiate intensely enough to make it essential for astronauts to be specially protected from them. The particles are trapped in these regions by the Earth's magnetic field.

Van de Graaff generator see ELECTROSTATIC GENERATOR.

Van der Meer, Simon (b. 1929)
Dutch physicist who received the 1984 Nobel Prize for Physics jointly with CARLO RUBBIA for their discovery of the charged W and the neutral Z particles of the WEAK NUCLEAR FORCE (see also SUBATOMIC PARTICLES).

Van der Waals, Johannes Diderik (1837-1923)
Dutch physicist who investigated the properties of real GASES. Noting that the KINETIC THEORY of gases assumed that the molecules had neither size nor interactive forces between them, in 1873 he proposed Van der Waals' Equation: $(P+a/V^2)(V-b)=RT$, (where P, V and T are pressure, volume and absolute temperature, R is the universal GAS constant, and a and b are constants whose values depend on the particular gas in question) in which allowance is made for both these factors. The weak attractive forces between molecules are therefore named Van der Waals forces. He received the 1910 Nobel Prize for Physics.

Van 't Hoff, Jacobus Henricus (1852-1911)
Dutch physical chemist awarded in 1901 the first Nobel Prize for Chemistry for his work laying the foundations of STEREOCHEMISTRY. (See also OSMOSIS.)

Van Vleck, John Hasbrouck (1899-1980)
US physicist who shared the 1977 Nobel Prize for Physics with P.W. ANDERSON and N.F. MOTT for his work on the QUANTUM THEORY of magnetic materials. He also contributed to LIGAND field theory.

Vane, Sir John R. (b. 1927)
British pharmacologist who shared the 1982 Nobel Prize for Physiology or Medicine with BERGSTROM and SAMUELSSON for control of PROSTAGLANDINS. Vane's research provided the means of measuring prostaglandin activity in tissue.

Vapor
The gaseous state of a substance (usually one that is solid or liquid at room temperature). An isothermal increase in pressure can convert a vapor to a liquid. To convert a solid or liquid to vapor, heat is needed to overcome the cohesive forces between molecules and allow them to escape.

Vapor pressure
The pressure exerted by a VAPOR in EQUILIBRIUM with its liquid or solid. In an enclosed space this occurs when equal numbers of MOLECULES are entering and leaving the vapor, which is then saturated. For a pure substance, the saturated vapor pressure (SVP) depends on the temperature. The BOILING POINT of a liquid is reached when the saturated vapor pressure equals the external pressure.

Variable stars
Stars that vary in brightness. There are two main categories. Extrinsic variables are those whose variation in apparent brightness is caused by an external condition, as in the case of eclipsing binaries (see DOUBLE STAR). Intrinsic variables vary in absolute brightness owing to physical changes within them. They may vary either regularly or irregularly: most intrinsic variables are pulsating or eruptive in nature. Pulsating variables, which vary in size, are the most common type of variable star: they include the RR Lyrae stars, with periods from 1.5h to a little over a day, W Virginis stars and RV Tauri stars (all three types appearing principally in GLOBULAR CLUSTERS); long period and semiregular variables, which are red giants, and the CEPHEID VARIABLES. Types of variable stars whose periods are known to have a relationship to their absolute brightness are especially important in that they can be used to determine large astronomical distances. Eruptive variables include flare stars (UV Ceti stars), faint red stars that suddenly flare up by one or two magnitudes in brightness once or twice a day. NOVAE and SUPERNOVAE involve much more extreme eruptions.

Variation
The genetic diversity found in all natural populations of organisms. Variation is the result of differing effects of MUTATION and RECOMBINATION. Genetic diversity is important because it provides variants, some of which may be more suited to prevailing conditions than others, thus providing raw material on which NATURAL SELECTION operates. (See also EVOLUTION.)

Varicella see CHICKENPOX.

Varices
Swellings; see VARICOSE VEINS.

Varicose veins
Veins in which the valves have ceased to prevent the backflow of blood, so they become swollen. They are most common in the superficial veins of the leg.

Varve
The layer of sediment deposited in the course of a single year, specifically in a lake formed of glacial meltwater. Characteristically, a varve has a SILT layer overlying a SAND layer. Study of varves is of great use in geological dating. (See also GLACIER; SEDIMENTATION.)

Vascular plants
Members of the PLANT kingdom in which vascular tissues such as XYLEM and PHLOEM are present and the plants are fully differentiated into roots, stems and leaves. Vascular plants include PTERIDO-

PHYTES, GYMNOSPERMS and ANGIOSPERMS. In all these, the sporophyte is either dominant and the gametophyte is inconspicuous or enclosed within the sporophyte; see ALTERNATION OF GENERATIONS. (♦ page 139)

Vascular system
Any system for the conduction of fluids around a living body. In plants it comprises the PHLOEM and XYLEM. In vertebrate animals it includes the CIRCULATORY SYSTEM, comprising BLOOD, ARTERIES, CAPILLARIES, VEINS and the HEART; the LYMPH vessels form a further subdivision. Its function is to deliver nutrients (including oxygen) to, and remove wastes from, all organs, and to transport HORMONES and the agents of body defense.

Vasectomy
Form of STERILIZATION in males in which the vas deferens on each side is cut and tied off to prevent SPERM from reaching the seminal vesicles and hence the urethra of the PENIS. It does not affect erection or ejaculation but does effect permanent sterilization.

Vasoconstriction
Narrowing of blood vessels, facilitating control of blood pressure and body temperature.

Vasodilation
Or vasodilatation, widening of blood vessels, facilitating control of blood pressure and body temperature.

Vasopressin (antidiuretic hormone, ADH)
HORMONE produced by the HYPOTHALAMUS and posterior PITUITARY GLAND, which is a mild vasoconstrictor, but primarily inhibits diuresis or loss of water in URINE. It is a vital link in the system for preserving the HOMEOSTASIS or constancy of body fluids. (♦ page 111)

Vector
A quantity having both magnitude and direction, unlike a scalar, which has only magnitude. One example of a vector quantity is VELOCITY.

Vega
Alpha Lyrae, the fourth brightest star in the night sky (apparent magnitude 0.03, absolute magnitude +0.5). It is 8pc distant, and 58 times as bright as the Sun. The InfraRed Astronomical Satellite (IRAS) discovered infrared radiation near Vega in 1983, suggesting that planetary material is orbiting the star. If this proves correct, it is the first evidence of a SOLAR SYSTEM forming apart from our own. (♦ page 228)

Vegetables
Any edible plant part that is not a fruit, cereal, herb, spice or beverage. Many vegetables are in fact FRUITS in the botanical sense (eg tomatoes, peppers, green beans and cucumbers). Others are the storage organs of BIENNIAL and PERENNIAL plants (eg potatoes, radishes, onions, carrots, beetroot), the leaves of plants (eg cabbage), enlarged buds (eg Brussels sprouts), enlarged stems (eg fennel, celery), flowerheads (eg broccoli, cauliflower, globe artichokes) or seeds (eg peas and beans).

Vegetative reproduction
The various forms of asexual reproduction seen in plants, but not including spore production. Forms of vegetative reproduction include the production of tubers, corms, bulbs, rhizomes, runners, bulbils (small bulbs developing asexually from flowers) and gemmae (clusters of cells produced by liverworts).

Vein
A mineral formation of far greater extent in two dimensions than in the third. Sheetlike fissure veins occur where fissures formed in the rock become filled with MINERAL. Ladder veins form in series of fractures in, eg, DIKES. Saddleveins are lens-shaped, concave below and convex above. Veins that contain economically important ORES are often termed lodes.

Veins
Thin-walled, collapsible vessels that return BLOOD to the HEART from the tissue CAPILLARIES. They contain valves that prevent backflow, especially in the legs in humans. Blood drains from the major veins into the inferior or superior VENA CAVA. Blood in veins is at low pressure and depends for its return to the heart on intermittent muscle compression, combined with valve action. (♦ page 28)

Veld (veldt)
Open GRASSLAND of South Africa, divided into three types: high veld, around 1500m above sea level, which is similar to the PRAIRIES; Cape middle veld, somewhat lower, covered with scrub and occasional low ridges of hills; and low veld, under about 750m above sea level.

Velocity
The rate at which the position of a body changes, expressed with respect to a given direction. Velocity is thus a VECTOR quantity, of which the corresponding scalar is speed: the rate of change in the position of a body without respect to direction. Translational velocity refers to movement through space; angular or rotational velocity to rotation about a given axis. (See also ACCELERATION.)

Vena cava
Two major veins in the human body. The superior vena cava is a vein collecting blood from the head, neck and arms, and delivering it to the right side of the heart. The inferior vena cava performs the same function with blood from the legs and abdomen. (♦ page 28)

Venera
Series of unmanned orbiters and landers launched by Soviets to explore VENUS, starting in 1961. Venera 3 in March 1966 was the first vehicle to land on another planet; Venera 9 was the first spaceprobe to transmit photographs of Venus back to Earth.

Venturi tube
Short, open-ended pipe with a central constriction, used for measuring the flow rate of a FLUID. The fluid velocity increases and its pressure drops in the constriction. The fluid velocity is calculated from the pressure difference between the center and ends of the tube.

Venus
The planet second from the Sun, about the same size as the Earth. Its face is completely obscured by dense clouds containing sulfuric acid, though the USSR's VENERA probes and the orbiter of the Venus PIONEER program in 1978 produced images of broadscale topography of the planet, revealing that two-thirds of the planet was covered by rolling plains, with larger craters. The American Pioneer Probe has indicated that Venus has active volcanoes. Venus revolves about the Sun at a mean distance of 0.72AU in 225 days, rotating on its axis in a retrograde direction (see RETROGRADE MOTION) in 243 days. Its diameter is 12.1Mm, its atmosphere is 97% carbon dioxide, and its surface temperature is about 750K. It has no moons, and almost certainly supports no life. (♦ page 134, 221)

Verd antique
Green mottled form of SERPENTINE used for indoor decoration; also a dark green PORPHYRY containing FELDSPAR. Verde antico is the green PATINA formed on bronze.

Verdigris
Blue-green powder, a basic copper (II) acetate, made by pickling copper in acetic acid and used as a pigment and a mordant in dyeing.

Vernal equinox see EQUINOXES.

Vernier scale
An auxiliary scale used in conjunction with the main scale on many instruments (in particular, the vernier caliper), allowing greater precision of reading. The vernier scale is graduated such that nine graduations on the vernier scale equal ten on the main scale. By observing which vernier graduation nearest the zero on the vernier scale coincides with a graduation on the main scale, the precision with which a reading can be made is improved by a factor of 10.

Verruca see WART.

Vertebrae
Small bones forming the backbone or spinal column, the central pillar of the SKELETON in VERTEBRATES. Vertebrae are specialized to provide the trunk with both flexibility and strength. In the neck, cervical vertebrae are small and their joints allow free movement of the head. The thoracic vertebrae provide the bases for the ribs. The lumbar spine consists of large vertebrae with long transverse processes that form the back of the ABDOMEN; the sacral and coccygeal vertebrae, which in humans are fused, link the spine with the bony PELVIS. Within the vertebrae there is a continuous canal through which the SPINAL CORD passes; between them run the segmental nerves. Around the spinal column are the powerful spinal muscles and ligaments. (♦ page 216)

Vertebrates
Members of the Subphylum Vertebrata, a division of the Phylum CHORDATA. It includes the MAMMALS, BIRDS, REPTILES, AMPHIBIANS and FISH. All have a vertebral column or SPINE, within which the spinal cord (see NERVOUS SYSTEM) is enclosed.

Vertigo
Sensation of rotation in space resulting from functional (spinning of head with sudden stop) or organic disorders of the balance system of the EAR or its central mechanisms. It commonly induces nausea or vomiting.

Very high frequency (VHF) see ELECTROMAGNETIC RADIATION; RADIO.

Vesalius, Andreas (1514-1564)
Flemish biologist regarded as a father of modern ANATOMY. Initially a Galenist he became, after considerable experience of dissection, one of the leading figures in the revolt against GALEN. In his most important work, *On the Structure of the Human Body* (1543), he described several organs for the first time.

Vestigial organ
An anatomical structure that is nonfunctional and frequently underdeveloped in a modern species but that represents the remnant of an organ that in the remote past was fully functional in an ancestor species, for example the vermiform APPENDIX in humans.

Veterinary medicine
The medical care of sick animals, sometimes including the delivery of their young.

Vibration
Periodic motion, such as that of a swinging PENDULUM or a struck TUNING FORK. The simplest and most regular type of vibration is SIMPLE HARMONIC MOTION. ENERGY from a vibration is propagated as a WAVE MOTION. Excess mechanical vibration, as with noise pollution, can do considerable damage to buildings.

Viking program
Series of US unmanned space probes designed to land on and study Mars. Viking I landed on 20 July 1976, sending back TV pictures; its soil-analysis experiments yielded results that initially suggested the presence of life, but probably represent only unusual chemical reactions. Viking 2 landed on 3 September 1976. Both craft consisted of a lander and an orbiter.

Vincent's angina
Sore throat due to infection of the PHARYNX with BACTERIA, commonly seen in undernourished or debilitated persons.

Vinyl compounds
Compounds containing the vinyl group, $CH_2{=}CH-$, formally derived from ethylene. They are polymerized (see POLYMERS) to form PLASTICS and RUBBERS. (See also PVC.)
Viral diseases
INFECTIOUS DISEASES due to VIRUSES. The common COLD, INFLUENZA, CHICKENPOX, MEASLES and RUBELLA are common in childhood, while lassa fever and YELLOW FEVER are important tropical virus diseases. Viruses may also cause specific organ disease such as HEPATITIS, MENINGITIS, ENCEPHALITIS, MYOCARDITIS and pericarditis. Most virus diseases are self-limited and mild; there are few specific drugs effective in cases of severe illness.
Virchow, Rudolf (1821-1902)
Pomeranian-born German pathologist whose most important work was to apply knowledge concerning the CELL to PATHOLOGY, in the course of which he was the first to document LEUKEMIA and EMBOLISM. He was also distinguished as an anthropologist and archeologist.
Virgo
The Virgin, a large constellation on the ECLIPTIC; the sixth sign of the ZODIAC. It contains the bright spectroscopic binary (see DOUBLE STAR) Spica and a cluster of galaxies.
Virtanen, Artturi Ilmari (1895-1973)
Finnish biochemist awarded the 1945 Nobel Prize for Chemistry for his work on winter SILAGE. He showed that keeping the silage acid ($pH<4$) stopped the FERMENTATION that would otherwise destroy it, without reducing its nutritive value or its palatability to animals (see pH).
Virus
A minute parasitic organism that can reproduce only inside the cell of its host. The virus particle consists of its genetic material, either DNA or RNA, enclosed by a protein coat or capsid. In the more complex viruses, such as the one that causes smallpox, the capsid is enclosed by a membranous envelope, usually derived from the cell membrane of the host. Viruses of this type are also more complex internally, their DNA being found in association with special core proteins.

Viruses replicate by invading host cells and taking over the cell's "machinery" for DNA replication, and by PROTEIN SYNTHESIS. The cell produces new copies of the virus's DNA (or RNA) and synthesizes its coat proteins. These constituents then assemble themselves into new viral particles and break out of the cell during lysis. All viruses cause diseases, and few living organisms are without their viral parasites. The viruses that infect bacteria are known as bacteriophages; some can carry fragments of bacterial genetic material from one cell to another, in a process called TRANSDUCTION.

Outside the host cell viruses have no metabolic functions, and in pure form they can be crystallized. The question of whether they are truly living organisms or not is debatable. It is believed that they are descended from more complex cells, and have undergone a process of degeneration that is common in PARASITES, losing most of their cellular structures and metabolic pathways. The fact that they have the same GENETIC CODE as cellular organisms supports this theory. Viruses could be descended from pathogenic bacteria, or they could be fragments of genetic material that have escaped from the genome of the host organism itself.

Some viruses are known to play a part in the initiation of tumors; see CANCER. Cells infected by virus express viral coat proteins on their surface before the new viral particles are released, and certain cells in the immune system recognize these foreign proteins and are programmed to destroy the cell; see IMMUNITY. (See also RNA VIRUSES, RETROVIRUSES.)
Viscosity
The property of a FLUID by which it resists shape change or relative motion within itself. All fluids are viscous, their viscosity arising from internal FRICTION between molecules which tends to oppose the development of velocity differences. The viscosity of liquids decreases as they are heated, but that of gases increases.
Vision
The special sense concerned with reception and interpretation of LIGHT stimuli reaching the EYE; the principal sense in humans. Light reaches the CORNEA and then passes through this, the aqueous humor, the lens and the vitreous humor before impinging on the RETINA. Here there are two basic types of receptor: rods concerned with light and dark distinction, and cones, with three subtypes corresponding to three primary visual colors: red, green and blue. Much of vision and most of the cones are located in the central area, the macula, of which the FOVEA is the central portion; gaze directed at objects brings their images into this area. When receptor cells are stimulated, impulses pass through two nerve cell relays in the retina before passing back toward the BRAIN in the optic nerve. Behind the eyes, information derived from left and right visual fields of either eye is collected together and passes back to the opposite cerebral hemisphere, which it reaches after one further relay. In the cortex are several areas concerned with visual perception and related phenomena. The basic receptor information is coded by nerve interconnections at the various relays in such a way that information about spatial interrelationships is derived with increasing specificity as higher levels are reached. Interference with any of the levels of the visual pathway may lead to visual symptoms and potentially to BLINDNESS. (♦ page 115, 213)
Visual display unit (VDU), or visual display terminal (VDT)
A CATHODE-RAY TUBE that will display the output from a COMPUTER. It is usually connected to a keyboard for entering and editing data.
Vitamins
Specific nutrient compounds that are required only in very small amounts but are essential for body growth and metabolism. Different species have different vitamin requirements. In humans, the vitamins are denoted by letters and are often divided into fat-soluble (A, D, E and K) and water-soluble (B and C) groups. Vitamin A, or retinol, is essential for the integrity of EPITHELIUM, and its deficiency causes SKIN, EYE and mucous membrane lesions; it is also the precursor for RHODOPSIN, the retinal pigment. Vitamin A excess causes an acute encephalopathy or chronic multisystem disease. Important members of the vitamin B group include thiamine (B_1), riboflavin (B_2), nicotinic acid, pyridoxine (B_6), folic acid and cyanocobalamin (B_{12}). Thiamine acts as a coenzyme in CARBOHYDRATE metabolism and its deficiency, seen in rice-eating populations and alcoholics, causes BERIBERI and a characteristic encephalopathy. Riboflavin is also a coenzyme, active in oxidation reactions; its deficiency causes epithelial lesions. Nicotinic acid and nicotinamide are coenzymes in carbohydrate metabolism; their deficiency occurs in millet- or maize-dependent populations and leads to PELLAGRA. Pyridoxine provides an enzyme important in energy storage, and its deficiency may cause nonspecific disease or ANEMIA. Folic acid is an essential cofactor in NUCLEIC ACID metabolism, and its deficiency, which is not uncommon in PREGNANCY and with certain DRUGS, causes a characteristic anemia. Cyanocobalamin is only needed in minute quantities and is stored by the body. It is essential for all cells, but the development of BLOOD cells and GASTROINTESTINAL-TRACT epithelium and NERVOUS SYSTEM function are particularly affected by its deficiency, which occurs in pernicious ANEMIA and among vegans. Pantothenic acid, biotin, choline, inositol and para-aminobenzoic acid are other members of the B group. Vitamin C, or ascorbic acid, is involved in many metabolic pathways and has an important role in healing, blood cell formation and bone and tissue growth; SCURVY is its deficiency disease. Vitamin D, or calciferol, is a crucial factor in CALCIUM metabolism, including the growth and structural maintenance of BONE; lack causes RICKETS, while overdosage also causes disease. Vitamin E, or tocopherol, appears to play a part in blood cell and nervous system tissues, but its deficiency is uncommon and its beneficial properties have probably been overstated. Vitamin K provides essential cofactors for production of certain CLOTTING factors in the LIVER; it is used to treat some clotting disorders, including that seen in premature infants. Vitamin A is derived from both animal and vegetable tissue and most B vitamins are found in green vegetables, though B_{12} is found only in animal food and yeast. Many fruits and vegetables, notably citrus fruit and kiwi fruit, are rich in vitamin C. Vegetables such as potatoes and cabbage are also a good source of vitamin C, but soaking or prolonged boiling destroys the vitamin. Vitamin D is found in animal tissues, COD LIVER OIL providing a rich source. It is also added to margarine, and is formed naturally in the skin when skin is exposed to sunlight. Vitamins E and K are found in most biological material. (♦ page 40, 233)
Vitriol, Oil of
Obsolete term for SULFURIC ACID.
Viviparous animals
Animals that retain their eggs and bear live young, the young being nourished by the mother while they are within her body. The young may be directly nourished within the mother's body by means of a PLACENTA or similar organ, as in the eutherian mammals, the major group of viviparous animals. (The marsupial mammals also nourish the embryo within the body in its early stages, but the later stages occur within a pouch where the young one is sustained by milk.) Other viviparous animals include some sharks and snakes, which have a placenta-like organ, and some teleost fish, whose young are suspended in a nutrient-rich liquid and have a long, branched, food-absorbing organ extending from the body. Although a distinction is usually made between viviparous and OVOVIVIPAROUS ANIMALS, some sharks show both methods, the young being sustained at first by the yolk of their large eggs and at a later stage by a placenta.
Vivisection
Strictly, the dissection of living animals, usually in the course of physiological or pathological research; however, the use of the term is often extended to cover all animal experimentation. Although the practice remains the subject of considerable controversy, it is doubtful whether research, particularly medical research, can be effectively carried on without some measure of vivisection.
VLSI (Very Large Scale Integration) see ELECTRONICS.
Volcanism (vulcanicity)
The processes whereby MAGMA, a complex of molten silicates containing water and other volatiles in solution, rises toward the Earth's surface, there forming IGNEOUS ROCKS. These may be extruded on the Earth's surface (see LAVA; VOLCANO) or intruded into subsurface rock layers as,

eg DIKES, SILLS and LACCOLITHS. (See also FUMAROLE; GEYSER; HOT SPRINGS.) (◆ page 142)

Volcano

Fissure or vent in the Earth's crust through which MAGMA and associated material may be extruded onto the surface. This may occur with explosive force. The extruded magma, or LAVA, solidifies in various forms soon after exposure to the atmosphere. In particular it does so around the vent, building up the characteristic volcanic cone, at the top of which is a crater containing the main vent. There may be subsidiary vents forming "parasitic cones" in the slopes of the main cone. If the volcano is dormant or extinct the vents may be blocked with a plug (or neck) of solidified lava. On occasion these are left standing after the original cone has been eroded away. Volcanoes may be classified according to the violence of their eruptions. In order of increasing violence the main types are: Hawaiian, Strombolian, Vulcanian, Vesuvian, Peléan. Volcanoes are generally restricted to belts of seismic activity, particularly active plate margins (see PLATE TECTONICS). At ocean ridges magma rises from deep in the mantle and is added to the receding edges of the plates (see SEA-FLOOR SPREADING). Such volcanoes have generally quiet eruptions and their lavas are mostly BASALT, eg those in Iceland and Hawaii. In MOUNTAIN regions, where plates are in collision, volatile matter ascends from the subducted edge of a plate, perhaps many kilometers below the surface, bursting through the overlying plate in a series of volcanoes. These volcanoes are violent and have a lava that is mostly RHYOLITE, more viscous and lower in silica than basalt. Volcanoes in the Andes and in the Cascade Ranges are of this type. (See also EARTHQUAKES.) (◆ page 142)

Volt (V)

The SI unit of electric POTENTIAL, potential difference, ELECTROMOTIVE FORCE, etc, defined such that ENERGY is dissipated at the rate of one WATT when a one-AMPERE current drops through a potential difference of one volt.

Volta, Alessandro Giuseppe Antonio Anastasio (1745-1827)

Italian physicist who invented the voltaic pile (the first BATTERY) and thus provided science with its earliest continuous source of electric current. Volta's invention (*c*1800) demonstrated that "animal electricity" could be produced using solely inanimate materials, thus ending a long dispute with the supporters of GALVANI's view that it was a special property of animal matter.

Voltmeter

An instrument used to estimate the difference in electrical potential between different points in a circuit. Most consist of an AMMETER connected in series with a high resistance and calibrated in volts. By OHM'S LAW the current flowing is proportional to the potential difference, though the instrument itself inevitably reduces the potential under test. Accurate determinations of potential difference must employ a POTENTIOMETER.

Volume

The amount of SPACE occupied by a three-dimensional object.

Volumetric analysis

Method of quantative chemical ANALYSIS in which quantities are measured in terms of volumes, either of solutions of gases, using apparatus such as the BURETTE, the pipette (a calibrated tube, filled by suction, capable of delivering a known volume of liquid), and the volumetric (calibrated) flask. The chief technique of volumetric analysis is TITRATION; also important is the measurement in a gas burette of the gas produced in a reaction, the weight of one reactant being known.

Vomiting

The return of food or other substances (eg blood) from the STOMACH. It occurs by reverse PERISTALSIS after closure of the pyloric SPHINCTER and opening of the esophago-gastric junction. It may be induced by DRUGS, MOTION SICKNESS, GASTROENTERITIS or other infection, UREMIA, stomach or pyloric disorders. Morning vomiting may be a feature of early PREGNANCY.

Von Neumann, John (1903-1957)

Hungarian-born US mathematician who contributed to QUANTUM MECHANICS, showing the equivalence of HEISENBERG's matrix mechanics and SCHRÖDINGER's wave mechanics. His most important work was to formulate game theory, especially the minimax theorem. He also devised high-speed computers that contributed to the US development of the HYDROGEN BOMB.

Vortex

A whirling mass of FLUID such as is seen in a TORNADO, a WHIRLPOOL, a smoke ring or water running out of a bath. The term is used in HYDRODYNAMICS for a portion of fluid in which the individual particles have circular motions. A vortex may be produced when two adjacent fluid streams have different velocities, or when a solid body moves through a fluid and vortex lines cannot begin or end inside the fluid.

Voyager

Two unmanned spaceprobes launched by the US in 1977 to reconnoiter the outer planets and their satellites. Voyager 1 was launched in September 1977, encountered Jupiter (March 1979) and Saturn (December 1980) before being committed to interstellar space. Voyager 2 was launched on a different trajectory in August 1977 to fly by Jupiter (July 1977), Saturn (August 1981), Uranus (January 1986) and Neptune (August 1989).

Vries, Hugo de (1848-1935)

Dutch botanist who rediscovered the Mendelian laws of inheritance (see HEREDITY) and applied them to C. DARWIN's theory of EVOLUTION in his *Mutation Theory* (1900-1903).

Vulcanization

Process for enhancing the durability of RUBBER by heating it with sulfur or sulfur compounds, usually in the presence of "accelerators" that speed the reaction. Vulcanization involves the creation of sulfur bridges between the long-chain rubber polymer molecules.

Vulva

The external genital area of female mammals.

Waals, Johannes Diderik Van der see VAN DER WAALS, JOHANNES DIDERIK.

Waders (shorebirds)
Name given to birds such as plovers, oystercatchers and avocets, which feed at the water's edge, usually on invertebrates. They generally have long legs, and some have elongated bills for probing deep into the sand or mud. Waders come from various different bird families in the Order Charadriiformes. Related birds that do not feed near water, such as the woodcock, may also be referred to as "waders".

Wafer
A thin slice of SILICON or other SEMICONDUCTOR material on which INTEGRATED CIRCUITS are formed. After manufacture, the wafer may be separated into individual CHIPS. Wafer scale integration is a technique to connect the separate chips on a wafer, rather than connecting them on an integrated circuit. No one has achieved this yet, because of the problems of ensuring that the large number of components work reliably.

Wagner Von Jauregg (or **Wagner-Jauregg**), **Julius** (1857-1940)
Austrian psychologist and neurologist awarded the 1927 Nobel Prize for Physiology or Medicine for his discovery that inoculation with MALARIA markedly helped sufferers from general PARESIS, hitherto fatal.

Waksman, Selman Abraham (1888-1973)
Russian-born US biochemist, microbiologist and soil scientist. His isolation of streptomycin, the first specific antibiotic (a term he coined) against TUBERCULOSIS, won him the 1952 Nobel Prize for Physiology or Medicine.

Wald, George (b. 1906)
US chemist and prominent pacifist whose work on the chemistry of VISION brought him a share, with H.K. HARTLINE and R.A. GRANIT, of the 1967 Nobel Prize for Physiology or Medicine.

Wallace, Alfred Russel (1823-1913)
British naturalist and socialist regarded as the father of ZOOGEOGRAPHY. His most striking work was his formulation, independently of C. DARWIN, of the theory of NATURAL SELECTION as a mechanism for the origin of species (see EVOLUTION). He and Darwin presented their results in a joint paper in 1858 before the Linnean Society.

Wallach, Otto (1847-1931)
Russian-born German experimental organic chemist awarded the 1910 Nobel Prize for Chemistry for his analysis of the structures of the TERPENES.

Walton, Ernest Thomas Sinton (b. 1903)
Irish nuclear physicist who shared with SIR JOHN COCKCROFT the 1951 Nobel Prize for Physics for their development of the first particle ACCELERATOR, using which they initiated the first nuclear FISSION reaction using nonradioactive substances.

Wankel engine
INTERNAL-COMBUSTION ENGINE that produces rotary motion directly. Invented by the German engineer Felix Wankel (1902-1987), who completed his first design in 1954, it is now widely used in automobiles and airplanes. A triangular rotor with spring-loaded sealing plates at its apexes rotates eccentrically inside a cylinder, while the three combustion chambers formed between the sides of the rotor and the walls of the cylinder successively draw in, compress and ignite a fuel-and-air mixture. The Wankel engine is simpler in principle, more efficient and more powerful weight for weight, but more difficult to cool, than a conventional reciprocating engine.

Warburg, Otto Heinrich (1883-1970)
German biochemist awarded the 1931 Nobel Prize for Physiology or Medicine for his work elucidating the chemistry of cell RESPIRATION.

Warfarin
A derivative of coumarin used as an ANTICOAGULANT in the treatment of heart disease, and in much higher doses to kill rodents by causing internal bleeding.

Warm-blooded animals
Animals that maintain constant internal temperature, more correctly known as HOMOIOTHERMS.

Warning coloration (aposematic coloration)
Striking patterns and colors that advertise the distastefulness or poisonous nature of an organism. Experiments have shown that after one or two unpleasant encounters with such organisms predatory animals learn to avoid them in future. Banded or spotted patterns of red-and-black or yellow-and-black are the predominant form of warning coloration, though plain red or even plain black are also used. The similarity between warning coloration patterns in widely different groups is no accident – the patterns serve to reinforce each other by a process known as Mullerian MIMICRY. Thus a newly evolving species that develops warning coloration will benefit a great deal if it produces a "traditional" color and pattern. Warning coloration is often exploited by other organisms that are not distasteful, venomous or toxic – they may carry warning coloration in imitation of a noxious model, and thus gain protection; this is known as Batesian mimicry.

Wart
Benign, scaly excrescence on the SKIN caused by a VIRUS; it may arise without warning and disappear equally suddenly. Verrucas are warts pushed into the soles of the feet by the weight of the body.

Wassermann test
Screening test for syphilis (see SEXUALLY TRANSMITTED DISEASES) based on a nonspecific serological reaction that is seen not only in syphilis but also in YAWS and diseases associated with immune disorders. More specific tests are available to discriminate between these.

Waste disposal
Disposal of the waste products of agricultural, industrial and domestic processes, where an unacceptable level of environmental POLLUTION would otherwise result. The most satisfactory waste-disposal methods are usually those that involve recycling, as in manuring fields with dung, reclaiming metals from scrap or pulping waste paper for remanufacture. Recycling, however, may be inconvenient, uneconomic or not yet technologically feasible. Many popular waste-disposal methods consequently represent either an exchange of one form of environmental pollution for another less troublesome form, at least in the short term – eg the dumping or burying of non-degradable garbage or toxic wastes – or a reduction of the rate at which pollutants accumulate – eg by compacting bulk wastes before dumping. Urban wastes are generally disposed of by means of dumping, sanitary landfill, incineration and sewage processing. Agricultural, mining and mineral-processing operations generate most solid wastes – and some of the most intractable waste-disposal problems: eg the "factory" farmer's problem of disposing of surplus organic wastes economically without resorting to incineration or dumping in rivers; the problems created by large mine dumps and open-cast excavations. Another increasingly important waste-disposal problem is presented by radioactive wastes. Those with a low level of RADIOACTIVITY can be safely packaged and buried; but "high-level" wastes, produced in the course of reprocessing the fuel elements of NUCLEAR REACTORS, constitute a permanent hazard. Even the practice of encasing these wastes in thick concrete and dumping them on the ocean bottom is considered by many environmentalists to be an inadequate long-term solution (see also NUCLEAR ENERGY).

Water (H_2O)
Colorless odorless liquid which, including that trapped as ICE in icecaps and glaciers, covers about 74% of the Earth's surface. Water is essential to LIFE, which began in the watery OCEANS; because of its unique chemical properties, it provides the medium for the reactions of the living CELL. Water is also the most precious natural resource for humans, which must be conserved and protected from POLLUTION. Chemically, water can be viewed variously as a covalent HYDRIDE, an OXIDE, or a HYDROXIDE. It is a good solvent for many substances, especially ionic and polar compounds; it is ionizing and itself ionizes to give a low concentration of hydroxide and hydrogen ions (see pH). It is thus both a weak ACID and a weak BASE, and conducts electricity. It is a good, though labile, LIGAND, forming HYDRATES. Water is a polar molecule, and shows anomalies due to HYDROGEN BONDING, including contraction when heated from 0°C to 4°C. Formed when hydrogen or volatile hydrides are burned in oxygen, water oxidizes reactive metals to their ions, and reduces fluorine and chlorine. It converts basic oxides to hydroxides, and acidic oxides to OXYACIDS. (See also DEHYDRATION; HARD WATER; HEAVY WATER; HYDROLYSIS; POLYWATER; STEAM.) mp 0°C, bp 100°C, triple point 0.01°C, sg 1.0. (◀ page 37)

Water cycle see HYDROLOGIC CYCLE.

Waterfall
A vertical fall of water where a river flows from hard rock to one more easily eroded (see EROSION), or where there has been a rise of the land relative to sea level or blockage of a river by a landslide. Largest in the world is one of the Angel Falls, Venezuela (815m).

Water gap
A short, narrow gorge cut through a ridge or region of high ground by a stream or river (see EROSION). If the river no longer passes through it, the gorge is termed a wind gap.

Water gas (blue gas, because of its blue flame)
A FUEL GAS consisting of HYDROGEN and CARBON monoxide, made by blowing steam (alternately with air) over red-hot coke. Enriched with petroleum hydrocarbons, it becomes carbureted water gas. (See also SYNTHESIS GAS.)

Water glass
Aqueous solution of sodium SILICATE (Na_2SiO_3) – concentrated, syrupy and alkaline – made by fusing sodium carbonate with silica. It is used to preserve eggs, as a cement, in ore flotation, and for water- and fireproofing.

Water pollution see POLLUTION; WATER.

Water power see HYDROELECTRICITY; TURBINE.

Watershed
The boundary separating the headwaters of a series of streams flowing down one side of a ridge from the headwaters of streams flowing down the other side.

Water softening see HARD WATER.

Waterspout
Effect of a rotating column of air, or TORNADO, as it passes over water. A funnel-like CLOUD of condensed water vapor extends from a parent cumulonimbus cloud to the water surface, where it is surrounded by a sheath of spray.

Water table see GROUNDWATER.

Watson, James Dewey (b. 1928)
US biochemist who shared with F.H.C. CRICK and M.H.F. WILKINS the 1962 Nobel Prize for Physiology or Medicine for his work with Crick establishing the "double helix" molecular model of DNA. His personalized account of the research, *The Double Helix* (1968), became a bestseller.

Watson, John Broadus (1878-1958)
US psychologist who founded BEHAVIORISM, a dominant school of US psychology from the 1920s to 1940s, and whose influence is still strong.

Watson-Watt, Sir Robert Alexander (1892-1973)
British physicist largely responsible for the development of RADAR, patenting his first "radiolocator" in 1919. He perfected his equipment and techniques from 1935 through the years of WWII, his radar being largely responsible for the British victory of the Battle of Britain.

Watt (W)
The unit of POWER in SI units, defined as the power dissipated when energy is utilized at a rate of one joule per second.

Watt, James (1736-1819)
Scottish engineer and inventor. His first major invention was a steam engine with a separate condenser and thus far greater efficiency. For the manufacture of such engines he entered partnership with JOHN ROEBUCK and later (1775), more successfully, with Matthew Boulton. Between 1775 and 1800 he invented the sun-and-planet gear wheel, the double-acting engine, a throttle valve, a pressure gauge and the centrifugal governor – as well as taking the first steps toward determining the chemical structure of water. He also coined the term HORSEPOWER and was a founder member of the LUNAR SOCIETY.

Waveguide
Means for channeling high-frequency ELECTROMAGNETIC RADIATION by confining it within a tube whose walls are made of a conducting material (typically metal). Waveguides find their greatest use in MICROWAVE technology.

Wavelength see WAVE MOTION.

Wave mechanics
Branch of QUANTUM MECHANICS developed by SCHRÖDINGER which considers MATTER rather in terms of its wavelike properties (see WAVE MOTION) than as systems of particles. Thus an orbital ELECTRON is treated as a 3-dimensional system of standing waves represented by a wave function. In accordance with the UNCERTAINTY PRINCIPLE, it is not possible to pinpoint both the instantaneous position and velocity of the electron; however, the square of the wave function yields a measure of the probability that the electron is at any given point in space-time. The pattern of such probabilities provides a model for the "shape" of the electron ORBITAL involved. Given wave functions can be obtained from the Schrödinger wave equation. Usually, and not unsurprisingly, this can only be solved for particular values of the ENERGY of the system concerned.

Wave motion
A collective motion of a material or extended object, in which each part of the material oscillates about its undisturbed position, but the oscillations at different places are so timed as to create an illusion of crests and troughs running right through the material. Familiar examples are furnished by surface waves on water, or transverse waves on a stretched rope; SOUND is carried through air by a wave motion in which the air molecules oscillate parallel to the direction of propagation (longitudinal wave motion), and LIGHT or RADIO waves involve ELECTROMAGNETIC FIELDS oscillating perpendicular to it (transverse wave motion). The maximum displacement of the material from the undisturbed position is the amplitude of the wave, the separation of successive crests, the wavelength, and the number of crests passing a given place each second, the frequency. The product of the wavelength and the frequency gives the velocity of propagation. Standing waves (apparently stationary waves, where the nodes and antinodes – points of zero and maximum amplitude – appear not to move) arise where identical waves traveling in opposite directions superpose. The characteristic properties of waves include propagation in straight lines; REFLECTION at plane surfaces; REFRACTION – a change in direction of a wave transmitted across a plane interface between two media; DIFFRACTION – diffuse SCATTERING by impenetrable objects of a size comparable with the wavelength which can cause waves to "bend" round corners; and INTERFERENCE – due to the superposition of waves. In destructive interference, the cancellation of one wave by another wave half a wavelength out of step (or phase) means that the crests of one wave fall on the troughs of the other; in constructive interference, the crests occur together and reinforce each other. If the wave velocity is the same for all wavelengths, then quite arbitrary forms of disturbance will travel as waves, and not simply regular successions of crests and troughs. When this is not the case, the wave is said to be dispersive and localized disturbances move at a speed (the group velocity) quite different from that of the individual crests, which can often be seen moving faster or slower within the disturbance "envelope", which becomes progressively broader as it moves. In transverse wave motion only, different polarizations of the wave are distinguished (see POLARIZED LIGHT). Waves carry ENERGY and MOMENTUM with them just like solid objects; the identity of the apparently irreconcilable wave and particle concepts of matter is a basic tenet of QUANTUM MECHANICS. (◆ page 21, 118, 225)

Wax
Moldable water-repellent solid. There are several entirely different kinds. Animal waxes were the first known – wool wax when purified yields LANOLIN; beeswax, from the honeycomb, is used for some candles aud as a sculpture medium (by carving or casting); spermaceti wax, from the sperm whale, has been used in ointments and cosmetics. Vegetable waxes, like animal waxes, are mixtures of ESTERS of long-chain ALCOHOLS and CARBOXYLIC ACIDS. Carnauba wax, from the leaves of a Brazilian palm tree, is hard and lustrous, and is used to make polishes; candelilla wax, from a wild Mexican rush, is similar but more resinous; Japan wax, the coating of sumac berries, is fatty and soft but tough and kneadable. Mineral waxes include montan wax, extracted from lignite (see COAL), bituminous and resinous; ozokerite, an absorbent hydrocarbon wax obtained from wax shales, and paraffin wax or petroleum wax, the most important wax commercially: it is obtained from the residues of PETROLEUM refining by solvent extraction, and is used to make candles, to coat paper products, in the electrical industry, to waterproof leather and textiles, etc. Various synthetic waxes are made for special uses.

Weak nuclear force
The force underlying the beta type of RADIOACTIVITY, and which is fundamental to the burning of stars like the SUN. At the quantum level it is described by the exchange of W and Z bosons. (See also SUBATOMIC PARTICLES; UNIFIED FIELD THEORY.) (◆ page 49, 147)

Weather
The hour-by-hour variations in the atmospheric conditions experienced at a given place. (See ATMOSPHERE; METEOROLOGY; WEATHER FORECASTING AND CONTROL.)

Weather forecasting and control
The practical application of the knowledge gained through the study of METEOROLOGY. Weather forecasting, organized nationally by government agencies such as the US National Weather Service, is coordinated internationally by the World Meteorological Organization (WMO). There are three basic stages: observation; analysis; and forecasting. Observation involves round-the-clock weather watching and the gathering of meteorological data by land stations, weather ships, and by using RADIOSONDES and weather SATELLITES. In analysis, this information is coordinated at national centers, and plotted in terms of ISOBARS, FRONTS, etc, on synoptic charts (weather maps). Then, in forecasting, predictions of the future weather pattern are made by the "synoptic method" (in which the forecaster applies his experience of the evolution of past weather patterns to the current situation) and by "numerical forecasting" (which treats the ATMOSPHERE as a fluid of variable density and seeks to use hydrodynamic equations to determine its future parameters). These methods yield short- and medium-term forecasts – up to four days ahead. Long-range forecasting, a recent development, depends additionally on the statistical analysis of past weather records in attempting to discern the future weather trends over the next month or season.

Weather control, or weather modification, is an altogether less reliable technology. Indeed, the natural variability of weather phenomena makes it difficult to assess the success of experimental procedures. To date, the best results have been obtained in the fields of CLOUD seeding and the dispersal of supercooled FOGS.

Weathering see EROSION.

Weber (Wb)
The unit of magnetic flux in SI units, defined such that an ELECTROMOTIVE FORCE of one VOLT is induced in a single coil when the flux changes in the coil at the rate of one weber per second. (See INDUCTION, ELECTROMAGNETIC.)

Weber, Wilhelm Eduard (1804-1891)
German physicist best known for his work with GAUSS on GEOMAGNETISM.

Weedkillers see HERBICIDES.

Wegener, Alfred Lothar (1880-1930)
German meteorologist, explorer and geologist. His *The Origin of Continents and Oceans* (1915) set forth "Wegener's hypothesis", the theory of CONTINENTAL DRIFT, whose developments in succeeding decades revolutionized our understanding of Earth (See also PLATE TECTONICS.)

Weight
The attractive FORCE experienced by an object in the presence of another massive body in accordance with the law of universal GRAVITATION. The weight of a body (measured in newtons) is given by the product of its MASS and the local ACCELERATION due to gravity (*g*). Weight differs from mass in being a VECTOR quantity.

Weinberg, Steven (b. 1933)
US physicist who received the 1979 Nobel Prize for Physics jointly with A. SALAM and S. GLASHOW for their theory which united the WEAK NUCLEAR FORCE and ELECTROMAGNETIC FORCE.

Weismann, August Friedrich Leopold (1834-1914)
German biologist regarded as a father of modern GENETICS for his demolition of the theory that

ACQUIRED CHARACTERISTICS could be inherited, and proposal that CHROMOSOMES are the basis of HEREDITY. He coupled this proposal with his belief in NATURAL SELECTION as the mechanism for EVOLUTION.

Welch, William Henry (1850-1934)
US pathologist and bacteriologist whose most significant achievements were in the field of medical education, playing a large part in the founding (1893) and development of the Johns Hopkins Medical School.

Well
Manmade hole in the ground used to tap water, gas or minerals from the Earth. Most modern wells are drilled and fitted with a lining, usually of steel, to forestall collapse. Though wells are sunk for NATURAL GAS and PETROLEUM oil, the commonest type yields water. Such wells may be horizontal or vertical, but all have their innermost end below the water table (see GROUNDWATER). If it is below the permanent water table (the lowest annual level of the water table) the well will yield water throughout the year. Most wells need to be pumped, but some operate under natural pressure (see ARTESIAN WELL).

Weller, Thomas Huckle (b. 1915)
US bacteriologist and virologist who shared with J.F. ENDERS and F.C. ROBBINS the 1954 Nobel Prize for Physiology or Medicine for their cultivation of POLIOMYELITIS virus in non-nerve tissues.

Wells, Horace (1815-1848)
US dentist and pioneer of surgical ANESTHESIA, using (largely without success) nitrous oxide (see NITROGEN).

Welwitschia
A species of plant found in the Namib Desert of Africa. Its leaves grow continuously, snaking across the desert sands. Reproductive organs are produced in cone-like bodies on short stalks from the center of the plant. Although usually classified with the GYMNOSPERMS, *Welwitschia* is not closely related to any other living plant. (◀ page 139)

Wen
Blocked SEBACEOUS GLAND, often over the scalp or forehead, which forms a cyst containing old sebum under the SKIN. It may become infected.

Werner, Abraham Gottlob (1750-1817)
Silesian-born mineralogist who taught for over 40 years at the Freiberg Mining Academy, disseminating the doctrines of NEPTUNISM.

Werner, Alfred (1866-1919)
French-born Swiss chemist awarded the 1913 Nobel Prize for Chemistry for his theory of coordination complexes. (See BOND, CHEMICAL; LIGANDS.)

Wertheimer, Max (1880-1943)
Czechoslovakian-born US psychologist; founder, with K. KOFFKA and W. KÖHLER, of the school of GESTALT PSYCHOLOGY.

Westerlies see PREVAILING WESTERLIES.

Weyl, Hermann (1885-1955)
German mathematician and mathematical physicist noted for his contributions to the theories of RELATIVITY and QUANTUM MECHANICS.

Wheatstone bridge
An electric circuit used for comparing or measuring RESISTANCE. Four resistors, including the unknown one, are connected in a square, with a BATTERY between one pair of diagonally opposite corners and a sensitive GALVANOMETER between the other. When no current flows through the meter, the products of opposite pairs of resistances are equal. Similar bridge circuits are used for IMPEDANCE measurement.

Whiplash injury
Impact injury in which the neck has been jerked back, causing pain, damage to vertebrae and, in severe cases, a broken neck. Common in people who have been in car crashes.

Whipple, George Hoyt (1878-1976)
US pathologist awarded the 1934 Nobel Prize for Physiology or Medicine for his discovery that feeding raw liver to anemic dogs improved their condition: the successful work of G. MINOT and W. MURPHY (who shared the award with him) in finding a treatment for pernicious ANEMIA sprang directly from this.

Whirlpool
A rotary current in water. Permanent whirlpools may arise in the ocean from the interactions of the TIDES (see OCEAN CURRENTS). They occur also in streams or rivers where two currents meet or the shape of the channel dictates. Short-lived whirlpools may be created by wind. (See also VORTEX.)

Whirlwind
Rotating column of air caused by a pocket of low atmospheric pressure formed – unlike a TORNADO – near ground level by surface heating. They are far less violent than tornadoes. Whirlwinds passing over dry, dusty country are sometimes called "dustdevils".

White corpuscles see BLOOD.

White dwarf see STAR.

White gold
An alloy of gold with nickel and sometimes other noble METALS, used in dentistry.

White lead see CERUSSITE.

Whooping cough (pertussis)
BACTERIAL DISEASE of children causing upper respiratory tract symptoms with a characteristic whoop or inspiratory noise due to INFLAMMATION of the LARYNX.

Wieland, Heinrich Otto (1877-1957)
German organic chemist noted for his work on STEROIDS, especially his research on the BILE acids, which brought him the 1927 Nobel Prize for Chemistry.

Wien, Wilhelm (1864-1928)
Prussian-born German physicist best known for his work on BLACKBODY RADIATION, work which was later to be a foundation stone for Planck's QUANTUM THEORY.

Wiener, Norbert (1894-1964)
US mathematician who created the discipline of CYBERNETICS. His major book is *Cybernetics: Or Control and Communication in the Animal and the Machine* (1948).

Wiesel, Torsten (b. 1924)
Swedish neurobiologist working in the US who shared the 1981 Nobel Prize for Physiology or Medicine with D. HUBEL for their work on recording from single cells, which revealed the distribution and structure of nerve cells in the brain.

Wigner, Eugene Paul (b. 1902)
Hungarian-born US physicist who shared with J.H.D. JENSEN and M.G. MAYER the 1963 Nobel Prize for Physics for his work in the field of nuclear physics. He also worked with E. FERMI on the MANHATTAN PROJECT, and received the 1960 Atoms for Peace Award.

Wilkins, Maurice Hugh Frederick (b. 1916)
British biophysicist who shared with F.H. CRICK and J.D. WATSON the 1962 Nobel Prize for Physiology or Medicine for his X-RAY DIFFRACTION studies of DNA, work vital to the determination by Crick and Watson of DNA's molecular structure.

Wilkinson, Sir Geoffrey (b. 1921)
British inorganic chemist awarded with ERNST OTTO FISCHER the 1973 Nobel Prize for Chemistry for their work on ORGANOMETALLIC COMPOUNDS.

Will-o'-the Wisp (Jack O'Lantern, ignis fatuus)
Light seen at night over marshes, caused by SPONTANEOUS COMBUSTION of METHANE produced by putrefying organic matter. Luring travelers into danger, it was popularly regarded as a wandering damned spirit bearing its own hell-fire.

Willstätter, Richard (1872-1942)
German chemist awarded the 1915 Nobel Prize for Chemistry for his studies of the structure of CHLOROPHYLL.

Wilson, Charles Thomson Rees (1869-1959)
British physicist awarded with A.H. COMPTON the 1927 Nobel Prize for Physics for his invention of the CLOUD CHAMBER (1911).

Wilson, Kenneth Geddes (b. 1936)
US physicist who won the 1982 Nobel Prize for Physics for his method of analyzing basic changes in matter that occur under the influence of pressure and temperature.

Wilson, Robert Woodrow (b. 1936)
US radio astronomer who shared the 1978 Nobel Prize for Physics with A. PENZIAS for their discovery of the cosmic microwave background radiation (see COSMOLOGY), and with P. KAPITZA.

Wind
Body of air moving relative to the Earth's surface. The world's major wind systems, or general winds, are set up to counter the equal heating of the Earth's surface and modified by the rotation of the Earth. Surface heating, at its greatest near the EQUATOR, creates an equatorial belt of low pressure (see DOLDRUMS) and a system of CONVECTION currents transporting heat toward the poles (see HORSE LATITUDES). The Earth's rotation deflects the currents of the N Hemisphere to the right and those of the S Hemisphere to the left of the directions in which they would otherwise blow, producing the NE and SE TRADE WINDS, the PREVAILING WESTERLIES and the Polar Easterlies. (Winds are named after the direction from which they blow.) Other factors influencing general wind patterns are the different rates of heating and cooling of land and sea and the seasonal variations in surface heating. Mixing of air along the boundary between the Westerlies and the Polar Easterlies – the polar front – causes depressions in which winds follow circular paths, counterclockwise in the N Hemisphere and clockwise in the S Hemisphere (see CYCLONE). Superimposed on the general wind systems are local winds – winds, such as the CHINOOKS, caused by temperature differentials associated with local topographical features such as mountains and coastal belts, or winds associated with certain CLOUD systems. (See also ATMOSPHERE; BEAUFORT SCALE; HURRICANE; JET STREAM; MONSOON; TORNADO; WEATHER FORECASTING AND CONTROL; WHIRLWIND.) (◀ page 25, 208)

Windaus, Adolf Otto Reinhold (1876-1959)
German organic chemist awarded the 1928 Nobel Prize for Chemistry for his work on the STEROLS, in the course of which he determined the structure of CHOLESTEROL and showed that ULTRAVIOLET RADIATION transforms ergosterol into VITAMIN D.

Wind gap see WATER GAP.

Windpipe
A common name for the TRACHEA.

Wind tunnel
Tunnel in which a controlled stream of air is produced in order to observe the effect on scale models or full-size components of airplanes, missiles, automobiles or such structures as bridges and skyscrapers. An important research tool in AERODYNAMICS, the wind tunnel enables a design to be accurately tested without the risks attached to full-scale trials. "Hypersonic" wind tunnels, operating on an impulse principle, can simulate the frictional effects of flight at over five times the speed of sound.

Wire chamber
A device for detecting charged SUBATOMIC PARTICLES. The basic principle is to collect electrons released when a charged particle ionizes a gas, on an electrified grid of wires. A number of different varieties exist, including multiwire proportional chambers, drift chambers and GEIGER COUNTERS.

Wishbone see COLLARBONE.

Withering, William (1741-1799)
British physician, mineralogist and botanist, who first made use of digitalis in the treatment of dropsy (see EDEMA). The mineral witherite, an ore of barium ($BaCO_3$), is named for him.

Wittig, Georg (1897-1987)
German chemist who shared the 1979 Nobel Prize for Chemistry with H.C. BROWN for devising new ways of making organic chemicals with boron- and phosphorus-containing compounds.

Wöhler, Friedrich (1800-1882)
German chemist who first synthesized an organic compound from inorganic material, UREA from ammonium cyanate (1828), and who, with LIEBIG, discovered the benzoyl radical.

Wolff, Caspar Friedrich (1733-1794)
German anatomist regarded as the father of modern EMBRYOLOGY for his demonstration that the organs and other bodily parts form from undifferentiated tissue in the fetus. It had earlier been thought that the fetus was a body in miniature.

Wolfram see TUNGSTEN.

Wolframite (iron manganese tungstate, $(Fe,Mn)WO_4$)
The chief ore of TUNGSTEN. It forms brown to black monoclinic crystals, and is found with TIN ores. China is the main producer.

Wollaston, William Hyde (1766-1828)
British chemist and physicist who, through the process he devised for isolating PLATINUM in pure and malleable form, founded the technology of powder METALLURGY. He also discovered PALLADIUM (1803) and RHODIUM (1804), was the first to observe the FRAUNHOFER LINES (1802), and invented the reflecting GONIOMETER (1809).

Womb see UTERUS.

Wood
The hard, fibrous tissue obtained from the trunks and branches of trees and shrubs. Woody tissue is also found in some herbaceous plants. Botanically, wood consists of the pith (remains of the primary growth), the water-conducting xylem vessels, the CAMBIUM (a band of living cells that divide to produce xylem and phloem); PHLOEM (vessels that conduct a sugary solution containing nutrients made in the leaves), and the bark. The new wood nearest the cambium is termed sapwood because it conducts water. However, the bulk of the trunk is heartwood in which the xylem is impregnated with LIGNIN, which gives the cells strength but stops them from conducting water. In temperate regions, a tree's age can be found by counting its ANNUAL RINGS. Commercially, wood is divided into HARDWOOD (from ANGIOSPERM trees) and SOFTWOOD (from GYMNOSPERMS).

Wood alcohol see FUEL ALCOHOL; METHANOL.

Woodward, Robert Burns (1917-1979)
US chemist who synthesized QUININE (1944), CHOLESTEROL (1951), CORTISONE (1951), STRYCHNINE (1954), LSD (1954), reserpine (1956), CHLOROPHYLL (1960) and VITAMIN B (1971). He received the 1965 Nobel Prize for Chemistry for his many organic syntheses.

Wordprocessor
A microcomputer with SOFTWARE for preparing, editing, storing and transmitting documents.

Work
Alternative name for ENERGY, used particularly in discussing mechanical processes. Work of one JOULE is done when a FORCE of one NEWTON acts through a distance of one METER.

Worms, True see ANNELIDS.

Wren, Sir Christopher (1632-1723)
English architect who had a brilliant early career as a mathematician and professor of astronomy at Oxford University (1661-73), and was a founder-member of the ROYAL SOCIETY OF LONDON (President, 1681-83). In 1663 he turned to architecture, and after the Great Fire of London (1666) he was appointed principal architect to rebuild London.

Wright, Sewall (b. 1889)
US geneticist, a founder of population GENETICS, the use of STATISTICS in the study of EVOLUTION.

Wrist
In humans, the flexible set of bones linking the end of the arm with the hand. The small bones of the wrist allow movement in various directions and provide basic maneuverability for the hand.

Wundt, Wilhelm (1832-1920)
German psychologist regarded as the father of experimental PSYCHOLOGY, opening the first psychological institute in 1879 and so ushering in modern psychology.

Wurtzite
The high-temperature (α) form of zinc sulfide (ZnS); a minor ore of ZINC which inverts to SPHALERITE. It forms brown-black pyramidal crystals in the hexagonal system, and occurs in Bolivia, Montana and Nevada.

Xanmorphyta see ALGAE.

Xenon (Xe)
One of the NOBLE GASES, used in flash lamps, electric-arc lamps, BUBBLE CHAMBERS and radiation counters; also to produce ANESTHESIA. It forms stable covalent compounds with fluorine and oxygen, with oxidation numbers +2, +4, +6 and +8. They are reactive and highly oxidizing, and unstable in acid solution. Most of the crystalline xenate (VIII) salts (XeO_6^{4-}) are stable. AW 131.3, mp −112°C, bp −108°C. (♦ page 130)

Xerophyte
A plant that has adaptations enabling it to live in dry habitats. These adaptations allow it to store water, cut down water loss by having a small leaf area, and recover after wilting. Many xerophytes have a thick waxy CUTICLE, and their STOMATA sunk in pits or furrows on the underside of the leaf to reduce transpiration. (See also SUCCULENTS.)

X-ray astronomy
The study of the X-RAYS emitted by celestial objects. As the Earth's atmosphere absorbs most X-radiation before it reaches the surface, observations are made from high altitude balloons, satellites and rockets. X-rays are normally emitted from regions of superhot gas; X-ray sources include the solar corona, solar flares, compact stars in binary systems, quasars and intergalactic gas.

X-ray diffraction
A technique for determining the structure of CRYSTALS through the way in which they scatter X-RAYS. Because of INTERFERENCE between the waves scattered by different ATOMS, the scattering occurs in directions characteristic of the spatial arrangement of the unit cells of the crystal, while the relative intensity of the different beams reflects the structure of the unit cell itself.

X-rays
Highly energetic, invisible ELECTROMAGNETIC RADIATION of wavelengths ranging between 0.1pm to 1nm. They are usually produced using an evacuated ELECTRON TUBE in which ELECTRONS are accelerated from a heated CATHODE toward a large tungsten or molybdenum ANODE by applying a POTENTIAL difference of perhaps 1MV. The electrons transfer their energy to the anode, which then emits X-ray PHOTONS. X-rays are detected using PHOSPHOR screens (as in medical fluoroscopy), with GEIGER and SCINTILLATION COUNTERS and on photographic plates. X-rays were discovered by RÖNTGEN in 1895, but because of their extremely short wavelength their wave nature was not firmly established until 1911, when VON LAUE demonstrated that they could be diffracted from crystal LATTICES. X-rays find wide use in medicine both for diagnosis and treatment (see RADIOLOGY) and in engineering where RADIOGRAPHS are used to show up minute defects in structural members. X-ray tubes must be shielded to prevent damage to living tissue. (♦ page 118)

Xylem
The main water-conducting tissue in higher plants; see WOOD; TREE. (♦ page 53)

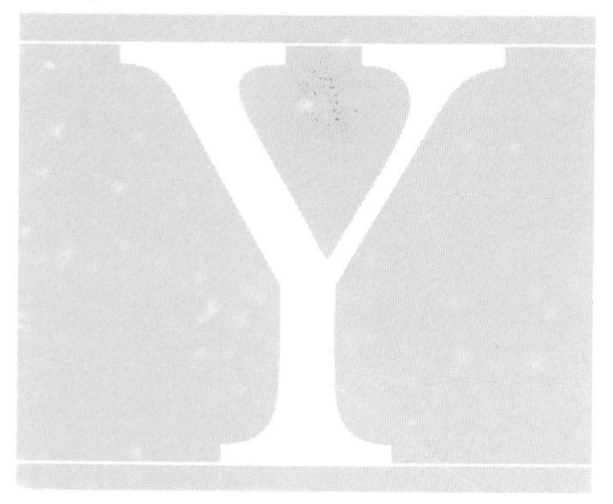

Yalow, Rosalyn Sussman (b. 1921)
US medical physicist who was awarded a half-share in the 1977 Nobel Prize for Physiology and Medicine (the other half was shared by R. GUILLEMIN and A.V. SCHALLY) for the development of radioimmunoassay techniques for peptide HORMONES.

Yang, Chen Ning (b. 1922)
Chinese-born US physicist who shared with TSUNG DAO LEE the 1957 Nobel Prize for Physics for their studies of violations of the conservation of PARITY.

Yard (yd)
Base unit of length in the US Customary and British Imperial unit systems. The International yard, defined as 0.9144m exactly, is generally used as a scientific standard. The yard of US commerce, defined as 3600/3937m, is marginally greater.

Yaws
A disease, caused by an organism related to that of syphilis (see SEXUALLY TRANSMITTED DISEASES), common in the tropics. It occurs often in children and consists of a local lesion on the limbs; there is also mild systemic disease. Chronic destructive lesions of SKIN, BONE and CARTILAGE may develop later. The WASSERMANN TEST is positive as in syphilis.

Year (yr)
Name of various units of time, all depending on the revolution of the Earth about the Sun. The sidereal year (365.25636 mean solar DAYS) is the average time the Earth takes to complete one revolution measured with respect to a fixed direction in space. The tropical year (365.24220 mean solar day), the year measured by the changing SEASONS, is that in which the mean longitude of the Sun moves through 360°. The anomalistic year (365.25964 mean solar days) is the average interval between successive terrestrial perihelions (see ORBIT). The civil year is a period of variable duration, usually 365 or 366 days (leap year), depending on the type of CALENDAR in use.

Yeasts
Single-celled FUNGI mostly belonging to the ASCOMYCETES. Many yeasts live on the surface of fruits in the wild, decomposing the fruit's sugars and causing a characteristic "bloom" on the skin. Some yeasts cause plant diseases, such as Dutch elm disease, while others cause diseases of the skin and mucous membranes in humans (see THRUSH). Certain "domesticated" yeasts are used in baking, brewing and wine-making. In the absence of sufficient oxygen, yeasts are capable of FERMENTATION, the decomposition of sugars to yield alcohol (ETHANOL) and carbon dioxide. Yeasts are also grown as a source of B-complex vitamins. Several fungi have both a yeast-like unicellular form and a more normal mycelial form (see MYCELIUM), and can switch between the two as circumstances demand.

Yellow fever
INFECTIOUS DISEASE caused by a VIRUS carried by MOSQUITOES of the genus *Aëdes* and occurring in tropical America and Africa. The disease consists of fever, headache, backache, prostration and vomiting of sudden onset. Protein loss in the urine, kidney failure, and liver disorder with JAUNDICE are also frequent. Hemorrhage from mucous membranes, especially in the gastro-intestinal tract, is also common. A moderate number of cases are fatal but a mild form of the disease is also recognized. VACCINATION to induce IMMUNITY is important and effective as no specific therapy is available; mosquito control provides a similarly important preventive measure.

Yerkes, Robert Mearns (1876-1956)
US pioneer of comparative (ie animal/human) PSYCHOLOGY and of intelligence testing. He initiated the first mass psychological testing program in WWI, involving nearly 1.75 million US army men, and in the 1920s and 1930s he was the foremost authority on PRIMATES.

Yersin, Alexandre Émile John (1863-1943)
Swiss bacteriologist who discovered, independently of KITASATO, the PLAGUE bacillus (1894), and developed a SERUM to combat it.

Young, Thomas (1773-1829)
British linguist, physician and physicist. His most significant achievement was in demonstrating optical INTERFERENCE, to resurrect the wave theory of LIGHT, which had been occulted by NEWTON's particle theory. He also suggested that the eye responded to mixtures of three primary colors (see VISION), and proposed the modulus of ELASTICITY known as Young's modulus (E – see MATERIALS, STRENGTH OF).

Ytterbium (Yb)
One of the LANTHANUM SERIES of elements. AW 173.0, mp 824°C, bp 1500°C, sg 6.972 (25°C). (♦ page 130)

Yttrium (Y)
Silvery white RARE-EARTH metal. It is used in ALLOYS and as a "getter" to help evacuate ELECTRON TUBES. A red PHOSPHOR (yttrium oxide or vanadate excited by europium) is used in color televisions, and yttrium-iron GARNETS are used in RADAR and MICROWAVE devices. AW 88.9, mp 1510°C, bp 3300°C, sg 4.46 (25°C). (♦ page 130)

Yukawa, Hideki (1907-1981)
Japanese physicist who postulated the meson (see SUBATOMIC PARTICLES) as the agent bonding the atomic nucleus together. In fact the mu-meson, discovered shortly afterwards (in 1936) by C.D. ANDERSON, does not fulfil this role, and Yukawa had to wait until C.F. POWELL discovered the pi-meson in 1947 for vindication of his theory. He received the 1949 Nobel Prize for Physics.

Zeeman, Pieter (1865-1943)
Dutch physicist who shared with H.A. LORENTZ the 1902 Nobel Prize for Physics for his discovery of the Zeeman effect. This involves the splitting of spectral lines (see SPECTROSCOPY) in a MAGNETIC FIELD.

Zenith
In astronomy, the point on the CELESTIAL SPHERE directly above an observer and exactly 90° from the celestial horizon. It is directly opposite to the NADIR.

Zeolites
Group of hydrated aluminosilicate minerals, found mostly in volcanic ROCKS and hydrothermal veins. Variable in form, they are light, with an open-framework structure that permits their use as molecular sieves and for ION EXCHANGE (especially for softening hard water). They undergo reversible DEHYDRATION.

Zernike, Frits (1888-1966)
Dutch physicist awarded the 1953 Nobel Prize for Physics for his development of phase-contrast microscopy, whereby CELLS may be studied without prior staining.

Zero-point energy
The kinetic ENERGY that, in accordance with the UNCERTAINTY PRINCIPLE, is retained by a substance at ABSOLUTE ZERO. Each of the component oscillators retains one half quantum ($h\nu/2$) of energy.

Ziegler, Karl (1898-1973)
German chemist who shared with G. NATTA the 1963 Nobel Prize for Chemistry for his work on ORGANOMETALLIC COMPOUNDS, in the course of which he dramatically advanced POLYMER science and technology.

Zinc (Zn)
Bluish white metal in Group IIb of the PERIODIC TAELE, an anomalous TRANSITION ELEMENT. It occurs naturally as SPHALERITE, SMITHSONITE, HEMIMORPHITE and WURTZITE, and is extracted by roasting to the oxide and reduction with carbon. It is used for GALVANIZING; as the cathode of dry cells, and in ALLOYS including BRASS. Zinc is a vital trace element, occurring in a number of enzymes. Chemically zinc is reactive, readily forming divalent ionic salts (Zn^{2+}), and zincates (ZnO_2^{2-}) in alkaline solution; it forms many stable LIGAND complexes. Zinc oxide and sulfide are used as white pigments. Zinc chloride is used as a flux, for fireproofing, in dentistry, and in the manufacture of BATTERIES and FUNGICIDES. AW 65.4, mp 419.6°C, bp 910°C, sg 7.133 (25°C). (♦ page 130)

Zinsser, Hans (1878-1940)
US bacteriologist who directed the development and production of vaccines to counter certain strains of TYPHUS.

Zircon ($ZrSiO_4$)
Hard SILICATE mineral, a major ore of ZIRCONIUM, of widespread occurrence. It forms prismatic crystals in the tetragonal system that when transparent are used as GEMS. They may be colorless, red, orange, yellow, green or blue, and have a high refractive index.

Zirconium (Zr)
A silvery white TRANSITION ELEMENT in Group IVb of the PERIODIC TABLE. It occurs naturally as baddeleyite (ZrO_2) and ZIRCON; the metal is extracted by reducing zirconium (IV) chloride with magnesium. It is corrosion-resistant at ordinary temperatures, owing to an inert oxide layer, but reactive at high temperatures. The metal is used in photographic flash bulbs and to clad uranium fuel elements in atomic reactors. The refractory oxide is used for ceramics, and other zirconium compounds are used in pharmaceuticals and as mordants in dyeing. AW 91.2, mp 1850°C, bp 4500°C, sg 6.49 (20°C). (♦ page 130)

Zodiac
The band of the heavens whose outer limits lie 9° on each side of the ECLIPTIC. The 12 main CONSTELLATIONS near the ecliptic, corresponding to the 12 signs of the zodiac, are ARIES; TAURUS; GEMINI; CANCER; LEO; VIRGO; LIBRA; SCORPIUS; SAGITTARIUS; CAPRICORNUS; AQUARIUS; PISCES. The orbits of all the planets except Pluto lie within the zodiac and their positions, as that of the Sun, are important in ASTROLOGY. The 12 signs are each equivalent to 30° of arc along the zodiac.

Zodiacal light
A hazy band of light, usually to be seen in the W after sunset or the E before sunrise, but in fact extending all along the ECLIPTIC. Associated is the Gegenschein, a circular patch of light on the ecliptic directly opposite to the Sun. Both phenomena are caused by the reflection of sunlight on clouds of interplanetary dust.

Zone refining
Purifying process, used particularly with METALS and SEMICONDUCTORS, in which a moving zone of an ingot of the impure material is melted, impurities tending to remain in the moving molten zone while the purified material is left behind. Similar processes are used for preparing large uniform semiconducting crystals and for distributing a desired impurity uniformly through a pure ingot.

Zoogeography
The study of the geographical distribution of animal species and populations; part of BIOGEOGRAPHY. Physical barriers, such as the sea, and mountain ranges, major climatic extremes, intense heat or cold, may prevent the spread of a species into new areas, or may separate two previously linked populations, allowing them to develop into distinct species. The presence of these barriers to movement and interbreeding, both now and in the past, are reflected in the distributions and adaptive radiations found today. The major zoogeographic regions of the world are the Ethiopian (sub-Saharan Africa); the Oriental (India and SE Asia); the Australasian (including Australia, New Guinea and New Zealand); the Neotropical (Central and South America), and the Holarctic (the whole northerly region, often divided into the Nearctic – North America – and the Palearctic – most of Eurasia with N Africa).

Zooids see BRYOZOA; ENTOPROCTA; HEMICHORDATA.

Zoology
The scientific study of animal life, including comparative ANATOMY and PHYSIOLOGY, EVOLUTION, GENETICS, EMBRYOLOGY, BIOCHEMISTRY, ANIMAL BEHAVIOR, ECOLOGY and TAXONOMY.

Zooplankton
The animal and animal-like elements of the PLANKTON, made up of small marine animals (eg arrow worms), the larvae of larger marine creatures (mainly mollusks and crustaceans, but also some fish larvae), and unicellular organisms. (♦ page 44)

Zsigmondy, Richard (1865-1929)
Austrian-born German chemist awarded the 1925 Nobel Prize for Chemistry for his work on COLLOIDS.

Zwicky, Fritz (1898-1974)
Swiss-born US astronomer and astrophysicist best known for his studies of SUPERNOVAE, which he showed to be quite distinct from, and much rarer than, NOVAE. He also did pioneering work on JET PROPULSION.

Zwitterion
Complex ION carrying both a positive and a negative charge in different parts (owing to both an acidic and a basic FUNCTIONAL GROUP being present), and hence neutral overall. They usually show TAUTOMERISM with an uncharged molecule; eg glycine:

$$H_2NCH_2COOH \rightleftharpoons {}^+H_3NCH_2COO^-.$$

Zygomycetes see FUNGI.

Zygote
The cell produced by the fusion of two GAMETES, which develops into a new individual. It contains the diploid chromosome number (see MEIOSIS). (♦ page 204)

Zymase
ENZYME complex found in YEASTS that catalyzes the alcoholic FERMENTATION of CARBOHYDRATES. It was first isolated by E. BUCHNER.

CREDITS

Photographs
12 Science Photo Library/Physics Department, Imperial College. 33 Chemical Designs Limited. 49 Science Photo Library/Lawrence Berkely Laboratory

Artists
Robert and Rhoda Burns, Kai Choi, Simon Driver, Chris Forsey, Alan Hollingbery, Kevin Maddison, Dave Mazierski, Colin Salmon, Mick Saunders

Typesetting
Peter and Robert MacDonald

Production
Joanna Turner